Essential Fish Biology

Essential Fish Biology

Diversity, Structure and Function

Derek Burton

Professor Emeritus, Department of Biology,
Memorial University of Newfoundland, Canada

Margaret Burton

Professor Emeritus, Department of Biology,
Memorial University of Newfoundland, Canada

OXFORD
UNIVERSITY PRESS

OXFORD
UNIVERSITY PRESS

Great Clarendon Street, Oxford, OX2 6DP,
United Kingdom

Oxford University Press is a department of the University of Oxford.
It furthers the University's objective of excellence in research, scholarship,
and education by publishing worldwide. Oxford is a registered trade mark of
Oxford University Press in the UK and in certain other countries

The front cover image shows a mormyrid, capable of electrolocation, and the back cover image
shows beach spawning capelin

First Edition published in 2018

Impression: 4

Published in the United States of America by Oxford University Press
198 Madison Avenue, New York, NY 10016, United States of America

British Library Cataloguing in Publication Data

Data available

Library of Congress Control Number: 2017943712

ISBN 978–0–19–878555–2 (hbk.)
ISBN 978–0–19–878556–9 (pbk.)

DOI 10.1093/oso/9780198785552.001.0001

Printed and bound by
CPI Group (UK) Ltd, Croydon, CR0 4YY

Preface

St. John's, with its historical associations in the Newfoundland fishery, has been a stimulating location in which to develop an overview of fish biology. Since the arrival of Europeans in Newfoundland in the fifteenth century, fishing has been a major economic activity in the province. More recently St. John's has become an important centre in the Northern Hemisphere for the study of fish biology and fisheries science. These studies involve a number of university departments and institutions as well as Newfoundland and Canadian government research units. This has produced a diversity of teaching and research programmes in fish biology, fisheries and aquaculture, attracting undergraduate and graduate students from within the province and internationally.

The aim of this book is to provide an introductory overview of the functional biology of fish and how it may be influenced by the contrasting habitat conditions within the aquatic environment. The book includes advances in comparative animal physiology made since the mid-twentieth century, which have greatly influenced our understanding of fish function or have generated questions on this subject, still to be resolved. Fish taxa represent the largest number of vertebrates, with over 30,000 extant species. Much of our knowledge, apart from taxonomy and habitat descriptions, has been based on relatively few of these species, usually those which live in fresh water and/or are of commercial interest. Unfortunately there has also been a tendency to base interpretation of fish physiology on that of mammalian systems, as well as to rely on a few type species of fish. We have attempted to redress the balance by using examples of fish from a wide range of species and habitats, emphasizing diversity as well as recognizing shared attributes with other vertebrates.

We have limited this book to a single volume appropriate to an introduction to the subject. We have also avoided a partitioning of chapters between a large number of specialist authors which has been a trend in the case of fish physiology texts. Although such 'sectionalized' books can be advantageous to advanced readers, students new to the subject can benefit from a more concise and consistent approach, written by fewer authors. Our objective is to provide a comprehensive introduction to each topic, together with detailed references leading to the roots of current knowledge. To provide useful background some topics are prefaced by a brief account of basic aspects of relevant more general cellular, anatomical or physiological biology. The book will be suitable for undergraduate and graduate students in fish biology, fisheries and aquaculture, as well as for more general readers interested in fish.

In a world in which security of food supplies will become an increasingly important issue, a sound development of aquaculture and sustainability of fisheries will be integral to planning. This implies a comprehensive understanding of factors influencing fish survival, including their basic physiology in their natural habitats and in captivity.

We wish to thank the many participants in informal discussions and more formal seminars in St. John's and at conferences elsewhere, often emphasizing how much remains to be researched in this subject as well as what has been accomplished. Members of the Biology Department of Memorial University have given courteous assistance and advice. We particularly thank Dr Paul Marino, Shena Quinton, Roy Ficken, Peter Earle and Kerri Green, as well as Mike Hatcher, Richard Hu, Julia Trahey and Graham Hillier at the Ocean Sciences Centre. We have benefited from the specific help and cooperation of Dr G. Allen, Dr A. Bal, Dr Vern Barber, Dr Eastman, Dr Fryer, the family of Dr Greenwood, Dr R. Khan, Prof Dr Schmitt and Dr Stevens. Many institutions have helped and cooperated with information retrieval, and we thank Joe Tomelleri for the 'slender chub' photograph, Dr Craig Purchase for the capelin photograph and Catherine Burton for the 'garden eels'. We appreciated Rudie Kuiter's advice and patience in accessing photographs of coral-reef fish. We thank Nancy Simmons for information on copyright. We are grateful to Ian Sherman, Lucy Nash and Bethany Kershaw of Oxford University Press for their help, guidance and patience. We are also grateful to Narmatha Vaithiyanathan for organizing the final assembly of the material. Finally, we acknowledge the many students whose curiosity and interesting questions over the years prompted us to develop this book which we hope will be helpful to many people. We welcome comments and corrections.

Contents

CHAPTER 1

Fish diversity

Classification and variety

1.1 Summary

Fish have a long evolutionary history and have successfully colonized a large range of habitats including deep oceans and shallow ponds.

Formal classifications of the very numerous fish species are subject to adjustment as new information accrues. The major divergences separate the extant jawless fish from the jawed fish, and 'cartilaginous' fish from those with bony skeletons. A few extant bony fish species are regarded as 'relict' (or primitive), remnants of ancient lineages.

Fish diversity parallels the numerous habitat niches they occupy, with size attained from minute to enormous, shapes from spherical to flat to elongated, and other attributes, both morphological and functional, enable 'lifestyles' mainly aquatic but, for a few, even terrestrial.

1.2 Introduction

Fish are aquatic vertebrates with gills and fins rather than lungs and legs, although some fish lack paired fins and others have lungs. In geologic time they preceded the other vertebrate groups which are derived from them.

Although a few fish spend quite a lot of time on land, almost all fish species are still recognizably characterized by their need to be in water, with very special adaptations to that particular environment with its density, low oxygen and high thermal capacity reducing rapid temperature changes. Many fish species are also high on the trophic level, being predaceous even as juveniles, and for many active carnivores this means rapid movement and a fairly complex sensory system. Fish represent huge diversity in shape, size, colour and habit. The distribution of particular genera may be widespread, either naturally (deep-sea grenadiers) or by introduction (trout, carp) or severely limited (desert pupfish, cave-fish), and the numbers of individuals in different species may also be extremely numerous (some marine species) or very few for species in specialized habitats such as caves.

Among the huge array of multicellular animals classified as such in the Animal Kingdom a dichotomy of structure is recognized by the number of embryonic cell layers, most having three, being therefore triploblasts (Appendix 1.1). In many animals including the vertebrates the middle embryonic cell layer develops a cavity (the coelom, Appendix 1.1) which allows gut movement. The formal classification of animals assigns each species to a series of groups or taxa, the largest or most inclusive being a Phylum. Each Phylum is usually divided into Classes, and each Class into Orders. Orders are subdivided into Families, each of which contains Genera. Any individual genus contains one or more species. Within this basic system may be many subunits; for example, there can be Subclasses, Infraclasses and so on to fit the complicated relationships within the groups. Where there are large numbers of similar species the term 'species flock' has been utilized; the definition of a species usually depends on morphological differences, and taxonomists set up a 'type' specimen for reference.

Structurally fish share with other vertebrates a basic body plan of bilateral symmetry, with a dorsal supporting structure the notochord (hence Phylum Chordata) as a precursor to the series of harder vertebrae. All vertebrates are metamerically segmented, which means the serial repetition of body

Essential Fish Biology: Diversity, Structure and Function. Derek Burton & Margaret Burton.
© Derek Burton & Margaret Burton 2018. Published 2018 by Oxford University Press.
DOI 10.1093/oso/9780198785552.001.0001

parts, a system which facilitates locomotion in the many animals which have this organization. The segmentation is not as obvious as in, for example, earthworms or millipedes, but its existence can be seen in the vertebrae, the 10, perhaps +1 (11.8), pairs of cranial nerves, the gills and gill supports (gill arches), the spinal nerves and many of the blood vessels supplying the body wall.

Physiologically fish share with other vertebrates a high degree of activity except under extreme conditions and a capacity to regulate the composition of body fluids, though temperature regulation is limited for most fish to seeking optimum conditions.

Currently fish are by far the most numerous vertebrates, both as individuals and species, representing roughly 50 per cent of all vertebrate species now alive (extant). It is generally agreed that there are in excess of 20,000 extant fish species; Nelson (1994) recognized about 25,000 species, and this number is augmented as new species are described, differentiated and/or discovered, for example from the deeper oceans or from big river systems such as the Amazon/Rio Negro complex. Thus Nelson (2006) reports nearly 28,000 species, Berra (2007) more than 29,000 and Nelson et al. (2016) state 32,000.

Fish are the oldest group of vertebrates, with an enormous diversity of first, jawless, and subsequently, jawed species showing adaptive radiation in aquatic habitats and prolific speciation as far back as the Ordovician Era for jawless fish (480–390 million years ago, mya). The Devonan Era (350–300 mya) indeed has been termed the 'Age of Fishes' with numerous species particularly associated with freshwater during that time period (Romer 1955). The jawless fish are now reduced to two relict groups (hagfish and lampreys), with a tendency towards a parasitic habit. Recently the hagfish have been detached from the vertebrate classification altogether (Nelson et al. 2016), and are regarded as a sister group.

1.3 Classification

The large number of fish species presents a challenge in identification and classification; the following provides a brief overview. To aid orientation, all species referred to are subsequently listed separately in the Index. Appendix 1.2 summarizes widely used classifications from the 1970s ('old') and the more recent ('new', Nelson 2006). The most recent radical revision (Nelson et al. 2016) is dealt with in the text.

As with other organisms, the species to which an individual is assigned is best characterized and named by a Latin-based binomial to avoid muddle (Appendix 1.3). Each species belongs to a genus (always capitalized) such as 'Gadus' for cod-like fish. The individual species also has a 'specific' name which, with the generic name, provides the 'binomial'. Thus Atlantic cod is 'Gadus morhua'. The naming convention is to italicize the constituents of a binomial, and if the genus is used repeatedly then it can be reduced to a capital, such as *G. morhua*. The strict naming convention, generally at first use in a publication, and notably in the title, is to include the name of the first person (Linnaeus and post-Linnaean) describing the species, immediately after the Latin binomial (Appendix 1.4).

For identification purposes, particularly for live fish, superficial characteristics may be very useful. Hence the presence of a small, but readily recognized, adipose fin between the dorsal fin and the tail may indicate a fish of the 'salmon–trout' grouping in comparison to perch in the same habitat although adipose fins are also found on some catfish and characins, as well as in five other orders (Reimchen and Temple 2004). Similarly, fish shape can be indicative; whereas the skewed eyes and flat shape of 'flatfish' together place these fish, the slim shape of eels and barracuda are only helpful. Similarly although several gill slits on each side, or ventrally, place fish in the 'shark–dogfish–ray' category, the presence of male claspers shared by this group and the chimaeras (rabbit/ratfish) has obvious limitations if a female is caught. Fortunately, however, the general appearance of both groups is so different from other living fish that confusion is actually most unlikely.

Because there are now so many different kinds of fish in addition to the very large number of extinct forms, formal classification of fish (by taxonomists) has presented a formidable task as new information becomes available. Revisions are ongoing at various levels; some species have undergone renaming or reclassification fairly often. Rainbow trout which is apparently more a salmon than a trout has shifted from *Salmo irideus* to *Salmo gairdneri* to *Oncorhynchus mykiss*, placing it with the Pacific salmon such as chum and sockeye. At another taxonomic level some of the 'primitive' fish like gar, the bowfin, the lungfish and the coelacanth may have been shifted from one major grouping to another, with the lungfish and coelacanth now placed closer to the terrestrial vertebrates than to most other fish. With this degree of change in groupings, any classification scheme may be regarded as temporary.

Some older nomenclatures have been incorporated in newer schemes but others have been lost, albeit sometimes temporarily. Thus 'Chondrostei' is retained for sturgeon-like fish as a Subclass but Holostei is no longer always applied as a formal taxon for *Amia* the bowfin (Nelson 2006), although still used and recognized informally and reinstated as a Subclass by Nelson et al. (2016). The taxon 'Isospondyli' (meaning similar vertebrae), used in the nineteenth century for soft-finned teleosts (such as herring, trout) which also shared some basic features like undifferentiated anterior vertebrae, air bladders (if present) linked to the gut (physostomatous), posteriorly placed pelvic fins, was superseded early in the twentieth century (Boulenger 1904), although still used by Gregory (1933, reprinted 1959).

During the latter half of the twentieth century, fish were classified as in Figure 1.1 and some of this nomenclature still stands, or is otherwise recognized as useful, but some recent changes to fish classification have been profound (Nelson 2006; Nelson et al. 2016; Figure 1.2).

The status of hagfish (1.3.1, Figure 1.2) is under consideration; Nelson et al. (2016) removed them from the vertebrates, a major adjustment. Hagfish ancestors are only represented so far by a single possible fossil hagfish (Bardack 1998), with a further type described more recently (Janvier and Sansom 2015). Even more difficult to assess is the situation with the conodonts and their possible relationship within the chordates (Aldridge and Donoghue 1998; Nelson et al. 2016). Only very recently have entire fossils of conodonts been available; previously these animals were known only from their teeth, which were however abundant and perhaps reminiscent of hagfish (Aldridge and Donoghue 1998).

Phylum Chordata	All animals with a notochord at some stage in their life cycle
Subphylum Vertebrata (Craniata)	All animals with a vertebral column
Superclass Agnatha	Jawless fish
Superclass Gnathostomata	Jawed vertebrates
Grade Pisces	Jawed fish
Class Chondrichthyes	Cartilaginous fish, but not sturgeons
Class Osteichthyes	Bony fish

Figure 1.1 Fish classification based on Nelson (1976).

1.3.1 Jawless fish

Most considerations of fish diversity divide fish into those with jaws and those without, the latter being 'Agnatha' (Figure 1.1), many of which are now extinct, with two living groups the hagfish and lampreys. In 2016 Nelson and colleagues separated the hagfish from the lampreys and all other extant fish (Figure 1.2) on the basis of lack of any vertebral elements (arcualia, 3.5.2) in hagfish. However as hagfish are noted for the modification of various structures, including eyes, the lack of vertebrae may be secondary.

Jawless fish, such as osteostracomorphs with bony head shields (Figure 1.3), have been found in deposits dating back to the Ordovician Era, with jawed fish gradually more prevalent into the Devonian. The term 'ostracoderms' was at one time a collective term used for early agnatha (Nelson

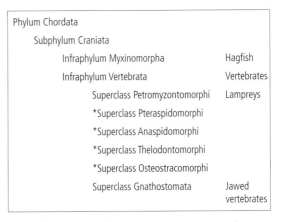

Figure 1.2 A new classification for hagfish and lampreys after Nelson et al. (2016).

*Extinct.

Figure 1.3 Extinct jawless fish: Left, an osteostracomorph with head shield and heterocercal tail. Right, an anaspidomorph with hypocercal tail. Both reproduced with permission, J.S. Nelson (1976), *Fishes of the World*, published by John Wiley and Sons. © 1976 by John Wiley and Sons Inc.

1976), and is still used informally for extinct agnatha whereas the term cyclostomes (literally 'round-mouth') is reserved for the persistent agnatha which lack exoskeletons. Cyclostomes lack cellular bone which was, however, present in some extinct agnatha; 'ostracoderms' typically had bony head shields, although the anaspids did not.

The early jawless fish had either heterocercal (longer upper lobe) or hypocercal (longer lower lobe) tails. The former were found in osteostracomorphs, the latter in anaspidomorphs which lacked the head shield (Figure 1.3).

Currently only a few genera of agnatha persist as the marine hagfish and lampreys which can be migratory (anadromous, with freshwater larvae) or freshwater (*Mordacia praecox*; Allen et al. 2002). The mouths are toothed in lampreys, and can attach to the skin of their prey; hagfish have teeth on the tongue and palate (Nelson 1976). Adult cyclostomes tend to attack jawed fish and whereas hagfish are regarded as 'scavengers', some lamprey species are considered true parasites, undesirable when they thrive on local commercially important lake or river teleosts. Hagfish are also known as slime-hags as they can secrete huge amount of mucus (Fudge et al. 2005).

Superficially, cyclostomes, which are elongated (Figures 1.4 and 1.5), are similar to eels in size and shape but have uncovered gill openings (7 bilateral in lampreys, 5–16 bilateral in the hagfish *Epitretus* reduced to a common aperture in the hagfish *Myxine*, generally one each side but two on the left side in *Notomyxine*). Cyclostomes do not have paired fins.

Larval lampreys are interesting. Termed 'ammocoetes', they are reminiscent of adult amphioxus, the protochordate (*Branchiostoma lanceolatum*), with bilateral symmetry and a filter-feeding pharynx.

With very specialized habits, for example parasitism, the cyclostomes are somewhat difficult to interpret morphologically and physiologically in relation to extant jawed fish. Atz (1972), reviewing Hardisty

and Potter (1971), nevertheless believed that the most interesting question about lampreys is to what extent they reflect the structure of the earliest vertebrates.

Atz (1972) observed that studies on lampreys tended to be skewed towards control of parasitic species, and hoped that a more balanced set of work would ensue. However, it is still difficult to incorporate information about the structure and function

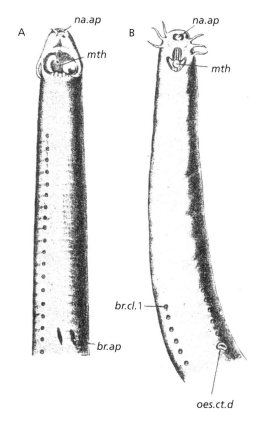

Figure 1.4 Hagfish. A. *Myxine* (the Atlantic hagfish) with two ventral gill openings (br.ap). B. *Epitretus* (*Bdellostoma*) with 6 (5–16 possible) bilateral openings (br.cl 1, etc.). Both after Bridge (1904), br.ap = left external branchial aperture, mth = mouth, na.ap = nasal aperture, oes.ct.d = oesophageal—cutaneous duct. The small openings on the left of A (as a row) are mucous glands.

of the extant jawless fish in comparison to the extant jawed fish; for example, the hagfish group has blood isosmotic (7.5.5) with seawater (Nelson 1976) and a very unusual blood system (Satchell 1986) with an additional, portal, heart (5.9) but no distinctive ammocoete larvae, all points distinguishing the hagfish markedly from the lampreys, not just from the extant jawed fish, contributing towards a justification for the recent detachment of hagfish from the vertebrates (Nelson et al. 2016).

Monographs and bibliographies on lampreys are available (Hardisty 2006; Hardisty and Potter 1971, 1982; Beamish et al. 1989). Recent texts on hagfish include Jørgensen et al. (1998) and Edwards and Gross (2015). Hardisty (1979) and Beamish et al. (1989) deal with cyclostomes (i.e. lampreys and hagfish).

The survival of the cyclostomes into modern time is attributed to their specialized niches, where they are not in direct competition with either the cartilaginous or bony fish. The lampreys actually spend much of their life cycle as the filter-feeding and benthic ammocoete stage, a phase quite different from the juvenile stages of either cartilaginous or bony fish (Ilves and Randall 2007).

1.3.2 Jawed fish

For taxonomic purposes, most extant jawed fish have, until fairly recently, been divided into bony or cartilaginous fish (Appendix 1.2a), but although this still works quite well, adjustments continue to be made. There is a small group of 'relict' fish (e.g. the coelacanth, lungfish, the bichir, etc.; see 1.3.6), which show a variety of 'primitive' (Janvier 2007) or plesiomorphic characteristics, indicating their status as remnants of very ancient lineages. Classifying the relict fish has apparently represented a challenge, as there has not always been a clear pattern in assigning them. While the Class Chondrichthyes is still

utilized for most cartilaginous fish (i.e. chimaeras (rat or rabbit fish) and the sharks, dogfish and rays), there has been (Nelson 1994, 2006) a re-evaluation of the other groupings so that many bony fish are now assigned to a new Class Actinopterygii with an additional Class Sarcopterygii (Appendix 1.2b) accommodating some, but not all, relict fish species.

It is convenient to think in terms of the 'old' classification with two Classes, Chondrichthyes (cartilaginous fish) and Osteichthyes (bony fish), and a newer classification with three Classes, Chondrichthyes (cartilaginous fish, no change from the old classification), Actinopterygii (most bony fish) and Sarcopterygii (bony fish species close to the terrestrial vertebrates).

Recently (Nelson et al. 2016) a further adjustment has been made, making both Sarcopterygii and Actinopterygii Subclasses of Class Osteichthyes (Figure 1.6).

The classification of cartilaginous fish has not undergone much change: cartilaginous fish are assigned to the Class Chondrichthyes in all schemes referred to (i.e. 'old', 'new' and Nelson et al. (2016)), although some fish like sturgeon with a large amount of cartilage are excluded from this category based on other features.

The survival of the cartilaginous fish, in spite of the success of the bony fish, has been attributed to their size and speed in open waters, habitats where manoeuvrability is not particularly advantageous (Ilves and Randall 2007). The fact that the young are quite large when first independent is thought to contribute to their persistent success.

1.3.3 Class Chondrichthyes and Class Osteichthyes

Class Chondrichthyes (old and new classifications) comprises fish with cartilaginous skeletons, and the following: soft fin-rays (unsegmented), no

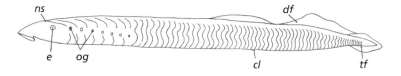

Figure 1.5 The cyclostome *Petyromyzon*, a lamprey, with seven external bilateral gill openings. After de Beer (1951). cl = urinogenital aperture, df = dorsal fin, e = eye, ns = nostril, og = openings of the gill pouches, tf = tail-fin.

Grade *Placodermomorphi

Grade Chondrichthyomorphi

    Class Chondrichthyes   Cartilaginous fish

Grade Teleostomi

    *Class Acanthodii

    Class Osteichthyes   Bony fish and tetrapods

        Subclass Sarcopterygii   Bony fish and tetrapods

            Infraclass Actinistia   Coelacanths

            Infraclass Dipnomorpha   Lungfish

            Infraclass Tetrapoda   Tetrapods

        Subclass Actinopterygii

            Infraclass Cladistia

                 Order Polypteriformes   Bichirs

            Infraclass Chondrostei

                 Order Acipenseriformes   Sturgeons, paddlefish

Unranked Neopterygii

            Infraclass Holostei   Gars, bowfin

            Division Ginglymodi   Gars

            Division Halecomorpha   Bowfin

            Division Teleostomorpha   Teleosts

Figure 1.6 A new classification for bony fish, after Nelson et al. 2016.

*Extinct.

swim bladder or lung; internal fertilization, male has claspers (Figures 1.8 and 1.9); high blood concentration of urea (a few exceptions); spiral valve in intestine; heterocercal tail. These fish are almost all marine and fairly large. The Class Chondrichthyes is routinely divided into two Subclasses, Holocephali and Elasmobranchi, with very unequal size (number of genera and species) composition (Figure 1.7). The elasmobranchs have special features like the ampullae of Lorenzini (12.9.3) which facilitate prey detection via electroreceptors, and some (electric rays) are capable of discharging high-voltage pulses (13.8).

The classification of bony fish has been subject to fairly frequent change and rearrangement. The 'bony fish', which may actually include fish with a large amount of cartilage but otherwise very different from the Chondrichthyes, were until recently all assigned to a single Class Osteichthyes, subdivided into four Subclasses (Table 1.1, Appendix 1.2a). Class Osteichthyes

Subclass Holocephali: top jaw fused to neurocranium (holostylic suspension, 3.5.4, Table 3.4), gill-cover/operculum, teeth as plates, no stomach, male has three claspers (two pelvic, one frontal/cephalic, (Fig. 1.8).

6 extant genera e.g. rabbitfish/ratfish (chimaeras). All marine, and medium-sized.

Subclass Elasmobranchi: top jaw not fused to neurocranium (amphistylic or hyostylic suspension, see Chapter 3, Table 3.4) and the following: 5–7 gill openings each side (Figs. 1.9, 1.10), spiracles (Fig. 1.10), numerous teeth; male has two pelvic claspers.(Fig. 1.9).

145 extant genera e.g. dogfish/sharks (Fig. 1.9), dorsally flattened skates and rays (Fig. 1.10).

Figure 1.7 The two Subclasses of the Class Chondrichthyes.

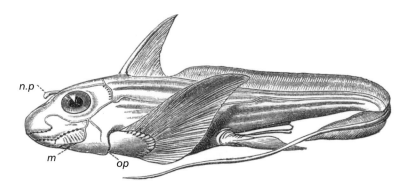

Figure 1.8 Male chimaera (*Chimaera monstrosa*). np = frontal (cephalic) clasper; m = mouth, op = operculum. Note pelvic claspers above pelvic fin. After Bridge 1904.

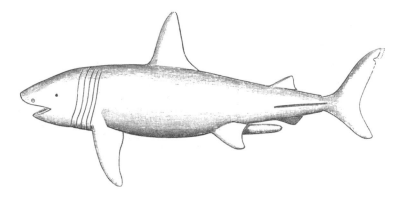

Figure 1.9 Male *Cetorhinus maximus*, the basking shark, showing gill slits and pelvic clasper beyond pelvic fin (after Bridge 1904).

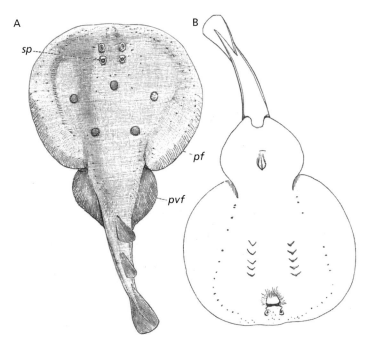

Figure 1.10 Female *Torpedo*, electric ray: A. (left) Dorsal view, sp = spiracle; pf = pectoral fin; pvf = pelvic fin. B. (right) Ventral view showing gill slits. (Both after Bridge 1904).

Table 1.1 Class Osteichthyes (after Nelson 1976).

Subclass	Known as	Extant fish habitat	Extant genera (#)
Dipneusti (Dipnoi)	Lungfish	Freshwater	3
Crossopterygii	Coelacanths	Marine	1
Brachiopterygii	Bichirs, reedfish	Freshwater	2
Actinopterygii	Rayfins	Varied	4100*

*approximate.

in the old classification were fish with (usually) bony skeletons, gill-covers and males lacking claspers.

Moyle and Cech (1996) retain the designation 'Class Osteichthyes' for all bony fish in preference to the newer classification (Nelson 1994) in which the two first Subclasses (Table 1.1, Appendix 1.2b) are assigned to a new Class Sarcopterygii, while the remaining two are subsumed into a new Class Actinopyterygii, retaining the name of the last subclass in the old classification, but having a wider composition.

1.3.4 Class Sarcopterygii

Fish of the Class Sarcopterygii, in the classification of Nelson (2006), have thick fleshy fins or limbs, with two Subclasses (Figure 1.11), and the currently unranked Tetrapodomorpha with all tetrapods. This classification reflects opinions on the closeness of the lobefins, represented by one extant genus *Latimeria* (the coelacanth) and the lungfish, the Australian *Neoceratodus*, the African *Protopterus* and the South American *Lepidosiren* to the tetrapods (including mammals).

Extant fish in this Class are regarded as relict or plesiomorphic, as are some of the fish in the new Class Actinopterygii (1.3.5).

Subclass Coelacanthimorpha (Actinistia) containing the extant coelacanth,

Subclass Dipnotetrapodomorpha = Sub-Class Dipneusti (Dipnoi), containing the lungfish.

Figure 1.11 The two subclasses of the Class Sarcopterygii (after Nelson 2006).

1.3.5 Class Actinopterygii

Class Actinopterygii in the classification of Nelson (2006) has two Subclasses (Figure 1.12, Appendix 1.2b).

1.3.6 The relict fish

The four non-teleost orders in Figure 1.12 are regarded as 'relict' fish, as are other non-teleost bony fish, the coelacanth and the lungfish. Whereas the non-teleost relict fish are characterized by retaining 'primitive' or plesiomorphoic features (Table 1.2), the teleost fish (1.3.7) can be characterized by a lack of certain features, like no spiral valves in the intestine, and the acquisition of particular traits including changes in the jaws, fins and scales.

Extant relict fish comprise the coelacanth and three genera of lungfish, the polypterids (bichir and reedfish), the chondrosteans (acipenseriform sturgeons and paddlefish) and the holosteans (semionotiform gar and amiiform bowfin). The relict fish are called 'archaic' in older literature (e.g. Hyman 1942) and the expression 'primitive' persists (McKenzie et al. 2007). Distinguishing the more 'primitive' or plesiomorphic relict bony fish are a varying set of features which may include complex scales of heavy and distinctive structure, lungs which may (lungfish) or may not (coelacanth) function in respiration, heterocercal tails and larvae with external gills (Table 1.2). Additionally, sturgeons of the Acipenseriformes may have a small spiracle, which is also present in the paddlefish *Polyodon* (Bridge 1904). It has been described in larval lungfish, is vestigial in the coelacanth but present (Fig. 1.15) and functional in *Polypterus* (Budgett 1903). The paired fin structure of the sarcopterygians is distinctive and regarded as presaging tetrapod limbs.

Subclass Chondrostei, with cartilage in the adult skeletons, and two extant orders, Order Polypteriformes, e.g. the bichir *Polypterus*, and Order Acipenseriformes e.g. sturgeon, paddlefish. The cartilage may undergo ossification in older fish.

Subclass Neopterygii, with teleosts in a separate Division, and two non-teleost orders, Order Semionotiformes e.g. *Lepisosteus*, the gar, and Order Amiiformes, i.e. *Amia calva*, the bowfin.

Figure 1.12 The two subclasses and non-teleost extant orders of the Class Actinopterygii (after Nelson 2006).

The continued survival of the primitive or relict bony fish may be attributed to their specialized niches and perhaps their capacity to tolerate hypoxia (Ilves and Randall 2007).

Distribution of relict fish

With the exception of *Latimeria*, the coelacanth, which is marine, all extant relict bony fish are anadromous (e.g. some sturgeon such as *Acipenser transmontanus*, Scott and Crossman 1973) or wholly freshwater (e.g. lake sturgeon *A. fulvescens*) in current distribution. The lungfish are in warm rivers or lakes and the two genera of Polypteriformes are in African rivers. Sturgeons (four genera) are found in Northern Europe and North America and/or the adjacent oceans. Paddlefish (two genera) are found (one species *Polyodon spathula*) in the Mississippi River (North America) historically as far north as the Ohio River and even Lake Eirie (Trautman 1957), and (one species *Psephurus gladius*) in the Yangtze River (China). The gar (one genus *Lepisosteus*, seven species) are found in America as far south as Costa Rica (Nelson 1976) and bowfin (*Amia calva*) are North American, presently restricted to the Eastern region, though fossils are reported from a much larger area including Europe and Asia. The coelacanth and lungfish, sometimes called lobefins, were previously far more abundant. Thomson (1969) provides an interesting summary of their prevalence in the fossil record. Today these fish survive as the coelacanth (Crossopterygii; one genus) and the lungfish (Dipnoi; three genera). In the Devonian, Thomson lists 27 crossopterygian genera and 14 in the later

Triassic. Similarly he lists 20 dipnoans (lungfish) in the Devonian and 5 in the Triassic.

Lungs and air bladders in relict fish

Lungs or air bladders occur in relict fish, indicating an ancient dependence on air breathing. The distinction between 'lungs' and air bladders depends on features such as their position and the degree of internal subdivision. The importance of these structures in gas exchange varies widely within the relict fish, some of which are found in warm freshwater which might be oxygen-depleted. The single 'lung' of the extant coelacanth *Latimeria* is fat-filled while some of the extinct species are reported to have had a 'calcified lung' (Brito et al. 2010). The lungfish have dorsal lungs, either bi-lobed or paired, with some subdivision. The bi-lobed single lung of *Neoceratodus* is derived from the right side (Thomson 1969). *Lepidosiren* and *Protopterus* have paired dorsal lungs (Romer 1955). While the sturgeons have a single undivided dorsal air bladder the gar and bowfin have a bi-lobed dorsal bladder with some internal division, and *Polypterus* has simple ventral lungs. Romer's (1955) view of the fish air bladder is that it derives from lungs; Ilves and Randall (2007) consider the discussion of which evolved first to be 'superfluous' in that initially the air bladder would have had both buoyancy and oxygen-storage capacity, as well as sound transmission.

The coelacanth

The coelacanth (*Latimeria chalumnae*) was discovered, as a single specimen, in 1938, off the coast of South Africa. The discovery was hailed as a 'living

Table 1.2 Plesiomorphic features of extant relict fish, Classes according to Nelson (2006).

Fish	'Lungs'	Ganoid scales	Heterocercal tail	Spiral valve	Larvae with external gills
Class Sarcopterygii					
Coelacanth (1 genus)	Vestigial			+	
Lungfish (3 genera)	Present			+	+
Class Actinopterygii					
Polypteriformes (2 genera)	Present	+		+	+
Acipenseriformes (6 genera)	Bladder		+	+	
Semionotiformes (1–2 genera)	Bi-lobed bladder	+	Abbreviate	+	
Amiiformes (1 genus)	Bi-lobed bladder			Vestigial	

fossil' as the group had been thought extinct since the Cretaceous. The Second World War (1939–45) precluded any active searches for more examples and it was not until 1951 that a small population was located off the Comores Islands, further north than the original discovery location. Recently additional populations, including one from Indonesian waters (a new species *L. menandoensis*) and others, off South Africa, have been reported. The extant coelacanths are remarkably different in several respects from all other living fish. This genus combines the outward appearance of a bony fish with heavy but flat scales and unusual fins, having peg-like supports terminating in spatula-like fins (Figure 1.13), with several interesting internal features including a non-respiratory lung used as a fat store and a hinged skull (with an 'intracranial joint', Thomson 1967, a characteristic of the crossopterygians). It has the largest known fish eggs (with internal fertilization), and high blood urea, a feature usually associated with chondrichthyans. It also has a rectal gland, a feature of extant chondrichthyans. The paired fins are somewhat reminiscent of *Neoceratodus*, the extant Australian lungfish (Figure 1.14). While the reproductive system and the high blood urea, and perhaps the structure of the pituitary (Lagios 1975), are similar to the elasmobranchs, the tail structure is not heterocercal but symmetrical, although unusual in comparison to all other extant forms, with a projecting central lobe. *Latimeria* has a spiral valve in the intestine, a feature it shares both with the Chondrichthyans and with all the other relict fish (Table 1.2) although that of *Amia calva* is vestigial. The spiracle in *Latimeria* is reduced and non-functional, a blind pouch (Thomson 1969).

Lungfish

Extant lungfish include three geographically disjunct genera, *Neoceratodus* (Figure 1.14) in Australia, *Lepidosiren* in South America and *Protopterus* in Africa. These freshwater fish have functional lungs and unusual elongated paired fins but, apart from the caudal fin, no other fins. They also have spiral valves in the intestine and the larvae of *Protopterus* and *Lepidosiren* have external gills (Thomson 1969). The scale structure varies from heavy (reminiscent of rhomboid scales, 2.4) for *Neoceratodus* to much more delicate for *Lepidosiren*. *Neoceratodus* (one species) is scarce with a very limited distribution, while *Lepidosiren* (one species) and *Protopterus* (four species) have a wider distribution and are more common.

Bichir and reedfish

Extant species of the Polypteriformes, the bichir *Polypterus* (ten species) and the reedfish *Calamoichthyes* (previously *Calamichthys*) or *Erpetoichthyes* (one species) are found in African rivers. These fish are elongated with ganoid scales (2.4). They have ventro-lateral bi-lobed lungs (6.6.1) opening to the foregut and a spiral valve in the intestine. Unlike the lungfish, the bichir has an anal fin and an array of dorsal finlets each with one anterior spine to which the fin-rays are attached; the fins are smaller in the reedfish (Figure 1.15). Like the lungfish *Protopterus* and *Lepidosiren*, the larvae have external gills. The bichir and reedfish were previously assigned to the Subclass Brachiopterygii, and the

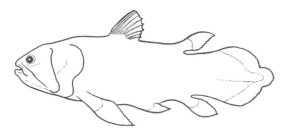

Figure 1.13 *Latimeria*, the coelacanth. Reproduced with permission, J.S. Nelson (1976). *Fishes of the World*, published by John Wiley and Sons. © 1976 by John Wiley and Sons Inc.

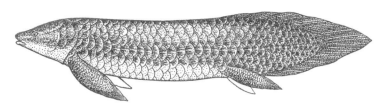

Figure 1.14 *Neoceratodus*, the Australian lungfish (after Bridge 1904).

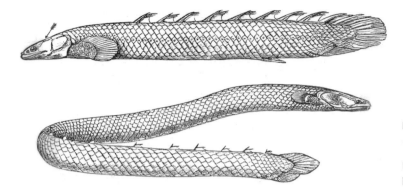

Figure 1.15 Polypteriformes: (top). *Polypterus*, the bichir. (bottom). *Calam(o)ichthyes*, the reedfish. Both after Bridge (1904). The arrow indicates the position of the left spiracle.

Order Polypteriformes, but Nelson (2006) placed them with the sturgeons and paddlefish in a Subclass Chondrostei (Appendix 1.2b). In 2016, Nelson and colleagues placed the Polypteriformes in an Infraclass Cladistia of the Subclass Actinopyterygii.

Chondrosteans

Extant chondrosteans (Subclass Chondrostei, previously a Superorder), so-called because they have a preponderance of cartilage in their endoskeleton, include the sturgeons and paddlefish (Figure 1.16). These also share the somewhat surprising feature of a heterocercal tail with *Lepisosteus* (previously *Lepidosteus*) the gar (Figure 1.17), which in turn shares the plesiomorphic ganoid scale characteristic with the Polypteriformes. Sturgeons, paddlefish and gar all have spiral valves in their intestines. A small spiracle, covered by the operculum, is present in most sturgeons and was reported for the paddlefish also, as well as in *Polypterus* (Bridge 1904).

Paddlefish currently occur in two widely separate locations, China and North America, and have considerable similarity in body form with a long flat rostrum (snout), hence the name. However, they are widely divergent in habit; the Chinese genus *Psepherus* has a protrusible mouth and is piscivorous (fish eating) whereas the North American *Polyodon* has a non-protrusible mouth, is planktivorous (feeds on small floating organisms, the plankton) and (6.5.11) ram ventilates (Berra 2007).

Holosteans

In some older classifications, such as in Pearson and Ball (1981), the gar and bowfin (Figure 1.17) were placed together in a Superorder Holostei and together were referred to as holosteans. The gar have also been placed in a Suborder Lepisostoidei, also known as 'Ginglymodi' (Nelson 1976). Nelson and co-workers (2016) now recognize holosteans as a taxon, ranked as an Infraclass (Figure 1.6).

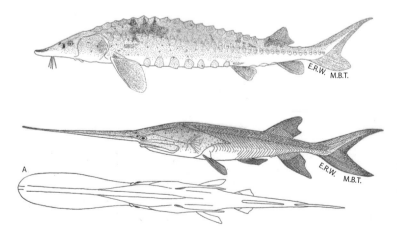

Figure 1.16 Chondrosteans: (top) *Acipenser fulvescens*, a sturgeon; (middle) *Polyodon*, the paddlefish, with ventral view (bottom). Reproduced, with permission, from M.B. Trautman (1957), *The Fishes of Ohio*, Ohio State University Press.

These 'relict' bony fish have a mixture of 'primitive' (plesiomorphic) features and more derived characteristics, the sum of which have tended towards their classification at the root of the bony fish divergence.

The gar (*Lepisosteus*), found in North American rivers, superficially resembles the pike, and is sometimes referred to as the gar-pike. It is externally distinguished by an 'abbreviate' (Scott and Crossman 1973) heterocercal tail and heavy (ganoid) scales (Figure 1.17). The upper jaw of the alligator gar (*L. spatula* or *Attractosteus spatula*) has a double row of teeth (Anon. 2012); these fish can grow to a large size (to 9 feet, approximately 3 m, Greenwood 1975), becoming formidable predators. Also found in North American rivers is *Amia calva*, the bowfin (Figure 1.17). It is usually placed close to the teleosts; it does not have a heterocercal tail but does have a vestigial spiral valve and a heavy skull structure.

1.3.7 Teleosts

The non-relict extant bony fish of the Class Actinopterygii are usually referred to as 'teleosts'. Modern fish assigned to this group have bony skeletons as adults and lack certain features (no heterocercal tails, no spiral valve in the intestine), although the considerable diversity in this group does not necessarily make instant recognition easy. The taxon has varied in the level (recently a 'Division' Teleostomorpha, Figure 1.6, and 'Subdivision' Teleostei, Figure 1.18) to which it is assigned, but the grouping itself remains stable. Fish in this group are all the extant jawed fish with the exception of the Chondrichthyes and the 'relict' fish (1.3.6). Exceptionally, the holostean *Amia* might be included as a teleost. Teleost fish are currently very successful, having undergone adaptive radiation within the oceans, the coastal seas and freshwater lakes and rivers, within caves and even venturing onto land in some specialized niches.

Particular taxa of the teleosts have specialized in certain habitats or have evolved novel physiological assets, thus the Ostariophysi and Clupeomorpha, which Nelson (2006) placed together in a new taxon, the Subdivision Ostarioclupeomorpha (Table 1.3), have superior hearing with links from the air bladder to the ear enabling sound amplification. However, the Ostariophysi are dominant in many freshwater systems whereas the Clupeomorpha are principally marine or anadromous. The deeper-bodied, highly coloured coral reef fish are usually euteleost acanthopterygians (Table 1.3, Figure 1.18). Various changes in the fin structure, from soft to hard fin-rays, from cycloid (round, smooth) scales to at least some ctenoid (toothed or spiny) scales (2.4), and changes in the skull and jaws are seen as trends, culminating in the Superorder Acanthopterygii (Figure 1.18), considered to be the most

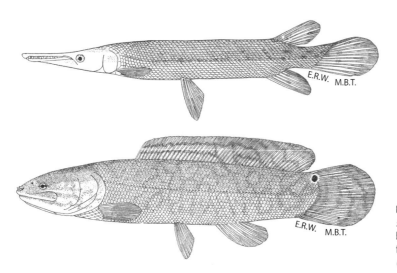

Figure 1.17 Holosteans: (top), *Lepisosteus spatula*, a gar; (bottom), *Amia calva*, the bowfin. Both reproduced, with permission, from M.B. Trautman (1957), *The Fishes of Ohio*, Ohio State University Press.

recent major fish divergence. Acanthopterygians are particularly successful in coastal marine habitats.

1.3.8 Teleost classification

Nelson (2006) separated the Division Teleostei into four Subdivisions: Osteoglossomorpha, Elopomorpha, Ostarioclupeomorpha and Euteleostei (Table 1.3).

Table 1.3 Numbers of genera and principal habitat of the teleost Subdivisions (after Nelson 2006).

Subdivision	Comprises	Habitat	Extant genera (#)
Osteoglossomorpha	Bony tongues, e.g. *Osteoglossum*, Mormyridae	Freshwater	26
Elopomorpha	Eels, tarpon	Marine, Catadromous	157
Ostarioclupeomorpha	Herrings, carp, characins, catfish, etc.	Pelagic oceanic, Anadromous, Freshwater	1080–1180 approx.
Euteleostei	*	Varied	2800 approx.

*A very large diversity of fish including salmon and trout, many coral reef fish (e.g. wrasses), coastal species (e.g. sculpins), commercially important marine fish including haddock and cod, extreme benthic fish such as anglers and flatfish, oceanic fish like *Mola mola*.

The number of divisions had not changed from previous schemes (e.g. Nelson 1976) but the composition is different; the Superorder Ostariophysi was moved from the Euteleostei to form the new Subdivision Ostarioclupeomorpha which include the species from the previous Infradivision Clupeomorpha. These four subdivisions are very unequal as to the number of genera (Table 1.3).

In 2016 Nelson and co-authors adjusted the teleost Subdivisions to Cohorts, retaining Osteoglossomorpha, Elopomorpha and Euteleostei, but adding the slickheads (previously in the Euteleostei) to the herring/carp group as Otocephala (Figure 1.18).

The Osteoglossomorpha

Subdivision (Nelson 2006) or Cohort (Nelson et al. 2016) Osteoglossomorpha (literally 'bony tongues') contains fish with teeth on the parasphenoid and tongue. This is an interesting strange freshwater group, including genera from South America, for example *Arapaima gigas* (Figure 1.19) and *Osteoglossum*, North America (*Hiodon*) and Africa (Family Mormyridae, elephantfishes). The mormyrids

Division Teleostomorpha	Examples
Subdivision Teleostei Cohort Osteoglossomorpha*	bony tongues
Cohort Elopomorpha*	eels
Cohort Otocephala	
Superorder Clupeomorpha	herrings
Superorder Alepocephali	slickheads
Superorder Ostariophysi	carp, catfish
Cohort Euteleostei	
Superorder Protacanthopterygii	salmon
Superorder Osmeromorpha	smelt
Superorder Ateleopodomorpha	jellynose fishes
Superorder Cyclosquamata	lancet fish
Superorder Scopelomorpha	lantern fish
Superorder Lamprimorpha	velifers, opahs
Superorder Paracanthopterygii	cod, anglers
Superorder Acanthopterygii	sticklebacks, flatfish

Figure 1.18 The teleosts, after Nelson et al. 2016, but *sequence reversed.

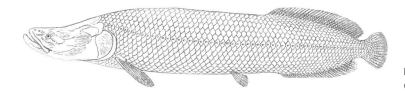

Figure 1.19 *Arapaima gigas*, an osteoglossomorph (after Bridge 1904).

inhabit turbid rivers and have remarkably large brains, and electrolocation (11.6.2, 12.9.1).

The Elopomorpha

Subdivision (Nelson 2006) or Cohort (Nelson et al. 2016) Elopomorpha comprises eels and tarpons which share the features of thin leaf-shaped leptocephalous larvae. This group is predominantly marine, but includes the migratory 'freshwater', actually catadromous, eels of the genus *Anguilla*.

The Ostarioclupeomorpha/Otocephala

Subdivision Ostarioclupeomorpha (Cohort Otocephala Nelson et al. 2016 with the slickheads Alepocephala) is a recent amalgamation of the herrings (which had their own Infradivision, Clupeomorpha) with the carp group (previously forming the Superorder Ostariophysi within the Subdivision Euteleostei). These fish may have superior hearing with morphological links between the air bladder and the ear.

The Ostariophysi

The ostariophysans, which comprise approximately 70 per cent of extant freshwater species, may dominate the freshwater habitat although some of the most 'primitive' (Series Anotophysi, based on three vertebrae contributing connections from the air bladder to the ear, e.g. *Gonorhynchus* and the milkfish *Chanos*) are marine. The remaining ostariophysans (Series Otophysi) all have four vertebrae contributing elements to a chain of Weberian ossicles linking the air bladder to the ear (12.8.4). Members of this group are important for pond- or tank-based culture and some are very popular aquarium fish, being semi-domesticated. The Otophysi are divided into four Orders (Table 1.4) which may be easily recognized; the Siluriformes (catfish) typically have barbels on the head

(hence their name). Some forms of catfish have heavy bony plates (armoured catfish) and others have unusual behaviour (upside-down catfish). Two of the orders contain electric fish (13.8). The Eurasian minnow (*Phoxinus phoxinus*) shows typical external features (Figure 1.20) of the cypriniform order.

Table 1.4 The Otophysi, ostariophysans with Weberian ossicles (based on Nelson et al. 2016).

Order	Examples	Features	Extant genera (#)
Cypriniformes	Carp, goldfish, *Phoxinus*	No teeth on jaws, pharyngeal teeth, herbivorous Freshwater, temperate, Eurasia, North America	489
Characiformes	Characins, piranha	Teeth on jaws, carnivorous, frugivorous Freshwater, tropical Africa, South America	520
Siluriformes	Catfish	Long barbels, usually no maxillary teeth Herbivorous, freshwater and a few marine Widespread in warm temperate and tropical regions At least one species 'electric'	490
Gymnotiformes	Knifefish Electric eel	Elongated freshwater, tropical Central and South America, some 'electric', e.g. electric eel *Electrophorus*	33

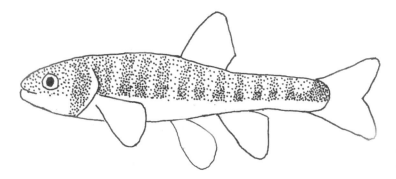

Figure 1.20 *Phoxinus phoxinus*, the Eurasian minnow, a cypriniform ostariophysan.

The Euteleostei

The Subdivision (or Cohort) Euteleostei is a large group which may include features such as spiny fins, anterior pelvic fins (i.e. close to the pectorals) and air bladders (if present) which are physoclystous (closed, 3.8.5). The Subdivision Euteleostei was divided (Nelson 2006) into nine Superorders (Table 1.5), including Ateleopodomorpha.

The Superorder Protacanthopterygii contains a number of commercially important fish such as the generally anadromous Atlantic (*Salmo salar*) and Pacific salmon (*Oncorhynchus* spp.), the charrs (*Salvelinus* spp), brown trout (*Salmo trutta*) and the coregonids (whitefish) as well as the marine capelin (*Mallotus villosus*) and anadromous smelt (*Osmerus*). Migration (9.14, 14.5.6) is a feature of the life history

for many of these fish although the migration may not occur for all fish, even in a species which is generally regarded as migratory. Conversely, pike (e.g. *Esox*) are strictly freshwater fish. The entire group is confined to cooler waters and charr are reckoned to be particularly susceptible to warming trends. Trout and salmon are popular subjects for aquaculture and introductions have been common, with disruption of native species in some instances.

The Superorder Stenopterygii (Table 1.5) is exclusively marine and includes some deep-sea species; these fish commonly have photophores (13.4.7, Plate 14), tiny luminescent spots (Figure 1.21). Stenopterygians may also have extreme lateral flattening and reflective plates (e.g. marine hatchetfish). This group has been eliminated from the most recent classification (Nelson et al. 2016), being transferred to the Superorder Osmeromorpha (Figure 1.23).

The Superorder Cyclosquamata inhabit marine or brackish water and are characterized by special gill arch structure. Some of them are synchronous hermaphrodites (9.4). This group includes the lancet fish, 'Bombay ducks' and lizard fish.

The Superorder Scopelomorpha used to incorporate genera now in the Cyclosquamata. In the diminished form, the Superorder is wholly marine, containing a host of small sardine-shaped mesopelagic fish with lateral and cephalic photophores, for example lantern fish Myctophidae (Figure 1.21).

The next two Superorders, Lampriomorpha and Polymixiomorpha, are small groups, the latter very small; both have been removed (Nelson 2006) from the Superorder Acanthopterygii. Their placing depended on interpretation of features such as the absence of 'true spines' in the fins and an unusual upper-jaw protrusion mechanism in the

Table 1.5 Numbers of genera and principal habitat of the Subdivision Euteleostei (after Nelson 2006), excluding 4 genera of the marine jelly nose fishes Ateleopodomorpha.

Superorder	Examples	Principal habitat	Extant genera (#)
Protacanthopterygii	Salmon, trout	Mostly freshwater	94
Stenopterygii	Hatchetfish	Marine	53
Cyclosquamata	Lancet fish, lizard fish	Marine or brackish	44
Scopelomorpha	Lantern fish	Marine	35
Lampriomorpha	Velifers, opahs	Marine	12
Polymixiomorpha	Beard fish	Marine	1
Paracanthopterygii	Cod, anglers	Mostly marine	270
Acanthopterygii	Wrasses, cichlids, Sticklebacks, flatfish	Predominantly coastal marine	2422

lampriomorphs, whereas the sole genus *Polymixia* (the beard fish) in the polymixiomorphs has what is described as a unique feature in positioning of the palato-premaxilliary ligament in the skull. Nelson et al. (2016) changed the status of the polymixi-omorphs, placing them in the paracanthopterygian group, with the comment 'few groups have shifted back and forth as frequently as this one'. These fish are not well known to most fish biologists.

The Superorder Paracanthopterygii is better known because the group incorporates the commercially important gadids (Family Gadidae; cod, haddock, pollack, etc.), the grenadiers (Family Macrouridae) found throughout the deeper oceanic regions and the anglers, unusual benthic fish with a long 'fishing lure', attached to their heads, that can be dangled in front of their very large mouths. This group, with several families, contains some deep-sea genera which have 'parasitic' males (9.3, 13.12) that become permanently attached to the larger females. The Superorder is overwhelmingly marine with the exception of the burbot (*Lota lota*) a cod-like gadid found in European freshwater.

The Superorder Acanthopterygii is a huge group with many species in the marine inshore, including coral reefs, and some in brackish and fresh water, the latter exemplified by the sticklebacks (Figure 1.22) (Family Gasterosteidae) and the perch (Family Percidae). The group is characterized by spiny rays in the dorsal, pelvic and anal fins, as well shown by the sculpins (Family Cottidae). The wrasses (Family Labridae) are typically tropical with many reef species, while the northernmost representative, the cunner (*Tautogolabrus adspersus*), having reached the shores of Newfoundland, becomes torpid during the winter. Other interesting members of this group are the notothenoids (Suborder Notothenoidea) prevalent in Antarctic waters, where members of the Family Channichthyidae have, remarkably, lost their haemoglobin (few or no red blood cells, 5.13), a situation which is presumed to reduce viscosity in a relative oxygen-rich but cold environment.

Often listed at the end of the many orders of the acanthopterygians are the economically important flatfish (Pleuronectiformes) and the puffers, boxfish, trunkfish and sunfish of the Order Tetraodontiformes. The sunfish (*Mola mola*) is interesting because it can become very large (3.3 m length, Hart 1973) on a surprising diet of jellyfish, is reputedly highly fecund and has the reputation of floating on its side, although this may be due to morbidity (Hart 1973). The warm-water puffers, boxfish and trunkfish, with some reef species, are much smaller but present a strange-looking array of rigid (boxfish, trunkfish) or inflatable (puffers) bodies, often with spines. Some species produce very toxic secretions which may cause fatalities if eaten by unwary diners (humans and presumably other fish too). One of the toxins, tetrodotoxin, has been useful in physiological investigations of neural function.

Recently (Nelson et al. 2016), the composition of the Euteleostei has been adjusted (Figure 1.23),

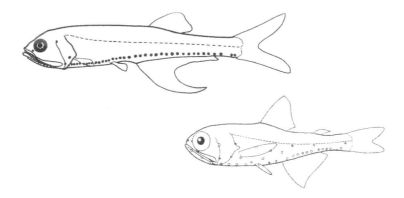

Figure 1.21 Left, *Bonapartia*, a bristlemouth* (A.S. Harold). Right, a myctophid. (J.R. Paxton and P.A. Hulley). Reproduced with permission. © The Food and Agriculture Organization of the United Nations, 1999, *FAO Species Identification Guide, Vol. 3*. The small circles (shaded or unshaded) represent photophores. *The classification has recently been changed (Nelson et al. 2016); the Family is Gonostomatidae, the Superorder is Osmeromorpha.

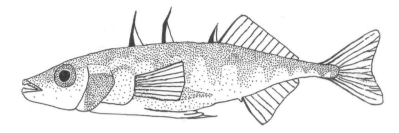

Figure 1.22 *Gasterosteus aculeatus*, the three-spined stickleback, an acanthopterygian teleost. Reproduced with permission. Original drawing D. Burton (1978). Melanophore distribution within the integumentary tissues of two teleost species, *Pseudopleuronectes americanus* and *Gasterosteus aculeatus* form leiurus, Canadian Journal of Zoology 56, 526–535. © Canadian Science Publishing or its Licensors.

Protacanthopterygii

Osmeromorpha, includes four orders, Argentiformes, Galaxiiformes, Osmeriformes, Stomiiformes

Ateleopodomorpha, jellynose fishes, 4 genera, previously in Superorder Acanthopterygii (Nelson 1976)

Cyclosquamata

Scopelomorpha

Lampriomorpha

Paracanthopterygii

Acanthopterygii

Figure 1.23 Superorders of the Euteleostei after Nelson et al. (2016).

retaining seven of the superorders of Nelson 2006, transferring one (Stenopterygii) to the Osmeromorpha which was added with other fish from the Protacanthopterygii now diminished. Polymixiomorpha was also removed, transferring to the Paracanthopterygii, whereas Ateleopodomorpha was included.

1.4 Variety in fish structure and function

Extant fish species have often been studied by concentrating on a few, easily captured or kept, species (15.6) used as 'types' and representative, which downplays differences and may emphasize similarities.

It is likely that of the many fish species known to exist some features (such as distribution, life cycles and physiology) of many of them are only understood at a rudimentary level. In the case of deep-sea fish, specimens obtained are perhaps singular, usually dead, and the opportunity to observe live fish very rare; when a live fish becomes available, as

occurred recently with *Dolichopteryx longipes* (Wagner et al. 2009), very interesting new data can be obtained, in this case the functioning of accessory eyes (12.3.8). The diversity of fish species is remarkable (Chapter 13) and includes a range of morphology and physiology, including extremes of size, ages attainable, habitats colonized, shape, colours and prevalence.

1.4.1 Diversity in size and longevity

Fish species vary in attained adult length from extremely small, for example about 10 mm for a goby species (Lourie and Randall 2003), to huge; over 2 m for *Arapaima gigas* in freshwater, 5 m for the freshwater catfish *Silurus glanis* (Giles 1994), 8 m for oarfish, 'king-of-the herring' (Nelson 1976) and over 10 m for basking sharks (Hart 1973), even up to 13.6 m (Nelson 1976). The largest of all (for extant species) is the whale shark (Nelson 1976) which may reach 18 m. The largest size, however, reached by many species is less than 20 cm. Some fish have a protracted growth and maturation phase, as occurs with elasmobranchs (9.15). In other cases early maturity may be possible but it does not preclude further somatic growth so that there is a whole range of sizes which can be 'adult', as occurs with halibut (*Hippoglossus hippoglossus*). This species and cod (*Gadus morhua*) are therefore said to have indeterminate growth, continuing to grow after sexual maturity, unlike birds and many mammals. Very short life cycles may be found with some South American 'annual' fish such as the cyprinodont *Rivulus stellifer* (Thomerson and Turner 1973), which completes the life cycle in a year, with drought-resistant eggs capable of initiating a new cycle when conditions become appropriate in the rainy season.

Very small fish species (of which there were 47 in Asian freshwater as described in 1993, (Kottelat et al. 2006) were termed 'miniature'. Little is currently known of the life cycle of these or the various 'miniature' fish more recently discovered (Kottelat et al. 2006) but Winterbottom and Southcott (2008) list an Australian pygmy goby (*Eviota sigillata*) as having a maximum recorded age of 59 days. Miniature fish, some of which are reproductively mature at a length of around (or less than) 10 mm, include gobies, a cyprinid *Paedocypris*, pygmy seahorses such as *Hippocampus denise* (Lourie and Randall 2003) and *Schindleria brevipinguis* (Watson and Walker 2004) and some apparently originate by paedogenesis, a process when sexual maturity occurs in a larval form (Abercrombie et al. 1961); see Appendix 1.5. Bennett and Conway (2010) amassed data on the minimum length at maturity, considering those maturing at sizes (standard length, 7.13.4) less than 20 mm at maturity, to be 'miniature'. On this basis these authors report only 7 miniature freshwater fish from North America, compared with 24 such species from Africa, 100 from South America and 50 from South-east Asia.

Age of fish can be known in captivity or deduced from ring counts from otoliths (7.13.5, 12.8.2) or scales (2.4) which grow by apposition, material being added annually in seasonally changing environments to the outer surface. If growth rate is known, then reasonable estimates of age can also be based on lengths attained. Using these criteria it is possible to conclude that many species can attain ages over 5 years in the wild, and unless/until there is size-selective intensive fishing, fish over 20 years in age are possible; wild *Sebastes* spp. are reported to attain 50–100 years (Leaman 1991).

1.4.2 Diversity in habitat

Water in which fish live can be warm or very cold, and fish have adapted to function well in a range of temperatures, though an optimum may be identifiable. Remarkable cold-water adaptations include a suite of features such as the presence of antifreeze proteins/glycoproteins either year round or seasonally (7.4, 8.5.4, 13.5.3), moderation of blood-cell number (5.2), aglomerular kidneys to conserve antifreeze (8.5.4) and extensive winter fasts for some fish (4.8), and the possibility of torpor (7.4). A fundamental distinction can be made between fish normally inhabiting fresh water and those in salt (usually marine) habitats, though some fish can migrate from one to the other (9.14, 14.5.6). Freshwater fish inhabit still and/or flowing water, still water provides fish habitats varying from huge lakes to small ponds. Flowing water includes springs and streams as well as small and large river systems with tidal estuaries at their mouths, where interchange with the marine habitat can occur.

Still water

Lake environments include those of the far north in Canada like Great Bear Lake and Great Slave Lake which have populations of coregonids and trout, and small lochs in Scotland with Arctic charr and powan (a coregonid). The meromictic (freshwater layered over saltwater) Lake Ogac in the East Canadian Arctic has a single species of fish, a landlocked population of cod (*Gadus morhua*), not, confusingly, *Gadus ogac*, a species which does occur, for example, off the Labrador coast. In contrast to the species-poor (depauperate) Northern Lakes are the African Great Lakes which were renowned for species diversity, notably hundreds of different cichlids in specialized niches. Unfortunately, introductions of Nile perch reduced the cichlid diversity considerably but it is still a species-rich environment. The North American 'Great Lakes', the linked systems from Lake Ontario through to Lake Superior, had more than 200 fish species (Hubbs and Lagler 1958), some of which have been negatively impacted, with local extinctions, due to overfishing and lamprey access. Lake Baikal, the seventh largest (in area), is 'the world's deepest lake' (Kozhov 1963), with a maximum depth of 1620 m. It has 50 fish species, with one endemic family, the Comephoridae, oily scaleless and viviparous cottids.

Smaller lakes and ponds occur in many regions under widely differing conditions of composition and temperature. Even in one Canadian province, Nova Scotia, a survey (Alexander et al. 1986) of 781 lakes with a surface area from 1 to 500 ha showed a pH varying from very acidic (pH 4) to very alkaline (pH 9.7), and an average of 6.2. The authors admit that the extremes might be questionable but one county had a mean of 5.5, still very acid, and vulnerable to acid rain. Of the 744 lakes surveyed for fish present a few (5 per cent) had no fish caught. The total number of species found was 37, some of which, for example winter flounder, were normally thought of as marine but capable of surviving varying salinity (i.e. euryhaline); the maximum number of species found in any one location was 12 (Kejimkujik Lake). One fish species, *Coregonus canadensis*, supposed to be present was not found; this species is listed as endangered and one local riverine population is believed to have been 'extirpated by acid rain'. These Canadian lakes were not very deep (mean maximum 8.2 m), overall had low levels of dissolved solids and were termed 'relatively unproductive'. An acid lake (pH4.4) studied earlier in Wisconsin (Jewell and Brown 1924) was reported to have four fish species: perch (*Perca flavescens*), catfish (*Ameiurus nebulosus*, more recently *Ictalurus nebulosus*), sunfish (*Lepomis incisor*, more recently *L. macrochirus*) and pike (*Esox lucius*), the first three being quite abundant in spite of the acidity.

Some smaller ponds may indeed not have acquired any fish species but if they do, the ephemeral nature of shallow ponds in warmer climates favours small fish with short life cycles and drought-resistant eggs.

Tropical peat swamps have recently been studied with particular reference to endemic species and miniature species which may be particularly prevalent in that environment (Kottelat et al. 2006).

Flowing water

Flowing water varies in bulk and depth seasonally and represents spawning sites for some migratory (anadromous) fish like salmon, as well as more permanent residence for species like catfish and perch, and many cyprinids and characins. In small fast-flowing streams, typically originating in mountain ranges, fish like cyprinid loaches may have clinging discs (e.g. modified pelvic fins) to avoid being swept downstream. In contrast to these often cold but well-oxygenated environments are the small, very warm springs in desert environments, a restricted habitat with rare species like the 'desert pupfish', some species and subspecies of *Cyprinodon*, now only found in desert springs and streams. Similar

difficult niches are presented by cave-waters, with species eking out an existence in the dark (13.5.1); such species may be blind and albino, and are prone to flash floods and insecure, scant food supply. Big river systems include the Mississippi in the United States, the Yangtze in China and the Amazon–Rio Negro complex in Brazil. Kramer et al. (1978) state that the 'Amazon fish fauna is richer than that of any other river system', with 1300 species recorded by 1967 (Roberts 1972). Fluctuations in water level are considerable and give concentrations of species at times of low water, with different feeding opportunities, including access to fruit and nuts. Changes in the direction of water flow can occur, with mixing of waters of very different composition. The River Negro as its name suggests is a 'blackwater' river; at its confluence with the Amazon it mixes its flow with 'whitewater'. Different species may only be able to survive in either black- or whitewater so changes in flow can be very significant in species survival. Blackwater can be exceedingly acid, as low as pH 2.8 (Henderson and Walker 1986). The composition of black- and whitewater is compared by Horbe and da Silva Santos (2009); they note that whitewater is distinguished by high pH (i.e. tends to be alkaline), with high calcium and magnesium.

Salt water

There are some extremely salty inland waters (e.g. the Dead Sea), which generally do not support any fish. Marine environments, with high sodium chloride levels (and total salt contents around 3.5 per cent, Schmidt-Nielsen 1990) range from coastal waters, reefs and rock pools to the multilayered oceanic waters, with fish specialized for the surface waters, mid-waters (mesopelagics like the lantern fish) and varying greater depths down to the sea bottom (including benthic species which are 'abyssal', Table 13.4). The salt (notably sodium chloride) levels may be reduced somewhat in coastal regions with varying dilution from river input. Conversely some mangrove regions can be hypersaline: off Belize, the killifish *Rivulus* (now *Kryptolebias*) *marmoratus* can survive in salinities of about 4.8 per cent (Frick and Wright 2002b).

As with swift streams, inshore fish may have developed clinging discs (*Cyclopterus lumpus*) to help keep stationary. Littoral species such as those found

in rock-pools have remarkable resistance to desiccation and abrasion as well as to salinity changes.
Gillichthys mirabilis is well-named; a goby from Californian shores, it can survive out of water for several
hours. Shallow coastal waters provide habitats for a
very large number of species, mainly acanthopterygians. Coral reef species are particularly numerous,
known for their bright colours and niche adaptations
including the clown-fish which hide among anemone
tentacles (13.3.3,13.13). Some reef fish have complex
group relationships which can involve sex reversal,
for example in response to death of dominant fish.

In open oceanic waters the surface may favour
pelagic species such as 'flying fish' (*Exocoetus*),
which have very long pectoral fins and can launch
into the air and glide several metres, apparently to
escape predators. The sunfish (*Mola mola*), which
feeds on jellyfish, occurs in open waters as well as
fast-swimmers like tuna and lamnid sharks, and
the slower planktonic feeders like the basking and
whale-sharks. Offshore surface waters are however
regarded as species-poor, particularly in comparison with inshore waters which benefit from input
of nutrients (and ensuring plankton blooms) from
estuaries and upwelling.

Lower in the water column are mesopelagic
forms like the myctophids (lantern fish) with small
size and photophores.

With increasing depth, light does not penetrate
and turbidity will affect the depth at which the environment is aphotic. Fish (and invertebrates) may
be red in the deeper photic areas and black in the
aphotic regions. In the very deep regions prey may
be very scarce and the fish species tend to be small
but with huge mouths and a capacity for abdominal expansion to accommodate a wide range of
prey size. Deep-water fish may have photophores
(13.4.7) and remarkable, even bizarre adaptations
like illuminated fishing lures (deep-sea anglers) as
well as 'parasitic males' (some deep-sea anglers,
13.12), thus optimizing the reproductive capacity,
as males attach to the female once found.

1.4.3 Diversity in shape

The typical fish shape is usually portrayed as
'streamlined' associated with swimming in the
water column. A lot of fish species, particularly

those that are pelagic in habit, have a torpedo
shape which minimizes drag, making them more
efficient and potentially fast swimmers. The herrings (clupeids) and the trout, as well as the dogfish
and sharks, have this sleek shape while the lamnid
sharks and scombrids (tuna, mackerel) have it in
association with a narrow caudal peduncle and a
lunate tail, a combination of shapes which is ultra
efficient for speed (3.8.2). Slower fish include those
that can ambush prey with a short burst of speed,
having a cigar shape like that of the gar or pike, and
a very large elongated mouth, well toothed.

Deep-sea fish (13.4.3) are noteworthy for diversity of shape and may also have extra large mouths.
The hammerhead sharks (*Sphyrna*) have moderate-
sized or extremely large lateral projections on each
side of the head (Figure 1.24) with an eye on the end
of each. The manta rays also have well-separated
lateral eyes (Scott and Scott 1988).

Unusual positions may be taken by some fish
(13.11). The shrimp-fish (Centriscidae) take on
an odd shape in life because they tend to swim
in groups with their heads down, vertical in the
water column, twisting and turning together. The
male Australian western carp gudgeon (*Carassiops
klunzingeri*) can also assume the vertical (head up
or down) when guarding eggs (Lake 1967), and the
coelacanth has been reported to do 'headstands'
(Balon et al. 1988). The ultimate strange orientation
may be that of the 'upside-down' catfish (*Synodontis
nigriventris*), which is noted for inverted swimming
(Blake and Chan 2007).

Flattened fish can be laterally flat like marine
hatchetfish, or the marine sunfish (*Mola mola*) and
many coral reef species in which presumably a flat
form facilitates narrow niche exploitation. Alternatively, other flattening can be regular dorso-ventral
(some elasmobranchs like skates and rays) and
a skewed dorso-ventral flattening as in 'flatfish'
(Pleuronectiformes) in which, at metamorphosis,
the individual fish lies on the right or left side and
the bottom eye migrates dorsally. Depending on the
family flatfish are generally dextral (right-side up)
or sinistral (left-side up) but one family (Psettodidae) shows no such bias.

Extreme elongation is typical of the cyclostomes
and eels but is also seen in other groups such as
the small group of gymnotiform ostariophysans

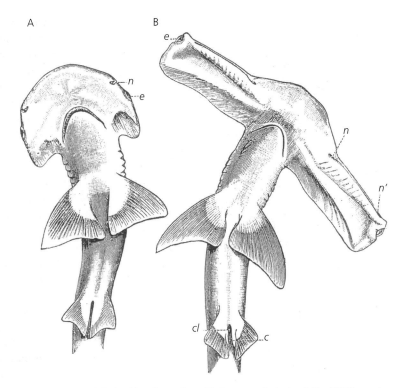

Figure 1.24 Ventral views of Hammerhead sharks: (A). *Sphyrna tiburo*, (B). *S. zygaena*. Both after Bridge (1904); c = clasper, cl = cloacal aperture, e = eye, n = nostril, n' = nasal groove.

(Nelson et al. 2016) including the 'electric eel' and some of the lampriomorphs. Some lampriomorphs are very elongated; the oarfish (Regalecidae) can attain 8 m (the longest bony fish) and have over 100 vertebrae (Nelson 2006). Fish elongation is discussed by Ward and Mehta (2010).

1.4.4 Diversity in colour

Coral reef fish, in particular, show a wide variety of colours, which may change as the fish ages. Yellow and blue predominate on the reefs while the lowest level of the photic zone in the sea has a number of red species; pelagic fish which swim near the water surface are often silvery. Many fish species are darker on the upper dorsal surface and lighter (white, grey or silver) beneath. Deepwater fish are usually dark all over (deep-sea anglers, grenadiers) in comparison to cave-fish which are usually both blind and albino. More information on the morphological bases for fish colour and the mechanisms for colour change is in 2.6–2.9 and Appendices 2.1–2.3. The autonomic nervous system's role in rapid colour change is explored in 11.10.6.

1.4.5 Diversity in numbers

Some fish species have occurred in very large numbers, documented as weight of fish removed (Table 1.6) and sold in some fisheries. Reports of huge numbers have occurred particularly often for pelagic species like herring and anchovy. The small schooling species come to attention when they are seen *en masse* from ships at sea, and it is not easy to gauge the accuracy of some of the reports. Unfortunately, masses of fish can attract extreme fishing effort from which some species seem very slow to recover, although other explanations for fishery failures have been given, particularly for pelagic species (Hart 1973). The advent of remote-sensing methods has made it easier to locate fish concentrations at sea, but translating concentrations into species numbers has never been particularly accurate. However, it is probably reasonable to conclude that medium-sized schooling fish like herring and mackerel have sometimes occurred in many millions, in the recent past, in the North Atlantic, and similar or larger numbers have occurred for sardine, anchovy and anchoveta in the Pacific. Catch data for California (Table 1.6) show very large values for the pelagic fish taken, huge amounts were taken off Peru (Idyll 1973). The situation with North Atlantic cod is that very large numbers were reported by early explorers, with enthusiastic stories of obtaining fish so readily that all the gear that was needed was lowered baskets (Harris 1990). Perhaps these men were confused by the presence of the much smaller capelin. Large numbers of cod were indeed present until recently (Harris 1990); although not quite as abundant as the 'basket' story suggests, there were large numbers caught on an annual basis, until the early 1990s. There are still thousands of cod in the North-west Atlantic but the population has been much reduced, although reports suggest that there are now many more large fish than in the recent past.

In the deep seas, a habitat fairly continuous over wide areas, it is common to find reports of grenadiers which must still be widely distributed, and as individual species, in spite of directed fisheries, may still occur in tens or hundreds of thousands. Lantern fish and other small 'forage' fish like capelin and the bristlemouths

(e.g. *Cyclothone*) may be very numerous. The latter stenopterygian species was said (Moyle and Cech 1996) to be probably the most numerous fish, and hence presumably the most numerous vertebrate. In contrast to the successful species, some of which are still withstanding high fishing pressure, a number of species may never have had high numbers, like the desert pupfish (13.5.2) and cave-fish (13.5.1) because their habitats are limited in scope. With the possibility of single or combined factors such as limited habitat or slow growth and late maturity, or extreme vulnerability to fishing practice, it is not surprising that the current trend is overwhelmingly towards loss of numbers within species and declining species diversity in many regions affected by increased human activity, the African Great Lakes, the American Great Lakes, the North Sea and the Aral Sea all having been negatively affected in recent decades (Chapter 16). Pauly et al. (2002) discussed the 'serial depletions' associated with fisheries, and the extended reach and scope of global effort on an industrial scale. Chapter 16 briefly considers the 'future for fish' in the present context; unfortunately the optimistic view of a superabundance of fish appears to have been wildly overoptimistic.

Appendix 1.1 The embryonic layers of animals

During early development (ontology) of animals, the fertilized egg (zygote) undergoes multiple cell divisions, with migration of cells and subsequent differentiation. The zygote becomes successively a blastula, an undifferentiated cell mass with a cavity (blastocoel) and then a gastrula, with cell layers becoming apparent. Animals are either derived from two cell layers (and are therefore diploblasts), the outer ectoderm and the inner endoderm (as with coelenterates; jellyfish, corals, etc.) or have an additional middle layer, the mesoderm, during development, as with most animals, which are triploblasts. Many triploblasts have a cavity, the coelom, within the mesoderm and are thus triploblastic coelomates. Triploblasts with no coelom (e.g. nematodes, the round-worms) are termed triploblastic

Table 1.6 Commercial catch data (landed lb) for pelagic fish off California (Bureau of Marine Fisheries 1929, 1941, 1942).

Year	Anchovy	Sardine	Herring	Mackerel*
1926	60, 157	286, 741, 250	453, 607	3, 623, 290
1927	368, 201	432, 275, 289	1, 168, 321	4, 740, 639
1936	195, 122	955, 525, 700	840, 530	100, 542, 214
1939	2, 147, 901	1, 160, 793, 581	302, 242	80, 909, 374
1940	6, 317, 797	905, 973, 403	453, 193	120, 503, 712

*Generally Pacific mackerel.

acoelomates. Vertebrates, including fish, are triploblastic coelomates. Developing embryos may have
a large yolk source, as with most fish, and the embryo lies on top of the yolk initially forming a blas-
todisc (a flattish array of cells rather than a sphere),
with an embryonic blood supply conveying nutrients from the yolk to the growing embryo.

Appendix 1.2 A summary of fish classification systems (jawed fish only)

Examples given are of extant fish

1.2a A summary of the 'old' fish classification system (based on Nelson prior to 2006)

CLASS CHONDRICHTHYES	cartilaginous fish
SUBCLASS HOLOCEPHALI	chimaeras (rabbit/ratfish)
SUBCLASS ELASMOBRANCHI	dogfish, sharks, rays, e.g. *Squalus, Cetorhinus, Raja*
CLASS OSTEICHTHYES	bony fish
SUBCLASS DIPNEUSTI/DIPNOI	lungfish *Neoceratodus, Protopterus, Lepidosiren*
SUBCLASS CROSSOPTYERYGII	coelacanth *Latimeria*
SUBCLASS BRACHIOPTERYGII	bichir, reedfish *Polypterus, Calamoichthys*
SUBCLASS ACTINOPTERYGII	
INFRACLASS CHONDROSTEI	sturgeons and paddlefish, e.g. *Acipenser, Polyodon*
INFRACLASS NEOPTERYGII	
DIVISION GINGLYMODI	garpike *Lepisosteus*
DIVISION HALECOSTOMI	
SUBDIVISION HALECOMORPHI	bowfin *Amia calva*
SUBDIVISION TELEOSTEI	
INFRADIVISION OSTEOGLOSSOMORPHA	bony tongues, e.g. *Arapaima, Hiodon,* Family Mormyridae
INFRADIVISION ELOPOMORPHA	tarpons and eels, e.g. *Anguilla*
INFRADIVISION CLUPEOMORPHA	Herrings, e.g. *Clupea*
INFRADIVISION EUTELEOSTEI	
SUPERORDER OSTARIOPHYSI	carp, catfish, characins, electric eel, e.g. *Cyprinus*
SUPERORDER PROTACANTHOPTERYGII	salmon, trout, capelin, smelt, whitefish, e.g. *Salmo*
SUPERORDER STENOPTERYGII	e.g. marine hatchetfish

SUPERORDER SCOPELOMORPHA	e.g. Family Myctophidae
SUPERORDER PARACANTHOPTERYGII	e.g. Family Gadidae and anglerfish
SUPERORDER ACANTHOPTERYGII	very large taxon*

*Includes:
Order Atheriniformes, e.g. flying fish, cyprinodonts (killifish), silversides
Order Lampridiformes, e.g. oarfish
Order Beryciformes, e.g. squirrelfish
Order Zeiformes, e.g. dories
Order Syngnathiformes, e.g. seahorses, sticklebacks
Order Scorpaeniformes, e.g. sculpins, snailfish
Order Perciformes, e.g. perch, grunts, icefish, tuna, blennies, gobies, cichlids, wrasses, parrotfish, surgeonfish, damselfish
Order Pleuronectiformes, flatfish
Order Tetraodontiformes, e.g. boxfish, sunfish (*Mola*)

1.2b A summary of the 'new' fish classification (based on Nelson 2006)

Note: there is also a more recent radical adjustment
of fish classification (Nelson et al. 2016).

CLASS CHONDRICHTHYES	cartilaginous fish
SUBCLASS HOLOCEPHALI	chimaeras (rabbit/ratfish)
SUBCLASS ELASMOBRANCHI	dogfish, sharks, rays, e.g. *Squalus, Cetorhinus, Raja*
CLASS SARCOPTERYGII	
SUBCLASS COELACANTHIMORPHA	coelacanth *Latimeria*
SUBCLASS DIPNOTETRAPODOMORPHA	lungfish *Neoceratodus, Protopterus, Lepidosiren*
CLASS ACTINOPYTERYGII	
SUBCLASS CHONDROSTEI	
Order Polypteriformes	Bichirs
Order Acipenseriformes	sturgeons, paddlefish
SUBCLASS NEOPTERYGII	
Order Semionotiformes	garpike *Lepiosteus*
Order Amiiformes	bowfin *Amia calva*
DIVISION TELEOSTEI	
SUBDIVISION OSTEOGLOSSOMORPHA	bony tongues, e.g. *Arapaima, Hiodon,* Family Mormyridae
SUBDIVISION ELOPOMORPHA	tarpons and eels, e.g. *Anguilla*
SUBDIVISION OSTARIOCLUPEOMORPHA	herrings and the carp/catfish group (ostariophysans)
SUBDIVISION EUTELEOSTEI†	
SUPERORDER PROTACANTHOPTERYGII	salmon, trout, capelin, smelt, whitefish, e.g. *Salmo*

SUPERORDER STENOPTERYGII   e.g. marine hatchetfish

SUPERORDER SCOPELOMORPHA   e.g. Family Myctophidae

SUPERORDER CYCLOSQUAMATA   lancetfish, lizardfish, 'Bombay ducks'

SUPERORDER PARACANTHOPTERYGII   e.g. Family Gadidae and anglerfish

SUPERORDER LAMPRIOMORPHA   opahs, velifers, oarfishes

SUPERORDER POYMIXIOMORPHA   *Polymixia*

SUPERORDER ACANTHOPTERYGII   very large taxon*

†Has also included the small group of jellynose fish as SuperorderAteleopodomorpha, sometimes placed with the Lampriomorpha
*Includes:
Order Atheriniformes, e.g. flying fish, cyprinodonts (killifish), silversides
Order Beryciformes, e.g. squirrelfish
Order Zeiformes, e.g. dories
Order Syngnathiformes, e.g. seahorses, sticklebacks
Order Scorpaeniformes, e.g. sculpins, snailfish
Order Perciformes, e.g. perch, grunts, icefish, tuna, blennies, gobies, cichlids, wrasses, parrotfish, surgeonfish, damselfish
Order Pleuronectiformes, flatfish
Order Tetraodontiformes, e.g. boxfish, sunfish (*Mola*)

Appendix 1.3 Examples of confusion in the use of common names

Confusion can easily occur if a common name is used for different species.

The name 'sunfish' is used for the oceanic *Mola mola* and for the freshwater *Lepomis* species, including *Lepomis gibbosus*, the pumpkinseed sunfish, *L.macrochirus*, the bluegill, *L.auritus*, the redbreast sunfish, *L.cyanellus*, the green sunfish, and *L.megalotis*, the longear sunfish.

The term 'trout' is used for members of the genera *Oncorhynchus*, *Salmo* and *Salvelinus*.

The term 'ratfish' is used for chimaeras (Holocephali) and 'rat-tail' for grenadiers (Macrouridae).

The term 'minnow' is applied to many different freshwater fish of small size, but which are generally cyprinids (Klosiewski and Lyons nd).

The term 'turbot' has been applied to two quite different flatfish: one primarily in the North-east Atlantic (*Scophthalmus maximus*) and one in the North-west Atlantic (*Reinnhardtius hippoglossoides*). The latter fish is also referred to as 'Greenland halibut', which, as it co-occurs with Atlantic halibut *Hippoglossus hippoglossus*, can also give rise to problems. Halibut and turbot, in their original designations, are high-value fish for consumption

and the use of the terms for species of different genera can mean buyers are not receiving what they thought they had paid for. The confusion has extended to difficulties when designating catch allowances and the status of a particular fish species as 'under-utilized'. There have been attempts to control the use of the term 'halibut' to protect the market for *Hippoglossus*. Thus the Food and Drug Administration of the United States has issued a ruling under the Code of Federal Regulations Title 21 (revised 1 April 2013) under Section 21CFR102.57 that the term 'halibut' may only be used for *Hippoglossus* species, and that the term 'Greenland turbot' is the common or usual name for *Reinhardtius hippoglossoides*. This ruling is not necessarily accepted, for example by Canadians. The controversy is discussed by Dyck et al. (2007) who refer to an attempt by Pacific halibut (*Hippoglossus stenolepis*) processors to control the name 'halibut' as far back as 1968.

The situation with *Reinhardtius* and *Hippoglossus* thus represents a considerable confusion mainly based on the common names 'turbot' and 'halibut'.

A similar problem may occur with the terms 'sole' and 'plaice' which represent desired food-fish off Europe. The most valued 'sole' is Dover sole, *Solea vulgaris*, but *Microstomas kitt*, the lemon sole, but of a different flatfish family, is also highly regarded in Europe. The 'true' plaice is *Pleuronectes platessa*, found in shallow areas off the European coast. The 'American plaice', *Hippoglossoides platessoides*, also occurs off Scotland where it does not grow very large and is known as the 'long rough dab'.

Deliberate confusion was associated with the use of 'rocksalmon' for dogfish marketed for human consumption in the United Kingdom.

Multiple common names may be used for the same fish; for example, *Pimephales promelas* can be the fathead minnow, Northern fathead minnow, blackhead minnow, tuffy minnow, fathead, blue-headed chub, crappie minnow, rosy-red minnow (for a particular colour-form or 'morph') (Danylchuk et al. 2011). *Glyptocephalus cynoglossus* is known as witch flounder, witch, greysole, gray sole, Craig fluke, pole flounder or flet in English Canada and *plie grise* in French Canada (Scott and Scott 1988).

Appendix 1.4 Further conventions and explanations in the use of naming taxonomy

The name of 'first' describers follows the Latin binomial, a tradition following the reorganization of taxonomy by Linnaeus, recognized by the initial L. Any plant or animal first systematized by Linnaeus is so designated by the initial L. after the binomial, otherwise the full surname is used, usually in brackets.

Linnaeus (Carl von Linné or Carolus Linnaeus), the Swedish botanist, was the instigator of a comprehensive system of biological classification, using the Latin binomial, previously used by, for example, Willoughby (1686) in fish classification. In the case of fish, Linnaeus had access to the work of his friend Petrus (Peter) Artedi who had died prematurely, this work being published posthumously in 1738.

Conventionally the attribution (name of first describer, Table 1.7) is given on first use in a publication, but this is often omitted; the use of the date is even rarer.

Many specific names are directly derived from Latin adjectives; for example 'auratus' = gold, as in goldfish, Carassius auratus 'minor' = small, as in the largescale tonguefish (sole à grand écailles), a bit conflicting, as the fish Symphurus is minor, which is small though the scales are relatively large; 'ferox' = fierce, as in longnose lancetfish, Alepisaurus ferox.

Distribution may also be indicated by the specific name, such as 'americanus' = American, as in Pseudopleuronectes americanus (winter flounder),

Hemitrypterus americanus (sea raven), 'atlanticus' = Atlantic (ocean), as in Lycodes atlanticus (Atlantic eelpout), 'marinus' = marine, as in Petromyzon marinus (marine lamprey).

Appendix 1.5 The concept of persistence of larval traits in an animal which is sexually mature

Over many years, zoologists realized that some animals resembled young or larval forms of other animal species, and where larvae strongly differed from adult forms this could give rise to a very different set of morphological characteristics in an evolutionary line. Perhaps the most noteworthy of this trend was the theorized derivation of chordates (including fish) from tunicate (ascidian) larvae (or 'tadpoles'*). Tunicates (seasquirts) tend to be sedentary but have motile larvae and it was deduced that larvae which had become sexually mature were the basis for evolution of the vertebrates. This view has since been challenged (see review by Lacalli 2005). In the Amphibia it is possible for the larva of the salamander to become sexually mature and then be known as an axolotl; it retains external gills and is aquatic and quite large. Humans are thought to retain some infantile characteristics such as a large head. These tendencies, for larval forms to 'become adult' or for adults to retain larval structures, which can lead to quite marked changes in body form, have acquired terminology:

Neoteny (adj. neotenic: persistence of a larval form; metamorphosis fails to occur (De Beer 1951)

Paedomorphosis (pedomorphosis): juvenile traits in adults

Paedogenesis (pedogenesis): reproduction in a larval form

Progenesis: used as in Paedocypris progenetica, indicates precocious sexual development

Many fish do not show a distinct metamorphosis, exceptions being flatfish in which an eye migrates and the fish lose their symmetry to lie on either the left or right** side and eels which have a leaf-shaped (leptocephalous) larva which changes to an elongated form, round in cross-section. In fish, precocious sexual

Table 1.7 Names of three extant fish.

Common name	Latin binomial	Named by	Short form attribution
Atlantic cod	Gadus morhua	Linnaeus 1758*	Gadus morhua L.
Winter flounder	Pseudopleuronectes americanus	Walbaum 1792*	Pseudopleuronectes americanus Walbaum
Coelacanth or Gombessa	Latimeria chalumnae	Smith 1940*	Latimeria chalumnae Smith

*These dates are given in Scott and Scott 1988.

development may lead to miniaturization; the very small fish species known as 'miniature' (1.4.1, 13.5.5) retain a fry-like size and skeletal structure and, at least in the case of *Schindleria*, a pronephric kidney (Nelson 1976).

* Not to be confused with frog tadpoles.

** 'Lefteye' flounders (e.g. Bothidae) have the left side up, and are termed 'sinistral' (Nelson 1976), Pleuronectidae are 'Righteye' flounders and therefore 'dextral'.

Bibliography

Early taxonomy

Broberg, G. (1985). Petrus Artedi in his Swedish context. In: S.O. Kullander and B. Fernholm (eds). *Fifth Congress of European Ichthyologists*. Stockholm: Swedish Museum of Natural History.

Recent taxonomy

Nelson, J.S. (1976–2006). *Fishes of the World*. New York, NY: John Wiley and Sons.

Nelson, J.S., Grande, T.C. and Wilson, M.V.H. (2016). *Fishes of the World*, 5th edn. Hoboken, NJ: John Wiley and Sons Inc.

Chondrichthyans

Carrier, J.C., Musick, J.A. and Heithaus, M.R. (eds) (2004, 2012). *Biology of Sharks and their Relatives*. Boca Raton, FL: CRC Press.

Hamlett, W.E. (ed.) (1999). *The Biology of Elasmobranch Fishes*. Baltimore, MD: Johns Hopkins University Press.

Relict fish

McKenzie, D.J., Farrell, A.P. and Brauner, C.J. (eds) (2007). *Fish Physiology*, Vol. 26, *Primitive Fishes* Amsterdam: Elsevier/Academic Press.

Thomson, K.S. (1991). *Living Fossil. The Story of the Coelacanth*. London: W.W. Norton and Co.

CHAPTER 2

The integument

The skin and associated structures

2.1 Summary

The fish integument consists of layers forming the skin; the cellular layering from the outside inwards is the epidermis and dermis, with an inner hypodermis. The composition, thickness and colour of the skin layers can vary seasonally and with reproductive status.

Most fish have scales and some have bony plates or scutes. The depth of insertion as well as the shape and composition of the scales shows considerable inter-specific variation. Scales may protrude through the skin or be covered by the epidermis.

The colour of fish skin is due to special cells (chromatophores) and/or other reflecting features. In some fish, the skin can change colour rapidly or slowly under the control of either or both the nervous and endocrine systems.

2.2 Introduction

The outer covering of any animal is termed an integument which is the skin in vertebrates. Fish skin usually has a layer of scales or bony plates/scutes embedded in it. However, scales may be much reduced and scutes, large bony plates, are only present in a few species. Fish integument, acting as the main boundary between the animal and its external environment, has many important functional roles including shape maintenance, mechanical protection, defence, communication and camouflage as well as gas exchange and ion regulation, a homeostatic role for some species. Esteban (2012) has reviewed literature on the roles of fish skin in immunity against infection and has considered it the largest immunologically active organ, although actual antibody production sites for skin are unknown.

Fish have a mucigenic skin in contrast to the keratinized skin of terrestrial vertebrates, although keratinized areas may occur in some fish. The fish integument is a dynamic structure, variously changing morphology, colour and secretions in response to stress, reproductive state, temperature and/or season and immediate habitat. The fish integument is penetrated by special sense organs such as the ampullae of Lorenzini of elasmobranchs (dogfish, sharks, rays) and canals of the lateral line system.

2.3 Fish skin structure

As with other vertebrates, fish skin (Figure 2.1) is composed of an outer epidermis (an epithelium, see Paulsen 1993) derived from the embryonic ectoderm (outer early developing layer) and an inner dermis, largely connective tissue (Paulsen 1993), developed from the embryonic mesoderm, the middle early developing layer (see Appendix 1.1). A thin non-cellular external cuticle, secreted by the epidermis, can be rubbed off on handling and is not usually seen in microscopic sections. Immediately next to the dermis may be subdermal fat in the hypodermis, beneath which are the muscle blocks (Figure 2.7) that enable movement. Scales may be present (Figure 2.1) or absent, for example in the Families Gasterosteidae (Figure 2.1), Galaxiidae and Comephoridae (Nelson 1976). Preparations for sectioning require chemical treatment which may introduce artefacts such as spaces between layers, between the epidermis and dermis or dermis and hypodermis. Unless special precautions are taken, fatty/oily tissue will also present as spaces.

Essential Fish Biology: Diversity, Structure and Function. Derek Burton & Margaret Burton.
© Derek Burton & Margaret Burton 2018. Published 2018 by Oxford University Press.
DOI 10.1093/oso/9780198785552.001.0001

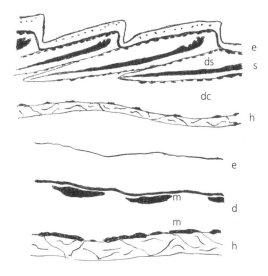

Figure 2.1 Vertical sections through fish skin, drawings based on light micrographs. (Above). Upper skin of winter flounder, *Pseudopleuronectes americanus*. (Below). Dorsolateral skin of three-spined stickleback, *Gasterosteus aculeatus*, Magnification for stickleback skin is about five times that of winter flounder. e = epidermis; d = dermis; dc = stratum compactum; ds = stratum spongiosum; h = hypodermis; m = melanophore; s = scale. Melanophores. In winter flounder skin show as dots in the epidermis and intermittent black areas subepdermally and above the hypodermis. The scales in flounder are ctenoid, i.e. toothed (2.4, Plate 3). Upper drawing based on previously published material, used with permission. D. Burton, 1978. Melanophore distribution within the integumentary tissues of two teleost species, *Pseudopleuronectes americanus* and *Gasterosteus aculeatus*, form *leiurus*. *Canadian Journal of Zoology* 56(4), 526–35. © Canadian Science Publishing or its licensors.

2.3.1 Epidermis

The epidermis covers the outer surface of the fish entirely in teleosts, forming a layer over the scales where they protrude through the dermis to lie horizontally or imbricated (sloped and overlapping like tiles) on the skin surface (Figure 2.1). However, in elasmobranchs, such as dogfish, which may have vertical scales, the epidermis does not cover the scale tips. In teleosts with deeply embedded scales the skin morphology can be quite complex as the epidermis follows the contours of overlapping scales (Figure 2.1). Conversely, fish with horizontal superficial scales (typically pelagic species) readily lose them (deciduous/caducous condition) on

handling and therefore skin preparation for microscopy may not show the scales at all.

The epidermis is technically an epithelium, separated from the dermis by a non-cellular basement membrane (Figures 2.2 and 2.3). Typically fish have a multilayered epidermis (stratified epithelium), the number of layers varying between species, and which may be reduced at times in the seasonal cycle (2.10). The dorsal epidermis is usually thicker in pelagic fish but in benthic species the ventral epidermis is often thicker (Bullock and Roberts 1975). The epidermis is not usually penetrated by blood vessels, thus in general any nutrients and chemical signals reach the epidermal cells by diffusion through the intercellular matrix (Van Oosten 1957), which presumably limits the depth of the epidermis. Access to the epidermis may be promoted by special regions in the basement membrane (see Figure 2.3).

An unusual adaptation has been described (Grizzle and Thiyagarajah 1987) for part of the skin of *Rivulus ocellatus marmoratus* (now known as *Kryptolebias marmoratus*). This fish can spend over a month out of water and apparently uses cutaneous respiration; its epidermis on the dorsal surface is supplied with capillaries which lie within 1 µm of the skin surface. A similar situation occurs in some gobies.

Most of the epidermis consists of cuboidal or ovoid Malpighian cells: those at the surface may be somewhat smaller but may be responsible for

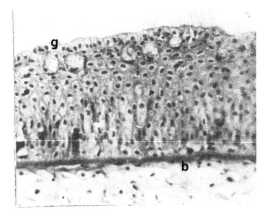

Figure 2.2 Light micrograph of lower epidermis of winter flounder, showing basement membrane (b) and goblet cells (g).

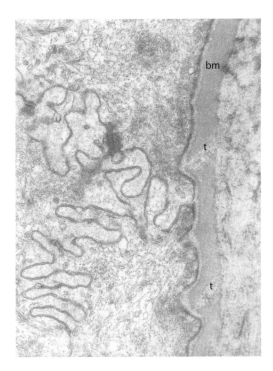

Figure 2.3 Electron micrograph (ultrathin section) showing basement membrane (bm) of winter flounder epidermis. Note special transmembrane regions (t) which may facilitate transport across the membrane (original magnification ×61,500).

secreting the mucoid cuticle. Scattered among these cells are specialized goblet cells, which secrete mucous, particularly prevalent at the surface. A few fish groups have club cells, which release a pheromone (14.7) on injury, alerting conspecifics to danger. Deeper in the multilayered epidermis may be eosinophilic granular cells implicated in immunity. Sacciform cells, with eosinophilic secretory material (Whitear 1986a), may produce toxins and be antiparasitic in, for example, salmonids (Pickering and Fletcher 1987). Ionocytes, mitochondria-rich cells similar to the chloride cells of fish gill, may be present amongst the outermost cells, and contribute to ionic regulation. Some fish have large epidermal melanophores, contributing to dark pigmentation. Sensory cells, and their supporting cells, including tastebuds, mechanosensory neuromasts (12.7.1) and tactile cells with their nerve fibres, are also found in the epidermis as well as free nerve endings (Whitear 1952, 1971a; Zaccone 1984) with unclear functions.

Epidermal ultrastructure also reveals Merkel cells (Lane and Whitear 1977; Lucarz and Brand 2007) in some fish species, which are innervated and may be mechanoreceptors (12.5.1).

Malphigian cells are so-called because they are mitotic and hence similar to the single layer of mitotic cells in mammalian skin which lies next to the basement membrane. However, in fish the mitotic capacity is retained in the multiple layers of the epidermis. In mammals, cells arising from the Malpighian layer become thinner and die as they reach the surface, becoming progressively keratinized with deposition of the relatively water-insoluble protein keratin but this process does not generally occur in fish which however have some keratinization in the cytoplasmic intermediate filaments. High concentrations of keratin have been described in a few specialized areas (e.g. lips) in some fish (Mittal and Whitear 1979; Zaccone 1979; Whitear 1986a,b) but it is not generally detected in most of the epidermis at the light microscope/histochemical level, in which case trans-skin diffusion (e.g. some gas exchange) is possible in contrast to its reduction or absence in highly keratinized tetrapods.

At the electron microscope (EM) level, intermediate filaments can be seen in the outer region of fish Malpighian cells and also form tonofilaments 7 to 12 nm thick, organized as clumps in desmosome regions at cell boundaries (Figure 2.4) and these may be keratin-rich as in other vertebrates. The desmosomes link cells, giving tensile strength. The protein keratin occupies an important position in the evolution of vertebrate skin, particularly in epidermal 'keratinization' integral in the transition from aquatic to terrestrial taxa. Padhi et al. (2006) present an overview of the keratin gene family in teleosts and other vertebrates. Keratin genes comprise 75 per cent of those for the structure of cellular intermediate filaments (Owens and Lane 2003). The human genome reveals up to 54 functional keratin genes (Moll et al. 2008) in comparison to 23 keratin genes in zebrafish (Padhi et al. 2006). Interestingly, two keratin genes, *K8* and *K18*, found in *Ciona* (a protochord tunicate) and expressed as simple epithelial keratins, are conserved in a wide range of species including teleosts (Schaffield et al. 2003). The indication is that epidermal intermediate filaments

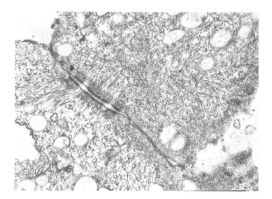

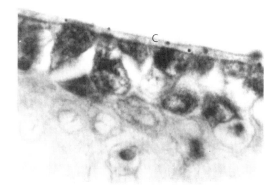

Figure 2.4 Electron micrograph (ultra-thin section) of winter flounder epidermis, showing desmosomes between adjoining Malphigian cells (original magnification ×75,000).

Figure 2.5 Light micrograph of section, showing a thin cuticle (c) lying on the epidermal surface (winter flounder stained with Alcian blue). Reproduced with permission, D. Burton, M.P. Burton and D.R. Idler, 1984. *Journal of Fish Biology* 25, 593–606.Epidermal condition in post-spawned winter flounder, *Pseudopleuronectes americanus* (Walbaum), maintained in the laboratory and after exposure to crude petroleum. Publishers John Wiley& Sons Ltd., © 1984, The Fisheries Society of the British Isles.

represent the evolutionary origins of 'keratinized' epidermis and hard keratin structures in terrestrial tetrapods, as well as the localized structures that resist abrasion in some fish.

Cuticle

The Malpighian cells of the epidermis may be strongly associated with, and produce, a 'cuticle', covering the surface (Whitear 1970). The cuticle, also referred to as a glycocalyx (Iger et al. 1988), is an acellular layer which may be seen overlying the epidermis (Figure 2.5). Shephard (1994) refers to it as a 'macromolecular gel'. The cuticle is mucoid, rich in mucopolysaccharides and may contain a variety of proteins, steroids such as cortisol (Easy and Ross 2006) and antimicrobial peptide (Cole et al. 1997). Shephard (1994) linked antibacterial qualities of the cuticle to the presence of lysozymes and/or complement.

The protein transferrin has been identified in the epidermal mucous of Atlantic salmon, *Salmo salar* (Raeder et al. 2007; Easy and Ross 2010) with an antibacterial role suggested. Lectins with potential antimicrobial activity also occur in skin mucous (reviewed by Ingram 1980 and Esteban 2012). Overall, fish epidermis is thus well adapted for a role as a primary contact in defence in an aqueous environment with a potential to be rich in pathogens. The cuticle although usually less than 1 μm in thickness (Shephard 1994) may be seasonally up to 60 μm thick in winter flounder (Burton and Fletcher 1983). Mucous in the cuticle is reportedly derived from the

surface layer Malpighian cells (Whitear 1986a) and may serve to reduce friction, drag and energy consumption when the fish swims, a concept reviewed by Shephard (1994).

Goblet cells

Goblet cells (mucous cells) are commonly found in the surface layers of fish epidermis (Figures 2.6 and 2.7). They are similar to the specialized mucous secreting cells found in vertebrate guts and gills. They are often elongated or goblet-shaped (hence their name) with a central area holding mucous globules; the nucleus is therefore eccentric. They may also be spherical or appear so in histological sections. Goblet cells are scattered amongst the epidermal surface cells as well as occurring in clusters or concentrations, for example in folds between scales. The EM (Figure 2.6) shows large, pale inclusion, the mucous globules. Numbers of goblet cells may increase in response to season, reproductive cycle or pollution (2.10).

Pleurocidin (an antibacterial peptide) has been described in skin mucous cells (Cole et al. 1997). Whitear (1970) suggested that the role of goblet cell mucous is distinct from that of the cuticle, providing emergency lubrication with indications that its discharge can displace the cuticle. Carp, *Cyprinus carpio*, subjected to one environmental stressor, exhausted

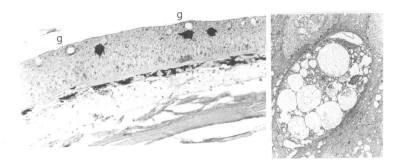

Figure 2.6 Goblet cells of winter flounder skin. Left: light micrograph of sectioned upper skin showing superficial goblet cells (g) in epidermis. Black pigmented epidermal and dermal melanophores are present. Right: electron micrograph (ultrathin section) of an epidermal goblet cell with mucous globules (original magnification ×9300).

its goblet cell supply within three days, new goblet cells appearing in eight days (Iger et al. 1988).

Club cells

These (Figure 2.7) tend to be larger than goblet cells, are club-shaped, have a central nucleus and no mucous granules and occur deeper than goblet cells in the epidermis. They are restricted in occurrence, having been described from freshwater ostariophysans, for example, cyprinids, silurids and also some marine fish.

If an ostariophysan is injured, damaged club cells release a pheromone, termed alarm substance or 'Schreckstoff' (von Frisch 1938, 1941a). This apparently warns conspecifics of danger, eliciting a fright reaction (Smith 1992), and the club cells are sometimes called 'alarm substance cells'. The presence of club cells does not necessarily mean the fish species will show a fright reaction; thus piranha have club cells but do not show a fright reaction (Irving 1996). It has been suggested (Irving 1996), following the work of Smith (1982), that alarm signalling is a secondary function and that the club cells are primarily concerned with antipathogenic secretions. Numbers of club cells vary interspecifically (Pfeiffer 1963) and are androgen-depressed (Irving 1996). The chemical nature of the Schreckstoff is still under discussion: a recent paper (Mathuru et al. 2012) found chondroitin fragments to be more active in zebrafish than hypoxanthine-3-N-oxide previously suggested as the pheromone.

The role of club cells in marine fish is not well understood; in Tetraodontiforms club cells have been

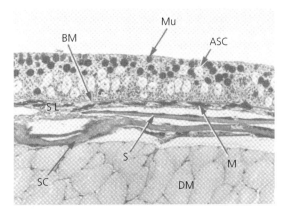

Figure 2.7 Light micrograph of a vertical section through the skin of a common shiner (*Notropis cornutus*). ASC = alarm substance (club) cells; BM = basement membrane; DM = dorsal musculature; M = melanophore; Mu = mucous (goblet) cell; S = scale; SC = stratum compactum; SL = stratum laxum (stratum compactum). After J.F. Smith (1976). Seasonal loss of alarm substance cells in North American cyprinoid fishes and its relation to abrasive spawning behavior. *Canadian Journal of Zoology* 54, 1172–82 ©Canadian Science Publishing or its licensors.

linked with secretion of defensive toxins (Thomson 1969) although sacciform cells have also been implicated.

Eosinophilic granular cells

Eosinophilic granular (EG) cells occur in highly variable numbers in deeper layers of the epidermis (Plate 1). They are particularly abundant in the Pleuronectidae (Roberts et al. 1972; Bullock and Roberts 1975) and occur in carp, *Cyprinus carpio*, in manured water (Iger et al. 1988). Their appearance is reminiscent of eosinophilic granulocytes (leucocytes) of the blood (5.13.2) and it is possible that is what they are, if eosinophils can cross the basement membrane (leucocytes can enter tissues by diapedesis, Paulsen 1993). However, Blackstock and Pickering (1980) believe that the EG cells are not 'transformed blood cells' but normal epidermal components. EG cells, as their name suggests, have a very high affinity for eosin and are readily detectable as rounded cells with bright red granules, in skin sections stained by routine haematoxylin and eosin procedures. The granules show as dense black structures in electron micrographs (Figure 2.8). The prevalence of granular cells in the upper epidermis of lampreys (Pfeiffer and Fletcher 1964) is interesting, but it is not possible to infer a direct relationship with the teleost EG cell, particularly as the lamprey cells can open to the exterior via a pore whereas although teleost (plaice) EG cells have been recorded as rupturing at the surface (Roberts and Bullock 1981), this was under very unusual circumstances (ultraviolet (UV) illumination). Also, Pfeiffer and Fletcher (1964) believed that the lamprey granular cells developed from 'club cells' in which case it seems unlikely that the club cells in lampreys are equivalent to those of teleosts.

Sacciform cells

These cells are relatively large and elongated. They are eosinophilic secretory cells with a basal nucleus and granular cytoplasm, and their secretion

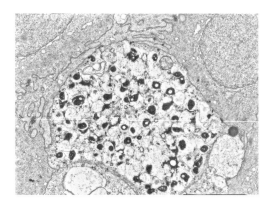

Figure 2.8 Electron micrograph (ultrathin section) of winter flounder epidermis showing an eosinophilic granular cell (original magnification ×24,000).

is contained in a large vacuole supplied by small cytoplasmic channels (Whitear 1986a; Fasulo et al. 1993). Sacciform cells are portrayed (Fasulo et al. 1993) with pores to the exterior surface. These cells have been reported from teleosts as well as chimaeras, rays and polypterids (Whitear 1986a). Sacciform cells increased when an ectoparasitic flagellate-infested immature brown trout and were concluded to be defensive against skin parasites (Pickering and Fletcher 1987). Fasulo et al. (1993) described 'bioactive peptides' and serotonin in sacciform secretions of two teleosts; tetrodotoxin (of puffer fish) may be localized in these cells (Whitear 1986a), features of these cells inviting comparison with the toxin-producing capacity of amphibian skin.

Ionocytes

Ionocytes (chloride cells) are not readily distinguished from Malphigian cells in routine preparations and may not be widespread. They have been described from sea bass, *Dicentrarchus labrax*, embryos (Sucré et al. 2010) and herring, *Clupea harengus*, larvae (Wales and Tytler 1996) and from the operculum of adult *Gillichthyes mirabilis* (Marshall and Nishioka 1980) using specialized stains appropriate for chloride cells (7.7). It is postulated that they precede (larvae) or augment the activity of gill chloride cells, assisting in ion regulation. At the EM level skin ionocytes would be characterized by a large number of mitochondria.

Epidermal melanophores

Epidermal melanophores, if present, contribute to dark colouration. In winter flounder these melanophores are different in shape and size from deeper dermal melanophores (Figure 2.9). As with the dermal melanophores, the cells contain dark-coloured inclusions termed melanosomes, packed with the pigment melanin (usually black) which can migrate within the cell. Aggregation of the melanosomes results in skin paling whereas their dispersion gives a darker appearance (2.6, 2.9).

2.3.2 Dermis

Whereas the epidermis is an epithelium, the dermis is basically composed of connective tissue with collagen fibres, which in vertebrates are expected to

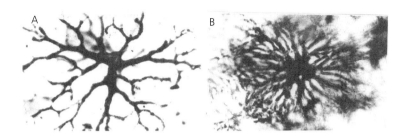

Figure 2.9. Melanophores of winter flounder skin, surface view from whole mounts of scales. A: Light micrograph, epidermal melanophore. B: Light micrograph, dermal melanophore.

be derived from individual cells termed fibroblasts, not readily distinguished in adult fish. Dermal fibrocytes (mature cells) of the minnow *Phoxinus* are figured by Whitear et al. (1980).

The dermis may be more complex in appearance than the epidermis. Scales, if present, may be embedded in it (Figure 2.1) and it is well-vascularized and contains myelinated (e.g. from the lateral line) and non-myelinated nerve fibres (e.g. to melanophores). The dermis acts as a strong flexible jacket with the scales and collagen fibres providing constraint.

The dermis can consist of two major layers, the stratum spongiosum and the stratum compactum (Bullock and Roberts 1975), as in winter flounder (Burton 1978), although in the stickleback, which has scutes rather than scales, the distinction between the layers is not so clear.

Stratum spongiosum

The outermost layer of the dermis is the stratum spongiosum or stratum laxum (Smith 1976); it has the scale bases within it and also contains chromatophores immediately beneath the basement membrane of the epidermis (Figure 2.1). The chromatophores can be of various types: melanophores (dark, Figure 2.9), erythrophores (red), xanthophores (orange or yellow, Plates 2 and 3) and iridophores (reflecting, Figure 2.14) and the recently described (Goda and Fujii 1995) cyanophores (blue). Chromatophores are the principal agents providing colour and colour change (2.6, 2.9). The stratum spongiosum has a less dense appearance than the stratum compactum beneath it.

Stratum compactum

The innermost layer of the dermis, the stratum compactum, contains strands of collagen organized in opposing spirals around the myotomal musculature of the body wall (bias sleeve, Whitear

1986b) so that the body of the fish is wrapped in a geodesic lattice giving support and strength while enabling bending and preventing wrinkling of the skin. This pattern can be discerned at the EM level (Figure 2.10). Whitear et al. (1980) presented an ultrastructural study suggesting an endothelial

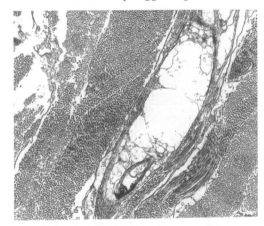

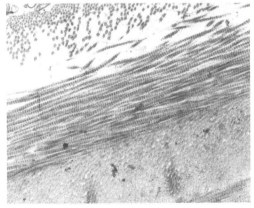

Figure 2.10 Electron micrographs (ultra-thin sections) showing collagen fibres in the stratum compactum of winter flounder skin. Note the different patterns (bias sleeve) of fibre orientation. Above: collagen fibres on each side of scale pocket (original magnification ×16,500). Below: higher magnification showing two sets of fibres at approximately 90° to each other (original magnification ×37,500).

layer on the inner surface of the stratum compactum. Flammang (2010) describes the stratum compactum of sharks, composed of five main layers, and the relationship to the radialis muscle, with the possibility that the combined effect is postural, stiffening the caudal fin. Ono (1982) found spray-type nerve endings in the compactum of pickerel (Esocidae) and suggested they were proprioceptors, supporting the possible postural function. A deep layer of chromatophores may occur in some species in the boundary region between the stratum compactum and the hypodermis (Figure 2.1).

2.3.3 Hypodermis

Beneath the dermis is the subcutis (Mittal and Banerjee 1975) or hypodermis (Figure 2.1) formed of loose connective tissue (Bullock and Roberts 1975) with adipose tissue representing a potential fat reserve which varies with the nutritional status of the fish (Maddock and Burton 1994). In histological sections the hypodermal fat may be extracted by the processing and the hypodermis itself may not be very obvious. The amount of hypodermal fat deposited varies not only with the individual but with the species; whereas aquatic mammals, for example, may have large subdermal deposits of fat as blubber, fat deposits in fish, principally acting as energy reserves, can occur in a number of different taxon-related sites such as the liver, the mesenteries and the hypodermis (7.12.2). Bullock and Roberts (1975) state that the hypodermis is flexible and allows movement between the dermis and the muscle (myotomes).

Yancey et al. (1989) described a gelatinous layer in the subcutaneous (i.e. hypodermal) region of mesopelagic teleosts, to which they attributed a possible buoyancy function (3.8.5).

2.4 Fish scales

The terminology for scales, the materials composing them and their evolutionary (phylogenetic) and embryonic (ontogenetic) history has become complex (Smith and Hall 1990; Richter and Smith 1995; Sire et al. 2009). Teleost scales may show annual rings, when growth is seasonal, and this can be used to obtain age of individuals (1.4.1, 3.3, 7.13.5).

Dogfish scales (Figure 2.11) strongly resemble teeth and are sometimes called 'dermal denticles'. This similarity has stimulated a number of hypotheses, with a relationship between scales and teeth (4.4.5) regarded as being due to evolutionary movement 'outside-in' or 'inside-out' or a modification of this directional migration (Huysenne et al 2009; Witten et al. 2014). The 'outside-in' hypothesis regards teeth to be derived from scales /odontodes which invaded the oral cavity, with the reverse movement envisaged for the 'inside-out' hypothesis.

The insertion of scales into the dermis is either vertical, as in dogfish, or angled, as in most teleosts. Whereas the insertion point in dogfish can be regarded as a socket, the insertion in teleosts is often referred to as a pocket (Bullock and Roberts 1975; Burton 1978; Whitear 1986b), although the overlap of some teleost scales also produces 'pockets' (Burton and Fletcher 1983).

Many fish have a more or less continuous covering of scales which may be deeply embedded (Figure 2.1) or superficial and easily detached, the condition known as deciduous or caducous, as found in pelagic fish like herring or mackerel. The development of the scales is discussed by Sire and Akimento (2004); the scale pattern which develops in teleosts is usually overlapping (imbricated) and predetermined, whereas that of elasmobranchs is neither, but may still have a regular appearance (Figure 2.11).

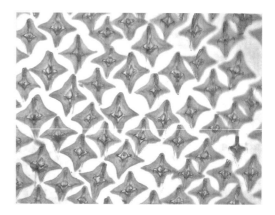

Figure 2.11 Light micrograph, surface view of dogfish skin showing regular pattern of scales (dermal denticles).

Scales are basically formed of a hard material like bone or dentine (Tables 2.1 and 2.2) covered by epidermis in teleosts but protruding through the epidermis in elasmobranchs. The relict fish species generally have more robust scales than the teleosts and the structure of scales in general has been the subject of much study and debate. Part of the interest in scale structure occurs because scales of extinct forms are available for comparison; scales are often readily available for study and they may reflect the individual history of a fish (age/spawning events). The considerable interest in these structures from morphologists, taxonomists and palaeontologists has produced a rather large number of sometimes confusing terms (Tables 2.1 and 2.2).

The supporting material of fish scales is a hard substance like bone or dentine with a matrix, expected to have some calcification and collagen fibres within it. It may contain channels (dentine, cosmine, Table 2.2) and the collagen fibres may be layered in a manner reminiscent of plywood. Some scales, particularly the large ones, associated with plesiomorphic (older lineages) forms may have several layers including a basal plate of bone, overlaid by, for example, cosmine or ganoine (a form of enamel). Sire et al. (2009) distinguish between 'hypermineralized' non-collagenous materials (such

Table 2.1 Scale types, occurrence and features.

Fish	Scale type	Alternate terms	Features of the scales
Elasmobranch	Dermal denticle	Placoid scale, odontoid	Very robust, like teeth, with enamel/enameloid
Teleost	Cycloid*	Elasmoid	May be very thin, 'acellular' bone?
	Ctenoid*	Elasmoid	With teeth
Coelacanths	Cosmoid		Multilayered, enamel/enameloid surface
	Modified cosmoid		Less robust, no enamel/enameloid
Gar	Ganoid		Multilayered, very hard, surface layers enamel/enameloid
Bichir	Ganoid		

*Cycloid and ctenoid can co-occur.

Table 2.2 Types of supporting material associated with scales and scutes, with layers of different material present in some relict fish and their forbears.

Name of material	Major features	Found in
Dentine	Channels present in matrix	Elasmobranchs
Cosmine	Branched channels in matrix	Coelacanths/ crossopterygians
Elasmodine	Matrix with a distinct array of collagen fibres similar to plywood structure (Sire et al. 2009) May be unmineralized. Variously termed lamellar dentine or bone	Teleosts, *Latimeria, Polypterus*
Enamel	Very hard, highly calcified	Elasmobranchs? Gar? Bichir?
Enameloid	Distinct patterns of crystal bundles (Richter and Smith 1995) Less ordered than enamel, not prismatic (Sire et al. 2009)	Elasmobranchs? Gar? Bichir?
Ganoin(e)	Layers*of enamel or enameloid	Gar, bichir
Hyaloin(e)	Very similar to ganoin but has a mesenchyme filled space separating it from the epidermis (Sire et al. 2009).	Catfish (over scutes)
Acellular bone	Matrix as in bone, but no cells present in the interior	Some teleosts
Isopedine	Dense bone (Greenwood 1975)	Coelacanths/ Crossopterygians Ganoid scale bases in Polypteriformes (bichirs), *Amia*

*But according to Richter and Smith (1995), ganoin(e) can be single-layered; they consider 'ganoine' to comprise either enamel or enameloid.

as enamel, ganoine and hyaloine) and plywood-like arrays of collagen fibres. However, the distinction between subsets of these materials is not always clear; hyaloine is said to be 'poorly understood' and its identification appears to depend on a 'mesenchyme-filled' space next to the epidermis as well as the presence of non-collagenous fibrils.

Scale patterns and types (squamation) can be used taxomically to differentiate species and this area of research specialism is termed lepidology (Jawad 2005).

A reduction in scale weight over evolutionary time is noted (Gemballa and Bartsch 2002), with ganoid scales replaced by elasmoid scales in the actinopterygians and cosmoid scales reduced as in *Latimeria*. Meunier (1984) prefers the dichotomous classification of scales into rhomboid or elasmoid, the latter being 'thin and flexible' with an 'imbricated fitting in the tegument'. Using this system, Meunier places scales of *Latimeria*, the lungfish, and *Amia* in the elasmoid category. As neither *Latimeria* nor *Neoceratodus* seem to have scales which fit readily into the 'thin, flexible' category, it is likely that this debate has not been settled, although the more recent work of Sire and Akimento (2004) and Sire et al. (2009) provides a further basis for discussion.

The tendency for bone reduction and the theme of 'plywood-' type tissues formed of layers of collagen rather than acellular bone as the main structure of scales in derived forms are important new issues. However, illustrations of scale structure may be simplified by omitting the surrounding dermis or overlying epidermis, both of which, being soft and easily abraded, can be lost in preparation and storage. This simplification may perhaps give the wrong impression of the scale's environment and functions *in situ*. The 'plywood' layering of collagen frequently and recently referred to in discussions of hard tsssues is reminiscent of the 'bias sleeve' arrangement of collagen in the stratum compactum (see Figure 2.10).

The first scales develop (in bony fish) at a time relevant to the ultimate size of the scales and the fish, with a species-specific minimum fish length when scale formation begins. The eel is exceptional, having late scale development (Sire and Arnulf 1990). In general, induction of scales occurs first when several scales appear at a species-specific site, often the caudal peduncle initially. In the cyprinodont *Rivulus marmoratus*, however, the first scales appear on the head (parietal region) at 8 days post-hatching with a caudal patch seen at 23 days and squamation complete after 6 weeks (Park and Lee 1988).

Elasmobranch scales are often referred to as dermal denticles, being morphologically similar to the jaw teeth: they are also known as 'placoid scales' or odontoids (Greenwood 1975). In their most typical form they are seen as numerous

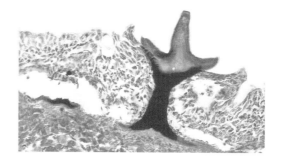

Figure 2.12 Light micrograph, vertical section through dogfish skin showing a scale (dermal denticle).

protrusions covering the surface of dogfish (Figures 2.11 and 2.12) and sharks, giving a rough surface. They are indeed like small teeth with their tips covered by enamel, but not by the epidermis. In skate they form groups of varying size on the dorsal surface, the larger scales forming long spines in some species. The covering of small scales typical of dogfish and sharks is thought to improve swimming performance (Raschi and Tabit 1992). The taxonomically related chimaeras have few scales.

Teleost scales vary in diameter from microscopic in eels to huge (about 5 cm diameter) in tarpon (Greenwood 1975). In many teleosts the scales are formed of a very thin tissue, originally thought to be bone, which, however, may be described as acellular, lacking cells and avascular (no blood vessels) with a tendency towards acellularity increasing in 'advanced' teleosts such as acanthopterygians. However, the Thunnidae are exceptions to this trend, perhaps explained by their high metabolic requirements (Meunier and Huyseune 1991). The lack of internal cells and blood vessels means that growth may occur by peripheral accretion with formation of annual growth rings, added to scale circumference, in fish with marked seasonality in their growth cycles. Scale rings may be used to estimate fish age (3.3, 7.13.5).

The concept of 'acellular bone' has recently been complicated/ superceded by further research on isopedine and elasmodine in extinct and extant species. These two tissues are characterized by parallel collagen fibres and the two types of tissue are considered distinct (Sire et al. 2009). The term 'elasmoid scales' is now being applied to scales of extant teleost species (Sire and Akimenko 2004), a condition regarded as highly derived.

In shape teleost scales are divided into two types: cycloid which are round and smooth (Plate 2) and ctenoid which are round and toothed or spiny (Figure 2.1, Plate 3) at the upper posterior end. Fish may have both types present, thus Scott and Scott (1988) report that pleuronectid flatfish can have ctenoid scales on the upper (ocular) surface and cycloid scales beneath (blind, lower surface), although it has also been reported (Bejda and Phelan 1998) that the distribution is sex-related, with upper and lower ctenoid scales in males. An extreme form of ctenoid scale, reported from the male darter *Percina*, develops during the reproductive season. These mid-ventral scales are very spiny and thought to be important for holding the female (Page 1976).

The relict (plesiomorphic) fish may have scale types not found in other extant groups.

The coelacanth (*Latimeria*) has modified cosmoid scales according to older sources. A typical cosmoid scale (Romer 1955) had basal layers of dense bone overlaid by spongy bone, with a surface layer of cosmine, similar to dentine with canals (but branched in cosmine) in it, and a thin layer of enamel. However, *Latimeria* does not show this typical cosmoid structure and Sire et al. (2009) have abandoned the 'cosmoid' attribution and consider the extant coelacanth's scales to be 'elasmoid', consisting of elasmodine overlaid with ganoin(e). The lack of typical bone distinguishes these scales from the ganoid scales of the gar.

The situation with the extant lungfish is not yet clear; of the three extant genera (*Neoceratodus, Lepidosiren* and *Protopterus*), *Neoceratodus* appears to have the heaviest scales, somewhat similar in external appearance to the large scales of *Polypterus*. Scales of extant lungfish have also been referred to as modified cosmoid, which may well be true in an evolutionary sense as early (Devonian) forms had cosmoid scales. However, Sire et al. (2009) state that cosmine is not present in any of the three modern genera and that the scales are 'typically elasmoid'.

The gar (*Lepisosteus* formerly *Lepidosteus*) and the bichir (*Polypterus*) are usually described as having ganoid scales, thick scales with a basal bone layer which was, in older sources, described as overlaid by cosmine in *Polypterus*, and a thick surface layer of ganoin, composed of many layers of enamel. These ganoid scales form a heavy covering, with some flexibility; the scales, which are roughly rhomboid (rhombic) in shape, in surface view may be linked with embedded pegs and fibres (see illustration in Greenwood 1975).

Sire et al. (2009) regard the gar scale as derived from the polypteroid type but do not use the term 'ganoid' for extant gar scales, preferring 'lepisosteoid'. The same authors do not use the term 'cosmine' for any layer between the bone and the capping tissues in the 'ganoid' or 'cosmoid' scales they figure, apparently regarding the term as confined to material with well-developed pore canal systems as found in some fossils. Sire et al. (2009) regard the rhombic shape as basic, featuring 'putative ancestral rhombic scales' in their figured evolutionary scheme of scale relationships.

The presence of isopedine sometimes as a very thin but apparently distinctive layer of parallel or twisted collagen fibres—for example between the dentine and the bone layers in Polypteriformes, but also described in *Amia* and sarcopterygians—has been useful in aiding identification of fossil scales. It is regarded as diagnostic of Polypteriformes if it occurs in a ganoid scale (Daget et al. 2001). However, there does not seem to be general agreement yet on the use of the term isopedine as distinct from elasmodine; if evolutionary origin of these determines the terminology it will be quite difficult to differentiate them for practical purposes. The reinterpretation of these terms is discussed in Sire et al. (2009).

2.5 Bony plates and scutes

Bony plates, with typical bone cells (osteocytes), which may be termed 'scutes' if capped with hyaloin(e)or ganoin(e) (Sire et al. 2009), are found in some groups, in which case the fish may be regarded as 'armoured', as in 'armoured catfish'. Greenwood (1975) provides a figure showing eight teleost species which can be considered 'armoured', with some of the armour more scale-like.

The presence of bony plates in some species seems to be dependent on the immediate environment of individual fish so that the number or size of bony plates is highly variable. This applies to fish in the Family Gasterosteidae (sticklebacks); Scott and

Crossman (1973) describe two forms of *Gasterosteus aculeatus*, the marine *trachurus* with large lateral plates, forming a 'complete series', and the freshwater *leirurus* with few or no lateral plates (Figure 1.22). Mattern (2006) reports that *Pungitius pungitius* has 5 to 13 lateral plates while *P. sinensis* has 28 to 35. Very large scutes are found on sturgeons (Acipenseridae) and contribute to their unusual appearance; however, the capping ganoin(e) found in fossil species is lost in extant forms. The extra weight of scutes must slow down fish bearing them and represents the reverse of the usual trend, which seems to have been toward weight reduction, termed a 'general regression of dermal ossification' (Meunier 1991).

2.6 Skin colour

Fish are noted for a huge diversity of colour and pattern, both of which are due to special structures in the skin. Many fish such as the gadids (cod, etc.), pelagic herring and the salmonids are grey or silvery in colour but there is an enormous range of patterns and colours among fish species, with small 'tropical' fish noted for bright colours such as yellow and red. Modifications of both patterns and colour of individuals are possible over time. Flatfish have patterns which can be complex (Figure 2.13) and which can change, becoming adapted to both the texture and colour of the current substrate. Flatfish patterns (which can change very quickly, Appendix 2.2) can include large or small, pale or dark spots and dark bars or the skin can appear uniformly dark or pale.

The colour of fish can be cryptic, providing concealment, or bright, communicating status to other fish. The range of colour between species is enormous but it can also differ within species, changing during maturation, or with habitat change on a long- or short-term basis. Thus stickleback males redden at the spawning season, salmon smolts become silvery losing their black parr marks as a preadaptation to marine habitat and flatfish can change colour and pattern in a few minutes. The skin colour may also change during ontogeny; larval colour is discussed for over 200 species of tropical teleosts (Baldwin 2013) and the need for more information emphasized; a particular problem is the loss of some pigmentation on attempting preservation. Colour patterns on fish skin can be extremely complex; patterns may be very labile, changing with stage of maturity or sex, with social dominance and with background. Conversely, some coral-reef fish are characterized by a stable though bright and distinctive pattern. Patterns may also mislead predators (Plates 4 to 7), with 'false eyes' in various positions, a situation seen with some reef species.

Fish-skin colouration is produced either by pigments, in specialized cells called chromatophores, which absorb different wavelengths of light (and reflect others producing the observed hue), or by intracellular or extracellular structural features such as fibrous layers and/or layers of guanine crystals (e.g. the stratum argenteum, Denton and Nicol 1966), producing irridescence due to light interference or other bright colours by light scattering. In some fish small reflecting plates are stacked in an

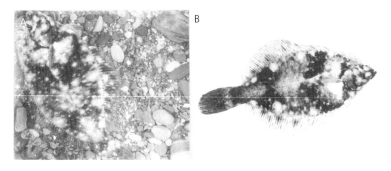

Figure 2.13 Photographs of winter flounder. A: Following adaptation to a pebble background. B: Newly transferred to a white background, showing the pattern before paling has occurred. Reproduced with permission, D. Burton 1979. Differential chromatic activity of melanophores in integumentary patterns of winter flounder (*Pseudopleuronectes americanus* Walbaum). *Canadian Journal of Zoology* 57, 650–7 © Canadian Science Publishing or its licensors.

orderly array in special chromatophores called iridophores (Figure 2.14). Skin colour often represents a visual summation of the chromatic effects of more than one morphological element. It is important to reconcile the appearance, if described in air, of the fish as it appears in water, viewed in the light available to conspecifics and potential prey and predators.

2.6.1 The physical basis for colour: chromatophores

Pigmented chromatophores contain single colours/ pigments (unlike some invertebrate chromatophores which may have more than one) and include brown or black melanophores (Figure 2.9). The pigments in fish chromatophores are segregated in cell inclusions, chromatosomes, individually termed by their contained pigment. With melanin-containing melanosome inclusions (Figure 2.14), the melanin is derived from the amino acid tyrosine. Another major group of fish chromatophores is carotenoid-containing and diet-dependent; the xanthophores (yellow or orange) and erythrophores (red). Guanine forms either white granular inclusions in leucophores, shaped like melanophores, which are reported from cyprinodonts (Fujii 1993) or parallel plates (Figure 2.14), responsible for the reflective or white surfaces of iridophores. The blue pigment of recently described cyanophores (Goda and Fujii 1995) has not been chemically characterized. Some of the pigments, with the notable exception

of melanin, do not preserve well, and therefore descriptions of fish colour using dead specimens may be unreliable.

2.6.2 Dark colouration

Dark colouration in fish is usually due to dermal melanophores with contribution from epidermal melanophores in some species. Both kinds of melanophores can show adaptive movement of the intracellular melanosomes, achieving gradations of darkening or paling at the microscopic and macroscopic level. Changes in individual melanophores are shown as a graded series in Figure 2.15. The melanosomes aggregate to achieve skin paling, or disperse to achieve darkening. Unlike cephalopod (octopus, etc.) chromatophores the cells do not expand or contract. Fish melanophores strongly resemble glial cells (e.g. astrocytes, 11.3.1) and have been termed paraneurons (Fujii 1993).

The control of melanosome movement (2.9 and Appendix 2.3) is a complex physiological situation; slow endocrine control has been observed in some elasmobranchs whereas teleosts may have slow endocrine control and/or swift colour change under neural control.

Although melanosome motility is not fully understood, it is accepted that it depends on mechano-chemical activity of microtubules, radiating from the melanophore centre to its periphery (Fujii 1993), mediated by changes in the level of a second messenger such as cyclic adenosine monophosphate

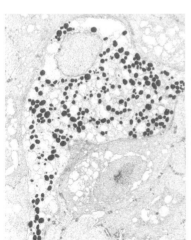 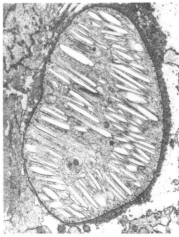

Figure 2.14 Electron micrographs (ultra-thin sections) of winter flounder dermis. Left: electron-dense melanosomes in a melanophore (original magnification ×20,400). Right: parallel plates in an iridophore (original magnification ×24,000).

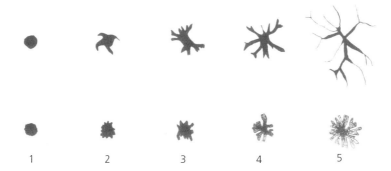

Figure 2.15 Drawing of apparent 'shape changes' associated with melanophores of winter flounder. Top row: epidermal melanophores; middle row: dermal melanophores; bottom row: melanophore index: 1 = fully aggregated, 5 = fully dispersed. Note: the melanophore itself does not change shape but movement of the internal melanosomes in the translucent cytoplasm produces an apparent shape change of the melanophore and skin colour change. Reproduced with permission. D. Burton 1979. Differential chromatic activity of melanophores in integumentary patterns of winter flounder (*Pseudopleuronectes americanus* Walbaum) *Canadian Journal of Zoology* 57, 650–7. © Canadian Science Publishing or its licensors.

(cAMP), hormonally (10.10) and/or neurally (11.10.6 and Appendix 2.3) controlled.

Many fish have a dark dorsal surface, and/or dark stripes, bars or spots, all of which are due to the discontinuous distribution of melanophores. Melanin pigment may also be found round the gut, notably in the peritoneum (4.4.6), and around blood vessels. Its function in these locations is not clear (13.4.7).

2.6.3 Red, orange and yellow colouration

Red, orange and yellow colours in fish are usually attributable to concentrations of dermal chromatophores such as xanthophores (yellow/orange) and erthythrophores (red). Male teleosts such as sticklebacks and salmonids often have increased red colouration prior to and during spawning, in association with higher androgen levels, as a secondary sexual characteristic. The carotenoid pigments of the xanthophores or erthyrophores are in discrete inclusions (xanthosomes, erythrosomes) which can aggregate or disperse (Hayashi et al. 1993; Murata and Fujii 2000).

2.6.4 Blue and green colouration

In fish, blue and green colours are not generally due to a specific pigment but to a physical light-reflecting quality of iridophores and/or other skin structure. Resulting irridescence may be perceived as green or blue, as with mackerel and salmonids and some smaller tropical fish. Goda and Fujii (1995) have described a blue pigment, which remains chemically undefined, in callionymids. The occurrence of 'extra-cellular scatter' producing blue colouration in vertebrates is discussed in Bagnara et al. (2007) in which it is attributed to 'ordered dermal collagen arrays'. Whereas blue fish are quite common, especially on coral reefs, green fish or green markings are relatively rare, unlike in amphibia and reptiles. *Amia calva* breeding males are described as having green fins and there is a green-sided darter *Diplesion* (*Etheostoma*) *blenniodes* (Forbes 1907). The breeding male rainbow darter (*Etheostoma caeruleum*) has an array of colours including green, particularly on the head (Plate VI, Scott and Crossman 1973, p. 730). Amongst marine fish there is a 'green moray' eel (*Gymnothorax funebris*) which also has a brown form (Stokes 1980) and a green 'queen parrotfish', *Scarus vetula*. The Australian surf parrotfish is also portrayed as green (Kuiter 1996) and other parrotfish may be blue-green or have green areas as with the head of male 'yellowhead' parrotfish (*Scarus spinus*), also Australian.

2.6.5 Silver and white colouration

Silver colouration may be due to iridophores. Denton and Nicol (1966) describe reflectivity in numerous fish species, with the presence of a special subdermal layer, the stratum argenteum, which contains numerous guanine crystals. In some fish, the peritoneum is silvery too.

White colouration in fish is due to leucophores or iridophores. Many fish have a pale or white ventral surface and white or pale spots on their upper surface, which may be masked by overlying melanophores if their melanosomes are dispersed.

2.7 Habitat and colour

Pelagic fish, which are usually silvery, may display countershading (Cott 1940) with a dark upper body camouflaging the fish when viewed from above against the darkness of the water or substrate, and a pale lower body blending with the bright water surface when viewed from below. Concealment can also result from 'disruptive colouration' (Cott 1940, Appendix 2.1, Plates 4 and 5) which makes the body outline difficult to distinguish as occurs with some flatfish patterning (Figure 2.13). False 'eyes' found in many tropical fish (Appendix 2.1, Plates 5 and 7) as a conspicuous body marking distant from the real eye, and eye stripes or body stripes (Appendix 2.1, Plates 4 to 7) may similarly mislead predators in their view of the size and orientation of potential prey. Male *Amia calva* develop a marked 'tailspot' during the breeding season (Reighard 1903); it seems unlikely that this functions to conceal or mislead, rather it could distinguish mature males from females. Johnsen (2014) has reviewed the principles involved in camouflage in 'featureless' pelagic environments.

Cave-dwelling fish (13.5.1) are usually blind and generally lack melanophores and the carotenoid-bearing chromatophores, that is, they are 'albino'. They may appear pink (Plate 13) because of the presence of haemoglobin. Deep-sea fish may also be blind or conversely have very large eyes but they are seldom, except for very deep and rare hadal forms (13.4.3) described as lacking colour; abyssal forms are usually black, indicating melanin presence. In the lowest region of the marine photic zone, where some blue light penetrates, fish (and other animals, e.g. crustacea) may be red (thus red-fish, *Sebastes*), presumably due to erythrophores, and probably concealing these fish from

predators in an environment only penetrated by longer wavelengths. Coral reef fish are typically very colourful (Plates 4 to 7), a situation presumably evoked by the complex ecosystem where conspecific and interspecific signalling occurs in conditions of bright light.

2.8 Abnormal skin colour

Under culture conditions some fish, notably flatfish, develop abnormal pigmentation: for flatfish this appears as more white than usual (hypomelanosis) on the upper surface, or more black than usual (hypermelanosis) on the lower surface. These abnormalities may be accompanied by incomplete eye migration and the resulting fish may not be readily marketable. Flatfish with partially or completely dark lower surfaces, termed ambicolouration with similar melanophores on both sides, have been reported from the wild in species which normally have a white lower surface. The 'turbot' or Greenland halibut (*Reinhardtius hippoglossoides*) always has both sides with melanophores, a derived state related to its pelagic habit. The dermal melanophores (Figure 2.16) on the two sides are morphologically quite different in this species (Burton 1988a).

2.9 Colour change

As many fish undergo colour change, colour itself is not necessarily a good taxonomic characteristic, although it can contribute to identification and also may be used with confidence for some coral-reef species.

Fish can change colour during their life cycle by changing numbers, pigment amount or types of chromatophores (2.6.1) or they may briefly adjust colour by intracellular changes in response to background or stress. Colour adjustment is due to neural or hormonal activity, the hormones include melanophore-stimulating hormone (MSH), melanin/melanophore-concentrating hormone (MCH), adrenalin, somatolactin, affecting different chromatophores, especially the melanophores. A role for somatolactin (10.12.8) has been suggested recently (Cánepa et al. 2006). It is reported that some fish (e.g. elasmobranchs) do not have neural regulation

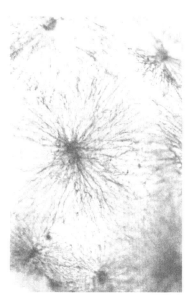

Figure 2.16 Light micrographs (surface view) of dermal melanophores from 'turbot' (*Reinhardtius hippoglossoides*). Left: from left side. Right: from right side. Reproduced with permission. D. Burton 1988a. Melanophore comparisons in different forms of ambicolouration in the flatfish *Pseudopleuronectes americanus* and *Reinhardtius hippoglossoides. Journal of Zoology, London*, 214, 353–60. John Wiley & Sons Ltd. © 1988 The Zoological Society of London.

of their chromatophores whereas others (flatfish) have both neural and humoral control with the former dominant.

Fish may be able to change colour over a long or very short time period; thus there may be a colour change before spawning or a quick change in response to background. Colour change in fish is reviewed by Fujii (2000) and Burton (2011) and the complex pattern adjustments and their control typical of flatfish are discussed by Burton (2010). The measurement of fish colour can present more complexity than that encountered with animals that have less subtlety in the integumentary pigmentation. Thus, bird or lepidopteran (e.g. butterfly) integuments can be measured on representative patches (Endler 1990) but single areas of fish skins can have several variable components (i.e. chromatophores) changing on short time scales so that measurement of individual units may be necessary.

In fish that show colour change it has been usual to distinguish 'morphological' colour change, in which chromatophore numbers and/or pigment amounts vary over a fairly long time period (weeks/months), from 'physiological' colour change. The latter is a more rapid adjustment in which changes within cells produce variation in colour perceived by masking of some of the pigment when chromatosomes migrate or producing changes in angle of reflecting surfaces. In physiological colour change, chromatosomes within chromatophores become aggregated or dispersed, notably in melanophores but also described from other chromatophores (e.g. xanthophores), whereas some guanine-bearing chromatophores (particularly iridophores of neon tetra) have been described (Yoshioka et al. 2011) as having platelets which can tilt like the slats of a Venetian blind.

The control of colour change and patterning is a very complex topic (see Appendices 2.2 and 2.3 for the details of one species). For fish with only hormonal control (e.g. elasmobranchs), removal of the pituitary (hypophysectomy) will eliminate the capacity to darken, dependent on the presence of MSH. However, fish with neural control of darkening and paling do have MSH which may affect melanophore condition, and it is likely that this hormone can also function to control xanthophore dispersion (Burton 1993). Hayashi et al. (1993) describe norepinephrine (noradrenalin), MCH and melatonin as having aggregating action on xanthophores, with melatonin the most effective.

Based on mammalian autonomic regulation (11.9), for many years it was postulated (erroneously) that fish with neural control of melanophores had dual (dineuronic) innervation via the antagonistic sympathetic and parasympathetic action.

Formerly it was believed (von Frisch 1911; Parker 1948) that whereas sympathetic stimulation controlled paling, parasympathetic stimulation caused darkening (i.e. 'double innervation'). However, such a dual innervation is no longer accepted. Instead, there is increasing support for only a single (mononeuronic) sympathetic innervation with different types of adrenoceptors for darkening (β) and paling (α) and different neurotransmitter concentration sensitivities (Mayo and Burton 1998a; Burton 2008, 2010).

Another mononeuronic mechanism has been reported for guppy, *Lebistes reticulatus*, and tilapia, *Sarotherodon* (*Oreochromis*) *niloticus*, in which ATP is released from adrenergic nerves as a pigment-dispersing cotransmitter along with adrenalin (Fujii and Miyashita 1976; Kumazawa and Fujii 1984; Kumazawa and Fujii 1986; Fujii and Oshima 1986). The ATP is subsequently inactivated more slowly than the catecholamine, thereby reversing the aggregating effect of the latter.

Melanophore responsiveness, in contrast to that of striated muscle, is not dependent on membrane depolarization (3.8.1, Appendix 11.1). This conclusion is based on *in vitro* ionic manipulation combined with exogenous catecholamine incubation (Fujii and Taguchi 1969) and on chemical sympathectomy with 6-hydroxydopamine followed by electrical stimulation or incubation with exogenous catecholamines (Mayo and Burton 1998b).

Colour change in response to background depends initially on the eyes perceiving the background type/colour. This perception incorporates the ratio (albedo) between incident light from above stimulating the lower part of the retina and reflected light from the background below the fish stimulating the upper part of the retina. Subsequent pathways include the optic nerve and the paling centre (medulla oblongata) in some fish, with pituitary involvement in others.

2.10 Skin variability

Skin structure for extant fish can vary considerably from the generalized picture often presented, including degree of scaliness as well as other cellular details (e.g. presence or absence of club-cells), which may secrete toxins rather than pheromones as is the usual expectation. Even within a taxon (e.g. at 'order' level) the skin may be very flexible or rigid as with the Tetraodontiformes which include boxfish (rigid) and puffers (inflatable). Considerable variation can also occur at the specific level in association with migrations, season and reproduction.

Individual fish can also show changes in their skin during their lifetimes which may be ontological (silvering in salmon smoltification), cyclical and/or sexual (including fish which naturally show sex reversals, as with some wrasse) or related to nutrition (hypodermal fat).

Cyclical changes in relation to summer/winter or to the spawning season may be obvious, such as changes in colour: lumpfish (*Cyclopterus lumpus*) males become red and females become blue, close to spawning.

There may be other changes which are more subtle but which can seriously affect the fish if it is handled.

Profound seasonal changes can occur in skin morphology, including sexual dimorphism in some teleosts. Sexual differences in epidermal thickness were briefly noted by Stoklosowa (1966) in wild migratory trout, *Salmo trutta*, during spawning. Epidermal thickening and increases in goblet cell numbers during the breeding season have also been reported in several species of cyprinoid fish (Smith 1976), winter flounder, *Pseudopleuronectes americanus* (Burton 1978), and stickleback, *Gasterosteus aculeatus* (Burton 1979), as well as a loss of alarm substance cells in cyprinoids (Smith 1976).

Seasonal changes have been studied in detail by regular sampling over whole-year cycles (Pickering 1977; Wilkins and Jancsar 1979; Burton and Fletcher 1983). In hatchery-reared Atlantic salmon, *Salmo salar*, precociously mature male parr can develop a thicker epidermis and more mucus cells than in immature fish (Wilkins and Jancsar 1979). In captive brown trout; *S.trutta* (Pickering 1977); and in a wild inshore population of winter flounder *P. americanus* (Burton and Fletcher 1983), the epidermis thickens during the period of gonad recrudescence, most particularly in males, and mucus cells increase in females. Presumably these epidermal changes are a preparation for contact or abrasion while spawning. Post-spawning fish (winter flounder and brown trout) undergo epidermal thinning.

Captivity can also influence seasonal changes in epidermal thickness which increases in tank-held *P. americanus* (Burton and Everard 1991a) relative to wild fish but it is unknown what regulatory factors are influenced by captivity.

Experimental treatment has implicated the sex steroids 11-ketotestosterone and testosterone (androgens) in stimulating epidermal thickening (Idler et al. 1961; McBride and Van Overbeeke 1971; Yamasaki 1972; Burton et al. 1985; Burton and Everard 1991b; Pottinger and Pickering 1985). A direct correlation between increasing natural plasmatic androgens at sexual maturity and epidermal thickness has been demonstrated in salmonids (Burton et al. 1985; Pottinger and Pickering 1985).

The dominant plasmatic androgens in male salmonids (Scott et al 1980; Kime and Manning 1982) and in pre-spawning male winter flounder (Campbell et al. 1976; Truscott et al. 1983) is 11-ketotesterone, but is testosterone in females. Androgen treatment can attenuate the post-spawning epidermal thinning in winter flounder (Burton and Everard 1991b).

Exposure to crude oil (Burton et al. 1985) inhibits epidermal cellular proliferation and cell elongation associated with pre-spawning epidermal thickening in *Salmo salar*. This inhibition during the spawning period is most likely systemic in origin, probably arising from oil-related suppression of plasmatic androgen levels (Burton et al 1985) rather than a direct effect on the skin. In *P. americanus*, superficial epidermal cells may increase mucigenesis directly in response to stressful environmental conditions such as exposure to oil (Burton et al. 1984).

2.11 Skin responses to heat and cold

Being adjacent to the surrounding water, it could be expected that skin and in particular its outer layer would encounter and be resistant to thermal changes. The presence of the cuticle and mucus, production of the latter able to respond very quickly to changes, whether thermal or chemical stress, may be supplemented by other changes.

Heat-shock protein was reported in brown trout epidermis in response to a sudden temperature rise (Burkhardt-Holm et al. 1998). Similarly, some fish

(e.g. winter flounder) living in cold waters are capable of antifreeze production in the skin (Murray et al. 2003). It seems that the special skin-type antifreeze protein is synthesized in the general epidermal cells ('pavement cells' = Malpighian cells), that it is active intracellularly but can be found in interstitial spaces where it is expected to slow ice-crystal formation.

Thermal adaptations, tolerance limits and behavioural responses to environmental extremes and the concept of 'stress' can be considered part of homeostasis (7.5).

Appendix 2.1 Some pattern types in reef fish (examples only, not comprehensive)

Based on illustrations in Stokes 1980 (S), Kaplan 1982 (Ka), Robins et al. 1986 (R), Kuiter 1996 (K).

MULTIPLE STRIPES (DISRUPTIVE)		
Pomacanthus imperator juvenile	emperor angelfish (K)	Plate 4
Antennarius striatus	striped anglerfish (K)	
Pterois spp.	Lionfish (K)	
Apogon limenus	Sydney cardinalfish (K)	
Scolopsis bilineata	monocle bream (K)	
Lutjanus sebae	red emperor (K)	
Chelmon rostratus	beaked coralfish (K)	Plate 5
Chaetodon tricinctus	three-band coralfish (K)	
Equetus punctatus	spotted drum (S)	
FALSE EYE		
Chaetodon capistratus	four-eye butterflyfish (S)	
Chaetodon unimaculatus	teardrop butterfly fish (K)	Plate 7
Chromis multinucleatus	brown chromis (R)	
Pomacentrus fuscus juvenile	dusky damselfish (K)	
Sparisoma aurofrenatum terminal male	redband parrotfish (S)	
Amblycirrhitus pinos	redspotted hawkfish (R)	
Apolemichthys trimaculatus juvenile	three-spot angelfish(K)	
Lepidotrigla papilio	southern spiny gurnard (K)	
Wetmorella spp.	possum-wrasses (K)	
Apogon limenus	Sydney cardinalfish (K)	

Parupeneus signatus	black-spot goatfish (K)	
Chelmon rostratus	beaked coralfish (K)	Plate 5
EYE STRIPE		
Siganus virgatus	double-barred rabbitfish (K)	
Parupeneus multifasciatus	half-and-half goatfish (K)	
Chelmon rostratus	beaked coralfish (K)	Plate 5
Chaetodon capistratus	four-eye butterflyfish (S)	
Chaetodon tricinctus	three-band coralfish (K)	
Chaetodon rafflesi	latticed butterflyfish (K)	Plate 6
Anisotremus virginicus	Porkfish (Ka)	

Appendix 2.2 Colour and pattern in a flatfish, winter flounder

Winter flounder individuals can vary in appearance from uniformly pale or dark or can show a surface pattern with dark bars, or spots (dark, white or orange).

1. The patterns can change in response to whether the fish are on sand, mud, pebbles or gravel.
2. The colour of each fish is dependent on different types of chromatophores, which include dark melanophores, yellow/orange xanthophores and white iridophores.
3. Local differences in chromatophore distribution and size partially account for the possibility of appearing spotted, barred, etc.
4. Some chromatophores can also undergo intracellular redistribution of pigment which can enhance the colour observable.
5. Melanophores are the dominant cell producing darkening, dark spots and dark bars by dispersion of the intracellular dark pigment, partly by masking other chromatophores.
6. Melanophore changes can be long term or short term.
7. Short-term changes of melanophores are controlled by the autonomic nervous system (11.10.6 and Appendix 2.3).

Appendix 2.3 Autonomic control of melanophores in winter flounder

By removing individual scales with its attached slip of skin, the responses of melanophores as part of this 'neuromelanophore preparation' can be observed microscopically and measured.

Darkening: Low concentrations of noradrenalin cause melanosome dispersion in the melanophore affected, by activation of beta-adrenoceptors on the melanophore membrane and raised cyclic AMP (cAMP) levels in the melanophore. The noradrenalin is secreted along the dermal sympathetic (spinal autonomic nerves, 11.9.1) nerves, not at a specific junction, and diffuses to the melanophores (field effect).

Paling: If the concentration of noradrenalin is locally high (values specific to different patternregions) then the melanophore alpha adrenoceptors are stimulated and override the beta adrenoceptors and reduce cAMP activity, the result being aggregation of the melanosomes producing paling.

Individual winter flounder melanophores, from different skin pattern areas, each have a specific range for the concentration of noradrenalin which produces darkening and the higher range which produces paling. Therefore a single physiological concentration can simultaneously produce opposite melanophore responses in different regions of the fish.

Bibliography

Bagnara, J.T. and Hadley, N.C. (1973). *Chromatophores and Color Change*. Englewood Cliffs, NJ: Prentice-Hall.

Bereiter-Hahn, J., Matoltsky, A.G. and Richards, K.S. (eds) (1986). *Biology of the Integument 2. Vertebrates*. Berlin: Springer-Verlag.

Bullock, A.M. and Roberts, R.J. (1975). The dermatology of marine fish. I. The normal integument. *Oceanography and Marine Biology, An Annual Review*, 13, 383–411.

Bullock, A.M. and Roberts, R.J. (1976). The dermatology of marine fish. II. Dermatopathology of the integument. *Oceanography and Marine Biology, An Annual Review*, 14, 227–46.

Fujii, R. (2000). The regulation of motile activity in fish chromatophores. *Pigment Cell Research*, 13, 300–19.

Roberts, R.J. and Bullock, A.M. (1978). Histochemistry and kinetics of the epidermis of some British teleost fishes. In R.I.C. Spearman and P.A. Riley (eds). *The Skin of Vertebrates*. Linnean Society Symposium series Number 9. London, New York: Academic Press. pp. 13–27.

Schliwa, M. (1986). Pigment cells. In: J. Bereiter-Hahn, A.G. Matoltsky, and K.S. Richards (eds). Biology of the Integument 2. Vertebrates. Berlin: Springer-Verlag, 1986, pp. 65–73.

Smith, D.J. and Smith, R.J.F. (1986). Effect of hormone treatment on alarm substance cell counts in the Pearl dace, *Semotilus margarita*. *Canadian Journal of Zoology*, 64, 291–5.

Spearman, R.I.C. (1973). *The Integument. A Textbook of Skin Biology*. Cambridge: Cambridge University Press.

Whitear, M. (1986a). Epidermis. In: J. Bereiter-Hahn, A.G. Matoltsky, and K.S. Richards (eds). *Biology of the Integument 2. Vertebrates*. Berlin: Springer-Verlag, 1986, pp. 8–38.

Whitear, M. (1986b). Dermis. In: J. Bereiter-Hahn, A.G. Matoltsky, and K.S. Richards (eds). *Biology of the Integument 2. Vertebrates*. Berlin: Springer-Verlag, 1986, pp. 39–64.

CHAPTER 3

The skeleton, support and movement

3.1 Summary

Special hard materials, typically cartilage and bone, form the base for skeletons in fish.

Scales and, rarely, scutes and bony plates represent an outer exoskeleton in most fish.

All extant fish have an internal endoskeleton composed of bone or cartilage and divided into distinct regions: a prominent axial skeleton (skull and vertebral column) and, except in agnatha, an appendicular skeleton supporting the paired fins as well as additional elements supporting the body wall and median fins.

Jawed fish have specialized elements originally associated with the gills which have become the lower and upper jaw plus mechanisms for supporting and enabling the jaw operation.

Movement (locomotion) in fish can depend on body-wall undulations and/or fin movements. The skeletal muscle system of fish generally shows a preponderance of anaerobic white muscle (for burst swimming) in the body wall with the lesser amount of red muscle important for sustained aerobic swimming. Braking, steering and buoyancy are all important features of fish locomotion.

3.2 Introduction

Animals have two basic types of skeleton, hard or the less familiar, which is 'softer' and based on enclosed fluid; a hydrostatic skeleton (Appendix 3.1) such as that formed by the major body cavity (e.g. the coelom) or the notochord (the precursor of the vertebral column).

The skeleton of animals has several functions including support and the enabling of movement by the provision of a firm basis for muscle attachment. Skeletons often also fulfil a protective role, surrounding soft organs.

Hard skeletons may be differentiated into external exoskeletons and internal endoskeletons. Hard skeletons are usually composed of small pieces linked by ligaments, tendons and muscle; this articulation allows some flexibility and more precise movement of appendages (e.g. limbs, fins) as well as the operation of jaws. Cailliet et al. (1986) provide a practical guide to the bones (osteology) and the skeletal muscles of fish which together can achieve movement.

Aquatic organisms such as fish derive considerable support from water buoyancy and hence the supporting role of the skeleton is substantially different from terrestrial forms.

Fish and all other vertebrates are bilaterally symmetrical and show metamerism (metameric segmentation, 1.2) in their organization. This means that the basic body plan allows for repetition of limbs (or fins) as well as the bending of the body in a series of waves controlled centrally from the brain. Differentiation of segments from the anterior to the posterior region is controlled embryonically by a genomic cluster known as the homeobox which has also been described from other bilaterally symmetrical animals (McGinnis et al. 1984) and, perhaps more surprisingly, from the echinoderms (Di Bernardo et al. 2000) which are radially symmetrical (basically 'circular') as adults. The original concept of metamerism required that each individual of a metamerically segmented species had the same number of segments but fish may show some variation (3.5.2).

Essential Fish Biology: Diversity, Structure and Function. Derek Burton & Margaret Burton.
© Derek Burton & Margaret Burton 2018. Published 2018 by Oxford University Press.
DOI 10.1093/oso/9780198785552.001.0001

The origin and evolution of the vertebrate skeleton has been discussed by Donoghue and Sansom (2002).

Early jawless fish (agnatha) found as fossils from ancient rocks included groups with massive exoskeletons, some with large headshields, which presumably would have slowed their movement and tended to keep them benthic. The exoskeletons of these forms may have evolved in response to large predatory invertebrates such as eurypterids, or possibly formed a reserve of calcium and phosphate (Eliot 2000). Extant fish have varying amounts of exoskeleton, in the form of bony plates and scutes (2.5), as well as scales which may be quite thin and fragile or thick and robust (2.4).

All fish have an internal endoskeleton which may be mainly cartilage or mainly bone (although somewhat contentious in its identity, 3.3) in the adult. Where bone-like material (rather than cartilage) is prevalent in the adult it is possible to determine that some bone is formed fairly directly, and usually superficially (dermal or membrane bone), whereas other deeper elements have a cartilage precursor which then undergoes ossification as 'cartilage bone' (Romer 1955) or endochondral bone (Hammond and Schulte-Merker 2009). It is usual to think of dermal bone as representing a former exoskeletal element which may have become deeper in evolutionary time; this view is partially reflected by Morriss-Kay (2001) who observes 'the vertebrate skull roof is a very ancient structure, forming a protective covering for the head in agnathan fossil fishes'. The endoskeleton itself may thus have superficial elements applied to it as well as having what can be thought of as 'original' components.

The endoskeleton of fish comprises two main elements: the axial skeleton (skull and vertebral column) and the appendicular skeleton supporting the appendages (fins).

Most extant fish have jaws (i.e. are gnathostomes, 'Gnathostomata', 1.3.2), derived from anterior gill supports and the relationship between the movable jaws and the skull, the 'jaw suspension', can be rigid or loose, the extremes providing (respectively) strength or manoeuverability.

Skulls can be very robust but more recent taxa have a lighter structure than the earlier forms. Unusual skull structures include those supporting very elongated heads and hinged skulls such as found in the coelacanth. Asymmetrical skulls support the heads of adult flatfish and slight asymmetry occurs in at least one cichlid, *Perissodus microlepis* (Takahashi and Koblmüller 2011).

Most extant fish have paired fins as well as median fins and a caudal (tail) fin, and the skeletal supports (the appendicular skeleton) underlying them reflect considerable variability including anterior positions of pelvic fins in some taxa.

In comparison with tetrapods, which have tiny ear ossicles derived from larger gill arch elements, some fish have an ossicle system linking the air (swim) bladder to the ear which assists hearing too. The Weberian ossicle system (which aids sound reception of freshwater fish like carp) is derived from parts of the anterior vertebrae (1.3.8, 3.5.2, 12.8.4).

3.3 Cartilage and bone

The hard tissues termed cartilage and bone are the basis for the skeleton in vertebrates, although the nature of material termed 'bone' in fish has generated debate.

These two tissues are often considered to be confined to the vertebrates (although cartilage has been described from several different major taxa of invertebrates, Hall 2005) and are characteristic of the group. Both are composed of specialized cells and the matrix they produce, a large mass of relatively amorphous and hard material forming the bulk of the supportive and protective structure in the individual animal. Although we think of these hard tissues primarily in association with the endoskeleton, bone particularly may also be a component of exoskeletal structures, though recent research (2.4) has reassessed the type of tissue forming some scales, suggesting that tissue previously called bone may not fit that categorization.

The difference between cartilage and bone lies in the composition of the matrix secreted by specialized cells. For cartilage, the cells are termed chondroblasts when young, chondrocytes when established. For bone, the young osteoblasts become osteocytes. In both cartilage and bone the cells lie in little spaces (Plate 8), lacunae. However, many fish (Table 3.1) do not have 'cellular' bone, unlike mammals in which mammalian osteocytes are linked by canaliculi (tiny channels) and mammalian bone is vascular.

The density of bone has recently been studied, with particular reference to birds (Dumont 2010) which are generally regarded as having a 'light-weight' skeleton. However, this author found that the bone tissue in birds is actually denser than that of rodents and bats; lightness of bird bones is achieved by the presence of air spaces. Similar studies for fish do not seem to be available but comparative studies emphasizing other features have been published: Witten and Huysseune (2009) conclude that most teleosts have a range of characteristics distinguishing their bone from that of mammals and these features include, for many species, the absence of osteocytes in mature bone ('acellular' bone, Table 3.1).

In cartilage the cells secrete a dense matrix containing protein (including collagen fibres) and a high percentage of an acid mucopolysaccharide (chondroitin sulphate) which can be differentiated from bone histochemically (Appendix 3.2), and as hyaline cartilage (Plate 8) is clear/translucent. Cartilage is not expected to resist tension and compression forces as well as bone but can be flexible. Though hard, it can be cut fairly readily, unless it becomes calcified. There are no channels linking the lacunae (Plate 8) and any nutrients must reach the cells through the matrix, and if it is fully calcified the cells may die. However, in elasmobranchs, regions of calcified cartilage may contain live chondrocytes (Moss 1977). Blocks of calcified cartilage with a distinct polygonal shape (Plate 9) may be found applied to, for example, the pectoral girdle; these blocks have been termed tesserae (Dean and Summers 2006). Bands of calcified cartilage are also found in elasmobranch vertebrae, radiating from the centra (Romer 1955, Figure 99A).

With bone, the cells secrete a very hard matrix containing collagen and impregnated with calcium hydroxyapatite—$Ca_{10}(PO_4)_6(OH)_2$—strongly resisting tension and compression forces. Traditionally, in vertebrate bone the osteocytes are described as being in linked lacunae and the matrix can be vascularized and/or have ready access to nutrients as it is channelized. However, many teleosts have bone which is considered 'acellular', lacking obvious osteocytes in the main matrix, although vascular channels have been described (Witten and Villwock 1997).

Characteristically bone can grow three-dimensionally and undergo complex remodelling, with new material added interstitially, within the bone, as well as by accretion on the surface, though Ham (1957) states that (human) bone grows only by surface addition. Conversely, a more recent text (Bloom and Fawcett 1975) states there is 'continual internal reconstruction' in mammalian bone. In some situations, as in scales from many teleost species, the interior of the scale, whether it be classified as bone or not, is acellular and growth can apparently only occur on the surface, by apposition, thus forming circuli, close together or further apart, producing the effect of 'annuli', distinctive annual rings, useful for ageing fish, if growth is highly seasonal (Nikolsky 1963). The criteria for ageing based on this phenomenon in Atlantic salmon were fully explored in an International Council for the Exploration of the Sea meeting (Anon. 1984) and a compilation of methods for 'age determination' in North-west Atlantic species was presented by Penttila and Dery (1988). In fish like salmon with extensive pre-spawning migrations and stress, scale resorption may occur, giving a spawning check mark on the scale (Greenwood 1975) which can also be a useful indicator of the fish's history.

Annual accretion patterns have been described for other bony structures like the cleithrum (of the pectoral girdle) and opercular bone (Nikolsky 1963), as well as ear otoliths (12.8.2), and may be used for determining age too; in fact, the otoliths may be preferred to scales as scales can be damaged and/or replaced and may therefore not reflect the entire life span. The otoliths (important for balance perception, 12.8.2, and not strictly part of the skeleton) are composed of calcium carbonate 'deposited on an organic matrix' (Gauldie 1993) and hence are neither cellular nor acellular bone; otoliths are not normally subject to resorption except in extreme conditions. Otoliths may occasionally appear 'abnormal', for example on one side (being chalky rather than semi-translucent in appearance, and also apparently friable), but this condition is usually fairly obvious and an alternate specimen can be selected.

Phelps et al. (2007) compared the use of various skeletal structures as well as the otoliths in estimating age for carp and concluded that although otoliths had been very useful in this role, pectoral fin-rays had advantages, including non-lethal sampling. A similar study (Sylvester and Berry 2006) for white sucker (*Catastomus comersonii*) did not recommend the use of scales for mature fish. Scales have not been found appropriate for older fish such as pike either, and a long-term cleithrum-based study was launched in the 1970s for the muskellunge *Esox masquinongy* (Casselman 2007).

Bone in non-scale regions, including the cleithrum (or part of it in *Caranx latus*, Bueno and Glowacki 2011) of many 'derived' (previously termed 'advanced') teleosts may also present as acellular (Moss 1961a, 1961b, 1963; Meunier and Huysseune 1991; Horton and Summers 2009) with a distinctive fibrous structure. Witten and Villwock (1997) quoting Moss (1961b) describe acellular bone in fish as being 'woven', with vascular channels. 'Woven bone', according to Currey (2002), is a term used as one of the two main bone types in mammals, based on the distribution of the enclosed collagen fibres. It is distinguished from lamellar bone in which a more regular layering of the linked osteocytes is observed, a condition not generally described for fish, although superficially resembling the annual rings.

Horton and Summers (2009) investigated the properties of acellular bone from a teleost (*Myoxocephalus plyacanthocedphalus*) and concluded that it did not have better stiffness than cellular bone, contrary to other studies. Cohen et al. (2012) have compared the features of cellular and acellular bone in two teleosts, observing that acellular bone occurs in more derived species and that the loss of osteocytes must confer some advantage. The species used in this study were carp (*Cyprinus carpio*) for cellular bone and tilapia (*Oreochromis aureus*) for acellular bone. However, any advantage to loss of cellularity remains unclear.

Witten and Villvock (1997) consider tetrapod bone to have two major functions: as an endoskeleton and as a source of calcium by resorption. The latter function may not be highly significant in fish for which calcium is readily available from the surrounding water and thus the role of resorption for fish skeletons would be primarily remodelling during growth, as described for *Oreochromis niloticus*.

In vertebrates with a high degree of remodelling of bone, old bone is removed by osteoclasts, described as multinucleate in mammals. The search for osteoclasts in fish led to the conclusion that they may appear sporadically, for example at times of growth, and that mononuclear osteoclasts occur (Witten et al. 2001). Witten and Huysseune (2009) have reviewed skeletal remodelling in teleosts. Particularly spectacular and rapid skeletal change occurs in the head (teeth, jaws, vomer, etc.) of some male salmon prior to breeding. These changes result in a prominent forward extension of the lower jaw, termed a kype. Tchernavin (1938) produced a detailed study of bone changes and tooth replacement in Atlantic salmon. Witten and Hall (2003) also studied this remodelling in Atlantic salmon which can survive spawning (as kelts), in which case the kype may be reabsorbed, a major physiological stress in post-spawned animals.

Some bone contains haematopoietic (blood-cell forming) tissue in bone marrow. This has not been reported for fish (Witten and Huysseune 2009) which have haematopoietic tissue elsewhere (e.g. the kidney, 5.13.6). Similarly, the complex Haversian systems found in some vertebrate bone, with concentric lamellae around blood vessels in the matrix, do not occur in most fish (Currey 2002). As accounts of vertebrate bone are often based on mammalian structure, the terminology applied can be a problem for clarity. Descriptive expressions (Table 3.1) applied to bone include 'Haversian', compact, cancellous, lamellar and acellular, as well as expressions referring to site of origin (dermal, membrane) or precursors (cartilage bone). Bone formation (ossification) accounts for mammals have often depended on detailed descriptions of long bone elongation, a situation not readily related to fish. Similarly descriptions of bone may be restricted to Haversian systems as with Rowett (1953).

3.4 The exoskeleton

Extant fish generally have scales (2.4) and some have more substantial external bony plates (2.5) forming an exoskeleton. The degree of protection afforded by these exoskeletal elements is very

Table 3.1 Terms applied to vertebrate bone.

Note that multiple terms can be used, e.g. acellular bone can be dermal; mammalian long bones are Haversian with zones of outer compact and inner cancellous bone and are overall 'cartilage bones' because they are formed on a cartilage model.

Type of bone	Characteristic	Taxon usually indicated	Applied to fish?
Haversian	Concentric lamellae round distinct vascular channels	Mammals, birds, some Reptiles	No
Lamellar	Layered	Vertebrates but not universal	Yes, for some bone
Cancellous/spongy	Multiple cavities	Mammals	Rare
Compact	Dense	Mammals and others	For some relicts, Fossils
Woven	Fibrous, not layered*	Mammals and others	Yes
Acellular	No obvious osteocytes	Fish	Yes, frequent
Hollow	Filled with 'marrow'	Mammals (long bones)	No
Hollow	Filled with air	Birds	No
Membrane/dermal	Site of origin (outer)	Vertebrates	Yes
Cartilage-bone	Site of origin (inner/cartilage)	Vertebrates	Yes
Endochondral	Bone formation site	Mammals but also applies to any bone formed in cartilage	Yes

*See Witten and Villwock (1997); Currey (2002).

variable as some fish are virtually scaleless (snailfish, Scott and Scott 1988) or have small scales only developed in older fish in fresh water (anguillid eels). The exoskeleton may form a complete or partial casing, helping to maintain shape, providing good or limited protection, but apart from contributing to streamlining in some forms, not otherwise directly important in locomotion.

3.5 The endoskeleton

All extant fish have an endoskeleton composed of cartilage or bone with larval cartilage generally replaced by bone in most non-chondrichthyans. The endoskeleton of vertebrates and thus fish, whether cartilaginous or bony, has a common basic morphology principally based on arrangements of the individual parts within either the central 'axial' skeleton or the appendicular skeleton which supports the appendages (fins or limbs). In addition, some teleosts have a complex and extensive array of long narrow bones including neural spines and ribs, forming an intricate support system, diagrams showing this can be found in Young (1950) and Caillet et al. (1986);

a similar complex arrangement of narrow bones is found in the bowfin, *Amia* (Figure 3.1).

3.5.1 The axial skeleton

Ontogenetically (during development) and phylogenetically (by descent), the axial skeleton has its origin in the dorsal flexible notochord. This rod has a fibrous sheath filled with large vacuolated cells and is characteristic of all members of the Phylum Chordata, in which the vertebrates are classified. The notochord persists throughout life in the Subphylum Cephalochordata (e.g. the lancelet, *Branchiostoma lanceolatum*, also known as amphioxus) but undergoes partial replacement in the vertebrates by the cartilaginous or bony elements of the vertebrae. However, in some relict fish such as the coelacanth and sturgeon a large portion of the notochord does persist and the sturgeon has no centra, which form a major part of the support system in the vertebrae of most fish.

The main role of the notochord is to provide a structure which resists shortening (compression) of the body when the body-wall muscles contract but permits bending during movement.

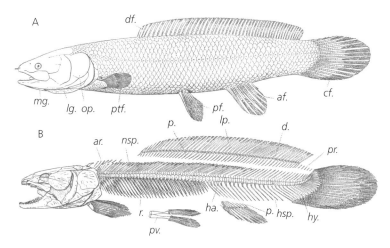

Figure 3.1 *Amia calva* lateral views, Goodrich (1909). Above (A): whole fish, after Brown Goode. Below (B): skeleton after Franque; af, anal fin; ar, anterior dorsal radials (*); cf, caudal fin; d, distal segment of dorsal radial; df, dorsal fin; ha, haemal arch; hsp, haemal spine; hy, hypural arches; lg, lateral gulars; lp, lepidotrichia; mg, median gular; nsp, neural spine; op, operculum; p, proximal segment of radial; pf, pelvic fin; pr, posterior dorsal radials; ptf, pectoral fin; pv, pelvic bone; r, pleural rib. *Goodrich questioned this nomenclature.

3.5.2 Vertebrae

The vertebrae also provide resistance to shortening but additionally allow flexibility to the fish and give protection to the dorsal nerve cord, as they surround it forming a neural arch, with small exit holes (foramena) where the spinal nerves emerge between the vertebrae (Figure 3.2). The vertebral column does not have a major 'girder role' in contrast to the tetrapod structure (Young 1957); support in fish is provided by water buoyancy. The major part of the bony vertebra, which encloses the notochord, is the centrum, and these centra form a linked series as the basis of the vertebral column. In early studies of vertebrae the concept of 'arcualia' was established; these were seen as arch components of pairs of cartilages per vertebra, also sometimes contributing to the centrum. Units of cartilage corresponding to arcualia are present in elasmobranchs (Figure 3.2) as two pairs per vertebra (basidorsal/basiventral and interdorsal/interventral) with a row of supradorsals present in, for example, *Scyllium* but not *Squalus* (Goodrich 1930a). The concept of arcualia has been challenged for tetrapods (Williams 1959) but is well established for fish (Schultze and Arratia 1986).

Vertebrae form a series from immediately behind the brain through to the tail and there may be a change in size posteriorly. Whereas anterior

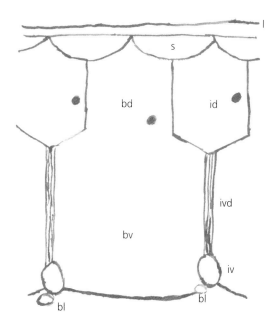

Figure 3.2 Dogfish vertebrae, diagrammatic lateral view based on *Scyllium* (after Bridge 1904 and Goodrich 1930). Blocks of cartilage are as follows: bd, basidorsal; bv, basiventral; id, interdorsal; iv, interventral (irregularly developed); ivd = intervertebral disc; s, supradorsal (not found in *Squalus*); l = a ligament. Segmental blood vessels (bl) emerge below the interventrals. The black circle in the interdorsal is the foramen of the spinal nerve dorsal root. The black circle in the basidorsal is the foramen of the spinal nerve ventral root.

vertebrae may have ventral ribs, the posterior (caudal) vertebrae have a ventral arch, the haemal arch (Figure 3.3) which encloses the caudal artery and vein, facilitating venous return (5.7). Although the difference between anterior and posterior fish vertebrae is well recognized in the absence or presence (respectively) of a haemal arch, it is also commonly considered that, apart from graded size variations, individual anterior or posterior vertebrae are similar. The vertebral column does not have the regional specializations associated with the terrestrial life in mammals. However, in the ostariophysi (e.g. roach *Rutilus*, minnow *Phoxinus* and goldfish *Carassius*) modifications of the first three or four vertebrae gave rise to a bilateral series of Weberian ossicles (see Greenwood 1975, Figure 71) which link to the ear and enhance hearing (12.8.4), and the vertebrae involved are distinctively different (Figure 3.4); see also Greenwood et al. 1966, Figure 8, showing the

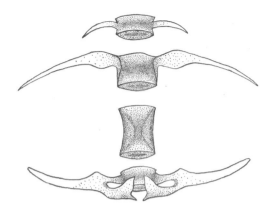

Figure 3.4 Anterior vertebrae, ventral view, of *Phoxinus phoxinus*, the Eurasian minnow, an ostariophysan. Vertebra 1 at top.

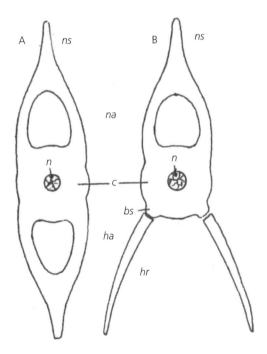

Figure 3.3 Sketches of transverse (cross-) sections through typical bony fish vertebrae, based on Kingsley (1926). A: posterior (caudal) vertebra, B: anterior vertebra: bs, basal stump; c, centrum enclosing remains of notochord (small circle); ha, haemal arch enclosing ventral blood vessels, hr, rib; na, neural arch enclosing nerve cord (large dorsal circle); ns, neural spine.

four anterior vertebrae of an anotophysan (*Gonorhynchus*). Other variations in the vertebral column include relatively fine neural and haemal spines (Chiasson 1974; Cailliet et al. 1986). Conversely some neural and haemal spines are very robust (Figures 3.1 and 3.5).

The vertebrae of terrestrial vertebrates, and some fish, have projections (the zygapophyses, Figure 3.5) which articulate and strengthen the vertebral column, preventing rotation between vertebrae. Similar structures are described as absent in a 'generalized' fish such as the tarpon (Rockwell et al. 1938) but may occur in very active bony fish (Bond 1996) and are figured in Cailliet et al. (1986). Parapophyses in elasmobranchs are ventrolateral projections (Bridge 1904).

Descriptions of vertebrae of fish and tetrapods have included the shape of the centrum surface (Table 3.2) forming the link to the vertebrae before and behind each of the individual units and the number of pieces seen during development of each vertebra. The intervertebral articulation may form a 'ball-and-socket joint' where a convex surface abuts a concave one (proceolous/opisthocoelous types, Table 3.2) but the fish situation is almost always amphicoelous (two concave surfaces with remains of notochord between them) 'which allows the column to resist longitudinal compression yet remain flexible' (Young 1950). An exception to the amphicoelous condition of fish is the gar *Lepidosteus* which is opisthocoelous (Goodrich 1930a).

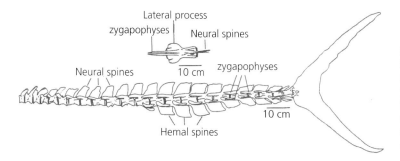

Lateral process
zygapophyses | Neural spines
Neural spines | zygapophyses
10 cm
10 cm
Hemal spines

Figure 3.5 The vertebral column (vertebrae 3 to 24) and lunate tail of Pacific blue marlin. Reproduced with permission. J. H. Hebrank, M.R. Hebrank, J.H. Long Jr., B.A. Block, S.A. Wainwright (1990). Backbone mechanics of the blue marlin *Makaira nigricans* (Pisces Istiophoridae). *Journal of Experimental Biology* 148, 449–59.

Table 3.2 The types of centrum (after Hyman 1942).

Name	Anterior surface	Posterior surface	Example
Amphicoelous	Concave	Concave	Elasmobranchs, teleosts
Procoelous	Concave	Convex	Frogs, alligators
Opisthocoelous	Convex	Concave	Gar, frogs, neck of ungulate mammals
Heterocoelous	Transverse saddle	Transverse saddle	Birds
Amphiplatyan	Flat	Flat	Mammals

The number of vertebrae differs between fish species (McDowall 2008) with values typically from 25 up to 50 within a taxon such as the Centrolophidae, and also shows some variation intraspecifically; that is, within a species (Videler 1993). Similarly, there may be considerable intraspecific variation in other skeletal elements (3.5.3). Hyman (1942) describes another interesting situation in which there may be two centra per single neural and/or haemal arch. This 'diplospondyly' occurs in *Amia* and in the tail of elasmobranchs where the doubling includes the arches.

Goodrich (1930a) and Young (1981) explore the fact that the vertebral centra can be regarded as intersegmental, with the centra alternating with muscle myomeres, giving two myomeres attached to each centrum and facilitating bending of the vertebral column. While the vertebrae may be associated with ribs and median fin (3.5.6) supports, more peripherally some teleosts can form intramuscular bones supporting the body wall.

3.5.3 The skull

The skull is usually regarded as an anterior development of the axial skeleton (but see Appendix 3.3, with a suggestion that at least part of the skull may derive from teeth). It consists of elements protecting the brain and anterior sense organs as well as the jaws and the branchial skeleton supporting the gills. It may appear to be quite simple, as in chondrichthyans, or very complex when a very large number of separate bones can be seen, as in teleosts. Initially, and in, for example, adult elasmobranchs, the skull has a cartilaginous neurocranium (Figure 3.6) around the brain with a series of capsules around or partially supporting the prominent cephalic sense organs—the olfactory organs, the eyes and the internal ears.

In bony fish the neurocranium becomes ossified, the centres of ossification in the cartilage are termed 'cartilage bones' and a number of outer dermal bones may be added to the structure as the fish develops (Appendix 3.4a). Hyman (1942) distinguishes a chondrocranium with cartilage bones and an outer dermatocranium, also termed the exocranium (Schultze 1993). In addition to the original

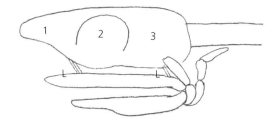

Figure 3.6 Dogfish anterior skeleton (modified from de Beer 1951) showing (chondro/neuro) cranium, with 'capsules' (1. Olfactory; 2. optic; 3. auditory), jaws and branchial supports. Note that ligaments (L) are important in securing the position of the upper jaw.

neurocranium (Figure 3.6) with its applied dermal bones (Appendix 3.4a), there may be elements of the gill supports which become incorporated in the skull (e.g. as jaws) and are termed the splanchnocranium (or branchiocranium, Gregory 1959). Gregory refers to the whole skull as the syncranium, comprising the neurocranium, 'the braincase' (plus applied dermal bones), and the branchiocranium.

The lateral region of the head and the gills may be protected by substantial opercular bones whereas the ventral surface can be covered by gular plates as in *Amia* and *Elops* (Goodrich 1930a).

The numerous components of bony fish skulls (Appendix 3.4 and Appendix 3.5) have formed the focus of much research, particularly as fossil forms of extinct ancient forms can be related to extant osteology, including that of tetrapods. Because bones are relatively durable it is their structure which has frequently formed the only information available on morphology of the many extinct species.

With such a mass of material available and potential long lineages, it has not always been possible to agree on homologies of individual ossified plates. Bones may have been named because they seem to be similar in position to named bones in extant tetrapods. Romer (1956) observed that, for actinopterygians, this practice, in which fish bones have 'customarily' been given 'familiar names', is 'little guarantee of true homology'. The situation has been seen as 'difficult' for lungfish (Kemp 1999), with interpretations and 'affiliations' of individual bones not consistent. Authors have resorted to alphabetical labelling of individual bones for

lungfish, presumably therefore avoiding premature conclusions on relationships with other lineages. Moreover, both in lungfish (Kemp 1999) and *Amia* (Jain 1985), considerable intraspecific variation is reported in bone number and conformation. In *Amia* the 'parietal' bone is usually bilateral with two bones but a single bone may occur.

From a tetrapod-based nomenclature, but with reference also to fish, Hyman (1942) summarized skull bones as forming a number of different groups. She tabulated cartilage bones as three categories: in the chondrocranium, associated with the sense organs, and as part of the jaws. She further subdivided each of these three types, with three major groups in the chondrocranium category: the occipitals, the posterior sphenoids and the orbitosphenethmoids. Hyman viewed skull dermal bones as being of three regions: part of the skull roof (which would include paired nasals, frontals, parietals and post-parietals); applied to the upper jaw (including the premaxilla and maxilla) or composing the palate (i.e. the roof of the oral cavity), including paired prevomers and pterygoids and the parasphenoid.

Gregory (1933, 1959, 2002) described individual skull elements (Appendices 3.4a, 3.4b) based on their derivation and position. A single diagram is not capable of showing all the skull elements in bony fish as bones can be obscured or deep, overlaid by more superficial elements. Gregory provides a series of useful diagrams from different fish and different aspects which can help understanding of this complex situation. Another series of figures, based on cod (de

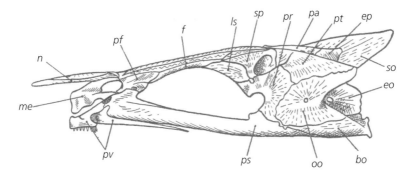

Figure 3.7 Neurocranium of cod (after de Beer 1951). Lettering: bo, basioccipital; eo, exoccipital; ep, epiotic; f, frontal; ls, laterosphenoid; me, mesethmoid; n, nasal; oo, opisthotic; pa, parietal; pf, prefrontal; pr, prootic; ps, parasphenoid; pt, post-temporal; pv, prevomer; so, supraoccipital; sp, sphenotic.

Beer 1951), illustrate the layering of the skull with the most internal set forming the neurocranium surrounding the brain (Figures 3.7, 3.10, 3.11).

3.5.4 Jaws and the branchial skeleton

Jaws are regarded as derived from the anterior elements of the branchial skeleton (Appendix 3.4b); that is, the rods which support the gill arches. The jaw support mechanism (suspension) may also derive from the branchial skeleton. The first gill or branchial arch (mandibular arch) on each side becomes the upper and lower jaw while the second arch (the hyoid) may contribute elements towards the jaw suspension (Table 3.3).

Table 3.3 The elements of branchial support based on dogfish (after Hyman 1942).

Arch	Element	Current function	Previous function?
Mandibular I	Pterygoquadrate	Upper jaw	Upper arch I
	Meckel's cartilage	Lower jaw	Lower arch I
Hyoid II	Hyomandibular	Supports posterior upper jaw and gills	pharyngobranchial and/or Epibranchial Arch II
	Ceratohyal	Jaw suspension, particularly lower jaw*	Ceratobranchial Arch II
	Basihyal	Supports lower jaw	Hypobranchial/Basibranchial Arch II
III–VII	Pharyngobranchial III–VII	Supports pharynx	
	Epibranchial III–VII	Supports pharynx, gills	
	Ceratobranchial III–VII	Supports pharynx, gills	
	Hypobranchial III–V/VI	Supports pharynx	
	Basibranchial III–V/VII**	Supports pharynx	

*Supports branchiostegal rays in teleosts (de Beer 1951).
 **Some fusion presumably but the structure underlies the length of the arch support. There is inconsistency between Hyman's text (two basibranchials) and diagram (three basibranchials).

The remaining branchial skeleton supports the gills on either side of the pharynx forming a structure collectively known as the 'branchial basket' or the visceral skeleton, forming a series of inclined/vertical structures on each side behind the main part of the skull in the pharynx wall. On each side the individual components form a line of struts from the ventral basibranchial, up through the hypobranchial, ceratobranchial, epibranchial and pharyngobranchial (Table 3.3), with some reduction posteriorly (e.g. only three pairs of hypobranchials; the text states three, but the diagram seems to show four, Hyman 1942). The ceratobranchials and epibranchials support the gills via gill rays. According to Gregory (1933) the major elements of the branchial supports are based on cartilage bone and dentigerous (tooth-bearing) plates attach to the dorsal part of the hypobranchials as well as the fourth pharyngobranchials. In the throat region of some fish (e.g. *Amia*) gular plates cover the ventral surface of the basibranchials.

The cartilaginous base for the upper jaw is the palato- or pterygoquadrate cartilage which can bear teeth. This cartilage may have an upward projection, the ascending or orbital process into the orbital (eye socket) region. The cartilaginous base for the lower jaw (the mandible) is 'Meckel's cartilage' as found in elasmobranchs, which can bear teeth. However, dermal bone elements, notably the dentaries, have become very important in the mandibles of teleosts and tetrapods.

Teeth (4.4.5) are found on the jaws and on other head regions including the roof of the mouth (e.g. the pterygoid), the tongue (for the Osteoglossomorphs, Lauder and Liem 1983, Figure 32) and the pharyngeal floor; in other words the basibranchial can be tooth-bearing or dentigerous. Stock (2001) states that 'bony fish may possess teeth on virtually any of the bones lining the oropharyngeal cavity'. Indeed, the branchial supports (Table 3.3) may form a structure termed the 'pharyngeal jaws' (Lauder and Leim 1983, Figure 32), highly important in many fish for holding prey (4.4.3).

The upper jaw of *Amia* (Figure 3.8) has applied dermal elements such as the premaxilla and maxilla and this anatomical development applies basically to many teleosts as well (Figure 3.9), but in some 'advanced' teleosts the maxilla becomes part of a mechanism enabling the premaxilla to move

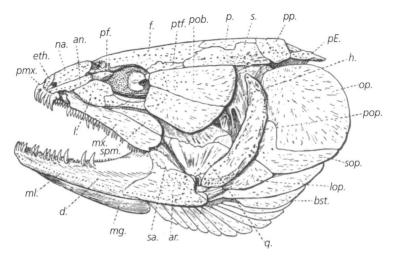

Figure 3.8 *Amia* skull: Note premaxilla (pmx) and maxilla (mx) forming a continuous structure on the upper jaw (after Goodrich 1958). The other lettering is as follows: an, adnasal; ar, articular; bst, branchiostegal ray; d, dentary; eth, mesethmoid; f, frontal; h, hyomandibular; iop, interopercular; l, lacri(y)mal; mg, median gular; ml, lateral line in mandible; na, nasal; op, opercular; p, parietal; pf, prefrontal; pob, post-orbital; pop, preopercular; pp, post-parietal; pt, post-temporal; ptf, postfrontal q, quadrate; s, pterotic; sa, suprangular; sop, subopercular; spm, supramaxilla.

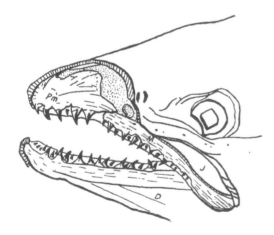

Figure 3.9 Salmon head showing premaxilla (Pm) and maxilla (M), similar in position to *Amia*. Reproduced with permission.V.V. Tchernavin (1938), *Transactions of the Zoological Society of London*, 24, 103–84.

forwards; that is, the premaxilla is protrusible, an important feature of a feeding movement (4.4.2). In such situations the maxilla is not part of the gape (Figure 3.10).

Protrusibility is apparently polyphyletic, having developed in different forms, including elasmobranchs, in which the jaws are quite mobile relative to the chondrocranium. Extreme active jaw protrusion (dependent on maxillary support) has been described for neotropical cichlids (Waltzek and Wainwright 2003) and in *Luciocephalus pulcher* (Lauder and Liem 1981). In both cases the protrusion exceeds 30 per cent of the head length. The mechanisms were reviewed by Motta (1984) and the possible functions were summarized by Lauder and Liem (1981) as well as by Motta (1984).

Some fish have fixed jaw protrusion, that is, the upper or lower jaw is extended permanently beyond the other; herring have a protruding lower jaw, whitefish a protruding upper jaw. Halfbeaks (Hemiramphinae) typically have a very long lower jaw relative to the upper while swordfish have extreme elongation of the upper jaw, the premaxillae and nasal bones forming a rostrum (Nelson 1976); this 'sword' can be used to slash and disable prey (Stillwell and Kohler 1985). A similar situation occurs with billfish such as marlins (Nelson 1976). Extreme elongation of both jaws occurs in needlefish (Belonidae).

Active jaw protrusion enables influence of the fish on surrounding water, producing a suction effect facilitating prey capture (Holzman et al. 2008).

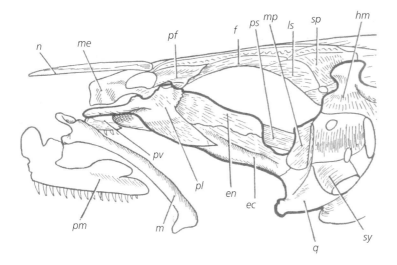

Figure 3.10 Cod skull, some bones removed: Maxilla (m) in supporting position relative to premaxilla (pm) (after de Beer 1951). Other lettering as follows: ec, ectopterygoid; en, endopterygoid; f, frontal; hm, hyomanibula; ls, laterosphenoid; me, mesethmoid; mp, metapterygoid; n, nasal; pf, prefrontal; pl, palatine; ps, parasphenoid; pv, prevomer; q, quadrate; sp, sphenotic; Sy, symplectic.

The combined effect of the pharyngeal jaws and protrusibilty, studied by Drucker and Jensen (1991), produced prey content separation ('winnowing').

The teleost lower jaws are complex in composition as compared to that of mammals which are formed of two simple 'dentaries' fused anteriorly at the symphysis. Bony fish lower jaws have bones from ossification at several sites in the embryonic cartilaginous palatopterygoquadrate bar, and a single site in 'Meckel's cartilage', forming a posterior articular bone. Dermal bones applied to the lower jaw in teleost fish are the dominant dentary and small posterior angular bone (Young 1981).

The upper jaw may be fused to the cranium which gives strength or may be attached in such a way as to optimize movement (as in protrusibility above) and increase the gape. The overall support of both jaws is referred to as a suspensorium and often incorporates elements of the hyoid arch. The terminology, however, and the interpretations of elements involved have been somewhat difficult (Hyman 1942), with a large body of previous literature. Table 3.4 shows some of the more usual categories used in the fairly recent past.

It has been usual to think of fish jaws as somewhat limited in action in comparison, for example, with mammals which can chew, using a side-to-side movement. However, studies of jaw protrusibility and other movements show that many fish can move their jaws in an antero-posterior motion and manipulate prey. Lauder (1979, 1980) describes a forward and backward chewing mechanism in 'primitive' fish such as the bichir (*Polypterus*), the gar (*Lepisosteus*) and the bowfin (*Amia*). Thomson (1969) provided a description of chewing in extant lungfish, with food chewed as it is pushed backwards and forwards between the tooth plates.

In elasmobranchs such as dogfish (Figure 3.6), the separate structures forming the basic skull (i.e. the neurocranium, capsules and jaws) may be quite clear, whereas in bony fish (Figures 3.7 and 3.8) there may be many additional structures (bones) overlying the neurocranium and the branchial arches may be condensed and partly covered by bones of the

Table 3.4 Relation of upper jaw to cranium.

Attachment	Termed	Example
Hyomandibular not involved, palatoquadrate of upper jaw articulated or fused to cranium	Autostylic*	Most vertebrates, but not most fish
Upper jaw fused to cranium	Holostylic†	Chimeras, lungfish**
Two-part attachment, posteriorly via hyoid	Amphistylic	Some elasmobranchs
Hyoid important for attachment of upper jaw	Methyostylic	Teleosts ('teleostomes')*
Hyomandibular crucial both jaws	Hyostylic	'Most modern fishes'**

*Hyman 1942.
**Romer 1955.
†Secondary autostylic, Walker and Homberger 1992.

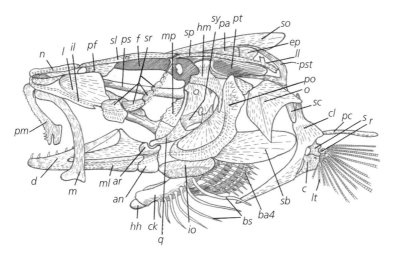

Figure 3.11 Cod skull and pectoral girdle, lateral view (after de Beer 1951). Note the supracleithrum (sc) which links the girdle, cleithrum (cl) to the skull; maxilla (m) in supporting position relative to premaxilla (pm). Other lettering as follows: an, angular; a, articular; ba4, fourth branchial arch; bo, basioccipital bs, branchiostegal rays; c, coracoid; d, dentary; ep, epiotic; f, frontal; hh, hypohyal; ck, ceratohyal; hm, hyomanibula; il, lateral line ossicle; io, interopercular; l, lachrymal; ll, infraorbital lateral line canal; lt, lepidotrichia; ml, mandibular lateral-line canal; mp, metapterygoid; n, nasal; o, opercular; pa, parietal; pc, postcleithrum; pf, prefrontal; po, preopercular; ps, parasphenoid; pst, post-temporal; pt, pterotic; q, quadrate; r, radials; s, scapular; sb, subopercular; sc, supracleithrum; sl, supraorbital lateral line canal; so, supraoccipital; sp, sphenotic; sr, suborbitals; sy, symplectic.

operculum, the 'gill cover' (Figures 3.8 and 3.11). In teleosts there may also be a link (the supracleithrum, Figure 3.11) between the skull and the pectoral girdle. Unusual structures in the skull include the crossopterygian intracranial hinge found in *Latimeria*, the extant coelacanth (Thomson 1966), which permits raising of the anterior region of the head on the hind part (1.3.6). This could facilitate prey capture.

3.5.5 The appendicular skeleton

The appendicular skeleton supports the fins, including the paired fins, the anterior pectorals and the basically posterior pelvics, which may in some derived fish such as acanthopterygians, have a more anterior position ('thoracic/jugular') just beneath the pectoral fins.

The paired fins depend on a system of structures secured within the main body of the fish linked distally to a series of long narrow fin-rays allowing the thin fins to project into the water.

The basic structures of the paired fin supports can be summarized as an internal 'girdle', with basal plates, linked to radials (smaller hard plates) and distally, projecting into the water, the long fin-rays

which can be generalized as dermal fin-rays (3.5.7) or 'dermotrichia' (Goodrich 1930a).

The paired fins have been studied with reference to tetrapod limbs and comparisons drawn between the larger, more internal-supporting structures and the girdles of tetrapods. The pectoral system of many fish (but not the Chondrichthyes, Romer 1955) includes dermal elements such as the cleithrum, a large structure which links the pectoral region to the cranium (Figure 3.11) but which does not, however, have an equivalent in extant tetrapods, in which it could limit the development of neck movement.

In some fish, including elasmobranchs and many teleosts, the 'girdles' are quite well developed in association with prominent paired fins. In such fish with well-developed paired fins there may be obvious large basal hard structures (such as the pelvic basipterygium, Figure 3.12) supporting the radials and fin-rays. Muscles can control fin angle and for teleosts, furling of the fin-rays. In some bony fish such as eels, however, the pectoral fins are small and pelvic fins absent. The loss of pelvic fins as 'reductive evolution' has been discussed recently by Yamanoue et al. (2010).

The elasmobranch pectoral and pelvic 'girdles' are, of course, cartilaginous and composed of elements which span the mid-line ventrally. The pectoral girdle of dogfish has two fused elements, meeting mid-ventrally at a concavity overlying the heart: the pericardial depression. On each side of the pericardial depression is a thick curved bar termed the coracoid portion and beyond this is a strongly curved scapular which tapers at its end. The fin skeleton itself links to the coracoid/scapular area on each side (Figure 3.12).

In a teleost fish such as *Pterois* (Figure 3.13) with well-developed paired fins the pectoral girdle has

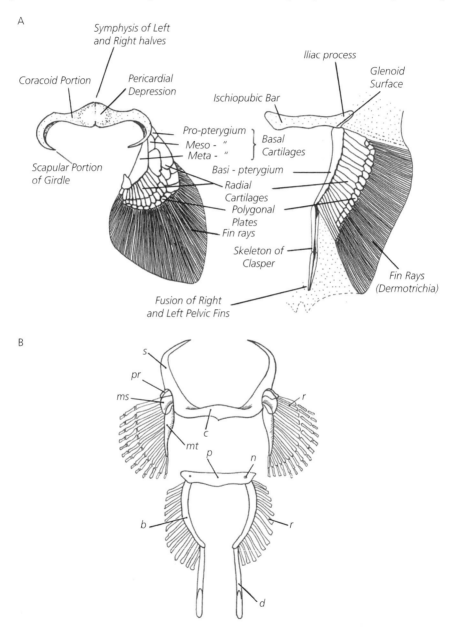

Figure 3.12 The pectoral and pelvic fin skeleton of dogfish (*Scyliorhinus/Scyllium*). A: viewed from above. Left: pectoral. Right: pelvic, after Wells (1961). B: top (pectoral), bottom (pelvic), viewed from below (de Beer 1951). Note the difference in representation of the radials; b, basipterygium; c, coracoids region; d, skeleton of clasper; ms, mesopterygium; mt, metapterygium; n, nerve foramen; p, pelvic cartilage; pr, propterygium; r, radials; s, scapular region.

many bony elements including (from dorsal to ventral) the posttemporal, supracleithrum, cleithrum, scapula(r) and coracoid, with the large cleithrum passing anteriorly and fused to its neighbour in the

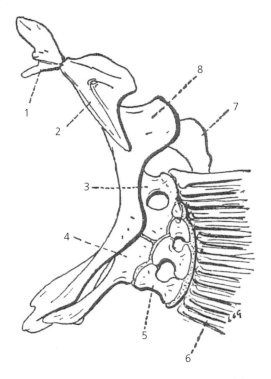

Figure 3.13 The pectoral girdle of a teleost *Pterois*, viewed from the side, after Goodrich (1909). 1, Post-temporal; 2, supracleithrum; 3, scapular; 4, coracoid; 5, radial, 6, lepidotrichia; 7, postcleithrum; 8, cleithrum.

mid-line. The posttemporal abuts the skull to which it is attached in (most) teleosts (Goodrich 1930a). Short plates termed radials, with a joint each end, link the fin-rays to the scapular/coracoid region.

The pelvic 'girdle' of dogfish (Figure 3.12), *Amia* and teleosts such as trout (Figure 3.14) is simpler than the pectoral girdle; it has no dermal elements. The dogfish has a single ischiopubic bar straddling the mid-line and articulating at its end with the basipterygium of the fin support on each side, at the glenoid surface (Wells 1961). A short projection, the iliac process, lies anterior to the glenoid surface (Figure 3.12A). In teleosts, the 'pelvic bone' (i.e. a bone formed by ossification of an ischiopubic equivalent) may not be present, and the radials are 'lost or fused' (Helfman et al. 2009).

The perch has paired pelvic plates (basipterygia) which may fuse at the mid-line (Chiasson 1974) and, as this is a derived teleost with relatively anterior pelvic fins, 'the anterior tips of the pelvic plates extend forward' between the fused cleithra of the pectoral girdle. The pelvic fin of perch has no radials linking it to the pelvic fin-rays, which link directly to the pelvic plates. Unlike in tetrapods, which transmit thrust to the vertebral column as a necessary feature of locomotion, the pelvic girdle of fish is not attached to the vertebral column. An extreme modification occurs with fish like *Cyclopterus lumpus* in which the united pelvic fins form a 'sucker' (Scott and Scott 1988). The structure and function of teleost pelvic fins have recently been reviewed by Yamanoue et al. (2010).

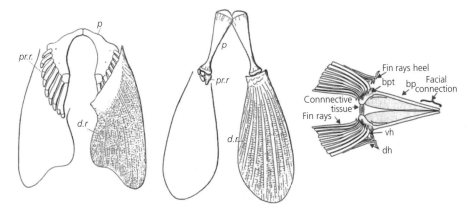

Figure 3.14 The pelvic region skeleton of bony fish, viewed from below. Left: *Acipenser*; middle *Amia*, after Goodrich (1901) see also Figure 3.1: dr, dermal rays; p. pelvic girdle, prr, preaxial radial. Right, trout; bp, semi-ossified basal plates; bpt, basipterygium; dh, dorsal hemitrichia; vh, ventral hemitrichia. Reproduced with permission.E.M. Standen (2010), Muscle activity and hydrodynamic function of pelvic fins in trout *(Oncorhynchus mykiss) Journal of Experimental Biology* 213, 631–41.

Male elasmobranchs and holocephans have paired claspers attached posteriorly to the pelvic complex (Figure 3.15). These structures are important in sperm transfer (9.5).

3.5.6 Median fin support

In addition to the paired fins, there may be several median fins, dorsally or ventrally in the midline, as well as a tail or caudal fin. The median fins have supports which can be described in the same terms as those for the paired fins, with reduced basal elements, radials and dermotrichia. Underlying the basal elements (Caillet et al. 1986) may be long pterygiophores, shown in teleosts as interdigitating with neural spines of the vertebrae. The dermotrichia supporting the fin-rays can be 'soft' or hard, forming spiny projections in some teleosts. The dermotrichia, which may be modified

scales (Goodrich 1930a; Young 1981), provide support within the fins and in teleost muscles which furl the fins are attached to them. The separation of the segmental supporting radials of the dorsal (except the adipose) and ventral fin bases from the vertebral column facilitates independent movement of the fins (Goodrich 1930a). The dorsal fins may be singular and extensive, forming a continuous structure along the back or divided into more than one, usually two, termed the anterior and posterior dorsal fins. The dorsal fin of the bichir (*Polypterus*) is formed of a series of 'finlets' (1.3.6, Figure 1.15). Ventral fins may include an anal fin, posterior to the anus. In live-bearers like *Xiphophorus* and *Poecilia* (9.5) the anal fin of the male is modified to form a gonopodium which accomplishes internal fertilization (Chambers 1987; Kwan et al. 2013).

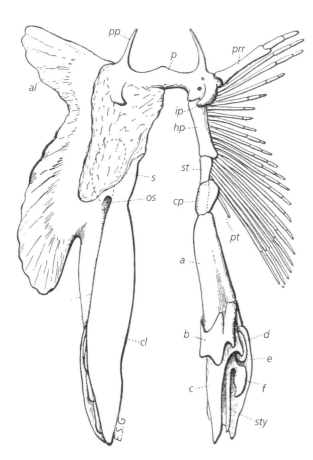

Figure 3.15 Pelvic complex of a male ray, view from above showing paired claspers, with skeleton dissected at right (after Goodrich 1909); a–f, cartilages of clasper; al, anterior lobe of pelvic fin; bp, basipterygium; cl, clasper; cp, covering plate; ip, iliac process; os, opening of sac; p, pelvic girdle; pp, prepubic process; prr, enlarged preaxial radial; pt, posterior preaxial radial; s, dotted line indicating ventral glandular sac; st, second segment of basipterygial axis; sty, styliform cartilage.

3.5.7 Fin-ray supports

The fin-ray supports (dermotrichia) may be segmented, soft or hard, forming spines as commonly occurs in acanthopterygians, which may have anterior spines in a fin, followed posteriorly by soft fin supports. The composition of the fin-rays is described as horny (similar to keratin, Bond 1996) or elastoidin (a form of collagen, Alexander 1970) in the ceratotrichia (dermal fin-rays, Wells 1961) of elasmobranchs with bone in lepidotrichia of teleosts (actinotrichia during development, Goodrich 1930a). The soft fin-rays of teleosts yield because they are segmented whereas the hard fin-rays or spines do not bend (Bond 1996). The segmented rays can also be branched. Lungfish (Dipnoi) have jointed, bony fin-rays termed camptotrichia (Goodrich 1930a).

3.5.8 Fin adjustment

Fin adjustment varies between taxa. Whereas the fin-rays of elasmobranchs enable some bending of the entire surface, the fins cannot be furled. In contrast, the support structure and complex musculature of teleosts may enable the furling of fins or generation of a series of waves for propulsion (Alexander 1970). The angle of the paired fins can be adjusted via joints of fin attachment to the girdles.

3.5.9 The adipose fin

The adipose fin, found in eight orders of bony fish (Reimchen and Temple 2004) including myctophids (1.4.2, Figure 1.21), does not have fin-rays; it is a small fatty projection arising dorsally just before the caudal fin or ventrally (e.g. in *Ichthyococcus*, Nelson 1976). This fin may have a hydrodynamic role in swimming (Reimchen and Temple 2004), mitigating or detecting turbulence; it is suggested the prevalence/persistence in several evolutionary lines underscores a positive role in spite of its small size.

3.5.10 Caudal fin support

The base of the caudal fin supporting the distal dermotrichia consists of hypurals, expanded processes of the haemal arch of the posterior vertebrae, and in teleosts a urostyle often supports the end of the notochord. In elasmobranchs, dermotrichia occur in both lobes of the heterocercal tail; the base of the upper lobe being supported by radials, the lower lobe is supported by hypurals (Goodrich 1930a).

The tail support and caudal fin is, in most fish, either obviously or covertly asymmetrical, with evidence of heterocercal ancestry. A true heterocercal tail has a larger upper lobe (Figures 1.3, 1.9, 3.16a) and provides posterior 'lift', counterbalancing the lift of pectoral fins. A reversed heterocercal (hypocercal) tail has a larger lower lobe as found in some extinct jawless fish, agnatha (pteraspids, anaspids) and the extinct reptilian ichthyosaurs (Young 1981). The elasmobranch selachians, for example dogfish, have clearly heterocercal tails both externally and internally (Figure 3.16a). *Amia* (Figure 3.1) and the teleosts have internal asymmetry (Figures 3.16b and 3.17) but most teleosts have external symmetry with equal sized lobes, a homocercal tail (Figures 3.16b and 3.17). Exceptions are male sword-tails (e.g. *Xiphophorus helleri*) which have an elongated lower lobe (the 'sword') of the caudal fin (Allen et al. 2002). The bichir *Polypterus* and the extant lungfish have a diphycercal tail, secondarily symmetrical as phylogenies show ancestral heterocercal tails (Hyman 1942). The coelacanth has a diphycercal tail (Figure 3.16c) with a terminal small additional lobe in most specimens.

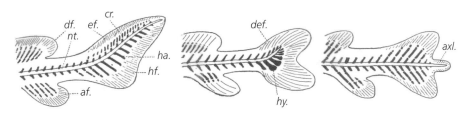

Figure 3.16 Diagrams (after Goodrich 1930) to show, left to right, a heterocercal tail (dogfish); a homocercal tail (teleosts), a diphycercal tail (coelacanth); af, anal fin; axl, axial lobe; cr, epichordal radial; df, dorsal fin; ef, epichordal lobe; hf, hypochordal lobe; def, dorsal lobe; nt, notochord.

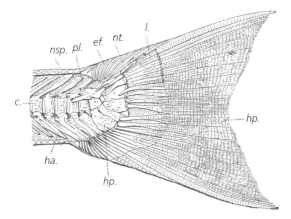

Figure 3.17 Teleost tail (*Salmo*, dissected); homocercal with external symmetry and internal asymmetry. After Goodrich 1909: c, centrum of caudal vertebra; ef, epichordal dermal ray; ha, haemal arch; hf, hypochordal; fhp, expanded hemal arch or hypural; l, dermal ray of opposite (right) side; nsp, neural spine; nt, upturned extremity of the notochord; pt, covering bony plate (modified neural arch, queried by Goodrich).

3.6 Trends in the fish skeleton

Fish are very ancient vertebrates, the fossil record for Agnatha (jawless fish) extends back to about 480 million years ago (mya; Janvier 1999). In general, too, it is elements of the skeleton that have survived, so the information is considerable if difficult, sometimes, to interpret. The relationships between bones through the years have not always been agreed upon and nomenclatures for some, such as those of lungfish, have resorted to numbers or letters (Kemp 1999) to avoid confusion or overly ambitious patterns of deduced relationships.

Reviewing the diverse structures of fish it has been possible, however, to detect a number of changes in some phylogenies, with taxa reflecting trends over evolutionary time. The changes include reduction in scale and skull thickness in bony fish (and therefore reduction in mass), modifications of vertebrae, girdles and fin supports, and considerable changes in the composition and support of jaws.

A change in structure from cartilage to bone has been a major trend as well as a subsequent reduction in bone mass, and a decrease in bone cellularity (Moss 1963; Meunier and Huysseune 1991; Horton and Summers 2009).

Reduction in scale thickness—Relict fish tend to have large heavy scales, such as the ganoid scales of gar or the bichirs and the prominent scales of lungfish and the coelacanth. By contrast, many teleost groups typically have very thin cycloid scales, though individually each scale may have a large area. The reduction in weight achieved by thinner scales can aid buoyancy and efficiency of movement.

Reduction in skull thickness—The bones of *Amia* (bowfin) skull form a solid casing round the brain whereas in *Gadus* (cod), a paracanthopterygian teleost, the separate skull bones may be so thin as to appear translucent. An *Amia* skull will be much heavier than a *Gadus* skull of the same length. The change in mass reflects a functional change from maximized protection to increased motility,

3.7 Articulation of the skeleton

The various elements of the endoskeleton are linked by a system of ligaments, tendons and muscles, which secure the skeleton, retain the abdominal organs and allow mobility. While some parts of the skeleton are fused or articulated to provide rigidity and protection, others allow for movement with the skeleton proving a firm base for muscle attachment and groups of antagonistic muscles (see 'antagonistic control' 11.4) operating mobile joints.

3.7.1 Joints

Where individual elements of the skeleton abut there will be a joint, which may allow movement of the elements relative to each other, or may be immobile as occurs with many of the skull components. Immobile joints are delimited by lines (sutures) which, with increasing age, may become obscure, and interpreting skull components may therefore be quite difficult with older animals. In association with feeding, however, there may be extreme mobility of the gape, to enable swallowing of very large prey (4.4.1). Hinging of the skull occurs in coelacanths (3.5.4) and some deep-sea fish have the capacity to bring the entire skull back over the vertebrae while the lower jaw remains depressed (4.4.1, 13.4.3).

Jointing of the vertebral column and of the fins may be more than a simple hinge but does not enable as much rotation as a tetrapod 'ball-and-socket' joint. The articulation of the paired fins to the girdles is usually via several basal elements rather than the single one of tetrapod limbs.

3.8 Locomotion

Most fish, being aquatic, move as adults from place to place by swimming; many fish species are capable of prolonged swimming for long distances (as in pre-spawning migrations) and many show very precise local movements to achieve feeding. Larvae may, however, go through a phase when passive floating occurs, and some fish are fairly sedentary, keeping stationary in burrows (see garden eels, 13.5.6.) or lying on the sea floor (e.g. most flatfish and anglers, 13.4.4).

Fish locomotion depends on muscle contraction with the endoskeleton operating as a firm base (compression strut, Young 1957; Thompson 1961) to prevent shortening of the main axis. Fish can swim for extended periods of time at suboptimal speeds or they can make rapid short movements. They also need to be able to stop abruptly (braking) and to steer. Some fish can even swim backwards (e.g. the eel, D'Août and Aerts 1999). Because there are so many fish shapes and sizes the mode of locomotion may vary considerably between species, and initially most studies were on medium-sized fish, easily held in captivity.

Fish movement has been the subject of intense scrutiny from the time of Gray (1933a,b,c, 1968) onwards, taking advantage of new technologies such as cine-cameras (see comments by Gillis 1996 and Lauder and Tytell 2004). A large literature has built up and two monographs (Blake 1983 and Videler 1993) have explored the details of fish swimming, including the biophysics of the medium in which fish move. Some fish (lungfish, coelacanth) operate their paired fins somewhat like the limbs of tetrapods and the relationship between the paired fins and tetrapod pentadactyl limbs has attracted much interest, including the basic work of Goodrich (1930a). Webb (1982) summarized major trends in the evolution of actinopterygian locomotion including increased ossification of the axial skeleton but

reduction in scale mass and armour, reduction in the number of vertebrae, development of the homocercal tail and a tendency to anterior pelvic fins.

Movement of fish is of two general types, either involving undulations of the main body and tail, or propulsion by fin movements (3.8.2). In either case, movements of the fin muscles are important for steering and stability. The jointing of the skeleton and number of muscles operating the joint determines the positioning and flexibility of the moveable parts.

In selachians (sharks and dogfish) the pectoral fins can be inclined but they have a wide base of attachment so the movement is quite limited. The rays (ceratotrichia) of the fins, particularly the pectoral and caudal fins, are fairly stiff but are linked to the radial cartilages by muscles which enable some bending and inclination of the fin margins.

In teleosts the basal elements of each fin are linked to multiple radials and the individual rays are attached one per radial, with a wide flexible membrane between the rays and two hinge joints enabling complex fin movements, each articulation having six muscles (except for catfish, Alexander 1970): multiple undulations as well as fin opening and closure are possible.

3.8.1 Skeletal (striated) muscles

The skeletal muscles hold the endoskeleton in place (with the tendons and ligaments) and also provide propulsion and operate stabilizers. It is usual to think of the body-wall muscles as originally composed of segmented muscle blocks, the myomeres (myotomes), divided by fibrous myocommata. The basic chordate structure is complicated in fish, however, in which the myocommata and the myomeres are bent in zig-zag formations, which is supposed to give a smooth bending action. Most of the muscle mass in fish is in the dorsal body wall where it can contract to bend the body. Exceptions occur with fish which have stiff bodies, such as many of the Tetraodontiformes (boxfish, etc.) which necessarily depend on fin movements for propulsion. Other fish which can bend their bodies may actually rely on fin movements for most propulsion unless stressed or lunge-feeding when they may utilize a very rapid body/tail bending as an escape or attack.

Protochordates like Amphioxus have simple muscle blocks (myotomes or myomeres), one pair per segment, with the myotomes enclosed by tissue septa (myocommata). In extant fish, however, the myotomes are not simple blocks: they have multiple cone shapes, with successive cones fitting into the adjacent myotomes and the individual myotomes therefore have the external appearance of extending across segments. These complex shapes show a variety of patterns if a transverse section is cut at successive levels across the trunk of a fish. The outer fibres within the myotome run horizontal to the body surface but deeper fibres are inclined, an arrangement thought to enable similar shortening of sarcomeres (the small units in the fibres) 'at different body flexures' (Johnston 1981a). However, Alexander (1970) states that 'the myotomes... have a very complicated shape which has never been explained satisfactorily'.

The body-wall muscles of extant fish are generally arranged in two layers: a thin strip of outer red muscle with an inner, very thick, white (or mosaic) muscle region (Plate 10) with a larger proportion of red muscle in active swimmers and very little or none in some of the more benthic species (Greer-Walker and Pull 1975). In very active fish like tuna there is also a region of internal red muscle (Figure 3.18) which functions, with a *rete* heat exchanger (5.10), to maintain high temperatures, these fish being heterotherms (7.4).

The red muscle contains a high proportion of myoglobin and has a rich blood supply (hence the colour). It is aerobic, with a high fat content, and is slow to fatigue, being used for slow, prolonged swimming (about 3 to 5 lengths/sec for teleosts, Alexander 1970). The individual fibres have a relatively small diameter and are classified as tonic or slow. White muscle by contrast does not have much myoglobin and is anaerobic, being glycogen-dependent. It is used for fast 'burst' speed, about two to three-times cruise speed, and is quick to fatigue; it can only maintain this speed for a few seconds (Alexander 1970). The individual fibres are large diameter, classified as fast 'twitch', similar to those of frog hind legs (e.g. the gastrocnemius muscle). The ability to initiate a fast response is also dependent on the special Mauthner fibres (11.6.6) in the spinal cord. These fibres have not been described in tetrapods except some urodele amphibia.

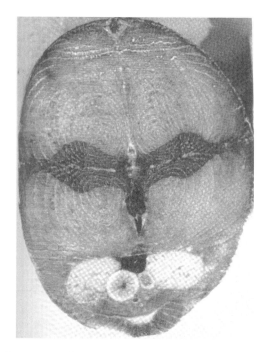

Figure 3.18 Body wall muscles in transverse (cross) section of a tuna. The internal dark area is red muscle. Reproduced with permission.E.D. Stevens, H.M. Lam and J. Kendall (1974), Vascular anatomy of the counter-current heat exchanger of skipjack tuna. *Journal of Experimental Biology* 61, 145–53.

Instead of 'pure' white muscle some fish have a mixture of fibre types which, microscopically, results in a mosaic effect and macroscopically, a pink colouration as in salmonids. Additionally, regions of small diameter intermediate-type and/or 'pink' fibres have also been figured (Johnston 1981; Johnston et al. 1975). Pink fibres occur in various teleosts, discussed by Mosse and Hudson (1977). Strong pelagic swimmers like herring and mackerel also have a relatively high proportion of red muscle in the outer layer. Both lamnid sharks and tuna also have a large block of red muscle close to the vertebral column in association with a heat-exchange mechanism (7.4, Figure 3.18). Fish which use slow regular fin movements for locomotion have red muscles operating these fins.

Fish skeletal muscle is very similar to that of mammals in the linear organization of the units termed sarcomeres (Rome and Sosnicki 1991; Burkholder and Lieber 2001) and the protein fibres (actin and myosin which slide past each other) forming the basis of the contractile mechanism

Table 3.5 Sarcomere lengths (relaxed) in μm for mammals, a bird and fish.

Species	Sarcomere length	(O) *optimum or (M)** maximum
Rabbit	2.27	(O)
Rat	2.4	(O)
Cat	2.43	(O)
Chicken	2.08	(O)
Carp (*Carassius* sp.)	2.11	(O)
Coalfish (*Gadus virens*)	1.82 (red)[†]	(M)
	1.85 (white)[†]	(M)

*Burkholder and Lieber 2001.
**Patterson and Goldspink 1972.
[†]red or white muscle.

(Huxley 1957; Connell and Howgate 1959). Information on the details of muscle contraction were obtained in early studies (e.g. Huxley 1957) by isolating individual muscle fibres (Szent-Györgi 1949) and stimulating their contractile action. More recent fish studies (e.g. Lieber et al. 1992) have used inferred sarcomere changes on live fish by measuring related parameters. Statements on fish sarcomere length have given values somewhat smaller than that of, for example, rabbit (Table 3.5). This value can be very important in setting up experiments measuring individual fibre contraction because a resting state (Szent-Györgi 1949) is desired at outset, and securing fibre ends to avoid undue stretching is also relevant. Johnston (1981a,b) has provided studies of fish muscle function and Lieber et al. (1992) have measured sarcomere length changes in swimming catfish with the interesting suggestion that the changes observed represented that white muscle was shortening and elongating 'during swimming powered by red muscle'. Phosphates (creatine phosphate or phosphocreatine and adenosine triphosphate, ATP) are expected to be the proximate source of energy for muscle contraction.

The control of striated muscle contraction is expected to occur via innervations of the neuromuscular junctions (NMJ) which are cholinergic (11.3.4). The muscle membrane is excitable (Appendix 11.1) and stimulation gives rise to depolarization and a muscle action potential leading to fibre contraction when actin and myosin slide past each other.

One of the phenomena associated with fish white muscle, in contrast to tetrapod muscle, is its capacity to apparently act as a protein store (7.12.3), with a reversible loss of structure, replaced as necessary by water (as a possible support) to give hydrated muscle (Love 1960). This condition, also referred to as jellied muscle (7.12.3, Appendix 7.4), may result in a very high degree of hydration (e.g. over 90 per cent, Patterson et al. 1974). Johnston (1981b) states that the loss of white muscle protein during starvation is both reversible and non-pathological. Black and Love (1986) studied the recovery of cod after starvation and confirmed that the high hydration could be reversed and that red muscle was conserved. Love (1970) associated utilization of white muscle with gonad growth. It is interesting that Connell and Howgate (1959) emphasized the instability of fish muscle relative to that of mammals; fish muscle can be readily degraded, a useful attribute if it is to be utilized as an energy source or source of amino acids for gonad growth. The loss of some protein can apparently be fairly well tolerated in an aquatic situation in which the muscles are not bearing weight (supportive) in the same sense as would occur with a terrestrial vertebrate. However, starved fish have decreased swimming endurance, with time and distance considerably reduced (Martinez et al. 2003).

3.8.2 Types of fish locomotion

Movements producing fish locomotion can usually be categorized as two types: 1. The body/tail bending (Body/Caudal Fin or BCF, Webb 1984) thought to typify early chordate movement, and 2. A variety of propulsive fin movements (Median Paired Fin or MPF, Webb 1984) which can be dependent on one fin such as the long anal fin of some knife fish (Superfamily Gymnotoidae of the Ostariophysi) or a combination of them, for example the pectoral, dorsal and caudal fin of the lumpfish (*Cyclopterus*). Many fish use both types of locomotion, with slow swimming based on fin movements while a rapid body/tail flip achieves a fast escape. Where body bending is the major propulsive mechanism it is common to classify it according to the number of waves passing along the body and this is also linked to the relative shape of the fish (Table 3.6): long, slim fish like anguillid eels can have two or more waves passing along the body simultaneously.

Table 3.6 Terminology of fish body/tail movement.

Description of movement	Name of movement	Example (Bond 1996)
More than one wavelength along body	Anguilliform	*Anguilla* (freshwater eel)
Half to one wavelength along body	Sub-carangiform	Trout, goldfish, cod
Half wavelength along body	Carangiform	Carangidae (jacks) *Scomber* (mackerel)
Caudal area only	Ostraciform	*Ostracion* (boxfish)
Caudal area only, lunate tail	Thunniform	Scombrids (tuna, billfish, Lamnid sharks

Thunniform locomotion is interesting because it has evolved in two very different groups, the scombrid teleosts, tuna, bill-fish (e.g. marlin) and lamnid sharks, which share other characteristics such as the development of heterothermy enabling maintenance of an internal environment warmer than the surrounding water. These fast, perpetually swimming (Alexander 1970) fish with a relatively high metabolic rate have a similar fusiform (torpedo-like) shape with a narrow caudal peduncle and a crescent-shaped or lunate tail. (Figure 3.5). The overall streamlined shape of the body reduces drag while the shape of the tail, described as having a high aspect ratio (obtained from span/width), is reported to give good lift relative to drag (Alexander 1970).

MPF locomotion, with fin propulsion paramount, is seen in many species, notably marine fish that feed inshore, in eel-grass, like seahorses (*Hippocampus* spp.), in kelp beds or on coral reefs, where slow movements of the fins and a short, deep body optimizes the ability to browse. It is necessary for these fish to make continual adjustment of position to allow for external water movements and also to allow for the jet of water emerging from the operculum which tends to propel them forwards.

The chondrichthyan skates and rays have greatly expanded pectoral fins which flap up and down to produce locomotion, while the elongated pectoral fins of the 'flying fish' (*Exocoetus*) enable short aerial glides apparently to escape predation. Although not the exclusive means of locomotion, the long dorsal fin of *Amia* (the bowfin) is utilized for propulsion. Electric fish such as the gymnotids use modified body muscle to generate electric fields (13.8.1) and have elongated median fins for slow swimming. The mormyrids, which use electrolocation, have fairly stiff bodies and MPF locomotion, thus optimizing their capacity to use their electric fields for navigation.

3.8.3 Fast-start

Recently interest has focused on 'fast-start' swimming used by predators for prey capture and for escape reactions of potential prey. These fast movements have been differentiated into 'C-starts' and 'S-starts', where 'C-starts' (previously L-starts, Domenici and Blake 1997) are used by predators and 'S-starts' as escape movements. Mauthner fibres have been traditionally associated with effective escape movements via tail-flips and the startle response (11.6.6), though Domenici and Blake (1997) have not recounted this association except for some work on goldfish (Canfield and Rose 1993). These authors believe, quoting Eaton et al. (1991), that C-starts depend on Mauthner fibre transmission. The C and S nomenclature depends on the typical body shape early in the reaction.

3.8.4 Steering, braking and stabilizers

Efficient swimming or maintenance of position ('holding station') requires ability to steer, to brake and to remain at a desired level in the water column. It is also necessary to counteract any tendency to lose orientation—that is, to turn over, to spin, to reverse unexpectedly or to be at any angle other than the best for optimal function. Steering, that is, directional swimming, can usually be achieved by movements of the paired fins or the caudal fin. Braking is achieved by movements of the pectoral fins and/or the caudal fin.

Stabilizers require the activation of 'righting' reflexes controlled via the ear when position changes are detected by the bi-lateral semi-circular canals (three each side in jawed fish) each in a different plane (12.8.1). In nautical terms there are three problems for keeping horizontal: yawing is a displacement so that the head swings to one side and then to the other repeatedly. Pitching occurs when the head moves up and down in the water column. Rolling describes a rotating movement of the entire body around its longitudinal axis. The righting reflexes counter the tendency to yaw, pitch and roll:

the median fins minimize rolling and yawing while the paired fins control pitch.

3.8.5 Buoyancy

The ability to remain in the water column, for any aquatic organism, can depend on shape and density. Dogfish and sharks have a tendency to sink, being denser than water; this is counteracted by the position of the pectoral fins which can give anterior lift, and the heterocercal tail which can give posterior lift. These functions have been discussed by Wilga and Lauder (2002). The occurrence of different tissue densities, with possible annual changes in, for example, fat deposition, is discussed by Brix et al. (2009). Pelagic notothenoids have lipid sacs (Plate 10) and a retained notochord (Eastman 1990). Yancey et al.

(1989) consider a possible buoyancy function for subcutaneous gelatinous tissue in mesopelagic fish.

Such gelatinous tissue, fatty livers which occur in chondrichthyans and some bony fish (e.g. cod), the lipid sacs and notochord in notothenioids (Plate 10), can provide some buoyancy but this is not capable of short-term regulation, which can occur with dorsal gas (swim) bladders found in many bony fish. The gas bladder may connect to the anterior gut, a condition called physostomatous (Figure 3.19). In this state, found in ostariophysans like goldfish, the pressure can be adjusted by gulping or expelling air. Such adjustment is not appropriate for deepwater fish and coming to the surface to gulp air is risky. Deep-sea fish and many others have a closed swim bladder (physoclistous) with pressure adjustments through the blood: gas (oxygen or nitrogen)

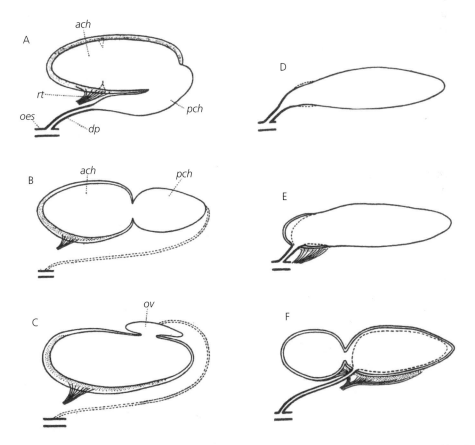

Figure 3.19 Swim bladders (from Goodrich 1930): A, hypothetical, physostomatous, with duct linked to gut, B, physoclistous (closed), with posterior chamber; C, physoclistous with oval; D, physostomatous, salmonid; E, physostomatous, esocid: F, physostomatous, cyprinoid. ach, anterior gas-secreting chamber; dp, ductus pneumaticus; oes, oesophagus; pch, posterior chamber; rt, *rete mirabile*.

is secreted or reabsorbed into the blood controlled through the autonomic system (11.10.4).

Although the gas in the swim bladder may be mostly nitrogen as has been reported for coregonids (Scholander and van Dam 1953), the gas enclosed and regulated can have a high level of oxygen, boosted in those fish with gas glands by secretion of lactic acid and subsequent local release of carbon dioxide, which, by the Root effect (6.5.7) causes release of oxygen (from haemoglobin) to the area. The regulation of gas secretion and loss or resorption occurs via the autonomic system (11.10.4), but from the point of use of the swim bladder for respiration there is also morphological variation in the blood supply and drainage. Physoclistous (closed) swim bladders commonly have a counter current *rete* (5.10, 6.5.6) in the gas gland, the red body or gland (Scholander 1954) in the anterior wall, which enables rapid gas release into the bladder from the blood. Gas absorption can be achieved in a posterior chamber, or smaller 'oval' (Figures 3.19 and 3.20). Loss of gas from the bladder may be prevented by the incorporation of relatively impermeable fibres or (guanine) crystals in the wall, giving many swim bladders a silvery appearance (Scholander 1954).

Some fish have a special gas gland for gas secretion, with a complex countercurrent blood supply (5.10) in a '*rete mirabile*' (Figure 3.20) in the red body (physostomatous fish) or red gland (physoclistous fish) of the swim bladder. In such a system, oxygen gas secretion can be promoted by lactic acid which

dissociates oxyhaemoglobin. Myctophids (lantern fish), paracanthopterygians (e.g. cod and angler fish) and those acanthopterygians that have swim bladders (flatfish do not) can have a special reabsorbent organ (the 'oval') in the swim bladder. This is in a posterior region of the bladder which can be closed off by a sphincter muscle when resorption is not required.

Adjustment of the swim-bladder gas content enables fish to keep stationary at different depths. Harden Jones and Schole (1985) reported that cod had slow gas secretion compared to gas absorption and they quoted Alexander (1972) who had felt that teleosts' capacity to adjust the swim bladder's effectiveness as a hydrostatic organ was not very impressive.

Not all pelagic bony fish have swim bladders and this may relate to the fact that soundwave reflection off the swim bladder facilitates echolocation by potential predators such as whales and dolphins: 'The swim bladder of a fish is the structure having the most influence on the echo intensity', state Au et al. (2007). This may explain the lack of a swim bladder in mackerel but its absence also promotes swift diving in pursuit of its prey; Schmidt-Nielsen (1990) outlines the problems of a soft-walled swim bladder being 'extremely sensitive to pressure changes'. Swim bladders are also reduced or absent in benthic fish such as flatfish, many very deep bathypelagic fish (Denton and Marshall 1958) and in fish which inhabit fast-moving streams, where they are apparently redundant.

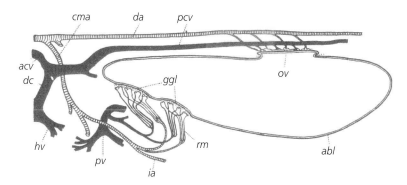

Figure 3.20 *Rete mirabile* of a teleost swim bladder, left side view, with oval (after Goodrich 1930). Veins black, arteries cross-lined. abl, bladder; acv, anterior cardinal vein; cma, coeliaco-mesenteric artery; da, dorsal aorta; dc, ductus Cuvieri; ggl, gas gland (red body); hv, hepatic vein; ia, intestinal artery; ov, oval; pcv, posterior cardinal; pv, portal vein; rm, *rete mirabile*.

3.8.6 Fish shape and swimming

Fast swimming is promoted by streamlining, in which case a smooth torpedo shape as found in tuna and lamnid sharks (thunniform swimmers) reduces turbulence. The smooth shape is improved if fins can be furled close to the body. A short, deep body promotes manoeuvrability, while the flattened body of many benthic species indicates very slow swimming unless the escape reaction is evoked. Some fish with many of the plesiomorphic features of the benthic habitat appear to have secondarily returned to a more pelagic habit, either making frequent excursions into the water column or spending the majority of their time off the bottom. The pleuronectid 'turbot' or Greenland halibut (*Reinhardtius hippoglossoides*) is such a fish. Although at first glance a typical asymmetrical flatfish, it is dark on both sides and viewed from the front shows more symmetry than the truly benthic pleuronectids, the mouth in particular is less skewed.

3.8.7 Movement out of water

'Flying' fish, the Exocoetidae, can use large, paired fins for short aerial glides (13.10.1) and the gastropelicids may use the pectoral fins actively for leaps (13.10.1).

Some fish, like the agnathan lampreys (Allen et al. 2002) and eels, can move on land. Any such fish (see 'terrestrial fish' 13.5.5) has effective aerial breathing (6.6) and can either use its body undulations (as with eel) or paired fin movements for terrestrial locomotion. Climbing perch (*Anabantus*) and the mud-skipper (*Periophthalmus*) both prop themselves up with their paired fins and can move on land, though not very fast. This type of movement seems to presage that of the true tetrapods, but is not on the direct phylogenetic line, being polyphyletic and to some extent opportunistic. The fish thought to be closest to the tetrapod ancestry, from the point of fin/limb morphology, are the lungfish and coelacanth which have well-differentiated paired fins, with supports more easily related to the tetrapod structure and hinging. Terrestrial locomotion obviously has advantages for freshwater or littoral fish in cases of drought or stranding, and provides opportunities for dispersal.

Temporarily stranded fish can make body movements (perhaps not directed) which may enable them to resubmerge. Fish like capelin (*Mallotus villosus*) or the grunion (*Leuresthes tenuis*) which have a beach-spawning strategy may be particularly vulnerable to stranding, and selection pressures may advantage those which can move best on shore, with the possibility of repeat spawning then available.

Appendix 3.1 The hydrostatic skeleton

A soft skeleton performing a support function is common in animals and is often referred to as a 'hydrostatic' skeleton. The support is generally provided by a fluid-filled structure such as the coelom, the major body cavity which develops within the embryonic mesoderm in triploblastic coelomates such as annelids, arthropods and vertebrates. However, the coelom does not have much of a support role in arthropods which have an exoskeleton. In annelids however (e.g. earthworms) the coelom has an important support role, also enabling muscular contraction to effect movement. In the annelids, which consist of a series of segments, the hydrostatic skeleton formed by the coelom, extends from the most anterior region right through to the last segments.

Vertebrates, such as fish, also have large coelom which provides some protection of abdominal organs but is not significant in providing a basis for locomotion. The vertebrate notochord however, which consists of large vacuolated cells surrounded by a fibrous sheath, can be regarded as 'hydrostatic' and can be important in locomotion, for example in larvae.

Appendix 3.2 Histochemical differentiation of cartilage and bone

Because the matrix of cartilage has a high carbohydrate content it is possible to use specific dyes to detect it and differentiate it from bone and other tissues at the microscopic or macroscopic level. At the microscopic level, Mallory's triple stain has been used; this procedure incorporates three dyes; acid fuschsin (red), aniline blue and orange G (Galigher and Kozloff, 1964), and the results show cartilage and fibrous connective tissue as blue, while muscle and bone stain red. The orange G stains erythrocytes. At the macroscopic level a complex procedure (Cailliet et al. 1986) allows staining of small intact skeletons, with alcian blue staining cartilage and alizarin red staining bone.

Appendix 3.3 The origin of bony skeletons: conodonts

The traditional view of the vertebrate skeleton relates the origin of the skull as a part of cephalization meaning elaboration of the anterior axial skeleton with deposition of cartilage and bone around the notochord. However, recently interest in the conodonts, an extinct jawed group, suggests that teeth became fused to form structures anteriorly like the ostracoderm headshields, and that at least the exoskeleton might therefore be derived from teeth. Knowledge of conodont morphology has burgeoned since more intact fossils have become available for analysis (see Donoghue et al. 2000), and other opinions may develop.

Appendix 3.4 Individual bones and their approximate position

Appendix 3.4a The neurocranium of bony fish (based on Gregory 1959).

Cartilage bones (deep)			Dermal bones (superficial)
Anterior (olfactory region)			
	Ethmoid	Covered dorsally by:	Rostral, postrostrals of relict fish; give rise to Mesethmoids
		Covered dorsolaterally by:	Nasal
		Covered ventrolaterally by:	Adnasal
		Covered ventrally by:	Vomer = prevomer of tetrapods
	Parethmoid	Covered dorsolaterally by:	Prefrontal
Orbital region			
			Sclerotics
			Circumorbitals (of which some, infraorbitals, represent the jugal and lacrimal of e.g. tarpon) see Gregory 1959; Schultze 1993
	Orbitosphenoid	Covered dorsally by:	Frontal
			'Postorbitals' not = tetrapod postorbital
Otic region			
	Sphenotic	Covered dorsolaterally by:	Dermosphenotic
	Pterotic	Covered laterodorsally by:	'Pterotic' = supratemporal of tetrapods
			Parietal
	Otic capsule		
	Prootic		
	Opisthotic		

	Exocciptial		
	Epiotic		May be covered by and fused with	'Epiotic' (superficial)
	Supraocciptial	May be covered by and fused with		'Supraocciptial'
				'Scale bone'
				Posttemporal (connects pectoral girdle to skull)
Basicranial region				
	'Basisphenoid'	Covered ventrally by		Parasphenoid = vomer of mammals
	Basioccipital	as above		as above

Appendix 3.4b The branchocranium of bony fish (based on Gregory 1933)

Cartilage bones (deep)			**Dermal bones (superficial)**
JAW region (First arch)			
Secondary upper jaw:			Premaxilla, supramaxillae, maxilla
Primary upper jaw: (Palatoquadrate)			
	Autopalatine	May be covered ventrally by:	'Dentigerous palatine I'
	Suprapterygoids		
	Metapterygoid		
	'Endopterygoid'	= pterygoid of tetrapods	
	'Pterygoid' = ectpoterygoid of terapods		
Primary lower jaw			
(Meckel's cartilage)			
	(Proximal) Articular	Covered inferiorly by:	Angular (generally reduced in teleosts)
		Covered laterally by:	Dermarticular (possibly fused with articular)
		Covered dorsally by:	Surangular (generally lost in teleosts)
		Covered medially by:	Prearticular
			Sesamoid articular ('tendon bone')
			mandibular adductor* insertion point
			Dentary
JAW SUPPORT (Second arch)			
Hyomandibular		Bearing posterolaterally:	Opercular
			Subopercular
		Covered laterally by:	Preopercular
Symplectic (distal hyomandibular)			
Interhyal			Interopercular

Cartilage bones (deep)		Dermal bones (superficial)
JAW support (second arch) continued		
Epihyal	Bearing dorsally	Dentigerous plate (i.e. toothed)
Ceratohyal	Bearing ventrally	Branchiostegal rays
Hypohyal		
Glossohyal	Covered dorsally by:	Dentigerous plate

*Muscle affects jaw closing

Appendix 3.5 Skull bones found in extant fish (compiled from Gregory 1933; de Beer 1951; Saunders and Manton 1951)

Name of bone	Bone type: C, cartilage; D, dermal	Location: N, neurocranium; S, splanchnocranium
Articular	C	S
Angular	D	S
Autopalatine	C or mixed	S
Basioccipital	C	N
Basisphenoid	C	N
Dentary	D	S
Ectopterygoid	D	S
Endopterygoid	D	S
Epiotic*	C	N
Ethmoid	C	N
Exoccipital	C	N
Frontal	D	N
Jugal**	D	N
Lacri(y)mal	D	N
Maxilla	D	S
Mesethmoid	C	N
Nasal	D	N
Opercular***	D	S
Opisthotic	C	N
Orbitosphenoid	C	N
Palatine (see autopalatine)		
(Palato)quadrate	C	S
Parasphenoid	D	N
Parietal	D	N
Prefrontal	Mixed	N
Premaxilla	D	S
Preopercular	D	S
Prevomer	D	N
Prootic	C	N
Pterotic	Mixed	N
Pterygoid	C	S
Rostrals	D	N
Sphenotic	Mixed	N
Suborbitals	D	N
Supraoccipital	C	N
Suprapterygoid	C	Uncertain
Symplectic	C	S
Vomer	D	N

*According to de Beer (1951).
**The jugal is described from some fish, e.g. Sarcopterygians (Young et al. 2013) but it is doubtfully present in *Amia*. The relationships of the sclerotic ring and the circumorbital bones may be problematic (Burrow et al. 2011).
***According to de Beer, the operculum consists of four bones, is an extension of the hyoid arch, ventrally the area was supported by gular plates, which in teleosts (Romer 1955) are reduced to branchiostegal rays.

Bibliography

Hanken, J. and Hall, B.K. (1993). *The Skull*. Chicago, IL: University of Chicago Press.

McGowan, C. (1999). *A Practical Guide to Vertebrate Mechanics*. Cambridge: Cambridge University Press.

Moss, M.L. (1961b). Studies of the acellular bone of teleost fish. 1. Morphological and systematic variations. *Acta Anatomica*, 46, 343–62.

Webb, P.W. (1998). Swimming. In: D.H. Evans (ed.). *The Physiology of Fishes*, 2nd edn. Boca Raton, FL: CRC Press, pp. 3–24.

CHAPTER 4

Food procurement and processing

4.1 Summary

Fish need an external source of food, which can be particulate or larger prey. Few fish are herbivores.

Procuring and processing the food requires a complex set of functional adaptations with procurement usually reliant on ingestion by the jawed mouth and subsequent passage through a series of chambers constituting the gut.

Food is rendered into absorbable small molecules by a series of mechanical and chemical digestions in the gut.

The details of the gut structure and function vary interspecifically and between major taxa; elasmobranchs and some relict fish have a spiral valve in the intestine and a distinct pancreas whereas teleosts have neither; the pancreatic tissue is diffuse.

Fish in their natural habitats may show feeding periodicity, with non-feeding periods being possible and extensive in certain species.

The nutritional requirements of fish have been the subject of considerable research effort with the aim of optimizing growth for aquaculture.

4.2 Introduction

As 'ingestive heterotrophs' animals need to obtain a source of organic nutrients, usually in the form of complex molecules in food. Vertebrates process the food through a gut which enables mechanical and chemical breakdown (digestion) of large structures, increasing the surface area of the material taken in (ingested). Mechanical digestion is followed by chemical digestion in the gut lumen, extracellularly. The small molecules produced by chemical digestion can then be absorbed and utilized. Although the food of vertebrates is quite variable, ranging from other animals to plants, the general pattern of food procurement and processing is very similar, relying on a sequence of 'prey' (animal or plant) detection, ingestion at the mouth, digestion and absorption by the gut and egestion of undigested material at the anus.

The majority of fish are carnivores, feeding on other animals, but some fish consume plankton (a mixture of floating organisms) and a small, but important, minority are herbivores feeding on plants. Some species are highly specialized in their food requirements and a few are parasitic (13.13). Food procurement in most fish depends on adaptations of the mouth and jaws combined with appropriate body movements. Although for most fish the food-processing system is very similar to the typical vertebrate gut, the presence or absence of constituents such as the stomach, and links to the air or swim bladder, as well as the presence of multiple pyloric caeca in most bony fish and the spiral valve in the intestine of cartilaginous fish, and of course pharyngeal clefts in the adult, differentiate the system from the mammalian gut. Feeding may have seasonal patterns and may change considerably throughout the life-history for those species which have continuous growth producing intraspecific large size ranges. As many fish begin active life as small larvae the initial size of food ingested is very small, and larval feeding has become an area of intense study for successful rearing of fish in aquaculture. Understanding of optimal nutritional content of food in captivity, relative to cost-effectiveness, has also become a very important field of study.

Essential Fish Biology: Diversity, Structure and Function. Derek Burton & Margaret Burton.
© Derek Burton & Margaret Burton 2018. Published 2018 by Oxford University Press.
DOI 10.1093/oso/9780198785552.001.0001

4.3 Food and feeding

Fish may be plant eaters (herbivores) or animal eaters (carnivores) or non-selective omnivores. The last category would include planktivores that may select (if at all) solely on the basis of size of prey (small) which is obtained by a form of filter-feeding. A recent investigation of hagfish feeding (Glover et al. 2011) shows that, as scavengers, they can absorb dissolved organic nutrients through their gills or skin directly from the body cavity of carcasses.

Only 5 per cent of extant fish species are herbivores (Tolentino-Pablico et al. 2008) and most of these are from one taxonomic group, the Ostariophysi, which also comprises the majority of freshwater fish, and includes the cyprinids (carp, etc.) and the silurids (catfish). *Ctenopharyngodon*, the grass carp, has even been maintained on lettuce, and carp have been very successfully pond-cultured for human consumption, using plant-based food, in both Asia and Europe, for many centuries. Herbivorous ostariophysans commonly have secondary pharyngeal jaws with tooth-plates adapted for crushing vegetation. One South American characin, the tambaqui or pacu (*Colossoma macropomum*) has a reputation for including Brazil nuts in its diet when the fish reaches a large enough size. Australian terapontids may also have 'terrestrial prey' in their diet (Davis et al. 2010), including fruit. Other fish taxa, which are marine and about 30 per cent of the herbivores, may graze seaweeds (Tolentino-Pablico et al. 2008) and mostly live on coral reefs. These marine herbivores can show species-specific preferences, specializing in green or brown algae, for example (Tolentino-Pablico et al. 2008).

Trophic (feeding) ecology, including dietary components analysed from stomach contents, has been an important area of study, enabling understanding of the complex interactions in a habitat. It is pertinent to realize that within a biological system the diet of individual fish changes with age, size and season. Indeed, young fish may occupy a completely different habitat and be part of a different community than the adults such as when mangrove swamps act as a 'nursery' for juveniles but the adults are in more open water. Similarly, larval fish may be floating in the plankton while adults are pelagic (active swimmers in the water column) or benthic (bottom-dwellers).

Winemiller (1990) presented a detailed study of interrelationships as 'food webs' for freshwater fish in South and Central America (swamp/stream), while Edgar and Shaw (1995) studied shallow (marine) water 'trophic pathways' in southern Australia, using gut analyses. Interestingly, of 5113 fish used, 720 were found to have 'empty guts'. Fryer (1959) constructed food webs for different habitat zones in Lake Nyasa (Africa), relatively closed systems. The complexity of food webs in the oceans and with change of season is difficult to comprehend or analyse.

4.3.1 Catching prey

Most fish are carnivores and some are highly selective in their prey, either by size or taxon. Carnivorous (predaceous) fish may be mainly fish-eating piscivores and either active chasers like tuna, salmon or Greenland halibut or more passive ambush species such as gar (Figure 1.17) and anglers (Figure 4.4), with large gapes as well as sharp teeth. Other carnivorous fish may concentrate on invertebrates and some fish like small cod may have a mixed diet of crustaceans (e.g. shrimp) and small fish. Flying insects may be caught, for example by trout or archer-fish (*Toxotes*, Figure 4.7). As fish grow they may adjust their preferred prey.

In the case of active live prey there are various strategies which enable final prey capture, one of which is suction (Roos et al. 2009) for the actual intake. Fish with suction feeding include teleosts and elasmobranchs, and it is achieved by rapid enlargement of the buccal cavity (Roos et al. 2009), usually by a coordinated set of actions with mouth opening, which include depression of the lower jaw and the hyoid as well as elevation of the neurocranium and 'abduction' of the suspensorium (the jaw suspension 3.5.4) and operculum (Van Wassenbergh et al. 2011). Syngnathids like pipefish, however, use a different mechanism (Van Wassenbergh et al. 2011) with cranial rotation (pivot feeding).

4.3.2 Piscivory, prey retention and prey defenses

Fish which are piscivorous, catching and swallowing other fish, must have a firm grip on the prey which may be assisted by pharyngeal tooth pads

(Figure 4.1). According to Nikolsky (1963) it is common for predatory fish to seize their (fish) prey in the middle of the body then release it and reorient to swallow it head first.

Some potential prey organisms have defence mechanisms which make it more difficult to swallow them; tetraodontiform puffer-fish can inflate, and may also be poisonous, producing tetr(a)odotoxin (Fuhrman 1967); related trunk fish can release ostracitoxin (Nelson 1976). Sticklebacks, with three ten-dorsal spines, are prey for some salmonids such as land-locked *Salmo salar* (Leggett and Power 1969). The sticklebacks (in this case *Gasterosteus*) are apparently 'actively' selected, as many as 27 being found in one stomach. The spines are obviously not an adequate defence against predation in this relationship, but there was no other fish prey available in this habitat.

Hoogland et al. (1957) studied the effectiveness of stickleback spines in relation to the predators *Perca* (perch) and *Esox* (pike), using *Gasterosteus* (three dorsal spines) and *Pygosteus* (ten dorsal spines) as

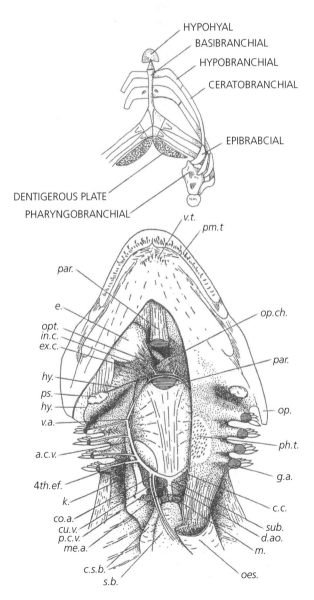

Figure 4.1 Additional teeth to retain prey: Above: 'Branchial apparatus' of perch showing dentigerous plates (tooth pads). Reproduced with permission: R.B. Chiasson (1974) Laboratory anatomy of the perch, © Wm.C. Brown Company Publishers, successor McGraw Hill Education. Below: Whiting with extra tooth patches in the buccal cavity, vomerine teeth and pharyngeal tooth pads, from Saunders and Manton (1949). a.c.v. anterior cardinal vein; c.s.b. lateral cornua of swim bladder; c.c. circus cephalicus; co.a coeliac artery; cu.v Cuvierian vein; d.ao. dorsal aorta; e eye; ex. c. external carotid; g.a. gill arch; hy. Hyoidean artery; in.c. internal carotid; k. kidney; m. Cut muscle which controls pharyngeal teeth; me.a. mesenteric artery; oes, oesophagus; op. Operculum; op.ch. optic nerves crossing; opt. ophthalmic artery; p.c.v. posterior cardinal vein; par. Cut parasphenoid; ph.t. pharyngeal teeth; pm.t. premaxillary teeth; ps pseudobranch; s.b. swim bladder; sub subclavian artery; vt, vomerine teeth; v.a. vertebral or posterior carotid artery; 4th ef. 4th efferent branchial artery.

potential prey. As described by these authors, when a stickleback is seized by a pike the stickleback can erect its spines and 'lock' them into this position. This reaction includes paired pelvic spines for *Gasterosteus*. Sticklebacks were observed to be regurgitated by pike and when a choice was available the pike showed a preference for minnows (*Phoxinus phoxinus*). However, this study used fairly small pike and perch and the authors quote a previous study (Frost 1954) which seems to show that large pike may prefer sticklebacks to minnows as prey. Perch which had swallowed sticklebacks in the later study (Hoogland et al. 1957) included one perch which was injured in the process with quite extensive bleeding.

4.3.3 Plankton feeders, microcarnivores and picking and scraping

Fish which are planktivorous as adults may strain small organisms from water by cruising with open mouths and may have simultaneous ram ventilation (6.5.11). Planktivorous fish such as basking shark and herring are characterized by long processes, gill rakers, as an array of processes extending from the outer surface of the gill arch (Figure 4.2), which partially cover the gill slits and so help to retain plankton as water passes from the pharynx across the gills ('filter-feeding'). However, the concept of passive, non-selective straining has been challenged (Langland and Nøst 1995), and these authors showed the same minimum prey size for fish species with considerable differences in gill raker spacing. The species used in this study were regarded as 'facultative planktivores', that is they were not highly specialized for small prey. O'Connell (1972) described Northern anchovy (*Engraulis mordax*) as alternating between filtering and specific 'biting' to obtain food and believed that biting activity could be more than 50 per cent of 'total feeding activity'. It was found that biting increased with the density of the prey.

Microcarnivores which pick up tiny prey, for example by suction, may also have long close (dense) gill rakers which prevent prey falling out of the pharynx, or the gill opening may be very small, thus achieving some prey retention. Carnivores have wider-spaced gill rakers than omnivores (Boyle and Horn 2006), and these authors were surprised to find high gill raker density associated with omnivory rather than with fish eating small invertebrates or detritus. Fryer (1959) described feeding behaviour of a cichlid genus *Lethrinops* which filters sand (Figure 4.3). Picking

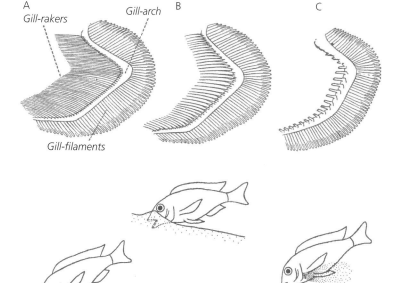

A Gill-rakers Gill-arch B C

Gill-filaments

Figure 4.2 Diagrams of gills showing gill arch and filaments. A = Allis shad (*Alosa alosa*). B = Twaite shad (*Alosa fallax*). C = Perch (*Perca fluviatilis*). After Greenwood (1975).

Figure 4.3 The African cichlid *Lethrinops* picks up a mouthful of sand plus prey and the sand is subsequently voided past the gill rakers. Reproduced with permission, G. Fryer (1959). The trophic interrelationships and ecology of some littoral communities of Lake Nyasa with especial reference to the fishes, and a discussion of the evolution of a group of rock-frequenting Cichlidae. *Proceedings of the Zoological Society of London* 132, 153–281 (*Journal of Zoology*). © John Wiley and Sons.

and scraping is a set of feeding behaviours from substrate. The role of 'trophic apparatus' in such specialized feeding behaviours is discussed by Hernandez et al. (2009), who examined 16 cyprinodonts. They differentiate such feeding from suction-feeding, describing 'picking and scraping' as needing fine control of the jaws but less force. Cichlids (Fryer and Iles 1972) include specialized 'pickers' and 'scrapers'. Horny lips (4.4.4) may also be used for 'scraping'.

4.3.4 Specialized diets, fin-biters, parasitism and cannibalism

Some fish have highly specialized diets, feeding on, for example, jellyfish (the sunfish *Mola mola*) or shelled molluscs (chimaeras) or corals (parrotfish). The wolffish (*Anarhichas*) may preferentially eat sea-urchins (Keats et al. 1986). African cichlids (Fryer and Iles 1972) have specialized to the point where several species, are described as scale-eaters (lepidophagy). Lepidophagous species include *Perrisodus microlepis* and four species of *Plecodus* (*P. paradoxus*, *P. straeleni*, *P. multidentatus* and *P. elaviae* from Lake Tanganyika and *Genyochromis mento* and *Corematodus shiranus* from Lake Malawi. *P. microlepis* has developed 'handedness', that is, individuals have mouths skewed to left or right which improves their capacity to evade detection as they attack from the appropriate side (Hori 1993; Palmer 2010). *Docimodus johnstoni* (Figure 4.11) is termed a 'fin-biter'. *Haplochromis* (*Dimidiochromis*) *compressiceps* is named the Malawi eye-biter, although it is unclear to what extent the name is justified in this case. Some African cichlids are described as paedophagous, eating the young of other cichlids, ramming mouth-brooders, for instance, so that the brood is dropped (McKaye and Kocher 1983; Ribbink and Ribbink 1997).

Cichlids are not the only taxon including 'fin-biters'. The well-known South American characin, the piranha (*Serrasalmus spilopleura*), with a popular reputation for fierce and collective carnivory, is also said to be a fin-biter, thus exploiting a 'renewable resource' (Northcote et al. 1987). Another 'piranha', the wimple piranha *Catoprion mento*, is reported to be a 'scale-feeding specialist'

(Janovetz 2005). Juvenile catfish of the Family Ariidae are reported to include scales in their diet whereas three genera of trichomycterids, the South and Central American catfish family which includes parasitic genera, are said to be specialist scale-eaters (Szelistowski 1989). One trichomycterid, *Ochmachanthus alternus*. was reported to be an 'obligate' mucus feeder (Winemiller and Yan 1989).

Parasitism of jawed fish by the jawless fish, notably most species of lampreys when adult, is a common problem in some freshwater fisheries (e.g. the North American Great Lakes). Adult lampreys can attach to other fish and cause considerable external damage.

Parasitism is rare as a means of nutrition for jawed fish and even where it has been described, the fish species involved may not fit the usual definition of parasitism (13.13). *Remora* species have a suction disc on the head and may attach to other fish, referred to as their 'hosts' (an implication of parasitism), but the food of *Remora* is described as parasitic copepods removed from the host's skin or food particles scavenged when the host is feeding (Scott and Scott 1988). 'Parasitic' male anglerfish are not true parasites as their relationship is intraspecific, the small male getting its nutrients from the female to which it is attached (13.4.4).

There are some small fish (*Vandellia* species) in the Amazon which are true parasites, that is they harm their host of a different species. These very small trichomycterid catfish, known locally as candirú, parasitize other fish by entering the gill area, for example; these fish also have a reputation for entering the orifices of mammals, including humans (13.13).

Smith and Reay (1991) provide a list of 106 teleost species for which cannibalism has been reported. In the case of gadids (e.g. cod) and sticklebacks (e.g. *Gasterosteus*), cannibalism, the consumption of conspecifics, seems to be well established, with multiple supporting references. Chondrichthyans also show cannibalism, perhaps best understood from the case of adelphagy or oophagy in which situation young or eggs may be devoured by siblings *in utero* (9.6).

Incipient parasitism may be seen in 'cleaner fish' (14.9.2).

4.4 The food-processing system

Structurally the food-processing system can be divided into a linear sequence, from the anterior mouth to the posterior anus. The mouth has associated jaws for prey capture, often with teeth in various locations including the buccal cavity. Digestion occurs extracellularly in vertebrates and there are several anatomical structures and chemicals for achieving this.

4.4.1 Mouth

The mouth of fish has enormous diversity; it may lack jaws (Agnatha) or have a variety of jaw actions, including protrusion. The gape may be large or small and, in some fish, at the end of a long tube, or positioned dorsally or ventrally. The jaws may be toothed or toothless. The teeth, if present, may be sharp or plate-like. Feeding may be fairly continuous with the mouth operational over extended periods, or it may be intermittent and highly selective. There are many different possible formats of mouth structure and action (Table 4.1).

The mouth can also be assessed as to its position and the size of the gape.

Large mouths enable ingestion of large prey or multiple small prey, while small mouths, as in trumpetfish, pipefish and cornetfish (Figure 4.4), are better for suction and precision. A long, terminal mouth is appropriate for fast predators such as barracuda while large wide mouths are found in slow ambush predators such as anglers which also have a 'lure' (esca) dangling over the mouth (Figure 4.4).

The seahorses and their kin (Family Syngnathidae) generally have a very long snout and small mouth. Recently, de Lussanet and Muller (2007) have described their size as inversely correlated ('the smaller your mouth, the longer your snout'), and refer to these fish as 'pipette feeders'

Table 4.1 Different fish feeding types, based on the mouth (after Nikolsky 1963).

Type	Description	Examples
Sucker	Jawless	Agnatha; hagfish, lampreys
Grasping	Predators, large mouth, plentiful sharp teeth	Pike-perch, pike, some catfish
Imbibing*	Bottom feeders; tube-like mouth, no jaw teeth	Mormyrids, bream. pipefish
Crushing	Feeding on invertebrates with hard exoskeleton (e.g. corals); sometimes with beaks, powerful teeth or plates	Trunkfish, skates
Planktophagic	Sieve plankton using gill rakers, no jaw teeth, large/medium immobile mouth	Herring, whitefish, some cyprinoids
Periphyton eating	Graze waterweeds with a lower slit-like hard-edged Mouth	Some cyprinoids

*Would be classified by many authors as 'sucking'.

(see also 'suction feeding', 'pivot feeding' and 'microcarnivores').

Ambush predators such as pike or gar (Figure 1.17) with huge, long mouths make a rapid dart ('C-start', 3.8.3) at their prey which they secure with sharp teeth. Relatively huge gapes are also associated with some deep-sea species including benthic anglers, usually with illuminated lures (13.4.4) positioned over the mouth (Figure 4.4), viper-fish, loose-jaws (Figure 4.5), swallowers and gulpers

Figure 4.4 Different-shaped mouths. Left: Small in syngnathids, e.g. a pipefish. Right: Large and wide in anglerfish. Both after Greenwood (1975).

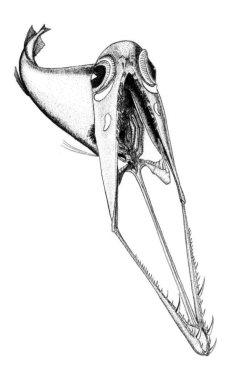

Figure 4.5 The loose-jaw *Malacosteus niger*, after Strenzke (1957), from K. Günther and K. Deckert (1950) Wunderweit der Tiefsee (p. 179) Herbig Verlagsbuchhandlung, Berlin.

(Figure 4.6). The loose-jaw, *Malacosteus niger*, (13.4.3 and Table 13.5) was predicted to feed on 'relatively large fish' (Scott and Scott 1988). It has a remarkably large gape but with huge gaps in the undersurface of the lower jaw region; hence apparently unsuitable for small prey. However, analysis of stomach

contents has shown a preponderance of copepods, which Sutton (2005) suggests is inappropriate for its jaw structure but possibly needed as source material for its visual pigment, and that these crustacean are selectively taken as well as intermittent large prey.

Deep-sea eels include the swallowers (*Saccopharynx*) and gulpers (*Eurypharynx*) with huge mouths (Figure 4.6), very small crania and minute eyes (Tchernavin 1947a). Tchernavin (1947b) notes that for *Eurypharynx*, 'a large part of the body (containing about 20 vertebrae) separates the cranium from the branchial apparatus'. It seems that both genera can flex this area backwards to increase the size of the gape, and *Saccopharynx* is apparently able to bend the vertebral column in this region to a right-angle, as depicted by Tchernavin (1947a,b). There is a photophore region at the end of the tail which may act as a lure, the implication being that the lure would be brought close to the mouth, in a manner similar to the angler-fish esca (Figure 4.4).

The position of the mouth reflects habitat as well as prey; a ventral mouth enables feeding off the bottom while cruising slowly above it, this mouth position is typical of elasmobranchs, particularly the skates and rays (Figure 1.10), which are slow, benthic feeders.

Dorsal gapes occur in marine anglers (Figure 4.4) as well as freshwater aerial feeders such as *Osteoglossum* (Figure 4.7), which can leap at aerial targets as well as feed in the water (Lowry et al. 2005), and the archer fish *Toxotes* (Figure 4.7), which eat insects brought down by spitting water at them (Temple

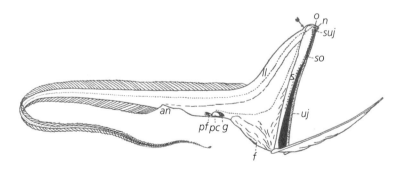

Figure 4.6 *Eurypharynx*, a deep-sea fish with a small head and huge gape. Reproduced with permission. V.V.Tchernavin (1947). Further notes on the structure of the bony fishes of the Order Lyomeri (*Eurypharynx*). *Journal of the Linnean Society of London, Zoology*, 41, 377–393. Publisher Wiley. The arrow indicates the posterior end of the cranium. an = anus; g = gills; ll = lateral line of the body; n = nostril; o = eye; pc = pericardium seen through the thin wall of the body; pf = pectoral fin; s = suspensorium mandibulae; so = infraorbital branch of the lateral line running along the bone acting as the upper jaw; suj = symphysis of the upper jaw; uj = upper jaw.

Figure 4.7 Dorsal gapes. Left: *Osteoglossum*, after Bridge (1904). Right: *Toxotes* (archerfish), after Greenwood (1975).

et al. 2010). These Asian/Australian fish can achieve ranges of up to three metres (Allen et al. 2002).

4.4.2 Jaws

With the exception of the Agnatha (lampreys and hagfish), all extant fish have jaws bordering the mouth. Movement of the jaws can increase the gape, or can optimize suction. In some fish the upper jaw or part of it can be moved forwards and downwards away from the skull; this protrusibility (3.5.4) facilitates prey capture or pick-up from the benthos. The upper jaw is attached to the cranium by fusion or by articulation (jointing) and ligaments, and the lower jaw links to the upper jaw with the aid of accessory bones and ligaments forming the 'suspensorium' (3.5.4). Both jaws are regarded as being derived from the anterior gill supports or 'arches' and the suspensorium may also be derived similarly, incorporating (depending on the fish taxon) elements of the second (hyoid) arch.

The jaws of elasmobranchs are formed from two cartilages on each side, the upper one attached to the cranium anteriorly by a ligament, which enables lowering and protrusion of the jaws away from the cranium (Figure 3.6). In bony fish the upper jaw is generally formed from two bones on each side, the anterior premaxilla and the posterior maxilla, both of which may border the gape and bear teeth (Figures 3.8, 3.9).

For some teleosts such as cod, the maxilla does not border the gape but is an important component of the jaw support (Figure 3.10). In other fish, for example cyprinids and many acanthopterygians, the maxilla becomes part of a complex sliding mechanism moving the premaxilla forward (Figure 4.8) to assist feeding, with some accompanying forward movement of the lower jaw. This protractile mechanism parallels the situation in elasmobranchs with both jaws protrusible.

4.4.3 Pharyngeal jaws and the tooth-bite apparatus

In addition to the principal (mandibular) jaws bordering the mouth there may be 'pharyngeal jaws', with the pharyngeal supports able to make complex movements securing or moving food, found according to Sanford and Lauder (1990), in all 'ray-finned' fishes, that is Actinopterygii including chondrosteans, holosteans and teleosts, therefore referred to as pharyngognaths (Liem and Greenwood 1981).

Yet another set of jaws has been described for osteoglossomorphs which possess three jaw systems (Figure 4.9), the third consisting of the 'tongue-bite apparatus' (TBA) now described also from

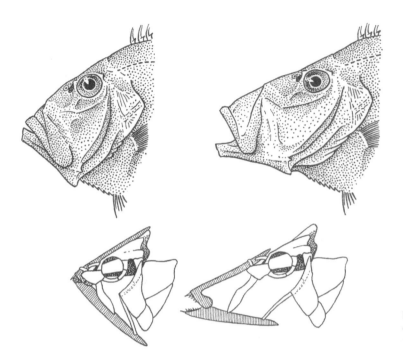

Figure 4.8 Protrusible jaws of John Dory (above), and large-mouthed wrasse (below). Both after Greenwood (1975).

salmonids (Camp et al. 2009). This system occurs between the anterior mandibular jaws and the pharyngeal jaws and may provide a formidable set of teeth (Figure 4.9) on the 'tongue' (basihyal) and the neurocranial bone above it. Camp et al. (2009) associate the TBA with distinctive prey-processing termed 'raking' (Konow et al. 2008) which also involves (retraction) movement of the pectoral girdle and achieves shearing; the process is differentiated from other food manipulations such as 'chews' and 'strikes' (Konow and Sanford 2008).

4.4.4 Lips

Soft tissue overlying the jaws and bordering the mouth may form distinct lips. 'Blubber lips' are hypertrophied structures found in Australian species of the genus *Hephaestus* (Allen et al. 2002; Morgan 2010). The blubber lips condition is found in some of the specimens and may be associated with a coarse substrate or perhaps surface feeding (Pusey et al. 2004). A cichlid (*Haplochromis euchilus*) with huge lips (Figure 4.11) is described by Fryer and Iles (1972). Large lips may have a role in gas exchange (6.6.4). Some fish (e.g. the cyprinids *Puntius sophore*, Tripathi and Mittal 2010, and *Cirrhinus*

mrigala, Yashpal et al. 2009) may have keratinized lips (2.3.1) and surface 'unculi', submicroscopic projections observable with the scanning electron microscope (Yashpal et al. 2009). It is suggested that the tough keratin in the unculi may have a role in scraping the substrate for food, including algal felts (Yashpal et al. 2009).

4.4.5 Teeth

Teeth used to crop, hold or crush food are found throughout the vertebrates and are usually associated with the jaws. In elasmobranchs such as the carnivorous sharks and dogfish, the teeth strongly resemble the very hard scales (dermal denticles, 2.4) covering the body surface, and may form multiple rows which migrate (Luer et al. 1990) to replace damaged rows on the front of the jaws. In bony fish there is a huge variety of dentition which may occur not only on the jaws but on the tongue, the roof of the mouth (palate) and the pharynx (3.5.4, 4.4.3). Cyprinids have no teeth on the jaws or in the buccal cavity but do have pharyngeal teeth. Teeth in the pharynx may be on tooth pads or as rows on the pharyngeal arches (Figure 4.1). Formulae have been applied to express number of teeth and numbers of these rows (Eastman and Underhill 1973; Bolotovsky and Levin 2011). Lauder and Leim (1983) show upper tooth plates in the protacanthopterygian *Novumbra*; this species has a small tooth plate fused to the third pharyngobranchial on each side, plus a more independent tooth plate adjacent but not fused to the somewhat reduced fourth pharyngobranchials. Lauder and Leim (1983) show teeth on the basihyal and basibranchials of pike (*Esox*), char (*Salvelinus*) and an osmerid (*Retropinna*), and also illustrate the complex musculature operating the pharyngeal jaws in a generalized euteleost.

Fish teeth are always hard and individually inflexible, though they may be hinged or depressible as in the pike, *Esox* (Andrew 1959), or the angler, *Lophius* (Scott and Scott 1988). They are described

as formed of dentine covered with enamel, like mammalian teeth (Andrew 1959), although the appearance of some teeth like those of wolffish suggests enamel is not always present. Replacement of teeth is serial in elasmobranchs and variable seasonally in, for example, the nurse shark (Luer et al. 1990) with a rate of 9 to 21 days per row in summer and 51 to 70 days per row in winter, but this occurs annually with the teleost wolffish which shed in the winter (Liao and Lucas 2000, quoting Jónsson 1982). Fish teeth are solid structures without a pulp cavity.

Unlike mammals which may be heterodont as in rodents, lagomorphs and primates, with teeth specialized for different biting or chewing functions (e.g. incisors, canines, molars), fish teeth do

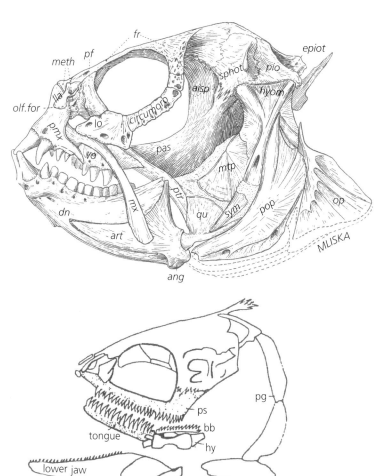

Figure 4.9 Teeth of (above) wolffish (Gregory 1933) and (below), the tongue-bite apparatus of an osteoglossomorph (Lauder and Leim 1983). Note teeth on dentary (dn), premaxilla (pmx) and vomer of the wolffish and on the lower jaw, tongue, parasphenoid (ps) and basibranchial (bb) of the oesteoglossomorph. Both reproduced with permission: W.K. Gregory (1933) Fish skulls: a study of the evolution of natural mechanisms. *Transactions of the American Philosophical Society* 23, 75–481. G.V. Lauder and K.F Liem (1983). The evolution and interrelationships of the actinopterygian fishes. *Bulletin of the Museum of Comparative Zoology*, 150, 95–197.

not usually show marked shape differences. However, the shape can change as the fish ages, such as in characiform fish, in which the larval teeth have single cusps and successive 'generations of replacement teeth' show increasing numbers of cusps (Stock 2001, after Roberts 1967). Fish dentition is therefore regarded as homodont even though considerable size differences can be present in an individual and illustrations of the wolffish indicate considerable shape differences, with canine-like teeth anteriorly and molar-type teeth posteriorly (Figure 4.9). *Cynoscion* and two other sciaenid genera, *Isopisthus* and *Macrodon* (Chao 1978), also have prominent 'canines' at the tip of the upper jaw. Taxonomists use a variety of terms to describe fish tooth shape. These include canine, viliform, molariform, cardiform and incisor (Helfman et al. 2009).

Mammals generally have a few well-differentiated jaw teeth, which can be expressed as a definite dental formula for any species. In general, dental formulae are not used for fish (except for pharyngeal teeth) as shape differences are exceptional. In fish it is usual to have many similar teeth on the upper and lower jaws; fish can also have teeth on the tongue and the roof of the mouth, on the vomer, palatine or parasphenoid bones. A group which is characterized by multiple tooth sites in the mouth/buccal cavity is the Osteoglossomorpha or 'bony tongues' (Figure 4.9). The daggertooth (*Anopterus pharaoh*) is an example of a fish with numerous teeth, with individual variation, also with multiple locations in the mouth. Some of the teeth are fixed while others can be depressed, facilitating food acquisition and retention. Hart (1973) describes the premaxillary teeth as 19 to 55 fixed on each side, with 5 to 16 depressible; the mandibular teeth as 12 to 17 fixed, 0 to 13 depressible and larger teeth on the roof of the mouth 8 to 10 fixed and 2 to 6 depressible.

Fish teeth can serve a number of different functions including cutting/cropping, holding and crushing.

Cropping or nipping vegetation or corals is accomplished by fused teeth, forming a 'beak', as in the parrotfish (Scaridae, Figure 4.10). To hold live prey and prevent it from escaping, many sharp pointed teeth work well; they may be angled or fold down (*Lophius*, *Esox*) to enable prey into the buccal cavity when flat but prevent it escaping when erect. Pharyngeal teeth can also hold prey, as with gadids which can take many small fish into the mouth and hold them posteriorly while taking in more prey.

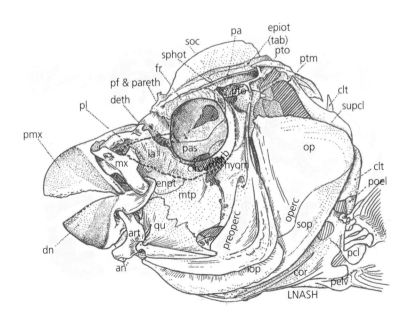

Figure 4.10 The 'beak' of a parrotfish. Note: the beak-like appearance of the premaxilla (pmx) and dentary (dn). Reproduced with permission: W.K. Gregory (1933). Fish skulls: a study of the evolution of natural mechanisms. *Transactions of the American Philosophical Society* 23, 75–481.

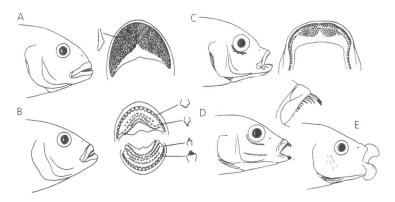

Figure 4.11 Cichlids from the African Great Lakes show enormous variety in diet and oral structure. Lateral view of heads, with upper dentition A–C. A, scale-eater *Corematodus shiranus*; B, fin-biter *Docimodus johnstoni*; C, scraper of rocks and leaves C*yathochromis obliquidens*; D, picker of small arthropods, *Labidochromis vellicans*; E, large sensitive lips for prey detection, *Haplochromis euchilus*. After Fryer and Iles (1972).

Crushing is utilized for hard-shelled invertebrates or vegetation using large strong oral teeth (chimaeras) for the former or pharyngeal teeth for the latter (cyprinids).

The variety of form and function for dentition within a fish family can be seen in cichlids (Figure 4.11).

Edentulous or edentate (toothless) fish include the sea-horses (*Hippocampus*) and pipe-fish, which suck in small prey by 'pivot feeding' a highly efficient very fast mode of prey capture (Van Wassenbergh et al. 2011); their capture time is five to seven milliseconds and depends on rapid head rotation and very good sight. Obligate planktivorous fish such as clupeids such as herring, and basking sharks, that have long anterior gill rakers which act as filters, may not have teeth (e.g. Clupeidae, 'teeth small or absent', Nelson 1976).

4.4.6 The gut

Following food procurement, the food is processed by passage through the gut (alimentary canal), as in most other animals. In fish and other vertebrates the food movement depends on smooth muscle in the gut wall for most of the gut length, and during the posterior progress of the food enzymes are added to the gut lumen to provide chemical digestion. This breaks down large food molecules to simpler ones which can then be absorbed, while undigestible material is voided at the anus. The gut is divided into three principal regions, foregut, midgut and hindgut, with regions demarcated arbitrarily for analytical purposes (Smith 1980). Basically, the foregut includes the oesophagus and stomach,

the midgut is the main intestine after the entry of the bile duct and the hindgut is the anal region. However, the gut structure varies considerably between taxa; the gut may lack a stomach, be coiled ('folded', Hale 1965) or straight (Hale 1965).

The gut has the usual vertebrate structure, in four major layers, the inner mucosa, the submucosa, the muscular coat, and the outermost serous coat (Andrew 1959) covered by the peritoneum (Figure 4.12). The mucosa includes the epithelium immediately lining the gut lumen; this is generally columnar and contains gland cells with, in some cases, more complex glands as invaginations. Under the epithelium, the submucosa is connective tissue and a '*muscularis mucosae*', smooth muscle cells enabling movement of the mucosa. The outer layer of smooth muscle has two sets of fibres, one circular in disposition and one longitudinal, enabling contraction of the gut lumen and therefore movement of the contents. The gut is innervated by the autonomic system (11.9.3, 11.10.1) via plexi; one in the submucosa (Meissner's/submucous plexus) and one between the two layers of the muscular coat (Auerbach's/myenteric plexus) or, according to Olsson and Holmgren (2001) in the generalized vertebrate model, with an additional plexus between the circular and longitudinal muscle (deep muscular plexus). The gut is well vascularized by a series of arteries coming off the dorsal aorta. The venous drainage from the absorptive regions of the gut (from the stomach posteriorly) is, as in all other vertebrates, directed to the hepatic portal vein and thence to the liver.

The peritoneum of some bony fish is blackened by melanization and a few fish are melanized

Nerve plexi	Gut layer from lumen to peritoneum		Major function
	Mucosa		Direct digestion, secretion, absorption
Submucosal (Meissner)	Submucosa + muscle fibres		Movement of the mucosa
Myenteric (Auerbach)	Muscular coat	Circular muscles	Peristalsis
Deep muscular		Longitudinal muscles	Peristalsis
	Serous, covered by peritoneum		

Figure 4.12 The major layers of the vertebrate gut wall, after Andrew (1959), Olsson and Holmgren (2001).

within the gut submucosa (Fishelson et al. 2012). The presence of such dark layers may provide a screen for bioluminescence (13.4.7) from prey in the gut which might attract potential predators (Fishelson et al. 2012). It is stated that closely related shallow-water species do not have this melanization (Fishelson et al. 2012), but whether this situation fits the difference between *Gadus ogac* (melanized) and *Gadus morhua* (non-melanized) is unclear. In the case of amphibia and reptiles with dark peritonea, the likely protection from deleterious ultraviolet radiation has been considered (Porter 1967). The same author concludes that the dark peritoneum does not act as a significant heat shield absorbing incident solar energy compared to skin and muscle heat absorption, with particular reference to desert reptiles; the possibility remains (though apparently not yet studied) that the dark peritoneum of fish conserves heat generated within the intestine, achieving a somewhat similar effect as described in heterothermic tuna with internal red muscle and countercurrent *retia* (5.10) with the internal temperature above ambient.

Food processing occurs in a sequence of specialized regions with the buccal cavity the most anterior. For many fish the buccal cavity is delimited anteriorly by jaw teeth and may also contain patches of teeth on the interior, particularly the upper region (palate). The buccal cavity may also be enlargeable which, in combination with a small mouth, promotes suction. The pharynx always contains gill slits on each side and these slits are bordered by the gills with gill rakers partially covering the exits. The rakers are particularly long in planktivores (Figure 4.2). The pharynx may also have tooth pads to help hold prey as it passes back or to crush vegetation for herbivores. The development of 'pharyngeal jaws' in cichlids has enabled a remarkable diversification of the buccal cavity in this taxon (Leim 1974) and the pharyngeal jaws have even recently been implicated in sound production (Rice and Lobel 2002) so the pharynx can be very complex with multiple roles. As Smith (1980) points out it is important that the feeding mechanism must not conflict with the respiratory role of the gills.

The oesophagus is a short link to the next region which is usually the stomach, a holding area for ingested material where mechanical digestion may occur and protein digestion can begin. Although the oesophagus is not generally expected to achieve more than transmission of food, Murray et al. (1994) reported some structural evidence supporting the possibility of pregastric 'digestion' in flatfish.

Mechanical digestion may occur in the stomach as occurs with 'the heavily muscularized cod stomach' (MacDonald and Waiwood 1982, quoting Tyler 1973). In contrast, the pout stomach is said to be primarily for storage being less muscular (MacDonald and Waiwood 1982). Proteolysis in the stomach occurs through the action of hydrochloric acid and protease (pepsin) as in other vertebrates. Smith (1980) describes the teleost stomach as taking three different forms; straight, U-shaped or Y-shaped with one long posterior caecum. The most posterior region of the stomach is the pylorus and many bony fish species have pyloric caeca attached to the gut just posterior to the pylorus; these may be few, several or very many. In the pleuroneciforms (flatfish) the number of caeca can be zero (soleids) to five to seven in witch flounder, *Glyptocephalus* (De Groot 1969). Kuperman and Kuz'mina (1994) compare the situation in pike, burbot and bream (Figure 4.13): burbot has many caeca; pike and bream have none. The function of the caeca has been debated (Buddington and Diamond 1986) and it is not clear how much food can enter them as they are often very narrow (Smith 1980). However, the pyloric caeca do contain digestive enzymes (Genicot et al. 1988) and it has been generally assumed they function to increase surface area for digestion (Smith 1980; Buddington and Diamond 1986).

Some fish lack stomachs and/or caeca; stomachless fish including some cyprinids and cyprinodonts and the labrid *Tautogolabrus adspersus* are herbivorous or omnivorous and lack pyloric caeca too. *Gasterosteus aculeatus* has a stomach (Figure 4.14) but lacks caeca (Hale 1965), as do pike (Figure 4.13).

The packing of the gut into the abdomen for the small stomach-less *Poecilia* involves complex looping (Figure 4.14) and is sometimes 'disarranged' by the presence, in the live-bearing females, of developing embryos (Hale 1965). Perhaps the lack of stomachs and/or caeca in small fish is related to the relative lack of available space in the abdominal cavity. The lack of caeca in pike, however, does not fit this explanation.

The principal gut region for digestion and absorption in the fish is the intestine as for other vertebrates; the anterior intestine is linked to pancreatic tissue and the bile duct. This gut region is variable in length, though the generalization that herbivores have a relatively longer intestine does not always apply (Smith 1980) but may hold within a taxonomic unit. De Groot (1969) reported considerable variation in the structure and length of the (inter-specific) flatfish digestive system. Allen et al. (2002) note that *Prototroctes maraena*, the Australian grayling, has a long 'intestinal tract' which they associate with herbivory, though this fish includes insects in its diet. Kramer and Bryant (1995a,b) studied

intestine length in a number of tropical fish species and found the relationship to diet to be quite complex. They found that omnivores commonly increased plant consumption with increasing body size, and that body mass was a better correlate with increased intestinal length than longer body length. Overall, however, they confirmed the general view that herbivores have longer intestines than omnivores and omnivores a longer intestine than carnivores of the same size. They warned that omnivores have considerable variation in food types and proportions (e.g. 4–60 per cent plant material) so that comparison of intestine length within the 'omnivore' category is difficult.

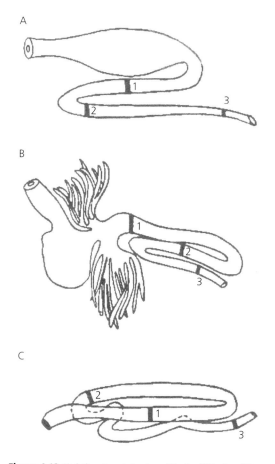

Figure 4.13 Variations in gut structure: (A) pike, (B) burbot, (C) bream. Reproduced with permission. B.I. Kuperman and V.V. Kuz'mina (1994). The ultrastructure of the intestinal epithelium in fishes with different types of feeding. *Journal of Fish Biology*, 44, 181–93. © The Fisheries Society of the British Isles.

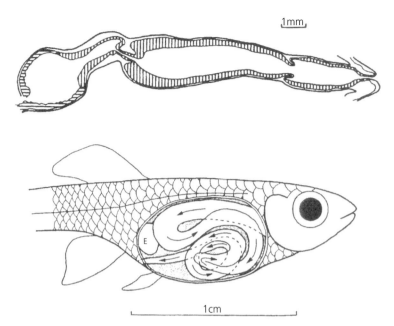

Figure 4.14 Above, the gut of *Gasterosteus* (anterior to the left). Below, looping of the gut in *Poecilia*. Reproduced with permission. P.A. Hale (1965). The morphology and histology of the digestive systems of two freshwater teleosts, *Poecilia reticulata* and *Gasterosteus aculeatus*. *Journal of Zoology* 146, 132–49. John Wiley & Sons Ltd.

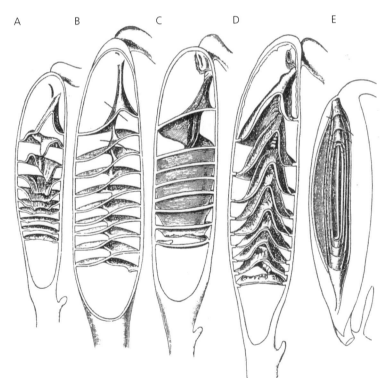

Figure 4.15 Spiral valves of chondrichthyans, after Bridge 1904. A–D, views of the opened intestine of *Raja* spp. E = opened intestine of *Sphyrna malleus*. Note rectal gland as small finger-like (blind) projection shown in A, C, D and E, posterior to the intestine.

The surface area of the intestine is increased by a 'spiral valve' in chondrichthyans (Figure 4.15) and other ancient lineages such as sturgeons (1.3.6). *Amia* has a rudimentary spiral valve, but a spiral valve is not found in teleosts which have villi instead; these are small processes protruding into the intestinal lumen similar to the mammalian situation except that there is no clear evidence for lymphatic 'lacteals' in fish villi.

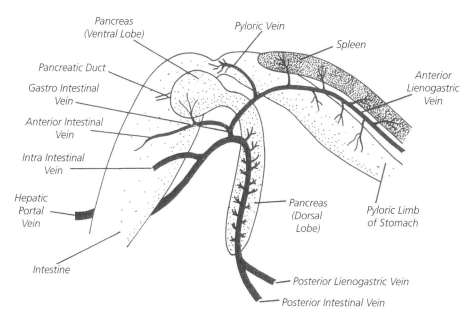

Figure 4.16 Pancreas of dogfish (*Scyliorhinus*), after Wells (1959).

Elasmobranchs have a distinct pancreas (Figure 4.16) and pancreatic duct, but the pancreatic tissue in most teleosts is diffuse, although the endocrine portion may be concentrated in Brockmann bodies (Wright et al. 2000). The gall bladder and bile duct is obvious in most fish and has the usual vertebrate position, emptying bile into the anterior intestine.

4.4.7 Digestion and digestive enzymes

Breakdown of food to provide readily absorbed small particles occurs in a series of phases. Most fish food is ingested as very large pieces including whole fish or invertebrates and the first phase of digestion may be purely mechanical, a crushing of the intake as can occur with pharyngeal jaws. Waves of contraction in the gut wall then pushes food down the gut where it is subjected to further physical breakdown due to local muscle action or to bile emulsifying fat to small droplets. As the food reduces to smaller particles, digestive enzymes secreted into the gut lumen or derived from microbial action convert macromolecules to smaller forms appropriate for absorption.

Understanding the details of enzymatic digestion, and making comparisons between fish species, have been complicated by the range of structures which may be secreting enzymes and the differences between regional demarcations, as well as the usually diffuse nature of the teleost pancreas, and the narrow lumen of the pyloric caeca in most teleosts. Studies of fish digestive enzymes have also been undertaken under varying temperatures, related to ambient or, less usefully, at levels above or below the conditions the fish would normally experience. The presence of microbial enzymes has also been a factor to consider in fish digestion, as has whether the enzymes extracted or *in situ* are endogenous (fish produced) or exogenous (microbial origin). Hidalgo et al. (1999) have listed some of the problems in making comparisons between species; they include a lack of 'uniformity' in gut tissue preparation in that attached glands may or may not be included, the fact that nutritional status may vary markedly, and finally the presence or absence of gut contents, whether the interior of the gut is washed or not.

Chemically the three major groups of potential food components can be described as protein, lipid ('fat') and carbohydrate. Each of these types are large molecules which can be rendered into smaller molecules through the hydrolytic actions of specific enzymes; proteases, lipases and carbohydrases, respectively.

The passage of food through the gut may be accompanied by the sequential effect of 'secretogogues' (also termed secretagogues), substances which stimulate the secretion of, for example, stomach acid or the extracellular enzymes which help to achieve chemical digestion. The effect of feeding and starvation on enzyme secretion was studied by Einarsson and co-workers (1996) in Atlantic salmon. These authors suggest that 'both synthesis and release' (of the enzymes studied, pepsin, trypsin and chymotrypsin) were controlled by cholecystokinin (CCK).

Proteases (proteolytic enzymes) act at amino acid-specific sites within (endopeptidases) or at the end (exopeptidases) of the amino acid chain forming a protein. The exopeptidases either cleave the terminal amino acid at the 'carboxyl' end of the protein, and are therefore carboxypeptidases, or at the other end, the amino terminal, and are aminopeptidases. The two amino acid dipeptides which may result from prior hydrolysis can be cleaved by dipeptidases in the gut lumen or it is feasible that these small molecules can be absorbed intact and cleaved intracellularly. Three major proteases found in vertebrate (including fish) guts are pepsin, trypsin and chymotrypsin; each of these is secreted extracellularly in a non-active precursor (zymogen form), pepsinogen, trypsinogen, chymotrypsinogen (proenzymes), which must be activated in the gut lumen. Hydrochloric acid can activate pepsinogen in the stomach, and once pepsin is present it may cause further activation (autoactivation). The release of both HCl and pepsinogen from the stomach wall into the stomach lumen are controlled by the hormone gastrin (10.12.4) and by the autonomic nervous system (11.9.3, 11.10.1). Activation of trypsinogen and chymotrypsinogen in the intestine may occur by the release of enterokinase from the intestinal wall and autoactivation may also occur to convert more precursor once the active form is present. Each of these three major proteases is an endopeptidase; pepsin cuts proteins at several locations, trypsin separates peptide bonds where the carboxyl terminal is derived from arginine or lysine (i.e. 'after' arginine or lysine) and chymotrypsin, which is less strict, functioning at a variety of locations, for example 'after' phenylalanine, tryptophan or tyrosine 'unless followed by proline' and, also but more

slowly, after other amino acids such as leucine and methionine.

Generalized accounts of chemical digestion in the fish gut begin with the stomach (if present). In the stomach, protein digestion can begin with pepsin. Pepsin works optimally in the low pH environment provided by hydrochloric acid secreted by the oxynticopeptic cells in the stomach wall which also release pepsin (Einarsson and Davies 1996). Subsequently, as food passes into the intestine it is acted upon by other enzymes, including trypsin and chymotrypsin, either secreted directly (as proenzymes) from the pancreas, as in elasmobranchs or, in teleosts from diffuse pancreatic cells located, for example, in (Genicot et al. 1988) or between (Overnell 1973) the pyloric caeca, or more rarely, a compact pancreas, as in the pike, *Esox lucius* (Bucke 1971). Less information is available for fish exopeptidases but 'carboxypeptidase H' has been described from anglerfish (Roth et al. 1991), and aminopeptidase activity was studied in stichaeids with different dietary habits (German et al. 2004). The intermediate product of protein digestion is small peptides (oligopeptides). The final product of protein digestion is expected to be amino acids or perhaps dipeptides.

Lipid (e.g. fat and wax esters) digestion may be achieved in two stages. Fat, as triglyceride, is emulsified by bile salts (synthesized by the liver, 7.11, and stored in the gall-bladder) at the beginning of the intestine to produce smaller droplets, a surfactant action. These provide a higher surface area on which lipases can act, hydrolyzing the fats to fatty acids and glycerol, which may aggregate to form tiny micelles which can move readily across the gut lumen to the epithelial surface where the constituents can be absorbed. The surfactant capacity of different gut regions of the gizzard shad (*Dorosoma cepedianum*) was studied by Smoot and Findlay (2000) who found activity similar to that of commercial surfactants (detergents) in the anterior intestine/caecal region.

Hagey et al. (2010) studied the diversity of bile salts in fish (and amphibia), recognizing three classes, based on C27 bile alcohols, C27 bile acids and C24 bile acids (i.e. 27 or 24 carbon atoms per molecule), with the bile alcohols usually found as sulphate esters. The principal bile salt in elasmobranchs was reported as a sulphated bile alcohol while the most usual bile 'salt' of bony fish was reported as based on C24 bile acids (Hagey et al. 2010) although cyprinids had a bile alcohol preponderant. A few teleost species had more than 5 per cent C27 bile acids in their bile while a combination of C24 bile acids and bile alcohols was reported from 37 species (notably Perciformes). The C24 bile acids are usually conjugated with taurine, for example as chenodeoxycholyltaurine (Hagey et al. 2010).

Fish may also digest wax esters by hydrolysis via a non-specific bile salt dependent lipase acting as a wax ester hydrolase (Patton et al. 1975; Bogevik et al. 2008), which yields free fatty acids and alcohols. This capacity is particularly relevant if a species has a high intake of crustaceans which have wax esters in their exoskeletons. Within the epithelial cells, lipids may be resynthesized to form small chylomicrons, spheres with a phospholipid/protein surface.

Carbohydrate digestion is important to the relatively few herbivorous fish and limited in others. Thus, most fish have weak starch-splitting (amylase) activity. Trout cannot utilize large amounts of carbohydrates in the diet and, in comparison to typical carnivorous fish like trout, carp have much higher amylase activity (Hofer and Sturmbauer 1985). Cellulases may also occur (Krogdahl et al. 2005) giving the capacity to utilize plant-derived cellulose. Papoutsoglou and Lyndon (2005) measured α-amylase and α-glucosidase enzymes activity in several teleosts including the herbivorous tilapia (*Oreochromis aureus*) and the more carnivorous salmonid *Oncorhynchus mykiss* (rainbow trout). Comparisons were based on a suite of temperatures and gut regions, and confirmed that the herbivore (tilapia) had much higher amylase activity than the trout. Total carbohydrase activity was also higher in tilapia than in trout (Papoutsoglou and Lyndon 2005). Because carbohydrate utilization would help reduce costs of rearing fish much effort has been put into promoting carbohydrate usage. One problem that has arisen is the inhibition of amylases by, for example, wheat flour (Hofer and Sturmbauer 1985).

Chitin is a 'mucopolysaccharide polymer' (Gutowska et al. 2004) which forms an important part of arthropod exoskeletons and hence may be ingested by predatory or planktivorous fish. Some freshwater predatory fish ingest large numbers of insects which would have a high chitin content but while chitinase is found in some fish guts, trout, which does naturally eat insects, cannot utilize effectively large amounts of chitin given in the diet (Lindsay 1984). Chitin is common in the marine ecosystems (Gutowska et al. 2004) and chitinase action would be advantageous to many fish. These authors measured chitinolytic activity in 13 fish species and concluded that the activity was higher in some gut tissues than in the gut contents suggesting intrinsic enzymes rather than microbial action. Lindasy and Gooday (1985) studied chitin digestion in cod and concluded that this fish had its own endogenous chitinolytic enzymes. However, Krogdahl et al. (2005) favour the idea that fish chitinases have an exogenous (microbial) origin and that cellulases (when present) have a similar source.

The final products of carbohydrate digestion are simple monosaccharide sugars (e.g. glucose) or acetyl glucosamine in the case of chitin. The role of carbohydrates in fish nutrition has been reviewed in Krogdahl et al. (2005).

4.4.8 Absorption

Absorption of the products of digestion (particularly amino acids, small peptides, fatty acids, glycerol, alcohols, sugars) occurs by passage through the columnar epithelium ('enterocytes') of the gut, principally of the intestinal wall, into the blood of the hepatic portal vein and thence to the liver (5.9). There is little evidence of a lymphatic uptake of lipids in contrast to the mammalian system with lacteals in the villi. Lipids, especially in the form of their constituents (notably fatty acids, glycerol, alcohol), are expected to be able to diffuse into the enterocytes but the products of protein and carbohydrate digestion require special carriers in the enterocyte membrane to enter the cells.

Each columnar epithelial cell (enterocyte) has a brush border with microvilli (Figure 4.17) at the apical end abutting the intestinal lumen. The microvilli contain embedded enzymes, peptidases which achieve further breakdown of small peptide chains (Tobey et al. 1985; Bai 1994) and also include special 'carriers' for sugars (monosaccharides), amino acids and small peptides (di- and tri-peptides, Randall et al. 2002) to enable ('transcellular') passage across the membrane. The lateral region of each enterocyte is linked to its immediate neighbour by

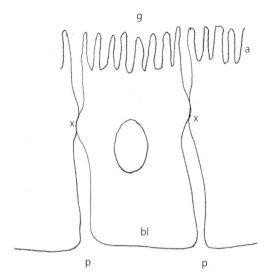

Figure 4.17 Diagrammatic representation of enterocytes showing their orientation with respect to the gut lumen. a = apical surface with microvilli (brush border); bl = basolateral surface; g = gut lumen; p = paracellular pathway, not generally available for absorption due to the presence of x = tight junction.

a 'tight junction' (Figure 4.17) and the base/basolateral surface of the enterocyte is expected to contain carriers to enable passage of monosaccharides and amino acids across the membrane.

Absorption of amino acids and monosaccharides may occur by paracellular diffusion (Figure 4.17) between the epithelial cells of the intestine but some accounts describe this route as unavailable because of the presence of tight junctions; however, other accounts do not exclude paracellular diffusion (Jutfelt 2006). Transcellular passage is promoted by active transport systems in the lumenal membrane of the intestinal epithelium, similar to those described for mammals. Stevens et al. (1984) described a 'multiplicity' of carriers for amino acids; Randall et al. (2002) list four types (e.g. one for aspartate/glutamate, another for the three dibasic amino acids, etc.) involved in Na+ linked active transport for mammals. Jutfelt (2006) has summarized the mechanisms involved with particular reference to salmonids. Hepher (1988) describes the active mechanisms as Na^+ dependent but adds that Cl^- may be important for marine fish. These active mechanisms enable uptake into the epithelial cells with special carrier proteins involved in exit from the basal membrane. Diffusion across the cells from the lumenal to the basal membranes is expected to achieve some of the transport. Lipid products (e.g. fatty acids, glycerol) are able to diffuse into the intestinal cells from the intestinal lumen and are predicted to reassemble into chylomicrons intracellularly before exocytosis at the basal membrane where they can access tissue fluid/blood for transport to the hepatic portal vein.

The absorption processes expected to occur, mainly based on mammalian studies, are summarized in Figure 4.18. In explanatory diagrams,

PROCESS	CARBOHYDRATES	PROTEINS	LIPIDS
Main digestive product (for fish)	Glucose	Small peptides	Fatty acids
		Amino acids	Glycerol
Entry point	Apical membrane		
Entry mechanism	Carriers (co-transport)	Diffusion . . . .
Intracellular transport	Diffusion ·		
Intracellular transformation		Cleaved* to amino acids	Recombined chylomicrons
Exit from cells	Carriers in baso-lateral membrane	Exocytosis
Uptake by tissue fluid/plasma			
Transport in blood to hepatic portal vein			
*Cleaved by intracellular enzymes			

Figure 4.18 Routes for absorption across gut epithelial cells (enterocytes), based on information/illustrations including 'comparative animal physiology' texts (e.g. Eckert/Randall et al. 2002) and a thesis (Jutfelt 2006).

individual cells are often shown involved with only one food group; it is not currently suggested that such specialization occurs. For monosaccharides, mammalian studies have followed fructose and galactose as well as glucose; although some fish are frugiverous (Davis et al. 2010) the amount of fructose/galactose normally in fish guts is likely to be small. Studies on fish enterocyte function have included Soengas and Moon (1998) on glucose transport and metabolism in isolated bullhead enterocytes, and research on lipids in caecal enterocytes (Oxley et al. 2005) and ultrastructural changes in char enterocytes fed high levels of linseed oil (Olsen et al. 1999).

4.4.9 Egestion

Egestion (evacuation) of undigested material occurs at the anus, directly to the exterior or via another chamber, the cloaca, as in elasmobranchs. It is important to note that technically, egestion is not excretion (8.2) though faeces may include some excretory material such as bile pigments.

4.5 Feeding periodicity

Fish may feed continuously if the food is particulate but every fish is limited by the space available to store ingested food and by the food-processing time, which can be temperature-dependent. Gastric evacuation times have been calculated for some species and, more rarely, estimates have been obtained for total evacuation, that is, time for a (marked or single) meal to pass through the gut (Smith 1980). Skipjack tuna are said to have much faster (twice as fast) passage rates than other fish (Smith 1980, quoting Magnuson 1969) and this may be due to the higher temperatures maintained internally for these heterothermic fish. Edgar and Shaw (1995) found 'empty guts' in 14 per cent of fish analysed in relatively warm waters off Australia; presumably clearance through the gut would be fairly rapid in these circumstances but complete cessation of feeding for extended periods (weeks/months) in cold water (4.8) or during migrations (e.g. pre- spawning migrations of salmonids) can occur. In the case of some salmonids the pre-spawning migration is long and the fish may be debilitated, not surviving

the combined effects of energy use during migration, the lack of food and the spawning activities. Although moribund fish may result, some species such as *Salmo salar*, show considerable variability in the outcome, with a proportion recovering, as 'kelts' and the possibility of a repeat spawn another year (9.14).

4.6 The control of feeding behaviour

Appetite control is quite complex, involving promotion or inhibition of feeding. Undoubtedly factors like temperature, season and reproductive status also have an effect. This subject was reviewed by Volkoff and Peter (2006) and by Volkoff et al. (2010); the many hormones likely to be involved being listed (10.12.5), although it is also emphasized that the control of food intake is still imperfectly understood. Currently hormones which influence fish feeding are divided into those that stimulate it (orexigenic) and those that inhibit it (anorexigenic). As with other endocrine studies, the fact that, experimentally, a putative hormone has an effect does not mean that it is naturally the controlling factor.

Forbes (2001) reviewed the affect of previous feeding on subsequent behaviour, particularly important for captive fish and aquaculture.

4.7 Gut activity and its control

The major gut functions in food processing depend on passage of food through the gut and secretion of enzymes and, for example, acid (stomach) and bile (into the anterior intestine) as the food passes along. Other less obvious gut functions, some not immediately concerned with food processing, have recently been reviewed by Grosell et al. (2010) and include the role of the gut in salt and water balance (Grosell 2010). The gut is also considered to have an important barrier function, resisting the uptake of pathogenic bacteria (Jutfelt 2006).

The movement of food through the gut of vertebrates is achieved by the action of smooth muscle in the wall of the stomach and intestine ('motility') with some striated muscle action enabling swallowing and movement in the oesophagus. The gut motility of non-mammalian vertebrates, as well as mammals, has been reviewed by Olsson and Holmgren (2001). A distinction can be drawn between the three major smooth muscle-driven events once food has been swallowed. These events are standing contractions (e.g. mixing food in the stomach), peristaltic contractions moving food along and sphincter contractions holding food in a compartment. The change in diameter of the gut is generally achieved by the contractions and relaxations of the smooth muscle layers (of the *muscularis mucosae*, 4.4.6), with contraction of the circular muscles simultaneous with the relaxation of the longitudinal muscles achieving a locally constricted diameter. Control of motility is complex and dependent on the presence of food, the autonomic nervous system (see enteric nervous system, 11.9.3) as well as specific peptides such as vasoactive intestinal polypeptide, VIP, and also by nitric oxide, NO (Olsson and Holmgren 2001).

For fish with stomachs, 'gastric receptive relaxation' occurs followed by, in sequence, stomach contractions and emptying.

Hormones, notably neuropeptides (Bjenning and Holmgren 1988), also play a role in both appetite and digestion, the latter through control of secretion (e.g. gastrin secreted in the stomach controls pepsin, 10.12.4). The role of secretin, a hormone originally described in mammals for the control of pancreatic enzymes is not very clear in fish although its presence has been reported; it now appears that secretin is just one of many similar hormones forming a 'secretin family' (Cardoso et al. 2010) some of which (e.g. VIP) may be very important in gut function.

The vertebrate gut is well vascularized with the large hepatic portal vein conducting absorbed nutrients direct to the liver from the gut. The mammalian small intestine has lacteals in the intestinal villi which, as part of the lymphatic system, absorb chylomicrons of absorbed fatty substances as a specialized route in parallel with the hepatic portal system. Such a lymphatic system has not been clearly identified for fish.

4.8 Temperature and feeding

Food processing increases the metabolic rate and when temperature rises the oxygen requirement also rises; the problems arising from this specific dynamic action or effect (SDA/SDE) can cause death of cultured fish if the feeding level is high. Any compensatory increased gill ventilation rate can increase oxygen demand, therefore producing more difficulty.

Keast (1968) studied winter feeding in Great Lakes fish and found feeding to be 'erratic': as the water warmed, different species commenced feeding at different temperatures; whereas some species were feeding at 6.5 °C, others did not start until 8.5 °C.

At the lower end of their temperature range fish may cease feeding for extended periods such as winter flounder (*Pseudopleuronectes americanus*) off Newfoundland (Fletcher and King 1978) and New Brunswick (McLeese and Moon 1989), which may have a winter fast of several months. During the winter fast the intestinal mass and surface area declined by about 50 per cent (McCleese and Moon 1989) but the length did not change. However, seasonal changes in intestinal length have been reported for Panamanian cichlids during the dry season (Townshend and Wootton 1987).

The effect of temperature on digestion was studied for bluefin tuna, *Thunnus thunnus*, by Carey et al. (1984). This large heterothermic teleost is capable of maintaining its viscera at temperatures higher than ambient, which is presumed to assist digestion and speed passage through the gut (Carey et al. 1984).

Papoutsoglou and Lyndon (2005) studied the effect of varying temperatures on carbohydrate digestion for some teleosts relevant to aquaculture, including salmonids and tilapia; these authors concluded that there were geographic differences in responses to temperature and also that temperature effects were species-specific.

4.9 Fish nutrition

Most animals, including fish, have fairly specific dietary requirements, including a range of amino acids (from protein and to form protein), fatty acids, carbohydrates and inorganic elements including iron and trace metals, as well as a range of small amounts of 'vitamins', organic molecules needed but not synthesized by the animal (Appendix 4.1). In the case of amino acids, some are considered 'essential' in that they must be obtained in the diet; the animal cannot synthesize them. Similarly, some inorganic components are 'essential' either in the diet or to be obtained directly, for example via the gills from the surrounding water. Thus vertebrates have a requirement for iron (e.g. to form haemoglobin) and zinc (to act in various metabolic pathways as a co-factor) as well as many other elements and organic molecules. For a fish to thrive, its intake, whether through the gut or gills or elsewhere, must include appropriate sources, either directly or indirectly (through synthesis) of all the components of its cells and extracellular fluid. Moreover, if it is to successfully grow and reproduce it has to ingest or absorb an excess of the necessary cellular components and energy sources.

Fish may have very particular species-specific requirements in their diet. It has been suggested that the loose-jaw *Malacosteus niger* (4.4.1) derives its special visual pigment from copepods (Sutton 2005). Other fish (e.g. salmonids) may derive carotenoid pigment in the muscle from consumed crustaceans too. Fish which are held in non-natural conditions may reveal other requirements for their diet and, as with mammals, may show that they cannot synthesize some amino acids or fatty acids and yet require them. Such 'essential' items must be taken into account in designing fish diets as well as any possible requirements for inorganic constituents (e.g. iron, calcium, zinc, etc.) and fish-relevant vitamins as well as the digestibility of any complex molecules included. Halver (1978) summarized vitamin requirements for some salmonids, with particular emphasis on the various B vitamins (thiamine, niacin, riboflavin, pyridoxin, etc.) and gave a tabulated list of vitamin deficiency syndromes for salmon, trout, carp and catfish. Some inorganic requirements (e.g. for Ca^{2+}) may be met by uptake from surrounding or engulfed water as suggested for winter flounder embryos (Fletcher and King 1978) and the same authors refer to previous studies showing both Ca^{2+} and Mg^{2+} obtained from water. Other inorganic ions (e.g. Cu^{2+} and Zn^{2+}) required as trace elements apparently need to be present in the food (Fletcher and King 1978) and may be stored in the liver for later release as required for gonad growth, for example.

Fish diets may be enhanced by varying the proportions of particular components and consideration of the presence of essential and non-essential amino acids as protein constituents, as well as the proportions of saturated and unsaturated lipids. The effect of PUFA (polyunsaturated fatty acids) has received particular attention (Thanuthong et al. 2011).

Fish held in pond or tank culture may be fed 'naturally' including (for piscivores) with chopped up 'trash fish' (low value for direct human consumption) or with specially formulated pellets which fish can learn to recognize as food. Pond fish nutrition is dealt with by Hepher (1988).

It has been necessary to carefully consider costs of high-protein diets as well as the capacity of particular species to digest individual components including the cheaper carbohydrates.

The increasing importance of aquaculture, as wild fish stocks decline, has led to specialized studies on fish nutrition (Cowey and Sargent 1979; Halver 1989), with the aim of maximizing growth. Traditional pond culture, particularly strong in Asia, within a conservative system, has existed for centuries and harvested herbivores or omnivores such as cyprinid carp. More recently, efforts have been directed towards the high-value piscivores like salmonids and finding a way to resolve the high costs of fish-based and pelleted diets. Most recently, genetically engineered fish have been developed to grow faster and to continue growth in cold weather.

Experiments involving fish feeding may require 'feeding to satiation' to make sure that all fish have accessed the food. If fish are not fed to satiation then a dominance effect is likely to occur when some fish do feed maximally and others get little or nothing. Experimental design which seeks information on levels of feeding may still need feeding to satiation, but with the frequency of feeding varied. Feeding to satiation is achieved when an excess of food is given, with some left uneaten (which usually has to be removed). Individual force-feeding may also be used to establish how much of a particular food a fish has had. This method is difficult, stressful for the fish and time-consuming.

Formulation of diets has been an important part of the aquaculture industry and there are many papers on topics like investigating diets for particular species, with particular attention paid to, for example, fatty acids and the analysis of the product for human consumption and health. In order to promote the pink/red colour of salmonids for increased marketability, pigment may be added to feed. Natural colour is derived from carotenoids in the prey (e.g. crustaceans such as shrimp, krill, etc.) but pigment/dyes used in the industry may be derived from other sources.

Unfortunately, high levels of feeding to maximize growth may give rise to problems with effluent.

Journals publishing papers on captive fish nutrition include *Aquaculture* and the *Journal of the World Aquaculture Society* and *Aquaculture Nutrition*. A valuable review of fish nutrition is provided in the book edited by Halver and Hardy (2002). Eddy and Handy (2012) discuss some aspects of the nutritional requirements for fish, emphasizing the lack of knowledge for many species. Appendix 4.1 summarizes some of the nutritional requirements expected for fish.

Appendix 4.1 Nutritional requirements for fish

The following summarizes some of the available information; as with other vertebrates, the diets may need to include particular constituents that the fish cannot synthesize.

Amino acids

There are 20 amino acids which can be constituents of animal or plant protein, forming long chains which may form sheets or be curled into compact structures with cross-linkages. These amino acids vary from the simple (glycine) to the more complex, such as tyrosine or tryptophan which incorporate a molecular ring structure. Two of the amino acids (cysteine and methionine) contain sulphur and may form links between amino acid chains, affecting protein shape. Amino acids are necessary in fish diets. Those termed essential cannot be synthesized in sufficient quantity to maintain health and therefore have to be obtained in the diet (normally from protein). Requirements for individual fish species have been discussed (Ketola 1982; Borlongan and Coloso 1993; Cowey 1994). Gaye-Siessegger et al. (2007) have reviewed the situation with levels of dietary

'non-essential' amino acids and the growth and metabolism of tilapia (*Oreochromis niloticus*).

The amino acids listed as being 'essential' for fish (Figure 4.19) resemble those for humans except that arginine is primarily regarded as essential only for pre-term or neonate humans (Wu et al. 2004). The requirements for the two pairs phenylalanine/tyrosine and methionine/cysteine are flexible in that they can be interconverted.

Lipids

Lipids comprise a complex set of water-insoluble molecules, including fats and oils (triglycerides) which consist of three fatty acids linked to glycerol. The fatty acids can vary in length and occurrence of double bonds, any double bond contributing to fluidity. Lipids vary according to the number and type of constituent fatty acids.

Lipids are basic components of cell membranes including myelin (11.3.5) and yolk. They can also form energy reserves (7.12.2) and contribute to buoyancy (3.8.5). The membrane viscosity of fish can change with environmental temperature and the composition of membranes may be adjusted (March 1993); dietary requirements for fish may thus vary with environmental conditions and for females transferring yolk constituents to maturing gonads.

A large amount of effort has recently been expended on understanding the nutritional role of lipids, particularly constituent fatty acids. Fatty acids can be divided into 'saturated' and 'unsaturated' and named for their length and the site of unsaturation (lack of hydrogen at a particular point in the chain, with a resulting double bond).

Polyunsaturated fatty acids (PUFA), with more than one double bond, have attracted particular attention, notably with regard to the subset of 'highly unsaturated' fatty acids (HUFA) which have 20 or more carbon atoms (Brett and Müller-Navarra 1997). Studies have included detailed analyses of fish and their food. These studies have indicated an array of 'essential fatty acids' needed in the diet of particular fish species.

Arginine, Histidine, Isoleucine, Leucine, Lysine, Threonine, Tryptophan, Valine

Phenylalanine and//orTyrosine,   Methionine and/or Cysteine,

Figure 4.19 Essential amino acids for fish.

Nomenclature for the fatty acids is often rendered as follows:

First: the number of carbons in the chain, e.g. 22 or 20

Second: the number of double bonds, e.g. 6n, 5n, 4n

Third: the position of the first double bond relative to the omega (ɯ) end of the chain, where the two ends are a COOH (alpha) grouping and the omega (terminal) CH3, e.g. −3, −6

The three essential long chain polyunsaturated fatty acids for vertebrates are (Sargent et al. 1999):

Docosahexaenoic acid: DHA, 22: 6n-3

Eicosapentaenoic acid: EPA, 20: 5n-3

Arachidonic acid: AA, 20: 4n-6

The ratios of fatty acids to each other are regarded as important (Sargent et al. 1999).

Essential metals

Metals can be toxic to fish but, usually in small quantities, they can also be necessary for fish function.

Various metals are regarded as important or probably important ('essential' in the diet) for fish:

Iron: for haemoglobin and cytochrome c oxidases formation

Copper: for cytochrome oxidase function

Zinc: in trace quantities for many metabolic processes, e.g. inhibits protein tyrosine phosphatizes, enzymes with diverse roles in cell function (Walton and Dixon 1993)

Cobalt: for the formation of cobalamin (a 'B' vitamin) necessary for haematogenesis (blood formation)

Nickel: present in some enzymes, but information on fish is lacking (Pyle and Couture 2011)

Selenium: necessary for the function of glutathione peroxidases (antioxidant enzymes)

Molybdenum: co-factor of at least seven enzymes but no studies have been presented for fish (Reid 2011)

Chromium: co-factor for insulin function.

Vitamins

Studies on organic molecules needed in the diet in small quantities (i.e. 'vitamins') have received some attention, notably in association with avoiding 'deficiency' situations where a suite of characteristics can be ascribed to lack of a nutrient. Thus lack of cobalamine (classified as a B vitamin) can result in a low blood count (anaemia). Vitamins, first characterized in mammals in relation to human nutrition, were originally so called because it was thought they invariably included an 'amine'. This is now known to be untrue but the other original designation, using the alphabet for naming, has persisted. Vitamins A and D are lipid-soluble; vitamins B (with numerous subsets) and C are water-soluble.

Vitamin A (a carotenoid) is needed for visual pigment (12.3.10) formation and, in fish, for some pink/red pigment formation, as in the main muscle mass of salmonids and the erythrophores/xanthophores of fish skin (2.6.3). Dietary carotenoids are obtained generally from crustacean in the wild or (presumably) directly from plants for herbivores.

The B vitamins, as described for humans (Tontisirin and Clugston 2004), include thiamine (B_1), riboflavin (B_2), pantothenic acid (B_5), pyridoxine (B_6) and cobalamine (B_{12}) The numerical designation developed over time and gaps indicate reassessment of the status of a substance previously thought to be a vitamin. Other B complex vitamins not included in the original numeration system are niacin, biotin and folic acid.

The various B vitamins have roles in cellular metabolism, for example acting as coenzyme components (niacin). Deficiencies of individual B vitamins are associated with recognizable problems in humans. Folic acid deficiency has been linked to nervous system development anomalies in humans like spina bifida (Erickson 2002).

Vitamin C (ascorbic acid) is not a 'vitamin' for many vertebrates as it can generally be synthesized except in mammals including primates, some bats and guinea pigs (Tontisirin and Clugston 2004). However, in spite of such statements, work on fish has shown a vitamin C requirement; there must be adequate vitamin C in the diet to prevent various structural problems during bone formation (Halver et al. 1969; Merchie et al. 1997; Darias et al. 2011).

D vitamins are variants of caliciferol, involved in bone formation; an overview of its role in fish is presented in Darias et al. (2011).

Vitamin E as tocopherols/tocotrienols function as antioxidants (Tontisirin and Clugston 2004) and their role in fish function is discussed by Hamre (2011).

Vitamin K (with quinone structure) is important for blood coagulation. Krossøy et al. (2011) review the requirements for this vitamin with particular reference to salmonids.

Halver (1978) presented an overview set of information on vitamins for fish, particularly B vitamins in association with deficiency conditions and there continues to be active research in this complex field.

Within the mammals where most vitamin work has concentrated there are considerable variations in vitamin requirements; for instance Vitamin C, necessary for humans is not essential for most other mammalian species. Fish, with a huge variation in species, may be also expected to have considerable variation in vitamin requirements.

Bibliography

Grosell, M., Farrell, A.P., and Brauner, C.J. (eds) (2010). *The Multifunctional Gut of Fish. Fish Physiology Vol. 30.* New York, NY: Academic Press.

CHAPTER 5

Transport: blood and circulation

5.1 Summary

All vertebrates have a closed blood system which is the principal route for transportation.

Fish generally have a single circulation, pumped by a linear connected series of chambers located ventrally and behind the gill region, and collectively forming the myogenic heart.

Fish hearts vary interspecifically and between major taxa: whereas teleosts have an anterior nonmyogenic bulbus arteriosus, elasmobranchs and relict fish have a valved conus arteriosus.

Adjustment of the circulation in fish may commonly occur through change in heart stroke volume and/or rate, as well as shunts. Hepatic and renal portal systems occur as do, more rarely, *retia* (dense networks of arterioles and venules) which expedite thermal or substance exchange

Arterial blood pressures are very low in some fish (e.g. the agnatha) and fairly high in others (e.g. salmonids).

Venous return depends on a variety of mechanisms and, perhaps surprisingly, the agnathan *Myxine* has an additional heart, the 'portal heart', whereas other fish may have other passive or active 'pumps'.

Elasmobranchs have large venous sinuses.

A secondary circulation, relatively cell-free, may occur, distinct from a lymphatic system.

Fish blood is different from that of mammals in several features including having nucleated red blood cells and thrombocytes rather than platelets. The plasma may contain antifreezes or osmotically active substances not seen in tetrapods and, seasonally, females may carry a yolk precursor.

5.2 Introduction

Active, large and/or complex animals use specialized systems to transport nutrients, dissolved gases like oxygen and carbon dioxide, and nitrogenous wastes, as well as having other functions, such as carriage of special phagocytic cells important in defence. Rapid carriage is usually achieved in a special fluid, like blood or lymph and, commonly, blood contains a respiratory pigment (haemoglobin in vertebrates and some invertebrates) which facilitates oxygen distribution. The transport system is generally either 'open' with a few blood vessels and many haemal spaces (as in arthropods such as insects) or 'closed' with the blood largely confined to vessels as in annelids (e.g. earthworms) and vertebrates. Pressure to achieve blood circulation can be provided by one or more muscular pumps ('hearts'), which may have intrinsic rhythms, modified by the nervous system. Movement of blood may also occur due to suction.

As in other vertebrates, the principle transport functions for fish are accomplished by the blood vascular system, powered by a contractile heart. In fish, the basically one-circuit system has a single passage of blood through the heart (although lungfish are exceptional), unlike that of terrestrial vertebrates which have progressively evolved a divided routing, with a double circulation in mammals (blood passing through the heart twice on its circuit through the body). Although it is common to find the typical (jawed) fish heart referred to as a two chamber structure because it has only one atrium and one ventricle, it actually has two other chambers, the posterior sinus venosus and an anterior conus arteriosus (chondrichthyans and relict fish) or bulbus arteriosus

Essential Fish Biology: Diversity, Structure and Function. Derek Burton & Margaret Burton.
© Derek Burton & Margaret Burton 2018. Published 2018 by Oxford University Press.
DOI 10.1093/oso/9780198785552.001.0001

(teleosts). In all vertebrates blood is pumped out of the heart into one or more aortas and arteries and returns to the heart via veins or venous sinuses.

With such a circulation, pressure drops at the gills and periphery can be considerable and venous return may depend on active or passive auxiliary devices.

However, some fish use other sites such as skin or buccal cavity or lungs for gas exchange and the vascular routing may be somewhat different, to the extent that the lungfish heart is 'septate' with partial division into two halves and an incipient double circulation.

The presence of a lymphatic system is not certain in fish but a secondary circulation for blood rather than lymph has been reported, although the system has restricted access by blood cells.

Change in the circulation can occur by shunts and changes in pressure, including pressure rise achieved by increased cardiac output through higher stroke volume rather than increased frequency. This contrasts with the situation in other vertebrates such as mammals, birds and amphibians, and some reptiles, in which heart rate tends to be adjusted rather than stroke volume (Shiels et al. 2006).There may also be auxiliary devices which boost pressure; these include the unique portal heart of the jawless hagfish (Johnsson and Axelsson 1996).

Nitric oxide (NO), produced from the amino acid arginine under the influence of the enzyme nitric oxide synthase (NOS) is an important modulator of mammalian/vertebrate cardiovascular physiology, its role recently discovered; it has local effects on cardiac muscle and vascular diameter.

The blood of fish is similar to that of other vertebrates; for the majority of species its major cellular constituents are red (erythrocytes) and white blood cells (leucocytes) in a fluid plasma. However, a family of Antarctic fish, the Channichthyidae (the icefish) are notable for having few, if any, erythrocytes (5.13).

5.3 Overview of the fish vascular system

Gill-breathing fish depend on blood passing to the gills from the short (thus keeping pressure high) ventral aorta, which supplies the gills via a series of paired afferent branchial arteries (Figure 6.8). The oxygenated blood from the gills drains into a series of paired efferent branchial arteries (Figure 6.8) which feed into the dorsal aorta, passing blood anteriorly to the brain via carotid arteries, and itself continuing

posteriorly, just beneath the vertebral column, until it reaches the caudal region. There this major artery becomes the caudal artery, alongside the caudal vein, both enclosed in the haemal arch (Figure 3.3) which helps to maintain blood pressure and venous return (5.7). Blood passes to the periphery via arteries from the dorsal aorta and caudal artery. In sequence, for example in dogfish, there are carotid arteries to the head, subclavian arteries to the pectoral fins (Figure 6.8), a gastric artery to the stomach and so on. Peripherally blood passes through a series of smaller vessels such as arterioles until the capillary bed is reached, where fluid containing oxygen and small nutrient molecules leaves the blood under pressure, rendering the blood remaining in further parts of the capillaries hyperosmotic simultaneously with a decrease in blood pressure as the blood passes through the small diameter capillaries (peripheral resistance). Fluid therefore tends to return to the blood by osmosis.

Details of the anatomy of the vascular system, either generalized or for particular species, are to be found in Goodrich (1930/1958), Hyman (1942), de Beer (1951), Romer (1955) and Young (1981).

5.4 The basic circulation for gill-breathing fish

The blood circulation pathway for fish fundamentally dependent on gills for gas exchange, is summarized in Figure 5.1. As a single circulation it is quite simple: muscular (smooth muscle in the walls) arteries conduct blood from the heart and thin-walled veins or sinuses return blood to the heart. The arterial walls in particular may also contain substantial amounts of elastic fibres which help to maintain pressure initially.

Blood is delivered to the tissues through a series of large diameter arteries, then smaller diameter arteries and even smaller arterioles. Finally, blood or, more accurately, its dissolved contents, reach the tissues through the very small diameter capillaries comprised basically of a single layer of (endothelial) cells.

The general structure of vertebrate arteries and veins, including fish (Hibiya 1982), is of layered tubes; the innermost forming the lining to the vessels' cavity (lumen) is the tunica intima composed of thin endothelial cells with some 'connective tissue' (collagen and elastic fibres) forming a 'tunica elastica'. The middle layer is the tunica media with

HEART	ARTERIES	GILLS	ARTERIES	CAPILLARIES VEINS	**HEART**

SV → A − V − CA/BA.→Ve Aorta→Afferents→. Efferents→D.Aorta→.PA →→→ ......→Cardinals* → SV

↓(LIVER)

→   Hepatic →

Figure 5.1 Generalized circulation for a gill-breathing fish. A = atrium; BA = bulbus arteriosus; CA = conus arteriosus; D = dorsal; PA = peripheral arteries; SV = sinus venosus; V = ventricle; Ve = ventral.
* Each of the two common cardinals = a ductus Cuvieri = a Cuvierian vein

variable (more in arteries) circular smooth muscle which constricts the vessel when it contracts. The outermost layer is the collagenous tunica externa.

5.5 The heart

The fish heart (Figure 5.2) comprises a linear series of four chambers situated ventrally, just posterior to the gills. From posterior to anterior these chambers are the sinus venosus, the atrium (auricle), the ventricle and either a bulbus arteriosus (teleosts) or a conus arteriosus (chondrichthyans).

The sequence of contractions starts posteriorly, moving forwards, and rhythmically repeated, with the possibility of rate adjustment via a pacemaker (5.5.5) and input from the autonomic nervous system (11.10.5). After contraction (systole) a relaxation period follows (diastole) before the next systolic sequence.

Blood returns to the thin-walled sinus venosus by two common cardinal veins (each a ductus Cuvieri or Cuvierian duct) or sinuses, one on each side and, posteriorly, hepatic veins from the liver. Each cardinal vein is fed by an anterior and posterior cardinal vein or sinus, draining blood from the head and the trunk respectively. Blood from the sinus venosus passes to the single atrium (auricle) which has a thicker wall than the sinus venosus. On contraction of the atrium, blood passes forwards to the thick-walled ventricle which then contracts, forcing the blood forward. The final chamber is the contractile conus arteriosus (chondrichthyans and 'relict' fish) or the non-contractile bulbus arteriosus (teleosts). The heart of the agnathan myxinoids lacks a conus or bulbus arteriosus and the atrium is to the left of the ventricle (Satchell 1986).

Icardo (2006) has reviewed the anterior chambers, the bulbus and conus arteriosus, and has named this region the heart outflow tract (OFT).

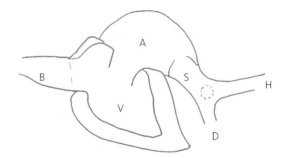

Figure 5.2 Generalized fish heart, based on information in various sources including Mott (1957), Randall (1968), Satchell (1991). A = atrium; B = bulbus arteriosus (teleost), conus arteriosus (elasmobranch); D = ductus Cuvieri; H = hepatic vein; S = sinus venosus; V = ventricle.

He suggests that the conus arteriosus persists in many teleosts.

5.5.1 Heart valves

Valves (preventing back-flow) occur between each main heart chamber; two sino-atrial (-auricular) valves between the sinus venosus and the atrium, and two atrio-ventricular valves between the atrium and the ventricle (Goodrich 1930). Valves are also described between the bulbus and the ventricle (Icardo 2006, quoting numerous authorities). The conus arteriosus is usually very well supplied with valves, which may occur in as many as six tiers in elasmobranchs while the elongated conus of the relict fish *Lepidosteus* has 'seven longitudinal and eight transverse rows of valves' (Goodrich 1930). Some teleosts such as *Tarpon* and *Albula* retain a small conus. A much reduced conus with, for example, two valves on a strip of muscular tissue is found in most other teleosts in addition to the bulbus. *Amia* has both a conus arteriosus with 12 valves (3 tiers of 4) and an anterior bulbus arteriosus (Goodrich 1930).

5.5.2 Septation

The heart of lungfish has 'partial septation' (Johansen 1968, 1970), with incomplete division allowing some degree of separation into two circuits with limited mixing in the heart, but two routes through the heart, left and right, one oxygenated via the gills, the other via the lung (see diagram in Johansen 1968). The coelacanth has no evidence of septation (Farmer 1997) and it is not reported for the other non-sarcopterygian relict fish such as *Polytpterus*, *Lepisosteus* and *Amia*.

5.5.3 Cardiac muscle

In many fish the sinus venosus is not muscular and in the teleost the bulbus arteriosus contains elastic tissue and smooth muscle rather than specialized cardiac muscle.

However, the atria and ventricles of fish hearts, as in other vertebrates, are mainly composed of distinct cardiac muscle which forms a series of linked units (myocytes, or more specifically cardiomyocytes).

Whereas in mammals the myocytes are separated by transverse intercalated discs (Figure 5.3) which may include different types of cell junction, in fish they have been described as linked by individual specialized junctions (Cobb 1974; Pieperhoff et al. 2009) which may be laterally placed (Figure 5.3). Fish myocytes are described as spindle-shaped rather than rectangular as in mammals (Satchell 1991; Pieperhoff et al. 2009). According to Satchell (1991), fish cardiac myocytes are generally shorter than those of mammals, the diameter much less and the 'myofibrils' (myofilaments) fewer.

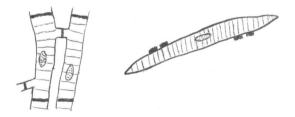

Figure 5.3 Cardiac muscle: Left, Mammalian, original drawing based on various sources including photomicrographs, Grove and Newell 1953 and Steen and Montagu 1959. Right, Fish, after Pieperhoff et al. 2009.

The myofilaments, similar to those found in skeletal muscle (3.8.1), are arranged within units called sarcomeres. Each sarcomere contains an array of thick and thin myofilaments, comprising the proteins myosin and actin respectively, and these can achieve length (and/or tension change) by a sliding mechanism. Unlike skeletal muscle however these cardiac sarcomeres change length rhythmically, under the overall influence of a pacemaker (5.5.5). Cardiac myocytes are essentially unicellular unlike skeletal muscle syncytia which have multiple nuclei and are formed by cellular fusion. Though mammalian cardiac myocytes are linked by the intercalated discs, each unit bears one or perhaps two nuclei (Bergmann et al. 2009) and is not itself syncytial.

5.5.4 Ventricle wall composition

The ventricle wall of the fish heart is thick and may have a complex structure, reviewed by Pieperhoff et al. (2009), and fish hearts can be categorized based on this structure, which may include layering, the development of a coronary blood supply, and colour (more or less myoglobin).

The cardiac muscle (myocardium) of the fish ventricle can be compact (dense) or spongy (trabecular). All fish have ventricles with an inner spongy myocardium with no coronary blood supply, and for some fish (e.g. flatfish, Santer and Greer Walker 1980; Icardo 2006) there are no additional layers. The spongy region depends, for nutrients and oxygen, on blood seeping into it from the heart lumen. The outer compact layer may be thick in very active fish like the tuna (Satchell 1991), and in such fish it contains a coronary blood supply.

Compilations of different fish heart types (e.g. Santer and Greer Walker 1980) show that elasmobranchs typically have a compact layer as well as the spongy layer whereas only about 20 per cent of the teleost species examined had the compact layer, and most of these (two eel species excepted) had a coronary supply. Species also vary in the amount of myoglobin present in the myocardium, resulting in 'red' or 'white' hearts (Driedzic and Stewart 1982; Driedzic 1983) in which 'red' hearts have much more myoglobin than 'white' hearts. The myoglobin apparently is significant in conditions of anoxia, as demonstrated in studies (Driedzic and Stewart

1982) of the red-hearted sea raven (*Hemitripterus americanus*) and sculpin (*Myoxocephalus octodecimspinosus*) in comparison to the white-hearted ocean pout (*Macrozoarces americanus*) and lumpfish (*Cyclopterus lumpus*). The relative size and shape of cardiac parts and microstructure is also interesting (Harrison et al. 1991), with active pelagic fish such as mackerel (*Scomber scombrus*) having a large 'pyramidal' ventricle as well as other features such as coronary arteries. The Antarctic notothenoid icefish (*Chaenocephalus aceratus*), which lacks erythrocytes and cardiac myoglobin, has a much larger ventricle than notothenoids with erythrocytes.

5.5.5 The pacemaker

The fish heart is actually capable of continuing contraction if it is removed from the fish and kept under suitable conditions, for instance in a saline. Such a heart is myogenic: it can contract without stimulus from the nervous system. All vertebrate hearts are myogenic but in order to continue beating outside the body, they may need oxygenation or tightly controlled temperature whereas the fish heart is not so condition-sensitive. The intrinsic (myogenic) contractions of the various regions of the heart may show a site-specific basic rate, that which is exhibited if the heart is removed from the body and kept in optimal conditions. Thus vertebrate cardiac myocytes tend to have an intrinsic rhythm and the rate of contraction of the cardiac chambers is controlled by the fastest myocytes and/or by a 'pacemaker' region. Contractions of the fish heart chambers are expected to be affected by a 'pacemaker' region, for example in the sinu-atrial junction, although Mott (1957) reported additional pacemaker areas; the whole of the sinus venosus for elasmobranchs and eels, a small ventral atrium region in teleosts, and in all fish studies the atrio-ventricular (AV) junction. Elasmobranchs appear to be very well endowed with pacemaker regions, with three areas shown by Mott (1957): the sinus, the A-V junction and one at the base of the truncus arteriosus.

Unlike in mammals (Solc 2007), there has been no firm evidence in fish for a special histologically differentiated cardiac conducting system (CCS) enabling rapid spread of the pacemaker effect, although recent work on zebrafish continues to explore the possibility and provides some support for a functional CCS (Sedmera et al. 2003). The generation of pacemaker rhythm is apparently associated with the phenomenon known as 'funny current' (Baruscotti et al. 2005; DiFrancesco 2010). 'Funny currents' are electrical currents generated during muscle contractions by special 'funny channels', limited to pacemaker regions, which have the capacity to initiate further electrical activity spontaneously, determining qualities of contractions (e.g. rates) for the other heart regions.

The rate of contraction can be modified; for example, increased by the action of spinal autonomic (sympathetic) fibres or adrenalin (11.10.5). Under exercise, trout showed an increase from 38 to 51 heart cycles per minute but in general, fish cardiac output increases via an increase in volume pumped (Satchell 1991) per cycle, that is by stroke volume.

5.5.6 The pericardium

The fish heart is enclosed in a membraneous sac, the pericardium, which, unlike the mammalian condition, has its outer surface linked to surrounding structures, like the pectoral girdle. Therefore, when the ventricle has contracted the partially anchored pericardium provides a suction effect, assisting filling of the atrium (Satchell 1991).

5.6 Blood pressure

Blood pressure is provided initially by the contraction phase (systole) of the heart but the single circuit system of fish shows a considerable pressure drop across the gills. Thus, Satchell (1991) reviews several sources and reports drops of around 25 per cent when comparing pressure values in the ventral aorta (emerging from the heart) and the dorsal aorta, after passage through the gills. Taking 'resting' rainbow trout as an example, mean blood pressures in the ventral aorta are reported as 5.17 kPa (Appendix 5.1) and in the dorsal aorta 4.13 kPa (Kiceniuk and Jones 1977). The same authors reported the low mean value of 0.19 kPa for the common cardinal vein. Satchell (1986) records dorsal aorta pressures (in cm H_2O, Appendix 5.1) as much lower in hagfish (7 cm) than in dogfish (22 cm) or some teleosts (cod 40 cm, rainbow trout 52 cm).

Stevens and Randall (1967) found the ventral aorta of rainbow trout to have 'resting' diastolic (relaxed) values of 32 mmHg (Torr, Appendix 5.1), rising to 40 mmHg in systole, while the dorsal aorta was less, with 25 mmHg pressure in diastole, rising to 29 mmHg in systole. These authors reported 6.9 mmHg in the subintestinal (hepatic portal) vein of rainbow trout; both sets of authors therefore show a large pressure drop in the periphery. Elasmobranchs have even lower (negative) pressures in the anterior and mid-region venous sinuses (Satchell 1991).

5.7 Venous return

The return of blood to the heart depends on a number of factors, some common to most fish. Anteriorly a valve on the anterior cardinal vessels prevents back-flow from the ductus Cuvieri on each side. Posteriorly the haemal arch (Figure 3.3) maintains venous pressure, which can be boosted by tail movements forcing blood forwards through the segmental veins into the haemal arch. Backflow is prevented by (ostial) valves in the segmental veins.

Some taxa, including the Ostariophysi and the eels, have an additional active pump, a minute two-chambered 'caudal heart', in the tail, which can beat continuously. A 'caudal heart' has also been described for two elasmobranch species whereas others have a different system, the 'caudal pump' in the tail (Satchell 1991). This system depends on two lateral sinuses, pressured alternately when the shark swims, being dependent on the bilateral bending movements along the body. The agnathan hagfish has a myogenic portal heart on the hepatic portal vein (Satchell 1986).

In elasmobranchs venous return occurs by means of veins and large sinuses in place of the major veins. Low (negative) pressure in the sinuses facilitates venous return into the central area.

Low or negative pressure in the flaccid sinus venosus assists filling during the heart's relaxation phase (diastole).

5.8 The circulation in air-breathing fish

Obligate air-breathing fish (6.6.6) which do not depend on gills for gas exchange may have a much reduced blood flow to the gill region, with blood shunted, for example through the fourth afferent artery, as in *Monopterus* (Hughes and Munshi 1979). Johansen (1968) provides a comparison of the circulation of air-breathing fish, with variants of the basic single circulation, or in the case of lungfish, an arrangement with 'two circuits' although there is some mixing in the heart which is not completely divided.

5.9 Portal systems

Portal systems (Hendersen and Daniel 1978) are a means to shunt venous blood to another region before return to the heart. All vertebrates have a hepatic portal vein (Figure 5.4), routing venous blood from the gut to the liver. This facilitates regulation of, for example, amino acids, lipid products and sugars absorbed from digested food. Hagfish (e.g. *Myxine*) have a 'portal heart' on the hepatic portal vein (Johnsson and Axelsson 1996) which has intrinsic contractile activity; this structure has not been reported in jawed fish, lampreys or any other vertebrates.

In the vertebrate kidney, two successive capillary systems occur, with blood from the glomerulus passing through peritubular capillaries via the efferent arteriole (Figure 8.3). Fish commonly have an additional portal system, via the renal portal vein (Figure 5.4), routing blood from the tail region through the kidneys. However, according to Hickman and Trump (1969), quoting Moore (1933), a standard renal portal system is not apparent in 'many freshwater teleosts'.

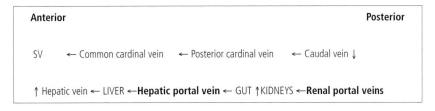

Figure 5.4 Schematic view of the hepatic portal and renal portal systems in fish SV = sinus venosus.

5.10 Retia

Retia (singular: *rete*) are complex networks of narrow blood vessels which may be visualized as dense (red) patches. They achieve rapid accumulation or exchange because they are small and thin-walled and very numerous.

Within the circulation, intense networks of blood vessels may occur; small arterioles and venules or afferent and efferent capillaries (Scholander 1954) organized in such a way as to be very close to each other with a countercurrent flow (6.5.6), thus maximizing exchange of oxygen for example, or lactic acid or heat. The term *rete* means 'net' and when first described the term '*rete mirabile*', usually translated as 'wonderful net', was applied. More recently the term has been anglicized so that the plural becomes 'retes'. *Retia* have been described in several locations for various species (and in, for example, mammals). The most spectacular may be those associated with deep-sea fish which have gas-exchange *retia* in their swim bladders, forming the 'red body' or gas gland (Figure 3.20 and Scholander 1954).

A choroid 'gland' associated with the retina of each eye (12.3.5) also has a *rete* (the choroid *rete*), maximizing oxygenation of this region, and this may actually predate the swim-bladder *retia* (Berenbrink et al. 2005). This choroidal *rete* 'concentrates oxygen for delivery to the retina' in teleosts (Gendron et al. 2011) and is not present in mammals. A choroid *rete* was not found in the eyes of several elasmobranchs (Wittenberg and Haedrich 1974).

For the few heterotherms like tuna which have internal red muscle acting to 'store' heat, a *rete* can assist the exchange process, keeping the internal organs warmer than the other tissues. Fudge and Stevens (1996) reviewed information available on the visceral *retia* of tuna and sharks, including a translation of very early papers on the subject.

5.11 The secondary circulation

The vertebrate transport system can be regarded as comprising a major (primary) blood vascular component backed up by tissue fluid drainage contributing volume to a secondary system, which in mammals, for example, does not contain erythrocytes, and connects to the venous system via a series of channels forming the 'lymphatic' system.

The presence of a lymphatic system in fish has not been fully established although Blaxter et al. (1971) referred to 'large lymph ducts' in some mesopelagic fish, but a parallel slow circulatory system with restricted access to blood cells has been described for teleosts, and may also occur in elasmobranchs (Satchell 1999). The anatomy and function of this system has been reviewed by Steffensen and Lomholt (1992) and Olson (1996); it is a high volume system linked (by arterio-arterial anastomoses) to the primary circulation through small twisted arteries from, for instance, the aorta and the efferent branchial (gill) arteries. Following these narrow entries, with microvilli, which prevent large-scale entry of blood cells, the subsequent vessels are larger, wide-bore arteries, a capillary network and a venous system which returns blood to veins of the primary system. The system has been compared to the lymphatic system of other vertebrates and may be a precursor of it.

Skov and Steffensen (2003) measured the volume of the primary (PC) and secondary circulation (SC) in cod and reported mean values of 3.4 and 1.7 ml/100 g respectively; that is, the PC had twice the volume.

5.12 Change in the circulation

Increased activity which in mammals, birds and amphibians may result in increased heart rate, in response to demand, leads in fish to increased stroke volume (Shiels et al. 2006). However, Thorarensen and co-workers (1996) had found with rainbow trout that a heart rate increase of 38 per cent and a stroke volume increase of 25 per cent contributed to improve cardiac output volume of blood expelled from the heart in unit time. Boutilier (1998) reports that 2–3 °C increased temperatures for cod resulted in a doubling of the heart rate. Vertebrate hearts, in general, respond to stretch with contraction in accordance with the Frank–Starling law (stroke volume of the ventricle increases as venous filling pressure rises with the larger volume of blood in diastole) developed during physiological studies centred on mammals, but the responses vary. Fish and some reptiles have cardiac myocytes that tolerate more stretch in comparison to the stretch-resistant cardiac myocytes of the other groups. Such considerable differences in function and the actual circulatory pathways may make it difficult to apply formulae achieved from mammalian work, for example, to fish studies. Thus, Metcalfe and Butler (1982) used the Fick method, developed for mammalian determinations, to measure the cardiac output of dogfish. In the Fick method cardiac output is estimated from the oxygen uptake per unit time at the respiratory surface, divided by the difference between the oxygen content of venous blood (e.g. fish ventral aorta) entering the respiratory surface and the oxygen content of arterial blood (e.g. fish dorsal aorta) leaving it. In the case of fish this method may not be reliable if there is also significant oxygen uptake through the skin or other respiratory surfaces (6.6), relative to the gills (Metcalfe and Butler 1982; Schmidt-Nielsen 1990). This may occur if catecholamines (10.12.6, 11.9.4) shunt some of the cardiac output away from the gill respiratory surface through non-respiratory vascular pathways (Hughes et al. 1982; Metcalfe and Butler 1982, 1986).

Decreased activity, contingent on drought or decreased temperature, may result in torpor, with minimal movement and circulation.

Under experimental conditions it can be demonstrated that changes in fish heart rate are under autonomic control (11.10.5). However, as shown by Shiels et al. (2006) earlier, changes in heart rate for fish are much less important than in other vertebrates for adjusting the blood circulation. The moderation of stroke volume that occurs in fish is linked to changed, and (myofilament) length-dependent Ca^{2+}, sensitivity in the myocytes.

The mammalian circulation is modulated by the action of the autonomic nervous system via the parasympathetic and sympathetic components (11.9), and by circulating factors/hormones like adrenalin which can affect heart rate and/or arterial diameter through vasoconstriction or vasodilation, but the role of some other potentially vasopressive factors in fish is not very clear (10.12.6). In mammals, heart slowing can be achieved through vagal (parasympathetic) fibres whereas acceleration can occur via sympathetic fibres or circulating catecholamines

(notably adrenalin). Autonomic modulation may occur in fish species (11.10.5) but there can be considerable interspecific variation in the physiological mechanisms. Moreover, the neuropeptide galanin, which has little effect on mammalian vasoconstriction, can achieve it in some elasmobranchs (Preston et al. 1995).

Nitric oxide (NO) also acts as a local vasodilator in vertebrates (Moncada and Higgs 2006) with immense medical importance and its action has been mainly studied in mammals, in which it may be evoked in response to, for example, acetylcholine (Huang et al. 1995). It has not been so thoroughly studied in fish but some recent papers have included fish species (Imbrogno et al. 2011; Angelone et al. 2012; Amelio et al. 2013).

It is expected that NO is produced locally by the action of an enzyme, nitric oxide synthase, on the amino acid arginine. NO has a very short half-life, acting briefly as a vasodilator (relaxing smooth muscle) close to the region where it is produced, for instance the endothelium of blood vessels.

5.13 Blood

It is generally understood that fish blood is very similar to the blood of other vertebrates, with pigmented and non-pigmented blood cells in an aqueous fluid. However, there are differences in the cell size and structure as well as the cell prevalence, with an extreme situation in Antarctic icefish which have very few or no red blood cells.

The composition of fish blood is determined following sampling, usually from a live animal (Appendix 5.2). Fish blood usually comprises large numbers of red cells (erythrocytes) and varying numbers of white cells (leucocytes) in a fluid plasma (Figures. 5.5, 5.6), with 'thrombocytes' less readily seen or distinguished. The packed cell volume (haematocrit) of fish varies from 47 per cent in salmon (similar to human values with 47 per cent haematocrit and about 5 million cells per cubic mm, Appendix 5.1), to 12 per cent in dogfish and 13 per cent or less in notothenoids, the Antarctic taxon which includes the icefish (Family Channichthyidae, e.g. *Chaenocephalus*). Icefish have few, if any, red blood cells, explicable in terms of high availability of oxygen and increased viscosity in a low-temperature environment; reduced blood-cell number would presumably assist blood flow. Blaxter et al. (1971) reported low haematocrits (5–9 per cent) in marine mesopelagic fish without swim bladders, in comparison to higher values (14–35 per cent) for mesopelagic fish with swim bladders and very high values (48–57 per cent) for some surface species.

The haematocrit values obtained can be affected by pre-sampling treatment of the fish (Lowe-Jinde and Nimi 1983), with a range of 35–41 per cent obtained for rainbow trout under different conditions of anaesthesia.

5.13.1 Red blood cells (erythrocytes)

Fish erythrocytes are nucleated, as for all other vertebrates, apart from mammals, although an exception is reported for *Maurolicus mülleri* (Wingstrand 1956) which has very small erythrocytes (maximum diameter 4–6 μm). Fish erythrocytes are usually seen as round/oval (Figure 5.5) or oval/ellipsoid (Figure 5.6) and generally larger than human red blood cells which have a maximum diameter of about 7–8 μm; teleost cells having a maximum diameter of 11–14 μm (e.g. Graham et al. 1985), and dogfish about 20 μm. Their principle function is the carriage of oxygen loosely bound to the cytoplasmic haemoglobin. For very active fish, haemoglobin levels can be roughly equivalent to the human value of 14–17 g/100 ml of blood; thus tuna have 13–18 g/100 ml blood, salmon have 10 g/100 ml, notothenoids, other than icefish, have 3–4 g/100 ml and dogfish a similar value. Although the oval erythrocyte shape is the conformation normally reported, Nilsson et al. (1995) found deformation in

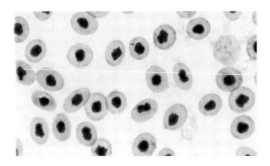

Figure 5.5 American plaice (*Hippoglossoides platessoides*) blood smear, Giemsa stain. Courtesy of Dr R.A. Khan.

Figure 5.6 Fish blood, showing (from left to right) a nucleated erythrocyte, leucocytes (a lymphocyte, granulocyte and a monocyte) and two thrombocytes. Composite sketch, based on various sources, including original slides, Catton (1951) for leucocytes and Shepro et al. (1966) for thrombocytes, Ardelli and Woo (2006) for all cells. Drawn to scale with erythrocyte maximum diameter 13 μm.

the gills, with extreme elongation occurring, a situation proposed to improve oxygen uptake. Erythrocyte shape change in capillaries was previously reported by Guest et al. (1963) for mammalian cells.

Erythrocytes also function in carbon-dioxide transport as in other vertebrates; a high concentration of the enzyme carbonic anhydrase in the erythrocytes promotes the formation of carbonic acid from carbon dioxide and water, in the peripheral tissues. The carbonic acid then dissociates to protons and bicarbonate; the protons become loosely bound to deoxyhaemoglobin and the bicarbonate diffuses into the plasma (Shier et al. 2002). Bicarbonate is expected to release carbon dioxide (the reverse reaction under the influence of carbonic anhydrase) at the gills.

Unstained erythrocytes do not individually appear red under the light microscope. They are eosinophilic however and thus appear red with eosin containing blood stains such as the usual 'Romanowsky' (Tavares-Dias 2006) procedures like 'Giemsa' and 'Wright's'. Many routine histological stains also contain eosin and therefore it is usual to see red or pink cytoplasm of erythrocytes in general preparations.

With the exception of Antarctic icefish, erythrocyte numbers far exceed the numbers of leucocytes and thrombocytes. For very active fish, the haematocrit (packed cell volume) is comparable to that of humans but as fish erythrocytes are generally larger than those of humans, the number of fish erythrocytes per unit volume is fewer (approximately 2–3 million cells μl^{-1}, Rough et al. 2005) than the value for humans (4–6 million cells/mm^3, Fox 2002) where the volumes measured, expressed in different systems, are equivalent (Appendix 5.1).

Mammalian blood, particularly after haemorrhage, may be seen to contain large numbers of immature erythrocytes, with no nucleus and a diffuse net-like structure (reticulum) in each cell, which used to be thought to represent the remains of a nucleus, before its final dissolution. However, the mammalian reticulum can now be related to ribosomal RNA (Hunt 1970) and reticulocytes have been described in fish blood (Groman 1982; Hibiya 1982); their equivalence to mammalian reticulocytes is difficult to explain in either case as fish erythrocytes retain their nucleus and fish 'reticulocytes' appear to have a distinct and granulated nucleus in a central location (Catton 1951); that is, the granular nucleus resembles a network only superficially, the network in mammalian reticulocytes being less dense and more peripheral. Erythrocytes develop from blast cells in special haematopoietic tissue (5.13.6).

5.13.2 White blood cells (leucocytes)

Fish leucocytes (Figure 5.6) are similar to those of other vertebrates, usually classified as several main types including lymphocytes, granulocytes (eosinophils and basophils), neutrophils and monocytes. Thrombocytes have also been described; their function is similar to mammalian platelets—they promote clotting (5.13.4). Enumeration of fish leucocytes (e.g. Engel et al. 1966) may categorize thrombocytes separately. Some leucocyte types, such as monocytes, are known to be phagocytic, but there is debate over the phagocytic capacity of thrombocytes (Tavares-Dias et al. 2007). Often described as round in blood smears, live leucocytes may be capable of movement via projected pseudopodia. Sizes reported are usually for the fixed rounded cells; Catton (1951) found that one type of leucocyte ('fine' granulocyte), measuring a maximum diameter of about 10 μm when live but inactive could extend to 24 μm when motile.

A coloured illustration of stained human blood cells, given in Bell et al. (1957), is useful for comparative purposes. Note, however, that the monocyte nucleus illustrated is lobed rather than bean- or kidney-shaped as described in some fish.

The classification of fish leucocytes has sometimes been controversial (Ellis 1977; Hibaya 1982) and numerous variants of the terminology have been listed (Zinkl et al. 1991). In general, blood leucocytes do conform quite well with the leucocyte categories of other vertebrates but Zinkl et al. (1991), who studied six fish species, could not find eosinophils or basophils in some species and found novel forms in goldfish. Eosinophils sometimes occur in extravascular locations like the epidermis (2.3.1); their absence in the circulation could be due to a seasonally or physiologically driven relocation event. Thus, some of the leucocytes undergo active migrations into tissues (diapedesis). Sizes of leucocytes are similar, proportionally, to those of mammals, with lymphocytes the smallest and monocytes the largest.

Lymphocytes, the smallest leucocytes (excluding thrombocytes) have a diameter of 6–9 μm in turbot (Burrows et al. 2001). These leucocytes are round, with a central round nucleus, and do not stain deeply with any of the usual blood stains (e.g. Giemsa).

Granulocytes are larger than lymphocytes and, as seen in smears, are round (diameter approximately 11 μm for turbot, Burrows et al. 2001) and have a lobed ('irregular') nucleus. The cytoplasm of neutrophils, basophils and eosinophils stains differentially, as the names suggest. With the usual blood stains neutrophil cytoplasm remains pale, while basophils become dark-blue/purple and eosinophils are red; the latter two types are sometimes collectively called 'heterophils'. The illustration in Yûki (1957) demonstrates the considerable difference in granulocytes under different staining regimes.

Monocytes may be the biggest fish leucocytes (though not in Radhakrishnan et al. 1976). They are usually described as rounded in the circulation, with a diameter of about 11.5 μm in turbot (Burrows et al. 2001) and are capable of phagocytosis. These cells are, however, not even mentioned by Hibiya (1982) and are regarded as immature cells in the circulation by Roberts and Ellis (2012), a stage prior to the mature cell's role in tissues as an active phagocyte. Monocytes are often described as having a kidney-shaped (reniform) or bean-shaped nucleus (Radhakrishnan et al. 1976; Evans et al. 1984; Ribeiro et al. 2007). Catton (1951) could not find monocytes in blood of trout, roach or perch.

Thrombocytes resemble small erythrocytes or lymphocytes in stained fish blood, taking a round form (Ardelli and Woo 2006) but they may also be ovoid (Shepro et al. 1966) and extended to a spindle-like form (Catton 1951; Shepro et al. 1966). Tavares-Dias (2006) recommends the use of alkaline toluidine blue rather than the usual blood stains to help distinguish these cells. Rowley et al. (1988) emphasized that reports of thrombocyte occurrence/prevalence had been subject to considerable confusion and hence required careful interpretation.

Leucocyte and thrombocyte numbers: According to texts of human physiology and anatomy (e.g. Shier et al. 2002) the 'normal' concentration of leucocytes (excluding platelets) for humans is approximately 5000–10,000/cubic mm of blood (mm^{-3}), and this rises in response to infection. The statement that 'fish have a far higher white cell count than man' (Satchell 1991) is interesting; Satchell quotes Mulcahy's value for pike as 100,000 mm^{-3}; the mean given by Mulcahy (1970) for 15 pike is actually 111,500/mm^3. For comparison, pre-treatment (i.e. before irradiation) leucocyte concentrations for *Lagodon rhomboides* were about 40,000 mm^{-3} and thrombocyte concentrations were about 70,000 mm^{-3} (Engel et al. 1966). Thomas et al. (1999) found seasonal variation in blood-cell counts for fathead minnows (*Pimephales promelas*) with leucocyte counts increasing threefold in August and September.

Although the comparability of fish leucocytes with those of other vertebrates, particularly mammals, has been queried (Ellis 1977), further work increased confidence that there was much similarity (Zinkl et al. 1991). Changes in fish leucocyte number and type have been widely studied in association with pollution problems and ecotoxicology, with reports of sharp decreases for leucocyte number (leucopenia) in experimental regimes with particular pollutants (Engel et al. 1966; Bendell-Young et al. 2000).

As for mammals, fish leucocytes are important in immunity (5.16), including antibody promotion and production (lymphocytes), and some (i.e. monocytes, granulocytes) are phagocytic, capable of engulfing particles, including bacteria.

Monocytes may be part of the reticuloendothelial system (RES), described as including monocytes and tissue macrophages (multinuclear phagocytes). Kupffer cells of the liver have been regarded as part of the RES too. Besides removing particulate matter (such as bacteria) from the circulation the RES has been described as important in iron recycling (Knutson and Wessling-Resnick 2003). Older views of the RES were summarized by Halpern (1959).

Leucocytes develop from blast cells (as do erythrocytes) in special haematopoetic tissue (5.13.6).

5.13.3 Plasma

The plasma is the fluid portion of the blood. The term serum is used for the fluid remaining after coagulation and removal of the cells and clot (fibrin, 5.13.4).

The composition of plasma may vary seasonally, diurnally or according to feeding status and reproductive condition, and the area/region of the circulation from which the blood is taken. The aqueous base contains ions as well as inorganic and organic molecules of a wide size range. In the case of ionic components, carbon dioxide can be carried as bicarbonate with venous blood generally having much higher values, and degraded amino acids can provide ammonium ions for excretion at the gill. The level of some ions, for example Na^+, K^+, Cl^-, $SO4^{2+}$, Mg^{2+}, Ca^{2+}, in the plasma, are well regulated, but the levels may rise or fall seasonally; for example, Na^+ and Cl^- rise in winter in winter flounder (Fletcher 1977), with highest values in January to April. Large molecules in the blood include proteins of various types, for example antibodies, fibrinogen, carrier (binding) proteins and for some cold-water fish, antifreezes such as antifreeze glycoprotein (7.4, 13.5.3), Seasonally, adult females may have large amounts of a yolk protein precursor called vitellogenin (9.10.2).

Depending on the feeding status of the fish and its food type and the site of the blood sampled there may be relatively large amounts of digestive products such as amino acids, fatty acids and glycerol, but this would be transient as these substances are regulated. The physiological status of the fish, whether reproductive, stressed, actively feeding, and so on will affect the levels of particular hormones in the plasma with, for instance, cortisol and adrenaline rising in stress. Some hormones including sex steroids such as testosterone may be carried in free or protein-bound form.

5.13.4 Coagulation

Vertebrate blood coagulates (clots) under conditions reflecting tissue damage or injury. The clots help to stabilize loss of blood from vessels. Clotting time is rapid and dependent on deposition of fibres.

The mechanism is similar to that of other vertebrates in which the protein fibrinogen, in the plasma, deposits fibrin fibres in response to injury, with a number of different participating factors (e.g. Ca^{2+}, thrombin), promoting fibrin formation and other coagulation events. In fish, thrombin action involves the thrombocytes rather than the much smaller platelets found in mammals. Vitamin K (Appendix 4.1) is important in the coagulation process, which involves vitamin K-dependent proteins (VDK) such as prothrombin (Factor II) as well as Factors VII, IX and X (Krossøy et al. 2011).

5.13.5 Osmotic fragility

Blood cells, particularly erythrocytes, may rupture or shrink under adverse osmotic conditions.

Tetrapod blood cells are prone to osmotic damage, but fish are more resistant.

Fish blood cells are unlike mammalian blood cells in their response to osmotic changes (7.5.7). Whereas mammalian erythrocytes lyse or shrink (crenation) respectively in response to hypotonic or hypertonic solutions of very moderate changes from the isotonic state, fish erythrocytes show considerable resistance to changes in their osmotic environment. Active volume regulation occurs with amino acid (taurine) adjustment (Garcia-Romeu et al. 1991), which has adaptive value particularly for estuarine or coastal fish with gill surfaces subject to varying salinities.

5.13.6 Blood-cell replenishment, recycling and haematopoiesis

Replenishment of blood cells occurs by recruitment from haematopoetic tissue.

Blood-cell formation or haem(at)opoiesis has been studied extensively in mammals. Mammalian red blood cells are enucleate (thus giving more available volume for haemoglobin) and mammalian erythrocytes have a short life in the circulation, expected to be a few months in humans. Old erythrocytes fragment and are phagocytosed, for example in the spleen and liver, where iron is retained as ferritin and haemosiderin and recycled. Sites of such phagocytosis may be referred to as part of the reticuloendothelial system also known as the mononuclear phagocyte system (Bloom and Fawcett 1975). A similar recycling is expected in other vertebrates, including fish. New erythrocytes are formed in various sites in mammals, including the bone marrow but in fish, which lack bone marrow, erythropoiesis (red blood-cell formation) occurs in sites like the anterior kidney (head kidney) or, less specifically in the interstitial tissue between nephrons in the aglomerular kidney of seahorses (Tsujii and Seno, 1990).

The formation of erythrocytes from erythroblasts (erythropoiesis) is well understood in mammals, following a set pathway with newly formed erythyrocytes released to the circulation from, for example, the spleen as recognizable 'reticulocytes' showing traces of nuclear structure. White blood cells are formed in various locations in mammals; these sites include lymph nodes and the spleen as well as the bone marrow. The lymphatic system in fish is not well understood; it is less obvious than in tetrapods with, for instance, no lacteals or major lymphatic vessels, although patches of 'lymphoid' tissue have been described in various locations including the spleen. Sites of leucocyte (leucopoiesis) and thrombocyte (thrombopoiesis) formation may be expected to occur in the spleen and anterior kidney as well as the thymus.

5.14 The spleen

The vertebrate spleen is a discrete structure, an organ important in blood-cell formation and a blood reservoir.

The spleen of vertebrates is usually a fairly obvious structure in the abdomen, appearing as an encapsulated reddish-brown organ close to the stomach in mammals. In fish the spleen is usually easily detected as a round (teleosts) or more elongated (elasmobranchs, Figure 4.16) reddish body although it is fairly small notably in teleosts, being less than 1 per cent body weight (Fänge and Nilsson 1985). It may be larger in chondrichthyans, of about 1 per cent body weight in *Chimaera* and 2.65–4 per cent body weight in *Mustelus*. The position of the spleen in whiting (*Gadus merlangus*) is illustrated by Saunders and Manton (1951), in the perch (*Perca flavescens*) by Chiasson (1974) and in the dogfish (*Scyliorhinus canicula*) by Wells (1961).

Splenic structure is traditionally described as consisting of small areas ('islands') of 'white pulp' surrounded by 'red pulp'. The white pulp contains many white blood cells, notably lymphocytes, whereas in the red pulp red blood cells predominate. Melanomacrophage centres (MMC) are also found (Rebok et al. 2011); these latter would be expected to phagocytose fragments of old erythrocytes and blood-borne particulates such as bacteria. The terms cortex and parenchyma are also applied to splenic structure, with the exterior cortex a 'reticulum' containing MMC, granulocytes and 'plasma cells', surrounding the parenchyma of red and white pulp beneath a sinus (Icardo et al. 2012).

Zapata (1980) reported erythropoiesis and thrombopoiesis in the red pulp of the elasmobranchs *Raja* and *Torpedo*. The overall functions of fish spleen are summarized by Spazier et al. 1992); they include 'blood-cell formation', 'blood-cell storage and release' and 'destruction of effete blood cells and foreign agents'. Spleen structure is liable to considerable changes under stress such as environmental changes including osmotic variations, exposure to toxic chemicals or drought (Spazier et al. 1992; Icardo et al. 2012).

5.15 The thymus

The vertebrate thymus is associated with the formation of white blood cells and is prominent in juvenile mammals, notably humans.

The fish thymus is not conspicuous and is not generally illustrated in manuals of fish anatomy. It has been described as occurring in the upper branchial chamber of teleosts, a pale mass under the operculum (Powell 2000). Mohammad et al. (2007) studied its structure in the lungfish (*Neoceratodus forsteri*). It is expected to be a source of lymphocytes.

5.16 The immune system

Vertebrates are expected to have a complex set of defence processes forming an immune system, activated in response to parasites or unusual environmental events. The blood system is a major participant in immunity, carrying the leucocytes and plasmatic proteins which achieve it.

Toxicology, with fish responding to various pollutants and pathogens and acting as indicators of environmental problems, has recently increased interest in fish resistance and the fish immune system, as has the increase in aquaculture globally. The possibility of vaccinating cultured stocks has been a commercially active situation. In intensive fish farming the possibility of rapidly spreading diseases, including infection of wild populations has encouraged further study of fish immunity. Some fish diseases have been associated with metazoan parasites, for example parasitic copepods or fish-lice, but many are fungal, bacterial or viral.

Vertebrate immunity is generally regarded as bipartite; innate and acquired. The possibility of acquired immunity is of interest in developing vaccines.

The fish immune system is generally expected to be very similar to that of other vertebrates so that knowledge of other vertebrate systems may assist in understanding fish responses to pathogens. Fish, however, do not have as well-developed a lymphatic system as that of tetrapods and fish vaccine development and application presents special problems including that of cost. Zapata et al. (2006) report several unresolved general problems in fish immunity with particular reference to ontogeny.

Non-specific, innate immunity depends on barriers to invading pathogens and also involves phagocytosis of 'foreign' particles such as fungal spores or bacteria.

Acquired immunity is a response to specific pathogens by multiplication of, for example, lymphocytes and antibody (complex proteins known as immunoglobulins) production by these cells. The term opsonins is reserved for proteins such as antibodies and complement (an array of plasma proteins, for instance (Silverthorn 2001), which coat pathogens, thus promoting them as targets for phagocytes. Antibodies can bind to viruses, for example, preventing them from penetrating host cells and lymphocytes can recognize and attack infected host cells and promote their destruction by apoptosis. The many complex mechanisms of the immune response, including inflammation, have been explored extensively for mammals, notably rats and humans; much less work has been done with fish, for which emphasis has usually been on high-value aquaculture subjects such as salmonids. A review of the fish immune system (Alvarez-Pellitero 2008) summarizes recent work with emphasis on teleosts and vaccination.

5.17 Conclusion

In 1957, Mott wrote, in the original two-volume set on the *Physiology of Fishes* (Brown 1957), that knowledge of fish circulation was 'fragmentary' and overly dependent on a few species. He considered the area to be difficult because of the aquatic environment. Since that time there have been numerous researchers contributing to the field, summarized from time to time in chapters in the continuing *Fish Physiology* series (Hoar and Randall 1969 onwards, e.g. Hoar et al. 1992). The criticism of narrowness in species studied may still have some point; salmonids seem to be popular for recent physiological studies, and they represent the more active fish. They are also individually quite variable (Thorarensen et al. 1996), which may give difficulties for interpretation. The common problem of applying mammalian methods to fish physiology has also appeared in fish circulation studies in which the use of Fick's principle to study fish heart function has been criticized (Metcalfe and Butler 1982). There have been, subsequently, some interesting comparisons (Thorarensen et al. 1996; Shiels et al. 2006) in which a possible 300 per cent increase in stroke volume for fish compares with around 30 per cent increases in stroke volume for other vertebrates (Shiels et al. 2006), but rainbow trout (Thorarensen et al. 1996) were only showing a 25 per cent increase in stroke volume during 'prolonged' swimming.

Appendix 5.1 Conventional units for blood system measurements

For measuring pressure: Traditionally, pressure was measured by the length of fluid (water or mercury) supported, with height of the column renedered metrically (i.e. cm). More recently the pressure may be reported as kilopascal (kPa).

1 kPa (kilopaschal) = 7.5 Torr. (= mm Hg = millimetres of mercury). 10 cm H_2O = 7.36 Torr.

For measuring volume: Volumes are usually given metrically.

Volumes for human blood have traditionally used the cubic millimetre (mm^3) or the expression 'per cubic millimetre' =/mm^3 or mm^{-3}, equivalent to 1 μl.

For measuring concentration, where the amount of solute dissolved in a volume is used, the information is commonly given directly as weight (mass) per unit volume or as a molar quantity (M) where M = molar, i.e. the molecular weight in gpl (grammes per litre).

Appendix 5.2 Blood-sampling techniques

The procedures for this have varied over time, but probably the best, avoiding contamination, is to remove a sample from the haemal arch region using a hypodermic needle. It may be advisable to use an anticoagulant (e.g. heparin) and the blood can subsequently be subsampled (aliqotted) and spun in a micro-centrifuge to separate cells and fluid. For microscopic examination, blood smears are prepared on microscope slides, using whole blood (i.e. not spun) and special blood stains such as Giemsa or Wright's (Humason 1967). The dry blood smear is 'fixed' for about two minutes by immersion in methanol, prior to staining.

To obtain a blood sample from a live fish the following process can be used. Note it may be necessary to obtain prior permission from the institutional Animal Care Committee or equivalent, depending on the rules/laws applicable for handling live vertebrates.

Anaesthesia

Blood sampling should be done with anaesthetized fish. Anaesthetics appropriate for aquatic animals (Bell 1964) include tricaine methane sulfonate (MS222) which may be buffered (Cho and Heath 2000), propoxate (Thienpont and Niemegeers 1964) and clove oil (Cho and Heath 2000). These substances can be dissolved in a small container of water at the appropriate temperature for the fish, and the fish then immersed for an appropriate time, but as clove oil has limited solubility in cold water (< 15 °C) it may have to be first dissolved in 95% ethanol (1:10 clove oil/ethanol). Concentrations appropriate will depend on the temperature; duration of immersion of the fish should be determined experimentally but should be about five to ten minutes, and be sufficient to have the animal quiet, determined by it gradually losing capacity to swim upright (loss of equilibrium, stage four anaesthesia) and to not react violently to handling.

A review of anaesthetics/sedatives is provided by Randall and Hoar (1971) and Bowker et al. (2015).

Blood removal

For MS222 at a temperature of 16–25 °C, small tropical fish like *Poecilia* would be subdued in 5 to 10 minutes with a concentration of about 0.2 gpl (0.12 g/600 ml Burton lab notes after Campbell and Idler 1976), and remain quiescent for about 10 minutes. A fish this size cannot be readily sampled and survive. The sampling is best done with heart puncture (into the ductus Cuvieri) by inserting a small capillary tube (with a heat-drawn tip) via the operculum.

For larger fish, at cooler temperatures (e.g. 5–10 °C), initial MS222 anaesthesia can be achieved with about 0.125 gpl (0.625 g/5l, Burton, laboratory notes) and sampling is done with a hypodermic needle and syringe. The size of the needle and syringe should be predetermined. It is advisable to practice with a freshly killed fish (using overanaesthesia) first.

On removal from the anaesthetic bath the fish should be lightly held with, for example, a wet paper towel. The needle and syringe(s) should be preprepared, rinsed with an anticoagulant (e.g. 0.15 M* sodium citrate (Shepro et al. 1966) or heparin (Hattingh 1975)) if necessary.

For symmetrical fish (the majority), the fish should be laid on its side or on its back and the needle angled to enter the haemal arch just in front of the caudal fin, ventrally. Once inserted, blood can be gently withdrawn. It is best to have a good mental 'picture' of the position of the haemal arch before commencing. For flatfish, the fish can be laid on its normal blind side and the haemal arch envisaged before starting. If the fish is to survive sampling, light pressure should be put on the needle insertion point when the needle is withdrawn, and pressure maintained for about a minute before returning the fish to water.

Cannulation of blood vessels (inserting a tube directly into the lumen, and stabilizing it there) a more intrusive, long-term procedure, is used to obtain serial samples (Wells et al. 1984).

Bibliography

Early material

Mott, J.C. (1957). The cardiovascular systems. In: M.E. Brown (ed.). *The Physiology of Fishes, Vol. I.* New York, NY: Academic Press, pp. 81–108.

G.H. Satchell has provided two comprehensive accounts of fish circulation (1971, 1991), both books published by Cambridge University Press.

Hunn's compilation is useful: Hunn, J.B. (1967). *Bibliography on the Blood Chemistry of Fishes*. Washington, DC: U.S. Department of the Interior, Fish and Wildlife Service. Bureau of Sport Fisheries and Wildlife. Research Report 72.

Later material

The Hoar and Randall (editors have changed over the years) Academic Press series, *Fish Physiology*, has the following relevant volumes:

Hoar, W.S. and Randall, D.J. (eds) (1970). *Fish Physiology, Vol. 4. The Nervous System, Circulation and Respiration.* New York, NY: Academic Press.

Randall, D.J. (1970). The Circulatory System. In: W.S. Hoar and D.J. Randall (eds) *Fish Physiology: Vol. 4*. New York, NY: Academic Press, pp. 133–72.

Riggs, A. (1970). Properties of Fish Hemoglobin. In: W.S. Hoar and D.J. Randall (eds) *Fish Physiology: Vol. 4*. New York, NY: Academic Press, pp. 209–52.

All chapters relevant, but note:

Farrell, A.P. and Jones, D.R. The heart. In: W.S. Hoar, D.J. Randall (eds) Fish Physiology: Vol. 12A: The Cardiovascular System. New York, NY: Academic Press, pp. 1–88.

Steffensen, J.F. and Lomholt, J.P. The secondary vascular system. In: W.S. Hoar, D.J. Randall, and A.P. Farrell (eds) Fish Physiology: Vol. 12A: The Cardiovascular System. New York, NY: Academic Press, pp. 188–217.

Hoar, W.S., Randall, D.J., and Farrell, A.P. (eds) (1992). Fish Physiology, Vol. 12A: The Cardiovascular System.

Hoar, W.S., Randall, D.J., and Farrell, A.P. (eds) (1992). Fish Physiology, Vol. 12B: The Cardiovascular System.

All chapters relevant, but note:

Fànge, R. Fish blood cells. In: W.S. Hoar, D.J. Randall, and A.P. Farrell (eds) *Fish Physiology: Vol. 12B: The Cardiovascular System*. New York, NY: Academic Press, pp. 1–54.

Richards, J.G., Farrell, A.P., and Brauner, C.J. (eds) (2009). *Fish Physiology, Vol 27: Hypoxia*. New York, NY: Academic Press.

The D.H. Evans (ed.) series *The Physiology of Fishes* also has relevant chapters:

Farrell, A.P. (1993). Cardiovascular system. In: D.H. Evans (ed.). *The Physiology of Fishes*, 1st edn. Boca Raton, FL: CRC Press, pp. 219–50.

Olson, K.R. (1998). The cardiovascular system. In: D.H. Evans (ed.). *The Physiology of Fishes*, 2nd edn. Boca Raton, FL: CRC Press, pp. 129–56.

Olson, K.R. and Farrell, A.P. (2005). The cardiovascular system. In: D.H. Evans (ed.). *The Physiology of Fishes*, 3rd edn. Boca Raton, FL: CRC Press, pp. 119–52.

Gamperl, A.K. and Shiels, H. (2013). Cardiovascular system. In: D.H. Evans (ed.). *The Physiology of Fishes*, 4th edn. Boca Raton, FL: CRC Press, pp. 33–80.

CHAPTER 6

Gas exchange

6.1 Summary

Fish need oxygen for aerobic respiration; they also need to excrete carbon dioxide and (usually) ammonia.

In general, gases are obtained from or voided into the surrounding water, but the amount of gas held by the water will be temperature-dependent.

The capacity to exchange gases with the environment is enhanced by the structure of the gas-exchange surface which is usually and principally that of the gill surface.

Most fish have an active gill ventilation mechanism.

Some fish, notably in warm fresh water, are air-breathing, with additional gas-exchange regions which can include the skin, the anterior or posterior gut and lungs.

6.2 Introduction

The concept of gas exchange in animals usually revolves around their ongoing need for oxygen if, as is usual, they are aerobic, and they continually generate of carbon dioxide and ammonia, which can be very damaging if retained. The term 'gas exchange' applies because, at least for oxygen and carbon dioxide, the mechanisms which achieve intake of oxygen and elimination of carbon-dioxide gas are interrelated and occur at the same site. Alternatively or additionally, the concept of exchange arises because oxygen is obtained from the environment, and carbon dioxide, at least, is eliminated directly to it.

Like all other vertebrates, fish are fundamentally aerobic, requiring oxygen for cellular respiration and having both carbon dioxide and ammonia as excretory products of metabolism. Because most fish are aquatic, they must, in general, obtain oxygen from water, and oxygen availability varies with water temperature; increased temperatures mean less dissolved oxygen, and there is much less oxygen per unit volume of water than there is in air at normal elevations hospitable to vertebrate life (Appendix 6.1). Moreover, to obtain oxygen from water requires the passage of relatively dense water across the respiratory surface, usually gills in fish, which means a large effort and a relatively high energy expenditure in comparison to the lung ventilation mechanisms in terrestrial vertebrates.

However, an advantage of gas exchange with an aqueous medium and a continuous clearance system is that not only carbon dioxide can be eliminated but ammonia too, in comparison to terrestrial vertebrates, for which excreting ammonia directly into air is not normally an option. Ammonia in the lungs, especially those that retain a residual volume (mammals) would be too toxic; terrestrial animals therefore convert ammonia into a less hazardous form (e.g. urea, uric acid) for routing out through the kidneys and ureters. Fish as juveniles or under unusual circumstances (e.g. burrowing) may resort to urea excretion.

Hence the gills in fish are termed 'multifunctional' (Evans et al. 2005), having a much wider range of functions than the tetrapod lung, including not only gas exchange for respiration and nitrogenous excretion but also osmoregulation and other facets of homeostasis such as acid-base regulation (7.5.3).

Gill ventilation mechanisms in fish usually involve active muscular pumps but some pelagic

Essential Fish Biology: Diversity, Structure and Function. Derek Burton & Margaret Burton.
© Derek Burton & Margaret Burton 2018. Published 2018 by Oxford University Press.
DOI 10.1093/oso/9780198785552.001.0001

fish use ram ventilation whereby they just swim through the water with open mouths.

Although the primary site of gas exchange in most fish is gills, there are accessory or auxiliary gas-exchange surfaces, including the skin, air bladder/lungs and parts of the posterior or anterior gut, notably the buccal cavity.

Most fish are aquatic and use water for gas exchange but some fish can be 'air breathers', an advantage if the habitat is subject to high temperatures and drought.

6.3 Major gases exchanged by fish

Oxygen required for tissue respiration is essential for all fish although requirements vary with season, temperature and activity levels. Carbon dioxide, a product of fish respiration, is eliminated at the gas-exchange surface, as is ammonia, a product of degradation of nitrogenous molecules. Oxygen is obtained from the environment via a special gas-exchange surface which in fish is usually the gills (6.5), though it may be other regions including skin, air bladder/lungs or regions of the gut (6.6). In most fish oxygen is then transported from the gas-exchange surface in the erythrocytes (5.13.1) loosely bound to haemoglobin (6.5.7, 6.5.8) and delivered to the tissues.

The reverse transportation occurs for carbon dioxide, formed in the tissues and transported in the blood, mainly as bicarbonate ion (6.5.7), synthesized in the erythrocytes under the influence of the enzyme carbonic anhydrase. At the gas-exchange surface carbon dioxide is released, for example from the bicarbonate, again under the catalytic effect of carbonic anhydrase, present in the erythrocytes and (with some dissension, see Weihrauch et al. 2009) perhaps in the tissues. Some elasmobranchs (Gilmour et al. 2001) may have plasmatic carbonic anhydrase. Some vertebrates are able to transport carbon dioxide in carbamino format, bound to haemoglobin, dependent on the availability of amino acid side chains for binding. It has been reported that teleosts have low levels of carbamino (carbamate, Jensen 2004) because the relevant amines are acetylated and/or there is competition with organic phosphates. In either case the binding site(s) may be blocked. According to Jensen (2004), some elasmobranchs have less amino-acid acetylation and therefore may have a bigger role for carbamino carriage of CO_2.

As ammonia is toxic (Ip and Chew 2010) in low concentrations, its transport and removal at the gas-exchange surface must be rapid and efficient. Earlier views of ammonia transport across the gills were summarized as a review by Wilkie (1997), who later (2002) described passage of ammonia 'across the branchial epithelium' as being accomplished by 'passive diffusion' for freshwater fish. The movement of ammonia has been thought of as commonly occurring in the ionic form (NH_4^+) and the formation of ammonium ion (NH_4^+) from gaseous ammonia (NH_3) is expected to be promoted at the gas-exchange surface by acidification linked to the release of protons (H^+) at the formation of bicarbonate (HCO_3^-) during the hydration of carbon dioxide.

However, the accounts of ammonia transport have recently been vigorously revised (Weihrauch et al. 2009) following the discovery of specific transport mechanisms including the occurrence of trans-membrane transporters (methylammonim/ammonium permeases and ammonium transporter proteins, Weihrauch et al. 2009). Ip and Chew (2010) suggest that passage of ammonia across cell membranes may occur via aquaporins (7.5.6). Revisions of ideas on ammonia excretion have included the realization that urea may be more often excreted in fish than was previously expected (Weihrauch et al. 2009, also 8.3.2).

6.4 The gas-exchange surface

Throughout the animal kingdom, gas-exchange surfaces have a set of characteristics which facilitate the rapid uptake (oxygen) or loss (carbon dioxide, ammonia) of gases by diffusion. A typical gas-exchange surface is thin, separating the blood or tissue fluid from the external environment by a few micrometres (μm). Most gas-exchange surfaces are also characterized by much folding, increasing the area for gas exchange. Gas-exchange surfaces are moist, and in terrestrial forms are usually inverted to protect the moist environment. Although external gills do occur in some larval forms (bichirs) a degree of protective inversion and covering is normal for fish. Generally, gas-exchange surfaces, particularly for large animals, have an accompanying ventilation mechanism, forcing air or water across the surface.

6.5 Gills

All fish have gills and can use them for gas exchange although for most fish the gills are of little use if the fish becomes stranded on land, because the gills collapse and their gas-exchange surfaces are not available to the air. Exceptional fish however, such as anguillid eels, can still use the gills in air, inflating the gill cavity with air 'about once a minute' (Schmidt-Nielsen 1990). However, for eels in air, most (about 66 per cent) of the gas-exchange role is taken by the skin, whereas in water this role is reduced to about 10 per cent (Schmidt-Nielsen 1990). Typical fish gills are a bilateral series, with highly folded, very thin epithelium supported by cartilage or bone (as gill arches with gill rays, Wells 1961) and permeated by a rich blood supply direct from the ventral aorta via the afferent branchial arteries. The gills are separated by clefts in the pharynx wall which enable water taken in at the mouth to flow across the gills. The occurrence of pharyngeal/gill clefts/slits is fundamental to the vertebrate taxon, at least during ontogeny (development) and is seen in protochordates like Amphioxus.

6.5.1 Gill filaments: primary and secondary lamellae

The gill tissue of fish is organized in a series of elongated gill filaments or primary lamellae supported by the gill arch (Figure 4.2), and approximately at right angles to it so that sections longitudinal to the filaments tend to cut the gill arch transversely (Figure 6.1). Each gill filament usually has additional folding as two (opposite) rows of secondary lamellae (Figures 6.2 and 6.5). Each gill arch normally supports two rows of primary filaments (the 'holobranch' Romer 1955), each row being a 'hemibranch'.

6.5.2 The pharyngeal clefts

The number of pharyngeal openings—clefts or slits—in extant jawed fish is usually expressed as five, with six or even seven each side in some elasmobranchs (hexanchiform sharks, e.g. *Hexanchus* with six, *Heptranchias* with seven) as well as an anterior spiracle (see also 6.5.3), taking the form of a modified gill slit, in chondrichthyans and sturgeon. The chimaeras and relict polypterids have only four gill slits (Romer 1955), most teleosts have five per side, but the deep-sea *Eurypharynx* has an extra slit each side (Tchernavin 1947b). The spiracle may have some gill tissue and be reduced relative to the more posterior gill slits or be well developed as in skate where it serves as an incurrent structure.

In counting gill-arches currently functioning as real gill supports morphologists may (or may not) take into account the evolutionary origins of the jaws and jaw support structures usually regarded as derived from the first and second gill arches on each side. Thus, counts can start at the jaws ('gill

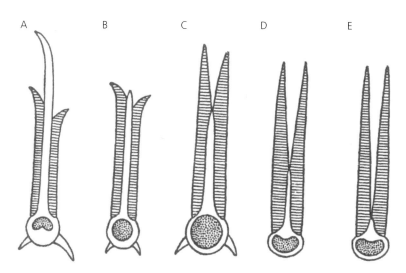

Figure 6.1 Cross-section of gill arches of A, an elasmobranch; B, chimaera; C, sturgeon; D and E, two bony fishes. showing longitudinal sections of filaments (after Greenwood 1975).

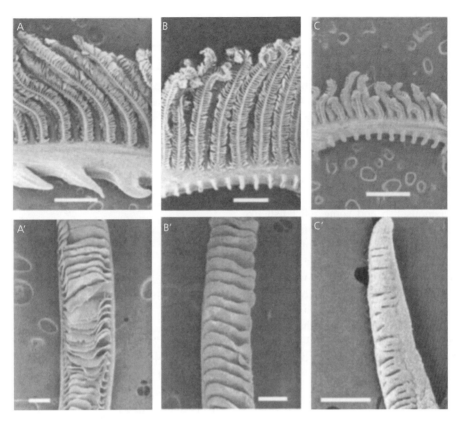

Figure 6.2 Comparison of gross morphological features of the gills in three sympatric gobies. Gill samples are from the third branchial arch. Top rows (A, B, C) are the branchial arches of individual species, and the representative filaments are shown below (A′, B′, C′). Scale bars (A, B, C) 1 mm; (A′, B′, C′) 200 µm. Reproduced from T.T. Gonzales, M. Katoh and A. Ishimatsu (2008), Respiratory vasculatures of the intertidal air-breathing eel goby, *Odontamblyopsis lacepedii* (Gobiidae: Amblyopinae). *Environmental Biology of Fishes*, volume 82, 341–51. With permission of Springer.

arch′ 1) or after the hyoid (phylogenetically 'gill arch' 2). The gill supports collectively may be referred to as the 'branchial basket', a flexible structure (Wells 1961) enabling ventilation movements.

Commonly, each gill arch has gill filaments on each side, the holobranch condition (Romer 1955); the last gill slit in most species has only one hemibranch anteriorly. Dogfish and sturgeon have a spiracle (closed in most adult teleosts, Goodrich 1930) with a reduced and atypical gill hemibranch (see pseudobranch, 6.5.3) anteriorly, and holobranchs for the remaining gill arches, nine hemibranchs each side. The lungfish by contrast may have fewer functional gills; *Protopterus* (the African lungfish) has no spiracular gill tissue, a somewhat reduced hemibranch on the anterior surface of gill slit 1, no gill tissue lining gill slit 2, or the anterior surface of gill slit 3 but a posterior hemibranch; gill slits 4 and 5

are fully furnished. Gill slit 5 for most fish, however, only has the anterior hemibranch. These variations are clearly depicted in Figure 219 of Romer (1955).

The gill arrangement is generally such that when water passes through the pharyngeal slits it has maximum exposure to gill tissue. In contrast to the African lungfish, however, with its gill surface reduction, the Australian lungfish (*Neoceratodus*) has a full complement of gill tissue for gill slits 2, 3 and 4 with a total of approximately 8.5 hemibranchs per side compared to approximately 5.3 per side for *Protopterus*; that is, each species includes a short hemibranch in its total. Teleosts (e.g. the perch *Perca*) have a hemibranch for the first and last gill slit and both hemibranchs for gill slits 2 to 4 with a total of 8 hemibranchs.

Most fish (Holocephali included) have a gill cover (operculum) on each side but the elasmobranch (sharks, rays) have a series of five to seven un-covered gill slits on each side (sharks, dogfish) or ventrally (the flattened skates and rays, Figure 6.3). The elasmobranch gills, though with no overall gill cover, have a septum on each gill arch which can partially occlude the gill slit (Figure 6.1A).

Synbranchids (swamp-eels) have a single gill opening on the ventral side of the head (Allen et al. 2002; Nelson et al. 2016).

6.5.3 Spiracle and pseudobranch

Most cartilaginous fish and some relict fish like sturgeons have an additional small, round opening, the spiracle on each side, generally anterior to the gills. This spiracle may have an active role in wa-ter flow across the gills as in skate, in which water is pumped into the dorsal spiracle (Figure 6.3) and out through the gills; the ventral mouth (Figure 6.3) is not appropriate for water intake in these species.

The spiracle may contain a gill-like structure, the pseudobranch, a structure that is deduced not

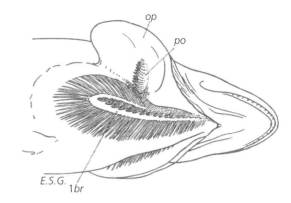

Figure 6.4 Location of pseudobranch of *Salmo salar*, after Goodrich 19300. 1br = first branchial arch; op = operculum (raised); po = pseudobranch.

to have a gas-exchange role because of the way its blood supply runs, being already fully equilibrated through the first gill in elasmobranchs (Satchell 1999). A pseudobranch also occurs within the oper-culum of many teleosts (Figure 6.4) but its function, though much studied, is not clear (Bridges et al. 1998; Mölich et al. 2009), though it may be involved with the blood supply to the eye, promoting oxygen de-livery to the retina via the choroid *rete* (5.10, 12.3.5).

6.5.4 Fine structure of the gills

The gas-exchange surface of each gill consists of two rows of thin filaments (primary lamellae), which

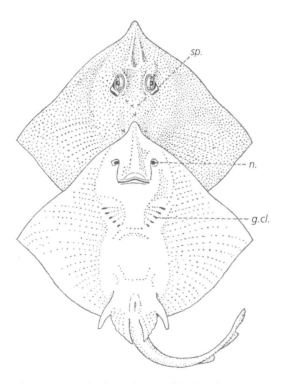

Figure 6.3 Dorsal and ventral aspects of the skate (from Greenwood 1975); g.cl = gill cleft; n = nostril; sp. = spiracle.

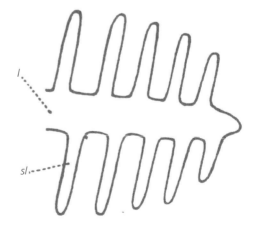

Figure 6.5 Disposition of secondary lamellae (sl) on a filament (l), after Goodrich (1930).

have secondary lamellae (also known as platelets, Gray 1954) on each side (Figure 6.5), increasing the surface area for diffusion. The secondary lamellae are more numerous in active fish; the frequency varying from 10 to 40 per mm of filament (Hughes 1982).

The microscopic structure of the gills shows a very thin epithelium ('pavement' epithelium) covering the lamellae. Particularly towards the base of the secondary lamellae may also be seen large chloride cells (Figure 6.7). These eosinophilic mitochondria-rich cells are considered to be involved in ion regulation (7.7, 8.3.7), an important feature of the gills being that they are a site vulnerable to osmotic flux with water and ion uptake or loss. Pillar cells (Wright 1973, Laurent 1982), pilaster cells (Goodrich 1930) help to support the lamellae (Figure 6.6) in conjunction with the blood passing through and the flow of blood can be controlled by sphincters in the vasculature. In the elasmobranchs the afferent branchial arterioles are linked by a capillary network within the filaments and each arteriole also links to the corpus cavernosum (Wright 1973) forming a reservoir of blood in spongy tissue between the filaments. The role of the corpus cavernosum is not clear (Metcalfe and Butler 1986). For a 'typical' fish (and *Latimeria*), but not lungfish, Hughes (1999) describes a gill circulation with arterio–venous anastomoses (omitting capillaries) which connect 'efferent lamellar arterioles' with the central venous sinus.

6.5.5 Blood supply

The blood supply to the gills occurs via the short ventral aorta and a series of paired afferent branchial arteries; five pairs in the dogfish (Figure

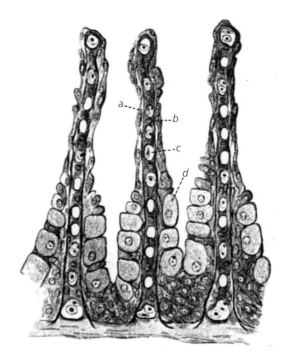

Figure 6.7 Chloride cells in fish gill epithelium. Reproduced with permission from A. Keys and E.N. Willmer (1932) 'Chloride secreting cells' in the gills of fishes, with special reference to the common eel. *Journal of Physiology* 76, 368–80. ©Wiley-Liss Inc. Note thin 'pavement' cells (a). b = pilaster cell; c = blood channel; d = chloride cell.

6.8), but with the first and second branching off a single 'innominate' artery on each side. In teleosts, there are four pairs of afferent branchial arteries.

The blood from the gills flows into a series of paired efferent branchial arteries en route to the dorsal aorta for distribution of oxygenated blood to the head and body. There are four pairs of efferent branchial arteries in the dogfish (Figure 6.8) and in teleosts (Figure 4.1), the latter having a bifurcation and reunification of the dorsal aorta, the *circulus cephalicus* (Figure 4.1), forming a loop above the gills which, in different species, receives all or some of the efferent branchial arteries (Goodrich 1930).

6.5.6 Countercurrent flow

Blood flow to the gills occurs under relatively high pressure; the blood passes very close to the surrounding water, separated from it only be the pavement cells of the gill lamellae. Van Dam (1938) is reported (Muir 1970) to have noticed that the water

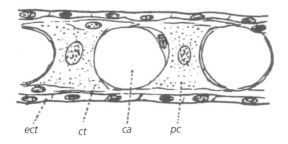

Figure 6.6 Pillar (pilaster) cells separate the secondary lamellar surfaces. Diagram after Goodrich (1930). ca = 'capillary'; ct = 'cutis'; ect = pavement cell; pc = pilaster cell.

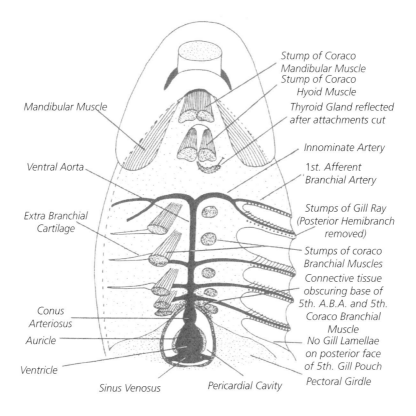

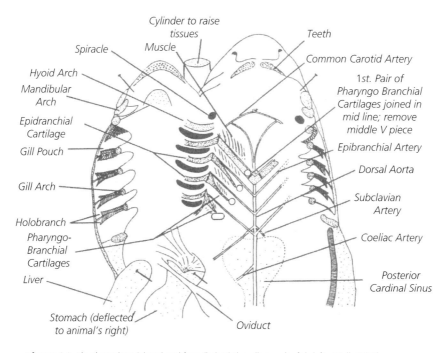

Figure 6.8 Blood supply to (above) and from (below) the gills in a dogfish (after Wells 1961).

flow across the gills ran against (counter to) the direction of (incoming) blood flow, thus improving the exchange of gases by maintaining a diffusion gradient (Figure 6.9). This 'countercurrent exchange' is regarded as important in non-respiratory contexts as well (e.g. the swim bladder *rete* 5.10, and Scholander 1954). In 1952, Hazelhoff and Evenhuis showed experimentally that the countercurrent ('opposite') flow of water past the gills of tench was much more effective in removing oxygen from water in comparison to parallel currents.

6.5.7 Oxygen gain and carbon dioxide loss

Oxygen is gained at the gills, from the water, by diffusion into the blood across the thin epithelial cells, and becomes loosely bound to haemoglobin in the erythrocytes to form oxyhaemoglobin (6.5.8). Carbon dioxide is lost to the water at the gills also by diffusion and with the intervention of the enzyme carbonic anhydrase (CA) which promotes the interconversion of carbon dioxide and carbonic acid. The CA is in the erythrocytes and may also be plasmatic in some elasmobranchs (Jensen 2004). However, it has been pointed out by Randall and associates (e.g. Lessard et al. 1995) that fish tend to build up lactic

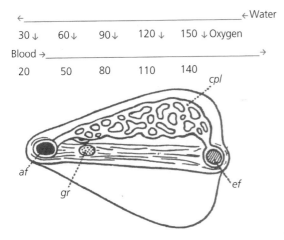

Figure 6.9 Composite diagram of the circulatory system in a fish gill lamella, showing the incoming (af) afferent vessel, linked to the 'capillaries' (cpl) with the efferent outflow (ef). Above are direction of blood and countercurrent water with oxygen values in mmHg. Gill diagram from Goodrich (1930). Oxygen values from Schmidt-Nielsen (1990).

acid in their white muscle and that additional acid (due to the action of plasmatic CA) would likely be disadvantageous. For vertebrates, in general increased levels of carbon dioxide (higher acidity) can reduce the affinity of haemoglobin for oxygen (Bohr effect) thus promoting the release of oxygen, for example, at the tissues. The reciprocal (Haldane) effect applies; deoxygenated blood carries more CO_2 (see Jensen 2004). In some teleosts there is a 'Root' effect, with increased acidity having more effect than in other vertebrates. Schmidt-Nielsen (1990) links the Root effect to oxygen secretion in the swim bladder. Interestingly, teleost blood may not achieve full saturation even in very high oxygen environments apparently because there can be two haemoglobins, one acid sensitive, one acid insensitive, the latter enabling oxygen carriage even when lactic acid levels are high following extreme muscle activity (Schmidt-Nielsen 1990).

The reverse diffusions occur at regions of low oxygenation, such as the peripheral capillaries, with carbon dioxide gained by the blood and oxygen lost to the tissues.

6.5.8 Respiratory pigments, haemoglobin and myoglobin

Many animals, including vertebrates, have pigments which promote the temporary attachment of oxygen. These pigments include haemoglobin and myoglobin found in most vertebrates. Both these pigments have a haem (heme) group which encloses a ferrous (iron) ion and are also linked to a protein globin sequence. The haem moiety is a porphyrin which resembles chlorophyl although the latter has magnesium, not iron, centrally. The porphyrin has a symmetrical structure with four sub-units around the central metal. Most vertebrates also have four of the porphyrin units in their haemoglobin (i.e. are tetrameric), although cyclostomes have only one when oxygenated but associating as dimmers or teramers when deoxygenating (de Souza and Borilla-Rodriguez 2007).

Whereas haemoglobin, whether vertebrate or invertebrate (e.g. in earthworms), is expected to occur in the circulation to assist oxygen uptake and transport, vertebrates also have a very similar molecule, myoglobin, which is concentrated in muscle, particularly the red muscle. Myoglobin is not as well studied as haemoglobin: its role appears to be acting as a reservoir for oxygen; it does not normally circulate.

Although haemoglobin has a very important role in oxygen dynamics there is an interesting exception for the Antarctic icefish, Family Channichthyidae (5.13), some of which lack red blood cells in the circulation. The trade-off appears to be a matter of obtaining an improved circulation under ultracold conditions, in which case erythrocytes might interfere with the circulation as cold temperatures increase plasma viscosity. The cold Antarctic water will hold more dissolved oxygen than warmer water, about twice the level for 30 °C (Appendix 6.1), and hence the icefish can cope with a reduced or absent respiratory pigment.

The presence of respiratory pigments like haemoglobin may increase the carrying capacity for oxygen by as much as tenfold or more, at 20 °C (Table 15.3 in Withers 1992); for cyclostomes, the increase is only about twofold, for chondrichyans about sevenfold and for osteichthyans up to thirtyfold (Withers 1992).

6.5.9 Gill area

Calculations of the gill area available for gas exchange have produced values much greater for pelagic fish like albacore (*Gymnosarda alleterata*) or mackerel (*Scomber scombrus*), in comparison to benthic species such as *Lophius*, the angler, or flatfish like *Pseudopleuronectes americanus* (Gray 1954). Hughes (1966) confirmed Gray's findings. Wegner et al. (2010) state that tunas have the largest relative gill area of any fish group. Large gill surface area is achieved by longer and increased numbers of gill filaments, and more numerous and longer secondary lamellae with a shape enabling tight packing of filaments (Wegner et al. 2010). Gislason and Thorarenson (2006) showed considerable differences for gill area between populations of char (*Salvelinus alpinus*).

6.5.10 Ventilation mechanisms and control

Rhythmical movements achieve passage of water across the gills with a control mechanism (respiratory centre) in the medulla oblongata of the brain (11.6.2) as indicated originally for goldfish (Adrian and Buytendijk 1931). The respiratory centre has a neurogenic drive (Rang and Dale 1991) as a central respiratory rhythm generator, RRG (Taylor et al. 2010), acting as a pacemaker. However, fish ventilation can be discontinuous (6.5.11) and highly modified (14.10. 2), and the pacemaker and its connections therefore capable of achieving intermittent action (in comparison with the pacemaker of the heart). Fong et al. (2009) discuss the central rhythm generator in vertebrates including factors which result in episodic breathing, referring to the rhythm generator as 'conditional'. The switch to ram ventilation (6.5.11) is presumed to occur via the responses of mechanoreceptors associated with the respiratory muscles and gill arches (Randall 1982, Taylor et al. 2010), and changes in rhythm generally are ascribed to arterial oxygen receptors on the gill arches (Taylor et al. 2010).

Oxygen is more difficult for aquatic animals to obtain compared with those on land; its presence or absence is the primary stimulus for ventilation (Gilmour 1998) rather than the carbon dioxide in air breathers (6.6).

For fish like lung-fish with two separate possibilities for obtaining oxygen (i.e. air or water) it has been suggested that there are two separate RRGs (Taylor et al. 2010).

For a gill-breathing fish there are usually a series of active events achieving intake and then pushing intake water across the gills. The pump mechanisms are rhythmic and continuous although the rate can be adjusted to metabolic need. Unfortunately, as need increases so does the cost of muscular contractions and the cost could approach 30 per cent of the total oxygen consumption (Schumann and Piiper 1966, quoted by Schmidt-Nielsen 1990, who questions this estimate). Limiting factors for gas exchange are generally considered to concern oxygen which is much less soluble in water than is carbon dioxide (Schmidt-Nielsen 1990).

Water routing for gill access depends, for most fish, on a one-way flow into the mouth across the buccopharyngeal cavity and through the pharyngeal clefts/slits. At that point the water may be in an opercular/branchial cavity, that is, under the bilateral gill covers or opercular (one each side) as in bony fish and Holocephali (chimaeras) or in individual parabranchial cavities as in elasmobranchs. These latter cavities can be controlled by individual septa at the end of each gill slit (Figure 6.1A). The usual ventilation pump system depends on bucco-pharyngeal muscle contractions. Overall the system has been described as two pumps; a bucco-pharyngeal force pump and a suction pump, a two-phase system for many fish but which is more accurately described as overlapping in action; two pumps slightly out of phase so that continuous water flow is possible. However, as Hughes (1960) described it for marine teleosts, having measured pressure changes, some fish have a dominant suction pump and others have a dominant force pump. Two types of 'chambers' are involved, the bucco-pharyngeal and the bi-lateral branchial chambers, with a one-way flow-through system, usually in at the mouth, across the gills via the gill (branchial/visceral) clefts or slits and out laterally on each side.

Recent analyses of fish ventilation have been due to Hughes and Shelton, whose work has been reviewed by Wegner and Graham (2010) and Randall (2014). As usually described, the overall ventilation mechanism works as follows: the initial (conventional 'start') phase begins (Figure 6.10) with the mouth opening, and the buccal and opercular (branchial) cavities (bony fish) or the buccal and parabranchial cavities (elasmobranchs) enlarging, thus drawing water in (suction pump) through the mouth, with the operculum or gill slits closed. The suction effect can be enhanced by the action of the ventral branchiostegal area ('throat' region) in bony fish, which may be substantial: there may be up to 51 pairs of supporting skeletal rays (Farina et al. (2015).

The next (second) phase occurs (Figure 6.10) with the mouth closing; the cavities contract (muscles in the walls contract), forcing water (force pump) across the gills, and the opercular flaps or gill-slits open to let the water out.

The mouth may not actually close in this cyclical activity, but the aperture can be periodically occluded by valves (Dahlgren 1898), described as 'velar folds' of the buccal mucosa (Saunders 1961).

The diagram (Figure 6.10) shows the oesophageal opening changing from open (a) to closed (b). However, Grove and Newell (1953) show it partly closed in both phases and Goodrich (1930) shows it open throughout. Villee et al.'s work (1984) show it as illustrated here. In humans (Shier et al. 2002), the anterior diameter of the oesophagus can be changed by external musculature, inferior constrictor muscles.

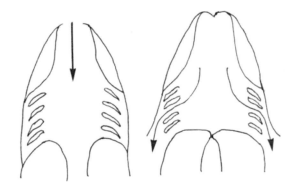

Figure 6.10 The two-phase ventilation mechanism, after Dahlgren (1898), Grove and Newell 1953, Villee et al. 1984.

The oesophagus can be very short in fish and posteriorly there may be a sphincter (Barrington 1957) which could prevent inspired water from entering the main gut.

6.5.11 Ram ventilation

Some very active fish just swim through the water with their mouths open and do not use pumps at this time. This system of ram ventilation has been described for tuna and mackerel which may be unable to use pump ventilation and are therefore obliged to swim continuously (Roberts 1975), making it difficult to hold some fish in small tanks. Obligatory ram ventilation avoids the potentially high cost incurred by active fish if using the usual muscular pumps of regular ventilatory mechanisms. As Steffensen (1985) puts it, these scombrids (now obligatory ram ventilators) have transferred the cost of oxygenation to their swimming muscles. Other fish that are in swift flowing currents could also avoid pumping by stationing themselves appropriately with open mouths (Steffensen 1985). Interestingly, the Mississippi paddlefish *Polyodon* is reported to use ram ventilation which fits well with its planktivory (Berra 2007). Although some pelagic fish must continuously swim in order to survive there is also evidence that some teleost species, such as trout or remoras (e.g. the sharksucker, *Echeneis naucrates*) can utilize either pumping or ram ventilation (Berra 2007). Steffensen (1985) found that a switch to ram ventilation in rainbow trout was accompanied by a drop in oxygen consumption.

Some active sharks from the Orders Myliobatiformes and Carchariformes also use ram ventilation, some intermittently, while other species are obligate ram ventilators. The latter include the very active lamnids, sphyrnids and carcharinids (Carlson et al. 2004).

Ram ventilation is discussed by Brainerd and Ferry-Graham (2006).

6.6 Air-breathing fish

The high volume of oxygen in air (Appendix 6.1) may be available to fish, though accessing it exposes fish to aerial predation and they need special adaptations to obtain it.

Most fish cannot survive out of water because their gills collapse in air, although anguillid eels have gill systems that can still function in air, though minimally, and the skin is also used for gas exchange, notably when these fish migrate across land. Similarly, some South American swamp-dwellers such as *Synbranchus* have stiff gills, allowing the gills to function, and the fish to migrate, on land (Eduardo et al. 1979). Fish like *Gillichthys mirabilis* (a Californian inshore goby) can live for hours out of water, just as can the few 'terrestrial' fish (13.5.5).

The ability to air-breathe using a variety of gas-exchange surfaces is so advantageous in hypoxic environments and seasonal drought that it has apparently evolved independently in many fish clades notably in warm-water fish (as oxygen solubility lessens as temperatures rise). Thus, although all fish have gills which can be used for gas exchange in water, many fish (particularly freshwater tropical species) have auxiliary gas-exchange systems enabling use of air. Risks attached to becoming air-breathing at the water surface include the vulnerability to aerial predators especially if visits to surface water are obligate.

Air-breathing fish are the subject of a monograph by Graham (1997) and a special journal issue (Lefevre et al. 2014); the concept of 'bimodal' breathing is presented in Fernandes et al. (2012). Overall about 450 fish species can access air-derived oxygen (Lefevere et al. 2014). Bimodal breathers can obtain oxygen from water or air, with some fish species capable of complete emergence while others only have brief surfacing behaviour.

Terrestrial species (13.5.5), which spend more time on land than in water, are rare. Fish subject to periodic drought conditions may actually aestivate, as occurs with the African lungfish *Protopterus* (Hochahka and Guppy 1987) and *Lepidogalaxias salamandroides* (Berra et al. 1989). In such cases metabolism will drop to a very low level but the requirement, source and routing of oxygenation is not clear (Berra et al. 1989).

6.6.1 Lungs

A few fish have what are regarded as genuine ventral lungs, with finely divided internal surfaces, connected to the external environment by a duct from the anterior gut. Lungfish, of course, are the prime examples of such fish; these fish live in warm fresh water in which oxygen solubility can be low; the ability to air-breathe is a definite advantage in such conditions. However, the belief that lungfish are obligate air-breathers has been challenged (Green et al. 2006) for *Protopterus aethiopicus* in Lake Baringo, Kenya, a body of water which is described as normoxic. Romer (1955) argued that, contrary to earlier beliefs, swim bladders were not 'progenitors of the lung'. Lungs, thought to have evolved in early bony fish, are found in *Polypterus*, which has paired ventral lungs. *Latimeria* retains a vestigial ventral 'lung', fat filled.

Schemes of evolutionary trends (Romer and Parsons 1986) show dichotomous lines of evolution (Figure 6.11) from a *Polypterus*-like basis, with one line proceeding to the fully developed tetrapod lung, and the other to the swim bladder, with this bladder still linked to the gut (physostomatous) or, ultimately, separate from the gut (physoclistous) and dependent on the blood for gas. Ilves and Randall (2007), however, consider the notion of which evolved first to be irrelevant as multiple functions could be expected.

6.6.2 The swim bladder used for respiration

The swim bladder (air bladder), a common dorsal structure in bony fish, provides an adjustable buoyancy system, a hydrostatic organ (3.8.5) but may also be an 'accessory air breathing organ' (Romer 1955) as in *Lepidosteus* and *Amia*. Adult *Arapaima gigas* (Figure 1.19) become dependent on the well-vascularized swim bladder (termed a lung by Hochacka et al. 1978), the gill lamellae atrophy as the fish grows (Fernandes et al. 2012; Ramos et al. 2013).

6.6.3 The gut used for respiration

Accessory gas-exchange areas used for air breathing include the pharyngeal epithelium as in the mud-skipper *Periopthalmus*, the buccal cavity epithelium as in the electric 'eel' *Electrophorus* (Kramer et al. 1978), parts of the anterior (stomach) or posterior (intestine) gut, which is described from several South American swamp-dwelling species (*Doras, Plecostomus, Hoplosternum, Ancistrus*) and from the Old World *Misgurnus*. Kramer et al. (1978) summarize the situation for Amazon-region catfish such as callichthyids using the intestine while loracariids use the stomach for air-breathing. Lefevre et al. (2014) comment on the question of balancing the two functions (respiration and digestion), which do seem somewhat incompatible, particularly when respiration is associated with the intestine.

The four-phase cycle of *Hoplosternum littorale*, which enables use of the intestine for breathing, is described by Jucá-Chagas and Boccardo (2006). As the air, in this case, is a flow-through system, taking surface air in at the mouth it is difficult at first consideration to reconcile this with feeding but the fish in this report were being fed and the description states that air is forced out of the anus as new air is taken in at the mouth. This suggests that the gut contains intermittent air bubbles and food boli.

6.6.4 The lips used for respiration

Some fish, notably South American characins, have a lower lip which becomes hyper-developed (Winemiller 1989) to aid air-breathing when water is hypoxic. The swollen lower lip of the tambaqui (*Colossoma macropomum*) and the matrinchã (*Brycon melanopterus*) is also described in terms of accessing the oxygen-rich upper layer of water (Almeida-Val 1999). Development of the swelling of the lower lip can occur very rapidly, and disappear quickly too; two to three hours has been recorded, according to Winemiller (1989) quoting Braum and Junk (1982).

Reid et al. (2003) quote Saint-Paul (1988) as describing the tambaqui's swollen lower lip acting as a scoop for the oxygen-rich surface water, directing it down past the gills. Swollen lips ('blubber lips') have also been reported in the Australian genus *Hephaestus*, grunters (Pusey et al. 2004; Morgan 2010), but this phenomenon, which is common in *H. tuliensis*, has been associated with feeding substrate, for example improved access between stones, rather than aerial respiration, although surface feeding has also been reported and in this situation aerial gas exchange might occur. Swollen extensible lips are also reported for an African cichlid, *Haplochromis euchilis* (Fryer and Iles 1972, see 4.4.4), but associated with feeding rather than gas exchange.

6.6.5 The skin used for respiration

Anguillid eels are the prime examples of fish able to use the general cutaneous surface for air-breathing when out of water. However, skin may also be used as an important gas-exchange surface by some fish when submerged. Steffensen et al. (1981) list several species for which cutaneous gas exchange is significant, including their principal study fish, the flatfish *Pleuronectes platessa* (plaice), which, when buried in sediment, may derive about 25 per cent of its oxygen via the skin. This source of oxygen becomes more important (rising to about 37 per cent) in conditions of lowered oxygen (hypoxia).

6.6.6 Facultative and obligate air-breathers

Broadly speaking, air-breathing is available to a surprisingly large number (450, Lefvre et al. 2014) of extant fish species, for which it may be facultative (most) or obligate (rare), the latter situation notably applying to those few species in which a period of obligate air-breathing enables survival. Air-breathing is generally a phenomenon of freshwater fish from warm/hot regions like the tropics, or of small inshore/littoral marine fish such as some gobies. For many air-breathing fish, access to the air means swimming to the water surface and partial emergence, taking air into the mouth or exposing the lower lip (Winemiller 1989), and is termed 'aquatic surface respiration' or ASR. This risky behaviour, which must occur at frequent intervals, exposes fish

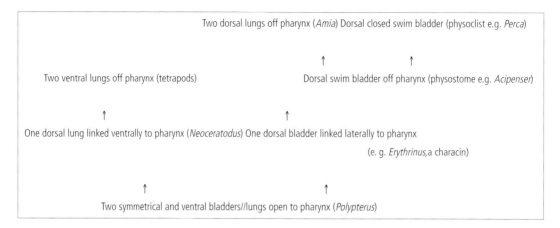

Figure 6.11 Scheme of evolution of lungs and swim bladder (after Romer and Parsons 1986).

to a new set of predators; however, as water levels are probably low at the same time, they might well be vulnerable to those predators anyway.

The blood supply and drainage associated with gas-exchange surfaces may be capable of adjustment with shunts (Hughes and Munshi 1979), particularly where fish can use alternate regions. Most fish use anterior structures such as gills, the branchial cavity or lungs for gas exchange in which case the anterior aorta provides the afferent vessels. If the swim (air) bladder is used then the coeliac artery is afferent.

The ultimate air-breather is perhaps *Protopterus*, the African lungfish, which has the reputation of being able to aestivate out of water in a cocoon for months (or even up to nine years, Hochachka and Guppy 1987) when drought occurs. The metabolism and hence the oxygen requirements are minimal in this situation.

6.7 Problems with oxygen and carbon dioxide levels in the aqueous environment

6.7.1 Hypoxia and hypercapnia

Warm water typically has less oxygen than cool water (Appendix 6.1, Tables 6.1, 6.2). Some aquatic systems have low oxygen (hypoxia) and/or high carbon dioxide (hypercapnia/hypercarbia), either as a fairly permanent situation or seasonally or sporadically. Such a situation may limit access by

fish (and other organisms) or elicit adaptations to the situation, as can notably occur with seasonal or sporadic hypoxia (Kramer et al. 1978) who review 'oxygen deficiency' in the Amazon.

In the oceans there are at least two large hypoxic regions (Herring 2002), in the northern Indian Ocean and the central east tropical Pacific. In each case the hypoxic area lies below a highly productive region, and excreta, detritus, etc., which descend become oxidized, thus reducing available oxygen for respiration of any organisms persisting in this region, the 'oxygen minimum layer' (OML). In the case of fish, some myctophids (lanternfish), the cod-let (*Bregmaceros nectabanus*) and some scopelarchids (pearlfish) are typical of the OML (Herring 2002).

Winter hypoxia is a recurring problem for ice-covered boreal lakes (Danylchuk and Tonn 2003). Depending on the population densities, which at high levels may exacerbate the problem, 'winterkill' may occur. As large-bodied species tend to have less tolerance for hypoxia, small species may dominate in lakes with this potential stressor (Danylchuk and Tonn 2003).

Both hypercapnia and hypoxia have been the subject of intense physiological studies from laboratories in Canada and elsewhere. As shown in Brauner et al. (2000), fish transportation (a necessary adjunct of some aquaculture enterprises) may be accompanied, particularly when in high density, by potentially damaging or lethal results even when water is oxygenated to compensate for predicted hypoxia. Hyperoxia (oxygen above ambient) and/or hypercapnia produce acidosis, which may elicit changes in gill structure (Brauner et al. 2000). Hypercapnia may also temporarily reduce blood flow to the gut (Farrell et al. 2001). Brauner and co-workers (2004) studied hypercapnia in an Amazon catfish (*Liposarcus pardalis*), comparing the acid-base regulation both intra- and extracellularly. They found that this catfish had excellent intracellular regulation but 'limited' extracellular control, thus regulating the cellular pH preferentially.

Somewhat the reverse of hypoxia occurs when environmental conditions increase atmospheric gas uptake by the gills, which may damage the fish especially if the environmental gas reaches 'supersaturation' levels.

6.7.2 Supersaturation

It has become recognized that fish and other aquatic animals can be afflicted with 'gas bubble disease' (GBD) analogous to the 'bends' which may occur with human divers. This condition in aquatic animals is related to situations where water acquires a 'supersaturated' status with atmospheric gases including nitrogen, sometimes due to extremely turbulent water as occurs below dams. It was also attributed to algal blooms (Woodbury 1942), in which case a higher oxygen level than usual was reported (algal blooms may be toxic to fish for a number of reasons, e.g. 'BOD', biological oxygen demand, 6.7.3). Various government reports, notably from the United States with respect to dams, have outlined occurrence of GBD (Ebel and Raymond 1976; Dawley 1996; Weitkamp 2000; Beeman et al. 2003), and the possibility that fish may avoid it by seeking deeper water if available. Nitrogen supersaturation may be physiologically more difficult for fish as they do not metabolize gaseous nitrogen. The reports generally agree that the small bubbles which form in the affected fish tissues, and may be observed as small pearl-like spherules in the fins, can cause death by blocking the circulation.

Supersaturation can also occur in tank systems, in which fish may be exposed to varying temperatures and cannot adjust their depth to any great degree. It is usual for water to be 'de-gased' by running down

specially designed columns where there is any perceived risk of GBD.

6.7.3 High oxygen demand

Where there are large numbers of organisms in the environment competition for oxygen can cause difficulties, for example, where there are 'algal blooms' or high levels of microorganisms as occurs in sewage. This biological oxygen demand may cause problems for fish, reducing oxygen availability. Run-off from agriculture or high levels of phosphates from detergents, which also gives rise to high levels of algal or micro-organism growth, termed eutrophication (16.6.3), has been particularly worrying in some highly developed regions.

Active fish, which tend to swim continuously for extended periods of time, may have extreme sensitivity to oxygen levels. Such fish would include the scombrids (mackerel, tuna) and the salmonids (salmon, trout). Some salmonids, such as char, are particularly sensitive to changes in water temperature and the accompanying shift in available oxygen.

Warm water has less dissolved oxygen than cooler water and saline water has less dissolved oxygen than fresh water. Schmidt-Nielsen (1990), using data from Krogh (1941), shows declining values for salt and freshwater oxygenation at temperatures from 0 °C to 30 °C, with 10.29 ml oxygen per litre at 0 °C and 6.57 ml/l at 20 °C in fresh water. Equivalent seawater oxygenation values are 7.97 ml/l at 0 °C and 5.31 ml/l at 20 °C. Thus, the amount of available oxygen at higher temperatures is considerably less than in cool water; char (considered to be glacial relicts) do not tolerate temperatures much above 20 °C although this may be due to susceptibility to proliferative kidney disease, PKD (Brown et al. 1991). Their natural environmental limits are such that they do not occur south of latitude 64°N, and their optimum temperature for growth is 12–16 °C (for Lake Windermere in the United Kingdom char, Swift 1964). Eggs of Canadian char are killed at temperatures exceeding about 8 °C (Scott and Crossman 1973). Rainbow trout by contrast tolerate higher temperatures, spawning at 10–15.5 °C (Scott and Crossman 1973), and they report that as long as water is well oxygenated these fish can do well in habitats exceeding 21 °C.

Appendix 6.1 Oxygen availability

The amount of oxygen which can dissolve in water (Tables 6.1, 6.2) depends on both temperature and other dissolved substances present. In particular, the amount of oxygen will vary inversely with temperature, warmer water holding less oxygen, and also inversely with, for example, ions such as chloride present. Therefore, sea water at a specified temperature will hold less oxygen than fresh water at the same temperature, and any water will hold more oxygen at lower temperatures:

Table 6.1 The comparative volumes of oxygen in water and air, under standard conditions, after Schmidt-Nielsen (1990). Oxygen concentration as litre per litre, at 15 °C.

Water	Air	Ratio, water to air
0.007 (i.e. 7 ml O_2 per litre)	0.209 (i.e. 209 ml O_2 per litre)	1:30

Table 6.2 Oxygen solubility in water under different temperatures and chlorinities (after Krogh 1941, Hoar 1983).

Chlorinity (g* per kg water)	Temperature, °C	Oxygen solubility at saturation (ml) per litre
0	0	10.29
0	30	5.57
20	0	7.97
20	30	4.46

*Total amount of chlorine, bromine and iodine. A value of 20 for chlorinity is the amount expected for full-strength seawater, which comprises about 3.5 per cent salts by weight (Schmidt-Nielsen 1990).

Bibliography

Brown, M.E. (ed.) (1957). *The Physiology of Fishes, Vol. I.* New York, NY: Academic Press.

Carter, G.S. Air breathing. In: Brown, M.E. (ed.) (1957). *The Physiology of Fishes, Vol. I.* New York, NY: Academic Press, Part 2, pp. 65–80.

Fry, F.E.J. The aquatic respiration of fish. In: Brown, M.E. (ed.) (1957). *The Physiology of Fishes, Vol. I.* New York, NY: Academic Press, Part 1, pp. 1–64.

Hoar, W.S. and Randall, D.J. (eds) (1970). *Fish Physiology, Vol. IV: The Nervous System, Circulation and Respiration.* New York, NY: Academic Press.

Hoar, W.S. and Randall, D.J. (eds) (1984). *Fish Physiology, Vol. 10A: Gills-anatomy, Gas Transfer, Acid-Base Regulation.* New York, NY: Academic Press.

Houlihan, D.F., Rankin, J.C., and Shuttleworth, T.J. (eds) (1982). *Gills.* Cambridge: Cambridge University Press.

Hughes, G.M. (1970). A comparative approach to fish respiration. *Experientia*, 26, 113–224.

Milsom, W.K. and Perry, S.F. (eds) (2012). New insights into structure/function relationships in fish gills. *Respiratory Physiology and Neurobiology*, 184, Issue 3, pp. 213–346.

Richards, J.G., Farrell, A.P., and Brauner, C.J. (eds) (2009). *Fish Physiology, Vol. 27: Hypoxia.* Amsterdam: Elsevier.

Wegner, N.C. and Graham, J.B. (2010). George Hughes and the history of fish ventilation: from Du Verney to the present. *Comparative Biochemistry and Physiology A*, 157: 1–6.

CHAPTER 7

Metabolism, homeostasis and growth

7.1 Summary

Like all other vertebrates, fish have a complex relationship with their environment with requirements for an external source of food and the ability to stabilize their internal conditions.

Most fish are not able to regulate their temperature actively but can regulate their ionic and osmotic composition, as well as the levels of metabolically important molecules like glucose.

Resistance to extremes of heat or cold may be aided by behavioural or metabolic adjustments including the production of antifreeze in cold conditions.

An important organ for many metabolic activities is the liver which may also act as an energy store.

Energy storage is also provided by the white muscle and the hypodermis as well as mesenteric fat.

Many fish have indeterminate growth and some species can reach very large sizes.

7.2 Introduction

Fish function under a very wide variety of conditions, with the main criterion being the ability to undergo all the necessary and complex interactions of a relatively large multicellular aquatic animal. Living organisms, including the tiny protozoans (*Amoeba*, etc.) and bacteria, have a basic set of characteristics which include the following: nutrition (Chapter 4), excretion (Chapter 8), respiration (Chapter 6), metabolism (current chapter), homeostasis (current chapter) and reproduction (Chapter 9), and sensitivity (Chapters 11 and 12), while most animals also show locomotion (Chapter 3), that is, they can move from place to place as juveniles if not adults (there are 'sessile' animals, e.g. corals). Growth (current chapter) is also a characteristic of living organisms though it is very limited in the microscopic forms.

Nutrition for animals involves heterotrophy, obtaining complex molecules from external sources (4.2), whereas excretion (8.4) often involves a set of special organs filtering potentially toxic wastes and eliminating them into the environment, which for aquatic species can occur through a gas-exchange surface, also available for obtaining oxygen for cellular respiration (6.4, 6.5).

Metabolism is the set of cellular processes which sustain the living organism by promoting the use or storage of molecules to maintain physical structure and activity. Metabolism is controlled in animals by the endocrine system (Chapter 10) and the nervous system (Chapter 11) and ultimately determined by the DNA inherited at each reproduction. It is controlled at the cellular level by enzymes and feedback in a modulated and stable internal environment, the process achieving this stability being homeostasis.

7.3 Fish metabolism

Metabolism is inherently a very complex subject as it comprises the myriad chemical processes necessary to sustain life. Most of what is understood of animal metabolism has been initially achieved through the biochemical study of mammals, particularly humans and rats. Comparative studies on other animals, including fish, demonstrate some general principles, including the need for

Essential Fish Biology: Diversity, Structure and Function. Derek Burton & Margaret Burton.
© Derek Burton & Margaret Burton 2018. Published 2018 by Oxford University Press.
DOI 10.1093/oso/9780198785552.001.0001

homeostasis (regulation of the internal environment) to provide a basis for efficient metabolism.

Although fish metabolic pathways are generally similar to those of mammals except in excretion (8.3), and the blood chemistry in part reflects that similarity, there are some marked differences in some facets of growth and osmoregulation as well as the role of white muscle in energy storage (7.12.3). Special differences too reflect yolk deposition as when females have high levels of yolk protein precursor (vitellogenin) in their blood. Elasmobranch metabolism is markedly different from that of teleosts, the former largely depending on ketone bodies and amino acids for oxidative fuels (Speers-Roesch and Treberg 2010).

Metabolism can be regarded as the sum of molecular dissolution (catabolism) and molecular construction (anabolism). Catabolic processes release energy (i.e. are exergonic) including heat, while anabolic processes use energy (i.e. are endergonic) often accessed via the conversion of adenosine triphosphate (ATP) to adenosine diphosphate (ADP). Fish like all other vertebrates are thought of as aerobic, using oxygen as a necessary part of their 'tissue respiration', the catabolism of food-derived molecules to provide energy, for example in ATP for anabolism. However, the white muscle which forms a major part of fish body mass for most species is anaerobic, which could mean interpretive problems when metabolism is measured during any short time interval as rate of oxygen consumption, even though oxygen debts incurred would be expected to be reflected eventually by increased oxygen use.

The study of metabolism often includes an estimate of 'metabolic rate' (MR) during low, moderate or high activity, in which MR may be measured in terms of oxygen use. For comparison between fish species or with other vertebrates a measure of 'resting' state MR may be attempted; this may be referred to as the 'standard' metabolic rate as compared to the basal metabolic rate of mammals (Hoar 1983). As the internal temperature of fish is not regulated (7.4) its MR reflects the environmental temperature.

Fish may show continuous activity (pelagic and shoaling species) or very little movement (benthic species like anglers, e.g. *Lophius*). The estimates of metabolic rate 'at rest' may therefore be a concept difficult to support, but possible with fish like trout

which show both quiescent and more active states, measurable in an enclosed system. Salmonids like trout do not have their major muscle mass as anaerobic white muscle so that measurement of metabolic rate is not complicated by the involvement of this tissue but anaerobosis may still occur for short periods of burst activity.

Ballantyne (1997) has discussed what he regards as the special case of elasmobranch metabolism, including their reported resistance to cancer. The unusual metabolism of elasmobranchs has been discussed by Speers-Roesch and Treberg (2010).

7.4 Temperature and metabolism

Most fish are termed poikilotherms, animals that do not regulate their body temperatures, as opposed to homeotherms such as mammals and birds, which maintain a body temperature around a set point. However, a few very active fish do maintain some parts of their body at higher temperatures than the rest, and thus are heterotherms.

Fish may be in fairly stable warm (coral reef) or cool (polar/benthic deep ocean) conditions, year round, or may endure considerable seasonal, or even daily, fluctuations particularly in shallow waters. In the case of seasonal changes in temperature fish, like invertebrates, may be capable of adjustments so that after days or weeks at a higher, or lower, temperature the fish may be operating at a different metabolic rate from that initially measured shortly after the change. It is possible that in such cases there are different isozymes (various forms of a single enzyme) appropriate for differing conditions (Shaklee et al. 1977; Lin and Somero 1995), with a delay before the new set of enzymes is synthesized. Adjustments to changes in environmental conditions may take place over a period of time and are referred to as acclimatization in the natural environment and acclimation under experiment in the laboratory.

Laboratory experiments comparing cellular function at different temperatures for a cool-water marine fish (winter flounder, *Pseudopleuronectes americanus*) with a freshwater tropical fish (sailfin molly, *Poecilia velifera*) found considerable differences in responses to noradrenalin (Burton 1988b). In this case, the comparisons did not utilize any acclimation but compared function under a range of temperatures for each species. Interestingly, in each case the optimum for the response time was higher than the normal temperature range, but much higher in the tropical fish (40 °C) than the cool-water fish (20 °C).

For many fish species living in surface waters, the metabolic rate (MR) varies considerably with external temperature and some fish species become torpid at low temperatures or very high temperatures when accompanied by drought (estivation). High-temperature responses may include production of heat-shock proteins (Hsps), shifts in sites of gas exchange as hypoxia increases, and air breathing may be an option (6.6). Hsps are expected to act as 'molecular chaperones' (chaperonins), protecting proteins from denaturation. The overall role of Hsps in stress responses (10.12.7) was reviewed by Feder and Hofmann (1999), though without specific reference to fish other than gobies like *Gillichthys* (Dietz and Somero 1992). This fish may be unable to escape the heating of shallow water but are well-adapted to such conditions including surviving in temperatures with 30 °C variation and are thus termed 'eurythermal'. These gobies were reported to show acclimatization with reference to the temperature at which an Hsp was induced.

Cold-water fish such as Antarctic notothenoids have a suite of adaptations including the production of antifreeze (glycoprotein/peptide) and a metabolic rate higher than would be expected for the temperature range (DeVries and Eastman 1981). Antifreeze production can also occur in Arctic and sub-Arctic species (Kao et al. 1986; Fletcher et al. 1989).

Low-temperature responses, which may include production of special antifreeze glycoproteins and antifreeze proteins (AFPs, reviewed for teleosts by Fletcher et al. 2001) in the liver for release to the blood or within, for example, the skin, or increased blood glycerol (Driedzic et al. 2006), enable fish to live in ice-laden seawater at a temperature of −1.9 °C in the case of Antarctic species (DeVries and Eastman 1981). Eastman et al. (1979) reported renal conservation of antifreeze glycoprotein (AFGP) in the aglomerular kidney of Antarctic eelpout. Other compensatory changes, for example, increased MR in cooler conditions to maintain activity, may occur during periods of acclimatization. Fish which live in highly variable conditions such as shore habitats and estuaries or shallow tropical seas or lakes are tolerant of a wide range of temperatures, that is, they are eurythermal, whereas fish in more stable conditions, such as the Northern lakes or the deeper sea habitats, are not so tolerant of temperature change (notably increase), that is, they are stenothermal. Some fish respond to extremes of temperature by entering condition of torpor. Cunners (*Tautogolabrus adspersus*) become torpid in the winter off Newfoundland (Green and Farwell 1971), and at the other extreme, in conditions of drought the African lungfish *Protopterus* may aestivate for years (Hochachka and Guppy 1987).

Low temperatures may produce localized mortality, 'winterkill'. This has been reported for cunners (Green 1974) and for pike (Casselman 1987).

Low-temperature-adapted fish in fresh water include coregonids (whitefish) and char; the Arctic char (*Salvelinus alpinus*) is circumpolar in distribution and grows well at 12–16 °C, with eggs having an optimal temperature < 8 °C (Giles 1994). The term 'glacial relicts' has been applied to both Arctic char and the powan (*Coregonus clupeioides* or *lavaretus*) which may have become 'land-locked' in deep, cold lakes and unable to migrate to sea for their adult phase, as many of the char and whitefish do. The burbot (*Lota lota*), an atypical gadid found in northerly fresh water, is reported to prefer water < 7 °C and to feed 'voraciously' in winter (Giles 1994), which contrasts with the euryhaline winter flounder of the North-west Atlantic, which has a winter fast off Newfoundland (4.8 and Fletcher and King 1978).

Some large marine fish are able to maintain parts of their bodies at higher than environmental temperatures and are termed heterotherms or regional endotherms, or red muscle endotherms (Sepulveda et al. 2005). The heightened temperatures are generally achieved by deep red muscle and/or *retia* (5.10). Fish with heterothermy include 13 species of tuna, some of the large lamnid sharks and one of the three thresher shark species (Katz 2002; Sepulveda et al. 2005).

High temperatures can give problems with oxygen consumption, especially in feeding fish. As with almost all other animals, fish are ingestive heterotrophs, requiring an external source of complex molecules. In order to survive, grow and reproduce, fish need to take in molecules either through food or via the gills as oxygen, and use these molecules to build fish- specific structures. Food is usually taken in as complex material which has to be rendered into a simpler format so that it can be absorbed into the fish plasma before restructuring can occur. The breakdown of food in the gut requires the production of many enzymes (4.4.7), a metabolically expensive process, and competing requirements for oxygen as temperature increases can give rise to problems (ascribed to specific dynamic action or effect SDA/SDE, 4.8) if captive fish are fed, for example, to satiation during such a change. In short, if a fish feeds to satiation at a high temperature the oxygen requirements may not be met and the fish may die, a situation that can develop in tank culture. Ballantyne (2016) briefly discusses SDA, observing that SDAs for both teleosts and elasmobranchs show about two to three times an increase over resting oxygen consumption, but the elasmobracnhs have lower rates overall, suggesting more efficient digestive processes in elasmobranchs.

7.5 Homeostasis and relations with the aquatic environment

The chemical internal environment of animals is regulated as part of homeostasis. In particular, ionic levels and pH (acidity), as well as circulating nutrients such as monosaccharides, are usually held close to a set point, though rapid, physiologically convenient, deviations may occur under special circumstances (as when frogs become hyperglycaemic when very cold, thus resisting freezing damage, Storey and Storey 1985). In vertebrates, ionic components of the tissue fluid and blood plasma are finely regulated, usually with hormonal control of specific constituents. Most fish need to actively regulate their ionic composition, as loss or uptake of water and ions, such as sodium (Na^+), calcium (Ca^{2+}) and chloride Cl^-, is a continuing problem, occurring at the gills, by intake at the mouth and to a lesser extent elsewhere.

7.5.1 The marine environment

Seawater, whether coastal or oceanic, is expected to have a high level of sodium chloride, and fairly high levels of some other inorganic ions such as

magnesium and sulphate, with an overall salinity of about 3.5 per cent (Schmidt-Nielsen 1990). The salinity varies somewhat, diminishing in regions close to large estuaries, and capable of being hypersaline (4.8 per cent) in mangrove regions (Frick and Wright 2002b). Tidal areas and the littoral (shore) zones are highly variable in the water composition and organisms living there are generally euryhaline, that is, capable of responding rapidly and appropriately to salinity changes (7.7). Organisms living in offshore habitats are more likely to be stenohaline with less capacity to adapt to changing salinity in the environment. Fish which migrate to or from the ocean, with a freshwater phase, may undergo a preparatory stage before the move is accomplished (see smoltification, 7.9).

Marine fish may need to counteract the tendency of ions such as Na^+ and Cl^- to load the internal environment, and the tendency of water to leave the fish. Marine elasmobranchs have a special mechanism to deal with the potential problems of water balance; they avert it by synthesizing and retaining particular organic molecules (7.5.9), thus altering the relationships of the body fluids relative to seawater. Marine teleosts have a different physiological mechanism (7.5.10).

7.5.2 The freshwater environment

The composition of fresh water varies considerably and different proportions of possible components will affect fish differentially. Water with high levels of small clay particles is common in some areas, and the particles may remain in suspension or be deposited as sediment. Even though water may be recognized as 'fresh', with the implication that it is appropriate for human consumption, its mineral composition and acidity may be quite variable even within a small geographic area. Fresh water is modified by the nature of the substrates in the region, thus calcareous rocks (chalk, limestone) contribute large quantities of Ca^{2+}, producing water known as 'hard'. Peat and/or decaying vegetation contributes organic acids, and water therefore with a low pH, popularly known as 'soft'. Acidity levels vary considerably; although it is usual to think of water as having a neutral pH of 7, fresh water is quite commonly acid or very acid, with pH values as low as

4.7 or even 4 (1.4.2, 7.5.3) still supporting some fish species such as brook trout (*Salvelinus fontinalis*, Alexander et al. 1986) or yellow perch (*Perca flavescens*, 1.4.2), though in general, high acidity is associated with depauperate ecosystems.

Freshwater fish tend to gain water (by osmosis, 7.5.5) and therefore need to void it. Freshwater fish deal with water which may be highly acidic, calcium-rich or calcium-poor, and fish in shallow water may need to cope with extremes of temperature and hypoxia (6.7).

7.5.3 Acid-base regulation

A particular facet of homeostatic ionic control is 'acid-base' regulation. Water is generally regarded as 'neutral' with a pH of 7, and body fluids including plasma could be expected to function best at neutrality, though local variations can be advantageous (6.5.7). Mammalian plasma is usually reported as having a pH just above 7 (Altman 1961). Seawater has a pH of somewhat greater than 7, for example in the range of 7.1 to 7.8 (Lorenzen 1971 quoted by Cushing and Walsh 1976). Conditions shifting pH in either direction from the near-neutral norm within organisms are termed either acidosis or alkalosis and can be regarded as deleterious (see hypercapnia, 6.7.1). However, active metabolism shifts pH as an ongoing process and the freshwater environment in particular can be prone to either stable extremes (particularly acidic) or temporary fluxes, after run-off or flood conditions. To some extent the bicarbonate ions and large molecular components of the blood can act as buffers (Appendix 7.1), thus modifying the effect of proton accumulation. Goss et al. (1992) discussed acid-base regulation in freshwater fish in the context of ion movement at the gills and the role of the chloride cells (7.7).

Lakes in individual counties in Nova Scotia were found to be generally acidic, but not in every case (Alexander et al. 1986). Of 45 lakes measured in Yarmouth County, 3 were neutral (pH 7) and 2 were alkaline (pH 7.5, 7.9). The lowest value reported in this county was pH 4 for Beaver Creek Lake, which even so had fish in it (yellow perch). However, highly acid fresh water is at risk of further acidification by environmental events such as 'acid rain'

and there has been considerable concern about this possibility.

'Blackwaters' are very common in 'humid equatorial' regions (Johnson 1968). Although often referred to collectively and generally considered to be acidic, exceptionally, the blackwater flows across limestone, moderating it towards neutrality. A comparison of the Amazonian blackwaters (e.g. the Rio Negro) and the contrasting white water of the Rio Solimôes is provided by Kramer et al. (1978). These two systems collectively form the Amazon River system with more fish species (approximately 1300 species by 1967) than any other river complex. The area is notable for extreme flooding with considerable ensuing variations in water conditions.

7.5.4 The effects of different salinities and osmotic movement of water

Fish and other organisms have to contend with different environmental salinities and the ensuing movement of ions down their concentration gradient and, in particular, the tendency of water to cross membranes from regions of high water concentration, for example fresh water, into an animal, or vice versa in salt water the tendency of water to leave and thus dehydrate an organism.

Studies of organism responses to changes in environmental salinity therefore include considerations of concentrations in surrounding water, and the responses vary if organisms have cell walls. Organisms like plants and bacteria have fairly rigid cell walls as opposed to the more pliable situation with animal cells, which generally only have a thin flexible and permeable cell membrane surrounding each cell. Organisms with cell walls in addition to the cell membrane can resist the osmotic effect of fresh water and avoid cell lysis due to swelling when water flows in due to osmosis. The maximum effect is turgid cells. In animals, however, unless osmoregulation occurs in the internal environment or at the cellular level, cells will burst due to osmotic swelling if surrounded by a dilute, solute-poor solution.

Studies of solutions relative to passage through membranes have generated various terms. The cell membrane is usually regarded as 'semipermeable', allowing passage of water and small ions and/or small molecules. If cells like mammalian erythrocytes are placed in a simple sodium chloride solution it is relatively easy to obtain different values which produce rupture (lysis), or no change in volume, or shrinkage (crenation). Thus, distilled water would be expected to produce lysis, 0.9% sodium chloride might not show a change in volume but 1% sodium chloride might show crenation. Schmidt-Nielsen (1990) crtiticizes the term '%' often used (as here) for different concentrations; for practical purposes, however, these terms can be widely useful and are generally taken to mean g/ml; that is, a 1% NaCl solution would be 1 g NaCl in 100 ml water. In such circumstances terminology applied is 'isotonic' for both cells and surrounding saline where there is no change in cell volume, while the terms 'hypotonic' and 'hypertonic' are used as follows: if the cells shrink the surrounding medium is said to be hypertonic to the cells; if the cells lyse the surrounding medium is said to be hypotonic to the cells. A similar set of terms apply to the usual more complex situation where a set of ions and other molecules exert an osmotic effect; the terms then are is(o)osmotic, hypoosmotic and hyperosmotic, each being used comparatively where two (sometimes complex) systems separated from each other by a semipermeable membrane, are compared.

The combined effect of different osmotically active particles such as ions or molecules are directly additive, that is, colligative and the 'osmotic pressure' exerted can be measured by depression of the freezing point and also expressed as 'osmolality' where one osmole (osM) produces a freezing point depression of 1.86 °C (Hoar 1983).

7.5.5 Fish and osmosis

The agnathan marine hagfish are apparently isoosmotic with seawater (Nelson 1976) by virtue of their ionic composition, as are many marine invertebrates, and will therefore not lose or gain water and not undergo active osmoregulation. The holocephalan *Chimaera* has been described as having a high concentration of plasmatic sodium chloride (Fänge and Fugelli 1963), which may eliminate the need for osmoregulation. Some fish (marine elasmobranchs) are approximately isoosmotic with seawater because their plasma includes high levels of

osmotically active, but not ionic, compounds (see osmoconformers, 7.5.8). Teleosts, whether marine or freshwater, and freshwater elasmobranchs have plasma which is not isoosmotic with the surrounding water and therefore need to react physiologically to continuous water influx (fresh water) or efflux (seawater).

7.5.6 Aquaporins: a route for water movement?

The usual view of aquatic life is that aquatic organisms are subject to osmotic movement of water as a passive situation which is then responded to, or which, in an evolutionary sense, has been partially resolved by adjustment of internal ions and molecules as a counteracting (7.5.9) situation. Recently, however, the movement of water has been assessed with reference to a group of molecules termed 'aquaporins' which may facilitate the transcellular movement of water as described by Kwong et al. (2013) for freshwater teleosts. Cutler and Cramb (2000) were interested in the situation with eels in the same context, and reported that water permeability was greater in freshwater- than seawater-acclimated eels, and that chloride cells (7.7) were fewer in freshwater-adapted animals. The same authors later (2002) were not able to explain why water transport across the gills is higher in freshwater eels as compared to seawater-acclimated eels. They suggest the possibility that water transport is a consequence of solute transport being a principal role of the aquaporin investigated.

7.5.7 Plasma regulation and the erythrocytes

Fish plasma constituents are regulated by a number of different processes similar to those of tetrapods but, unlike avian or mammalian erythrocytes, it is possible to demonstrate that some fish species have erythrocytes capable of resistance and/or volume regulation in osmotic challenge. Any tendency to shrink or swell may be mitigated, an important feature where blood cells are exposed to varying osmotic conditions at the gills. Thus, fish erythrocytes are more robust than those of some other vertebrates facing an osmotic challenge, an advantage for blood cells passing through the thin surface of the gills.

The osmotic fragility (5.13.5) for erythrocytes of five Amazonian fish species was measured by Kim and Isaacks (1978); they found two species (*Arapaima gigas* and *Electrophorus electricus*) susceptible to haemolysis but the arawama (*Osteoglossum bicirrhosum*) and the armoured catfish *Pterygoplichthys* were somewhat resistant, while the lungfish, *Lepidosiren paradoxa*, was very resistant. Garcia-Romeu et al. (1991) described the response of trout erythrocytes to osmotic swelling; the recovery of normal volume was accompanied by a loss of amino acids, notably taurine. The subject of animal cell volume regulation has been reviewed by Lang and co-workers (1998). For practical purposes it is often assumed that fish can cope well, with minimal osmoregulation necessary, in an environment roughly equivalent to one-third seawater (see Appendix 7.2) where seawater has a total 'salinity' of approximately 35 g/kg of water of which by far the highest non-aqueous constituents are sodium and chloride (Vernberg and Vernberg 1972). Prosser (1952) considered seawater to have a salinity equivalent to 3.4% NaCl and a chlorinity of 1.9%. For experimental purposes, where there is a need to keep fish tissue viable, special saline solutions may be used to minimize morbidity or for longer-term culture (Appendix 7.3).

The set-point levels of plasma ionic constituents may, however, vary seasonally as contributors, for example, to antifreeze conditions, with, Na^+ and Cl^-, for example, rising in the coldest season (Fletcher 1977) which should depress the freezing point. Similarly plasma glycerol may rise at such times, for instance in smelt (Raymond and Driedzic 1997).

Although plasmatic constituents can be expected to include a regulated amount of particular common ions like the monovalent Na^+ and Cl^- as well as the divalent Ca^{2+}, Mg^{2+} and SO_4^{2-} there have been reports of considerable variation, for example of SO_4^{2-} which may be utilized as a preferential osmolyte by freshwater eel (Nakada et al. 2005; Kato et al. 2009).

7.5.8 Osmoconformers and osmoregulators

Aquatic organisms, including fish, tend to be in one of two groups; osmoconformers which have body fluids, etc., similar in osmotic pressure to their environment which applies to some marine fish (elasmobranchs), and osmoregulators, which in fresh water maintain an internal environment with osmotic pressure higher than that of their environment or, which in seawater teleosts, have a plasmatic composition lower in salinity and osmotic pressure.

The osmotic pressure of the internal environment may be raised by the presence of organic molecules, osmolytes (7.5.9) maintained high by synthesis and retention. Yancey (2001, 2005) has reviewed organic osmolytes which may be regulated to maintain cell volume.

Most chondrichthyans, such as sharks and the like, maintain their blood at a high osmotic pressure with osmolytes such as urea at high levels, and this obviates the need for intense osmoregulation, cutting down ionic flux at the gills. Such fish are osmoconformers, but Treberg et al. (2006) refer to marine elasmobranchs as osmoconforming/osmoregulatory.

Most bony fish (e.g. teleosts), however, have blood plasma at osmotic levels approximating to one-third that of seawater but much higher than fresh water. Fish which live in fresh water therefore have a tendency to continually take in water through osmosis at the gills whereas bony fish in seawater have an opposing tendency to lose water. Freshwater fish are necessarily osmoregulators and so are marine teleosts, although whereas the former needs to lose incoming water the latter need to acquire it.

7.5.9 Osmolytes and counteraction

In comparison to bony fish and a few freshwater rays the marine chondrichthyans, which maintain high plasmatic levels of urea, require 'counteracting osmolytes', usually methylamines, typically trimethylamine oxide (TMAO) or, in the case of the holocephalan *Callorhincus millii* (elephant fish) probably betaine (the methylamine glycine betaine, Bedford et al. 1998). The methylamines are counteracting/compatible osmolytes in that they can protect proteins including enzymes from the destabilizing effect of urea. Yancey et al. (2002) reported increasingly high levels of TMAO with increasing water depths for marine fish and invertebrates.

This situation was concluded to be protective of the deleterious effects of high pressures on proteins.

7.5.10 Marine teleosts and osmoregulation

Marine teleosts have a potential continual water deficit and are expected to drink seawater in order to compensate for water osmotically lost at the gills. Excess salts must then be removed, for example at the gills or, particularly for some divalent ions, elsewhere such as at the kidney. Chloride cells (7.7) on the gills or operculum of marine fish have effectively reverse polarity to those functioning to import salts in freshwater environments. Marshall (1995) diagrams the two situations. While Cl^- accumulates and is subsequently removed from chloride cells through apical anion channels, Na^+ is reported to be removed paracellularly, for example between a chloride cell and a pavement cell.

7.5.11 The coelacanths' osmotic relations

The coelacanth has high urea levels (Pickford and Grant 1967) and TMAO present, at approximately a 3:1 ratio urea/TMAO in blood, and approximately 4:3 in muscle (Robertson 1975 quoting Lutz and Robertson 1971, and Griffith et al. 1974). Thus, the coelacanth resembles the marine chondrichthyans, avoiding osmotic difficulties rather than having to acquire water against an adverse osmotic gradient.

7.5.12 Freshwater fish and osmoregulation

The few chondrichthyans which are freshwater, such as *Potamotrygon* spp. or euryhaline, for example the Atlantic stingray, *Dasyatis sabina* (Janech et al. 2006), and the bull-shark *Carcharinus leucas* (Reilly et al. 2011) have osmoregulatory mechanisms somewhat similar to freshwater teleosts. *Potamotrygon* has low urea levels and the Atlantic stingray can markedly increase its kidneys' glomerular filtration rate and urine output under low salinity conditions (Janech et al. 2006).

Whereas marine teleosts have a potential continual water deficit, freshwater fish need to continually remove water from their bodies. Loss or gain of water other than at the gills is not so problematic; the skin is in general waterproof although some loss

or gain can occur with feeding, notably in the buccal region. Just as water loss or gain needs continual adjustments also salts/ions can be lost or gained, and in fresh water, ionic composition and organic acid presence may vary considerably in different water systems. In the Amazon, at Manaus, 'black' water (from the large Rio Negro) and white waters merge, and the ensuing complex of environments may limit distribution of some species. However, individual fish or species may tolerate short exposure to differing osmotic conditions quite well and can exude mucus to limit rapid ionic uptake or loss. In species which have a migratory habit there may be short periods of adjustment (acclimatization), enabling morphological and/or physiological changes preparatory to movement. Passive influx and efflux at the gills (and elsewhere) will depend on the relative concentrations in the environment and the blood/tissue fluid, as well as permeability of the contact tissues. Both freshwater and more saline environments will provide inbalance and concentration gradients with entry or loss of water and salts as ions, including calcium Ca^{2+}, sodium Na^+, potassium K^+, and chloride Cl^-. Both water and inorganic ions will be gained by feeding. Influx of 'salts' and water will be offset by subsequent elimination.

7.6 Mechanisms for water and ionic regulation in freshwater fish

Removal of water by freshwater fish, which are continuously at risk of plasmatic dilution by osmotic uptake of water at the gills, is effected via the kidneys, reported to produce a copious dilute urine, measured in lampreys by micropuncture of the kidney tubule (Logan et al. 1980), with the output estimated as 337 ml/kg/day, in comparison to values of about 1–11 ml/hr/kg reported for teleosts (8.5.3).

Passive loss of salts at the gills needs to be counteracted by active uptake by, for example, the chloride cells (mitochondria-rich cells) which (7.7) may have a very complex role in homeostasis.

7.7 Chloride cells

The chloride cells, are also known as mitochondrial-rich (MR) cells or ionocytes (2.3.1). These cells are particularly prevalent on the gills being present in freshwater and marine fish. They are found in both elasmobranchs and teleosts and are reputed capable of exporting or importing specific ions with the use of ATP to help effect ion transport. These cells can be detected at the base of the gill secondary lamellae (Figures 6.7 and 7.1); the cells are large and eosinophilic, with plentiful mitochondria and a characteristic structure including an apical pit (Karnaky 1986). The presence of these cells has been known for decades; their mechanism of action has been studied and their capacity for increase under environmental change has been reported; the increase is believed to impede respiratory gas exchange (Perry 1997). Perry also considers that the chloride cells of freshwater fish import Ca^{2+} and Cl^- while the thin gill 'pavement cells' may be capable of taking in Na^+ via a proton pump. The chloride cells are also believed to have an important role in acid-base regulation (7.5.3) with

a Cl^-/HCO_3^- exchange system (Perry 1997). Hulbert et al. (1978) summarized the established view of chloride cells as occurring in two forms, 'inactive' in fresh water and 'active' in seawater, with the implication that one was transformed to the other in changed environmental conditions, with a change in the 'directional pumping rates of Na^+ and K^+'. These authors point out, following Fenwick (1976), that Ca^{2+} influx increases in fresh water in comparison to seawater-adapted fish (in this case, anguillid eels).

7.8 Marine chondrichthyans and ionic regulation

Marine chondrichthyans (the majority of this taxon), which counteract the possibility of osmotic water loss by maintaining their plasma at high osmotic levels via urea and TMAO, may still have a problem with passive gain of salts, producing a need to export salts. Such removal can occur at various locations, which may include the gills, the kidney and the rectal gland (8.4.4) or rectal gland equivalent in holocephalans (Hyodo et al. 2007).

7.9 Euryhaline and stenohaline fish

Animals which are euryhaline tolerate or actively accommodate varying salinities. The term 'stenohaline' refers, in contrast, to animals with poor tolerance of salinity change. If indeed chloride cells operate one way in seawater and in the other direction in fresh water, the question of reversing polarity or recruiting different MR cells may be relevant, especially with fish living in environments undergoing frequent and rapid changes.

Fish which are euryhaline include those normally inhabiting estuaries where ionic composition of the environment fluctuates daily. Other fish which over a lifetime are capable of surviving varying salinities include those which are migratory, and such fish may make physiological and morphological adjustments to either fresh or salt water, as their migrations take them into different conditions. Some migrating fish undergo a complex set of changes; this is true of salmonids which undergo 'smoltification' in preparation for movement from fresh to salt water—changes include silvering of the integument

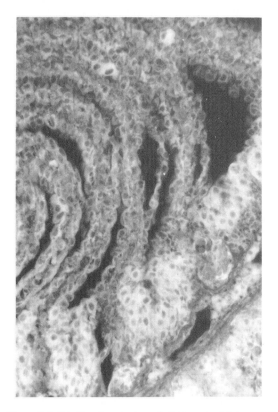

Figure 7.1 Chloride cells at the base of gill filaments of skate, visualized by an indirect immunofluorescent procedure (Appendix 10.1b) using an antibody to 1α hydroxycorticosterone binding protein.

and a loss of 'parr' marks (dark lateral blotches). Smolts may wait in estuaries for a completion of the osmoregulatory adjustments, for example changes to chloride cells.

Studies of the euryhaline bull shark *Carcharhinus leucas* show upregulation of particular branchial ion transporters in fresh water; the upregulated Na^+/K^+-ATPase pump and the Na^+/H^+ exchanger are believed to accomplish Na^+ uptake (Reilly et al. 2011).

7.10 Control of osmoregulation

Hormonal control of osmoregulation (10.11.3, 10.12.1) in fish has been studied intermittently with candidates for action including prolactin and cortisol and, more recently, growth hormone and insulin-like growth factor I (McCormick 2001). In mammals the central control of osmoregulation occurs via the hypothalamus in the brain; osmotic changes affect the release of ADH (antidiuretic hormone or vasopressin) from the 'posterior lobe' (neurohypophysis) of the pituitary gland. Thus a salty meal will stimulate release (to the blood) of ADH, and ADH will increase water retention by the kidney. Fish have similar hormones, small peptides secreted by the neurohypophysis, but their action is not yet clear; the situation in fish is complicated by the multiple sites of osmoregulation, including the gills (not present, of course, in mammals). A role for angiotensin has recently been discussed (8. 5.1).

7.11 The liver

The liver is a large organ in the anterior abdominal cavity of all vertebrates. It has multiple functions (Figure 7.3) and is a major site of metabolism with the hepatic portal vein bringing molecules absorbed from the gut direct to the liver. Plasmatic glucose is regulated though it may rise in stress (Fletcher 1975) or in response to cold temperatures (Umminger 1971); it can also be stored as glycogen in the liver or muscle. Liver glycogen undergoes glycogenolysis with release of glucose under hormonal influence (Moon et al. 1999) Plasmatic levels of individual amino acids are also regulated, with the liver playing a major role, with excess amino acids in bony fish deaminated to produce ammonia

as an end-point, rather than urea. Urea is however specifically synthesized and retained in the plasma in chondrichthyans and the coelacanth (7.5.8, 7.5.9, 7.5.11). Unlike in some mammals in which a fatty liver can be a pathological condition, fish species can have a normally fatty or oily liver, providing energy storage and buoyancy. Chondrichthyans, particularly holocephalans (chimaeras), have oily livers as do some flatfish (halibut) and gadids (cod). Seasonally, the liver may synthesize antifreeze protein or glycoprotein and females synthesize vitellogenin, the yolk precursor, at this site.

Liver size has been reckoned, in teleosts, to be approximately 2 per cent of body weight (Gingerich 1982, quoted by Dethloff and Schmitt 2000) in mature fish, but the hepatosomatic index/liver condition index (HSI/LCI, the liver weight expressed as a percentage of body weight) varies with feeding level and stress (Plante et al. 2003), and may be larger for fish like cod (Lambert and Dutil 1997; Levesque et al. 2005) or chimaeras (Oguri 1978), in which it is a lipid storage site. Lambert and Dutil (1997) reported the LCI for wild cod as varying between 1 per cent and 8 per cent.

Biochemically, the multiple functions of a generalized vertebrate liver have been listed by Schär et al. (1985) as being segregated into two main groups; one set (periportal) morphologically close to the hepatic portal vein, the other (perivenous) near the ultimate venous drainage (Table 7.1). In trout there was some evidence of this differential activity, with glucogen and phosphorylase mainly periportal and two other enzymes perivenous. However, these authors found that in trout some enzymes

Table 7.1. Two biochemical sets of vertebrate liver parenchyma (after Schär et al. 1985).

Periportal area	Perivenous area
Oxidative energy metabolism B oxidation (e.g. fatty acid metabolism)	Glycolysis
Amino acid catabolism	
Ureagenesis from amino acids*	Ureagenesis from ammonia*
Gluconeogenesis	Lipogenesis
Bile acid synthesis	Biotransformation**

*Urea formation would not be a major activity in most teleosts (8.3.2)
**A very general and inclusive term, often applied to detoxification processes

reflecting different biochemical pathways were not segregated.

The liver is a major site of detoxification, capable of increasing particular enzymes, for example aryl hydrocarbon hydroxylase (AHH), a microsomal cytochrome P450 and mixed function oxidase (MFO) inducible by petroleum, in response to pollution. Increased levels of such enzymes or 'biomarkers', including forms of cytochrome P450 such as ethoxy resorufin-O-deethylase (EROD) have been used to monitor environmental contamination (Arinç et al. 2000; Whyte and Tillitt 2000). The liver is also the site of steroids' modification, for example increasing their water-solubility as glucuronides (under catalysis by glucuronidase enzymes) preparatory to excretion. The teleost liver participates in detoxification of metals such as silver (Ag) via the induction of metallothionein, which forms a complex with silver; each molecule being capable of binding 12 silver atoms (Wood 2011).

The vertebrate liver is the site of haemoglobin breakdown with iron retained as haemosiderin or ferritin in the liver (and elsewhere, e.g. kidney, spleen) and bile pigments formed as excretory products, stored and subsequently released via the gall bladder. Concentrations of pigments may accumulate in the liver and elsewhere (spleen, kidney) forming melanophore–macrophage centres (MMC); these may contain iron pigments (haemosiderin/ferritin) as well as melanin and lipofuchsin.

Bile salts derived from cholesterol (Hagey et al. 2010) and synthesized by the liver also pass to the gall bladder and periodically enter the duodenum in the bile as an important factor in fat digestion (4.4.7). Phagocytic Kupffer cells (part of the reticuloendothelial system RES) lining vertebrate liver sinusoids are expected to play a role in removing particulate matter from the blood and have been reported to play a role in destroying old erythrocytes (Kalashnikova 2000). However, some authors report no sign of Kupffer cells in some fish species such as cod (Seternes et al. 2001).

The anatomy of the liver facilitates its numerous roles, with a very efficient blood supply; the hepatic portal vein direct from the gut and the hepatic artery from the dorsal aorta. Sinusoids, which may be lined with Kupffer cells, both supply and drain the liver, with blood returning to the sinus venosus via the short hepatic vein. The principal liver cells, the hepatocytes ('parenchyma', Schär et al. 1985), may be multifunctional, with no obvious specialization, unlike the metabolic zonation described for mammals (Buhler et al. 1992). However, some authors have described metabolic zonation in fish; Schär et al. (1985) did find metabolic zonation in rainbow trout, and regional differences were described for *Myxine glutinosa*, an agnathan fish (Foster et al. 1993). Mommsen and Walsh (1991) found two subpopulations of hepatic cells in the toadfish (*Opsanus beta*), an unusual fish metabolically in that it can be ureogenic (8.3.2). Individual vertebrate hepatocytes may be arrayed in cords, with one surface against a sinusoid and the opposite one draining into the biliary system via bile cananiculi (Figure 7.2), as in the mammalian system (Ham 1957, Figure 416) except that some mammals, including humans, have a more obvious arrangement of hepatocyte cords in lobules around the blood vessels with a regular and repeating pattern. However, Schär et al. (1985) consider the fish liver (as found in rainbow trout) to be quite different from the mammalian (which itself is not interspecifically identical in structure). These authors describe mammalian parenchyma (hepatocytes) as being plate-like, differing from the trout 'reticulum of tubuli', with four to six cells grouped around the bile canaliculi. These authors also refer to hepatocytes as being zoned within an acinus (an expression more usual for example, for glands like the pancreas) and perhaps conflicting with the tubular concept. Hampton et al. (1985) concur about tubules; the fish (rainbow trout) liver structure is seen as tubular, with cells arranged around the bile canaliculi as the lumen of the tube. In fish there are apparently no clear lobules arranged around the larger blood vessels (as in mammals).

Historically the anatomical arrangement ('architecture', Schär et al. 1985) of liver cells (e.g. in mammals) has been subject to some argument and may be seen differently if the amount of blood in the individual vessels varies. The liver is also a very large structure in some fish and there may yet be further descriptions of functional or even anatomical specialization.

The bile cananiculi collect the bile pigments resulting from haemoglobin breakdown and bile salts for fat emulsification, passing these to the gall

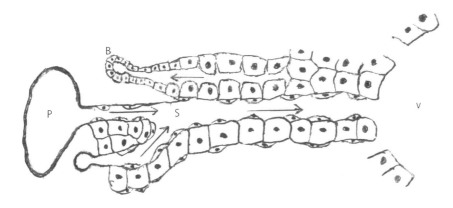

Figure 7.2 Diagram of vertebrate liver structure (after Ham 1957); B = bile duct; P = portal vein; S = sinusoid; V = central vein.

Metabolic type	Molecules affected	End-point
REGULATION	glucose, amino acids	homeostasis
SYNTHESIS		
bile salts	taurocholate	digestion
plasma proteins	globulins	lipid transport
	albumin	OP raised
hormone(s)	IGF	growth
seasonal		
antifreeze	AFP, AFGP, glycerol	freezing point depression
vitellogenesis	yolk components	eggs
occasional	MFOs	detoxification
CONJUGATION	steroids	elimination
CATABOLISM	amino acids	NH_3 formation
OSMOREGULATION	urea	OP raised
	TMAO	proteins protected
STORAGE	glycogen	source of glucose
	lipids	buoyancy
	ferritin from HB,,	Iron for HB synthesis

Figure 7.3 Major liver functions. AFP = antifreeze protein; AFGP = antifreeze glycoprotein; IGF = insulin-like growth factor; MFO = mixed function oxidase; OP = osmotic pressure; TMAO = trimethyl amine oxide.

bladder partially embedded in the liver, prior to periodic release via the bile duct to the duodenum. Some fish livers include areas of pancreatic tissue which may make it more difficult to obtain 'pure' cell cultures of hepatocytes.

7.12 Energy storage

Fish may utilize carbohydrates, or lipids, or even proteins as sources of energy, for metabolism and specifically for growth of muscle, bone or cartilage and gonads (often seasonally) and other internal organs in proportion to overall size.

7.12.1 Carbohydrates

As in other vertebrates, the liver (7.11) can be a carbohydrate store, with glycogen deposited in the hepatocytes and glucose released to the blood as needed. Blood glucose levels are hormonally regulated (10.12.1, 10.12.7) but their measurement may be complicated by the tendency to vary with stress levels (Specker and Schreck 1980; Bracewell et al. 2004).

7.12.2 Lipids

Whereas mammalian fat storage in special adipose tissue has been fairly well studied, with a differentiation into brown fat and white fat, in both cases involving special cells called adipocytes, the situation with fat storage in fish is less clear. Brown fat, used for thermogenesis in mammals, unsurprisingly has not been described in fish (Flynn et al. 2009). Fish lipids may be deposited in mesenteries round the gut or be located subcutaneously beneath the dermis in the hypodermis (Maddock and Burton 1994); the red muscle also contains a lot of lipid. Some fish livers normally contain a lot of fat (7.11). Not all fish have each of these sites for fat storage; mesenteric (coelomic) fat is not seen in elasmobranchs or many teleosts (e.g. gadids or pleuronectids) and is principally described from freshwater fish (Nikolsky 1963) such as perch and cyprinoids. Fatty livers occur in chimaeras, cod and halibut (7.11).

Hepatic fat is located within the hepatocytes whereas hypodermal and mesenteric fat occurs in special adipocytes. Red muscle lipid can be found as droplets closely associated with the mitochondria (Lin et al. 1974). Lipid levels may vary seasonally and/or with reproductive (Ribeiro et al. 2007) or feeding status (Maddock and Burton 1994). Nikolsky (1963) reports a scale used by Prozorovskaia (1952) for the Caspian roach, which reflects the amount of fat on the intestine ranging from Unit 0 (no fat) through to Unit 5 with fat covering the intestine. Nikolsky (1963) also comments on Hjort's (1914) scale for herring and methods for determining fat levels in fish. Hjort (1914) found these to vary

(as an average) seasonally from about 7.5 per cent to 13.7 per cent. The percentages quoted by Nikolsky for commercial fish vary from averages of 0.2 per cent in haddock to 23 per cent in anchovy. He does not comment on the oilfish, *Ruvettus pretiosus*, a widely occurring large fish, found in the Mediterranean, the Indian Ocean and elsewhere, and which has an even higher fat content including deposits in the skull (Bone 1972), interesting because the coelacanth which co-occurs with oilfish off the Comores also has fatty deposits round the brain. The more general occurrence of perimeningeal adipose tissue around fish central nervous systems is discussed in 11.6.1 The Comephoridae, a family endemic to Lake Baikal, are also referred to as 'oilfishes', with bodies 'usually high in fat content' (Nelson 1976).

7.12.3 Muscle

An unusual feature for some fish is their periodic utilization of white muscle protein, which may subsequently have very high water content giving 'jellied muscle' (3.8.1, Appendix 7.4). Love (1960) described wild cod as showing increased muscle hydration annually, rising from approximately 80 per cent to 83 per cent water. He also described the wild 'jelly cat' (*Lycichthys denticulatus*) alternatively known as the Northern wolffish (*Anarhichas denticulatus*) as naturally having muscle hydration from 82.5 per cent to 95.4 per cent (Love and Lavéty 1977), in which case the white muscle has large peri-tissue spaces but little sign of muscle disorganization in contrast to what is seen with starved cod. These authors suggested that watery muscle can be a feature of deep-water fish, compensating for lack of a swim bladder. Hochachka and Guppy (1987) discuss the utilization of white muscle. Whether periodic utilization of white muscle protein compromises locomotion is not clear; perhaps there is an excess of this tissue so that movement is not adversely affected if some is lost to maintain other tissues. Jellied muscle has not been reported from many taxa and may not occur in freshwater fish (Appendix 7.4) except migrating salmonids. Close to spawning, after a lengthy non-feeding migration they may have lost a substantial proportion of their mass, and if they survive, are 'kelts', relatively emaciated, that is, overall in poor condition. Pearcey (1961) observed that winter flounder off Connecticut tended to have soft muscle in the winter prior to spawning, after a non-feeding period, a situation somewhat similar to that of pre-spawning migratory salmonids. Templeman and Andrews (1956) found jellied muscle in large American plaice and considered this was due to the gonads 'having priority in use of protein' and with no evidence of parasitic involvement.

Some jellied or soft muscle reports, particularly for tuna, associate the state with the presence of kudoid parasites (Abe and Maehara 2013). A 'milky' condition described from some American Pacific-coast fish was ascribed also to myxosporidian parasites (Patshnik and Groninger 1964) and differentiated from the 'jellied' muscle of some flatfish.

7.13 Growth

Growth has been studied at the individual level when it deals with increased size presumably achieved via mitosis and/or increased volume of individual cells, with some deposition of extracellular material, and also at the population level when it deals with the capacity of a population to increase or maintain stability particularly under fishing pressure. The relationship between the capacity of a population to reproduce and individual growth is certainly bound up with the ability of individuals to access sufficient food to both grow to a size when reproduction can occur and the ability to maintain gamete development through to actual deposition/fertilization.

Because of the important commercial implications, fish growth has been widely studied and growth parameters defined, discussed and analysed. Many fisheries scientists have devoted their working lives to understanding individual and population changes with time, developing 'growth curves' which might help further understanding. A review of work up to the 1980s was undertaken by Weatherley and Gill (1987), with a chapter by Casselman. Comments on this overview were provided by Dabrowski (1990). Since the 1980s a series of fisheries collapses occurred (e.g. North Atlantic cod off Newfoundland, under a fishing moratorium since 1992 and 1993, see Myers et al. 1997, for other species see Hutchings 2000). The rather optimistic views of some fisheries scientists on the resilience

of population to exploitation, and the capacity to recover by 'compensation' with flexible reproduction and maximum growth has needed careful revision (Appendix 16.1).

7.13.1 The growth potential of individuals

The role of hormones in growth (10.12.3) is complex and includes pituitary (growth hormone), thyroid (thyroxin) and liver-sourced insulin-like growth factor (IGF).

The enormous range in adult fish sizes possible (1.4.1) encompasses some very small gobies, blennies, guppies and tetras, including popular freshwater aquarium fish, which become sexually mature at small sizes, for example 2 cm approximate length. Such small fish may show little or no further growth after sexual maturity; that is, they have 'determinate growth' and an expected fairly uniform size attained by adults for each species in this category. The extremely large fish include some freshwater osteoglossomorphs such as the South American *Arapaima gigas*, which reach about 4.5 m length (Nelson 1976), some catfish (the European wels *Silurus glanis* can attain over 3 m), and the huge planktivorous basking and whale sharks, the latter capable of exceeding 15 m length, as 'the world's largest fish'. Oarfish (*Regalecus*), rarely seen, very long, marine fish which live in deep water, have been recorded as 8 m length (Nelson 1976) and are thus the longest bony fish. These fish wash ashore occasionally and popular reports suggest they may exceed 8 m length.

These big fish species may not attain sexual maturity for some years and can continue growth after becoming adult. Larger fish thus very often have indeterminate growth, with no definite end-point, though growth may slow when the fish becomes very large. Large flatfish such as halibut have been reported as sexually mature at quite small lengths; female 41–45 cm, male 66–70 cm, and also as large fish; male 116–120 cm, female 176–180 cm (Kohler 1967). It is likely that these fish do not reproduce every year (9.15, 9.16), and that they have intermittent somatic growth.

Mommsen (2001) believes that most fish species show indeterminate growth and that their use of 'functional muscle for storage' sets fish apart from

Figure 7.4 Large cod, historically found off Newfoundland. Photographed in 1901 by Robert Holloway.

all other vertebrates with 'the possible exception' of some marine mammals.

The contribution of large females to the sustainability of a population may be much more important than previously allowed (Xu et al. 2013). The opposite view was indeed held on occasion; for example, large females close to spawning might have inferior-quality flesh (see 'jellied muscle' 7.12.3, Appendix 7.4) and their destruction could be advocated (Templeman and Andrews 1956). With continued pressure on populations large fish such as those seen in Figure 7.4 may have been effectively eradicated, in the case of cod a few large females were reported by May (1967); for 37, 9- to 10-year-olds caught on the southern Grand Bank off Newfoundland only 6 were 1 m or more in length. The largest cod on record weighed 95.9 kg (254 lb) and was caught off Massachusetts in 1895 (Scott and Scott 1988), much larger than the fished average weight

of < 4.5 kg reported in the same source. However, recently, since the cod moratorium off Newfoundland, there have been occasional records of very large fish, a 75 lb fish off Newfoundland and a 103 lb fish off Newfoundland and fishers in the 2016 restricted food fishery verbally report many large fish as compared to the recent past.

Some fish species such as some salmonids may have dwarf forms as land-locked populations, and may also have dwarf males which remain as freshwater residents ultimately in competition with large sea-run males. Similarly, and more extreme, dwarf males are the only form of male for some deep-sea anglers, and in some species the small males become permanently attached and thus 'parasitic' on the larger females (9.3, 13.4.4). Recently Rodriguez et al. (2012) have described a new species of *Loricaria*, notable for its small adult size, termed 'diminutive' in comparison with other species of this catfish genus. Similarly a 'pygmy' seahorse *Hippocampus bargibanti* was described in 1969, and, recently (Lourie and Randall 2003) it has become apparent that there are more 'pygmy' seahorse species. These pygmy seahorses can be little over a centimetre in length as adults, similar to the tiny goby *Trimmatom nanus* (Winterbottom and Emery 1981). The goby described as the smallest fish or vertebrate by Nelson (1976) is *Pandaka pygmaea*, which matures at 6 mm. Other species claimed to be exceptionally small include *Schindleria brevipinguis* and *Paedocypris* spp. (Kottelat et al. 2006). These 'miniature' fishes have characteristics (e.g. a functional pronephros, 8.5 and Nelson 1976) of larval fish and are hence termed paedomorphic.

7.13.2 Control of growth

Control of growth is a very complex topic. Embryonic growth is to a large extent predetermined but may be diminished if access to yolk is restricted (Gray 1928); the larva in this situation is just smaller than usual. In post-larval life the positive growth of the fish requires an adequate food supply as well as hormones (10.12.3) such as 'growth hormone' from the pituitary and thyroid hormone, with the possible involvement of anabolic steroids like testosterone. In the case of the pituitary it is known that vertebrates in general have a growth hormone which can be extracted from the anterior region of the gland, excess of which in humans causes acromegaly in which height and bone growth occurs beyond normal limits. In fish additional growth hormone can be genetically engineered which may induce quicker growth than normal provided additional feed is available. In vertebrates a thyroid hormone, thyroxin, T_4 (10.11.4), is associated with successful early development and metamorphosis, lack in young humans producing unsatisfactory height and mental retardation (cretinism). Growth hormone and thyroid hormone are expected to act synergistically on growth but Rousseaux et al. (2002) found evidence for negative feedback, with thyroid hormone controlling growth hormone in eels, that is, increased thyroid hormone decreased pituitary and blood levels of growth hormone.

Growth in vertebrates is expected to include increased cell number through mitoses, or in the gonads through both mitoses and meiosis. However, increased muscle mass may involve recruitment of extant satellite cells (Koumans et al. 1991) rather than production of new cells, and a similar mechanism may occur in fat deposits where lipocytes become active. The occurrence of 'jellied muscle' (Appendix 7.4) can confuse the understanding of growth, as extreme hydration can temporarily add to mass (as with terminal phases of oogenesis). Growth of certain organs may not occur at the same rate, thus the brain may not become proportionally bigger as length increases. In general it is not expected that vertebral number will increase in adults, following the general rule that metameric segments do not change in number with age once the larval somites have been developed. Because some anatomical characters are indeed stable or 'fixed' it is possible to use some 'countable traits' taxonomically. Such features, termed 'meristics' include fin-ray counts, particularly useful because they can be achieved without dissection, on a live or dead fish.

Secondary growth as a repair process enables replacement of parts lost through accident, but fish do not seem to have as much capacity for replacement as some amphibia and reptiles, some species of which can replace tails or even limbs. However, scale replacement is reported to occur as 'regenerated' scales (Casselman 1987) and tooth replacement occurs in elasmobranchs. The periodic growth of gonads is remarkably fast and substantial in some fish taxa, and with hydration of late-stage oocytes, may produce very large changes in ovarian mass relative to body mass.

7.13.3 Growth rate

The capacity of some fish species to grow very large has been associated with questions as to how fast can such growth occur and how can sound data be accumulated to find answers.

Growth rate data have been of considerable interest because they underlie the capacity of fish to withstand fishing pressure, to be good aquaculture subjects, and also because it is somewhat different from that of land vertebrates. Some of the assumptions developed with mammals and birds, which generally have determinate growth with a predictable and uniform size for adults and a known and very limited reproductive output, do not hold for fish. Even fish specialists have drawn conclusions which may not be tenable, and extrapolating from one set of data to a stock or population or species may be very unwise. Thus, because some fish have strongly periodic feeding and/or reproduction with the possibility of extended periods of starvation, loss of weight is recognized as a normal situation, particularly after spawning of large numbers of eggs, as occurs with many species. Length loss, discounted by some, except in special cases like eel metamorphosis (e.g. Brown 1957) may occur ('degrowth', 7.13.4).

Growth curves of individuals and of populations have been calculated and form the basis for much fisheries work. At the individual level growth (measured as weight or length change) is plotted against time (e.g. body length against age, von Bertalanffy 1957).

Population changes, as increase, decline or stabilizing at a maximum (dependent on the carrying capacity, K, of the environment), are less easy to measure, requiring representative and repeated subsampling. If it is desired to know whether the sampled fish are adults or pre-adults then it is essential to differentiate post-spawned fish from 'immature' fish which have never been reproductive. One serious problem has been changes in gear and search techniques which may enable catches far above what used to be possible so that the notion of 'catch-per-unit-effort' (CPUE) has changed with time and an over-optimistic view of population size can result. One set of data (Beverton and Holt or Ricker curves, Eberhardt 1977) has been aimed at plotting numbers or biomass of 'recruits' against stock.

Hopkins (1992) reviewed the three typical routes for measuring fish growth: 'absolute' (g/day), 'relative' (% increase in body weight) and 'specific' (%/day). The term 'instantaneous' is used alternatively for 'specific'.

Handling individuals for measurement may stress them and produce suboptimal growth.

Growth rates of young fish (e.g. salmonids) show the trade-off when yolk is absorbed and transferred into larval tissue (Gray 1928). For comparative purposes, the specific growth rates were calculated for different phases of the life history enabling quantification of the change in tissue accumulation per unit time, usually based on weight changes. Initial growth rates are expected to be very high, given optimum conditions. Maximum instantaneous growth rate (G_{max}) has been used by some scientists (Shepherd and Cushing 1980; Cushing 1995); its incorporation in predictive models to estimate survival to recruitment is discussed by Rutherford et al. (2003).

According to Sparre and Venema (1998), the work of Pütter (1920) forms the basis for developing growth models, including the von Bertalanffy (1957) curves for individual growth of organisms varying from invertebrates through mammals, although as recognized by Sparre and Venema (1998), individual arthropods (which moult) grow in sudden spurts giving a stepwise effect, although a smoother curve as a cohort. The desire to generate analysable smooth curves for use, for example, in fisheries or aquaculture predictions has generated some criticisms particularly of both 'ends' of the curves. Thus the idea that at the outset (beginning of the curve) a fish has zero length is biologically untenable with 'no meaning' (Sparre and Venema 1998), whereas Knight (1968) is not comfortable with use of the von Bertalanffy curve's indication that growth ceases. Recently a spate of errors has been identified (Garcia-Berthou et al. 2012), which according to these authors has produced several wrong identifications of the lowest point of seasonal growth.

The capacity of fish populations to increase or to maintain stability has been primarily studied from the fisheries viewpoint. Often the measurements of population has been rendered as 'biomass' of a stock, and the relationship between stock and 'recruitment' has been sought, where recruitment is the attainment of a fishable size. The age or size at first reproduction may be factored into predictions, and the capacity of individuals to grow quickly to reach reproductive status is obviously important. Unfortunately expectations were raised that a small spawning stock could be sufficient to regenerate a fishery, based on the ideas that more young would survive (compensation, Appendix 16.1). Unfortunately too, some of the optimistic expectations for resilient populations were based on atypical species such as semelparous migratory salmonids or the false trevally, *Lactarius lactarius* (Pauly 1980).

Eberhardt (1977) comments that some of the stock/recruitment curves are 'not realistic' and also applies the term 'fictitious'. However, these treatments of data have been widely used though they may now need careful reconsideration. Some of the characteristics of population analysis have been discussed by Pauly (1993) and Cushing (1995). The stock–recruitment relationship was reviewed by Sparre and Venema for the Food and Agriculture Organization of the United Nations in 1998, with some consideration of curves developed previously, for example by Ricker (1954) and Beverton and Holt (1957). Ricker's work was very influential; it considered reproduction, both invertebrate (fruit-fly, water-flea and starfish) as well as fish (three species of Pacific salmon, herring and haddock). Ricker considered that removal of adults could increase reproduction, an optimistic view that may have been inherent in overexploitation.

The output of a fishery may have been measured in local units of weight (e.g. pounds) and volume such as barrels and crans (Waterman 2001) and even hectolitres (Hjort 1914); this sometimes makes it difficult to make comparisons as to mass or individuals removed.

7.13.4 Size measurements

Measurement of size usually includes both length and mass (weight), with differing procedures standardized. Length measurements are usually metric, and of three basic types: fork length (FL), total length (TL) and standard length (SL). In most cases it is accepted that the anterior measurement will be taken from the most anterior part of the head. However, the posterior measurement can be to the fork of the tail giving 'fork length', the most posterior portion of the caudal fin (total length) or the caudal peduncle (standard length), the most posterior part of the mid-body before the caudal fin, or more technically 'the posterior margin of the last whole vertebral centrum' (Scott and Scott 1988).

As some fish, notably chondrichthyans (sharks, etc.) do not have 'forks' in the caudal fin and/or there may be fraying of the caudal fin, the use of 'fork length' may be limited. Loss of the finer, longer caudal fin-rays may also compromise the use of the

total length. Mass/weight is usually achieved in metric units and may vary by being taken from active live fish which makes accuracy difficult, anaesthetized live fish (which can be gently dried with, for example, damp paper towel), or dead fish which can be weighed whole or gutted, with further variants such as separate weighing of the gonads. Mass, of course, varies with feeding status. Individual fish living in highly seasonal conditions with little or no feeding for extended 'winter' periods lose weight and even length ('degrowth', Calow and Woollhead 1977) during the winter fast. The relationship between mass and length can be expressed as 'condition factor' (CF) often (and apparently erroneously) attributed as first used by Fulton (Nash et al. 2006). CF may be modified, for example by using gutted weight (± gonads). Traditionally CF is calculated as $W.100/L^3$ and, depending on the species it can be determined what range of values represents high or low condition during the life and seasonal cycles. Thus a CF value of unity for cod represents a well-fed fish, but in the wild, healthy individuals may have a CF of 0.8 or thereabouts.

7.13.5 Fish ages

The age of fish (1.4.1) can be known from captive specimens and information on growth rates can be established under controlled conditions. Information on age of some wild fish can be obtained with some degree of confidence from examining, for example, the otoliths (Halliday 1970) or scales of fish (1.4.1, 2.4, 3.3 12.8.2) with seasonal growth, in which case annual growth rings may occur, as these structures grow by accretion rather than interstitially. Casselman (1987) reviewed practices utilized in determining age and growth and Phelps et al. (2007) presented a study of different methods for age determination based on common carp.

Traditionally carp are expected to be able to attain decades, and fish which attain large sizes in cold water (e.g. cod) may also have reached ages of more than 20 years (May 1967). Carey and Judge (2000) presented a compilation of data for vertebrates in which carefully referenced information suggests that for fish, wild sturgeon reach the greatest (reasonably well authenticated) ages of more than 100 years.

The oldest (captive) carp in this dataset was 45 years old, though popular reports for (captive) Japanese koi carp give far greater life spans, based on local (family-based) recollections of the history of individual fish.

7.14 Conclusion

Initial growth of young fish will usually depend on maternally derived yolk, and subsequently on the capacity to feed optimally. Conversion of nutrients to somatic (body) tissue may occur in competition with gonad growth once the fish reaches puberty. Also, as the fish grows it may change its diet, for example, from small crustaceans to fish, including congenors, and this diet shift may entail other changes. As some fish become larger they may shift depth or even migrate from salt- to fresh water or vice versa. Such a drastic environmental change may be accompanied by changes in the skin, the gills and osmoregulation as occurs with salmonid smolts.

Maximizing growth is an aim of the aquaculture industry and diets are formulated to achieve this economically. Some fish have limited capacity to digest particular potentially nutritive components and this has to be factored into feeding regimens.

Just as growth is a controlled accretion of material, regulated in fish to the extent that a species-specific maximum size may be approached, the survival of the individual is dependent on stabilization of the internal environment of the fish by a series of regulatory systems with some degree of tolerance for variation in the composition and temperature of the surrounding water.

Depending on the body of water accessed by the fish, individuals can also move to or from conditions appropriate for their maximal survival. As movement itself may be very strenuous or stressful, especially if upstream or between salinities, some fish which undertake long migrations do so only occasionally, perhaps once only (9.14).

Appendix 7.1 Buffers

The concept of buffering refers to a number of different chemicals which dampen the effect of proton (H^+) accumulation. In a natural system such as vertebrate plasma, carbon dioxide can form the highly dissociated (i.e. 'weak') carbonic acid and a number of molecules (e.g. proteins including haemoglobin) can sequester the protons, thus stabilizing the tendency to become acidic. A range of buffering molecules, including those used by Good et al. (1966), have been developed, one of which is HEPES used by Higuchi et al. (2012) for eel testis culture in combination with the use of a commercially available culture medium. Ideally, buffers should be relatively inert except in their buffering role, and they should certainly be non-toxic. Cacodylate used for buffering some solutions (e.g. fixatives) used in electron microscopy contains arsenic and is therefore completely unsuitable for physiological work on live material. The Tris (Sigma) system can be used to buffer between pH 7 and 9 and is a popular buffer for biochemical procedures. Phosphate buffered saline is used for immunochemistry.

Chelating agents such as EDTA (ethylenediamine tetra-acetic acid) can be used effectively to remove ions such as calcium from a water-based medium, and may thus be used to modify a system, but the viability of, for example, cells or tissues with added EDTA could be problematic.

Appendix 7.2 The use of artificial media post-hypophysectomy

In order to study the role of individual pituitary hormones, removal of the pituitary (hypophysectomy) has been used followed by administration of putative or known pituitary hormones. Freshwater fish only survive hypophysectomy if they are put in appropriate salines after the operation, as their capacity to osmoregulate is impaired. Ball (1965) used dilute seawater (roughly one-third as 12 ppt, i.e. 12 parts per thousand salinity) for freshwater *Poecilia*, and Komourdjian and Idler (1977) used 11 ppt for *Salmo gairdneri*. Grau et al. (1984), however, used about twice that salinity (26 ppt artificial seawater) for *Fundulus heteroclitus*, a euryhaline brackish or freshwater species. Winter flounder (*Pseudopleuronectes americanus*) can be held in seawater post-hypophysectomy (Burton 1981).

Appendix 7.3 Fish salines

Simple or complex formulations of salines, originally based on a sodium chloride solution, have been used to bathe fish tissues during experimental investigations.

Specially formulated 'salines' can be used for short-term experiments (hours). It is generally supposed that fish salines do not need the addition of glucose or oxygen, however Burton and O'Driscoll (1992) did add glucose to fish skin incubating media. The composition of fish salines may generally reflect the various 'Ringer's solutions' that were developed for mammals and then modified for amphibia. 'Physiological salines' are often referred to by the name of the person who first developed them (e.g. Ringer) and changes acknowledged as, for example, 'modified Krebs'. Physiological salines, including 'Cortland' and 'Hanks', useful for freshwater teleosts, were reviewed by Wolf (1963) and 'balanced salt solutions' for fishes were tabulated by Randall and Hoar (1971).

It is interesting that fish cells (e.g. neuro-melanophore preparations, see Appendix 2.3) can respond differentially, and reflect natural function, to the simple substitution of Na^+ for K^+ or vice versa in a saline (Burton and O'Driscoll 1992).

For longer term culture (e.g. days), the composition of bathing fluid may be more carefully composed with frequent changes and aeration and adjustment of pH. Higuchi et al. (2012) thus describe the culture medium for Japanese eel testis, which had a number of amino acids and insulin as supplements in addition to a buffer (HEPES, Appendix 7.1) and adjustment of pH to 7.4.

Appendix 7.4 Jellied muscle

Some marine fish are prone to fluctuations in the degree of hydration of their white muscle, which has been reported from time to time as 'jellied muscle'. Fish with jellied muscle are regarded as inferior for human consumption and may be discarded following identification in a catch, 'firm' muscle being preferred. Physiologically it is not clear whether fish with jellied muscle can move as fast as 'normal' fish, and the reversibility of the condition is assumed but not demonstrated.

Iteroparous fish prone to jellied muscle to-date seem to be exclusively marine and include flatfish, wolffish and cod, all fish with commercial importance for which the condition would be noticed and recorded.

The level of hydration which merits the term 'jellied muscle' has not been defined but seems to be in excess of 80 per cent water. The periodicity is related to the spawning season or the pre-spawning migration.

Anadromous fish, such as salmonids, which may have a large proportion of semelparity, are noted for their poor muscle condition at spawning and those that survive, as kelts, may take years to recover.

Some marine (Pacific) tuna and swordfish seem prone to infestation with myxosporean parasites which result in soft muscle. These parasites have not been reported from fish in the North Atlantic which may show seasonally variable muscle hydration, which is currently deduced to be a natural, physiological adjustment to changing feeding and reproductive demands.

Flatfish (Pleuronectiformes) which may have jellied muscle include:

American plaice, the north-west Atlantic form of *Hippoglossoides platessoides* (Templeman and Andrews 1956). The Scottish, much smaller form, termed locally 'long rough dab', may not show this phenomenon.

Alaskan halibut *Hippoglossus stenolepis* (International Pacific Halibut Commission 2012) have been reported as 'mushy'.

Winter flounder (*Pseudopleuronectes americanus*) off Connecticut (Pearcey 1961) and off New Brunswick (McLeese and Moon 1989), associated with the winter fast.

Smooth flounder (*Liopsetta putnami*) of the western Atlantic (Scott and Scott 1988).

Dover sole (*Microstomus pacificus*) off the United States west coast (Fisher et al. 1985). Jellied condition associated with deeper water?

Other taxa with jellied muscle are cod (Gadiformes) *Gadus callarias/morhua* (see Love 1960; Love and Lavéty 1977) and wolffish (Anarhichadidae) *Lycichthys denticulatus* (Love and Lavéty 1977).

Bibliography

Casselman, J.M. (1990). Growth and relative size of calcified structures of fish. *Transactions of the American Fisheries Society*, 119, 673–88.

Heisler, N. (1984). Acid-base regulation in fishes. In W.S. Hoar and D.J. Randall (eds). *Fish Physiology, Vol. 10A. Gills Anatomy, Gas Transfer and Acid-base Regulation* New York, NY: Academic Press, pp. 315–40.

Weatherley, A.H. and Gill, H.S. (1987). *The Biology of Fish Growth*. New York, NY: Academic Press.

Wood, C.M., Farrell, A.P., and Brauner, C.J. (eds) (2011). *Fish Physiology, Vol. 31A: Homeostasis and Toxicology of Essential Metals*. New York, NY: Academic Press.

Wood, C.M., Farrell, A.P., and Brauner, C.J. (eds) (2011). *Fish Physiology, Vol. 31B: Homeostasis and Toxicology of Non-Essential Metals*. New York, NY: Academic Press.

CHAPTER 8

Excretion

8.1 Summary

Excretion by fish principally involves removal of metabolic wastes into the surrounding water.

Fish are able to eliminate both nitrogenous wastes as ammonia and carbon dioxide via the gills leaving less of an excretory role for the kidneys which, however, may eliminate divalent ions acquired, for example, through the gills.

Elasmobranchs have a small rectal gland which may also eliminate excess ions.

The skin may have a minor role in excretion and the liver has a major one in that it processes certain metabolites to expedite elimination.

Recently aquatic animals like fish have had to deal with the effects of numerous unusual chemicals in their environment, many derived from industrial processes and with structures not normally encountered in the natural environment.

8.2 Introduction

Excretion has been defined as the removal of metabolic wastes; this distinguishes it from egestion, the removal of undigested material from the gut. However, some metabolic 'wastes' such as bile pigments may end up in the faeces so the distinction is a fine one.

The major materials excreted by vertebrates are carbon dioxide and water (as a result of cell respiration) and simple nitrogen-containing molecules (as a result of protein and nucleotide metabolism). The end-points of protein or nucleotide catabolism vary somewhat according to taxon throughout the animal kingdom, but in aquatic vertebrates the main nitrogenous molecule excreted is commonly ammonia gas NH_3 (or ion NH_4^+), while in terrestrial vertebrates it is either urea (mammals) or uric acid (birds). Urea and trimethylamine oxide (TMAO) may be conserved or excreted in small amounts in fish. Creatine and creatinine from creatine phosphate (3.8.1) are also excreted, usually in small amounts. Excess inorganic ions, water and complex organic molecules such as steroid conjugates and bile pigments are also routinely excreted.

Fish may be exposed to particulates such as silt, with the possibility of high levels of toxic substances such as metals. Some waters contain high levels of dissolved xenobiotics, 'foreign' molecules, due to human activity. In the case of either particulates or dissolved toxic substances penetrating the tissues of the fish there may be either tissue storage or excretion, with some molecules inducing higher levels of certain enzymes (e.g. in the liver) which may facilitate elimination. Inshore and freshwater fish are particularly prone to such potentially toxic situations. Pollutants may actually overwhelm the capacity of the fish to withstand them and severe morphological changes may occur, particularly in structures most exposed, including excretory tissues and organs (Hinton et al. 1987).

The principal organs involved in excretion in fish are the gills, kidney and liver.

It is usual to associate nitrogenous excretion with the kidney in vertebrates but aquatic vertebrates have the option of voiding ammonia into the water, via the gills in the case of fish. The kidney tubules in fish, therefore, may have a minor role in nitrogenous excretion but other important roles in ion and water excretion. The anterior part of the elongated fish kidney may have non-excretory functions such as haematopoiesis (blood-cell formation) not present in the compressed metanephric kidney of mammals and other amniotes (i.e. vertebrates with an embryonic membrane known as the 'amnion', not present in fish). The anterior 'head-kidney' of teleosts may also incorporate endocrine tissue (10.11.3) which is separate in elasmobranchs and in other vertebrates.

The gills, with very large surface areas in direct contact with the water (6.5.9), can void large quantities of ammonia and carbon dioxide. Depending on its permeability the skin may also be capable of excretion via gaseous diffusion (2.3.1, 8.4.2).

The liver is directly involved in transforming many molecules (7.11), parts of which may be retained while other parts are excreted, either via the bile or the urine. The liver is particularly important in processing amino acids as well as haemoglobin and non-water-soluble molecules like steroids, and potentially poisonous molecules inadvertently acquired through ingestion or via the gills. Such molecules taken up from the environment include crude petroleum products which have been part of the natural environment during the evolution of fish, and xenobiotics, the result of human activity such as mining, farming and manufacturing processes. In general, the liver can detoxify by modifying molecules; for example, transforming some into more readily excreted water-soluble forms.

8.3 Pathways for excretion

The necessity for excretion of some molecules (or elements) rests on their harmfulness if retained, as a trade-off, against the cost of eliminating them. Regulation of the internal environment to enable optimum chemical actions maintaining life may require a very narrow range of pH with membrane organization actively excluding ions or free molecules that might adversely perturb metabolic processes. Transformation of some waste and/or excess molecules to forms that can be excreted may require expenditure of metabolic energy.

8.3.1 Amino acids

As ingestive heterotrophs, fish take in food which may be rich in protein. When the protein is digested preliminary to absorption and assimilation,

it forms a quantity of amino acids, of which there are 20 different types, with varying characteristics. Different amounts of these will be needed as bases for forming fish molecules, such as structural proteins or hormones. Excess may be transformable to other amino acids or may be catabolized, releasing energy for anabolism. During metabolism ammonia may be formed and as it is toxic it must either be quickly excreted or transformed (using more energy) to a less toxic molecule such as urea (NH_2CONH_2). Although it is generally expected that, at least in most teleosts, ammonia would be an important vehicle for nitrogenous excretion, and that the excretion site would be the gills, Table X in Hickman and Trump (1969) shows a substantial amount of ammonia in the urine of *Lophius piscatorius*, a marine teleost.

Bony fish do not usually synthesize urea but elasmobranchs do, though it is not commonly excreted but conserved, boosting the osmotic pressure of the plasma in an osmoregulatory role (7.5.8), reducing the possibility of osmotic loss of water. Urea is synthesized from ammonia via the ornithine cycle involving five enzymes (Figure 8.1), with an additional input from glutamine possible via the 'accessory' enzyme glutamine synthetase (Wright et al. 1995). The glutamine molecule formed from the amino acid glutamic acid and ammonia can be a means to transport ammonia 'in a non-toxic form' (Yudkin and Offord 1975). Although the ornithine cycle is not generally associated with teleosts, Wright et al. (1995) found the associated enzymes in young rainbow trout and observe that the capacity to evoke these enzymes could be important to teleosts in stress conditions such as alkaline pH or 'air exposure'.

8.3.2 Ureotely and ammonotely

Huggins et al. (1969) described the relationship of urea to fish in terms of urea synthesis (ureogenic fish), urea excretion (ureotelic fish) and/or urea utilized to achieve 'osmotic equilibrium' (ureosmotic fish). Ureotely is rare in teleost fish, having been described in species with a burrowing habit (*Opsanus tau*), habitats in alkaline water (Lake Magadi tilapia, Walsh et al. 2001) or poor-quality water (some catfish, e.g. *Hetropnesustes fossilis*, Saha and Ratha

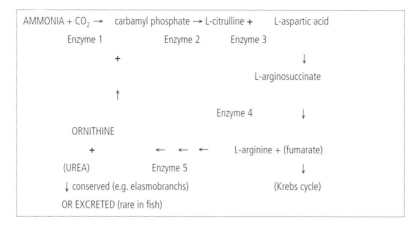

Figure 8.1 Simplified ornithine cycle, with 5 enzymes: 1 = carbamyl synthetase; 2 = ornithine carbamyl transferase; 3 = arginiosuccinate synthetase; 4 = argininosuccinate lyase; 5 = arginase. Based on Hoar (1966) and Yudkin and Offord (1975). In this cycle, ornithine plus carbamyl phosphate produces citrulline which, with aspartic acid, produces arginosuccinate. The arginosuccinate splits to form arginine and fumarate, the latter entering the Krebs cycle (and thus contributing some potential energy), the former (arginine) forming ornithine (to re-enter the cycle) and urea, to be conserved (8.3.2) in some fish or excreted (8.3.2) in some fish and tetrapods).

2007). Wright et al. (1995) also described ureotely in young rainbow trout and believed that the capacity to express ornithine cycle enzymes may be widespread in teleosts, and evoked under appropriate (e.g. high ammonia) conditions. Engin and Carter (2001) quoted by Cutler and Cramb (2002) report anguillid eels as excreting '23–42% of their nitrogenous waste as urea'.

Most teleosts are ammonotelic, excreting excess nitrogenous material via ammonia at the gills.

Fish that are capable of spending some time on land may utilize ureotelism (Frick and Wright 2002a) but this does not occur with the mangrove killifish *Kryptolebias*, formerly *Rivulus marmoratus*, which is noteworthy for being able to spend extended periods of time (weeks) out of water, but tends to resort to damp places including the interior of logs (Taylor et al. 2008).

The situation in elasmobranchs is interesting as marine elasmobranchs maintain high plasmatic urea. It is not clear however to what extent, how and where, they may excrete excess nitrogenous material, whether as ammonia at the gills or urea at the gills/kidney. Conservation of urea at the kidney level in the marine skate *Raja erinacea* is described by Morgan et al. (2003). The nephrons in elasmobranchs are looped with a countercurrent (6.5.6) disposition, facilitating passive resorption but there may also be more active resorption with urea transporters (Morgan et al. 2003). The freshwater stingray *Potamotrygon motoro* (Ip et al. 2009) could upregulate glutamine synthetase in response to increased salinity but could not synthesize urea.

8.3.3 Creatine

Creatine phosphate is found in vertebrate muscles (3.8.1) where it can be utilized as an energy source, particularly a source of phosphate for ATP synthesis. Excess creatine may be excreted in the urine, where reports showed it may form more than 50 per cent of the total nitrogen (N) excreted in fish (Whitledge and Dugdale 1972). These authors investigated the Peruvian anchoveta (*Engraulis ringens*) and found ammonia one-half of the N in the tank water, with creatine slightly less, and a small amount of urea. Creatine may also be excreted as creatinine (Danulat and Hochachka 1989).

8.3.4 Haemoglobin

Effete red blood cells are destroyed and the iron retrieved and recycled (5.13.6, 7.11), while bile pigments formed from the haem moiety (8.4.3) are eliminated from the liver via the bile into the intestine.

8.3.5 Steroids

Excess steroids in the circulation may be processed in the liver to water-soluble glucuronides (7.11, 8.4.3).

8.3.6 Water

In the aquatic environment fish have evolved and adapted to a range of salinities and face different problems in fresh water and seawater or in the rarer situation where they face daily changes in salinity. Thus, for freshwater fish, water taken in at the gills by osmosis needs to be eliminated; as this water has not been part of the metabolic process it is not strictly part of excretion but osmoregulation (7.5), whereas water produced by cellular respiration may be retained or eliminated as an excretory product.

8.3.7 Inorganic ions

In seawater, inorganic monovalent ions such as Na^+, K^+, Cl^-, and the divalent ions Mg^{2+} and SO_4^{2-} may be taken in at the gills or with the food. Their elimination may be regarded as part of osmoregulation rather than excretion, although once processed intracellularly (e.g. in active transport) the term 'excretion' could be applicable. Ions may be actively eliminated at the gills (via mitochondria-rich chloride cells, 7.7) or in the urine (divalent ions) or via the intestine (O'Grady and Maniak 2005) or the rectal gland (or equivalent) of chondrichthyans and the coelacanth.

8.3.8 Particulate materials

In some invertebrates relatively inert particulate matter has been reported as 'parked', stored, for example, in the integument and that may happen in some vertebrates too, although the best comparison may be with the deposition of arsenic in mammalian hair. Sites of deposition, for example of resistant bacteria, in fish, may include melanomacrophage centres of the spleen, kidney and liver (Agius and Roberts 2003). Metal-rich granules (MRGs) have been reported to be stored by killifish (*Fundulus heteroclitus*) living in polluted water (Goto and Wallace 2010).

8.3.9 Metals

Mining and industrial processes may release metals and other by-products into the water, notably 'tailings' ponds, with subsequent more general leaching into the surrounding water, including river estuaries. Bioaccumulation of metals may occur in fish, with particular concerns about human consumption of fish with raised mercury, cadmium or other highly toxic minerals. Silver may be eliminated via bile or urine (Wood et al. 2010). Recently the effect of nanoparticles of silver has been a concern. It was not known how much penetration would occur with these, which are increasingly used industrially (Scown et al. 2010). These authors have found some concentration of silver nanoparticles after experimental exposure of trout, with some accumulation in the gill and liver.

The elimination of both non-essential and excess essential metals has recently (Wood et al. 2011a,b) been reviewed. The induction of metallothioneins in the liver and/or kidney may be very important in binding metals and providing homeostasis and/or detoxification.

8.3.10 Xenobiotics

Xenobiotics, 'foreign' chemicals (Christiansen et al. 1996), are released into the environment by a number of industrial processes and by pharmaceutical-related molecules in human waste or farm effluents which may contain antibiotics and pesticides as well as fertilizers. Silva and Martinez (2007) compared populations of *Astyanax altiparanae* living in a 'clean' reference environment and a polluted urban site. They found detrimental changes in the fish kidney, notably the interrenal cells, in fish from the urban site. They did not however attempt to analyse the pollutants in the urban site.

Fish are particularly vulnerable to contamination from human (anthropogenic) activity, including release of wood-pulp products into fresh water and the potential long-term effects on their metabolism with the possibility that they cannot adequately excrete molecules like phytestrogens which resemble and mimic reproductively active steroids, for example oestrogens such as oestradiol, and which may enter the environment in physiologically significant quantities due to human activity such as milling. Canadian researchers have been particularly active in studying the problems with mill effluents (van den Heuvel 2010; Scott et al. 2011; Kovacs et al. 2013). Guerrero-Estévez and López-López (2016) reviewed endocrine disruption in viviparous Goodeinae, endemic in the Mexican Central Plateau, a region affected by 'intensive agricultural activities and pesticide use'.

Endocrine disruption may also occur when tamoxifen and other highly potent pharmaceuticals enter the freshwater systems. Tamoxifen has been in widespread use to block natural oestrogens in human reproductive cancers; it has been reported to negatively impact fish reproduction (Maradonna et al. 2009). More recently, aromatase inhibitors have been introduced for cancer treatment, and these may have fairly long half-lives *in vitro*; if they accumulate in fresh water they may be expected to be endocrine disruptors (blocking formation of oestradiol) for fish; their excretory pathways will presumably depend on their resistance to catabolism. Xenobiotics are mostly excreted via urine or bile: Christiansen et al. (1996) found that *Boreogadus saida*, a gadid with aglomerular (8.5.1, 8.5.2) kidneys, excreted a xenobiotic solely via the bile in comparison with *Gadus morhua* which utilized the kidney.

8.4 Sites for excretion

8.4.1 The gill surface

Fish can utilize their gills, with a continuous flow-through system, to void ammonia, which would be toxic in more enclosed structures like lungs. Thus the gills, as gas-exchange organs, can excrete carbon dioxide and ammonia; in the latter capacity many fish are therefore 'ammonotelic'. A few burrow-dwelling fish may have problems with ammonia build-up and therefore have an alternate system. Toadfish (*Opsanus beta*) is known to synthesize less toxic urea in crowded conditions. It voids the urea via the gills and the 'body surface', not the urine (Wood et al. 1995).

Ammonia and carbon-dioxide excretion may occur at the gills by passive diffusion, but in the case of carbon dioxide it reaches the gills as bicarbonate ion, and dissociation is accelerated by the enzyme carbonic anhydrase in the erythrocytes. Ammonia may also be actively eliminated through a Na^+/NH_4 exchange, as described for cultured gills of trout (Tsui et al. 2009). Wilkie (1997) discussed different possible mechanisms for ammonia elimination, comparing freshwater and marine fish, and Nakada et al. (2005) suggest that glycoproteins may take an active role in ammonia excretion at the gills although recognizing that passive diffusion is 'generally thought to be the mechanism of transport' for ammonia.

Chloride cells of the gills are also important in ion regulation including ion excretion (7.7).

8.4.2 The skin

The excretory function of the vertebrate skin has not been much studied, except for its occasional, intermittent role in gas exchange in fish like eels and the toadfish. In humans, ion loss can occur with sweat and it has been recognized that unusual inorganic elements may be voided via that route too; thus arsenic and heavy metals may be detected in sweat or hair. In fish, copious mucus secretion (from gills or skin) may act as a vehicle for excreta. The role of fish mucus has been reviewed by Shephard (1994); this author considers gas exchange and the loss of unspecified heavy metals as well as aluminium. The amphibious killifish *Kryptolebias* (formerly *Rivulus*) *marmoratus* may shift its route for nitrogenous excretion to the posterior end (skin and kidneys) when out of water (Frick and Wright 2002b).

8.4.3 Excretion via the liver, bile and intestine

The liver has functions in excretion and detoxification (7.11); its products can be eliminated via the bile and the intestine or via the blood and urine. O'Grady and Maniak (2005) described potassium ions as being actively secreted by the apical membrane of intestinal cells in *Pseudopleuronectes americanus* (winter flounder). Some fish have a rectal gland off the posterior intestine which may be excretory (8.4.4).

In ureogenic and ureotelic fish the liver is the major site of ornithine cycle activity, producing urea for release to the blood. Extrahepatic ornithine cycle enzymes have also been reported from the kidney and muscle, for example in the holocephalan *Callorhincus milii* (Takagi et al. 2012).

Very small 'foreign' particles in the circulation may be phagocytosed by Kupffer cells lining sinusoids in the liver; it is not clear what happens to the phagocytosed particles. Kupffer cells, originally described in mammals, are now recognized to be macrophages (Seternes et al. 2001). Their occurrence in fish may not be universal, indeed they may not occur in the cod liver: in cod, endocardial (i.e. heart) epithelial cells were shown to be important in removing some materials from the circulation.

The liver also participates in erythrocyte breakdown and haemoglobin degradation, with the recovery and storage of the integral iron moiety. Iron is stored as ferritin or haemosiderin in the liver and elsewhere (spleen/kidney). Residual parts of the haemoglobin form bile pigments (bilirubin, biliverdin) eliminated via the bile.

Steroids may be conjugated, in the liver, to glucuronides, which are water-soluble and this can facilitate their excretion in the urine or bile. Clearance of steroids by such mechanisms, as well as protein binding in the circulation allow for regulation of the sexual cycles.

Liver cells, notably the hepatocytes forming the bulk of the organ (7.11), can also respond to environmental changes, such as increases in crude petroleum and its constituents. A typical response is the induction of more enzymes of the cytochrome P450 class, which will assist detoxification and elimination of the products to the bile, or to the urine via the blood.

Numerous experiments and observations of polluted sites have described the role of liver enzymes in modifying xenobiotics, non-natural molecules introduced to the environment through human activity. Unfortunately some xenobiotics inhibit cytochrome P450 activity as appears to be the case with trinitrotoluene (TNT) which has entered the Mediterranean and Adriatic ecosystems during recent conflicts (Della Torre et al. 2006).

8.4.4 The rectal gland

The rectal gland is a small finger-like (digitiform) blind-ending structure attached to the rectum in many elasmobranchs (Figure 4.15) and the coelacanth. In at least two elasmobranch genera (*Hexanchus* and *Heptanchus*) the gland has 'several lobes' (Fänge and Fugelli 1962) while the chimaeras (Holocephali), a sister group to the elasmobranchs may have '10 tubular structures' in the posterior intestinal wall (Hyodo et al. 2007).

Burger and Hess (1960) and Burger (1962, 1965) found that the elasmobranch rectal gland secreted a concentrated sodium chloride solution but the overall role of this structure has not been well understood, because fish can osmoregulate without it (Haywood 1974). However it does, for example under the influence of certain peptides, that is vasoactive intestinal peptide, VIP, and cardiac natriuretic peptide, CNP (Epstein and Silva 2005) or cyclic AMP (Silva et al. 2012), have the capacity to variably secrete/excrete chloride ion (Silva et al. 2012). It is now considered to function in ion (notably Na^+, Cl^-) excretion, and it is contractile (Evans and Piermarini 2001). Epstein and Silva (2005) briefly reviewed some aspects of rectal gland function, emphasizing its specialization in chloride secretion, with product passing to the large intestine. Silva et al. (2012) described the presence of aquaporin molecules (7.5.6) in the rectal gland of *Squalus acanthias*. The aquaporins were expected to enable water movement necessary for chloride secretion.

8.4.5 The kidney

The kidney, which is the primary site for nitrogenous excretion in terrestrial vertebrates, has a different role in fish, in which the gills tend to be the major site for such excretion (6.3). The kidneys in fish, however, are large, long structures and their functions have evolved over a very long time period, being strongly associated with functions other than excretion; the posterior part of the kidney is regarded as primarily osmoregulatory via kidney tubules (nephrons) while the anterior part may not have nephrons. In its capacity as an osmoregulator the kidney will excrete salts notably in marine fish and water in freshwater fish. The kidney structure is interspecifically variable, being highly linked with the reproductive system in some fish, while the kidney tubules may have very large or no Malpighian/renal corpuscles (8.5.2).

8.5 The structure and function of the fish kidney

Fish kidneys are typically paired and lie on either side of the posterior cardinal veins which drain them. The kidneys are overlain by the peritoneum, the gut and the reproductive organs, and may not be evident at first on dissection. The anterior head kidney is primarily composed of haematopoietic tissue, blood vessels and endocrine tissue ('interrenal' and chromaffin, 10.11.3) in extant teleosts.

Classical zoologists and anatomists (e.g. Goodrich 1930b; Romer 1955) spent long hours carefully studying and debating the development of vertebrate kidneys and their relationships to the urinogenital systems. In some fish (e.g. elasmobranchs) there are clearly close relationships between the kidney and the reproductive system, seen during development and subsequently. Classical zoologists postulated an ancestral archinephros with segmentally repeated coelomoducts and even nephridia (similar to these structures in invertebrates). The coelomoducts carried gametes and were open into the coelom anteriorly, as were the nephridia. Each coelomoduct emptied into the same duct as the nephridium, which bore and processed excretory products. With evolutionary time the archinephros was seen to have become an anterior pronephros and a long mesonephros, which structurally persists in extant jawless fish. The kidney of extant jawed fish is called mesonephric while the compressed kidney of mammals which has no evident segmentation is a metanephros. All vertebrate kidneys now have a major component of 'kidney tubules' or nephrons which bear little obvious resemblance to nephridia, and have no cilia as usually presented, though Ojeda et al. (2006) write of 'multiciliated' cells in the second part of the proximal convoluted tubule of the African lungfish. Coelomoducts persist, for example in female elasmobranchs (and humans) as a single pair of oviducts. The sperm carrying vas (ductus) deferens in male elasmobranchs which has a series of several vasa efferentia (thus recalling segmentation) leading into it on each side, is separate from the urinary duct.

The fish kidney is long—much longer—relative to body size, than the compact kidney of other vertebrates. The essentially paired kidneys, with varying degrees of fusion particularly posteriorly; *Amia* has extensive posterior fusion (Youson and Butler 1988) and they lie dorsally, and typically, on each side of the aorta and posterior cardinal vessels. The kidneys receive blood from the dorsal aorta via short renal arteries, and also from the two renal portal veins (not present in e.g. mammals), one on the outer side of each kidney (5.9 and Figure 5.4). Blood drains to the posterior cardinal system. The kidneys lie beneath the peritoneum and generally extend caudally from just posterior to the heart to close to the rectum. At dissection the kidneys may be obscured by the gut and the swim bladder when present. They may be overlain by the gonads and, in the case of male elasmobranchs, the reproductive ducts may run across or even through the kidney (Figure 8.2). The anterior part of the kidney (the head-kidney) may be narrower than other regions or the kidney may be fairly uniform in width (Figure 8.2).

As viewed during dissection the kidney tissue is dark and loose; it is not clearly encapsulated. The pigmentation is due to deposits of lipofuchsin, with possible contributions of melanin and iron stores (ferritin/haemosiderin).

8.5.1 Kidney microstructure

Vertebrate kidneys contain a series of tubules (nephrons) with a highly defined structure, often typified by the mammalian (human) pattern. However, fish kidneys are markedly different from the mammalian organ both macroscopically and in nephron structure and presence. The anterior 'head-kidney' of teleosts does not contain nephrons; it consists of haematopoietic tissue, with concentrations of endocrine tissues (chromaffin and corticosteroid) around the veins. This tissue is separated away from the kidneys in elasmobranchs but is still very close to them. The chromaffin tissue, so-called because of its affinity for potassium dichromate (10.11.3), forms a series of suprarenal glands in elasmobranchs and can secrete adrenalin; the tissue is equivalent to that of the adrenal medulla in tetrapods. Tissue functionally equivalent to the tetrapod adrenal cortex forms a series of distinct interrenal glands in elasmobranchs (the number and size being highly variable) and the tissue, in all fish, is called 'interrenal'.

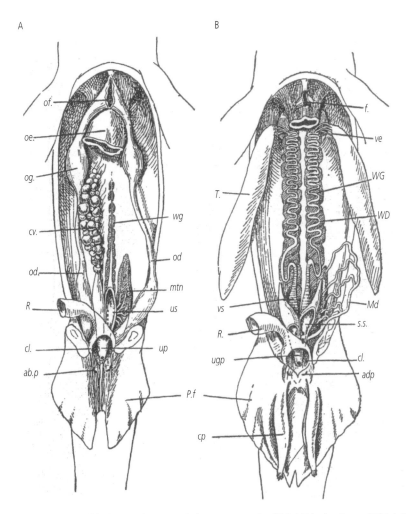

Figure 8.2 Ventral view of dissected dogfish showing the urinogenital systems. From Goodriich 1930, after Bourne 1902. A. Female. B. Male. Abp. abdominal pore; cl. cloaca; cp. clasper; f. vestige of oviduct in male; Md. ducts of posterior region of kidney; od. Müllerian oviduct; oe. cut end of oesophagus; og. oviducal gland; ov. ovary; Pf. pelvic fin; R. rectum; s.s. sperm-sac; T. testis; up. urinary papilla (female); ugp. urinogenital papilla (male); us. urinary sinus; ve. vasa efferentia; vs. vesicular seminalis; Wd. Wolffian mesonephric duct; WG. Wolffian gland or mesonephros. Note the male kidney (mesonephros) seems to be much more extensive than the female kidney.

Some teleosts may bear obvious 'corpuscles of Stannius' on the kidney surface (10.11.8). These corpuscles were first described in sturgeon by Stannius in 1839 (Nadkarni and Gorbman 1966) and were thought to be adrenal gland homologues. Rasquin (1956) states that the most usual number of corpuscles is two but salmonids have 6 to 14 and *Amia* as many as 50. The current view of these small, round, pale structures is that they are involved in calcium regulation, synthesizing the hypocalcaemic hormone stanniocalcin (STC-1, Greenwood et al. 2009), thus promoting calcium excretion, and are not adrenal gland homologues.

Yet another specialized cell-type, the juxtaglomerular (JG) cell, is found in vertebrate kidneys, including fish classified as 'aglomerular' (Oguri 1989). The JG cells are expected to be close to the anterior region of the nephron, to contain identifiable specific granules (Capréol and Sutherland 1968) and to secrete the enzyme renin which cleaves angiotensinogen in the blood to release angiotensin, and subsequently aldosterone, an important

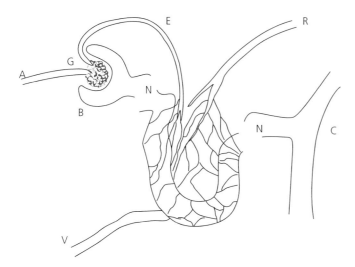

Figure 8.3 Diagram summarizing the blood supply to, from, and round a generalized glomerular teleost nephron. A. afferent arteriole; B. Bowman's capsule; C. collecting duct; E. efferent arteriole, G. glomerulus; N. nephron tubule; R. vessel incoming to kidney from the renal portal system; V. outgoing venule to the renal vein.

mineralocorticoid, at least in mammals. However, aldosterone may not be active in fish (McCormick et al. 2008) and the angiotensin may be directly important in osmoregulation (7.5, 7.10), for example in the elasmobranchs with effects on the gills and rectal gland (Hazon et al. 1999), and drinking (the dipsogenic effect, Anderson et al. 2001).

8.5.2 The nephron

Fish nephrons (kidney tubules) generally have a Malphighian (renal) corpuscle with a lumen passing into a thin neck which may be ciliated, and is not present in mammals. The corpuscle (Figures 8.3, 8.4) consists of a glomerulus, usually described as a cluster of small blood vessels 'the size of capillaries' (Romer 1955), partly enclosed by a Bowman's capsule, as in tetrapods. The corpuscle is the site of 'ultrafiltration' (achieved by high blood pressure) into the lumen of the capsule, with fluid and small molecules from the blood in the glomerulus passing into the lumen of the Bowman's capsule, from which the tubule arises. From the capsule lumen, fluid passes successively into the tubule's proximal and distal (convoluted) regions, with (unlike mammals) no loop of Henle between the two (Figures 8.3, 8.4), although other loops are described for elasmobranchs (Gambaryan 1994 and Morgan et al. 2003). The distal region drains into the collecting ducts, which link to the kidney duct.

The structure of the nephron varies considerably between fish taxa, some marine teleosts even lack glomeruli (8.5.4). The 'neck' linking the corpuscle to the proximal tubule, first region, is relatively longer in elasmobranchs (Romer 1955). As shown by Romer too, elasmobranch nephrons are long compared to teleosts. Indeed, Morgan et al. (2003) describe the elasmobranch nephron as having five loops; with a countercurrent arrangement, fewer loops are shown (Figure 8.4) by Gambaryan (1994). The corpuscles of both cyclostomes (agnatha) and elasmobranchs are very large, and useful for sampling the filtered fluid (7.6). The function of the corpuscle when present, as it is in most vertebrates, is that it allows retention of the blood cells and large molecules during urine formation by ultrafiltration.

Hickman and Trump (1969), who used the southern flounder (*Paralichthys lethostigma*) as their model, dropped the term 'convoluted' (previously used by, for example, Romer 1955) from the description of the proximal and distal parts of the nephron, but divided the proximal tubule into first (PI) and second (PII) regions. Cilia occur in the neck and in the distal tubule and the tubule is also lined by a brush border.

The structure of the nephron ('tubule') in *Amia* was described by Youson and Butler (1988). This nephron has a renal corpuscle with glomerulus and Bowman's capsule comprising two layers; the visceral podocyte layer (8.5.3) and the parietal

A   B

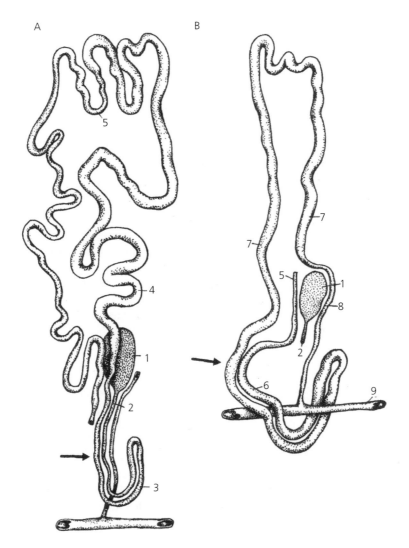

Figure 8.4 Scheme showing the arrangement of a typical elasmobranch nephron, based on microdissection. A = proximal and B = distal parts of the nephron. Arrows indicate proximal and distal countercurrent systems. 1 = glomerulus (enclosed in capsule?); 2 = neck segment; 3,4,5 = proximal segments I, II and III; 6, 7 early and late (distal?) segments; 8 = connecting segment; 9 = collecting duct. Reproduced with permission, S.P. Gambaryan (1994) *Journal of Morphology* 219, 319–39. ©1994 Wiley-Liss. Inc.

layer which continues on into the tubule region. No juxtaglomerular 'apparatus' (8.5.1) was found. The actual tubule had a series of regions: neck, first and second proximal, first and second distal and collecting. Mitochondria-rich cells, thought to be very similar to gill chloride cells, formed the second distal segment, while the first distal segment also had a lot of mitochondria and 'deep infolding of the plasma membrane' (i.e. a brush border with microvilli) which these authors considered indicated sites for ion transport.

Gambaryan (1994) microdissected nephrons from elasmobranchs (Figure 8.4) and teleosts (Figure 8.5), using dilute hydrochloric acid. His measurements show the nephron length in the cottid (*Cottus gobio*) to be more than twice that of some other teleosts such as herring (*Clupea harengus*) and (*Gadus morhua*).

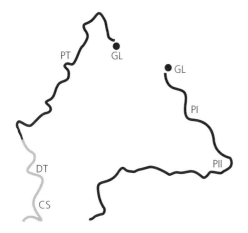

Figure 8.5 Dissected nephrons from two freshwater teleosts: (left) bream (*Abramis brama*) and (right) salmon (*Salmo salar*). Reproduced with permission, S.P. Gambaryan (1994). Microdissectional investigation of the nephrons in some fishes, amphibians and reptiles inhabiting different environments, *Journal of Morphology*, 219, 319–39.© 1994 Wiley-Liss, Inc. CS = connecting segment (to collecting duct); DT = distal tubule; Gl = glomerulus in the renal corpuscle; PT = proximal tubule; PI first part of proximal tubule; PII second pert of proximal tubule.

8.5.3 Nephron function

Items to be excreted enter the tubule either at the corpuscle during ultrafiltration, when water and other small molecules and/or ions enter the Bowman's capsule lumen under pressure, or pass through the cells of the proximal or distal convoluted tubule either actively, with the use of ATP, or by passive diffusion.

Ultrafiltration at the corpuscle depends on the structure of special cells called podocytes, forming slits round the corpuscle (Bloom and Fawcett 1975, Figures 31.7 and 31.8).

The blood supply to the nephron (Figure 8.3) comprises an afferent arteriolar input to the glomerulus, at high pressure from the dorsal aorta via short renal arteries. Both the input to and the output from the glomerulus is arteriolar. The output efferent arteriole gives rise to a capillary network round the nephron where there is also input from the renal portal vein (not found in mammals). The renal portal vein derives from the caudal vein, giving an opportunity for the venous blood from the tail (which can be highly muscular and active in locomotion, as in tail-flips) to disgorge excess compounds like

creatine. The blood supply round the nephron as opposed to the glomerulus gives an opportunity for resorbtion from or secretion into the lumen; in elasmobranchs the nephrons partly occupy a 'sinus zone', in which the tubules 'mix freely in large blood sinuses' (Lacy and Reale 1985).

During passage along the nephron the lumen contents can be modified, either by resorption, secretion or diffusion via the nephron wall (see the very impressive diagram summarizing these activities in Hickman and Trump 1969, Figure 48). The inner surface of the nephron lumen is expected to have a 'brush border', the surface area being increased by microvilli. The resulting urine may not have a high concentration of nitrogenous compounds as nitrogenous excretion is generally achieved elsewhere, for example, ammonia through the gills. Creatine levels voided may exceed those of urea in active fish (Whitledge and Dugdale 1972). Urea is actually conserved (resorbed) in most elasmobranchs, an exception being the freshwater *Potamotrygon*.

Excess divalent ions (Mg^{2+}, Ca^{2+}, $SO4^{2-}$) are voided in the urine, notably of marine fish (Table X in Hickman and Trump 1969). Sulphate transporters (Slc13a1 and Slc26a1) have been described in the apical and basolateral membranes of proximal tubule (PT) cells of freshwater eel, *Anguilla japonicus* (Nakada et al. 2005), enabling reabsorption: reduction in quantity of these transporters occurs after transfer of fish to seawater. Kato et al. (2009) identified a different sulphate transporter (mfSle26a6A) in the PT brush borders of the puffer fish *Takifugu obscurus*, which is predicted to be the major secretor/excretor route for sulphate in marine teleosts.

The process of ultrafiltration retains cells and large molecules but small proteins like antifreeze protein (AFP) may end up in the urine. This protects the urine from freezing but is expensive in terms of replacement and some cold-water fish, notably Antarctic notothenoids, have aglomerular (8.5.4) kidneys, which conserve AFP.

The composition and volume of fish urine is expected to vary considerably between taxa; it is predicted to be more copious in freshwater fish. The urine flow rates for freshwater fish are about 1–10 ml/hr/kg (Hickman and Trump 1969) with high rates (11 ml/hr/kg) for goldfish (*Carassius auratus*) at 18–23 °C and lower rates for the catfish

(*Catastomus commersonii*) at 4–18 °C (1–4 ml/hr/kg). The role of special water channels, aquaporins, in fish kidneys is not yet clear but may be important in expediting resorption for marine fish (Cutler et al. 2007). The urine flow rates for glomerular marine fish and euryhaline fish in seawater are low, about 1 ml/hr/kg (Cutler et al. 2007).

Urine composition may also vary annually/seasonally. There is likely to be steroids and steroid-conjugates released at the reproductive season, acting as pheromones.

Overall excretory products may vary seasonally too. Estuarine fish will respond to changes in salinity as will freshwater fish under drought or flood conditions. Freshwater fish will also respond to conserve or eliminate particular ions; the geologic/mineral base for some waters is highly variable; thus 'black' and 'white' waters in Brazil may mix at confluences or even flow backwards in drought. Calcium-rich fresh waters have a high pH in comparison to acidic waters of boggy areas which may be quite close. Fish may need acclimatization periods to adjust to these varying conditions if they cannot otherwise avoid the changes; they may also respond to external conditions including temperature changes by having internal changes in the regulated levels for particular compounds or ions; thus plasmatic levels for Na^+ may increase in a cold season (Fletcher 1977).

8.5.4 The aglomerular kidney

Regarded initially as unusual, it is now being suggested that marine teleosts in general might benefit from reduction in the glomerulus, to conserve water (McDonald and Grosell 2006). Ozaka et al. (2009) consider that aglomerular kidneys are more common than formerly reported. Previously stated to be a relatively rare situation; about 30, mostly marine, teleost species were termed aglomerular (Beyenbach 2004). Poor development of glomeruli has been associated with retention of antifreeze (Eastman et al. 1987). These authors used the term 'pauciglomerular', derived from Marshall and Smith (1930) for some boreal fish with small corpuscles.

In those fish which are aglomerular, urine formation is expected to occur by secretion into the nephron lumen.

8.6 Verifying regulation of, and pathways for, excretion

The regulation and routing of excretion in fish is not easy to study. Urine can be collected but this is difficult. Plasmatic levels of particular substances may indicate regulation but not routing. Analyses of changes in surrounding water does not provide information about routing either unless restraint and a split chamber (with head on one side and tail in the other) is used, in which case gills versus urine can be established as sites for excretion of particular products. Such a 'split box' is described by Thomas and Rice (1981), using anaesthesia and a 'rubber dam' to achieve a study of excretion routes for aromatic hydrocarbons in char.

Transport of material en route for excretion is generally dependent on carriage in the blood plasma, where it would be dissolved. Gases such as ammonia (NH_3) or carbon dioxide (CO_2) can be transported as dissolved gas or as ions (NH_4^+, HCO_3^-). Schmidt-Nielsen (1990) points out that ammonia is highly water-soluble and diffuses very rapidly so that it can be lost readily from fish surfaces into water. In conditions where fish are not surrounded by large volumes of water, ammonia can build up and become toxic. In these circumstances the ammonia may be detoxified; in the case of air-exposed marble goby, *Oxyeleotris marmoratus*, by conversion to glutamine (Jow et al. 1999). Toadfish (*Opsanus tau*) in burrows can achieve ammonia detoxification by conversion to urea (Wood et al. 1995) prior to excretion via, for example, the gills or skin.

Water itself may need to be excreted and its excretion is dependent on the osmoregulatory system (7.5) and the prevalence of water excretion depends on the salinity of the environment; it is expected that active excretion of water in the urine is very important for freshwater fish.

Large molecules like steroids and some xenobiotics are frequently modified by the liver prior to excretion via various routes and inorganic ions may depend for their regulation/excretion on special mitochondria-rich ('chloride') cells primarily found on the gills, though the kidney may be the export route for divalent ions.

8.7 The control/regulation of excretion

Various control mechanisms can be expected to operate in excretory homeostasis. Antagonistic pairs of hormones may control calcium levels (10.12.1). Natriuretic (10.12.1) peptides including atrial (cardiac) natriuretic peptide (ANP/CNP) may influence loss of sodium at the gills and via the rectal gland (Loretz and Pollina 2000). Mineralocorticoids have been investigated and the occurrence in fish of the potent tetrapod hormone aldosterone has not yet been established (Colombe et al. 2000; Sturm et al. 2005) and the role of steroids in excretory regulation in fish is not well understood though it is recognized that fish-specific non-mammalian steroids, such as 1-α hydroxy corticosterone found in elasmobranchs (Idler and Truscott 1972) may be important. McCormick et al. (2008), revisiting some of the questions about hormonal control of excretion in fish, found no evidence for aldosterone involvement (8.5.1) and queried whether there is another steroid involved.

Bibliography

Black, V.S. (1957). Excretion and osmoregulation. In: M.E. Brown (ed.). *The Physiology of Fishes, Vol. 1: Metabolism.* New York, NY: Academic Press, pp. 163–205.

Forster, R.P. and Goldstein, L. (1969). Formation of excretory products. In: W.S. Hoar and D.J. Randall (eds). *Fish Physiology, Vol. I: Excretion, Ionic Regulation, and Metabolism.* New York, NY: Academic Press, pp. 313–450.

Wright, P.A. and Anderson, P.M. (2001). *Nitrogen Excretion. Fish Physiology, Vol. 20.* San Diego, CA: Academic Press.

CHAPTER 9

Reproduction

9.1 Summary

Fish have an enormous range of reproductive outputs, from a few progeny to millions. The life cycles may be brief or very long with repeated, but possibly intermittent, reproduction over many years.

The production of gametes though similar to the process in mammals differs for females in that the early stages of oogenesis can be repeated during the adult stage.

Egg production in fish normally involves deposition of a large amount of yolk.

Some fish species, notably chondrichthyans, have internal fertilization.

Reproductive behaviours include migration, beach-spawning and mouth-brooding.

9.2 Introduction

Of all the vertebrate groups fish have the greatest variability in reproductive output and other features such as periodicity, parental care and reproductive sites. Thus egg size may be huge or very small, Females may be live bearers or deposit eggs, nests may be built and guarded or millions of eggs may be deposited in open waters. Sexual dimorphism (9.3) may lie in different colouration or size, and sex reversal may occur. Gametes may be developed over a long period (years) or the whole life cycle may be completed within a year. Some fish migrate very long distances and spawn just once (semelparity) whereas other fish have the capacity for many reproductions over several years (iteroparity). The divergence of chondrichthyans and teleosts has associated radical differences in reproduction, both in the structure of the gonads and their ducts as well as the reproductive process.

As reproductive capacity is important to maintain fish populations and withstand fishing pressure, there has been a lot of effort directed at understanding fish reproduction, its control and variability. Because of the applied aspect, however, data collected from wild fish have sometimes concentrated on fish of economic importance whereas experimental information may have been gathered from small freshwater fish easy to keep in aquaria. Therefore, in spite of the huge amount of information collected there are still areas including periodicity and individual output variability which are not well understood, even for commercially important species.

In the juvenile phase, gonads contain gamete precursor cells, such as primordial germ cells and/or cells arrested in early stages of pre-meiosis, fish enter into adulthood as they begin to develop their first crop of gametes, a process of gametogenesis (9.8).

After the first reproduction, fish may be able to repeat the process, following a period of recovery. Such iteroparity with repeat reproductions, particularly from year to year, is usual. Exceptions are the semelparous fish like annual fish (9.8, 9.15) and fish which undergo long migrations (9.14).

If a fish is in a post-reproductive stage with small gonads it may be referred to as 'immature' and this can give rise to confusion with fish that have not yet been through a reproduction. When categorizing fish it may be important to double-check the status of fish categorized as 'immature'; size of the fish, and size of the gonads and/or gonad wall thickness can be helpful in distinguishing fish which have never reproduced from adult fish which are between

Essential Fish Biology: Diversity, Structure and Function. Derek Burton & Margaret Burton.
© Derek Burton & Margaret Burton 2018. Published 2018 by Oxford University Press.
DOI 10.1093/oso/9780198785552.001.0001

reproductive phases or which feasibly have entered a senile non-reproductive state, which, however, is rarely reported even from large, old fish (Rideout and Burton 2000). Post-reproductive fish are sometimes called 'resting' and there may be signs of recent reproduction such as residual eggs or sperm, in which case the term 'post-spawning' (9.13) may be appropriate. Macroscopic inspection of the gonads can stage the fish (Appendix 9.1, Tables 9.3–9.6), assigning fish to categories: immature, maturing, etc.

It is usual for fish to have separate sexes, in which case they are termed gonochoristic and the gonads are either ovaries producing eggs or testes producing sperm. Some fish, however, are hermaphrodite, developing both ovaries and testes (9.4). Sex determination in fish (9.4) is thus more labile than in mammals.

The process of gamete (egg or sperm) production is termed gametogenesis (9.8), spermatogenesis (9.9) for male sperm production and oogenesis (9.10) for female egg production. Gametogenesis, as in other animals, goes through a pre-set sequence of cell divisions (Appendix 9.7), first mitoses to multiply numbers of cells and then meiosis which reduces (halves) chromosome number in preparation for the fusion of one egg with one sperm producing a doubling of chromosomes, back to the original number, but with a blended genetic composition.

Although the process of gametogenesis has been conserved in the major events occurring and is very similar throughout the animal kingdom, particularly for males, for females there may be some very important differences in the timing and in the presence of much, little or no yolk. Most fish share with other non-mammalian vertebrates and monotremes the need to deposit a lot of yolk in the developing eggs. Also, the large number of eggs produced by some fish means that production of new potential eggs (as oogonia, 9.10) through multiple mitoses should occur before each reproduction unlike the situation in, for example, mammals which apparently rely on a store of immature cells formed mitotically very early in the life cycle.

9.3 Sexual dimorphism

Some taxa have males with much smaller size than females. This trend is already apparent in flatfish such as winter flounder, in which males can undergo puberty at a smaller size than females. In some deep-sea

anglers the reproductive male is much reduced in structure. The small, free-swimming male can become permanently attached, at the mouth, to the female. The male remains small, and there is fusion to the female, with transfer of nutrients from female to male. The male is therefore termed 'parasitic', although that term is usually reserved for interspecific relationships where harm is done to the host (13.13). Some salmonids (*Oncorhynchus*) may have small precocious males (sneakers or jacks) which, as satellites to a spawning pair, may dart in to fertilize eggs as they are deposited (Foote et al. 1997; Young et al. 2013).

External differences between males and females may arise just before spawning or may be more permanent, as in maximum size achieved or the presence of specific structures such as elasmobranch or holocephalan claspers (Figures 1.7, 1.8, 1.9, 9.5). Colour change prior to spawning, with reversion afterwards, is quite common. Thus male sticklebacks, male salmon and male lumpfish (*Cyclopterus*) become reddened; male capelin become silver with longitudinal spawning ridges. The colour changes allow for appropriate recognition signals prior to courtship and the capelin spawning ridges help to hold a female in position before egg deposition. Prespawning salmonid changes may include a snout elongation, resulting in the hooked kype, which occurs in Pacific salmon (*Oncorhynchus* spp.), being particularly pronounced in male chum, coho, sockeye and pink salmon, with the latter also having a large hump developing dorsally (Hart 1973).

Bothids (flatfish) may have very different (wider spaced) eyes in males and other external differences besides size may permanently differentiate the adult sexes. Erlandsson and Ribbink (1997) have reviewed sexual size dimorphism in the cichlids. They are able to makes generalized statements: that where females are bigger than males (at the same age) the species with one exception are mouth-brooders (Appendix 9.2, Table 9.7) whereas where males are larger than females substratum brooding occurs.

9.4 Hermaphroditism, sex determination and manipulation

Hermaphroditism, with individuals developing both ovaries and testes (reviewed by Sadovy and Shapiro 1987), is relatively rare, occurring synchronously (simultaneously) in some deep-sea fish (Greenwood 1975) and sequentially in some Acanthopterygian coral reef species including species in the Labridae, Serranidae and Pomacentridae. The Pomacentridae (clown/anenome fish) are males first; that is, protandrous (Iwata et al. 2008), whereas labrids which change sex are commonly protogynous (female before male) but with one labrid species (*Halichoestes trimaculatus*) capable of either protandry or protogyny (Kuwamura et al. 2007). Sex changes occur in response to group social interactions and dominance hierarchies. Two fish species, *Rivulus* (now *Kryptolebias*) *marmoratus* and *K. hermpahroditus*, have been described as achieving internal self-fertilization (Earley et al. 2012). An earlier announcement of this unusual condition, unknown in any other vertebrate, was made by Harrington (1961), and the possibility of self-fertilization in deep-sea fish with ovotestes was briefly discussed by Marshall (1971).

Vertebrate eggs, after deposition, may be subject to conditions, notably temperatures higher or lower than 'normal' which affect the ratio of males/females developing from genetically equivalent material (Conover and Heins 1987; Schwarzkopf and Brooks 1987). In turtles, males may be more prevalent in the upper or lower temperature ranges (Schwarzkopf and Brooks 1987). Aquaculturists may be interested in using such environmental parameters to control sex of developing eggs and, for example, promoting female production (gynogenesis) in sturgeons for caviar production and, for maximizing growth, may wish to suppress gonad development by inducing chromosome abnormalities like triploidy. Steroid treatment can influence sex ratios too (Flynn and Benfey 2007).

Sex determination can thus be affected by environment, including temperature and steroids or steroid mimics in the water, but any reports on unusual sex ratios markedly different from the 'expected' 1:1 may be problematic if sampling of wild fish, for instance, has accessed pre-reproductive sexually segregated groups. For those fish which are 'thermosensitive' (Baroiller and D'Cotta 2001), the majority have increased numbers of males with higher temperatures although the reverse situation has been reported for two species (*Dicentrarchus labrax* and the catfish *Ictalurus punctatus*). Anomalous results reported include the situation with rosy barb

(*Puntius conchonius*). When reared at about 27 °C (Çek et al. 2001) the fish were strongly female sex-skewed (statistically significant) but this had not been seen previously with other workers rearing the same species at 25 °C. Wild carp from an Oklahoma lake were reported (Mauck and Summerfelt 1971) to have a slightly female-skewed composition (57 per cent) and a non-significant difference (male-skewed: 1.31 males per female) was found for *Micropogonius furneri* off Rio de Janiero (Vicentini and Araüjo 2003), a similar (but female-skewed; 1.23 females per male) ratio as reported for swordfish (*Xiphias gladius*) from the Eastern Mediterranean (Aliçh et al. 2012). An example of environmental pollution apparently affecting sex ratios is given by Larsson and Förlin (2002) with proximity to an active pulp mill being associated with male bias for an eelpout (*Zoarces viviparus*). Fish may be particularly prone to 'endocrine disruption' and individual species (e.g. brown trout in Ireland, Tarrant et al. 2005) can be used as 'sentinels' to monitor possible contamination of water.

9.5 The male reproductive system

The testes are usually paired and elongated though they are triangular in flatfish (Figure 9.1 left, and Figure 9.2) and convoluted in cod (Figure 9.1). The mesentery suspending the testes is termed the mesorchium.

Expected to be at their largest before the reproductive season, the testes may comprise 10 per cent (gonadosomatic index, GSI = 10, Figure 9.3) or more of the wet weight of the whole fish. Surprisingly, however, the highest GSI may occur prior to the final stages of spermatogenesis (sperm formation), actually months before reproduction (Figure 9.3) so it may be unsafe to use GSI profiles to determine the exact time of reproduction unless a sharp drop in GSI is found as seen for winter flounder between May and July, off Newfoundland, (Figure 9.3) when spawning occurs.

In chondrichthyes (dogfish, rays) the testes and kidneys form a linked urinogenital system; several ducts (vasa efferentia) pass on each side from the testes to a vas deferens (sperm duct) ventral to, and partially embedded in, the paired kidneys. The vas deferens conducts sperm to the cloaca for transfer during copulation. The internal structure

Figure 9.1 Dissected males showing the gonads, Left: winter flounder, one testis *in situ* (below), the other laid on the fish (above). Right: cod with testes removed and laid by the fish. Cod is presented in a smaller scale than flounder.

Figure 9.2 Dissected non-reproductive (see 9.16) male winter flounder showing both testes, one *in situ*, the other laid on the fish.

The lobule walls are expected to support groups (nests) of spermatogonia, the 'stem cells' for sperm production (Figure 9.11), and, depending on the seasonality of the particular species, may be in active spermatogenesis or have completed seasonal sperm production. The structure may change considerably during spermatogenesis, with cysts (9.5.1 and Figure 9.10) transiently present. Immediately prior to reproduction the testis may appear to be a fairly fragile, easily ruptured, container of sperm. At this point there is little evidence of ongoing spermatogenesis and the next season's output relies on some residual spermatogonia. A 'staging' system exists for the macroscopic appearance of bony fish testes (Appendix 9.1).

Chondrichthyans have internal fertilization, the males have external claspers (Figures 1.8, 1.9, 3.15, 9.5), used in copulation; a pair as extensions of the pelvic fins, and holocephalans (chimaeras) have an additional cephalic clasper (i.e. on the head). This structure (Figures 1.8, 9.5) is also known as a frontal tentaculum, and there are additionally two

of the testes may be 'ampullar' (Dodd 1983), a term also used for a swelling on the vas deferens of some elasmobranchs. In bony fish the testes are detached from the urinary system with a sperm duct per testis and most have a relatively simple lobular structure (9.5.1 and Figure 9.4) although tubules are described in some fish (Grier and Lo Nostro 2000, 9.5.1).

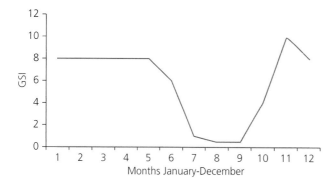

Figure 9.3 Gonadosomatic indices (GSI = gonad weight/body weight × 100) plotted against time (months) for winter flounder *Pseudopleuronectes americanus*. Based on original data and Burton and Idler (1984).

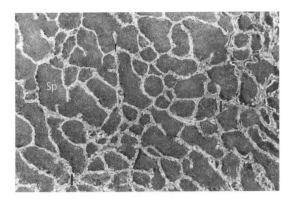

Figure 9.4 Testis structure of bony fish. Lobules (l) of winter flounder testis, containing spermatozoa (sp) showing androgens associated with lobule boundaries, Indirect antibody method (Appendix 10.1b) using anti-11-ketotestosterone as the first antibody.

prepelvic tentacula as patches on the skin (Figure 9.5). These structures have dermal denticles on them, assisting contact during copulation.

The pelvic claspers are grooved and one is used to transfer sperm to the female. The sperm are flushed through the groove with seawater from the siphons (Gilbert and Heath 1972), which are two muscular sacs. These lie subcutaneously in the abdominal wall at the level of the pelvic fins (*Squalus acanthias*) or from that point as far forward as the pectoral fins (*Mustelus canis*), in which case the siphons are very long. A few bony fish species such as poeciliids (guppies and mollies) and

ocean pout also have internal fertilization with deposition of sperm into the female's reproductive tract; male poeciliids have modified anal fins forming gonopodia for copulation/intromission. In both chondrichthyans and poeciliids sperm may be grouped in 'packages' for transfer. These aggregations of sperm are termed spermatophores or spermatozeugmata. In the basking shark *Cetorhinus maximus* the spermatophores, packets of sperm each about 2 to 3 cm in diameter, are formed in the enlarged vas deferens (Dodd 1983). Harder (1975) describes Leydig's gland, between the upper part of the vas deferens and the kidney in elasmobranchs, as involved in spermatophore production, as well as the secretion of fluid. In the Poeciliidae, groups of sperm form spermatozeugmata (Harder 1975) in which the sperm flagella are rolled into a spiral and the heads of sperm group together as a transferable unit.

In chondrichthyans there may be a swelling (seminal vesicle) of the vas deferens posteriorly, and Dodd (1983) reports a large ampulla, similarly an enlargement of the vas deferens, in basking shark (*Cetorhinus maximus*), containing a large volume ('gallons') of fluid. Accessory structures are rarely described in teleosts although the glands of blennies (Patzner and Seiwald 1985) are an exception and male sticklebacks (*Gasterosteus aculeatus*) are reported to use a kidney secretion ('spiggin' Rushbrook et al. 2007) for nest glue.

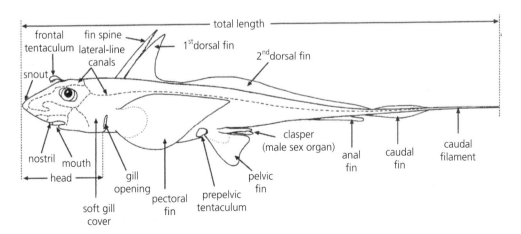

Figure 9.5 A male chimaera showing tentacula and claspers. Reproduced with permission. L.J.V. Compagno (1999). *The Food and Agriculture Organization of the United Nations, FAO species guide for identification, Vol. 3.*

9.5.1 Testis microstructure

Each testis is surrounded by a tunica albuginea, connective tissue covered by coelomic epithelium (peritoneum), and is subdivided into more or less permanent smaller chambers variously called lobules (most bony fish, Figure 9.4) or tubules (a few bony fish, e.g. viviparous species) or ampullae (chondrichthyans). Between the lobules, tubules or ampullae there will be some connective tissue, blood vessels, etc. Depending on the season the lobules, tubules and ampullae will contain developing or fully formed sperm, and the developing sperm may be in smaller groups, within cysts (Figure 9.10).

Connective tissue between the lobules or tubules in bony fish is similar to that of mammals and is called interstitial tissue; it is expected to contain steroidogenic (steroid-secreting) cells, usually termed Leydig cells. However, Leydig cells, steroidogenic (synthesizing androgens such as testosterone) in tetrapods, may not be obvious in fish. Androgens (e.g. 11-ketotestosterone) are synthesized and circulate in fish even if their exact site of formation is not always certain in males. Figure 9.4 shows association of androgens with lobule boundaries and the 'lobule boundary cells' may be Leydig cell homologues.

Terminology has been contentious and homology arguable, however the role and equivalence of lobule boundary cells (Lofts and Bern 1972) was a matter for vigorous debate. Exceptionally clusters of cells like Leydig cells have been reported for some teleosts such as *Gobius paganellus* (Stanley et al. 1965). The presence of Leydig cells in elasmobranchs has also been controversial (Dodd 1983); their steroidogenic role might be taken by the ampullar cells also termed 'Sertoli' cells. The presence and role of Sertoli cells in fish has also been a matter for debate. Sertoli ('sustentacular') cells, which are quite obvious in mammals, are also seen, as 'nurse' cells, in some invertebrates. In mammals the Sertoli cells are large, stretching from the basement membrane to the lumen of seminiferous tubules and form a distinct and apparently permanent feature. The mammalian Sertoli cell has several roles including secretion of androgen binding protein and inhibin (9.18), a paracrine (10.2), as well as forming a route for transfer of nutrients to the developing gametes (hence 'nurse' cells). It is possible that cells surrounding the cysts of teleosts may be Sertoli cell homologues.

9.5.2 Spermiation and spawning

Release of sperm (spermiation) to the sperm ducts may result in release of sperm from the gonopore if the fish is handled or slight pressure is applied. In natural spawning, however, the release is stimulated by the proximity of females, the presence of pheromones (chemicals released to the water which influence behaviour of conspecifics, 2. 3.1, 14.7) and often, a simple or elaborate courtship with visual stimuli.

Pheromonal stimulation of spawning in freshwater has been studied (Stacey and Cardwell 1995) and various conjugated or non-conjugated steroids have been implicated. It is presumably important that the pheromones would be species-specific, however, and the relative universality of some steroids might be a problem.

Although the final stimulation for spawning may be pheromones and courtship, for fish that live in a seasonally changing environment external cues (9.18) such as day length or temperature may be important to enable synchrony of gamete formation (gametogenesis) before release cues can operate. However, in the case of some species, like winter flounder, the males have completed spermatogenesis many months before the females are spawning-ready (Burton and Idler 1984), a strategy that will enable males to respond promptly.

9.6 The female reproductive system

Ovaries may be single (some elasmobranchs) or paired (most teleosts) and elongated or more spherical. The ovaries may also be partially fused posteriorly, as in cod. The ovarian output may be huge and the GSIs may approach 30 per cent, particularly when hydration of the oocytes (9.10.7 and Figure 9.6) occurs. The mesentery suspending each ovary is termed the mesovarium (Figure 9.8).

The ovaries of chondrichthyans and some teleosts (e.g. protacanthopterygian salmonids) are essentially 'naked' (gymnovarian), that is they have no thick muscular wall surrounding the oocytes and no attached duct (Figures 9.7b, 9.8). The products of

Figure 9.6 Female cod dissected to show ovaries. April, spawning ready with some hydrated oocytes (H). L = liver; O = ovary.

oogenesis, as in mammals, erupt (ovulate) through the thin ovarian epithelium into the body cavity where they may enter a separate open-ended duct (oviduct) before fertilization in the duct or externally in the water. The major groups of paracanthopterygians (cod, haddock, etc.) and acanthopterygians (flatfish, sticklebacks, labrids and many more), however, have a muscular wall enclosing the ovary (cystovarian); the ovarian tissues line a large lumen (Figure 9.7a) into which ovulation occurs, and the lumen communicates directly with the oviduct. The cystovarian condition is usually obvious macroscopically.

Macroscopic inspection of ovaries as well as their size can usually give considerable information about the stage of development and, as for males, for some bony fish species, particularly those of commercial importance, a staging system exists (Appendix 9.1). This may include size of the ovary, colour and consistency. As many fish produce large eggs, macroscopic inspection of the ovary also gives information on the closeness to ovulation based on the size of individual oocytes and their colour or transparency. Low-power microscopy with a dissecting microscope, following chemical treatment (Stockard's solution Goetz and Bergman 1978, Appendix 9.5), but no sectioning, can also show migration of the nucleus close to ovulation. Fleming and Ng (1987) preferred formalin to Stockard's for measuring salmonid eggs.

Microstructure of the ovary comprises lamellae (Figures 9.7, 9.15) to which the oogonia and oocytes are attached.

Steroidogenesis is expected to occur in small follicular cells which may be seen around developing oocytes. Each oocyte is surrounded by a follicle which may be clearly differentiated into an outer thecal layer (fairly flat cells) associated with testosterone production and an inner granulosa layer of cuboidal cells which are expected to synthesize oestrogens from thecal testosterone during oogenesis, with progesterone(s) synthesized later.

Apart from the presence or absence of distinct oviducts there may also be other structures associated with the ovary, notably fat bodies in sturgeons, similar to those of amphibians. Oviducts of fish with internal fertilization (principally chondrichthyans) accompanied by production of egg cases (skate) or retention of young post-fertilization, may be complex, for example forming shell glands or uteri respectively. The shell glands are seen as paired swellings of the oviducts and the uteri may even have a form of placentation providing maternal nutrients to the developing young. Young developing *in utero* may, alternatively, obtain nutrients by devouring other young (adelphagy) or eggs (oophagy), both situations having been reported from elasmobranchs (Gilmore 1993).

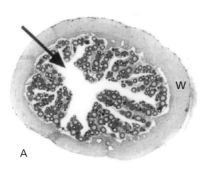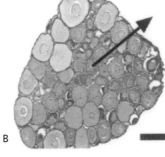

Figure 9.7 Sections through a cystovarian ovary of winter flounder (A) and gymnovarian ovary of a salmonid (B). Arrows indicate direction of ovulation; W = ovarian wall. After Burton, M.P.M. (2008). Scale bar = 250 μm.

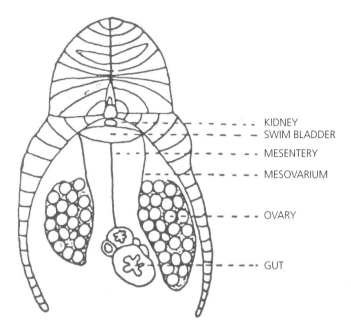

KIDNEY
SWIM BLADDER
MESENTERY
MESOVARIUM

OVARY

GUT

Figure 9.8 Transverse section through a female rainbow trout showing the gymnovarian ovaries. The body cavity has been opened mid-ventrally. After Scott (1962).

9.6.1 Vivipary, ovovivipary and ovipary

If it is established that following internal fertilization a female has a process like placentation in the reproductive system the species is said to show vivipary (live bearer, with nutrients derived from the mother post-fertilization i.e. matrotrophy). Vivipary (viviparity) in fish is reviewed by Wourms (1981) and Wourms et al. (1988).

Ovoviviparity describes retention of the developing eggs and larvae after internal fertilization, but with no direct maternal connection (i.e. no placentation), and is also termed aplacental vivipary (Conrath 2005). As discussed by MacFarlane and Bowers (1995), ovoviviparity can include a situation in which some nutrition is provided by the female later in gestation, apparently by uptake of serum proteins after fertilization. The source of nutrition in retained young for which there is no placentation is quite variable; it includes adelphagy and oophagy (9.6) as well as lecithotrophy (yolk dependence).

Ovipary, deposition of yolky eggs after internal fertilization, technically, as oocytes, before external fertilization, is by far the commonest situation.

Elasmobranchs never have a large output—that is, they have a low fecundity (9.10.6)—and have a variety of deposition strategies including ovipary

after internal fertilization for skates and chimaeras which have individual horny cases, popularly named 'mermaid's purses' (see illustration in Greenwood 1975), for each egg laid. Sharks and dogfish have either ovipary, ovovivipary or vivipary.

For bony fish the most common situation is ovipary with a large number of yolky eggs (thousands to millions) produced for external fertilization with or without parental care. These numerous eggs vary in size between species, with diameters of a few millimetres (salmonids) to much less. Large bony fish like cod *Gadus morhua* (May 1967) or sunfish *Mola mola* (Hart 1973) have been reported as having millions of eggs per female per season, though in the case of cod it is only the very large females that have over a million eggs and the situation for sunfish is not clear for the Atlantic (Scott and Scott 1988).

The coelacanth is ovoviviparous, with huge eggs (9 cm diameter, Balon 1984) retained and a few young produced.

9.7 Genital pores

Following spermiation, release of sperm to the ducts, the sperm are deposited, via a single sperm duct (vas deferens) per testis, either close to the female during courtship or mass spawning (most bony fish) or within the female during copulation (a few bony fish and chondrichthyans). In teleosts the sperm ducts unite posteriorly to empty into a single genital opening, the gonopore, separate from the anus and the opening of the excretory ducts. Lungfish, the coelacanth and elasmobranchs have a cloaca into which the reproductive and excretory ducts empty as well as the anus. Chimaera females have two separate reproductive external orifices. The relict fish, *Polypterus, Acipenser, Lepidosteus* and *Amia*, have a urinogenital sinus, opening as a 'median pore' behind the anus (Goodrich 1930).

9.8 Gametogenesis

Vertebrates reproduce sexually, with the process of gamete formation (gametogenesis) resembling, in its overall pattern, that of other animals (Figure 9.9). However, for fish the details of spermatogenesis in males and oogenesis in females have some important differences from other vertebrates, reflecting the non-terrestrial environment in which reproduction occurs and the high output for many fish. Such differences include somewhat ambiguous support structures and steroidogenic cells (9.5.1) in male fish with clonal separation in cysts (9.9.1 and Figure 9.10), and for the female's developing egg; one or more specific points (micropyles 9.10.4 and Figure 9.16) for sperm entry, and as differences from mammals, production of oogonia seasonally as well as yolky inclusions.

Gametogenesis can be rapid, particularly in males (winter flounder males complete it in two to three months, Burton and Idler 1984) and in annual fish (9.15) described from warm freshwater in South America, with the entire life cycle completed in a year (Thomerson and Turner 1973) or less (Park and Lee 1988).

However, for some females, particularly those with large eggs like elasmobranchs and those from cooler waters (winter flounder), the cycle may take years (Dunn 1970) in which 'group synchrony' (9.10.5) occurs, with several groups of oocytes at different stages; the next largest being capable of 'promotion' with growth, etc. after the current year's group (a set of large, yolky oocytes) completes ovulation.

As is usual in sexual reproduction, development of fish gametes occurs through a series of cell divisions: mitoses followed by meiosis. The gametogenic process in fish thus parallels that in other animals, with meiosis or reduction division ensuring that the chromosome number in the gametes is reduced from the diploid (two set, $2n$) number, in the parent, to a haploid (one set, n) number in the gamete. Initial mitoses establish a large base of cells '-gonia' (either spermatogonia or oogonia, Figure 9.9). When these cells enter meiosis they become '-cytes' (either spermatocytes or oocytes, Figure 9.9). For details of cell division see Appendix 9.7.

9.9 Spermatogenesis

Spermatogenesis, production of sperm, follows a pattern similar to that of other vertebrates, but sperm output may entail a fairly continuous process with most or all stages always present or, for highly seasonal fish, there may be a succession of stages and at some times of the year sperm are not being actively formed. Even for highly seasonal fish spermatogenesis may be completed well before spawning (winter flounder) or continue until very close to spawning (some salmonids). Fish spermatogenesis has recently been reviewed by Schulz et al. (2010).

9.9.1 Different types of spermatogenesis

In bony fish the lobular testis structure (Figure 9.4) is usual, but some fish, such as small tropical fish, may have a tubular testis (Grier and Lo Nostro 2000). The tubular testis occurs in poeciliids with internal fertilization and shows a process of continual spermatogenesis whereas lobular testes can show a succession of dominance culminating in a testis with sperm and a few spermatogonia, but no ongoing spermatogenesis. In cross-section lobules have a clear boundary, formed by 'lobule boundary cells' but no distinct lumen. They have spermatogonia round the periphery and, depending on the stage of spermatogenesis, may be filled with cysts (Figure 9.10) at similar or different stages of spermatogenesis. Each cyst contains cells at the same stage of development. The cysts develop to contain clones of a primary spermatogonium, one cyst per primary spermatogonium. The cyst is contained within a structure formed by 'cyst cells' which may be Sertoli

Spermatogenesis	Ploidy	Oogenesis
"Primordial germ cells"	2n	"Primordial germ cells"
Mitoses		Mitoses
(Primary) **Spermatogonia**		**Oogonia**
Mitoses		Mitoses*
(Secondary) **Spermatogonia**		**Oogonium**
>3Mitoses		Growth, Meiosis initiated, then arrested
(Secondary) **Spermatogonium**		(Primary) **Oocyte**
Meiosis initiated		Growth
(Primary) **Spermatocyte**		Vitellogenesis
Meiosis I completed	n	Meiosis resumed
(Secondary) **Spermatocytes**	n	Meiosis I completed
Meiosis II completed		(Secondary) **Oocyte** + polar body**
Spermatids		Ovulation
Spermiogenesis		Fertilization
Spermatozoa		Meiosis II completed
		Ovum + polar body

Figure 9.9 Cell divisions and cell nomenclature in fish gametogenesis.
* Assumed, rarely witnessed
**May subsequently divide. 2n = diploid; n = haploid.

cell homologues. The cysts are transient structures and the lobules eventually become filled with their products, sperm. Stanley et al. (1965) apparently use the term 'follicle' as synonymous with 'cyst'.

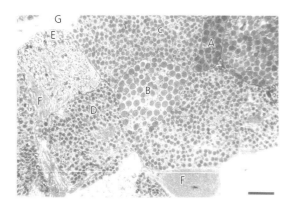

Figure 9.10 Cysts in a teleost testis. Reproduced with permission. C.M. Morrison (1990). *Histology of the Atlantic Cod* Gadus morhua; *An Atlas. Part Three: Reproductive Tract.* Canadian Special Publications of Fisheries and Aquatic Sciences 110, Department of Fisheries and Oceans Canada. A = spermatogonia; B = primary spermatocytes; C = secondary spermatocytes; D = spermatids; E = heads of spermatozoa; F = flagella = flagella of spermatozoa; G = interstitial tissue. Scale bar = 20 μm.

Dodd (1983) describes the elasmobranch testes as containing structures termed ampullae; each of these has a lumen and in cross-section resembles a mammalian seminiferous tubule. These ampullae are closed initially but become patent (open, with a lumen) and link to the sperm ducts. The ampullae develop in zones, arranged linearly, with the earliest stages of spermatogenesis furthest from the point where the sperm are released. The spermatogenesis occurs in clonal cysts (similar to those of bony fish), and the elasmobranch cysts are referred to as 'spermatocysts' (Roosen-Runge 1977). These spermatocysts are arranged one by one around the ampullary lumen. Cyst development (Wourms 1977) is described as starting with a bicellular condition, one germ cell (spermatogonium) and one support cell.

9.9.2 Spermatogonia and mitosis

Spermatogonia are fairly conspicuous round cells with a large nucleus and a prominent nucleolus; this description seems to be accepted now but earlier studies (e.g. Craig-Bennett 1931) showed smaller, less distinct, rectangular epithelial cells as spermatogonia and the cells now usually regarded as spermatogonia were termed spermatocytes. This view was probably associated with the idea of a distinctive 'germinal epithelium' composed of a separate cell line, 'primordial germ cells' expected to undergo meiosis, with component cells being basic spermatogonia in destiny. However, Roosen-Runge (1977) believes the term 'germinal epithelium' is not useful; Morrison (1990) states that male teleosts do not have a permanent germinal epithelium. The view of what would be now termed spermatogonia as spermatocytes (Craig-Bennett 1931) shifts the terminology but, as the final product of mitotic proliferation of the spermatogonia transform to spermatocytes, the adjustment of terms is a matter of potential; secondary spermatogonia eventually become spermatocytes without further division. Unless there is considerable cytoplasmic growth between divisions, the last spermatogonium of the series are most likely to be smaller than the primary spermatogonium before the muliple mitotic divisions occur so that, on that basis, Craig-Bennett's view shows a spermatocyte as much larger than would be expected.

Spermatogonia can be detected in juveniles and post-spawned fish, as well as during active spermatogenesis; they form the reservoir of cells from which sperm can be formed seasonally or more continuously. Primary spermatogonia (average diameter 13.4 µm for goldfish, Billard et al. 1974) initiate spermatogenesis by undergoing a series of mitotic divisions producing smaller secondary spermatogonia (average diameter 7 µm for goldfish) which are expected to bediploid with two chromosome sets per cell, although some fish, for example salmonids, are reported to show signs of polyploidy with, for example, evidence of four chromosome sets and ongoing diploidization (Rees 1964; Wolfe 2001; Ching et al. 2010).

According to Roosen-Runge (1977), the number of spermatogonial divisions will vary from species to species and is species-specific, but the minimal activity reported is 4 mitotic cycles to produce 16 secondary spermatogonia that will enter meiosis as primary spermatocytes (Figure 9.11).

9.9.3 Spermatocytes and meiosis

After the mitotic phase the spermatogonia become primary spermatocytes by starting the last set of divisions (meiosis) producing four spermatids per

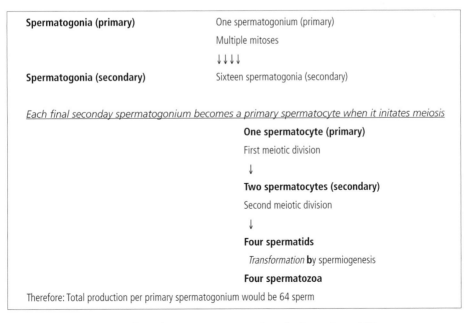

Figure 9.11 Minimal production (numbers of sperm) per primary spermatogonium, after Roosen-Runge 1977.

primary spermatocyte. The final two divisions are meiotic (halving the chromosome sets) with an intermediate production of two secondary spermatocytes before the spermatids are formed (Figures 9.9, 9.11). During these final stages, chromosomes become very obvious in appropriately stained (Appendix 9.6) preparations. During meiosis there may appear very distinct cells looking as if they have 'half an anaphase', perhaps due to a pause when cells have cytoplasmic division but retain the chromosome cluster (sometimes described as like parasols or umbrellas) at the original pole site. However, this stage is actually regarded as 'syn(i)zesis' when primary spermatocytes are in a prolonged zygotene of meiosis I (Appendix 9.7), with chromosomes aggregated close to one place of the nuclear membrane (Potter and Robinson 1991). This stage has also been termed 'bouquet', and ascribed to attachment of the ends of chromosome (telomeres) to the nuclear envelope, thought to be important in bringing homologous pairs of chromosome into close contact. The existence of the synzesis or bouquet arrangement

has been remarked on for over 100 years (e.g. McClung 1905), but its significance has only recently been explored, in a variety of organisms (Bass 2003; Harper et al. 2004; Tomita and Cooper 2007).

The diameter of the secondary spermatocytes/ spermatids is approximately half that of the secondary spermatogonia; the spermatocyte diameter in goldfish is about 3.6 μm (Billard et al. 1974).

9.9.4 Spermatids and sperm formation: spermiogenesis

Finally, each spermatid becomes transformed (spermiogenesis) without further division, to a spermatozoon with a flagellum (typically 50 μm long for acanthopterygian teleosts, Ginzburg 1972) for movement but most sperm do not move until activated by contact with a stimulating condition, for example water.

Most sperm (of those described) have only one flagellum, but a bi-flagellated condition occurs in some species such as pout (Yao et al. 1995) and

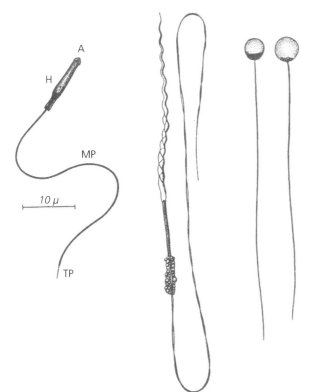

Figure 9.12 Spermatozoa from different taxa (Ginzburg 1972). From left to right: sturgeon, dogfish (*Squalus*), bowfin (*Amia*) and carp. Scale (bar = 10 μm) only applies to sturgeon. A = acrosome; H = head and middle piece (nuclear material and mitochondria); MP = principal piece of tail; TP = end piece.

lungfish (Ginzburg 1972). Whereas elasmobranchs, chimaeras, sturgeon, lungfish and the coelacanth have an anterior acrosome (Ginzburg 1972) for egg penetration, teleost sperm (and bowfin) generally lack this structure (Figure 9.12) which is not necessary as teleost eggs typically have a micropyle (9.10.4, Figure 9.16) for sperm access.

The final phases of sperm formation are associated with condensation of nuclear material, and in some fish a change to predominance of protamines rather than histones (Kennedy and Davies 1980) in the sperm head. The spermatids and sperm are strongly basophilic in contrast to the earlier stages and very small in nuclear and cell diameter (in goldfish 2.3 μm, Billard et al. 1974) with very little peri-nuclear cytoplasm.

9.10 Oogenesis

Fish oogenesis differs from that of, for example, mammals in that it is commonly an ongoing process with new oogonia being produced in adults, and for some fish, a very large potential output. Mammals have an output limited very early in the life cycle, with a finite number of oogonia produced early with no subsequent additions.

Moreover, unlike mammals, but similar to other non-mammalian vertebrates, fish eggs are yolky and the process of yolk deposition is an important phase of oogenesis. The resultant fish eggs can be huge (9 cm diameter; coelacanths, Balon 1991), moderate sized (*c.* 5 mm diameter, salmonids) or small (*c.* 1 mm diameter, anchovy, Jones et al. 1978). Most fish eggs are spheres so their size can be expressed by the diameter but eggs of *Engraulis japonicus* are ellipsoids (Yoneda et al. 2013).

9.10.1 Stages of oogenesis

Teleost oogenesis has recently been reviewed by Lubzens et al. (2010). The sequence of cell divisions is the same as for spermatogenesis, with oogonia

Duration of stage	Name of stage	Name of cell (singular)
	PUBERTY OR RECRUDESCENCE	
Brief	Mitotic proliferation	OOGONIUM
Long/very long	Meiotic prophase I/arrest	PRIMARY OOCYTE
Long	Primary growth	Small immature
Long		Large immature (perinucleolar)
	VITELLOGENESIS	
Short	Endogenous vitellogenesis	Late perinucleolar (plus CA*)
	Secondary growth	
Long	Exogenous vitellogenesis	Vitellogenic (yolk deposited)
	MATURATION	
Brief	Nucleus migrates to periphery	
Brief	Resumption of meiosis (GVBD*)	
Very brief	Completion of meiosis I	SECONDARY OOCYTE (main product)
		plus a polar body (1st), which may divide
Brief	Hydration and ovulation	
Very brief	Completion of meiosis II	OVUM (main product)
		plus a polar body (2nd)

Figure 9.13 Progression of oogenic stages.
* CA = cortical alveoli, GVBD = germinal vesicle breakdown = nuclear membrane disappears/end of prophase I.

and oocytes (Figures 9.9 and 9.13), but the product is a much larger cell than at the outset, with considerable accretion of cytoplasmic nutrients (yolk) in a growth phase which may result in very large ovaries relative to body weight, and a GSI of 25 per cent or more. Rapid uptake of water (hydration) into yolky oocytes immediately prior to deposition (9.10.7) can further increase both volume and weight (mass) of the ovaries so that GSIs can exceed 30 per cent.

Growth phases may be referred to as 'primary' and 'secondary' (Figure 9.13). Primary growth occurs in the pre-vitellogenic stages and secondary growth occurs during vitellogenesis, the deposition of yolk.

Cytologically, the different stages of oogenesis (Figures 9.13, 9.14) may be distinguished by the affinity for different stains and changes in size, the nucleoli and cytoplasmic inclusions. The cytoplasm of early-stage oocytes is basophilic, staining with haematoxylin and thus darker than other stages. The nucleus tends to present as a pale area with prominent nucleoli, particularly in early stages. It is unusual to see mitotic proliferation of the oogonia or meiosis. The latter begins at oocyte formation but stops (meiotic arrest) in early prophase (Figure 9.13). Deposition of yolk into the oocyte occurs during a very protracted meiosis when the actual process of division is suspended, a condition known as meiotic arrest.

As in other vertebrates, completion of the first and second meiotic divisions can be activated following ovulation by contact with sperm at fertilization. The divisions are unequal cytoplasmically with only one yolky cell of the two cells formed after the divisions. The tiny product which is non-yolky, with very little cytoplasm, is termed a polar body and it contains one, essentially discard, set of chromosomes. As there are two sequential divisions the discards are termed the first and second polar

bodies. The first polar body may complete meiosis, forming two more tiny cells which may temporarily adhere to the large, yolky cell.

Meiosis only resumes close to ovulation, with preparatory movement of the nucleus to the periphery, a stage which can be seen by stripping 'eggs' (by applying light pressure on the abdomen) and treating them with Stockard's solution (Appendix 9.5) which allows visualization of the nucleus/germinal vesicle (Goetz and Bergman 1978).

9.10.2 Vitellogenesis: yolk formation

The accretion of nutrients in eggs of animals is a very important phenomenon. All non-viviparous vertebrates undergo this process to a marked degree, with the accumulation of nutrients as yolk, a complex substance incorporating lipid, protein and phosphate. Much of the research into yolk deposition for vertebrates depended originally on studies using, for example, frogs or chickens. It was established that yolk formation in the oocyte (vitellogenesis) depends on the synthesis of vitellogenin (VG, Vtg) in the liver (also 'vitellogenesis'), with subsequent release to the circulation and eventual uptake by developing oocytes, where it is transformed into yolk. The circulating VG is a very large molecule, referred to as a glycolipoprotein; the protein component has many amino acids—1233 in zebrafish (Wang et al. 2000). Most vertebrate VGs contain a 'phosvitin' domain but that is lacking in zebrafish. The role of phosvitin is discussed by Finn (2007). Another component of VG, produced in the oocyte after uptake, is lipovitellin, discussed by Thompson and Banaszak (2002).

VG production is under the control of oestradiol (a steroid hormone, Appendix 9.3), itself under the

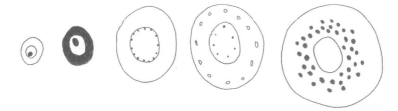

Figure 9.14 Stages of oogenesis. From left to right: an oogonium with prominent nucleolus; a small immature oocyte with basophilic cytoplasm; a large immature oocyte with multiple peripheral nucleoli; an oocyte with cortical alveoli; an oocyte with yolk globules.

control of pituitary derived gonadotropin (9.18, 10.12.2). Uptake of the VG occurs through the outer layers of the developing oocyte for transformation intracellularly into the yolk inclusions. Prior to this growth phase, sometimes called exogenous vitellogenesis (Figure 9.13), there is a preparatory stage called endogenous vitellogenesis during which a ring of 'cortical alveoli' become evident in the peripheral cytoplasm (Figures 9.13, 9.14). The cortical alveoli contribute, much later, to the cortical reaction (Kinsey et al. 2007) which occurs at fertilization, preventing polyspermy (Ginzburg 1961; Laale 1980).

Dense, basophilic cytoplasmic structures also obvious in some oocytes, before exogenous vitellogenesis is established, have been termed the 'yolk nuclei' or 'Balbiani bodies' (Wallace and Selman

1981): they may form a peri-nuclear ring (cod) or a more localized apparently single structure, like a nucleus in microscopic appearance but with no obvious enclosing membrane. Their presence may signal preparation for vitellogenesis, but their function remains a 'mystery' (Zelazowska et al. 2007). When exogenous yolk becomes deposited it takes on species-specific format as described by various authors (Table 9.1), distributed through the cytoplasm in solid or fluid form, the latter enclosed in spheres called yolk globules, which may gradually aggregate in one mass (Wallace and Selman 1981).

9.10.3 Layers around developing oocytes

During meiotic arrest and while undergoing vitellogenesis, the teleost oocyte can be seen, in stained sections, to be surrounded by layers forming a 'follicle', which is left behind in the ovary after ovulation, to form a post-ovulatory follicle (POF). These POFs (Figure 9.15), although homologous with those of mammals, do not generally have the same role; in mammals they can form the steroidogenic corpora lutea maintaining a pregnancy but in fish they may be transient or persist as very small structures.

The follicle which surrounds each developing oocyte is cellular and may be distinguished as double, an inner granulosa of small cuboidal cells and an outer theca, with flatter cells. It is generally expected (based on a body of work with yolk-bearing vertebrates including amphibia) that these two layers are steroidogenic, with testosterone synthesized

Table 9.1 Yolk types and terms in teleost fish.

Oocyte stage	Yolk stage	Fish species	Author(s)
Primary oocyte, endogenous vitellogenesis	Yolk vesicles, precursor to cortical alveoli	*Fundulus heteroclitus*	Wallace and Selman 1981
Primary oocyte, exogenous vitellogenesis	Yolk droplets, later a homogeneous mass	*Gadus morhua*	Morrison 1990
	Yolk granules	*Melanogrammus aeglefinus*	Robb 1982
	Yolk granules (translation)	Several, including *Clupea harengus, Coegonus albula, Esox lucius, Cyrpinus carpio, Perca fluviatilis*	Ginzburg 1972
	Yolk globules	*Pleuronectes platessa*	Barr 1963
	Yolk droplets then yolk globules	*Limanda limanda*	Htun-Han 1978
Primary vitellogenesis	Yolk droplets (thin zona radiate, scattered vacuoles)	*Mullus surmuletus*	N'Da and Déniel 1993
Secondary vitellogenesis	Yolk droplets dividing cytoplasm into two concentric rings (follicle distinct)		
Tertiary vitellogenesis	Yolk globules		

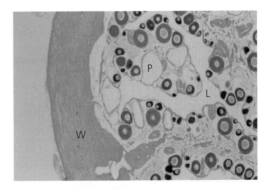

Figure 9.15 Ovarian section (winter flounder), showing post-ovulatory follicles (P), visible as circular empty spaces, oocytes are dark circles enclosing lighter circles (nuclei). The cystovarian ovary clearly shows the ovarian wall (W) with lamellae surrounding a lumen (L).

in the theca converted to oestradiol in the granulosa, the conversion being catalysed by the enzyme aromatase. The oestradiol when released to the blood will promote vitellogenin synthesis in the liver. The vitellogenin released to the blood from the liver is expected to be taken up through the follicle to be incorporated in the yolk of the growing oocyte.

Channels in the outer layer of the oocyte itself (facilitating uptake of vitellogenin) give the appearance of striations and this layer may therefore be called the 'zona radiata' (Table 9.2). As the oocyte becomes bigger and ovulation approaches, the follicle may change its mode of steroidogenesis to produce progesterones as maturation inducing steroids, MIS (Scott and Canario 1990), and structural changes in the outer layers of the oocyte may produce more elaborate and resistant or adhesive structures preparatory to egg deposition. The basic outer layer may then be called a chorion, but the nomenclature has become complex (Table 9.2).

Structures surrounding the developing oocyte/egg have been given a number of different names, especially as the appearance changes during oogenesis; in general terms they may be called membranes or envelopes, while the term 'follicle' is reserved for the cell layers external to the developing oocyte. Laale (1980) describes the acquisition of radial structures by the outermost oocyte membrane (the pellucid or vitelline membrane, also called the oolemma) to form the zona radiata. This surrounds the primary oocyte during the secondary growth phase and has a complex structure mediating the uptake of vitellogenin. The zona radiata

is eosinophilic (i.e. is red in standard preparations with haematoxylin and eosin 'H and E', Appendix 9.6) and the distinct radial striations reflect the close relationship with the granulosa layer of the follicle: the granulosa cells have cytoplasmic extensions penetrating the zona radiata during vitellogenesis and these are withdrawn during ovulation (Guraya 1986). An alternate term, 'chorion' (Guraya 1986), for the outer layer of the oocyte is probably best reserved for the later structure formed around the egg itself, the egg membrane. In the final stages of maturation the zona radiata changes, and the development of the egg membrane occurs.

9.10.4 The egg membrane or chorion

Towards the end of vitellogenesis structural changes occur in the oocyte periphery and eventually a relatively impervious egg envelope or membrane is formed; this is usually called a chorion. Oocyte and egg-enclosing layer terminology, as used by several authors, is summarized in Table 9.2.

The egg membrane or chorion varies considerably between species: demersal eggs may have multiple layers with adhesive structures (discussed in Laale 1980). Capelin (*Mallotus villosus*) has a two-layered structure, a double oocyte membrane, DOM (Flynn and Burton 2003, Table 9.2). Post-fertilization, the egg membrane(s) will become hard or sticky or both. In all cases they will become relatively impermeable, to sperm and to water. Hardening is discussed by Davenport et al. (1986) and is actually not dependent directly on fertilization but is linked

Table 9.2 Terminology for non-cellular layers enclosing oocytes and eggs.

Structure enclosed	Structure enclosing (i.e. envelope or membrane)	Description	Species	Author (s)
Oocyte (primary)	Plasma membrane	Very thin	All	
	= Oolemma = Vitelline membrane			Laale 1980
Oocyte (vitellogenic)	Zona radiata	Radial striations		Laale 1980
	Zona pellucia			Guraya 1986
	Chorion			
Oocyte (late vitellogensis)	Double oocyte membrane	Distinctly two layers, just before ovulation	*Mallotus villosus*	Flynn and Burton 2003
Egg	Chorion			Laale 1980

to the cortical reaction (9.10.2) and the presence of calcium ions.

Prior to fertilization, during the final stages of oocyte formation, the zona radiata of most bony fish (i.e. most teleosts) may have developed one or more micropyles for sperm entry (Figure 9.16). Immediately after fertilization, the micropyle(s) will become blocked to prevent polyspermy and to hinder osmotic or bacterial problems.

9.10.5 Synchrony or asynchrony in oogenesis

Marza (1938), in analysing animal oogenesis, noted three types: asynchronous, synchronous and group synchronous. As the names suggest, with synchrony the oocytes develop simultaneously and are ovulated and deposited as a single event or in very close (hours/days) events. This may be rare in fish but does apply to fish like semelparous salmonids.

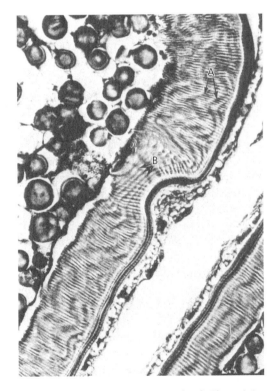

Figure 9.16 Micropyle in cod oocyte. Reproduced with permission. C.M. Morrison (1990). *Histology of the Atlantic Cod* Gadus morhua; *An Atlas. Part Three: Reproductive Tract.* Canadian Special Publications of Fisheries and Aquatic Sciences 110, Department of Fisheries and Oceans Canada. A = zona radiate; B = micropyle region.

For true synchrony one would not see any immature non-vitellogenic oocytes in the ovary once vitellogenesis was established. However, as semelparity in fish may be a fairly recent adaptation contingent on a long migration, it is feasible that some oocytes, laggards, would be seen, even though they would have no chance of recruitment later.

In fish, asynchrony is typical of small and/or warm-water fish which can breed over an extended period, and which do not have the abdominal space or the nutritional storage capacity to hold reserves of large yolky oocytes. However, the large pelagic billfishes (swordfish, marlin, etc.) show asynchrony (Arocha 2002). Group synchrony occurs in large marine fish from cool waters that have restricted spawning periods, timed to take advantage of spring plankton blooms for larval feeding (9.12). In these fish, a subset of oocytes undergoes vitellogenesis and these may either be spawned as essentially a single event or subdivided into batches, and spawned over a period of weeks (multiple, serial or batch-spawning). As the spawning season approaches it is possible to see, in group-synchronous fish, the large vitellogenic oocytes flanked by much smaller oocytes, which would develop the following season if conditions were appropriate.

9.10.6 Fecundity and atresia

Fecundity, the number of eggs produced per female per reproductive season (or by prediction, per life time), varies from very few eggs (the coelacanth and chondrichyans and most small teleosts) to millions (large marine teleosts). Fecundity for gravid swordfish (seven females) was calculated to be from 995,000 to 4.3 million (Poisson and Fauvel 2009), and Gilmore (1993) states fecundities of up to 16 million for large scombrids and billfishes. For fish with group synchrony it is possible to achieve estimates of egg number shortly before spawning by counting vitellogenic oocytes. As swordfish are said to have asynchronous development and multiple spawning events seasonally (Arocha 2002), fecundity estimates may have been particularly difficult. To achieve egg counts it has been usual to preserve the ovaries in Gilson's fluid (Appendix 9.5) which allows separation of the individual eggs, though the fluid as originally conceived is very toxic, with mercuric chloride, as well as being unsuitable for late-stage (hydrated) eggs. Lowerre-Barbieri and Barbieri (1993) proposed a two-step method they considered an improvement.

A complication of fecundity estimates is the occurrence of atresia, when developing oocytes may be absorbed, a downregulation of potential eggs. Similarly, though even less well known, there is the possibility of upregulation where immature oocytes may be recruited into vitellogenesis at a fairly late stage of the annual cycle (Urban and Alheit 1988). Fish that have this degree of flexibility can add to their output if feeding is good, so that like 'asynchronous' fish their spawning pattern is regarded as 'indeterminate'. Fish with synchronous oogenesis would be 'determinate spawners' and group-synchronous fish, although basically determinate, may have some upregulation possible. Evidence of atresia (downregulation) and late recruitment (upregulation) may be obtained by careful histological examination of ovaries. Scott (1962) reported that atresia increased with starvation; there is therefore evidence that resorption is selective.

Complete resorption and late abandonment of reproduction has been reported occasionally for some fish species, but non-reproduction is probably more often achieved by a 'decision' earlier in the annual cycle, in which case vitellogenesis does not begin (9.16).

9.10.7 Terminal phases of oogenesis

After the secondary growth phase, with yolk deposition, the oocyte can resume meiosis to form one haploid ovum with three very small polar bodies (Figure 9.13). Other processes, including hydration (9.6, 9.10.1Figures 9.6, 9.13), may accompany this final stage.

Endocrine control (9.18, 10.12.2) of these final stages of oogenesis has been associated with increased levels of progesterones in comparison to earlier stages under oestrogen control. However, Truscott et al. (1992) found little evidence for progesterones' involvement in winter flounder maturation; they concluded that pregnenolone was more likely involved. Various facets of endocrine involvement in fish reproduction have recently (2010) been presented in *General and Comparative Endocrinology* (Vol. 165, issue 3).

The cessation of meiotic arrest is linked to 'germinal vesicle breakdown' (GVBD), an expression derived from older terminology in which the nucleus was called a germinal vesicle, as advanced (post-prophase) stages of cell division will be accompanied by dissolution of the nuclear membrane. In large oocytes with very large nuclei it may be possible to see the nucleus in each cell, without the use of a microscope, or minimally, with a low-power microscope after treatment with fixatives. Therefore as oogenesis advances to ovulation, a concurrent situation, with resumption of prophase and meiosis proceeding, may be tracked: GVBD will reflect the end of prophase. Just before GVBD the nucleus (germinal vesicle) migrates to the periphery of the oocyte, close to the micropylar area, ready for sperm penetration.

Hydration, the uptake of water into the cytoplasm of maturing oocytes, results in oocytes known as 'hyaline' (Htun-Han 1978). It occurs (osmotically) as a result of protein cleavage during the final maturation process (Finn et al. 2000). The hydrated oocytes can be seen macroscopically in the ovary, with clear cytoplasm (Figure 9.6), quite distinct from non-hydrated cells. Their occurrence can be associated with GVBD (Scott and Canario 1990). In some fish, for example in cod, which is a batch spawner, vitellogenic oocytes do not all become hydrated at the same time (Morrison 1990). In other fish such as winter flounder, the entire output becomes hydrated simultaneously, increasing the volume and the mass of the ovary by a considerable amount. Hyaline oocytes are quite distinctive in sections (Htun-Han 1978), they tend to take on a star shape after processing, presumably an osmotic phenomenon similar to the crenation (shrivelling) response of mammalian red blood cells to a hypertonic environment. In section, the cytoplasm of hyaline oocytes is also clearer in appearance than that of non-hydrated oocytes. Hydration at this stage may be necessary to achieve ovulation; this and later hydration (see Pankhurst 1985) may also contribute to buoyancy, important for marine pelagic eggs which may also rely on oil droplets for flotation. Marine teleosts from four species with pelagic eggs showed a high degree of hydration at ovulation (Finn et al. 2000), increasing from 58 per cent to 68 per cent water pre-ovulation to over 90 per cent. These values were obtained after biopsy so no direct environmental uptake of water was included. According to Fabra et al. (2005) studying marine fish, the hydration process is aquaporin-mediated (7.5.6).

9.10.8 Ovulation

As with other vertebrates, emergence of the oocyte from its surrounding follicle (ovulation) is preparatory to fertilization. In fish, ovulation occurs either into the body cavity if the ovary is 'naked' (i.e. gymnovarian), or into the lumen of the enclosed ovary with a muscular wall (cystovarian) typical of many teleosts. Gymnovarian animals have open-ended oviducts into which the mature oocytes should pass. Ovulation can be induced hormonally (10.12.2).

9.10.9 Egg structure

The final product of oogenesis, eggs (ova) vary considerably between species, in size and enclosing structures. They may also show some variation in size intraspecifically although individuals usually have eggs which are more similar (Springate and Bromage 1985) but size may change for the same female as the season progresses, decreasing in yellowtail flounder (Manning and Crim 1998). The amount of yolk deposited is related to the length of time the enclosed larva will be dependent on it before hatching. Some larvae, for example halibut, hatch when relatively undeveloped (altricial), whereas salmonid fry emerge from the sediment fully eyed (more precocious) and with a small attached 'yolk sac' which is rapidly absorbed.

Descriptions of the stages of oogenesis and the resulting eggs of different species have led to some confusion in terminology, for inclusions called yolk (Table 9.1) and for covering layers or membranes (Table 9.2). For eggs that are deposited the structure will reflect whether it is buoyant and floats (halibut), or whether it is buried in sediment (salmonids or capelin), stuck to vegetation (goldfish), placed in nests and guarded (sticklebacks, lumpfish) or provided with an extra protective layer, an egg case (skate). The stickiness of some eggs produces difficulties in culture and 'unsticking' processes have been developed (Kujawa et al. 2010).

9.11 Fertilization

Fertilization is achieved when one (or, rarely, more) sperm enter the micropyle(s) (Figure 9.16), preparatory to fusion with the egg nucleus to form the diploid zygote. Sperm entry may actually stimulate the final meiotic division, and also leads to processes excluding any more sperm from entering. Post-fertilization, with nuclear fusion, the zygote can start mitotic divisions forming successively, blastula, gastrula and embryo (Appendix 1.1). For oviparous females the developing embryo will hatch at a species-specific stage of development, and development rate will be temperature-dependent, giving the phenomenon of 'degree-days', it being possible to calculate when hatching will occur (Appendix 9.4).

9.12 Egg and larval development

Most fish have eggs well provided with yolk and development of a fertilized egg draws on this nutrient source (i.e. lecithotrophy), which in its diminished form may still be available post-hatch, as a 'yolk sac' ventrally placed and provided with a special vitelline circulation, and at this point the stage is an eleutheroembryo (Morrison 1993). Once the yolk is used up the larva depends on its capacity to catch prey, and the mouth must be functional at this stage. The entire developmental period may be intensely temperature dependent, and a calculation 'degree-days' (9.11, Appendix 9.4) applicable to obtain time to hatch. Some relatively cold-water fish (e.g. salmonids) with eggs deposited in the autumn may take a very long time to reach hatch, with the timing such that the larvae can take advantage of spring plankton blooms (9.10.5).

Initial post-fertilization development follows the usual stages of lecithotrophic (yolky) eggs. The fertilized eggs become asymmetrical, with yolk at one end (the vegetal pole) and the developing embryo at the other end (the animal pole). The animal pole and the vegetal pole are determined by the position of the micropyle, with the animal pole close to it and the yolk at the vegetal pole furthest from it. The cell divisions occurring post-fertilization produce a blastula and then a gastrula (Appendix 1.1), with the larva gradually taking form subsequently. Hatch may take place when the larva is still relatively undeveloped, as occurs with halibut, or the larva may be well developed with functional eyes and a fairly well-developed gut, soon capable of 'first-feeding' (salmonids). Crawford (1986) describes development sequences for the flounder *Rhombosolea tapirina* at approximately 11–14 °C (Figure 9.17).

Some larvae are very unlike their parents, the slim, leaf-like 'leptocephalous larvae' of eels are particularly different from the parent, and the larvae of *Idiacanthus*, a deep-sea fish, have unusual stalked eyes, projecting at right-angles from the head (Nikol'skii 1961). The adult does not have projecting eyes. Where the larva is very different from the adult the change to the adult form may be referred to as 'metamorphosis' as with the pleuronectiforms (flatfish), in which the symmetrical larvae go through a phase when one eye migrates to pair with the other on either the right or left side, and the adult tends to become benthic, lying on the 'blind' side.

Description and identification of eggs and larval fish have been very important for both basic and applied research, particularly for marine fish (e.g. Saville 1964; Macer 1967; Fives 1970; Nichols 1971; Fritzche 1978; Hardy 1978a,b; Johnson 1978; Jones et al. 1978; Martin and Drewry 1978; Fahay 1983), with the possibilities of young forms of many different species in the plankton. A number of publications have dealt with differentiation and success of the successive stages, with much attention given to salmonids (e.g. Battle 1944; Petersen et al. 1977; Velsen 1980; Srivastava and Brown 1991). Many larvae are planktonic but some, including the salmonids, may start off in a gravel bed and 'emerge' to swim up into the water column. Emerging salmonids are often referred to as 'alevins' but young fish are generally referred to as 'fry'. Successful completion of larval

development and survival through the very hazardous phases of the fry and small juvenile stages for fish with no parental care indicate that very few fertilized eggs produce an adult. The Canadian Department of Fisheries and Oceans have issued a poster suggesting that 'one in a million' are successful in the case of cod.

Osse and van den Boogaart (1999) have pointed out that marine teleost larvae tend to be much smaller (on average 10 times smaller by weight) than freshwater teleost larvae, and the survival of the former much less, approximately 44 times fewer, presumably linked to the lower weight.

9.13 Post-reproduction

After spawning (ovulation/spermation), the gonads may enter a resting stage ('post-spawning'), during which time any unspawned gametes may be removed by phagocytes. However, 'residual eggs' may persist for some time, as may post-ovulatory follicles, POFs (Figure 9.15). The presence of residual gametes and/or POFs are certain signs of recent reproduction and may be used to distinguish adult ('mature') fish from large fish which have never reproduced ('immatures'). The term 'immature' can be confusing as it may also be applied to the state of a gonad or fish that is currently not in active reproduction.

Parents care for eggs or young in many fish species (14.8.6). It is expensive in terms of lost foraging for the parent involved and few young are produced in these circumstances, but the costs, though presumably high in terms of future reproduction, must be advantageous in terms of survival for the current young. The cost of parental care is discussed by Morley and Balshine (2003). Taborsky (1984) even describes 'brood-care helpers' in the cichlid *Lamprologus bircardi*.

Newly hatched	Floating passively, yolk sac up	Neither mouth nor jaws seen
3 days post-hatch	Yolk sac mainly resorbed	
5 days post-hatch		Mouth, jaws, gut, anus functional

Figure 9.17 Early stages of development in *Rhombosolea tapirina*, based on Crawford 1986.

An unusual egg-deposition strategy is that of the cyprinid bitterlings (*Rhodeus* and *Acanthorhodeus*) which place their eggs into the mantle cavity of bivalves, using a long ovipositor which grows shortly beforehand (Nikol'skii 1961). Nikol'skii also reports a similar adaptation for a liparid, *Careproctus*, which deposits eggs in the carapace of a crab.

9.14 Reproductive migrations and reproductive mortality

In terms of cost to parents, the ultimate cost is death which occurs in semelparous fish after a lengthy reproductive migration. Thus, some fish which migrate long distances to reproduce do not survive, particularly in cases where the migration involves a change in salinity as occurs with some salmonids and freshwater eels. In some fish species then, the life cycle incorporates just a single reproduction migration per individual, the adult dying immediately after reproduction. This is the expected outcome for most Pacific salmon, *Oncorhynchus*, species (except rainbow trout) which are anadromous, breeding in fresh water and returning to sea after a fresh-water development period when fry (alevins) become parr, then adjust to increased salinity as silvery smolts before an intense feeding period for one or more years at sea. *Lovettia*, an Australian short-lived 'annual' species, is also anadromous (Allen et al. 2002), spawning in fresh water. Freshwater eels such as *Anguilla anguilla* make the reverse migration; that is, they are catadromous. The adults reproduce at sea and the thin, leaf-like leptocephalous larvae migrate back to freshwater streams where they feed and grow to elongated adults prior to migrating back to sea to reproduce and die. The Australian barramundi (*Lates calcarifer*), also catadromous, makes a shorter migration, spawning in estuaries (Allen et al. 2002). Amphidromous fish have also been described, in which case adults (e.g. the Tasmanian smelt, *Retropinna tasmanica*, and some galaxiids) migrate from the sea to spawn in rivers and the larvae are returned to the sea by the natural water movement (Allen et al. 2002). Anadromous Atlantic salmon (*Salmo salar*) are capable of more than one migration; they may survive their first spawning as somewhat emaciated fish termed kelts.

Diadromy is the general term for fish which migrate to or from fresh water to reproduce. This type of reproductive cycle presents particular problems for conservation (McDowall 1999) but provides a separation of reproductive adults valuable for quantification.

There may be a high death rate for many fish species post-spawning and various possibilities for surviving even after long migrations, or an admixture of non-migratory subsets within a population as occurs with some small males which remain resident in fresh water to await the arrival of migrating females, as occurs with some Pacific salmon. Species may have facultative anadromy even amongst the larger individuals; a portion of the population can migrate. Some fish species may also have a complex situation where the post-spawning deaths are skewed to one sex; capelin (*Mallotus villosus*) can spawn high on shore and males may make several spawning attempts whereas it is likely that females only make one. The chances of males surviving are therefore not good anyway whereas females can and do survive (Burton and Flynn 1998). However, it has also been recently reported that neither male or female capelin from offshore spawning groups (Barents Sea) survive (Christiansen et al. 2008) and are essentially semelparous, which may be associated with long migrations. Other hazardous spawning behaviour (e.g. grunion, *Leuresthes tenuis*, off the Californian coast which beach-spawn) may be associated with high mortality rates of the adults but with good egg and larval survival. Another beach-spawner, an osmerid (like the capelin), is the surf smelt (*Hypomesus pretiosus*) which spawns year round off the Pacific North-West (Hart 1973; Moulton and Penttila 2001) with no reports in these sources of excessive mortality. Tewksbury and Conover (1987) discussed the advantage of intertidal egg deposition in the Atlantic silverside (*Menidia menidia*) and concluded that it protected eggs from aquatic predators. Martin (2015) has written extensively on beach-spawning.

Many fish which are capable of repeat spawning, that is, they are iteroparous, from year to year also undergo some pre-reproductive migrations, this includes movement of iteroparous salmonids and cod: the distances moved may be slight or considerable. Russian-Norwegian cod make a well-recognized movement south to spawn. In the case of individual bodies of fresh water, lakes linked to streams and rivers, iteroparous salmonids such as brook trout (*Salvelinus fontinalis*) will move pre-spawning to deposit eggs on appropriate substrate.

The ability of some fish to return to their place of origin (e.g. the natal river) has attracted particular interest as a 'homing phenomenon'. The mechanisms which enable this seem to include chemosensory receptors (14.7). Migration and navigation may involve magnetoreception (12.10). Cues determining migration paths for the small gobioid *Sicyopterus stimpsoni* are discussed in Leonard et al. (2012). These fish, which include waterfall scaling in their movement, apparently do not 'home' to their natal stream and have weak trail-following capacity; rather, they depend on high levels of organic growth as indicators for direction finding. Brönmark et al. (2014) have reviewed the migratory phenomenon in freshwater fishes, broadening the consideration of this behaviour beyond the 'dominance' of salmonid studies, and including new data on seasonal movement of a cyprinid, the roach *Rutilus rutilus*.

9.15 Spawning and reproductive patterns

The life cycle of individual fish species varies from the short fast one year of 'annual' fish to the capacity to reach very large sizes and the possibility of starting reproduction when quite small and continuing through to reproduce when large and old, for example halibut and cod.

Annual fish such as the cyprinodonts *Rivulus* and *Cynolebias* (Breder and Rosen 1966; Thomerson and Turner 1973) live in hot climates where the shallow fresh water they inhabit can dry up and kill the adults, though *Kryptolebias* (formerly *Rivulus*) *marmoratus* can seek terrestrial refuge as an 'amphibious' fish (Taylor et al. 2008, 13.5.4, 13.5.5). Reproduction before the dry season produces desiccation-resistant eggs which can develop when water returns to the area. Obligate annual fish have been described (Thomerson and Turner 1973) but some 'annual' fish can survive for longer than a year in captivity (Breder and Rosen 1966). In contrast, many fish reach large sizes and have indeterminate growth with few reports of senescence.

Fish like cod and halibut start reproduction when relatively small, for example 50 cm or less length for both male and female cod, but continue reproduction when more than 1-m long, with high fecundity. Both male and female halibut may start reproduction at around 50 cm length too and have been reported as 'mature' as well as 'immature' when over 1-m long (Kohler 1967).

For oviparous species, females may deposit eggs as a single event during the reproductive season or the eggs may be laid as a series of batches (9.10.5), a situation which may be more stressful for the female but which spreads the risk to the eggs and larvae. For any one female that is a batch spawner, the interval between egg deposits may be regular, or not, and the quality or size of the eggs may change with time. Also, for each individual female the timing of reproduction may be similar from year to year, or not. For large iteroparous fish the pattern of oogenesis often takes the form of group synchrony and in cold-water fish the time of development for each individual oocyte may be three years (Dunn 1970). Including with this the possibility of links between nutritional status and fecundity, the output from year to year may be influenced by mitotic events which occurred more than one year earlier.

Elasmobranchs may be slow to reproduce, adulthood being attained at large sizes (they may take two decades to mature, Gilmore 1993) with a very long gestation period (approximately six months to two years for different species, Compagno 1990) for live bearers, and the possibility of an extended 'rest' period between reproductions (Conrath 2005). Output is low in general, a matter for concern when fisheries continually remove the reproductive fish.

9.16 Reproductive omission: skipped spawning

Some females, and indeed some males, which have the potential for repeat spawning (or repeat reproduction) each year may not do so. For long-lived species it is now known that a proportion of the adults may not reproduce in any one year (Rideout et al. 2005). Reproductive frequency is an important facet of the life history, especially for potentially long-lived species, and it is important that it should be taken into account, especially for intensively fished species. The idea that females of some freshwater species do not reproduce every year is well established (Miller and Kennedy 1948; Kennedy 1953; Nikloskii 1971); as is the slow pace of reproduction in elasmobranchs; the idea that annual reproduction may not occur for marine teleosts previously thought of as highly fecund is now gaining acceptance. As it is likely a matter of irregular reproduction, periodically excluding individual males or females from participation, it is not surprising that the facts have been difficult to establish. Reports for female Chondrostei (sturgeon, paddlefish) indicate they are incapable of annual reproduction. Berra (2007) suggests that female *Polyodon* (the North American paddlefish) spawn every two years. It is very important to understand the stages of reproduction (Appendix 9.1, Tables 9.3–9.6) to interpret participation.

9.17 Reproductive success

In Darwinian terms fitness can be measured by survival of the young. Adaptations in fish which achieve enhanced survival vary from huge output (but with no parental care) to much smaller output with care, including retention of fertilized eggs (vivipary, ovovipary, 9.6.1). Added to this is the possibility of a long reproductive period and repeated reproductions. Some fish have potentially very long lives and late maturity. There are even fish that can start reproduction quite young and when small, with a potential for large size and considerably increased output with age (halibut), a situation unlike mammals which have a relatively fixed ultimate size (determinate growth). Moreover, there may be several variants in fish reproduction within a species; salmonids are well known for such variation, which includes the possibility of small, resident males intervening at the mating of an anadromous pair (9.3). Variability in some species includes the capacity to switch sex (9.4) and there have even been reports of parthenogenesis in which all female populations of *Poecilia formosa* (apparently originally derived by hybridization) rely on sperm from congenors (either of the two 'parent' species) to stimulate egg development but not contribute chromosomes; the species is ameiotic and thus retains diploidy (Tiedemann et al. 2005).

The production of all female progeny by such gynogenesis has been an aim of, for example, sturgeon culture and can be achieved by prior treatment of sperm so that they activate but do not contribute genetically (Flynn et al. 2006).

Because of the potentially huge output of some fish there may have been an overoptimistic view of their capacity to withstand fishing pressure. Individual fish are capable of suspending reproductive activity in unfavourable conditions but this can provide considerable fluctuations in output. If estimations of production solely rely on the concept of 'spawning biomass' as equivalent to numbers of adult fish, without correcting for such downregulation, there can be problems with sustainability.

The capacity to survive and repeat spawn may be particularly advantageous if increased size (particularly volume) facilitates increased fecundity. Bagenal (1965) studied long rough dab (*Hippoglossoides platessoides*) off Scotland and reported a variation of output from 9700 eggs for a 12.2 cm female to 251,300 for a 30 cm fish. This is a similar pattern seen for the increase in fecundity possible for cod, which at first reproduction at about 50 cm length can produce about half a million eggs in comparison to the 12 million eggs reported for a large female over 1 metre long (May 1967). These large increases would certainly explain the possible advantage of omitting an annual reproduction (9.16) to survive for a future output and suggest that large females may be particularly important in maintaining a population; their capture may lead to a disproportionate effect on a stock.

9.18 Factors controlling or influencing reproduction

Achieving reproduction depends on input externally to provide appropriate timing (e.g. indicating season, coordinating spawning events so that sperm are available to fertilize eggs) and, for individuals, requires a series of 'switches' determining beginning and progress of gametogenesis without severely compromising the state of that fish so that it can successfully complete reproduction.

The nature of the external cues (Figure 9.18) is known better than the internal cues connecting energy/nutritional status to reproductive output. The endocrine factors enabling gametogenesis and final maturation are fairly well known, particularly for later stages of the process (e.g. ovulation). Neural input for reproductive behaviour (10.11.1, 14.8) is recognized with the very important role of the hypothalamus. The possible roles of pheromones (14.7), chemical factors released by individuals to achieve coordinated behaviour such as spawning, have been studied for some species of fish.

Studies directed at understanding reproduction at the individual and population level for vertebrates have been crucial to agriculture, aquaculture and facets of medicine. Hormones (Chapter 10), particularly sex steroids (androgens, oestrogens and progestogens) and gonadotropins have been intensively studied in vertebrates and applied to a variety of fish, notably those of commercial importance. Overall the view is that, as for other vertebrates, fish reproduction is controlled via the hypothalamic–pituitary–gonadal (HPG) axis and recently, 'kisspeptin' systems in the brain have been linked to reproduction with the possibility that kisspeptins have a control function (Akazome et al. 2010; Escobar et al. 2013). At the steroid level, 'StAR' (Steroidogenic Acute Regulatory) protein can modulate steroidogenesis (Bauer et al. 2000), and at the cellular level, cyclins (Becker et al. 2000) will presumably be involved in the onset, pausing or restarting of phases of gametogenesis; a cyclin is described, for frogs (Ihara et al. 1998), as

an essential part of MPF (maturation-promoting factor) involved in resumption of meiosis seen as GVBD (germinal vesicle breakdown).

The sex steroids synthesized either in the testes or ovaries have a number of decisive roles. In the case of males, the major androgen in fish is 11-ketotestosterone (not found in mammals). It may be transported in the plasma bound to protein and can evoke/maintain secondary sexual characteristics such as changes in skin colour and thickness (2.10). Testosterone is synthesized in the follicles of females and, under the influence of gonadotropins, can be converted to the oestrogen oestradiol, then released into the plasma where it targets the liver cells (hepatocytes). These in turn synthesize vitellogenin (9.10.2) for release to the plasma and uptake by developing follicles to form yolk. Males injected with oestradiol will actually synthesize vitellogenin (Chester Jones et al. 1972). Towards the end of oogenesis the follicles synthesize progestogens rather than oestradiol as part of the final maturation (Figure 9.18).

The gonadotropins (10.12.2) in vertebrates have usually been described in relation to mammalian reproduction, with two separate hormones recognized, one known now as follicle-stimulating hormone (FSH) used to be also termed ICSH (interstitial cell-stimulating hormone) with reference to the interstitial androgenic cells (now usually called Leydig cells) of the testis. The other gonadotropin

	External cues	Internal cues	Specific Regulatory factors
	Temperature	Level of energy stores,	Inferred or known
	Water quality	circulating nutrients	
	Light		
	"Stress" may promote or delay		
Gametogenesis			Cyclins, GnRHs →GTHs
Mitoses			GnRH→FSH- like GTH
Initiation Meiosis I			Not known; kisspeptin?
Arrest			Not known
Vitellogenesis			GTHs→(testosterone→estradiol)
Resumption			GTH II (LH- like) ↑MIS→↑MPF
Ovulation			

Figure 9.18 Control of reproduction in iteroparous, oviparous fish. GnRH = gonadotrop(h)in releasing hormone; GTH = gonadotrop(h)in; FSH = follicle-stimulating hormone; LH = luteinizing hormone; MIS = maturation inducing steroid, e.g. 17α-hydroxy, 20βdihydroprogesterone (i.e. 17α 20βDHP/DHP, Jalabert and Finet 1986); MPF = maturation-promoting factor (Cdc2 + cyclin B in frog; Ihara et al. 1998).

was named LH, luteinizing hormone, because of its role in late stages of mammalian oogenesis when the postovulatory follicle became the corpus luteum and could maintain mammalian pregnancy. It is somewhat difficult to compare the role of these two hormones, which are believed to occur in fish, with the mammalian situation. The obvious similarities would be in a relationship between the gonadotropins perhaps best called I (FSH-like) and II (LH-like) and steroid control (steroid synthesis/release/regulation), although Jalabert (2008) reports acceptance of the mammalian terms. Most fish are expected to be capable of mitotic renewal of oocytes by producing oogonia. This does not occur in adult mammals; the processes controlling these mitoses is not well understood.

Another set of controlling factors may be found in activin and inhibin, acting locally in the gonads. These factors are regarded as paracrines (9.5.1, 10.2) with activin enhancing oocyte maturation but not inducing it (Petrino et al. 2007). Similarly, inhibin can inhibit oocyte maturation by blocking the action of stimulatory hormones. The role of inhibin in male fish has not received as much attention. Poon et al. (2009) summarize studies on these paracrines, with the views that activin is antagonized by inhibin and that follistatin is activin's binding protein. It has been found that in mammals, activin stimulates FSH but not LH whereas inhibin suppresses FSH but not LH. If there is a similar system in fish it seems that activin would promote early gametogenesis while inhibin would suppress it.

Any diagrams or schematics attempting to depict the control of vertebrate or fish reproduction have tended to be bewilderingly complex; Figure 9.18 summarizes some of the factors which are known or believed to stimulate various phases of gametogenesis for females. Inhibition, the control of meiotic arrest and omission of an annual cycle or downregulation (fewer gametes produced relative to inhibitory states like poor nutrition), is not figured; the physiology of these negative states is not well understood, although the phenomenon of 'apoptosis' (programmed cell death) could help underlie the progress or induction of atresia (resorbtion of oocytes).

Cyclins control the cell cycles and cell division in organisms including yeasts through to vertebrates which have various 'checkpoints' (Becker et al. 2000); these determine whether an individual cell proceeds into division. It may be helpful to think of the whole set of events leading to successful reproduction as passing through a series of checkpoints, with pause possible, particularly for females, at various stages. The cyclins may be synthesized or activated in response to steroid presence, and the steroids can be evoked by gonadotropins (10.12.2).

Presumably the exact nature and timing of the checkpoints is species-specific. It is important to remember that many of the regulatory studies on vertebrate reproduction have depended on frogs (e.g. for cyclins) or mammals (rats/humans), and this influences available detail for particular stages, while other information on the cell cycle has been derived from fruit-flies and much of the knowledge has proved widely applicable. However, there are some areas that cannot be extrapolated. Female mammals do not have post-embryonic mitoses of oogonia and the information about the recurrent mitoses in oviparous fish is scarce. In immature goldfish cyclin (B) was not detected in the oocytes (unlike the frog *Xenopus*), but this cyclin was synthesized in the oocytes following the addition of the steroid $17\alpha\ 20\beta$DHP (Ihara et al. 1998), the hormone recognized as maturation-inducing for fish, though a trihydroxy equivalent ($\sim17\alpha20\beta$THP/20beta-S/THP, Appendix 9.3) has also been recognized (King et al. 1997). The activity of the cyclin is expected to depend on its link with a cyclin-dependent kinase (cdk, also known as 'cdc' for cell division control), and the cyclin or linked cyclin (cyclin/cdk) when present may be inactivated or activated by degrees of phosphorylation, giving the possibility of control via cyclin synthesis and, subsequently, phosphorylation/dephosphosphorylation.

External physical factors (e.g. temperature, light including the lunar cycle) and/or stress (Sumpter et al. 1987; Schreck 2010) and internal nutritional factors (dependence on prey or food abundance and quality, competition) and various interactions including courtship and size hierarchies have also been factored in to the complex sets of agents which can impede or achieve reproduction. Most information has been sought using small (goldfish, zebrafish) or commercially important fish (cod, flatfish, trout), and there is much more to be known, including the control of early phases of gamete formation and gamete resorbtion. The onset of first reproduction (puberty) has also been of interest.

Palstra et al. (2014), working with sole (*Solea solea*), believe that these fish are under photothermal control and that a cold period is 'permissive' for the action of FSH and presumably therefore the beginning of gametogenesis, with LH increasing subsequently with temperature rise.

Adverse temperatures have been suggested as causing resorbtion of oocytes in Greenland halibut (Fedorov 1971) with 'failure to spawn'. Potts et al. (2014) have described temperature effects on reproduction of resident marine fishes (sea-bream, *Diplodus sargus capensis*) in relation to global warming.

The timing of particular events in gametogenesis may be set in each species to maximize reproductive success. One important endpoint would be achieving eventual hatch so that the young can feed well (9.10.5, 9.12). Some fish are cued to lunar cycles; this applies to reef fish such as the rabbit fish, *Siganus argenteus* (Rahman et al. 2003). In the Siganidae the synchronous spawning is apparently achieved by stimulation of gonadotropin 'associated with a particular lunar phase' (Takemura et al. 2004), and melatonin (10.11.2) levels (amongst other factors) respond to night brightness (Takemura et al. 2010) and are presumed to stimulate or inhibit the production/release of gonadotropins in the HPG axis.

The role of pheromones in fish reproduction has been recognized through studies of the smaller oviparous species, notably female goldfish which release two different pheromone types around spawning. The first 'periovulatory' pheromone effect is due to a steroid mixture (e.g. androstenedione and sulphated or non-sulphated progestogens released at the gills or in the urine), the second signals the postovulatory stage and is a mixture of (e.g. two prostaglandins released in the urine) (Stacey 2003; Stacey et al. 2003). These pheromones can induce males to respond appropriately; for example, to synchronize gamete maturation and invoke courtship/reproductive behaviour (Stacey 2003). In female zebrafish pheromones may suppress reproduction in neighbouring females (Gerlach 2006).

The view that female fish are in general very fecund has generated an approach to fishing that assumes easy replacement for fish removed.

Unfortunately this optimistic view of the effect of removal, that it would be compensated for by less competition promoting higher survivals, has not withstood high fishing pressures. In particular, the systematic removal of large fish from the population has apparently undercut the possibility of sustaining populations (Xu et al. 2013).

Fish are also prone to interference with their reproductive physiology by phytoestrogens and pharmaceuticals in waste water entering the ecosystem (8.3.10), a matter for concern and investigation (Brown et al. 2014).

Appendix 9.1 Staging systems for fish reproduction

Table 9.3 Staging for female long rough dab (American plaice) *Hippoglossoides platessoides* off Scotland (after Bagenal 1957).

Stage number	Ovarian description	Bagenal's opinion	Comment
1	Small, translucent, pink	Fish would not spawn next season	Young fish, had never spawned
2	Large opaque, 'eggs' Visible and opaque	Ripening	
3	Some translucent: 'eggs'	Close to spawning	
4	Blood-shot, empty, flacced	Spent	May include non-reproductive Adults (i.e. 4_1)
4_1	Not given	'Immature' but had apparently spawned	
4_2	Not given	'Maturing' recovering from spent	

Table 9.4 Maturity stages generalized from Russian sources (Nikolsky 1963) applied to both sexes.

Stage Number	Meaning
I	Young fish, prior to reproduction
II	Quiescent
III	Ripening
IV	Ripe
V	Actively reproducing
VI	Post-reproductive

Table 9.5 North-east Fisheries Center (USA) maturity stages (Burnett et al. 1989).

Stage	Meaning	Application
I	Immature	Both sexes
D	Developing	Both sexes
R	Ripe	Both sexes
E	Eyed	Applies only to oviparous (e.g. redfish) females (larvae being eyed)
U	Ripe and running	Males and oviparous females with external fertilization
S	Spent	Both sexes
T	Resting	Both sexes

Table 9.6 Department of Fisheries and Oceans (Canada) maturity stages (Templeman et al. 1978).

Stage	Meaning	Application
Imm	Immature	Both sexes
Spent L.	Spawned last year (L = last)	Both sexes
Mat P	Spermatogenesis for next season (P = present)	Males only
Mat A–P	Oogenic for next season, large (yolky) oocytes Visible but not translucent (i.e. not very close to ovulation)	Females only
Mat B–P	As Mat A–P but up to 50% of the yoky oocytes translucent	Females only
Mat C–P	As Mat B–P but more than 50% of the yolky oocytes translucent, Perhaps also spawning ('ripe')	Females only
Partly spent P	Partially spawned out, some milt or eggs remaining	Both sexes
Spent P Mat N	Spawning completed present year, some evidence of recrudescence for next year	Both sexes
Mat N	Developing from immature for first reproductive cycle	Both sexes

Appendix 9.2 Families with mouth-brooding

Table 9.7 Families with mouth-brooding.

Family	Common name	Type of brooding	Reference
Osteoglossidae	Bony tongues	Guarding or ♀ mouth-brooder	Blumer 1982
Cyclopteridae	Lumpfish	Guarding or ♂ mouth-brooder	Blumer 1982
Opisthognathidae	Jawfishes	Commonly ♂ mouth brooder	Blumer 1982 Nelson 1976
Ariidae	Sea catfishes	♂ or ♀ mouth-brooder	Blumer 1982
Bagridae	Bagrid catfishes	Biparental mouth-brooding (2 *Phyllonemus* spp.) Or ♂ mouth-brooding (2 *Lophiobagrus* spp.)	Ochi et al. 2002
Apogonidae	Cardinalfish	♂ mouth-brooding	Fishelson et al. 2004
*Belontiidae/ Osphronemidae	Gouramis Fighting fish	Guarding or ♂ mouth-brooding	Schindler and Schmidt 2008
Luciocephalidae	Pikehead	Oral but sex of brooder n/a	Blumer 1982
Cichlidae	Cichlids	Guarding or mouth-brooding	
		♂, ♀ or biparental	Grüter and Taborsky 2004

*These two families have undergone changes in attributions. Osphronemidae, which only had one genus previously (e.g. Nelson 1976) appears to have superceded Belontiidae. n/a = not available.

Appendix 9.3 Steroids known to be or possibly involved in fish reproduction

Steroids have a characteristic four-ring structure and various possible side-chains.

Androgens (basically thought of as 'male' hormones) have 19 carbon atoms, hence termed C_{19} steroids. Androgens may be conjugated (to form a glucuronide or sulphated form) or bound to a protein (e.g. androgen-binding protein in the plasma) so that androgens may be referred to as free or bound and the concept of 'total androgens' arises to include the bound reserve. Testosterone, not the major male hormone for fish, is found in females as a precursor to oestradiol; 11-ketotestosterone is the principal male hormone for fish. Other C_{19} steroids found in vertebrates include androstenedione, dehydroepiandrosterone and androsterone.

Oestrogens (one of two major female hormone groups) have 18 carbon atoms, hence termed C_{18} steroids. Oestrogens (also termed estrogens) can be conjugated, as with androgens, or protein-bound (Ozon 1972b).

Oestradiol (also known as E_2) is regarded as the crucial oestrogen for fish, formed in the outer layers of maturing ovarian follicles and responsible for invoking vitellogenesis. Other oestrogens found in fish are oestrone, identified in the ovary of some fish, both elasmobranchs and teleost (Ozon 1972) and oestriol, identified in the ovary of fewer fish than oestrone, but found in both elasmobranchs and teleost (Ozon 1972).

Progestogens are the female steroid hormones associated with late stages of vertebrate oocyte maturation (and pregnancy for mammals) and have a 21-carbon atom structure. An important progestogen in fish is 17α 20β dihydroxy progesterone (DHP), the principal maturation-inducing steroid (MIS). 17α20β trihydroxy progesterone (20beta-S)[*], the MIS for the teleost family Scianidae, may also increase sperm motility in pleuronectids such as Southern flounder (Tubbs et al. 2011).

[*] also known as 17 alpha, 20 beta, 21 trihydroxy-4-pregen-3-one or 20 beta-dihydro-11-deoxycortisol

Appendix 9.4 The concept of degree days

Warmer temperatures tend to accelerate biological processes like plant growth and larval development. By multiplying time to hatch (number of days to hatch) by the temperature (keeping this constant) the product (as 'degree days') can be used to calculate development times at different temperatures within a fairly narrow limit. Practically this involves dividing the calculated degree days value for the species by the ambient temperature at the time of development to gain time to hatch. This applies to various poikilotherms including fish and invertebrates such as nematodes

(Westcott and Burrows 1991; Baird et al. 2002; Chezik et al. 2014). In practical terms, time to hatch may be taken for a percentage, for example 50 per cent hatch.

Appendix 9.5 Fluids for egg treatment

In order to preserve, count or evaluate 'egg' status, particular fluids are useful. Stockard's (9.6, 9.10.1) is used for 'clearing' eggs so that the interior is seen, in which case the position of the nucleus (germinal vesicle) can be checked and closeness to final maturation gauged. Stockard's solution is 50 ml formalin (37% formaldehyde), 40 ml glacial acetic acid, 60 ml glycerin and 850 ml water (Moulton and Penttila 2001), which may be substituted, for example by using Cortland's solution instead of pure water (Goetz and Bergman 1978). Cortland's is a balanced salt solution (BSS) which minimizes osmotic disruption; three such BSSs used for fish, Earle's, Hank's and Cortland's, are tabulated by Wolf and Quimby (1969).

Gilson's fluid (GF) is used to preserve and separate eggs (9.10.6) for counting (i.e. fecundity estimation), however it contains mercuric chloride and does not preserve late-stage (hydrated) oocytes well, and is associated with some shrinkage, and thus formalin may be preferred (Fleming and Ng 1987; Lowerre-Barbieri and Barbieri 1993). A modified GF, previously used quite widely, is 100 ml of 60% ethanol (or methanol), 15 ml nitric acid, 18 ml glacial acetic acid and 20 g mercuric chloride (Lowerre-Barbieri and Barbieri 1993).

Appendix 9.6 Staining sections with haematoxylin and eosin

Mounted sections on microscope slides, cut at about 7 μm, from fixed and wax-embedded tissue, can be stained, for greater contrast with an array of different stains, the most common being haematoxylin, counterstained subsequently with eosin, therefore known as H and E. A common procedure for this is as follows:

Dewax and hydrate

Immerse sections in haematoxylin for five to ten mins

Rinse in water, 'blue' using an alkaline solution

Differentiate (remove stain) if necessary with acid alcohol; and reblue.

Dehydrate with an alcohol series, counterstain with alcoholic eosin (about 30 secs).

Rinse in ethanol.

Clear in, for example, xylene.

Mount.

Appendix 9.7 Cell division

Every live cell (with the exception of mammalian erthyrocytes) contains at least one chromosome set, and when the cell divides it is crucial that each daughter cell has its appropriate set of chromosomes. This is achieved by doubling of the chromosomes before division and then separation of the doubled material in an orderly fashion with the aid of internal cell structures. This procedure is mitosis when it occurs to achieve growth or asexual reproduction. When sexual reproduction occurs, two gametes fuse to form the zygote, and to avoid an accumulation of chromosomes, each gamete undergoes a special reduction division or meiosis.

Cell division, whether mitosis or meiosis, follows a common procedure, doubling of each chromosome to form two chromatids before cell division occurs. At the beginning of cell division the chromosomes condense (becoming visible under the microscope after appropriate staining). In the live dividing cell a set of movements within the cell then occur. These movements achieve separation so that each daughter cell has equivalent chromosomal material.

The movements of the chromosomes to achieve this occurs by the actions of a special structure called a spindle in animals. The spindle is only formed during cell division and disassembles afterwards.

For growth, each cell division procedes by mitosis and produces two near-identical cells, with the same number of chromosomes as each other and as the parent cell. The two chromatids of each chromosome lie parallel to each other and attached to each other by a centromere. The two chromatids of each chromosome are then separated and one of each pair passes to one of the two 'poles' of the cell.

There it is regarded as a chromosome and it can, after cytoplasmic separation, resume its normal activity after decondensation. The chromatid movement is achieved by microtubules of the spindle which can attach to the centromere of the chromatids.

Stages of mitosis

Prophase: chromosome condensation, dissolution of the nuclear membrane

Metaphase: spindle formation, pairs of chromatids attach to the spindle at its centre (the equator)

Anaphase: chromatids separate and one of each pair travels to a different pole

Telophase: the chromatids are now chromosomes which decondense and a nuclear membrane forms around each of the two chromosome clusters at the two poles.

Cytokinesis: the cytoplasm divides (usually equally) so that each nucleus has half the parent cell's cytoplasm and the same number of chromosomes as the parent

Meiosis

In gonial (oogonia, spermatogonia) cells which form the gametes a more complex process achieves cell division, known as meiosis. This produces four cells in two consecutive cell divisions know as meiosis I and II. As many organisms have two sets of chromosomes (paternal and maternal) sexual reproduction, in which two gametes fuse to produce a zygote, could produce unsupportable increases in chromosomes. Meiosis, a reduction division, produces gametes with one-half the chromosomes, as a complete set.

Meiosis I

In meiosis I the prophase has series of movements of the double chromosomes so that before division each individual chromosome is coupled with its equivalent. This brings each of the pair into very close contact and they can exchange material, 'reshuffling' the composition of each chromatid.

Meiosis 1 proceeds to separate the chromosome pairs so that a complete set passes to each spindle pole. However, this separation, for females, may be long delayed so that the cell remains arrested in prophase 1, until fertilization occurs. Presumably this delay allows chromosomes to remain active (precondensation) and the cytoplasm to grow, as it does, very considerably, in fish. Mammals may have meiotic arrest in diplotene, a situation during which very little activity is expected within the oocyte, whereas the arrested fish oocyte may have a very active phase with yolk accumulation (vitellogenesis).

When fertilization occurs the oocyte resumes prophase and passes through the other stages to achieve separation of the homologous paris of chromosomes and then in meiosis II separation of the chromatids. The cytoplasm divides unequally and only one of the four products survives as the ovum or egg. The rest of the nuclear material, with a minute amount of cytoplasm, is discarded as polar bodies. In contrast, the division of the spermatocyte produces four sperm. The prophase of meiosis I has several stages, as usually described, but such stages are not routinely seen in fish oocytes, and it is possible that meiotic arrest in fish is earlier than is described for mammals

Stages of meiosis I prophase

Leptotene: chromosomes partly condensed.

Zygotene: chromosomes begin to form pairs (bivalents), attached by synapsis, which eventually are described as tetrads (four chromatids). The close contact of the bivalents enables interchange of genetic material as 'cross-over'. During this phase, chromosome may cluster in one place, close to the nuclear membrane and attached by their telomeres, a sub-stage known as synzesis or bouquet formation.

Pachytene: further condensation.

Diplotene: some separation of the bivalents, with points of attachment seen as chiasmata.

Diakinesis: separation of bivalent chromosomes completed; nuclear membrane and nucleoli disperse.

At cytokinesis, in females most of the cytoplasm remains with one of the two products, the putative ovum. The other product is very small and is the first polar body (which may subsequently divide).

Meiosis II

Meisosis II resembles mitosis but again the cytoplasm does not divide equally in females; most of it remains with the ovum, and a very small second polar body is extruded.

Bibliography

Potts, G.W. and Wootton, R. (eds) (1984). *Fish Reproduction: Strategies and Tactics*. London: Academic Press.

Wootton, R.J. and Smith, C. (2014). *Reproductive Biology of Teleost Fishes*. Chichester: Wiley-Blackwell.

For physiology of fish reproduction:

Special issue: Reproductive Physiology of Fishes, *Journal of Fish Biology*, 2010, 76, Issue 1.

Norris, D.O. and Lopez, K.H. (eds) (2011). *Hormones and Reproduction of Vertebrates, Vol. 1: Fishes*. London: Elsevier, Academic Press.

Proceedings of the International Symposia on the '*Reproductive Physiology of Fish*' for example:

1. Paimpont, France (1977). *Annales de biologie animale, biochimie, biophysique* 18 (4).
2. Wageningen, Netherlands (1982). Waginingen: Centre for Agricultural Publishing and Documentation.
3. St. John's, Newfoundland (1987). St. John's: Third fish symposium 1987.
4. Norwich, UK (1991). Sheffield: FishSymp 91.
5. Austin, Texas (1995). Austin: Fish Symposium 95.
6. Bergen, Norway (1999). Bergen: Institute of Marine Research and University of Bergen.
7. St. Malo (2007). *Cybium* 32 (2 suppl.).
8. Cochin, India (2011) *Indian Journal of Science and Technology* 4

For pheromones, see Stacey, N and Sorensen, P. (2005), Reproductive pheromones. In K.A. Sloman, R.W. Wilson and S. Balshine (eds.) *Fish Physiology Vol. 24 Behaviour and Physiology of Fish* Amsterdam: Elsevier, 359–412.

Special issues on migration:
Canadian Journal of Zoology, 2014, 92, No. 6.
Journal of Fish Biology, 2012, Vol. 82(2).

CHAPTER 10

Integration and control: hormones

10.1 Summary

Within the circulation and tissues of vertebrates there are many individual molecules, acting as hormones, which can have profound effects on function, usually (but not always) transient effects which are reversible. The source of hormones in fish may be in diffuse tissues or well-defined glands.

Recognized hormones in fish have usually been sought after their presence was noted in mammals.

However, there are some functions in fish that would not be relevant in mammals and vice versa; mammals do not have adjustable chromatophores, and the presence of melanin-concentrating hormone (MCH) and melanophore-stimulating hormone (MSH) in mammals does not relate to colour change. Similarly, prolactin-like molecules in fish cannot be involved in lactation.

Whereas some hormones, notably some steroids, are identical in mammals and fish, the latter have steroids that are not found in mammals. Complex polypeptide hormones may be similar in both taxa but not identical, though shared sequences may enable activity across the taxa.

The multiplicity of sites for hormone secretion and actions present a challenge for understanding their importance to fish function especially as some hormones may also act as pheromones and some neurotransmitters may also act as hormones.

10.2 Introduction

Hormones are molecules which are synthesized in special endocrine or neural cells, then released to the tissue fluid or circulation, where they target other cells, bind to them and cause specific responses. Endocrine cells may form obvious clusters as endocrine glands or be more diffuse.

Hormones have been described in many taxa of the Animal Kingdom and are particularly relevant in complex multicellular forms with a well-developed circulation. They have been intensely studied in mammals because of their importance to the medical sciences. The emphasis has been on the effects of too little or too much with respect to individual factors, notoriously hormones like thyroid hormone which affects metabolism and mental activity, or the pituitary growth hormone which in excess can produce gigantism.

Vertebrates control function either through the nervous system (neural control, Chapter 11) or through hormones released to the circulation (endocrine or humoral control). The term endocrine means 'secreting within' and refers to hormones as secreted into the blood directly, not through a duct. Hence endocrine glands are also referred to as 'ductless glands' in contrast to exocrine glands.

The two control systems in vertebrates are linked via the hypothalamus and the pituitary gland, with a series of release and release-inhibiting hormones modifying the circulation of other hormones. Also, although it is generally agreed that hormones affect their target at some distance from their source, some hormones are paracrine (acting close to the source). Moreover, some hormones are chemically very similar to neurotransmitters (e.g. adrenalin/noradrenalin) and some hormones are 'neurohormones', secreted by neural tissue. However, most hormones are secreted by glandular cells, occurring in clusters and well vascularized. Some endocrine production occurs in very well-defined separate glands such as the thyroid and the pituitary.

Essential Fish Biology: Diversity, Structure and Function. Derek Burton & Margaret Burton.
© Derek Burton & Margaret Burton 2018. Published 2018 by Oxford University Press.
DOI 10.1093/oso/9780198785552.001.0001

10.3 Hormone nomenclature and classification

The nomenclature of fish hormones generally follows that of mammals and may reflect the source or the probable or known function. Hormone nomenclature, starting with the first hormone described, secretin, tends to have the suffix '-in', as with secretin, gastrin, prolactin, thyroxin, insulin, ghrelin. When hormones controlling other hormones were recognized in the pituitary, they were generally given the longer suffix '-trophin' as with gonadotrophin, or as an adjective '-trophic', as in adrenocorticotrophic hormone, ACTH for brevity, the latter controlling secretions of the adrenal cortex.

It is quite common for hormones to have several alternate names and at least one acronym, which is usually based on a sequence of initials representing the longer name; for example, thyroid stimulating hormone (TSH) is also termed thyrotrop(h)in. Common hormone acronyms are listed in Appendix 10.2.

In addition to classifying hormones by their chemical structure (steroids, polypeptides etc., 10.4), hormones can be differentiated into major classes by solubility; those that are lipid-soluble and hence can enter cells, and those that are lipid-insoluble and render their effects by external links to their target cells. The specificity of hormone action is regarded as relying on the presence of particular molecular receptors within the cells for lipid-soluble hormones and on the cell membrane for lipid-insoluble hormones and probably too for some lipid-soluble hormones that can penetrate the cell membrane but whose fast action suggests a shorter routing.

Lipid-soluble hormones comprise a major group; steroids including 'sex' steroids, corticosteroids as well as the non-steroid thyroid hormones, tri-iodothyronine (T_3) and thyroxine, tetra-iodothyronine (T_4). Lipid-insoluble hormones are amines, peptides and polypeptides including high molecular-weight proteins.

The association of just one hormone with one cell type was reflected in the nomenclature but that tenet may be less secure now: some endocrines are synthesized and released in a precursor form (prohormone) which may actually split to form several hormones. A good example of this is proopiomelanocortin (POMC, 10.8) of the pituitary which, as its name suggests, can be a precursor for several hormones.

On first encounter it may not be clear that a newly described substance in the circulation, or in an extract, is indeed a hormone, even though it shows consistent physiological effects if administered. It often happens that a possible hormone is called a 'factor' or a similar less exact term. Even when the putative hormone is accepted as a hormone, the term 'factor' may persist. In addition, when a hormone is first described its role may be unclear and the name applied may not, eventually, be appropriate. Examples of this in fish are MCH, melanin (melanophore)-concentrating hormone, which may be involved in appetite control (Volkoff et al. 2010) though named in relation to colour change and MSH, melanophore-stimulating hormone, which may be controlling xanthophores rather than melanosome movement in some species such as winter flounder, *Pseudopleuronectes americanus*.

10.4 The chemical structure of hormones

Chemically, hormones can be steroids or a catecholamine or based on a single amino acid (e.g. tryptophan or tyrosine) or a polypeptide chain composed of a few or many amino acids (Table 10.1). The largest polypeptides are generally regarded as proteins and the protein hormones may share sequences with other hormones (forming a 'family') such as the growth hormone (GH), prolactin (PRL) and somatolactin (SL) cluster, which form a 'superfamily' with the cytokines (Sakamoto and McCormick 2006). Gonadotropins (GtHs) and TSH (10.11.1, Table 10.2) have an α and a β sub-unit (i.e. are 'heterodimers', Appendix 10.1c), with the vertebrate scenario of an identical α subunit in the different GtHs such as LH (luteinizing hormone) and FSH (follicle-stimulating hormone).

Sequencing the entire molecule for individual fish species has produced some interesting comparisons, with elasmobranch GH more similar to tetrapods than the teleost GHs from bonito and salmon (Yamaguchi et al. 1989). Some of the protein hormones (e.g. TSH, GtHs) with subunits (Table 10.1) are also glycosylated, that is, they are glycoproteins, and therefore can be located histochemically

Table 10.1 Chemical basis of hormones in fish.

Chemical format	Hormone type	Example	Major source region
Steroid	Sex steroid	Testosterone	Gonads
	Corticosteroid	1-α hydroxycorticosterone	Interrenal
Catecholamine	Catecholamine	Adrenalin	Chromaffin tissue
Linked aminoacid	Iodinated tyrosine	Thyroxin	Thyroid
	Modified 5OHT* Based on tryptophan	Melatonin	Pineal
Polypeptide	Short[†]	AVT*	Pituitary PN*
	Medium length[†]	MSH*	Pituitary PI*
		Gastrin	Stomach
Protein		Insulin	Pancreas
	Long[†]	GH*	Pituitary PA*
		Prolactin	Pituitary PA
	2 sub-units, glycosylated	GtH*, TSH*	Pituitary PA

[†]Short = 3–10 aminoacids. Medium length = 11–99 aminoacids. Long = > 100 aminoacids.
*AVT = arginine vasotocin; GH = growth hormone; GtH = gonadotropin; MSH = melanophore stimulating hormone; 5OHT = 5hydroxytryptamine; PA = pars anterior; PI = pars intermedia; PN = pars nervosa; TSH = thyrotropin

via Schiff's reagent (Appendix 10.1). Pearse (1969) considered that many of the different cell types associated with peptide hormone secretion (in vertebrates) were part of a common embryological descent. He termed these cells APUD from their propensity to handle amines through uptake and decarboxylation (Amine content and/or Amine Precursor Uptake and Decarboxylation). This concept was reviewed by Goldsworthy et al. (1981) and more recently revisited by Boyd (2001) and Blackmore and co-workers (2001).

10.5 Hormone candidates

Traditionally, since hormones were first described, a candidate hormone is expected to meet certain criteria. These include identifying the source (x) of the putative hormone (y) and finding more of y in the venous outflow from x than the arterial inflow to x.

It is also expected that extirpation of x will lead to physiological effects reversible by injection of y. Obviously if the source x is a complex gland its extirpation (ablation) can either produce multiple physiological problems or even death. Moreover, some endocrine tissue is diffuse, not forming a distinct gland and its complete surgical elimination may be impossible. In spite of these difficulties many vertebrate hormones are recognized and well characterized in mammals; their occurrence or precise function in fish may not be so well understood (see 'substance P' in the review by Volkoff et al. (2005) and melanin-concentrating hormone (MCH) reviewed by Baker (1994)).

Once a hormone's presence is suspected it may be possible to localize its site of synthesis and/or its target by the use of antibodies and immuno-cytochemistry (Appendix 10.1.a), sometimes preceded by the identification of enzymes expected to be in the synthetic pathway. Such enzymes as 3βhydroxysteroid dehydrogenase (as an indicator of steroid synthesis) and enzymes in the melatonin synthetic pathway (Falcón et al. 2010) have been useful in finding putative locations of intracellular production of these hormones. Localization of a hormone can occur by extraction of tissue in comparison to control tissues (i.e. the putative hormone or the enzymes associated with its synthesis would be found in one region but not in another) or by sectioning tissue (Appendix 10.1) and using dyes or chemical reactions to show likely cellular sources of the hormone.

10.6 Characteristics of hormone control

Endocrine control tends to be slower and more sustained in action than neural control. The persistence of hormones in the circulation and their clearance rate may be affected by the presence of plasmatic binding proteins (e.g. with androgens) or the possibility of partial inactivation by conjugation (e.g. in the liver), as occurs with steroids, though conjugated steroids released (e.g. in the urine) may act as pheromones.

Hormones can achieve homeostasis or produce a unidirectional effect, not readily reversed, promoting cell division and/or long-term intracellular changes as occur in growth and reproduction.

Stimulation → *hypothalamus*↓

        TRH → *adenohypophysis* *↓

              TSH → *thyroid* ↓

                     T_4 → somatic target

Figure 10.1 A chain of hormones (underline) secreted sequentially from sites of endocrine tissue (italics). *of the anterior pituitary; TRH = thyrotropin-releasing hormone TSH = thyrotropin (thyroid-stimulating hormone) T_4 = thyroxin

Individual hormones may function to control the release of other hormones with as many as three hormones linked in a chain of action (Figure 10.1), with a hypothalamic hormone stimulating a pituitary hormone which controls the synthesis/release of a third hormone. Hormones may also act as antagonistic pairs as occurs with some hypothalamic hormones where there can be 'releasing' hormones and 'release-inhibiting' hormones (Table 10.2). Dual antagonistic control can also achieve homeostasis as with the regulation of plasma levels of glucose by insulin and glucagon. Hormones can be regulated by feedback, either negative or positive. Negative feedback provides homeostasis with downregulation of synthesis/release in response to high circulating levels of a hormone or a molecule whose presence it has stimulated. Positive feedback which leads to upregulation and increased output is associated with reproduction and growth where, at least in the short-term, accretion of cells or their products should occur.

Because so much research on vertebrate hormones has arisen from medical work the study of fish hormones has been greatly influenced by medical endocrinology. As the evolutionary history has fish hormones generally expected to pre-date mammalian hormones, there is the possibility of misunderstanding the function of apparently homologous hormones, as may occur with prolactin (10.7).

10.7 Fish hormones in relation to mammalian hormones

Hormones described from both tetrapods and fish may be identical, as with some steroids, or similar in both structure and effective action. Moreover, a fish or mammalian peptide hormone may act in non-source (heterologous) species very well, an example being salmon calcitonin which can affect calcium levels in mammals.

Hormones described as functioning in one way may turn out to probably have another, primary, function *in vivo* prolactin, involved in mammalian lactation, is a puzzle: structurally similar prolactins are present in fish and mammals but with necessarily very different functions; the function in fish is not yet clear though many different suggestions have been made including osmoregulation (Sakamoto and McCormick 2006) and some facets of reproduction (Whittington and Wilson (2013). The similar somatolactin, first described by Rand-Weaver et al. (1991a), may be involved in colour change (Pandolfi et al. 2009).

10.8 Stored and bound hormones, prohormones

Some of the steroids synthesized and active in fish are identical to those of other vertebrates, for example testosterone, but although testosterone occurs in fish it may not be the principally active steroid but can be present as a precursor to another steroid, specific to fish. Thus the predominant male steroid in fish may be 11-ketotestosterone. Steroids may also be conjugated, for example to glucuronides or sulphate, which increases the range of complexity in analysing their presence and clearance. In transmission, steroids may also be linked to plasmatic binding proteins giving the concept of free or bound fractions as well as the sum: the 'total' value. Bound steroids can act as a source or reservoir of active molecules close to their target cells.

Stored forms of hormones can also be detected close to their secretion point, as prohormones, or bound to protein or at the ends of special axons (10.9). Some polypeptides are synthesized as very large prohormones: pro-opio-melanocortin (POMC, 10.3) found in the vertebrate pituitary (10.11.1), can be cleaved to produce endorphins, adrenocorticotrophic hormone (ACTH) and forms of MSH. ACTH, itself a 39 amino acid polypeptide, can cleave to form αMSH (13 amino acids) and CLIP (16 amino acids) which is a corticotrophin-like intermediate lobe peptide (Goldsworthy et al. 1981, after Nakanishi et al. 1979).

The major thyroid hormone (T_4) is secreted into thyroid follicles where it is stored as a 'colloid' thyroglobulin, with the T_4 bound to protein. T_4 released to the circulation is transformed to T_3, active in the periphery. The presence of large amounts of stored hormone can be misleading because there may not have been much release to the circulation. Hoar (1957) and Leatherland (1994) figured different stages of activity for thyroid gland follicles.

10.9 Neurosecretion

Although it has been customary to regard hormones as acting at a distance from their secretion site it is now realized that some act very close to their origin. Vertebrate neurohormones are in this category. Associated with the hypothalamus and pituitary, neurohormones are secreted by special neurons with axons passing to the pituitary. Neurosecretions forming the neurohormones pass along the interior of the axon before release to the circulation. The role of these hypothalamic hormones in mammals is significant in smooth muscle contraction in females and in osmoregulation. The equivalent role in fish is more general as presently understood, possibly affecting blood pressure through vasoconstriction as well as osmoregulation through cell permeability changes. Hypothalamic neurosecretion is also highly important in control of other hormones (10.11).

10.10 Interaction of hormones with their targets

In order to achieve their role, hormones react with specific target cells, recognized by the presence in the target of specific receptor molecules which may be on the cell surface or within the cell, the latter meaning that the hormone has to penetrate the cell membrane.

Thus, the operation of hormones is expected to be through a 'messenger' system at the cellular/ intracellular level with specific molecular receptors binding to the individual hormone molecules. For peptides and proteins the receptors are on the cell surface and binding inaugurates a response via a 'second messenger' such as cyclic adenosine monophosphate, cAMP (the hormone being the 'first messenger'), which may take some time to accomplish. Steroid hormones, being lipid-soluble can enter the cell to bind intracellularly. The location of steroid receptors, however, may not necessarily be internal; for some steroids, external receptors may occur. The delay to response will vary with hormone type as does clearance rate. The response tends to be extended and slow in comparison to neural control. Though often reversible, the effect in the case of reproduction is likely to be unidirectional and not readily reversed.

Hormone action can be subject to 'endocrine disruption' (via EDCs, endocrine-disrupting chemicals) when environmental levels of either introduced or naturally occurring hormone mimics affect the receptor sites. This can occur when EDCs bind to receptors, thus either blocking or enhancing the system. Sex steroids are particularly vulnerable to endocrine disruption, such as when waste water contains high levels of synthetic steroids used in contraception or phytoestrogens occur in pulp-mill effluents (8.3.10, 9.18).

10.11 Endocrine organs, glands, tissues and cells

10.11.1 The pituitary gland and hypothalamus

The most complex source of hormones in fish, as currently understood, is the ventral region of the brain forming the hypothalamus (11.6.2) and the projection from it, the pituitary gland (hypophysis). The hypothalamus exercises control over the secretions of the pituitary by a series of peptide hormones (also known as 'factors') which act to stimulate or inhibit release at the pituitary sites (Table 10.2). These high-order controlling peptides may be carried close to the pituitary cells by special neurosecretory axons. Unlike mammals there is no hypothalamic–hypophyseal portal system (a short vascular route for control). Also, some peptide hormones are produced by hypothalamic

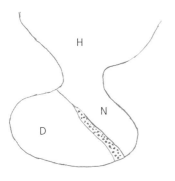 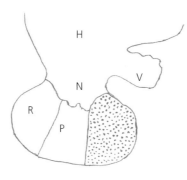

Figure 10.2 Diagrammatic saggital section through (left) a mammalian (human) pituitary and (right) a teleost (flounder) pituitary. D = pars distalis of the adenohypophysis; H = hypothalamus (of brain); N = neurohypophysis; P = proximal pars distalis; R = rostral pars distalis; V = saccus vasculosus. Stippled = pars intermedia. Anterior to the left for both diagrams.

neurosecretory cells for direct action at the periphery after storage and release in the pituitary. MCH is a hypothalamic hormone capable of appetite control (10.3) but originally named for its role in colour change (Baker 1994).

The pituitary gland projects ventrally from the hypothalamus of the brain and together these two areas are of major importance in producing and controlling hormones. The pituitary gland of vertebrates has been intensively studied. Regions recognized in mammals (Figure 10.2) include the anterior pituitary (pars anterior, PA, as part of the adenohypophysis, AH), the posterior pituitary (or neurohypophysis, NH) and the pars intermedia (PI) which, confusingly, is posterior in fish while the small neurohypophysis is intermediate in position (Figure 10.2). In some teleosts the pars anterior is divided into a rostral pars distalis, RPD, and a proximal pars distalis, PPD (Figures 10.2, 10.3). Initially based on embryonic origins the pituitary was thought of as neurohypophysis plus adenohypophysis, the latter including the pars intermedia.

It is relatively easy to remove the pituitary (hypophysectomy) from anaesthetized fish and, provided they are subsequently held in appropriate media (to compensate for ionic disturbances, Appendix 7.2), the fish can be used for experimental administration of individual hormones. Such experiments have aided study of the physiological effects of particular extracts or purified hormones. A comprehensive review of studies on the pituitary gland of fishes was produced by Pickford and Atz (1957). Since that time a number of new ideas, including the presence and activity of hypophyseal-releasing hormones (Table 10.2) have changed our understanding of this system.

Table 10.2 Vertebrate hypothalamic hormones controlling pituitary hormones.

Hormone name	Acronym	Function
		Reproduction
Gonadotrop(h)in releasing h.* and/or	GnRH	Controls s/r** of gonadotropins (GtHs)
Luteinizing h. releasing h.	LHRH	Controls s/r of luteinizing h. (LH)
Follicle-stimulating h. releasing h.	FSHRH	Controls s/r of follicle stimulating h. (FSH)
		Growth
Growth h. releasing h. also known as Somatotropin releasing h.	GHRH	Controls s/r of growth h. (GH) also known as somatotropin
Growth h. release-inhibiting h. also known as somatotropin release-inhibiting factor or somatostatin	SRIF	Prevents release of GH
		Metabolism/Homeostasis
Adrenocorticotrop(h)ic releasing h. also known as corticotrophin-releasing factor	CRF	Controls s/r of adrenocorticotrop(h)ic h. (ACTH)
Prolactin-releasing h.[†]	PRLRH	Controls s/r of prolactin (PRL)
Thyroid-stimulating h. releasing h.	TSHRH	Controls s/r of thyroid-stimulating h. (TSH)
		Colour control
Melanophore-stimulating h. releasing h.	MSHRH	Controls s/r of melanophore-stimulating h. (MSH)

*= hormone.
**s/r = synthesis and/or release.
[†]Not much known about this; see Lagerström et al. (2005).

The neurohypophysis of mammals is the source of two octa/nonapeptide hormones (formed by hypothalamic neurosecretion); antidiuretic hormone (ADH), also termed arginine vasopressin (AVP), important for osmoregulation of terrestrial terapods, and oxytocin, important in female uterine contractions and lactation (Withers 1992), none of which are directly relevant to most fish. Fish (and non-mammalian vertebrates generally) have a similar peptide, however, arginine vasotocin (AVT), whose functions have been said to include osmoregulation, modification of the circulation and facets of reproduction (Balment et al. 2006).

The pituitary gland and hypothalamus together have a role in endocrine production which impinge on or control many different functions including growth, reproduction, homeostasis and stress responses. Tables 10.2 and 10.3 list major hormones generally recognized as synthesized and released in this region. Some potentially active substances (e.g. CLIP) extracted from this region have yet to be established as bona fide hormones released to the circulation or acting as controlling hormones between the hypothalamus and pituitary.

The prior study of mammals has obviously produced some terminological problems when homologies are considered but tissue has changed position over evolutionary time (e.g. the 'pars intermedia' of mammals is posterior in fish). Other problems of interpretation arise because fish pituitaries may change composition seasonally and may change shape as the fish ages. Also, many fish are reported to have a prominent saccus vasculosus (SV/V, Figure 10.2) very close to the pituitary but not usually contiguous with it (11.6.2), with the exception of *Polypterus* in which the SV links directly with the neurohypopysis. The SV can be seen macroscopically in flatfish as a triangular red structure posterior to the pituitary, and microscopically the SV has very distinctive ('crown' or 'coronet') cells lining large spaces. It occurs in some fish (elasmobranchs, eels, flatfish) but not others (*Poecelia*), and its function is not known but possible roles are discussed in 11.6.2.

Over the past approximately 80 years many cell types have been described from the vertebrate pituitary (Table 10.3) and positions were taken on nomenclature and function. At one point it was

Table 10.3 The localization of major pituitary hormones in teleost fish (see also Appendix 10.1, a-c).

Hormone	Pituitary location[*]	Cell type	Cell characteristic
Growth	PPD	Somatotrop(h)s	Strong affinity for Orange G, i.e. are acidophils
Prolactin	RPD	Lactotrop(h)s	Weak affinity for Orange G
Gonadotrop(h)in	PPD/RPD	Gonadotrop(h)s	Strong affinity for PAS[**]
Adrenocorticotrophic	PPD	Corticotrop(h)s	Chromophobe[***]
Thyroid stimulating	PPD	Thyrotrop(h)s	Strong affinity for PAS
Melanophore stimulating	PI	MSH cells	Affinity for Pb haematoxylin, PAS negative[†]
Somatolactin	PI	SL cells	Some species PAS positive Chromophobe in trout[††]
Endorphin	PL/PD	nd	Bound antibody[****]

[*]Acronyms as in Figure 10.3.
[**]PAS = periodic-acid-Schiff procedure, which determines the presence of carbohydrate, i.e. the endocrines secreted are glucoproteins.
[***]does not stain.
[****]van den Burg et al. 2001.
[†]van Eys 1980.
[††]Rand-Weaver et al. 1991b. nd = not determined
Note: Other cell types reviewed in Ball and Baker (1969), Holmes and Ball (1974).

insisted that one cell type would only secrete one hormone, but this idea is shaken by the concepts introduced with POMC, with potential cleavage to produce three or more active constituents with very different targets. It has also been accepted procedure to classify cells by their staining affinities; quite reasonably, a glycoprotein hormone (such as TSH) would be expected to be associated with cells staining with periodic-acid-Schiff (PAS) which shows up carbohydrates. More recently, with access to specific antibodies capable of being labelled (tagged) individually, for example with a dye or as part of a linked and amplified chain (Appendix 10.1b), it has been possible to detect the likely presence of individual hormones especially if they are complex polypeptides and antigenic. It is more difficult to use these techniques for hormones which are not very antigenic or which share amino acid sequences with other hormones or are readily lost during processing, but antibodies can be raised to modified hormones or specific amino acid sequences (Appendix 10.1b), and recognition of the unmodified form can be achieved if it has not been lost during tissue processing.

Chemically all of the pituitary hormones are simple or complex peptides with some hormones very similar to each other, which may make immunochemical localization and/or separation challenging. Thus prolactin, growth hormone and somatolactin are similar long-chain peptides whereas TSH and the gonadotropins all have two subunits with the α-subunit identical within a species; in addition, these hormones are glycosylated, that is they are glycoproteins (hence PAS positive, Table 10.3) after post-translational modification.

Some vertebrate pituitaries have a blood portal system directly linking the hypothalamus to the pituitary; this hypothalamic–hypophyseal portal system enables direct and rapid passage of controlling hormones from the hypothalamus to the pituitary. The alternate or additional link is through the neurosecretions (neurosecretory substance, NSS) of axons with cell bodies in the hypothalamus and axons reaching down into the pituitary. As there are so many fish species it was a major effort to study these links but from those species which were studied it seems generally unlikely that there is a pituitary portal system in teleosts, which will therefore rely on the general circulation or neurosecretions.

The known pituitary and hypothalamic hormones encompass many different control functions, justifying the previous notion of the pituitary as the 'master gland'. When pituitary extracts are made for analysis there may be subtle but important changes which affect function, and identifying the naturally occurring polypeptide may be quite difficult. Comparisons with the better-known mammalian systems have been crucial but sometimes misleading; for example, somatolactin (SL) secreted by the fish pars intermedia has not been found in tetrapods (Sasano et al. 2012).

However, it is well established that the pituitary and hypothalamic hormones together control growth (growth hormone and its antagonistic hypothalamic controlling pair), metabolism and metamorphosis (TSH and its hypothalamic control), reproduction (probably two gonadotropins, GtHs, and their hypothalamic control), homeostasis (prolactin, adrenocorticotrophic hormone, ACTH,

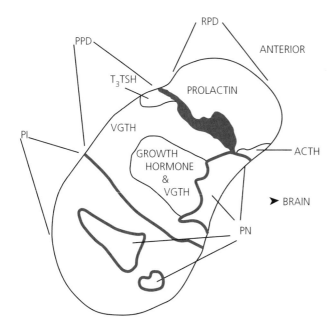

Figure 10.3 Parasaggital (longitudinal) section of winter flounder (*Pseudopleuronectes americanus*) pituitary at the spawning season. Drawing based on slides stained with PAS Orange G and immunofluorescent procedures (Appendix 10.1b) using first antibodies raised to fish hormone extracts. PI = pars intermedia; PN = pars nervosa; PPD = proximal pars distalis; RPD = rostral pars distalis; ACTH = adrenocorticotrophic hormone; T$_3$TSH = a tri-iodothyronine stimulating hormone; VGTH = a PAS positive region with gonadotrophic activity.

and arginine vasotocin, AVT) and also partake in stress responses through ACTH and endorphins. MSH, melanophore-stimulating hormone (and possibly SL, Sasano et al. 2012) has a role in skin colour and colour change in some fish.

10.11.2 The pineal organ

Like the pituitary gland, found ventrally attached to the brain, the pineal organ/gland is also attached to the brain, but on the dorsal surface. In contrast to the pituitary gland which secretes many hormones (Tables 10.1 and 10.2), the pineal is simpler in structure, and is generally only associated with the secretion of one hormone melatonin, also secreted or stored in the retinas and elsewhere in the brain. One view of the pineal is that it is a non-image forming photodetector (Bowmaker and Wagner 2004), an interpretation that tends to discount any endocrine role.

An overview of the pineal organ of vertebrates is provided by Kappers (1971), at which time the role of the pineal as a photoreceptor system was becoming recognized as supplemented by a secretory role. Kappers lists the three main cell types of the pineal (the epiphysis); the photoreceptors (sensory receptors, Chapter 12), supporting cells and sensory nerve cells. Martini (1971), in the same symposium, points out that the pineal had not been much studied previously and that the usual procedures for studying ablation and subsequent treatment with extracts had been unsuccessful because of the peculiar innervations of the pineal system.

Unlike the fish pituitary, which is easy to locate and offers many opportunities for study, the fish pineal can be difficult to find and study. The pineal of some vertebrates presents as a fairly obvious structure (the median dorsal third 'eye', also referred to as the parietal eye or the epiphyseal complex (Hoffman 1970), part of which may be seen by inspection of the dorsal head surface (as with some lizards and the rhynchocephalian reptile *Sphenodon*). Romer (1955) suggests that ancestral vertebrates may have had two additional 'median' eyes, one of which became the pineal and one the parapineal organ, the latter forming a small but recognizable extra eye in *Sphenodon* and some lizards. The pineal organ in extant species arises dorsally (11.6.2, Figure 11.5), generally

anterior to the optic tectum (Romer 1955), and may overlie the telencephalon (Bowmaker and Wagner 2004)) and/or terminates dorsally close to the skull or lies within a foramen (tunnel) or fossa (depression) of the skull. A pineal has been described in at least 79 teleost species though its position in some 'within the skull' (Ekström and Meissl 1997) suggests difficulty in visualization. Matty (1985) states that 'very little is known about the pineal of elasmobranchs' although some recent advances have been reported (Carrera et al. 2006). Of the relict fish the extant lungfish do not have a pineal foramen, but this was present in some fossil forms such as the early Devonian *Dipnorhynchus kurikae* (Campbell and Barwick 1985), though absent in the late Devonian *Rhinodipterus* (Clement 2012). However, interpretation of the fossil structures is complicated by reports of skull anomalies (Kemp 1999b), with 'abnormal skull roofing patterns' reported; the holotype of the Devonian *Tarachomylax oepiki* had a definite pineal opening whereas another specimen had none. The pineal of *Acipenser baeri*, a sturgeon, has been described by Yáñez and Anadón (1998) as having a long stalk, projecting forwards from the optic tectum, so that the pineal lies above the anterior telencephalon (11.6.2), in a 'parietal window'.

The fish pineal is expected to be photosensitive and secrete melatonin, suppressed by light, as in other vertebrates; however, detecting a pineal or studying it has been unrewarding in understanding the details of function. Melatonin is N-acetyl 5-methoxytryptamine, derived from 5-hydroxytryptamine (serotonin); it is expected to be secreted in response to change in light conditions and to enable seasonal cues to be effective, for example in controlling reproductive cycles. It may also be secreted by the retina or other regions such as the gut (Ekström and Meissl 1997). Histologically it is common to find reports of rod-like photoreceptors containing visual pigment in the fish pineal. It is interesting that melatonin production, indicated, for example, by the presence of appropriate enzymes, can occur both in the retina of the eye and the photoreceptors of the pineal gland. Enzymes important in melatonin synthesis have been reported in a few teleosts (e.g. pike, salmon). In addition, cultured pineal cells from many more species have been reported to produce melatonin.

Rüdeberg (1971) states that there is great structural variation in the teleost pineal organs, and contrasts those organs that are clearly photosensory with those which appear to be primarily vascular/glandular. More recent views do not preclude the pineal as having all these attributes (Ekström and Meissl 1997), with photosensitive cells capable of melatonin synthesis. The pineal organ is basically epithelium, which may be folded, surrounding a central lumen, and is referred to as a 'vesicle' (or 'end-vesicle', Bowmaker and Wagner 2004). The neural connections of the teleost pineal, most of which consist of non-myelinated axons may be very complex and some of the brain areas innervated can also receive input from the retina. The sturgeon has similarly complex brain links (Yáñez and Anadón 1998).

Matty (1985) reports the bony fish pineal as anatomically differentiated by the overlying structures. In this categorization the fish pineal is overlaid by opaque structures (i.e. there is no pineal foramen but the pineal organ may lie in an indentation or fossa of the skull) or by a 'transparent window' through which light access to the pineal can be modulated by chromatophore activity (in which case the skull is presumably penetrated by a foramen). Breder and Rasquin (1950) originated this classification, listing species of each of the categories: fish showing reaction to light via an 'exposed pineal area' included *Brachydanio rerio*, the zebrafish. Fish with chromatophores controlling light access included *Synodus synodus*, *Hemirhamphus brasiliensis* and *Sphyraena barracuda*. Fish with covered pineal regions, said to be 'light-negative', included *Astyanax mexicanus* and *Ameirus nebulosus*. However, these authors could not place some fish in any of these categories; for example, some fish with covered pineals were still classed as light-sensitive.

It can be suspected that effectiveness of the pineal in response to light, diurnal behaviour and seasonality would be limited to fish from cool, shallow water in cases where seasonal changes are important for migration or reproduction, for instance. However, a pineal was described in deep-sea fish *Bathylagus wesethi* and *Nezumia liolepis* (McNulty 1976) and in 'cave-fish' *Typhlichthyes subterraneus* (McNulty 1978a) and *Chologaster agassizi* (McNulty 1978b), as well as the blind goby *Typhlogobius*

californiensis (McNulty 1978c). Holmgren (1959, quoted by Hoffman 1970) studied the occurrence of 'pineal spots' (equivalent to Matty's 'transparent window' in oceanic fish) and found them more prevalent in deep-sea species, which he believed might relate to vertical migration.

The pineal organ of teleosts was reviewed by Ekström and Meissl (1997); these authors conclude that there is an imbalance between considerable knowledge of certain aspects of the fish pineal and scant understanding of its function which is concluded to be as a component of a 'circadian system', that is, it is related to patterns of time, season and daily rhythms. According to these authors, in the absence of a regular day/night cycle the pineal is able to maintain a circadian (e.g. approximately 24-hour) activity, with higher levels of melatonin in the time equating to the 'dark' period. The same study emphasizes that some teleosts can rely on the pineal to 'entrain' (i.e. adjust) a cycle to the local day/night situation.

In mammals, melatonin has been described as anti-gonadotrophic (Silman et al. 1979), which suggests that it suppresses reproduction in low light conditions, for example the northern winter. The functions of melatonin are still incompletely understood in fish but its cyclical activity appears to impact on the production of brain and pituitary hormones affecting feeding, growth, reproduction and immunity (Falcón et al. 2010) and inhibits stress-related corticosteroid secretion (Herrero et al. 2007; Azpeleta et al. 2010). It is reported to induce a sleep-like condition in zebrafish (Zhdanova et al. 2001).

10.11.3 The adrenal gland homologues

In mammals the adrenal gland has two different components: the outer adrenal cortex and the inner medulla. The adrenal cortex secretes steroids, mineralocorticoids such as aldosterone (regulating ions), and glucocorticoids such as cortisol (increasing plasma glucose as part of a stress reaction). The inner adrenal is also important in stress responses; it is a modified sympathetic ganglion (11.9) which secretes and releases noradrenalin (norepinephrin) in response to stress.

The presence of noradrenalin (and adrenalin) can be identified by the reaction with potassium

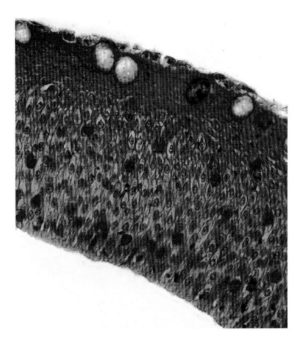

Plate 1. Eosinophilic granular cells (red) in the epidermis of winter flounder (*Pseudopleuronectes americanus*). Section (approximately 5–6 μm thick) stained with haematoxylin and eosin. (see page 31)

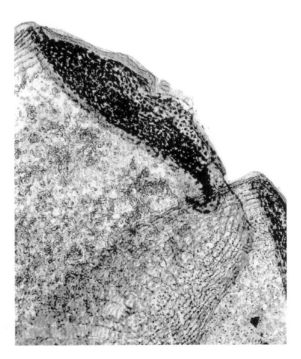

Plate 2. Cycloid scales, live and mounted in glycerine, each with a skin slip showing orange xanthophores and black melanophores, from winter flounder (*Pseudopleuronectes americanus*). (see pages 33 and 37)

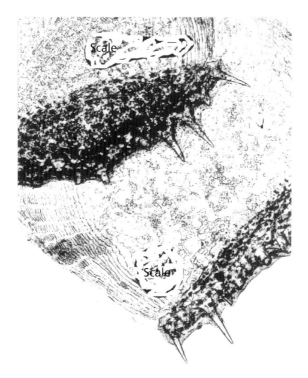

Plate 3. Ctenoid (toothed) scales, live and mounted in glycerine, each with a skin slip showing orange xanthophores and black melanophores, from winter flounder (*Pseudopleuronectes americanus*). (see pages 28, 33, and 37)

Plate 4. *Pomacanthus imperator* (juvenile). © Rudie Kuiter. (see pages 38, 41, 44, and 267)

Plate 5. *Chelmon rostratus* © Rudie Kuiter. (see pages 38, 41, 44, 45 and 267)

Plate 6. *Chaetodon rafflesi* © Rudie Kuiter. (see pages 38, 41, and 267)

Plate 7. *Chaetodon unimaculatus* © Rudie Kuiter. (see pages 38, 41, 44, and 267)

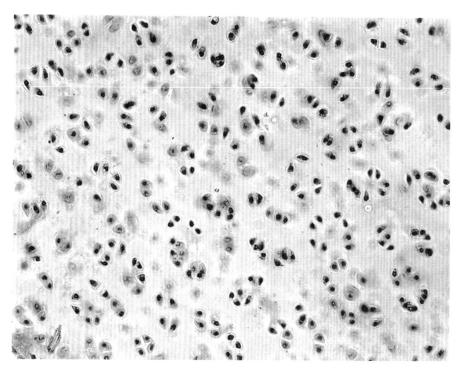

Plate 8. Hand-cut section of hyaline cartilage from the pectoral girdle of a skate (*Raja* sp). Stained with haematoxylin and eosin to show chondrocytes in lacunae. (see pages 48 and 49)

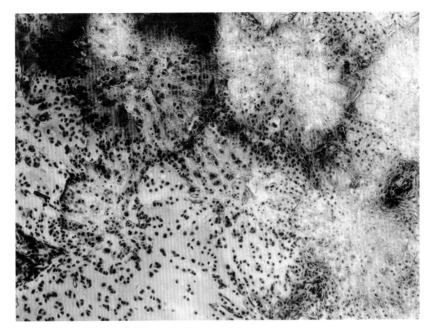

Plate 9. Hand-cut section of tesserae from the pectoral girdle of a skate (*Raja* sp.). Stained with haematoxylin and eosin. (see page 49)

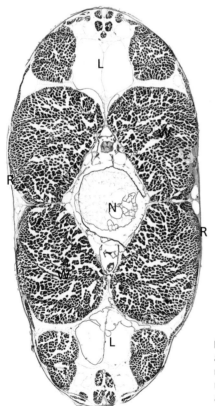

Plate 10. Transverse (cross-) section of the posterior (tail) region of an Antarctic notothenoid, *Pleurogramma antarcticum*, showing major muscle blocks white muscle myotomes (W), small thin red muscle regions (R) external to the white muscle, the central notochord (N) and lipid sacs (L). © Prof J.T. Eastman. (see pages 66 and 69)

Plate 11. Capelin (*Mallotus villosus*) spawning on a Newfoundland beach. © Dr Craig Purchase. (see page 289)

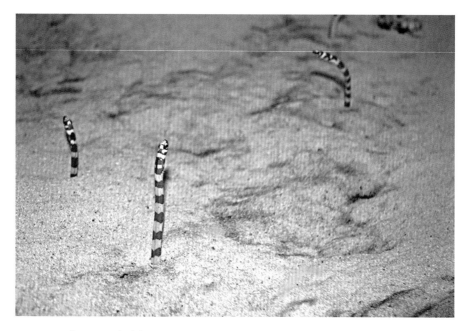

Plate 12. Splendid garden eels (*Gorgasia preclara*). © Catherine Burton. (see page 275)

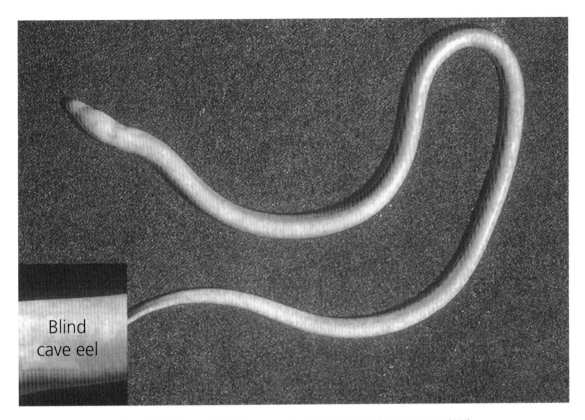

Plate 13. Blind cave eel (*Ophisternon candidum*). © Dr G.R. Allen. (see pages 273 and 275)

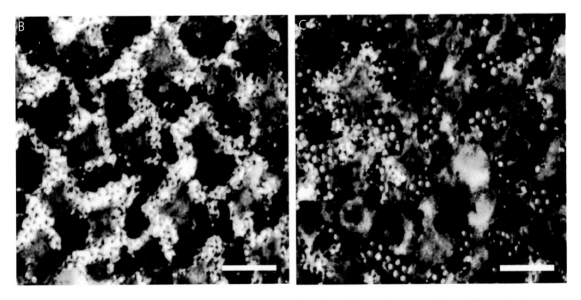

Plate 14. Skin patches bearing photophores, from a pygmy shark (*Squaliolus aliae*). B. Before melatonin stimulation. C. After melatonin stimulation. Reproduced with permission from J.M. Claes, H.-C. Ho and J. Mallefet, Control of luminescence from pygmy shark (*Squaliolus aliae*) photophores, *The Journal of Experimental Biology* 215, 1691–1699. © 2012 Published by the Company of Biologists. Scale bar = 250 μm. (see pages 271 and 272)

Plate 15. Banded rainbowfish (*Melanotaenia trifasciata*). © Dr G.R. Allen. (see page 267)

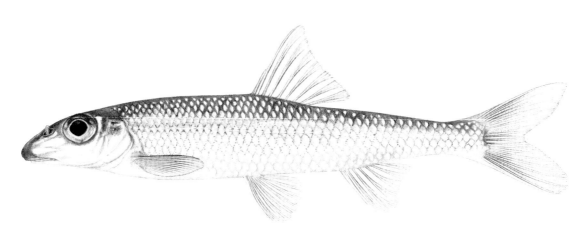

Plate 16. Slender chub (*Erimystax cahni*). © Joe Tomelleri. (see page 313)

dichromate producing brown/orange colouration due to the formation of noradrenochromes (and adrenochromes) which can persist during subsequent histological treatment. Any tissue that reacts this way is termed 'chromaffin' (affinity for chromate) and it is usual to introduce the reaction early after 'fixation' (Appendix 10.1). This post-chroming has been used in fish and identifies two different sites in chondrichthyans (elasmobranchs such as skates and dogfish) and teleosts. In, for example, skate, suprarenal glands of imprecise number and shape can be seen on the kidney; they are composed of chromaffin tissue. In teleosts chromaffin tissue cannot be distinguished macroscopically but is scattered within the anterior kidney (head-kidney) close to branches of the posterior cardinal vein. Chromaffin cells are also found in sympathetic ganglia (11.9).

The adrenal cortex homologue in fish is identifiable as interrenal glands (between the kidneys and macroscopically identifiable as small yellow patches) in elasmobranchs and interrenal tissue in the head-kidney of teleosts, also close to the posterior cardinal vein branches. Aldosterone has been sought in fish but it has been difficult to find, although reported in salmonids. In elasmobranchs, a corticosteroid, 1α hydroxycorticosterone, was identified by Idler and Truscott (1966). This steroid is reported as a mineralocorticoid with some glucocorticoid capacity (Carroll et al. 2008).

Cortisol is released on stress in actinopterygians (Barton 2002). The role of mineralocorticoids in regulating ions may actually be taken by some non-steroid hormones, including perhaps prolactin which can affect plasmatic sodium levels. The osmoregulatory needs and ionic losses at the gills in non-terrestrial vertebrates presents a very different problem set from that in tetrapods and it should not be surprising that some homologies do not give outcomes based on very simple predictions.

10.11.4 The thyroid gland

Hoar (1957) briefly reviews knowledge of the fish thyroid to that time. A longer and comprehensive overview of 'fish thyroidology' was provided by Leatherland (1994).

The thyroid gland of vertebrates occurs anteriorly and ventrally, below the head region. It usually forms an easily recognized discrete structure although this is not the case for teleosts in which thyroid follicles are more diffuse.

The thyroid synthesizes and secretes thyroxin, tetra-iodothyronine (T_4), in spherical follicles, storing T_4 within a central colloid. Once released to the circulation T_4 may be transformed to T_3 (triiodothyronine). Both T_3 and T_4 are expected to affect metabolism in all vertebrates, so that an excess of either hormone is associated with hyperactivity. Pioneer studies of the teleost thyroid showed seasonal changes (Barrington and Matty 1954) and also linked it to the protochordate and agnathan endostyle from which it was deduced to have evolved (Barrington and Sage 1963, 1966). An endostyle is seen in larval lampreys (ammocoetes) and 'transformation' of the larval endostyle to the adult thyroid has been described (Wright and Youson 1980).

Reviews of thyroid function in fish, notably teleosts, include peripheral metabolism (Eales 1985), the effect of 'nutritional state' (Eales 1988), the measurement of thyroid status (Eales and Brown 1993) and its response to the polluted environment (Brown et al. 2004).

Elasmobranchs have a very distinct thyroid gland lying ventral to the ventral aorta (Figure 10.4) where it bifurcates to form the two most anterior afferent branchial arteries. In dogfish, the thyroid is carrot shaped with the thicker part anterior and its axis parallel to the length of the fish. In skate, the thyroid lies along the surface of the arterial bifurcation, that is it lies at right angles to that of the dogfish. The surface of the thyroid has a distinctive semi-translucent granular appearance as seen in dissection. Microscopically the gland is composed of numerous round follicles containing more or less colloid, an acellular mass which stains in routine histological preparations.

The thyroid in teleosts is not readily seen macroscopically; it is diffuse, within connective tissue approximating to the general area near the ventral aorta. Geven et al. (2007) describe thyroid follicles as generally occurring below the pharynx but list several other possible locations including the heart and kidney. Difficulty in locating thyroid tissue in some teleosts is illustrated by the comprehensive search described in Twelves et al. (1975), who removed several pieces of the mouth floor for examination

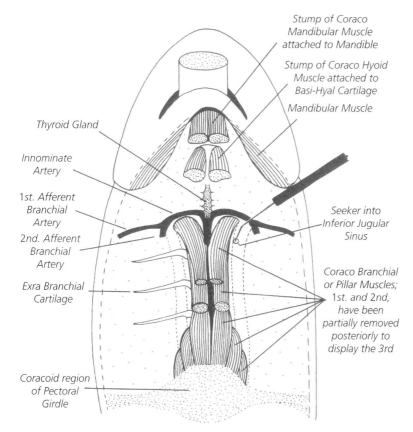

Stump of Coraco
Mandibular Muscle
attached to Mandible

Stump of Coraco Hyoid
Muscle attached to
Basi-Hyal Cartilage

Mandibular Muscle

Thyroid Gland

Innominate
Artery

1st. Afferent
Branchial
Artery

2nd. Afferent
Branchial
Artery

Exra Branchial
Cartilage

Seeker into
Inferior Jugular
Sinus

Coraco Branchial
or Pillar Muscles;
1st. and 2nd,
have been
partially removed
posteriorly to
display the 3rd

Coracoid region
of Pectoral
Girdle

Figure 10.4 The thyroid gland of a
dogfish *in situ*. Ventral skin and muscle
cut to show the thyroid lying on the
bifurcation of the first afferent arteries
(after Wells 1961).

of two Antarctic fish. However, Hoar (1957) figures a clear depiction of thyroid tissue round the ventral aorta and its anterior branches.

10.11.5 The gonads

Vertebrate gonads include special endocrine cells synthesizing steroids, notably sex steroids, principally androgens in the male and androgens, oestrogens and progestogens in the female.

Androgens in mammals are synthesized (steroidogenesis) by the interstitial (Leydig) cells (9.5.1) of the testis. Equivalent cells have been described in fish but are not always easy to locate and it is possible that 'lobule boundary cells' (9.5.1) may have this steroidogenic role. The major sex steroid in male fish is 11-ketotestosterone, first described (Idler et al.) in 1960. Testosterone identical to that of mammals is also found in the circulation of males

and females, and is regarded as the precursor for both 11-ketotestosterone in males and the major oestrogen (oestradiol) in females.

For females, follicles surrounding developing oocytes are the source of oestrogens and in mammals pregnancy is initially maintained by progesterone secreted from each persistent post-ovulatory follicle (POF) which forms a corpus luteum. In fish, which are usually oviparous, POFs do not have this role but progesterones functioning as maturation-inducing steroids (MIS) associated with final ovulation/final maturation, are thought to be produced by late-stage pre-ovulatory follicles.

Steroidogenic pathways have been well studied; the presence of the enzyme 3β hydroxysteroid dehydrogenase (3βHSD) in tissue is used to indicate sex-steroid synthesis. The conversion of testosterone to oestradiol in the follicles is achieved by the aromatase enzyme system.

10.11.6 The urohypophysis

In the caudal region of many, if not all, fish is a gland reminiscent of the hypophysis (pituitary) but with no obvious mammalian equivalent. It lies close to the spinal cord and has been carefully studied as a source of hormones. A variety of physiologically active substances have been extracted and characterized; they include a group of peptides termed 'urotensins'. This region is known as the urohypophysis or the caudal neurosecretory system (Bern and Takasugi 1962; Fridberg and Bern 1968). Richards et al. (2012) report a high level of gene expression for the hypocalcaemic hormone stanniocalcin (see corpuscles of Stannius 10.11.8) in the caudal neurosecretory system of the flounder *Platichthys flesus*.

10.11.7 The gut and pancreas

A source of hormones controlling metabolism and some aspects of digestion, as well as appetite, occurs in the anterior gut and pancreatic tissue of vertebrates. Fish with stomachs may secrete gastrin and a number of other peptide 'hormones' have been described from vertebrate stomachs and the intestine. The role of some of these peptides is not clear however, and opinion has shifted over time so that names may have been adjusted, for example gastric inhibitory polypeptide (GIP) may be better termed glucose-dependent insulinotropic peptide. Vasoactive intestinal peptide (VIP) has been accorded a number of functions in vertebrates. The concept of 'vasoactive peptides' discussed by Sander and Huggins (1972) was then a very wide one, including peptides which shared the capacity to affect the vascular system but which on reflection was found to be a relatively minor role in the effects elicited by some.

Ghrelin, which has been found in various gut regions including the stomach (Table 10.4), may have an important role in energy homeostasis and appetite control; currently much of the work on this hormone and leptin concentrates on obesity control in humans (Klok et al. 2007).

Table 10.4 lists some of the gut-sourced peptides with endocrine activity; their individual roles and importance may still need clarification for different species. Goldsworthy et al. (1981) point out that although numerous active peptides have been extracted from the gut wall their functional significance has not been established in many cases. The situation with teleosts is complicated by various interspecific structural variations, including the absence of a stomach in some, the presence of a few or many pyloric caeca and the lack of a separate pancreas for many (though 'Brockmann bodies' of pancreatic endocrine tissue occur in some teleosts, 10.11.7 and Wright et al. 2000). Identification of individual peptides has been complicated by the shared amino acid sequences which may occur, for example between gastrin and CCK, cholecystokinin (Goldsworthy et al. 1981). In such a situation, tests which rely on antibodies may not differentiate two similar peptides, thus reported in conjunction; for example gastrin/CCK as reported by Aldman et al. (1989). Further information is attained by injection of separate synthetic peptides and noting responses of various gut regions.

Anderson (2015) provides a comprehensive list of 'gastroentero-pancreatic hormones' of elasmobranchs and their physiological activity, which includes effects on the rectal gland.

Some peptides with potential or established hormonal status are associated with the nerve plexi of the gut and have therefore been termed 'neuropeptides' (Table 10.4).

The vertebrate pancreas has two main functions, one is exocrine digestive enzyme production (4.4.7). The other involves patches of endocrine tissue, termed 'islets of Langerhans', with three separate cell types, each producing either insulin, glucagon

Table 10.4 Major gastrointestinal endocrine peptides of elasmobranchs and teleosts.

Source region:	Stomach[*]	Intestine	Nerve plexi[**]	Pancreas
	Gastrin	Secretin	Bombesin-like	Insulin
	Ghrelin	GIP[†]	Substance P-like	Glucagon
		VIP[†]	VIP-like	Somatostatin
		CCK[†]		

[*]Stomach not present in all teleosts (4.4.6).
[†]CCK = cholecystokinin. GIP = gastro-inhibitory peptide. VIP = vasoactive intestinal peptide.
[**]from Bjenning and Holmgren showing the three most commonly found.

(important in glucose regulation) or somatostatin (regulating growth), also secreted in the hypothalamus. Elasmobranchs share with mammals the presence of a separate and distinct pancreas but many teleosts have diffuse pancreatic tissue in mesenteries close to the duodenum, and this distribution makes functional studies difficult. Some teleost, however, have a concentration of the pancreatic islet tissue as 'Brockmann bodies' (Yang and Wright 1995; Slack 1995; Rusakov et al. 2000), which makes extraction and analysis much easier.

10.11.8 Corpuscles of Stannius

Small, white, round bodies on the ventral kidney surface may occur in teleosts as sources of ion-regulating factors such as hypocalcin (also known as stanniocalcin, STC-1, Greenwood et al. 2009) and teleocalcin (Verbost et al. 1993). The corpuscles of Stannius have been described in various teleosts including salmonids, eels and goldfish (Flik et al. 1989; Verbost et al. 1993) and also the holostean *Amia calva* (Verbost et al. 1993). The corpuscles can be selectively removed as an operation (stanniectomy) under anaesthetic and the effect of removal studied. Recent studies on the stanniocalcin receptors have used material from flounder *Platichthys flesus* (Greenwood et al. 2009) and rainbow trout, *Oncorhynchus mykiss* (Richards et al. 2012).

10.11.9 Ultimobranchial organs, bodies or glands

A structure termed 'ultimobranchial' has been described in teleosts (e.g. salmonids) in the septum between the heart and the main abdominal cavity. According to Matty (1985) it develops embryonically in the last branchial pouch of each side before migrating to overlie the pericardium, but is unpaired in elasmobranchs.

It produces calcitonin which is hypocalcaemic, that is it lowers blood calcium levels. Its importance in relation to calcium regulation in fish and the possible co-occurrence of another hypocalcaemic factor (the hypocalcaemic factor of the corpuscles of Stannius) is not clearly understood.

Pang (1971, 1973) summarized the knowledge to that time, reporting (1971) ultimobranchial bodies

in 'sharks, bony fish and lungfish'. Sasayama et al. (1999) explain that the ultimobranchial glands may be difficult to localize and they produce guidance in the form of diagrams and immunohistochemistry. They regard calcitonin as important in goldfish and eels and, indeed, 'in all jawed vertebrates'. Salmon calcitonin injected into elasmobranchs had no effect on magnesium levels (Srivastav et al. 2003) but lowered blood calcium and raised blood phosphate in the stingray (Srivastav et al. 1998).

Yamane (1981) reported sexual differences in the ultrastructure in the ultimobranchial glands of adult Japanese eel, finding females to show more activity than males, in accordance with higher blood calcitonin levels in females.

10.12 Endocrine roles in controlling specific functions

Within vertebrates it is expected that hormones will be involved in a number of different activities achieving either homeostasis or long-term changes involving growth and reproduction. Hormones are also important in modulating digestion and food intake. In addition, hormones in fish can control colour and colour change either in conjunction with the nervous system or without it.

10.12.1 Hormones and homeostasis

Homeostasis in aquatic vertebrates with gills may mean a continuous adjustment of ions lost or gained.

Some fish are euryhaline, able to move between salt and fresh water or to tolerate tidal changes close to estuaries. In fresh water too there may be considerable variation in acidity and in calcium levels. The extent to which hormones are involved in particular ion regulation is not always clear. Removal of the pituitary (hypophysectomy), for freshwater fish, is not fatal if fish are then kept in an appropriate salinity (e.g. approximately one-third seawater). If hypophysecytomized fish are placed in fresh water they will lose sodium and this loss can be counteracted by prolactin injections.

Calcium levels in natural water vary from 32 mmoles per kg (mm/kg) in seawater to 5 mmoles in hard water and 0.22 in soft lake water

(Schmidt-Nielsen 1990), hence it is not surprising that salmonids, which are capable of extensive migrations (9.14), appear to be well provided with calcium-regulating hormones (10.11.8, 10.11.9). In mammals calcium 'upregulation' is prevalent and relies on the parathyroid hormone; in fish calcitonin and, more importantly, stanniocalcin (STC-1) are presumed to help regulate calcium levels as hypocalcaemic agents, that is, lowering blood calcium. Prolactin and somatolactin may be the principle hypercalcaemic factors in fish (Kaneko and Hirano 1993) but as access to calcium is generally regarded as producing high levels of calcium (hypercalcaemia) in fish rather than low levels, it is expected (e.g. Janz 2000) that hypocalcaemic factors, such as STC-1, are more important in calcium regulation. Devlin et al. (1996) have implicated involvement of the saccus vasculosus (10.11.1, 11.6.2) in calcium regulation as they have described the presence of a parathyroid hormone related protein in it, for the marine teleost, the sea bream *Sparus aurata*.

Hormones may be involved in fish osmoregulation but establishing the relative importance of, for example, corticosteroids has been difficult. Fish (notably teleosts) tend to lose water osmotically in seawater and gain it in fresh water, though marine elasmobranchs have blood which is osmotically very close to seawater and therefore avoid water loss (7.5.1, 7.5.8). In seawater, teleosts drink to gain water (and excrete excess ions like Na^+) whereas in fresh water they void gained water in the urine (8.5.3). Preparation for seawater migration in salmonids (smoltification) may include a role for growth hormone (Benedet et al. 2005).

In mammals, water content of the plasma is affected by the antidiuretic hormone (ADH) of the neurohypophysis, a hormone which controls water loss in the urine. Fish can only suffer dehydration (osmotically in seawater) if water is lost at the gills in excess of that recoverable by drinking. Moreover, individual fish blood cells can osmoregulate; this does not occur in mammals which are very dependent on central osmoregulation to protect individual cells from shrinking or swelling and bursting.

So-called chloride cells (7.7), particularly prevalent on fish gills, can actively pump ions and may be reversible if conditions change; that is, ions may be recovered or removed from the surrounding water.

The extent to which hormones modify such functions or permeability is not well understood. Mineralocorticoids secreted by the interrenal tissue may be active in this way and may also affect the function of the rectal gland (8.4.4) of elasmobranchs. It would be expected that ACTH from the pituitary would control the secretion of mineralocorticoids, but as their precise identity and role is still not clear any chain of action remains conjectural.

Natriuretic peptides (8.7), for example those found in fish hearts (Loretz and Pollina 2000), may have a role in 'sodium extrusion', notably at the gills and/or the rectal gland of the elasmobranchs.

Vertebrates have a complex set of conversions which comprise the renin–angiotensin system (RAS), which ultimately and notably in mammals (and clinically for humans) can be important in maintaining blood pressure. The renin component (an enzyme) is produced in juxtaglomerular cells of the kidney and previous to the action of ACE (angiotensinogen converting enzyme, notably in mammalian lungs) assists in cleaving angiotensinogen to the active angiotensin. The RAS may have a role in controlling water intake by drinking in fish. Components of the RAS have been described in fish via the effects of specific enzyme inhibitors (Anderson et al. 2001) with a 'dipsogenic' effect, a role in controlling drinking in response to different salinities.

As in mammals, glucose regulation is achieved through the dual antagonistic hormones insulin and glucagon, secreted by pancreatic tissue either in a separate pancreas (Chondrichthyans) or as smaller 'Brockman bodies' or islets embedded in the gut (Norris 1980). Insulin lowers blood glucose (hypoglycaemic) whereas glucagon raises it (hyperglycaemic).

In mammals, roles for energy homeostasis including appetite control, are ascribed to ghrelin and leptin amongst other hormones. Leptin is associated with adipocytes and ghrelin in secreted in various regions, notably in parts of the gut. Appetite control in fish is reviewed in Volkoff et al. (2010).

10.12.2 Hormones and reproduction

The role of hormones in fish reproduction is known to be fundamental to gonad growth and gamete maturation and deposition (9.18). It is expected that the main control mechanism for reproduction in fish is the hypophyseal–pituitary–gonadal axis with input from melatonin (for seasonal influence) and perhaps kisspeptin as a switch in the brain (9.18) as well as StAR (steroid acute regulatory) protein for fine control of steroidogenesis (9.18).

Sex steroids such as androgens, oestrogens and progestogens (Appendix 9.3) control fish reproduction at the gonadal level. Their induction and release is understood to be influenced by pituitary release of gonadotrop(h)ins; the number of these has been subject for much discussion in fish (Jalabert 2008). In mammals, the classical view is that there are two, one known as FSH (follicle-stimulating hormone for females) or (the same molecule) ICSH (interstitial cell-stimulating hormone for males), involved in stimulating early gametogenesis; the other LH (luteinizing hormone based on female effects) important in maturation and maintenance of post-ovulatory follicles in, for example, mammals. Mammalian gonadotropin can achieve dramatic effects in fish, ovulation/final maturation can be induced by injections of crude pituitary extract too, by human chorionic gonadotropin (HCG) or by LHRH. In fish, the debate over the presence and significance of one, two or more gonadotropins is discussed by Jalabert (2008). In general the factors controlling onset of repeat gametogenesis are not well understood; there has been more interest in late stages (maturation) of gametogenesis and ovulation in females. Thus Jalabert restricts some of his observations to the gonadotropin known as luteinizing hormone (LH), usually associated with maturation (and definitely so in mammals). The use of 'gonadotropin I and II' is still confusing; that there is more than one seems established but the equivalences are still unsatisfactory.

Some sex steroids found in fish, such as testosterone and oestradiol, in plasma and in steroidogenic cells, are identical to those of mammals. Thus testosterone occurs in the circulation of both male and females. In females, testosterone synthesized in the outer cells of the follicles in early oogenesis is the precursor molecule for oestradiol synthesized in the cells of the inner granulosa layer of the individual oocyte follicles (9.10.3). However, the principle androgen in male fish has been identified as 11-ketotestosterone (11KT), originally described in salmon by Idler et al. (1960, 1961). 11KT can influence many facets of reproduction including secondary sexual characteristics like epidermal changes (2.10).

In late oogenesis, fish synthesize progestogens (forms of progesterone, C21 steroids) which are involved in the final maturation, ovulation and spawning. Progesterones present as maturation-inducing steroids (MIS) and are identified in the follicle and the circulation.

A role for the pineal and melatonin in determining phases of reproductive activity is suspected but not understood.

A relationship between nutritional status and reproduction occurs but is not well understood at the endocrine level. Some fish, like winter flounder and cod, are capable of omitting reproduction (Burton and Idler 1984; Rideout et al. 2005) which may occur in relation to a poor nutritional status (i.e. low condition at a 'critical period' in the annual cycle, Burton 1991, 1994). Female fish are also capable of downregulating gamete numbers by atresia if nutrition is poor and even upregulating gamete number and 'fast-tracking' oocytes if condition is good. The whole interrelationship of nutrition and reproduction is complicated by extensive periods of low feeding or cessation of feeding during cold winter months for some fish. In long-lived marine fish, gametogenesis, particularly in females, is very long although it may have rapid phases, and can withstand periods of poor nutrition, relying on stored lipid or protein to maintain the gametes. Large-sized marine fish may be able to maintain condition and/or acquire stores by periods of intense feeding.

10.12.3 Hormones and growth

Both hypothalamic–pituitary hormones and peripheral hormones have been described as influencing somatic growth. In this context the major hypothalamic–pituitary axes would be those controlling growth hormone (GH) and thyroid-stimulating hormone (TSH), the latter promoting thyroid activity with the secretion of thyroxine (T_4) enabling increase in its peripheral product (T_3) (see Figure 10.1). There are hypothalamic-releasing and release-inhibiting hormones for both GH and TSH. The pituitary synthesis and release of growth hormone may be seasonally variable; some fish do not grow over cold winter months and in fact may lose both mass and length.

Surprisingly GH may increase in (experimental) starvation, which may be related to the need for 'mobilization of fatty acids and glycerol' (Sumpter et al. 1991). Interaction between GH and T_4 may occur, with GH promoting conversion of T_4 to T_3 (de Luze and Leloup 1984).

Growth hormone may have many roles, being 'highly pleiotropic', notably in salmonids (Benedet et al. 2005). Chemically it resembles both prolactin and somatolactin. GH may mediate its peripheral effect through liver-synthesized somatomedin C (Insulin-like Growth Factor, IGF 1). Because of the intense commercial interest in rapid fish growth for aquaculture, attempts to boost effective GH have included the production of transgenic fish (Du et al. 1992). Somatolactin found in the pars intermedia of the fish pituitary may promote reproductive growth (Benedet et al. 2008). Company et al. (2001) discussed the role of GH and SL in fish growth and adiposity.

The thyroid hormones are expected to control metamorphosis, such as occurs in eels and flatfish.

Leptin, described mainly from mammals, is secreted from white fat cells (adipocytes) in proportion to their mass/status which provides a measure of lipid storage. Its prevalence and possible role in fish has proved difficult to clarify.

10.12.4 Hormones and digestion

The possible presence of non-neural control of digestion led to the initial discovery of hormones; the first hormone described by Bayliss and Starling in 1902 (Bell et al 1957) was secretin in mammals, controlling the secretion of pancreatic enzymes. Gastrin, released from the gastric mucosa, and cholecystokinin (CCK) have also been described from mammals, as well as several other peptides like vasoactive intestinal peptide (VIP) and gastro-inhibitory peptide (GIP). Besides controlling release of digestive enzymes, stomach acid and bile, gut hormones may increase or inhibit gut motility.

According to Hadley (1996), gastrointestinal hormones are secreted by clear cells termed entero-chromaffin, argentophil or argentaffin, the last two

terms based on affinity for silver stains. Structurally he divides the hormones into two 'families': either gastrins and CCKs or secretin, vasoactive intestinal peptide (VIP) and GIP.

Gastrin has several actions, the primary function has been regarded as stimulation of acid (HCl) secretion in the stomach. CCK controls the release of bile. As some fish do not have stomachs and bony fish have diffuse pancreatic tissue, the search for fish hormones influencing digestion continues to reflect diversity and complexity. Fish gastrin has been reported from the stomach and intestine (Noaillac-Depeyre and Hollande 1981) and CCK-producing cells are described as having wide distribution in the gut of fish as well as the nervous system (Volkoff and Peter 2006). Fish CCK affects digestion and food intake. Secretin has been reported from some fish (Norris 1980).

Gut motility may be affected by bombesin and gastrin-releasing peptide (secreted widely in the gut) which can also increase blood flow to the gut and inhibit food intake (Volkoff et al. 2005).

10.12.5 Hormones and appetite

A large number of endocrine factors have been listed as affecting appetite (Volkoff et al. 2010). Those stimulating feeding (orexogenic) include GH, ghrelin and neuropeptide Y while those inhibiting feeding (anorexogenic) are more numerous and include MCH and melatonin as well as CCK.

10.12.6 Hormones and the vascular system

Circulating adrenalin tends to cause vasoconstriction and thus raises blood pressure in vertebrates.

Other vasoactive substances are substance P (a tachykinin, Volkoff et al. 2005) and the vasoactive intestinal peptide (Table 10.4). Whereas tachykinins are expected to increase blood pressure through vasoconstriction, bradykinins would lower blood pressure through vasodilation. Although bradykinin (as a small peptide) is well characterized in mammals, the presence and role in fish of such a peptide is less certain (Conlon 1999). The renin–angiotensin system has a role in tetrapod blood pressure and may have a similar vasopressive effect in fish (Anderson et al. 2001).

10.12.7 Hormones and stress

A response to the external environment may occur in such a way as to constitute a 'stress response', a suite of resulting changes which involve the blood vascular system, the immune system, ventilation rate, blood glucose levels and even perhaps a change in skin colour. At the cellular level, proteins capable of protecting other proteins from denaturation may be induced (7.4). Particular hormones may be secreted or released under stress conditions but as sampling from live fish to measure hormone levels may itself be stressful, obtaining clear information on endocrine stress responses can prove problematic.

Sage (1971) found a stimulation of thyroid activity in stressed *Poecilia*. Sumpter et al. (1985) reported elevation of MSH and endorphin in stressed trout, although plasmatic α-MSH was not elevated by moderate stress but raised in more severe stress (Sumpter et al. 1986) and plasmatic cortisol levels increased after a short delay, with a lag behind increased ACTH levels. Pickering (1989) looked for, and found, increased cortisol levels in stressed salmonids. Cortisol release is indeed expected to be under the control of ACTH (Conde-Sieira et al. 2013) and there is a role for glucose in the regulation of the cortisol secretion.

Wendelaar Bonga (1997) and Barton (2002) reviewed the stress responses in fish, including the role of endocrines.

Vertebrate stress responses are associated, initially, with two sets of hormones: catecholamines (adrenalin, noradrenalin) and corticosteroids (e.g. cortisol). Thus, acute or sudden stressors evoke quick release of adrenalin/noradrenalin which may increase heart rate and cause vasoconstriction temporarily. The release of corticosteroids, notably cortisol (hydrocortisone), follows. Cortisol is known to have multiple effects in humans, with gluconeogenesis an important result of increased levels of this hormone (Khani and Tayek 2001). The consequent increase in blood glucose levels may be used as a measure of stress but caution is necessary (Martínez-Porchas et al. 2009); nevertheless these authors tabulate post-stress blood glucose increases from 12 different fish species and 10 different sources.

In fish, melatonin may suppress corticosteroid secretion (Herrero et al. 2007; Azpeleta et al. 2010).

Stressed fish may show colour changes, for example rapid paling can occur with increased circulating adrenalin.

10.12.8 Hormones and colour change

Responses of fish to environment include hue changes to background colour by many. Flatfish are particularly noted for a capacity to rapidly pale or darken and change skin patterns, these rapid changes are generally regarded as mediated by the nervous system (2.9, 11.10.6) or release of adrenalin. However, fish can also have a slow, hormonally controlled response to light and background, which may be mediated by at least two pituitary/hypothalamic hormones, MSH and MCH, although the primary role of MCH is not clear (2.9) and may be related to appetite control (10.12.5). Some fish (e.g. elasmobranchs) only have the slow endocrine control of skin colour. The endocrine response includes increase or decrease of chromatophore number which can occur over a long time period (weeks); this kind of response is referred to as morphological colour change.

It has recently been suggested (Sasano et al. 2012) that somatolactin has a role in colour regulation.

10.13 Overview

The study of fish hormones has been greatly influenced by the growth of aquaculture. The emphasis recently has been on promoting growth and controlling phases of reproduction, and although much progress has been made there are still many areas which need further exploration. The possibility of applying information obtained from mammals to fish and vice versa has been basic to the directions of much research. Recently, a setback to application has occurred with the withdrawal of fish calcitonin from some medical use (long-term treatment for osteoporosis) due to increased risk of cancer (press release from the European Medicines Agency, 20 July 2012). In fish hormone research, the similarity of many fish hormones to mammalian forms, or even the exactly identical chemical structure in the case of some steroids, may have obscured the interpretation of functions. Thus, although fish hormones may be identical in the case of sex steroids there are also fish-specific sex steroids.

With the similarity of many of their hormones to other vertebrate systems fish are susceptible to endocrine disruption by medically derived/pharmaceutical pollutants (8.3.10, 9.18, 10.10) as well as other industrial effluent.

Appendix 10.1 Localization of endocrine cells

Endocrine cells can be located through histochemical stains and antibodies as well as by the identification of particular enzymes or specific RNA. Ultrastructure (e.g. smooth endoplasmic reticulum) has also been used to support the likely identity of steroidogenic cells in conjunction with the presence of steroidogenic enzymes.

Appendix 10.1a Histochemical reagents and procedures used to help identify endocrine cells

The processes of histochemistry are applied (usually) to thin sections of chemically fixed organs/tissues, cut on a special knife-holding machine (microtome) which can be set to cut thicknesses of from about 5 microns (μm) to about 100 μm. In order to render the tissue/organ firm enough to slice, the tissue is supported (embedded) in wax, and in order to make this possible the tissue/organ has to be dehydrated using a series of alcohols (usually ethanol; aqueous through to absolute) to remove water gently, then placed in a solvent (e.g. xylene, toluene) that is compatible with both ethanol and wax.

- Chemical fixation stabilizes cells for subsequent processing, preventing deterioration and loss of some contents.
- Chemical fixatives include buffered formalin and special mixtures such as Bouin's, Orth's and Zenker's fixatives (Humason 1967). Orth's fixative contains potassium dichromate and is appropriate for showing the presence of adrenalin and/or noradrenalin in tissues.
- Lipids are difficult to fix *in situ* as they tend to leach out if subsequently run through an ethanol series. It will most likely be necessary to use frozen material cut on a special 'freezing microtome' or cryostat for investigating the presence of lipids or ethanol-soluble substances such as steroids. Alternatively, prolonged post-chroming with potassium dichromate after fixation can be used to stabilize lipids.

Following the preparation of thin slices by a microtome, usually in the form of a long 'ribbon' of consecutive sections, the sections are mounted on a glass slide, using a thin smear of glycerin–albumen to enable adhesion, with addition of a little water and heat (a special hotplate) so that the sectioned material spreads slightly (after some cutting compression). The sections on the slides are then dried and are ready for further processing: staining.

Before staining, the wax has to be removed (e.g. by xylene), rinsed in ethanol and then rehydrated through a series of increasingly aqueous ethanols to water.

Useful stains to distinguish particular endocrine cells include 'periodic-acid-Schiff' (PAS) and Orange G.

PAS determines the presence of carbohydrates such as glycoprotein. This procedure (Galigher and Kozloff 1964; Pearse 1968; Humason 1967) uses Schiff's reagent which distinguishes carbohydrates by a bright magenta colouration, following release of aldehydes by prior oxidation with periodic acid. The reagent itself is unstable and should be freshly prepared and/or checked with a positive control, a material known to contain an appropriate carbohydrate. PAS will show the presence of the pituitary glycoprotein hormones (GTHs, TSH) by magenta staining in particular regions.

To determine the presence of 'acidophils', counterstaining with eosin or preferably Orange G reveals prolactin and growth hormone cells in the pituitary; with Orange G, prolactin shows pale orange/yellow and growth hormone orange. If PAS is used first its magenta colour does not contrast well with eosin but it does show well with Orange G.

The cytoplasm of many cells, including erythrocytes is acidophilic, that is, it binds acidic stains such as eosin or Orange G. These standard stains therefore do not specifically identify endocrine cells but help to determine probably location.

Thus, if an appropriate Orange G solution (e.g. 1%, 0.5% phosphotungstic acid, 1% acetic acid, Pearse 1968) is used, prolactin cells will show as pale orange and growth hormone cells as brighter orange in thin pituitary sections, compared to magenta of glycoprotein hormones prior stained with PAS.

Appendix 10.1b Localization of endocrine cells using antibodies

Particular amino acid sequences specifically associated with individual hormones can be located in cells using antibodies that may be commercially available or produced in small amounts with the appropriate permits, if sufficient pure sequences are available for injections over time appropriate to invoke antibodies. There are several different processes used to locate hormones using antibodies. In each case an antibody is 'labelled' or tagged with a dye that can be subsequently detected. The tag may be fluorescent, such as FITC (fluorescein–isothyocy-ante), detected using an epifluorescent microscope equipped with the relevant filters.

Because of the expense of producing specific antibodies it has been usual to use a 'string' of sequential antibodies with the second antibody tagged and reacting to the first antibody's protein structure. This is expected to be both cheaper, easier and enhanced in effect.

An example of this indirect procedure is shown here:

Dewax two consecutive slides, rehydrate through ethanol to water or buffered saline (phosphate buffered saline PBS). Incubation can be carried out with a minimum of fluid by putting individual slides in a petri dish.

Slide 1

Incubate with first antibody. Rinse with PBS.

Incubate with second antibody. Rinse with PBS.

Counterstain, for example with Evan's blue if FITC tag used.

Mount in, for example, glycerine and examine under microscope.

Slide 2 (control)

Omit antibody incubations. Use counterstain only.

First antibody: Antibody raised in rabbit. Antigen 'fish growth hormone', the antibody itself is a rabbit gammaglobulin. Antibody is thus rabbit anti-fish growth hormone.

Second antibody tagged with FITC: Antibody raised in goat. Antigen 'rabbit gammaglobulin' Antibody is thus goat anti-rabbit gammaglobulin and the only sites it binds to will be where the first antibody bound to growth hormone; there should be no other rabbit gammaglobulin on the slide.

Appendix 10.1c Specific problems with individual hormone types and localization

Hormones may share sequences, making identification through the use of antibodies difficult depending on the region of the hormone which the antibody locates. An example of this problem is the situation with TSH and GTHs, which share a near identical α sub-unit, shown below as aaaa representing a near identical sequence of aminoacids; the beta sub-unit in each case, represented as bbbbb . . . and βββββ . . . is, however, different. An antibody raised to the alpha sub-unit will not distinguish the hormones.

TSH: aaaaaaaaaaaaaaaa	GTH: aaaaaaaaaaaaaaaaa
bbbbbbbbbbbbbbbbbbbbbbbbbbbbbbbbbbb bbbbbb	ββββββββββββββββββββββββββββββββββ ββββββββ

Steroids present particular problems for cellular localization: (a) because they are soluble in ethanols and (b) because they are not antigenic and (c) because they may be very similar structurally and therefore antibodies may not be molecule-specific.

Loss in ethanol (a) can be avoided by using frozen sections and/or by stabilizing steroids chemically (e.g. using digitonin, Burton and Idler 1987); (b) has been addressed commercially by conjugating any individual steroid (e.g. to bovine serum albumen), which makes the molecule antigenic; (c) differentiating similar steroids requires careful attention to possible cross-reactions, for example by using consecutive sections and different antibodies. A reaction to testosterone antibody in one set of three sections could then be compared to the next three sections' reaction to 11-keto-testosterone antibody. The synthetic pathways, of course, might mean that both steroids were present anyway as part of the process.

Appendix 10.2 Hormone acronyms

Note: h. = hormone

ACTH: adrenocorticotrophic hormone

ADH: antidiuretic hormone

AVP. arginine vasopressin

AVT: arginine vasotocin

CCK: cholecystokinin

CLIP: corticotrophin-like intermediate lobe peptide

CRF: corticotrophin releasing factor

E$_2$: oestradiol

FSH: follicle-stimulating hormone

FSHRH: follicle-stimulating hormone releasing hormone

11KT: 11-ketotestesterone

LH: luteinizing hormone

LHRH: luteinizing hormone releasing hormone

MCH: melanin- (or melanophore-) concentrating h.

MIS: maturational inducing steroids

MSH: melanophore-stimulating hormone

MSHRH: melanophore-stimulating hormone releasing h.

PRL: prolactin

SL: somatolactin

SRIF: somatotropin release-inhibiting hormone

GH: growth hormone

GnRH: gonadotrophin-releasing hormone

GtH, GTH: gonadotrop(h)in

HCG: human chorionic gonadotrophin

ICSH: interstitial cell stimulating hormone

T$_3$: tri-iodothyronin

T$_4$: tetra-iodothyronine, thyroxin

TSH: thyroid-stimulating hormone

TSHRH: thyroid-stimulating hormone releasing h.

VIP: vasoactive peptide

Bibliography

Goldsworthy, G.J. Robinson, J., and Mordue, W. (1981). *Endocrinology*. New York, NY: John Wiley and Sons.

Holmes, R.L. and Ball, J.N. (1974). *The Pituitary Gland. A Comparative Account*. Cambridge: Cambridge University Press.

Norris, D.O. (1980). *Vertebrate Endocrinology*. Philadelphia, PA: Lea and Febiger.

Sherwood, N.M. and Hew, C.L. (eds) (1994). *Molecular Endocrinology of Fish, Vol. XIII: In Fish Physiology*. New York, NY: Academic Press.

Schreck, C.B., Tort, L., Farrell, A.P., and Brauner, C.J. (eds) (2016). *Fish Physiology Vol. 35, Biology of Stress in Fish*. Amsterdam: Elsevier.

Abstracts of the four-yearly International Symposia on Fish Endocrinology may be available online.

CHAPTER 11

Integration and control: the nervous system

11.1 Summary

Vertebrate nervous systems share anatomical and physiological features enabling rapid response to events in the external and internal environments.

The central nervous system, the anterior brain and the dorsal nerve cord enable collection and processing of information based on incoming data via sensory nerves. Executive responses occur via motor nerves.

Collation of information enables learning and modification of responses as an ongoing process.

Control of internal organs and some aspects of skin function is enabled by the autonomic nervous system.

11.2 Introduction

The evolution of muscular activity, characteristic of the animal kingdom, is paralleled by the co-evolution of the nervous system, with sensory and control elements. Integrative functions associated with such systems are achieved by neurons, cells with elongated processes for transmission of information, but which also form circuitry for complex behavioural activity. The capacity of fish for gathering information and for complicated behaviour indicates that a high degree of organizational and functional complexity of the nervous system has evolved in these taxa.

In all animals sensations perceived by the sensory cells/organs (Chapter 12) enable appropriate responses from effectors (muscles/glands) and in fish, as well as some arthropods, squid and some reptiles, special pigmented cells (chromatophores, 2.6.1, 2.9). The nervous system links sensory structures (receptors) to effectors.

Fish nervous systems are pivotal in our understanding of general vertebrate neural organization, but the diversity of fish species is also enormous and is reflected in their morphological and functional neural variation. The nervous systems of extant taxa should not be interpreted solely in terms of evolutionary linearity but also as demonstrations of adaptational diversity in major groups which are widely separated taxonomically.

Fossil agnathans (jawless fish) such as cephalaspids and anaspids from about 400 million years ago most likely included ancestral forms of the extant cyclostomes. Impressions in their crania (Stensio 1958, cited by Young 1981) indicate a brain like that of extant lampreys, with a pineal and vertebrate organs of special sense (e.g. eyes). They appeared to be detritus feeders (Young 1981) in contrast to adults of extant cyclostomes, with their sucking mouths and parasitic or scavenging life which may be considered either specialized or degenerate, with neural implications.

Major gnathostome taxa, including Chondrichthyans and Osteichthyes, have no currently known or recognized obvious fossil connecting links with their probable agnathan ancestry (Gilbert 1993) and have undergone extensive adaptive radiation. This is reflected in the neural and behavioural adaptations to the extensive range of habitats found in freshwater and in coastal and deeper seawater. The relative size of the regions and centres of the brain in different fish species is considered indicative of their importance and of the related sensations (Kotrschal et al. 1998) in the various habitats, although functional details are often incompletely understood.

Essential Fish Biology: Diversity, Structure and Function. Derek Burton & Margaret Burton.
© Derek Burton & Margaret Burton 2018. Published 2018 by Oxford University Press.
DOI 10.1093/oso/9780198785552.001.0001

Information gathering and its neural processing in fish involve remarkable adaptations in different habitats with physical conditions ranging from the dark deep oceans to the turbid waters of some African rivers and the bright warm and shallow coral reefs. The nervous system provides access to information on the environment (Chapter 12) as well as controlling some of the activity of internal organs, and locomotion. The fish brain provides a series of centres for learning, comparison of sensory inputs from different sources, reflex control of, for example, balance and swimming and control of rhythmic activities such as respiratory pumps as well as the complex behaviours of shoaling/schooling, prey detection and feeding, and the set of reproductive behaviours which may include migrations, courtship and nest-building.

The diversity of habitats within the aquatic environment is exploited by fish, often within a web of complex predator–prey relationships, displaying a variety of locomotory and behavioural responses, which can be very rapid in attack, defence or escape. Hence fish may have a special tract of very large-diameter axons in the nerve cord (Mauthner fibres 11.6.6), enabling quick reactions. Although considerable variations in the fish nervous system must occur these are not well understood except perhaps in the context of special senses like electrolocation (12. 9.1) and even so, the anatomical variations are probably better known than the details of function.

The fish nervous system achieves rapid communication and coordination in comparison to the endocrine system of multiple hormones (Chapter 10), which provides a slower, more sustained set of links and controls via the blood system. In contrast, the neurons of the nervous system are specialized for rapid conduction of signals which coordinate more precise activity.

11.3 Neural cells

Vertebrate neural cells include neuroglia ('glia') as well as the conducting neurons, both possessing a perikaryon (soma), a cell body with a nucleus, and projecting processes.

11.3.1 Neuroglia

Neuroglial cells occur as three types within the neural tissue; astrocytes, oligodendroglia and microglia.

Both glial cells and neurons, as well as the pigmented melanophores (2.6.1), share a large degree of morphological similarity and develop embryologically from neural crest cells. The neural crest (reviewed by Le Dourain and Kalcheim 1999) is a transient structure along the early embryonic anterior–posterior axis. It releases cells, possessing considerable developmental plasticity, which migrate to target sites to differentiate into a variety of cell types. The extent to which neural crest precursors of the cell types are segregated is yet to be determined. Many neural crest progenitor cells are pluripotential; others appear to be more restricted in potential. There are indications that glial cell development depends on paracrine signals from recently formed neurons (LeDourain and Kalcheim 1999). Each retains the capacity for continued mitosis which contrasts with neurons. Astrocytes and oligodendroglia probably provide structural support for the neurons (Gardner 1975) but there is increasing evidence for their involvement in various types of neurophysiological activity including synaptic transmission (11.3.4).

Recent work (Verkhratsky et al. 2002; Scemes and Giaume 2006) supports the concept that astrocytes have a special form of intracellular excitability quite different from that associated with the membrane depolarization (Appendix 11.1) associated with neurons and skeletal muscle. It is thought that calcium is crucial in astrocyte intracellular excitability and that both the endoplasmic reticulum and mitochondria are intimately involved.

The astrocytes are the largest glia and are star-shaped; one of the processes, the end foot, can be applied to a capillary, contributing to a 'neurovascular unit' (Tam and Watts 2010). Cell-to-cell signalling using adenosine triphosphate (ATP) can occur between neurons, astrocytes and capillaries. Astrocytes release neuromodulators such as ATP and cytokines which can change neuronal communication, including information flow in brain neural networks (Burnstock 2007, 2009). In the teleost *Astatotilapia burtoni*, tight junction and desmosome interconnections between astrocytes have been described using immunohistochemistry (Appendix 10.1b) and freeze-fracture electron microscopy. In the optic tract of this species astrocytes can be involved in nerve regeneration, possibly by glia–neuronal interaction affecting the micro-environment of growing neurons (Mack and Wolberg 2006).

Oligodendroglia are myelin-forming cells (Figure 11.2) found in rows along the nerve fibres in the white matter of the central nervous system (CNS). Their processes are shorter and fewer than in astrocytes. Microglia have only a few short processes and are smaller than both astrocytes and oligodendroglia. They have an immune function; it is reported that neural injury can trigger their division and they become phagocytic (Gardner 1975; Burnstock 2009).

11.3.2 Neurons

Morphologically, fish neurons display the general characteristics of those in vertebrates. The perikaryon (cell body) has the main intracellular organelles around the nucleus. The elongated processes give neurons diverse and highly irregular shapes. These processes include dendrites, increasing the surface of the perikaryon for neuronal interaction, a long dendron (Grove and Newell 1953; Eccles 1992), conducting impulses towards the perikaryon of some neurons, and an axon conducting impulses away from the perikaryon of some neurons. Neurons with repeatedly branching processes (as occurs with many dendrites) are appropriate for estimating their fractal dimension as a quantitative index of morphological complexity, represented by the space occupied (Isaera et al. 2004).

Axons (and dendrons) may be very long (constituting nerve fibres), in long-distance signalling neurons, or short in local signalling neurons, or absent in neurons such as retinal amacrine cells (Figure 12.2) possessing only dendrites.

Some of the diversity in fish neurons, as in those of other vertebrates, is apparent in the examples in Figure 11.1. Simplistically, there are three basic types: motor neurons to effectors (Figure 11.1A), sensory neurons from receptors (Figure 11.1B and C) and interneurons or internuncial neurons (Figure 11.1D) which connect other neurons and can form intricate networks or circuits in parts of the brain facilitating behavioural complexity. Another classification is based on neuron morphology, describing them as unipolar, one process which divides into a peripheral and a central branch (Figure 11.1B), bipolar (Figure 11.1 C), two processes, peripheral and central, and multipolar (Figure 11.1A and D), one axon and many dendrites.

Sensory neurons respond to environmental (external and internal) stimuli usually after transduction to electrical signals by sensory receptors (Chapter 12); motor neurons control effectors such as muscle fibres and chromatophores. Interneurons, found in the brain and spinal cord, provide links; they may have long, short or no axons. The sensory neurons conduct centrally towards the central nervous system (CNS: brain and spinal cord). Although the peripheral long process of the sensory neuron has been called a 'dendron' (Grove and Newell 1953; Eccles 1992), it has recently also been referred to as an axon. Axons of motor neurons conduct signals (impulses) away from the perikarya which lie within the central nervous system, those of sensory

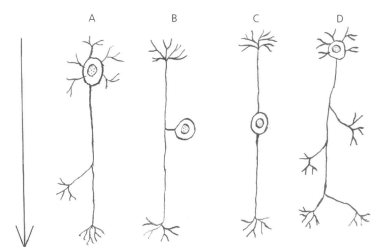

Figure 11.1 Examples of vertebrate neuron types. A = motor neuron (multipolar); B = sensory neuron (unipolar) from a spinal nerve; C = sensory neuron (bipolar) from a cranial sensory nerve; D = Interneuron (multipolar). The circles are perikarya with nuclei; the axons of A and D have collateral branches. The arrow indicates the direction of conduction.

neurons being outside it, for example, in the dorsal root ganglia (Figure 11.4) of the spinal nerves or closely associated with sense organs in the cranial region. Interneurons can relay impulses to variable numbers of motor neurons depending on their number of processes and their degree of branching. This can enable widespread motor responses to local sensory stimulation.

11.3.3 The neuron membrane and conduction

Animal cells and tissue fluids include a variety of electrically charged ions such as Na^+ and K^+. Collectively such ions represent an internal source of electrical energy which may be used physiologically as in processes such as neural conduction. Such processes can be facilitated by cellular membranes which are able to maintain electrical gradients, and change them in some cases (excitable membranes, Appendix 11.1), with important roles in information biotransmission. Cells possess an electrical transmembrane gradient between the inside and outside, that is, a membrane potential with a negative internal voltage (up to about −100 mv); thus cell membranes are electrically polarized.

The membrane surrounding neurons and at least some muscle cells is different from that of most other cells (Appendix 11.1). It has special voltage-sensitive molecular 'gates', probably membrane proteins which can open or close pores very quickly by conformational change in response to specific anions (Na^+, K^+). Any cell membrane operates to exclude some ions from the cell via a pump mechanism and the combination of ion distribution and sudden movements of ions provides a messaging system (action potentials) with changes moving along the neuron until a link (synapse) is reached. At the link, small amounts of a chemical may be released to stimulate the next stage of the pathway. Transmission in such an 'excitable' membrane (Appendix 11.1) and myelin insulation (11.3.5) to provide rapid transit are the basis for fast conduction by neurons.

11.3.4 Synapses

Transmission between neurons occurs at specialized 'physiological junctions' known as synapses. A synapse is a space between neurons, with pre- and post-synaptic membranes and, more recently, a third glial element has been implicated, giving the term 'the tripartite synapse' (Araque 1999; Volterra et al. 2002). Dendrites, which are branched, provide an increased surface area for contact with large numbers of pre-synaptic axons, making synapses sites of integration as well as transmission. Most synapses are chemical synapses, signals being transmitted across them by a neurotransmitter ('chemical messenger') such as acetylcholine released in response to an inflow of calcium ions during depolarization (Appendix 11.1) by an arriving action potential. Astrocyte (glia) end feet can occur around synapses and they may channel neurotransmitter diffusion between neurons (Gardner 1975).

Chemical synapses possess neurotransmitter receptors on their post-synaptic membrane. Such 'molecular receptors' may be ionotropic, responding directly to the neurotransmitter (e.g. acetylcholine) by changing the state of the gated channels and their current flow to generate a post-synaptic potential (a voltage change in a post-synaptic membrane receiving a synaptic signal) which will initiate appropriate activity. Alternatively, chemical synapses may be metabotropic, reacting to a neurotransmitter such as noradrenalin, activating secondary intracellular messengers, commonly cAMP (cyclic adenosine monophosphate) with the end-result of opening a gated channel or facilitating another response (e.g. melanosome movement, 2.9).

The neurotransmitter acetylcholine (Ach) is important in the CNS for transmission between neurons and also operates at the localized and specialized neuromuscular junctions (NMJ) for striated muscle (which are thus cholinergic) achieving rapid contraction for locomotion where precise control is necessary. By contrast, another neurotransmitter, noradrenalin, operates in autonomic (11.9) components of the peripheral nervous system where it may have a longer effect on, for example, the viscera. Noradrenalin is released along the lengths of adrenergic fibres associated with smooth muscle (Randall et al. 1997); a dermal plexus 'field effect' has been observed in adrenergic melanophore control (Mayo and Burton 1998b; Burton 2010).

Other neurotransmitters or 'neuromodulators' (a neurotransmitter which alters the synaptic responsiveness to another neurotransmitter) have been described. Burnstock (2009) lists 22 neuromodulators including co-transmitters that can modify synaptic responses. Electrotonic, or electrical synapses, can also occur. They allow electrical current to flow directly through protein channels known as connexons; conduction across electrical synapses is reported to be faster than that across chemical synapses.

11.3.5 Myelination

Vertebrate axons and dendrons may be medullated/myelinated (Figure 11.2), with additional fatty insulating layers of myelin around them. However, this is not the case for cyclostomes (Bernstein 1970; Young 1981). Myelin is composed of layers of phospholipid. In the peripheral nervous system these are products of Schwann cells which may also be regarded as a form of neuroglial cell (LeDouarin and Kalcheim 1999). The Schwann cells wrap around axons and dendrons to form a myelin sheath of spiralling plasma membrane layers, with the cytoplasm and nucleus squeezed to its outer region. Within the vertebrate (including jawed fish) spinal cord and brain, myelination involves oligodendroglia cells. The neuroglia in cyclostomes do not include oligodendroglia (Bernstein 1970).

The myelin sheath acts as an insulator around the axon or dendron, but between adjacent Schwann cells, at intervals generally 1 to 2 mm long, are nodes of Ranvier, gaps of a few micrometres, in the myelin sheath. Membrane electrical change for the myelinated neuron is restricted to the nodes of Ranvier, a saltatory conduction (from node to node), resulting in increased velocity in comparison to non-myelinated neurons. In saltatory conduction impulses 'jump' from node to node because transmembrane ionic current flow and action potentials are limited to the nodes (Figure 11.3) thereby reducing conduction time.

In nodes of Ranvier the exposed neural membrane includes clusters of Na^+ channels which are absent from the internodal axon membrane (Waxman 1997). Depolarization (Appendix 11.1) of a node generates ionic currents through its Na^+ channels to complete local circuits inside and outside the internode, which excite the next node membrane.

A difference in myelin morphology at the nodes of Ranvier has been observed between elasmobranchs and teleosts. The myelin sheath in elasmobranchs tapers towards the node but is smoothly rounded, like that of mammals, at the teleost node (Ranvier 1872 quoted by Thomas and Young 1949).

The morphogenesis of myelinated fibres in vertebrates appears to involve active participation of both glial cells and neurons. Waxman (1997) has reviewed various types of evidence suggesting that physiological contact with glia is involved in nodal aggregation of sodium channels in the developing myelinated nerve fibre. Such evidence also indicates, locally, axons can specify the number of myelin lamellae formed by a single oligodendroglial cell, as well as influence internode length. The number of lamellae is important since an optimum myelin thickness, relative to diameter, maximizes conduction velocity (Rushton 1951). Burnstock (2007) discusses the regulation of axon interactions with Schwann cells and oligodendroglia and the signalling and receptor systems which may be involved.

Thomas (1956) states that in *Salmo trutta*, myelination in the lateral line nerve starts shortly before hatching, and the number of myelinated fibres increases throughout life (from 10 to 1372 between fish lengths of 1.5 cm to 54.5 cm), the larger fibres

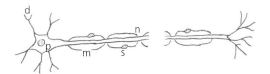

Figure 11.2 Myelinated multipolar neuron. d = dendrite; m = myelin sheath; n = node of Ranvier; s = Schwann cell nucleus; p = nucleated perikaryon. The long process to the right of the perikaryon is the axon.

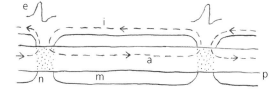

Figure 11.3 Process of saltatory conduction. a = axoplasm; i = internodal current circuit; m = myelin sheath; n = node of Ranvier; p = plasma membrane of axon; e = action potential (spike) localized to node of Ranvier. The stippling represents clusters of Na^+ channels.

acquiring their myelin before the smaller. The diameter of the largest fibres also increases with fish length but at a decreasing rate, which may reflect surrounding mechanically restrictive action (Young 1945; Cragg 1955). In *Salmo trutta* and *Raia clavata* the thickness of the myelin sheath for any particular axon diameter is constant, and not influenced by fish size, at comparable sites on the lateral line nerve (Thomas 1955).

In fish, internodes may be several times longer than those reported for mammals and amphibian (Thomas and Young 1949 quoting Ranvier 1872; Key and Retzius 1876). In the elasmobranch *Raia clavata* and the teleost *Conger conger*, internodes are up to 8 mm long in larger fish and a minimum of 0.5 mm; each internode can include several Schwann cell nuclei (Thomas and Young 1949). Conduction in non-myelinated axons is considerably slower than in those which are myelinated. Increased axonal diameter also speeds up conduction.

Myelination facilitates increased activity levels and faster neural coordination compared with cyclostomes. The evolution of a myelination process is remarkable in that it involves the development of a type of glial cell with a role which may be solely ancilliary to another type of cell. While this relationship benefits neuronal function, the glial cells may only gain holistically from the resulting increased survival of the fish. The absence of myelin in extant cyclostomes invites questions about whether it occurred in ancestral agnatha but was then lost during the evolution of modern parasitic species, whilst persisting in gnathostomes.

11.3.6 The neuropil

A neuropil (i.e. a network) of densely interlacing nerve fibres may characterize neural areas associated with integrative activity and information processing. The morphology of nerve cells with their processes facilitates complexity in neuropil neuronal circuitry, but often the intercellular relationships between afferent (incoming sensory) and efferent (outgoing, executive, motor) neurons is unclear. Most invertebrate ganglia possess a neuropil core with surrounding neuron perikarya (Simmons and Young 1999). The mammalian cerebral cortex, associated with complex behaviour, is largely neuropil in which numerous synapses may facilitate interaction between specific afferent pathways and information stores. A neuropil is a feature of the optic tectum (Healey 1957 and 11.6.2) of teleost fish with afferents from various sensory regions in the brain (Wullimann 1998). The mammalian cerebral cortex and teleost optic tectum are laminated, possessing layers of neuropil; up to 15 in teleosts, 6 in mammals.

11.4 Antagonistic control

In order to achieve fine control, many neuroeffector systems (i.e. the nervous system and its innervated effectors) operate with two features, one of which may be promotional, increasing activity, the other counteracting it. At the cellular level this can mean involvement of excitatory and inhibitory neurons and neurotransmitters. At the vascular level it can mean vasoconstriction and vasodilation. At the muscular level, in order to achieve locomotion there can be sets of muscles on two sides of a skeletal component with opposing effects, one set contracting while the other relaxes and vice versa. For smooth muscle, the anatomical arrangement of two sets of muscles can achieve constriction or widening of the gut or ducts (Appendix 11.2).

Neural mechanisms are integral to many of the systems of antagonistic control which have essential roles in animal functioning. Such antagonism involves an active response to counteract another physiological response and encompasses a diverse range of systems. These systems include macro-responses of tissues/organs and micro-responses at cellular/subcellular levels, both controlled neurally and/or hormonally (Chapter 10).

Antagonistic regulation of movement can result from the anatomical arrangement of muscles so that their contraction produces opposite somatic responses only by excitation of their respective motor nerves. Thus the erector and depressor muscles on opposite sides of fish fin-rays can change fin shape and area in some species. Another example is provided by the circular sphincter and radial muscles in the iris of the Mediterranean star-gazer (*Uranoscopus scaber*) which control pupil size (11.10.7) and probably involve neural control. In both examples one muscle contracts, counteracting the effect of the other during its relaxation to reverse the result of its previous contraction.

In other systems antagonistic responses at the cellular level can depend on counteractive intracellular processes initiated by contrasting excitatory and inhibitory neurotransmitters and/or hormones, as in the case of adrenergic and cholinergic control of smooth muscle responses (Appendix 11.2). Although the responses are often thought of as adrenergically controlled contraction and cholinergic relaxation, other responses may occur in non-mammalian vertebrates (Jensen and Holmgren 1994). Also, the same neurotransmitter at high or low concentration can evoke counteracting excitatory and inhibiting responses respectively in a single effector, through antagonistic molecular receptors, such as described for teleost melanophores (2.9, 11.10.6, Appendix 2.3).

11.5 The organization of the fish nervous system

There is a well-established literature on the anatomy and physiology of fish nervous systems and extensive activity in the subject up to the mid-twentieth century, including fundamental translated material, which has been reviewed by Healey (1957).

The linearity of the fish nervous system, with a succession of morphological and functional regions in the brain and repetitive innervated body segments, has been used to help interpret the evolutionary development of the nervous system associated with tetrapod anatomy.

The brain and spinal cord are collectively known as the central nervous system (CNS); the cranial and spinal nerves (11.8) together with the autonomic nervous system (ANS, 11.9) comprise the peripheral nervous system.

The direction of information flow (e.g. to (afferent/sensory) or from (efferent/motor) the CNS) is based on the morphology and functioning of individual nerve processes (fibres), each part of a single neuron, and on the point of origin of any stimulus. Efferent nerve cell perikarya are located in the CNS or in ganglia; for example, as a linked series (the sympathetic chain) on each side of the spinal cord. Afferent perikarya occur in ganglia external to the CNS.

The sensory fibres conduct from receptors (sensory cells/tissues/organs, Chapter 12) receiving information from the external environment, from exteroreceptors (e.g. the ear, lateral line), or from sensory receptors receiving information from the internal environment, from enteroreceptors (e.g. stretch receptors/proprioceptors, some forms of chemoreceptor).

Motor fibres evoking locomotion involving skeletal (striated) muscle (3.8.1), may be termed 'somatic'; these fibres also regulate support through controlling muscle tone. Other motor fibres, in the involuntary ANS, innervate smooth muscle of the viscera (internal organs) and ducts, glands, skin melanophores, and the cardiovascular system. Any efferent pathway between the CNS and skeletal muscle consists of a single neuron, contrasting with two neurons generally in efferent ANS pathways and more in enteric innervation (11.9.3).

Fish typically display what is regarded as the basic segmental vertebrate arrangement with a linear sequence of brain regions (Table 11.1) serviced by a set of 10 (or possibly 11) paired specialized cranial nerves (Table 11.3) with fibres running from and/or to the brain. Posteriorly these are symmetrical, more obviously metameric, spinal nerves from the spinal cord, one pair per segment, innervating skeletal (myotomal) muscles (3.8.1), the fin musculature and the skin, also containing autonomic fibres. Each spinal nerve has a ventral root, carrying motor fibres with perikarya in the CNS, and a dorsal root carrying sensory fibres with their perikarya located in a dorsal root ganglion external to the CNS (Figure 11.4). These ventral and dorsal roots join, except in lamprey (Goodrich 1930), to form 'mixed' segmental nerves carrying both motor and sensory fibres. A pathway of neurons linking specific sensory receptors and effectors producing an autonomic response, is termed a reflex arc.

The additional, less obvious ANS links the internal organs and the CNS and has both excitatory and inhibitory efferent nerves. The outflow of autonomic motor neurons from the spinal cord occurs in the ventral roots of spinal nerves. However, the tissues and organ systems under autonomic control do not have a segmental organization; the anatomical organization of the autonomic nerves provides diffuse patterns of innervation (11.9) appropriate for locomotion and rhythmic pumps for respiration, as well as achieving orientation and special behaviours, memory and learning.

The cranial or spinal nerves both contain many individual nerve 'fibres' (axons/dendrons). These fibres may all be similar as in some cranial nerves, conducting towards the CNS (sensory) or away from the CNS (motor), or the nerve may contain both motor and sensory fibres and is therefore 'mixed' as in the spinal nerves and some cranial nerves (Table 11.3).

In contrast with the fish spinal nerves the cranial nerves are specialized, each pair having a unique function whereas the spinal nerves tend to be more similar to each other, particularly in relation to the repetitive myotomal innervation. The spinal nerves in this role contain the peripheral part of the somatic nervous system controlling locomotion.

11.6 The central nervous system

The neural tissue of the CNS has a surrounding meningeal epithelial covering whilst internally there are centrally located and extensive fluid-filled cavities: the ventricles in the brain (Figure 11.5) and a central canal in the spinal cord (Figure 11.6).

11.6.1 The meninges

The vertebrate CNS is covered by one or more fairly thin epithelia, each of which is a meninx (plural: meninges). In mammals there are three layers: the outer fibrous dura mater, a middle arachnoid layer and an inner pia mater. Sagemehl (1884), Sterzi (1901) and Kappers (1926) identified only a single tissue, the meninx primitiva, around the central nervous system of fish. Kappers considered this to represent the origin of the dura, arachnoid and pia mater, the meninges of 'higher' vertebrates. However, Kappers did identify in *Lophius piscatorius* an additional fibrous differentiating layer similar to the dura mater of early embryos of humans. An adipose perimeningeal tissue lies peripherally to the meninx primitiva (Sagemehl 1884; Sterzi 1901; Kappers 1926) and Kappers considered this to have a shock-absorbing role in the moveable enclosure between the spinal cord and vertebrae.

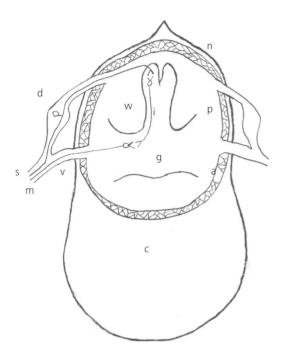

Figure 11.4 Transverse (cross-) section of a teleost vertebra with spinal cord demonstrating spinal (segmental) nerve origins and a somatic reflex arc. a = adipose perimeningeal tissue; c = centrum; d = dorsal root (with ganglion) of spinal nerve; g = grey matter; i = interneuron; m = motor neuron; n = neural arch; p = meninx primitiva; s = sensory neuron; v = ventral root of spinal nerve; w = white matter. s + m = a mixed spinal nerve (based on Burton 1964a).

the distribution of their target tissues and organs. Sensory afferent fibres from the viscera reach the CNS through the dorsal root of segmental nerves and their perikarya are in the dorsal root ganglia, like those of somatic afferent fibres (Goodrich 1930). In mammals it is possible to distinguish three parts of the ANS; the sympathetic, with a mid-region (thoraco-lumbar) component, the parasympathetic, with cranio-sacral components in the head and posterior areas and the enteric nervous system of the gut. The division into sympathetic and parasympathetic is not obvious in fish which may, however, have some features of both components. The CNS, as in other vertebrates, has extreme anterior development, associated with cephalization, or formation of a distinctive head from the anterior body segments. A succession of brain regions occur, with special sense organs and centres enabling

11.6.2 The organization of the brain

Although there is considerable variation in the relative size of fish brain regions in association with the dominance of special senses there is a particular succession of these regions from anterior to posterior which is regarded as standard (Figure 11.5 and Table 11.1). During embryonic development three regions are distinguished: the anterior fore-brain or prosencephalon; the mid-brain or mesencephalon and the hind-brain or rhombencephalon. The prosencephalon develops into the anterior telencephalon and the posterior diencephalon which, as the 'between-brain', links to the mid-brain. Traditionally the fore-brain is associated with smell, whereas the diencephalon and mid-brain are associated with vision. The hind-brain is associated with hearing and balance and the lateral line system. Various regions of the brain are linked internally by nerve fibres forming tracts; Wullimann (1998) reviews information about these nerve tracts in fish brains. Specialized tracts called commissures link right and left regions.

In addition to the major regions (seen as swellings) associated with smell, vision and hearing/balance there are two small out-growths from the brain surface, one mid-dorsal as the pineal (10.11.2) and one mid-ventral as the hypophysis/pituitary (10.11.1), both of which have endocrine functions, linking the two control systems (neural and endocrine). The pineal can be difficult to locate and, being attenuated, can easily be destroyed on dissection. The pituitary, however, is ventral to the hypothalamus close to the crossover of the optic nerves (optic chiasma) and is relatively easy to find on dissection or surgery.

As well as brain regions associated with sensation there are centres associated with motor responses including the generation of rhythms. Respiratory rhythms are controlled from the medulla oblongata of the hind-brain whereas the cerebellum achieves fine control of fish balance within the often labile water column with motor output to the myotomal muscles in response to sensory input from the semi-circular canals of the ear.

In fish the medulla oblongata and the tegmentum (floor of the mid-brain) are together known as the brain stem, the usage of this term differing from that for the human brain in which the optic tectum and diencephalon are also included (Wullimann 1998).

Fore-brain

The fore-brain or prosencephalon includes the telencephalon and diencephalon (Table 11.1) and is relatively large in elasmobranchs, their olfactory sense (smell) being particularly important in the search for food, but Wullimann (1998) reviews research on other possible sensory roles. In the teleosts, based on experiments on goldfish, *Carassius* (Kumakura 1927; Janzen 1933; Healey 1957), the telencephalon

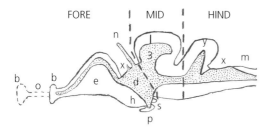

Figure 11.5 Plan of longitudinal section through a generalized teleost brain demonstrating the main regions and ventricle system. Note the relative sizes of the regions may vary *considerably* between taxa, as also the length of the olfactory tracts. b = olfactory bulb; d = thalamus; e = telencephalon; h = hypothalamus; l = optic lobe; m = medulla oblongata; n = pineal; o = olfactory tract; p = pituitary; s = saccus vasculosus; x = choroid plexus (anterior and posterior); y = cerebellum enclosing the 4th ventricle. Stippling represents the ventricle system; 3 = third ventricle (after Healey 1957).

Table 11.1 Fish brain regions.

Major region	Associated special sense	Sub-regions
Prosencephalon (Fore-brain)	Smell	Telencephalon (cerebral hemispheres)
	Vision	Diencephalon, Thalamus, hypothalamus
Mesencephalon (Mid-brain)	Vision	Optic lobes: dorsal tectum and ventral tegmentum
Rhombencephalon (Hind-brain)	Auditory Acoustico-lateralis	Metencephalon (cerebellum) Myelencephalon (medulla oblongata)

has a regulatory role over other sub-regions of the teleost brain. In European minnow (*Phoxinus*), the fore-brain has a role in the acceptance of newcomers to schooling (Berwein 1941; Healey 1957). Other parts of the brain are involved in conditioned responses to colours or shapes (Nolte 1933, cited by Healey 1957).

The telencephalon of teleosts is atypical of vertebrates as the cerebral hemispheres lack lateral ventricles (Young 1981), although these are present in cyclostomes and elasmobranchs. Also, there is no development of a cerebral cortex, with its neuropil, in each of these three groups (Healey 1957), there being significant development of a cortex only in the lungfish (Bull 1957; Young 1981). In teleosts, the roof of the cerebral hemispheres (telencephalon) is entirely a thin ependymal membrane (Kappers et al. 1936; Young 1981), that is, a cuboidal/columnar epithelium with no neurons. This condition is accompanied by the entire location of the cerebral 'grey matter' beneath the ventricles, and contrasts with the thickened roof in elasmobranchs. The ventral area receives nerves from the olfactory tracts, dorsally optic information is received through the diencephalon, the roof of which is evaginated, forming the eyes and the pineal body, sometimes called the third eye as it is light-sensitive.

The pineal body structure and size is quite variable (10.11.2) but generally contains photoreceptor cells (pinealocytes), sensitive to light intensity which may influence diurnal cycles, including skin colour (2.9) as well as seasonal reproductive cycles (10.11.2). The pineal body is regarded as both a light receptor and an endocrine gland (Fenwick 1970). Production of melatonin (10.11.2), the pineal hormone, is generally high in dark conditions and low in light, but according to some authors, including Kulczykowska et al. (2001), there can be an intrinsic cycle reflecting a diurnal rhythm in circulating melatonin. Where this occurs it is taken as an indicator of an intrapineal oscillator though an external (brain/central) oscillator is also suggested (Ekström and Meissl 1997). Pineal excitation is transmitted to ganglion cells, which show electrical activity in the dark (Fujii and Oshima 1986); the expectation is that afferent fibres of pineal 'ganglion' cells pass to other parts of the brain; 'nuclei'

such as the suprachiasmatic nucleus (SCN) in the hypothalamus which in mammals particularly is associated with control of rhythm (Dibner et al. 2010) and which is still present in blind cavefish (Wullimann 1998).

The hypothalamus (ventral diencephalon) has regulatory functions involving internal systems, including circulation and the endocrine system. It is as well developed as in mammals and a conspicuous neurohypophyseal tract runs to the pituitary. Incoming fibres are from the telencephalon as well as the gustatory (taste) and acoustico-lateralis (lateral line) systems. Outgoing fibres run both anteriorly and posteriorly, the latter to the medulla oblongata.

In relict fish, as well as chondrichthyans and many teleosts, the saccus vasculosus (10.11, Figures 10.2 and 11.5) is attached posteriorly to the hypothalamus. It has folded, vascularized walls, and mainly consists of large coronet cells with apical 'hairs' and is often described as having substantial amounts of carbohydrates (acid mucopolysaccharide, Jansen and van de Kamer 1961, or glycogen, Sundararaj and Prasad 1964) associated with it. Its function is not understood but at least two different roles have been suggested; that it is sensory or that it is secretory (Nishiyama 1963). A role in resorption or 'monitoring' of cerebral fluid has been suggested (Lanzing and van Lennep 1971) and, in contrast Devlin et al. (1996) report the presence of parathyroid hormone-related protein. The old idea (Dammerman 1910, quoted by Te Winkel 1935) that the saccus vasculosus responds to water pressure ('a peculiar organ for detecting water pressure', Sasayama 1999) has not been thoroughly investigated but is to some extent supported by its occurrence and structure: it is well developed in many marine or migratory teleosts (e.g. *Anguilla*, Holmes and Ball 1974) but absent or reduced in some freshwater teleosts such as the cypriniformes; it has never been found in some taxa, notably Atheriniformes and Cyprinodontiformes (e.g. *Poecilia*, Holmes and Ball 1974) which live in fresh or brackish water (Tsuneki 1992).

Langecker and Longley (1993) report two blind catfish species, *Trogloglanis pattersoni* and *Satan eurystomus*, from Texas artesian water, both of which show a very well-developed 'hypertrophied' saccus vasculosus (in comparison to the epigean catfish

Ictalurus punctatus) and which live in conditions of high hydrostatic pressure.

However, analyses or deductions based on size of the saccus vasculosus may need extra care in view of Nishiyama's (1963) findings that the overall shape (notably length and width) and extent of this structure shows considerable intraspecific (i.e. individual) variability as well as having an interspecific variety in typical shape ranging from rhombic and cone-shaped to dumbbell-shaped.

The thalamus, dorsal to the hypothalamus, receives incoming information from the retina, optic tectum, cerebellum and spinal cord and functions as a relay station for sensory signals.

Mid-brain

The mesencephalon (mid-brain) is the dominant region in cyclostomes, elasmobranchs and teleosts. The optic lobes receive the optic tracts ('nerves', 11.8 and Table 11.3) from the eyes. In cyclostomes the relationship between optic tract and optic lobe is ipsilateral (i.e. the same side, Greenwood 1975). In elasmobranchs and teleosts there is a junction (chiasma) of the two optic tracts below the brain, the neural input from each eye being to the contralateral (opposite) optic lobe (Grove and Newell 1953). The optic nerves, which cross (decussate) below the brain, end in the roof (optic tectum) which has a complex pattern of layers (laminae) of neurons and neuropil. Healey (1957) and Wullimann (1998) emphasize that in fish the optic tectum is concerned not only with vision but also with correlation of sensory impressions, body position and movement, the mesencephalon exhibiting functions associated with the tetrapod cortex. In this context it is interesting that in gymnotoids and mormyrids the optic tectum also receives input from the centres for smell, hearing and taste as well as electrosensory information. Neural input from non-visual sources is relayed to the tectum via centres such as the torus semicircularis, a sensory processing site in the fish mid-brain (Wullimann 1998). Efferent neurons run to the medulla oblongata or the spinal cord. The mid-brain controls much of fish behaviour and learning. Like the cerebral cortex of mammals, it regulates spinal centres controlling movement. There is a spatial relationship between cells in the retina and those of the tectum of fish, as

in mammals (reviewed by Healey 1957). The organization of the tectum is appropriate for identifying objects and coordinating suitable motor responses. The structural similarities of the fish optic tectum and mammalian cerebral cortex suggest that the anatomical similarities may correlate with analogous physiological functions (Healey 1957). The involvement of different regions of the brain in fish and mammals in cognitive development may have originated as sensory and behavioural adaptations to distinct aquatic and terrestrial attributes requiring different neural region dominance in these taxa. Experiments involving the control of colour change in minnow, *Phoxinus* (11.10.6), suggest that nerve pathways from the retina to the autonomic nervous system do not involve the optic tectum (reviewed by Healey 1957). The tegmentum or base of the mid-brain has major motor centres but their regulation by the tectum appears to be necessary for coordinated movement (Young 1981).

Hind-brain

Within the rhombencephalon (hind-brain) the cerebellum is very large in both elasmobranchs and teleosts, projecting under the mid-brain in the latter group. It has a sensory input from the ears and lateral lines and an output to motor centres. A stratum of flask-shaped neurons, the Purkinje cells, occurs in the cortex of the cerebellum. These neurons possess large perikarya with an elaborate system of dendrites and an axon which runs to the centre of the cerebellum to cerebellar 'nuclei' with an output to motor nuclei and to the thalamus. The cerebellum functions to maintain balance which can involve precise responses to strong lateral forces generated by moving water in order to remain upright and can be well developed in shallow water species. The cerebellum regulation involves a complex system of excitation and inhibition of motor control (Young 1978) of, for example, different fins, the angle of which can be altered, or in the case of teleosts, the area can be changed as the fins can be furled.

An important variable in analysis of interspecific variation in the size of the cerebellum is the level of active swimming; the cerebellum is larger in active swimmers like *Thunnus* (Young 1981). Exceptionally, in the Mormyridae, which can electrolocate (12.9.1), the cerebellum is so large it covers the rest of the brain (von der Emde 1998), even though these fish are slow swimmers. This enlargement is caused by hypertrophy of part of the cerebellum, the valvular cerebella, innervated by the anterior lateral line root (Bell 1981) and related to perceiving electrical signals involved in direction finding (12.9.1). In the chondrichthyans the dorsal wall of the rhombencephalon has prominent thickenings associated with electrosensation in these fish (Kotrschal et al. 1998).

The medulla oblongata merges with the spinal cord and is clearly derived from it. Sensory and motorneurons of the cranial nerves V to X (Table 11.3) occur in this region which is largely a relay station between the brain's higher centres and the spinal cord (Healey 1957). Tetrapods differ, with their tracts from higher centres running through the medulla without interruption. The medulla is the site of the respiratory centre which controls respiratory movements of the pharyngeal floor and, in teleosts, the rhythmic movement of the operculum, enabling ventilation of the gills (6.5.10). This respiratory centre is affected by anaesthestics only at their highest sub-lethal concentrations. It was demonstrated by Adrian and Buytendijk (1931) in their pioneering recordings of electrical activity in the brain that spontaneous activity in respiratory centres of goldfish correlates with its respiratory rhythm. The medulla oblongata of the Eurasian minnow (*Phoxinus*) has a paling centre controlling rapid colour changes (von Frisch 1911).

11.6.3 The ventricles and cerebro-spinal fluid

Within the neural tissues of the brain, vertebrates have an interconnected cavity system of ventricles (Figure 11.5), filled with a clear 'cerebro-spinal fluid', which extends through the length of the brain and continues into the spinal cord as the central canal. Two lateral ventricles (absent in teleosts, Young 1981) connecting with a third (IIIrd) ventricle occur in the fore-brain. An 'iter', or cerebral aqueduct, in the mid-brain, connects the third ventricle with a fourth (IVth) ventricle in the medulla oblongata. Mammalian cerebrospinal fluid has few/no cells, little protein and an ionic composition different from that of blood plasma (Gardner 1975) but this fluid in fish requires further study. The ventricles,

and also the spinal cord central canal, are lined by an epithelium known as the ependyma. An anterior and a posterior 'choroid plexus' in the roof of the fore- and hind-brain (Figure 11.5) are thin, folded, vascularized structures which project into the IIIrd and IVth ventricles respectively.

11.6.4 The blood–brain barrier

The choroid plexi (Figure 11.5) of the brain ventricles achieve a filtration of substances entering neural tissue; this system is referred to collectively as the blood–brain barrier (BBB), and it excludes, for example, red blood cells and most white blood cells (in mammals) from entering neural tissue. In fish, however, large numbers of lymphocytes and macrophages have been reported in white matter of the spinal cord (Dowding and Scholes 1993). The BBB has the capacity for selective transport of some solutes, there being 'differential entry' (Bernstein and Streicher 1965).

The endothelial (lining) cells of the blood vessels and endfeet of neuroglial cells (astrocytes) form a neurovascular unit (Tam and Watts 2010) which contribute to the BBB, selectively transporting substances, there being differential entry from the blood to the brain. Based on anion uptake by the brain of goldfish Bernstein and Streicher (1965) suggested that BBB transportation processes differ in teleosts and mammals but the general view of cerebrospinal fluid is that it is formed at the choroid plexi and passes through neural tissue, draining to the venous system at the meninges. In the tilapia, a cichlid teleost, *Oreochromis* (*Sarotherodon*) *mossambicus*, electron microscopy and immunohistochemical studies (Dowding and Scholes 1993) indicate that leucocytes (lymphocytes and macrophages) in the white matter of the CNS of fish are similar to those in fish blood and that they are more numerous than in neural tissue of mammals. The macrophages are reported to engulf old myelin, a role performed by 'microglia' in mammals.

11.6.5 Brain size and complexity

Some fish do not have a finite growth pattern, and the brain in the largest fish may not fill the brain cavity; it seems that the brain does not grow proportionally with the soma. In carp, 'the larger the fish, the proportionally smaller its brain' (Kotrschal et al. 1998), which fits to some extent with the concept that neurons do not divide. Indeed, the relatively small size of brains of, for example, adult coelacanths has been remarked upon. Northcutt et al. (1978) reported a likely range for the brain weight of less than 2 g for a coelacanth with total body weight of approximately 30 kg. The cranial cavity may contain fatty material round the brain; the composition of this material has been reported for *Hoplostethus mediterraneus* (Sargent et al. 1983); Kotrschal et al. (1998) call the equivalent tissue in carp 'lymphatic fatty tissue'.

Huxley (1863, quoted by Striedter 2004) observed that fish brains are very small, and within extant fish taxa lungfish are noted for small simple brains whereas the Mormyridae, a family of electric fish which use their electrogenic capacity to electrolocate have very large brains, with a particularly well-developed cerebellum (11.6.2). Striedter (2004), quoting from Nilsson (1996), states that the brain of mormyrids uses relatively more oxygen than that of humans; 60 per cent of consumption compared to 20 per cent for humans and 2–8 per cent in other vertebrates.

In relation to body size the brain in elasmobranchs is generally larger than in teleosts, that of cyclostomes being the smallest (Ebbeson and Northcutt 1976 cited by Young 1981). The size of the elasmobranch brain with its large olfactory bulbs and cerebral hemispheres may be associated with the neural processing due to high levels of both olfactory sensitivity and electrosensitivity to bioelectric fields (12.9.3), which facilitates food location in the potentially huge spaces of their marine environment. The cerebral hemispheres with their thick roof contrast with those of teleosts which are 'everted' with a thin roof, but neither has a complex cerebral cortex. More is known about teleost behaviour than that of elasmobranchs. In both groups the optic lobes have neuropil development which, with its complex neural circuitry, provides a capacity for memory and learning systems as well as for migratory, schooling, courtship and nesting behaviour.

A recent evaluation of head size in teleosts has suggested that the overall small body size during early stages of development has necessitated eversion rather than evagination of the telencephalon (Striedter and Northcutt 2006).

Variation in fish brain structure and relative complexity is reviewed in Kotrschal et al. (1998).

11.6.6 The spinal cord

The spinal cord in cyclostomes is morphologically simpler than in elasmobranchs and teleosts. In transverse section it is flattened and is not differentiated into grey and white matter. In elasmobranchs and teleosts the cord is tubular and there is differentiation into inner 'grey' matter, including perikarya, and outer 'white' matter, with tracts of myelinated fibres. In all three taxa there is a central canal (containing cerebro-spinal fluid) and commissures which are groups of fibres decussating (crossing) the spinal cord. A transverse section of teleost spinal cord (Figure 11.6) shows the outer white matter and the inner grey matter, representing a reversal of these regions in the brain.

The inner grey matter, with two dorsal and two ventral 'horns', includes interneurons and the nucleated perikarya of motor neurons. The teleost dorsal horns of the grey matter are closer together than in mammals and are larger relative to the spinal cord diameter towards its cephalic end, being almost filled by the substantia gelatinosa Rolandi which contains numerous small neurons. Knowledge of the dorsal horns, particularly the substantia gelatinosa Rolandi and its neurons, is poor. The gelatinous nature may be due to a dense network of dendrites (Keenan 1928), and according to Carleton and Leach (1949) the mammalian corpora gelatinosa Rolandi are rich in glial cells. The dorsal horns of the minnow *Phoxinus* are surrounded by fine myelinated fibres. Within the spinal cord the grey matter provides a basis for dorsal root sensory neurons to synapse with interneurons forming reflex arc pathways through local motor neurons (Figure 11.4), and with interneurons having more distant connections including the brain. The ventral horns include perikarya of somatic motor neurons. Axons from dorsomedial somatic motor neuron perikarya in the ventral horns innervate the myotomal musculature, more ventral motor neurons in these horns innervate the fins (Kappers et al. 1936).

The white matter of fish is less extensive than in tetrapods with their complex limb movements to

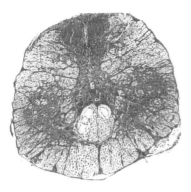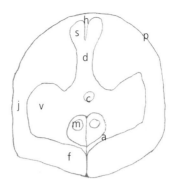

Figure 11.6 Transverse (cross-) sections of a teleost (minnow *Phoxinus phoxinus*) spinal cord. Left: photomicrograph, myelin preserved by post-chroming; section stained with Mallory trichrome (see Burton 1964a). Note the presence of an additional Mauthner fibre (normally there are only two). Right: diagram of the cross-section, showing distribution of tissues and tracts (nomenclature after Edinger 1899; Keenan 1928; Kappers et al. 1936; Berkowitz 1956). a= accessory commissure; c = central canal; d= corpus commune posterius; f = ventral funiculus; h = dorsal funiculus; j = ventrolateral column; m = Mauthner fibre; p = meninx primitiva; s = substantia gelatinosa Rolandi; v = ventral horn (dark ventral region). The dorsal horns (dark dorsal region) include the substantia gelatinosa Rolandi.

coordinate, this difference being particularly noticeable around the dorsal horns of the grey matter.

Discrete neural tracts are identifiable within the white matter, each distinguished by the range of diameters of its axons (Kappers et al. 1936 and Ramón-y-Cajal 1952). The ventral funiculus (Edinger 1899) or medial longitudinal fasciculus (Berkowitz 1956) between the ventral horns of grey matter contains many descending motor-coordinating myelinated fibres which synapse in the ventral horns with motor neuron perikarya of the spinal nerves. In teleosts these fibres typically include the Muller fibres with overall diameters between 11.1 μm and 22.5 μm in *Ameiurus* (Graham and O'Leary 1941) and 8 μm and 22 μm in *Phoxinus* (Burton 1969), and the bilaterally paired larger Mauthner fibres (Figure 11.6), between 25 μm and 40 μm overall diameter in *Ameiurus* (Graham and O'Leary 1941) and 32 μm and 60 μm in *Phoxinus* (Burton 1969). Mauthner fibres have been described in teleosts (and some urodele amphibians). They are also present in larval lampreys (Young 1981) but they are not found in elasmobranchs (Aronson 1963 cited by Diamond 1971). They are smaller in fish which are eel-shaped (anguilliform) or benthic as well as in fast-moving scombroids (Young 1981). Similar 'giant', but unmyelinated, fibres also occur in some invertebrates such as cephalopods, enabling rapid conduction for startle responses. In teleosts and amphibia, the bilateral pair of Mauthner fibres have perikarya

each with two large dendrites in the nucleus motorius tegmenti of the medulla oblongata (Bartelmez 1915), and their wide axons decussate (cross over) to pass down opposite sides of the spinal cord. They have numerous short collateral branches which synapse in the ventral horns with segmental motor neurons to the myotomal muscles. The Mauthner cell bodies have inputs from the lateral line and vestibular acoustico system. The size of these neurons has facilitated experiments, mainly on goldfish and zebrafish, on the causal relationship between a specific behaviour pattern and neuronal activity, such studies being rare in vertebrates. As in invertebrates, the giant fibre system of fish mediates startle responses although lumpfish (*Cyclopterus lumpus*), which appears to lack Mauthner neurons, is able to display rapid startle responses (Hale 2000). Parker (1908) demonstrated that slight leaps forward of the squeteague (*Cynoscion*) in response to tapping an aquarium cease if the input from the lateral line and vestibular acoustic system is blocked experimentally.

Stimulation of the Mauthner cells excites contralateral myotomal contraction to bend the body away from the stimulus and achieve tail-flips to escape attacks by predators. The synaptic input from the VIIIth cranial nerve is both chemical and electrotonic (Diamond 1968). The electrotonic synapses have a tight junction ultrastructure and appear to transmit weaker signals. The Mauthner cells are a component of short delay pathways with a conduction velocity of 85 m/s in goldfish. The action potentials on their decussated axons simultaneously excite ipsilateral (same side) spinal motor neurons and inhibit those which are contralateral (opposite side), possibly through inhibitory neurons activated electrotonically from the Mauthner axon.

The occurrence of giant fibre systems, including those of invertebrates, has facilitated physiological studies of the nervous system, particularly conduction and the nature of the nerve impulse because in some, electrodes can be inserted within the fibres for 'intracellular' recording.

11.6.7 The posterior spinal cord

In Actinopterygians (the taxon for 'rayfins' which includes teleosts), the spinal cord ends in a caudal neurosecretory system, the urophysis (Fridberg and Bern 1968), the function of which, as its name implies, has been compared with the pituitary (the hypophysis). However, although physiologically active substances have been extracted from the urophysis (10.11.6), its significance in the overall function of fish has not been clearly demonstrated.

11.7 The peripheral nervous system

The peripheral nervous system, including cranial and spinal nerves, provides links between the CNS and other systems, enabling control of locomotion, operation of ducts, movement in viscera, changes in the cardiovascular system and, for some fish, change in chromatophores (Table 11.2).

11.8 The spinal and cranial nerves

Each myotomal segment has a symmetrical pair of mixed spinal nerves arising from the spinal cord and innervating the myotomal muscles and skin. These spinal nerves are largely repetitive in structure and function and contain the peripheral portion of the somatic nervous system controlling locomotion by striated muscles of the myotomes and the fins. Also, peripheral fibres of the autonomic nervous system, innervating the non-striated (smooth) muscle of the internal organs as well as chromatophores of the skin, have associations

Table 11.2 Systems controlled via peripheral nerves (not comprehensive).

System	Effector	Function
Somatic motor	Striated muscle	Locomotion, feeding
Sensory	Exteroreceptors	Information gathering relative to the environment
	Enteroreceptors	Information gathering in the internal environment
Autonomic	Smooth muscle	Gut movement, duct control Vascular diameter change
	Cardiac muscle	Changes in heart action
	Melanophores	Skin darkening or paling

within the spinal nerves. The origin of each spinal nerve from the spinal cord possesses a dorsal root and a ventral root (Figure 11.4), the dorsal root bearing sensory fibres and the ventral root motor fibres. The dorsal and ventral roots join to form the mixed spinal nerves with sensory neurons conducting towards the spinal cord and the motor neurons conducting away from it.

In fish there are 10 or possibly 11 (including the terminal nerve, nerve zero) pairs of cranial nerves directly associated with the brain (Table 11.3). In contrast to the spinal nerves, each pair of cranial nerves is specialized, being linked to a specific sensory system, to muscles of the eyeballs or to muscles of the jaws and sense organs. Each cranial nerve has either a sensory, motor or a mixed function.

The cranial nerves may be understood in relation to the increasing complexity and specialization of the head region; the evolution from jawless agnatha to the gnathostomata with biting jaws was a major change, enabling vertebrates to become more active carnivorous feeders.

It is generally considered that chordate ancestors had a terminal mouth and a pharynx with a series of gill slits and gill arches with cartilaginous supporting rods. As the mouth became used for grasping the arch supporting the first gill slit became bent over into the position of the vertebrate jaws. Thus in jawed fish the head has evolved a gradual modification of the anterior metameres of the chordate body during a process of cephalization.

Consequently the cranial nerves represent a series of segmental nerves with dorsal and ventral

Table 11.3 Fish cranial nerves.

	Name	Direction of conduction	Termed	Main role
0	Terminal	To brain		Under investigation
I	Olfactory	To brain	Special	Enables smell
II	Optic	To brain	Special	Enables sight
III	Oculomotor	From brain	Motor	Eye movement
IV	Trochlear	From brain	Motor	Eye movement
V	Trigeminal	To/from brain	Mixed	Sensation and Jaw movement
VI	Abducens	From brain	Motor	Eye movement
VII	Facial	To/from brain	Mixed	Like V
VIII	Auditory	To brain	Sensory	Enables ear function
IX	Glossopharyngeal	To/from brain	Mixed	Branchial movements and taste
X	Vagus	To/from brain	Mixed	Branchial movements and lateral line function

roots associated with the head somites which can be recognized during development (Balfour 1878 cited by Young 1981). In the head region the dorsal and ventral roots of the cranial nerves separate; a primitive condition in the cephalochordate *Branchiostoma* (*Amphioxus*), and agnatha (jawless fish), is retained. However, the olfactory and optic nerves are special cases (Table 11.3) not conforming to this pattern. Olfactory nerve fibres have their perikarya at the periphery with their inner sides forming the fibres to the brain which is inconsistent with the proposed dorsal root origin of cranial sensory nerves. These nerves may have pre-existed the development of a segmental structure (Young 1981). The optic nerve is an 'optic tract' between two parts of the central nervous system, based on its embryogenesis, rather than a peripheral nerve. The fifth, seventh, ninth and tenth cranial nerves are mixed, with segmental

sensory fibres and motor output for anterior ventral non-myotomal muscles (Table 11.3) comparable to what is considered an ancestral chordate condition (Young 1981). Many fish have a complex lateral line system over the head and flanks of the body, with neuromasts which detect water low-frequency vibrations (12.7), with an innervation which has been associated with cranial nerves V, VII, IX and X, with the possible involvement of VIII (Young 1981 and 12.7.1). Northcutt and others (2000) studied the possible contributions of these cranial nerves to the lateral line innervations in the channel catfish *Ictalurus punctatus*.

The number of fish cranial nerves is fewer than that in tetrapods with their additional accessory (XI) and hypoglossal (XII) nerves. These respectively innervate some neck muscles and the tetrapod tongue which is muscular; both features (neck, mobile tongue) related to the evolution of terrestrial vertebrates. The vagus nerve in fish is compounded of a 'vagus-accessorius series' of nerves from several segments which become separated into the tetrapod nerves XI and XII (Young 1981).

11.8.1 The terminal nerve (nerve zero)

The existence and possible function of an additional anterior cranial nerve are issues for conjecture. A 'terminal nerve', cranial nerve zero, was described over 100 years ago, in association with the olfactory tract as a supernumerary nerve in smooth dogfish, *Galeus canis* (Fritsch 1878), and as a new nerve in *Protopterus annectens* (Pinkus 1894), both cited by Wirsig-Wiechmann (2004). There is an extensive literature about this nerve including a review by Goodrich (1930) and special journal issues edited by Demski and Schwanzel-Fukuda (1987) and Wirsig-Wiechmann (2004). Individual accounts include those by Demski and Northcutt (1983), Műnz and Claas (1987), Fujita et al. (1991), Whitlock (2004) and Abbe and Oka (2007). The function of this nerve is incompletely understood but suggestions include chemosensory (Demski and Northcutt 1983) and neuromodulatory (Abbe and Oka 2007) roles. In fish it has clusters of perikarya in the olfactory bulb (12.4.7), sending their processes peripherally to the olfactory epithelium and centrally to the telencephalon and the retina. Gonadotrophin-releasing hormone (GnRH), other reproductive hormones and neuroactive peptides occur in the terminal nerve and they may have neuromodulatory roles in the physiology of the retina and olfactory epithelium during reproductive behaviour. The possibility of this neuromodulatory role is based on electrophysiological, immunohistochemical and biochemical data (Abbe and Oka 2007). The terminal nerve displays pacemaker activity, probably resulting from intrinsic properties of membrane ionic channels, the frequency of which may be changed by temperature, photoperiod and physiological conditions regulating the release of terminal nerve GnRH. This released GnRH could modulate neuronal activity in different brain regions motivating sexual and other behaviours.

11.9 The autonomic nervous system

This component of the vertebrate peripheral nervous system has controlling roles involving internal organs such as the gut, the blood system, some glands, as well as skin chromatophores (in teleost fish and in reptiles) and also the iris of the eye in some cases. Smooth muscle associated with the tubular blood vessels, gut and ducts (e.g. ureters) as well as with the urinary and swim bladders are innervated by the autonomic nervous system (ANS), as is the cardiac muscle of the heart. Autonomic neurotransmitter release from membrane-bound vesicles occurs from large varicosities or swellings along the endings of the innervating axons (Burnstock and Costa 1975); these axons therefore have a beaded appearance. Autonomic control is mediated through molecular receptors distributed on the effector system.

The nerves of the system run from the CNS to each effector area via a specific pathway, each typically involving two neurons with an intervening ganglion outside the CNS. This ganglion contains the perikaryon of the second neuron, which is 'post-ganglionic' and non-myelinated, the first neuron being 'pre-ganglionic' and myelinated. In many vertebrates there are two chains of 'sympathetic' ganglia close to the spinal cord as well as less obvious more peripheral ganglia. The ganglia enable impulses to pass from the myelinated pre-ganglionic fibres to non-myelinated post-ganglionic

fibres. This arrangement can provide independent, localized innervation to the separate viscera but also offers the possibility of anatomical amplification. This is relevant when pre-ganglionic fibres from only two to three spinal nerves, as in the minnow *Phoxinus phoxinus* (von Frisch 1911), after passing along the sympathetic chain, then synapse with post-ganglionic neurons innervating the entire dorsolateral skin, providing the capacity for simultaneous responses of melanophores over a very large area.

Vertebrate sympathetic ganglia are anatomically complex and incompletely understood (Gardner 1975). A pre-ganglionic fibre may have multisynaptic connections within a ganglion or may run through one ganglion to synapse in another, more anterior or posterior. Gardner (1975) states that sympathetic ganglia can also include chromaffin cells which contain catecholamines and may function as interneurons.

It has been recognized since Langley (1898) used the term 'autonomic nervous system' that, in mammals, it can be subdivided anatomically into sympathetic, parasympathetic and enteric components. In mammals, the sympathetic component has 'outflows' of pre-ganglionic neurons from the spinal cord in the thoracolumbar vertebral region synapsing (in a sympathetic ganglion close to the vertebrae) with post-ganglionic neurons. The post-ganglionic fibres then pass to the effectors (smooth muscle of the blood vessels, etc.) via individual special or segmental nerves. The parasympathetic outflows (of mammals) are cranial and sacral, the pre-ganglionic neurons synapsing (sometimes in readily recognizable ganglia) with post-ganglionic neurons distal to the spinal cord and sometimes on the innervated tissue (Donald 1998). Pre-ganglionic neurons are cholinergic (with acetyl choline as neurotransmitter); post-ganglionic parasympathetic neurons are also cholinergic whereas post-ganglionic sympathetic neurons are adrenergic (with noradrenalin and adrenalin as neurotransmitters). In mammals, while the sympathetic and parasympathetic nerves provide a direct link between some internal organs and the central nervous system, the gut has an additional 'enteric system' consisting of intrinsic nerve cell networks (basically two extensive plexi) with extrinsic autonomic innervation

CNS

EFFECTORS

SC

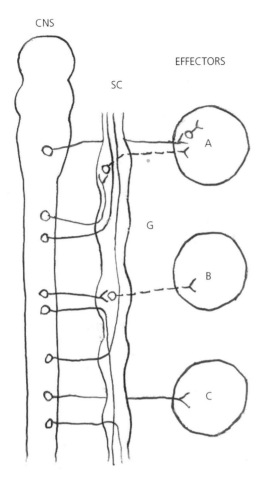

G

A

B

C

Figure 11.7 Schematic representation of types of teleost autonomic innervations. This schematization is conceptual, emphasizing relationships between pre-ganglionic and post-ganglionic neurons rather than a comprehensive anatomical illustration. A = innervations of heart, swim bladder and gut, from cranial and spinal autonomic outflows. B = innervations of smooth muscle, melanophores, some glands, from spinal autonomic outflows. C = innervations of chromaffin tissue from spinal autonomic outflows. CNS = central nervous system. G = sympathetic ganglion. SC = sympathetic chain. Based on accounts by Young (1931a, 1933a, 1981), Nilsson and Holmgren (1993), and Nilsson (2011). Cholinergic neurons are represented as complete lines, adrenergic neurons as broken lines.

from the CNS through the vagus nerve (cranial X) and the splanchnic nerves from the sympathetic chain.

An autonomic nervous system developed at an early stage in vertebrate evolution, with a simple organization in cyclostomes (agnatha) and greater complexity in elasmobranchs and teleosts,

as illustrated by Nilsson and Holmgren (1993) and Nilsson (2011). In more active fish there is a greater need to regulate blood flow as required by the various organs and tissues, consistent with the capability of the heart. A simplified conceptual schematic representation of the teleost autonomic nervous system is provided in Figure 11.7.

The division in mammals into sympathetic (thoracolumbar) and parasympathetic (craniosacral) outflow from the CNS is not appropriate in fish which do not have homologous regionalization of the body. Instead, Young (1931a) and Nilsson (1983) have proposed a simple differentiation into spinal and cranial autonomic systems (in addition to the enteric system), circumventing the uncertainty about a posterior 'parasympathetic' outflow in fish and non-mammalian tetrapods. However, the cranial outflow is still regarded as 'parasympathetic' in teleosts by some authors such as Seth and Axelsson (2010) and Mann and others (2010). The spinal autonomic system involves neuronal pathways through spinal nerves and the sympathetic chain (Figures 11.7 and 11.8); the cranial autonomic system pathways being in cranial nerves and with ganglia near, or on, the target effectors (Figure 11.7).

Teleosts have their sympathetic ganglia in well-developed paravertebral sympathetic chains similar to those of tetrapods. They are, however, unique in that they continue into the head (Burnstock 1969), their ganglia connecting with the vagus, glossopharyngeal, facial and trigeminal nerves. In some teleosts (Gibbins 1994) they fuse posteriorly forming a single chain; commissures can also occur between the two chains. Pre-ganglionic fibres to the head have outflows which are posterior, from the spinal cord, and run anteriorly through the sympathetic chains (Figure 11.7) to the cranial sympathetic ganglia which are found only in teleosts and holosteans (*Amia*). Nilsson (1984) suggests that the vasculature of the head and gills may be innervated by spinal autonomic neurons. In addition to the white *rami communicantes* routes for pre-ganglionic fibres from their spinal outflows in teleosts, there are also grey *rami communicantes* (Figure 11.8) through which some post-ganglionic fibres return to the spinal nerve to supply the skin and chromatophores (Young 1981). In the case of the cranial sympathetic ganglia there are no white rami but only the grey

A

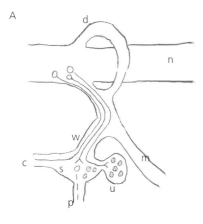

B

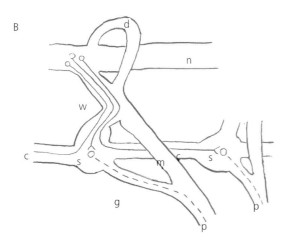

Figure 11.8 Relationships between spinal autonomic outflow nerve fibres and sympathetic ganglia. A. elasmobranchs. B. teleosts. c = sympathetic chain; d = dorsal root of spinal nerve; g = grey ramus communicans; m = spinal nerve; n = spinal cord; p = postglangionic fibre (to skin, melanophores, vasculature); u = suprarenal gland; s = sympathetic ganglion; w = white ramus communicans. Stippled circles represent chromaffin cells. Complete lines are pre-ganglionic neurons, broken lines are post-ganglionic neurons (based on accounts by Young 1931a, 1933a, 1981; Nilsson and Holmgren 1993).

rami communicantes carrying post-ganglionic fibres to nerves V, VII, IX and X, pre-ganglionic fibres routes being through white *rami communicantes* of spinal autonomic outflows (Nilsson and Holmgren 1993). Sympathetic fibres from the chain form anterior and posterior splanchnic nerves (Burnstock 1969). The posterior splanchnic nerve originates directly from a posterior sympathetic ganglion, while the anterior splanchnic nerve originates from

the large coeliac ganglion of the chain (Nilsson and Holmgren 1993).

The ANS of 'lower' vertebrates may now be regarded as having three major components: spinal autonomic (sympathetic), cranial autonomic (parasympathetic) and enteric.

11.9.1 The spinal autonomic system

Cyclostomes do not have sympathetic chains, but spinal autonomic outflows occur in ventral spinal roots in the myxiniformes and in dorsal and ventral roots in the petromyzontiforms. There are no segmental sympathetic ganglia but there are diffuse neuronal perikarya associated with some of the blood vessels which may be primitive sympathetic ganglia (Burnstock 1969). In elasmobranchs, continuous sympathetic chains are absent; segmental sympathetic ganglia are present, but not in the head. These receive pre-ganglionic fibres from spinal nerves through white *rami communicantes* (Figure 11.8) but there are no grey *rami communicantes* (Young 1933a), which are found in teleosts and tetrapods. Connections between adjacent ganglia are irregular, as also are transverse commissures between them on opposite sides of the fish. The ganglia are associated with clusters of chromaffin cells known as suprarenal glands or tissue (Young 1933a) which are equivalent to the medulla of the mammalian adrenal gland. Elasmobranch post-ganglionic fibres follow segmental blood vessels or are included in the splanchnic nerves to the viscera (Nilsson 1983).

11.9.2 The cranial autonomic system

Cranial autonomic outflows, which could be considered as a stage of parasympathetic development, have been identified in some or all of cranial nerves III, VII, IX and X in the major groups of fish. Those in the vagus (X) occur in cyclostomes, elasmobranchs and teleosts, innervating the head except in myxiniformes (Nilsson 1983). They also innervate the gut in elasmobranchs and teleosts, the gall bladder in myxinoformes and the swim bladder in teleosts. Since spinal autonomic fibres are also included in the teleost vagus it is also known as the vago-sympathetic nerve trunk (Young 1931a;

Nilsson 1983). Pre-ganglionic fibres occur in the oculomotor nerve of elasmobranchs and teleosts, synapsing in an anterior ciliary ganglion with post-ganglionic fibres innervating the eyeball. Cranial outflows in the facial (VII) and glossopharyngeal (IX) nerves have been identified in cyclostomes and elasmobranchs but their roles are not clear. The apparent evolutionary dichotomy in the extent of parasympathetic development in fish and tetrapods can be interpreted in terms of different energy requirements for their levels of homeostasis, as well as for the degree of support from the surrounding medium and ease of movement, in aquatic and terrestrial habitats.

11.9.3 The enteric autonomic system

This system controls gut motility via smooth muscle contraction and relaxation, and may evoke glandular secretions. The vertebrate enteric system consists of the myenteric (Auerbach's) plexus, between the longitudinal and circular layers of gut smooth muscle (4.4.6 and Figure 4.12), and a submucous (Meissner's) plexus in the connective tissue beneath the gut epithelium (Bell et al. 1957). Both of these intrinsic plexi are present in the enteric system of elasmobranchs and teleosts (Nilsson and Holmgren 1993), with the addition of a 'deep muscle plexus' within the circular muscle (hence also intrinsic) close to the submucosa of a generalized vertebrate (Olsson and Holmgren 2001). However, Burnstock (1959), working with brown trout, shows a somewhat different interpretation of the deeper plexi; he portrays a submucous plexus close to the deep circular muscle as well as a subepithelial plexus. This system links to the CNS through the extrinsic vagus (cranial X) and splanchnic nerves. In most fish, the myenteric plexus has perikarya and nerve terminals; the submucosal plexus lacks perikarya ('unlike the Meissner's plexus of mammals', Burnstock 1959) and is more sparsely innervated (Domeneghini et al. 2000).

Domeneghini et al. (1999, 2000) regard the gut as two 'neural compartments' in eel and sturgeon, with cholinergic innervation in the myenteric plexus from oesophagus to distal intestine, and adrenergic innervation only in the myenteric plexus of the intestine.

In fish, the possible roles of sensory neurons, extrinsic and intrinsic stimuli in reflexes, and of local stimulation in gastrointestinal activity, are not well understood, but Grove and Holmgren (1992a, b) reported experiments involving extrinsic nerve section and internal mechanical stimuli.

The enteric system is regarded as the most primitive component of the ANS, being present in non-vertebrate chordates lacking the other autonomic components (Gibbins 1994).

11.9.4 Neurotransmitters and receptors in the autonomic nervous system

Although the classical model for autonomic neurotransmitters in vertebrates is based on acetylcholine for pre-ganglionic neurons as well as cranial autonomic (parasympathetic) post-ganglionic neurons and on noradrenalin/adrenalin for post-ganglionic spinal autonomic (sympathetic) neurons (Figure 11.7), it is now recognized that the purine ATP and one or more neuropeptides may also be present with the classical neurotransmitters (Furness and Costa 1987; Gibbins 1989). Fish post-ganglionic adrenergic neurons usually contain a mixture of adrenalin and noradrenalin but it is not clear how extensively they co-function with neuropeptides. Acetylcholine can be identified by the presence of the enzyme choline acetyltransferase (ChAT), which has been identified in the heart, swim bladder and spleen in some teleosts and may be associated with their cranial autonomic post-ganglionic innervation. It has been suggested (Domeneghini et al. 2000) that in the enteric nervous system of eel the adrenergic component uses nitric oxide, substance P and calcitonin gene-related peptide as neuromodulators. Also, the release of these is possibly modulated in turn by the cholinergic component. Fish intestinal and gill innervations can also include 5-hydroxytryptamine (serotonin). The neuropeptides bombesin, neuropeptide Y, tachykinins and vasoactive intestinal peptide (van Noorden and Patent 1980; Nilsson and Holmgren 1993; Donald 1998) have also been identified in autonomically innervated tissues in the gut and pancreas. More information is needed about the roles of the different neurotransmitters in the enteric division of the fish autonomic nervous system.

Chromaffin cells, found either in sympathetic ganglia or as chromaffin tissue (principally in the head-kidney of teleosts and suprarenal glands of elasmobranchs, 10.11.3), have a high content of catecholamines, and may contain 5-hydroxytryptamine. The biology of these cells in fish is reviewed by Nilsson and Holmgren (1993), Sauter (1994) and Donald (1998). Chromaffin cells release their catecholamines locally or to the circulation, in response to pre-ganglionic cholinergic nerves, often closely associated with these cells (Nilsson et al. 1976). They may, by their release to the circulation, have an important role in cardiovascular regulation in those fish without the direct adrenergic innervations of the heart found in many teleosts. The term chromaffin originates from the histological application of dichromates (10.11.3).

Cell receptors are molecules with the sole function of being a cellular recognition site for a specific neurotransmitter or hormone which can bind to the site, thereby activating it. In the autonomic nervous control an adenyl cyclase component on effector adrenoceptors is activated specifically by catecholamines leading to a reaction sequence producing the effector response. Both the activation of intracellular secondary messengers and electrical changes in the excitable membranes of smooth and cardiac muscle cells can be involved in the response.

11.10 Individual effector control systems

A recent review of the comparative physiology of the ANS (Holmgren and Olsson 2011a) includes information on the effector control systems including those of the eye, the gut and the circulation.

The internal organs and systems controlled by the ANS include structures operating via muscle (smooth or cardiac), glands or chromatophores. In some circumstances antagonistic control occurs. The control of organs like the pancreas and rectal gland may occur directly (through innervations) or indirectly through catecholamines from chromaffin tissue or other humoral agents.

In fish, a range of different antagonistic autonomic control systems exists which produce an initial effector response and its reversal. Such antagonism can result from adrenergic and cholinergic neurons or from cholinergic neurons and chromaffin cells. Alternatively, antagonism can involve different adrenoceptor types, mediating opposite responses to either low or high catecholamine concentration on target cell plasma membranes (Wahlquist and Nilsson 1981; Mayo and Burton 1998a; Burton 2008). Also, purinergic inhibition of catecholamine action by an ATP neurotransmitter or cotransmitter has been proposed in some cases (Furness and Costa 1987; Gibbins 1989; Burnstock 2007). Compared with somatic peripheral regulation there is greater diversity in autonomic regulatory processes which require the following individual descriptions.

11.10.1 The gut

ANS control of gut function (movement and secretion) operates through the intrinsic enteric nervous system within the gut wall and extrinsic innervation via the vagus and splanchnic nerves.

Cholinergic innervations from the vagus nerve and adrenergic innervations from the splanchnic nerve(s) provide extrinsic control to the gut muscle. Contractions of the stomach, but not of the intestine, are evoked by electrical stimulation of the vagus of the angler fish *Lophius piscatorius* (Young 1936). There is interspecific variability in the response to stimulating the vagus and splanchnic innervations of the intestine (Nilsson and Holmgren 1993). As extrinsic control of the gut is ultimately attained through stimulating the intrinsic enteric system, with its own range of neurotransmitters and receptors, this could contribute to the gut's variable responsiveness to electrical stimulation in different regions and between species. Interspecific differences in gut muscle physiology and hormonal sensitivity would add to such variability. Also, these neural systems are involved in regulating the gastrointestinal blood flow. Based on surgical and pharmacological experiments on rainbow trout (*Oncorhynchus mykiss*), Seth and Axelsson (2010) concluded that the extrinsic vagus and splanchnic innervation is important in regulating gastrointestinal blood flow under normal conditions and the intrinsic enteric component more important in controlling increased blood flow after a meal. Olsson and Holmgren (2001) have reviewed autonomic control of gut motility in vertebrates including fish.

11.10.2 Spleen, pancreas and rectal gland

These three structures may be innervated by the ANS, but in the case of lungfish spleen the presence of direct sympathetic innervations is doubtful (Abrahamsson et al. 1979). The fish spleen (5.14) is expected to have a contractile mechanism which would add blood cells to the circulation as with mammals, but the contractions may be evoked under stress and a humoral response when catecholamines are released to the circulation. The pancreas of fish (4.4.7), which may occur as diffuse tissues or concentrated as Brockmann bodies (van Noorden and Patent 1980), may be innervated via the vagus and splanchnic nerves, with the individual fibres possibly producing vasoactive intestinal peptide (VIP) as a neurotransmitter. The rectal gland, present in elasmobranchs and the coelacanth (8.4.4), is also associated with VIP activity and other humoral agents (Holmgren and Olsson 2011b) with some autonomic input.

11.10.3 Urinogenital system

Relatively little is known about the autonomic innervation of the gonads, kidney and bladder of fish. Such work as there is has been reviewed by Nilsson and Holmgren (1993) and Donald (1998). The posterior splanchnic nerve from posterior sympathetic ganglia innervates the bladder and gonads. Adrenergic innervations of kidney preglomerular arterioles has also been reported (Elger et al. 1984) in rainbow trout (*Salmo gairdneri*). There is also evidence of non-adrenergic neurotransmitters such as acetyl choline, 5-hydroxytryptamine, nitric oxide and neuropeptides. Nilsson and Holmgren (1993) describe a ganglion on the ureter in *Gadus* and suggest this may be an indication of a primitive sacral parasympathetic component. There has been a debate over whether the nerve to the ureter, gonad and urinary bladder may represent equivalence to a mammalian sacral parasympathetic component. However, Young (1936, 1981) did not find anything in *Uranoscopus* corresponding to a pelvic nerve/sacral parasympathetic outflow in terapods.

11.10.4 Swim bladder

Teleosts which are physoclystous (with a closed swim bladder) have an antagonistic cholinergic

and adrenergic regulation of gas secretion and resorption respectively. The secretion of gas, mainly oxygen or nitrogen in deep-sea fish around 2000–3000 m (Hughes 1963), by increasing the activity of gas gland cells, thereby inflating the swim bladder, is controlled by cranial autonomic cholinergic fibres in the vagus nerve. A spinal autonomic pathway synapsing with anterior splanchnic post-ganglionic adrenergic neurons, regulates swim-bladder blood flow and resorption of gas. There is evidence (Nilsson and Holmgren 1993; Donald 1998) that swim-bladder intrinsic neurons also include neuropeptide- and nitric-oxide releasing fibres.

11.10.5 The cardiovascular system

The autonomic control of fish cardiovascular systems shows considerable interspecific diversity, from its presumed absence in hagfish to an antagonistic control described in some teleosts (Sandblom and Axelsson 2011).

Heart rate can be influenced by the ANS. There is a vagal inhibitory cholinergic innervation of the heart, except that it is excitatory in the petromyzontiformes. In teleosts, except pleuronectids (reviewed by Farrell 1993; Donald 1998), excitatory neural control of the heart is provided by spinal autonomic pre-ganglionic neurons synapsing in the sympathetic chain with adrenergic post-ganglionic neurons of the vagosympathetic trunk and/or anterior spinal nerves (Holmgren 1977; Nilsson 1983). Mann et al. (2010) have used pharmacological and surgical methods and digital videomicroscopical monitoring of cardiac activity to study cardiac rhythm in unanaesthetised larval zebrafish (*Danio rerio*), immobilized in agarose to avoid effects of anaesthesia on autonomic function. They concluded that heart responses of these fish to repeated mild electrical stimulation were physiologically similar to those in mammals to sudden startle stimulation, with a transient bradycardia (slowing of heart beat) under vagal control followed by tachycardia (increased heart beat) under sympathetic control. The cholinergic inhibitory, and adrenergic excitatory, post-ganglionic innervations can modulate the intrinsic rhythm of the cardiac pacemaker which is a ring of specialized myocardial cells of the sinus venosus (reviewed by Farrell 1993; Olson 1998). In the myxiniformes, intrinsic cardiac chromaffin cells have a regulatory role in heart function (Bloom et al. 1961; Axelsson et al. 1990), although these cells are not associated with an extrinsic innervation.

Vasoconstriction and vasodilation (vasomotor control) of blood vessels may be achieved by the ANS including the indirect action of chromaffin cells.

Blood flow to and from the gills via the afferent and efferent arteries may be controlled autonomically but the autonomic branchial vasomotor regulation in elasmobranchs is a matter of conjecture (Metcalfe and Butler 1984). In *Scyliorhinus canicula*, branchial vasculature control may be through catecholamines released from chromaffin cells (Davies and Rankin 1973). In teleosts, the spinal autonomic outflow is involved in regulating vasoconstriction within the branchial vasculature and viscera.

In cod, *Gadus morhua*, the anterior and posterior splanchnic nerves include spinal autonomic vasomotor fibres (Nilsson 1976; Uematsu et al. 1989). In pharmacological experiments on this species, the β-adrenoceptor agonist isoproterenol constricts the vasculature at high concentrations but causes its dilation at lower concentrations (Wahlquist and Nilsson 1981). Thus, antagonistic control may be achievable through one adrenergic pathway, two sets of innervation (adrenergic and cholinergic) not being necessary.

The fairly recent disclosure of nitric oxide as a local factor affecting vascular diameter requires its synthesis and local release presumably under ANS control.

11.10.6 Skin colour change

Some fish can display colour changes due to the activity of chromatophores (2.9) controlled by hormones and, in teleosts, the autonomic nervous system (reviewed by Burton 2011). There is considerable interspecific variation in the degree of neural control of the melanophores (black or brown chromatophores). Von Frisch (1911) demonstrated by electrical stimulation and surgery the existence of a paling centre in the medulla oblongata and a spinal pigmentomotor tract in the European minnow *Phoxinus phoxinus*. Stimulation of this paling centre aggregates melanophore pigments causing paling of the skin. The spinal pigmentomotor tract from the medulla oblongata is located dorso-medially, which includes

the corpora gelatinosa Rolandi (Figure 11.6) and dorsal area of the corpus commune posterius (Burton 1964a, b) both surrounded by fine nerve fibres, which appears to support the view (Kappers et al. 1936) that teleost visceral efferent fibres may originate at the base of the dorsal horns. There is localized spinal outflow into the sympathetic chain and postganglionic fibres reach a dermal plexus via the spinal nerves (von Frisch 1911, reviewed by Grove 1994 and Burton 2011). Dermal melanophores generally occur close to this adrenergic neural plexus. There is evidence (Whitear 1952; Gray 1955, 1956; Burton 1966, 1969) that neurotransmitter release occurs along the nerve endings rather than at neuro-effector junctions.

In some species there may also be epidermal melanophores, their control depending on the 'field effect' of neurotransmitter diffusion from the dermal plexus (Mayo and Burton 1998b; Burton 2010). Antagonistic adrenoceptors on both epidermal and dermal melanophores mediate graded intracellular pigment migrations and of the resultant skin colour rather than all-or-none changes (2.9). Iwata and Fukuda (1973) have proposed a hierarchy of centres in the brain exciting and inhibiting the paling centre in the medulla of crucian carp (*Carassius carassius*).

11.10.7 Pupillary size changes

The control of the pupillary diameter in fish has interested researchers into the ANS because of the antagonistic sympathetic and parasympathetic neurons involved in the mammalian iris regulation. In mammals radial muscle in the iris contracts to dilate the pupil in response to stimulation of the sympathetic, adrenergic neurons; iris sphincter (circular) muscle fibres contract to close the pupil under control of parasympathetic neurons in the oculomotor nerve (Table 11.3).

Alternatively to control of pupil size, light incident on the retinal photoreceptors in teleosts may be regulated by the 'photomechanical' pigment movement of intrinsic cells (12.3.11) and a few teleost show pupillary size changes. Amongst teleosts, the pupil of isolated eyes of eels closes in light and opens in darkness (Young 1981), the iris sphincter muscle responding directly to light. Another teleost, the Mediterranean stargazer (*Uranoscopus scaber*), has antagonistic neural control of the iris sphincter and dilator muscles (11.4), but this control is opposite to that in mammals. Thus, adrenergic spinal autonomic activity closes the pupil by contraction of the iris sphincter muscle; cholinergic oculomotor activity evokes dilation by contraction of the radial muscle of the iris (Young 1931b, 1933c, 1981).

The elasmobranch (selachian) iris is directly sensitive to light intensity. Although contraction of the selachian iris (via the sphincter) is not under nervous control, dilation (contraction of the dilator pupillae) is, via cholinergic fibres from the oculomotor nerve (Young 1933b; Neuhuber and Schrödl 2011).

11.11 Learned behaviour

In contrast with behaviour which is inherited and relatively fixed (as is colour change), fish may show some variability in behaviour, modified by experience (14.4). Thus fish may be able to 'home' to a natal stream based on early experience of the sensory system, with chemical composition of the environment 'imprinting' and determining future migratory patterns. Adult fish may also learn to respond to, or avoid, particular visual (or other sensory) cues. Captured fish, subsequently released, may avoid the place of capture for months (Beukema 1970). The capacity for memory and modification of behaviour (learning)—cognitive functions—may be associated with the optic lobes' neuropil rather than the cerebrum but recent work links goldfish learning, and modification of the startle response, to the telencephalon in goldfish (Portavella and Vargas 2005; Collins and Waldeck 2006). Recent studies of fish cognition and behaviour are reviewed in Brown et al. (2011). Aspects of fish cognitive behaviour are considered in Chapter 14.

11.12 Overview

The vertebrate nervous system is difficult to comprehend and study. The material in this chapter is, to a large extent, based on extrapolation, as is work on the mammalian brain, which has used rats and even squid axons to achieve partial understanding of function. The recent interest in an additional cranial nerve, necessitating the terminology 'cranial nerve zero' demonstrates that some

of our established assumptions can have uncertain foundations.

Appendix 11.1 Excitable membranes

Like all other animal cells, neurons have a membrane resting potential (i.e. are polarized) in which the interior has a negative voltage between −20 mV and −100 mV (varying between cell types) compared with that of the extracellular fluid. A membrane Na^+/K^+ pump (a membrane molecular system which actively transports sodium and potassium ions against their concentration gradients) contributes to the membrane potential by transporting more Na^+ ions out than K^+ ions inwards.

Excitable cells can 'depolarize' their membrane, changing the charge in such a way that changes are generated and transmitted (conducted) as 'action potentials', also referred to as signals or impulses.

The action potential (Hodgkin and Huxley 1939) is a transient change in the membrane potential of electrically excitable cells such as neurons and muscle fibres, due to the presence of voltage-sensitive (i.e. 'gated') Na^+ and K^+ ionic channels which probably include membrane proteins capable of opening and closing pores by conformational changes. An excitable membrane is stimulated if an applied or intrinsic current reaches a threshold voltage; then the excitable membrane 'depolarizes' (reduction/reversal of membrane potential) due to the opening of the voltage-sensitive Na^+ channels allowing an inflow of Na^+ ions into the cell at that location. The voltage-sensitive K^+ channels respond to the voltage changes after the Na^+ channels do, and open only as the action potential reaches its maximum, producing an outward flow of K^+ ions to return the changed membrane potential back towards its resting value when the K^+ channels close. The Na^+/K^+ pump restores the resting concentrations of the two ions and the resting potential too. Local currents generated by the action potential initiate the opening of the next Na^+ gates conducting the electrical charges along the membrane, normally in the direction away from the point of origin of the initial depolarization because the gates remain non-operational (refractory) for a few milliseconds after their activity.

Thus, neuron signal conduction depends on excitable neuronal cell membranes with a capacity to undergo small changes in transmembrane electrical voltage, produced by ionic currents through gated sodium (Na^+) and potassium (K^+) channels. This effect can be propagated along the processes of the neuron from the point of origin. These action potentials therefore represent the signals which transmit information, notably along axons and dendrons. Stimulus strength is represented by the frequency of action potentials.

Appendix 11.2 Neural control of muscles

Neural control of muscles distinguishes between those muscles (striated, skeletal, voluntary) that are involved in locomotion (3.8.1), cardiac muscle of the heart (5.5.3) and an array of involuntary smooth muscles which control gut movement and ducts or tubes including blood vessels.

All vertebrate muscles depend on the sliding action of actin and myosin (3.8.1) to shorten and then relax back to their resting tension and/or length.

The initiation of contraction depends on a series of events triggered by elements of the nervous system or by an intrinsic rhythmic cycle as occurs with the heart and gut and, to some extent, with ventilation (6.5.10, 11.6.2).

Striated muscle is controlled by fast motor fibres in cranial and spinal nerves whereas smooth muscle action is initiated or modified by small diameter autonomic nerves. Cardiac muscle (5.5.3) may be influenced by sympathetic fibres of the ANS which can increase rate or vagal ANS fibres which may slow activity (11.10.5).

Smooth muscle action has been reviewed by Webb (2003).

Bibliography

Burrows, M. (1984). *Mechanisms of Integration in the Nervous System*. Cambridge: Company of Biologists Ltd.

Katz, B. (1966). *Nerve, Muscle and Synapse*. New York, NY: McGraw Hill.

Kuno, M. (1995). *The Synapse: Function, Plasticity and Neurotrophism*, Oxford: Oxford University Press, Oxford.

Nilsson, S. and Holmgren, S. (1994). *Comparative Physiology and Evolution of the Autonomic Nervous System*. Chur, Switzerland: Harwood.

Wirsig-Weichmann, C. (ed.) (2004). Special issue: Anatomy and function of the nervous terminalis—Part 1. *Microscopy Research and Technique*, 65, 1–32.

CHAPTER 12

Perception and sensation

Sensory cells, organs and systems

12.1 Summary

Most fish have excellent perception of their environment via vision, taste and smell, with other senses including hearing and detection of pressure and position and, for some fish, electrical field detection.

The relative importance of different senses to different fish species has traditional emphasized sight in teleosts and smell in elasmobranchs but this view has not been sustained recently for elasmobranchs, in which sight may be very good.

Vision in fish is achieved with focus dependent on lens movement, backwards in the case of teleosts, forwards in the case of elamobranchs. Pupil size adjustment occurs with some fish. The cornea is not important for focus, it merely transmits light to the retina, generally but not always, through the lens. Judged by various parameters (cone diversity, cone mosaic, colour patterns and recognition) colour vision is probably excellent in many teleosts and may occur in elasmobranchs.

Chemoreception, besides being concentrated anteriorly in the nose and mouth, may also occur at other points on the body surface.

Pressure differences detection occurs through mechanoreceptors, including those organized into the special lateral line system and those in the inner ear which enable balance detection through a bilateral set of semicircular canals. Although no ear is externally visible and there is no 'middle ear', other regions adjacent to the semicircular canals can achieve hearing, being reactive to particular sounds such as grunts and whistles.

The capacity to be aware of changes in self-generated electrical fields or fields associated with muscle or nerve activity of other organisms occurs in some fish species, including elasmobranchs and some teleosts.

Other sensory systems such as magnetoreception, pain and thermal perception are not well understood.

12.2 Introduction

Animals require an ongoing stream of information about their environment, achieved via sensory cells, organs and systems.

In fish the major features of the vertebrate sensory systems are already well developed, although in the case of 'ears' not externally evident. As in tetrapods, the essential components of the system are the primary receptor cells (sensory neurons), for example olfactory bipolar cells, or secondary receptor cells (specialized epithelial cells), such as retinal rods and cones. These have the capacity for transforming (transducing) specific sensory stimuli into receptor potentials—change in receptor cell membrane potential (11.3.3, Appendix 11.1)—which initiates nerve impulses to the brain in neuronal receptor cells or, in the case of epithelial receptor cells, releases synaptic neurotransmitter to modulate sensory neuronal activity. A range of receptors individually sensitive to different stimuli can, with the integrative capacity of special areas of the brain (11.6.2), give impressions of the environment, including visual images, sound or even electrical characteristics of nearby objects. Perception of local environmental characteristics including their changes (e.g. dark/light; warmer/cooler), and the presence or absence of potential mates, prey and predators, depends upon peripherally located ('extero-') sensory receptors which may be aggregated within sense organs. Internal ('intero-') sensory receptors provide feedback information from organs and tissues essential for their control.

Essential Fish Biology: Diversity, Structure and Function. Derek Burton & Margaret Burton.
© Derek Burton & Margaret Burton 2018. Published 2018 by Oxford University Press.
DOI 10.1093/oso/9780198785552.001.0001

Animal sensory systems may be classified according to the modality, or type of stimulus, to which their receptors respond. Photoreceptors (12.3) respond to light. Chemoreceptors (12.4) are stimulated by specific molecules or ions, and include the senses of smell and taste as well as monitoring CO_2 and O_2 levels. Mechanoreceptors (12.5) are responsive to movements or pressures (12.7) and include the senses of balance (12.6), hearing (12.8) and touch and the monitoring of internal muscle tension (12.5.3). Sensory modalities also include temperature (12.11, 14.12) and electrical fields (12.9), detectable by special electroreceptors only (as far as is known) by some fish. Nociceptors are receptors of noxious stimuli such as may cause pain (12.11, 14.11.1) and can be responsive to excessive stimulation of another modality, for instance high temperature or loud sound resulting in tissue damage.

Although receptors respond to a specific sensory type attributes of the nerve impulses transmitted to the central nervous system (CNS, 11.3) do not indicate the stimulus type; this is determined by the area of the CNS receiving the input. Variations in stimulus intensity evoke differences in the frequency of nerve impulses in the sensory neurons. Receptors may be tonic (12.7.2, 12.9.3) with no decline (or a slow one) in nerve impulse frequency, when a stimulus is maintained, or they may be phasic (12.9.3), with a rapid decline ('adaptation'). A sensory stimulus in some systems (12.3.10) may inhibit impulse frequency.

Individual sensory systems within taxa can also show evolutionary adaptations to widely differing habitat conditions associated with the aquatic environment, and in extreme conditions may be enhanced or absent.

12.3 Photoreceptors: vision

Photoreceptors are concentrated in special organs (the pineal gland and the bilateral eyes) which contain one or more light-sensitive visual pigments, but nervous tissue of the CNS and even the skin of cyclostomes can also be sensitive to light.

The photo-environments of fish species represent an extremely wide range of diversity. Individual marine species can be adapted to contrasting light conditions in the photic zonation of the vertical realms in the ocean environment (13.4.2, Table 13.4).

Shallow-water species may actually be in relatively dark conditions if concealed in caves or rock formations, or burrows in silt, mud or sand. Habitats of freshwater species range from shallow, clear water to silt-laden rivers and estuaries, or the complete darkness of some cave systems.

Both the light intensity and wavelength (colour) are influenced by the depth of the water. Intensity can range from the full strength of the tropical sunlight near water surfaces to darkness of caves or darkness illuminated with erratic flashes from luminescent organisms in the ocean depths. Infrared, red and yellow wavelengths are rapidly absorbed during light transmission through water. Thus the long red and yellow wavelengths do not penetrate to the depths reached by blue and green light in 'pure' fresh water and salt water. However, the transmission depth of the individual wavelengths especially short wavelengths (i.e. blue), can be influenced considerably by the turbidity of the water; the penetration of blue and green wavelengths being truncated in coastal and estuarine waters. The apparent colour of an object perceived by a fish will be affected by the range of wavelengths illuminating the object and reflected by it. This may be quite different than its colour when viewed by a human in air.

12.3.1 The pineal

The pineal organ or gland which is situated on the mid-dorsal brain above the diencephalon (11.6.2, Figure 11.5) has been regarded as a third eye in some vertebrates. According to Fenwick (1970) there are two prevalent ideas about it as a photoreceptor system in fish; one dismissing the pineal as vestigial, while the other concentrates on its current functions, which are still not clear.

The pinealocytes (photoreceptors) of the pineal gland possess visual pigments somewhat different in characteristics (i.e. 'spectrally distinct') from the optic visual pigments in the same species (Bowmaker and Wagner 2004). It is not expected that the pineal of modern fish is image-forming, even though some pineal organs (e.g. in cyclostomes) have a lens-like structure above them (Fenwick 1970). The pinealocytes are fairly similar to the rods and cones of the optic retina (Figure 12.2) possessing an outer segment and cell body/perikaryon

(Fujii and Oshima 1986; Ekström and Meissl 1997). The outer segment has lamellae containing visual pigment (Ekström and Meissl 1997) as in rods and cones of the bilateral eyes (Figure 12.2).

The pineal is undoubtedly capable in fish of detecting light intensity; in one cavefish species, young fish (larvae) can detect shadows via the pineal (Yoshizawa and Jeffery 2008). It is possible that the most significant role of the modern fish pineal is the response to seasonal/diurnal light levels with release of the hormone melatonin in dark conditions (10.11.2, 11.6.2).

12.3.2 The eyes

As in vertebrates, generally fish eyes consist of an optical system which projects an image of the field of view onto the photoreceptors of a retina. Many fish, notably teleosts in shallow water, show complex colour change (2.9, Appendices 2.2, 2.3) and responses to colour and pattern of conspecifics, strongly indicating good or excellent colour vision, achieved by a complex array of different photoreceptors.

The bilateral eyes of elasmobranchs and teleosts are comparable in basic organization, which is like that of other vertebrates. Amongst cyclostomes (extant Agnatha) the eyes of lampreys are similar to other vertebrates but they also possess skin photoreceptors. Hagfish eyes appear to be degenerate, being embedded in adipose tissue and without a cornea, lens or iris (Hawryshyn 1998); their skin photoreceptors can detect changes in light intensity.

12.3.3 The eyeball optical system: biocamera

The eyeball optical system can be considered as a 'biocamera', in which the eyeballs can be moved to follow events in the environment, and light passes through a cornea via the pupil of the iris, with focus on the retina generally dependent on a moveable lens behind the iris.

12.3.4 Eyeball movement

The fish eyeball is moved within its orbit by six extrinsic muscles, attached to the sclera (Figure 12.1), a gnathostome characteristic. These muscles are the inferior rectus, anterior rectus, superior rectus, posterior rectus, inferior oblique and superior oblique. The activity of these muscles is coordinated through cranial nerves III, IV and VI (Table 11.3) to produce synchronous movements in both orbits, but in some species, as in *Blennius* and the Australian sandlance (*Limnichthyes fasciatus*), there can be independent eye movements.

12.3.5 Eyeball structure

The eyeball (Figure 12.1) is enclosed in the tough, opaque, flexible sclera continuous with an anterior transparent cornea. The sclera can include encircling cartilage which, together with the internal ocular pressure, resists changes in the shape of the eyeball. The sclera covers other layers (e.g. argentea, choroid and the 'pigmented epithelium') which absorb stray light, or reflect it (as a tapetum,

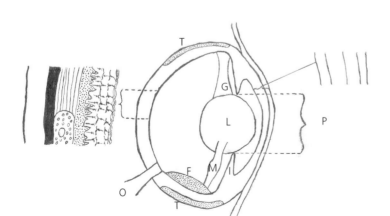

Figure 12.1 Vertical section of a generalized teleost eye (centre), with (left) layers of eyeball wall and retina, and (right) layers of the cornea. Main diagram (eyeball): F = falciform process; G = suspensory ligament; I = iris; L = lens; O = optic tract; P = pupil; M = retractor muscle; T = sclerotic cartilage. Left diagram, layers from left to right: sclera, argentum, choroid, pigmented epithelium, retina, nerve fibres. Right diagram, layers from left to right: cornea consists of autochthonous, sclera and dermal layers, conjunctiva. Based on Walls (1942), Brett (1957) and Young (1981).

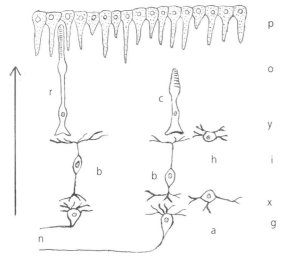

Figure 12.2 The pigmented epithelium and representative cells in the layers of the fish retina. a = amacrine cell; b = bipolar cell; c = cone; g = ganglion cell layer; h = horizontal cell; i = inner nuclear layer; n = optic tract; o = outer nuclear layer; p = pigmented epithelium; r = rod; x = inner plexiform layer; y = outer plexiform layer. Arrow = direction of light. Melanin migration occurs within the long processes of the pigmented epithelial cells in species lacking pupillary control of light intensity at the retina. Illumination extends the myoid in the rods but contracts it in the cones (12.3.11). For further details see Franz (1913), Brett (1957), Kandel et al. (1991) and Hawryshyn (1998).

12.3.12) depending on incident light intensity. The choroid nourishes the retina and contains blood vessels; the ventral falciform process (Figure 12.1) may also have a nutritive role (Young 1981). The choroid in most teleosts, but not in elasmobranchs, includes a *rete mirabile* (5.10) releasing oxygen to supply the retinal tissues (Wittenberg and Haedrich 1974). The photoreceptors occur within the retina (12.3.9, 12.3.10, 12.3.11) which lines the back and sides of the hollow eyeball. Most of the interior of the eye is filled with fluid; in mammals this is the 'aqueous humor' between the cornea and lens and a more viscous 'vitreous humor' between the lens and retina, terms not universally used for fish eyes.

The typical fish cornea (Figure 12.1) does not have high refractive power with a focusing role, unlike that of mammals, because it is covered by water and not air. The fish lens consists of very dense proteins and has a high refractive power and it alone is usually responsible for focusing.

12.3.6 The iris and pupil

The iris is a continuation of the choroid and has a central opening or pupil which regulates light entry into the eye, like a camera aperture. There is little capacity for change in pupil size in most teleosts but the iris of elasmobranchs is directly sensitive to light intensity (11.10.7) with reduction in pupil size in bright light.

In fish, pupil shape can produce an 'aphakic space' (with light bypassing the lens) which can be either localized between the lens and iris or under the lens (termed circumlental), the pupil being wider than the lens (Walls 1942; Munk and Frederiksen 1974; Sivak 1978). The significance of the aphakic space is not yet clear; Sivak (1978) suggests that it facilitates accommodation (i.e. lens movements), but other suggestions have included increased light gathering capacity in deep-sea fish or extension of the binocular field.

Those elasmobranchs that inhabit shallower waters display more extensive dilation and constriction of their pupils than deeper species. Elasmobranch pupil size is related to the amount of light entering the eye (Murphy and Howland 1991; Warrant and Locket 2004; Lisney and Collin 2007). The piked dogfish (*Squalus acanthias*) moves between shallow inshore water and deeper continental shelves to 1460 m, deeper than some 'deepwater' species (Compagno et al. 2005). In this shark, the pupil area as a percentage of eye area increases by 12.4 per cent in darkness compared with light (Kajiura 2010). In elasmobranchs, constriction depends on contraction of the sphincter iridis muscle in direct response to light (11.10.7) and it possesses a light-sensitive pigment (Steinach 1890 reviewed by Kuchnow 1971).

There is interspecific variation in the extent of pupillary responsiveness in teleosts (11.10.7). Muscle fibres have been identified in the iris of some teleosts (Steinach 1890; Grynfellt 1909 reviewed by Kuchnow 1971). In most of the teleost eyes that have been studied, for example trout, the pupil is round and little or no change in size has been observed (Bateson 1889, 1890; Young 1981), the quantity of light reaching the photoreceptors being controlled by pigment photomechanical movements round the actual photoreceptors (12.3.11). Extensive pupillary changes in diameter (11.10.7) have been recorded in *Uranoscopus scaber* and *Lophius piscatorius* (Beer 1894; Young 1931b, 1933c) and in the pearlfish, *Encheliophis jordani* (Walls 1942), and plainfin midshipman, *Porichthys notatus* (Douglas et al. 1998a). Autonomic innervation and control of pupil size (11.10.7) have been observed in some teleosts (Nilsson 1980; Rubin and Nolte 1981). Direct iris responsiveness to light occurs in other species, such as eels (11.10.7). It has been suggested that the sphincter photosensitive pigment involved may be the same as the retinal visual pigment since maximum values for the iris sphincter muscle action spectrum for contraction and the rhodopsin (a retinal visual pigment) absorption maximum in eel are correlated (Seliger 1962).

Partial covering of the pupil by a 'semicircular flap' from the upper edge of the iris, in daylight and its retraction at night occurs in turbot and brill (Bateson 1889, 1890). This contrasts with the occurrence of protective adnexa or eyelids in some elasmobranchs (Hueter et al. 2001).

12.3.7 Image formation

Image formation depends on light being focused on the retina by the lens and/or cornea. The quality of the visual image perceived by the retina will be affected by the ability of the eye to focus (accommodation) on near and distant objects. In tetrapods, accommodation depends on changes in lens shape but, apparently because of the higher density of their lens, fish may focus by changing the distance of the lens from the retina. In elasmobranchs, the lens can be moved forward by contraction of a *protractor lentis* muscle to focus on distant objects; in teleosts, accommodation involves contraction of a *retractor lentis* muscle (Figure 12.1) pulling the lens closer to the retina (Romer 1955). Elasmobranchs

may close their pupils to produce 'pinhole' apertures to form sharp retinal images of both near and distant objects, a stenopeic condition enhancing accommodation. Some teleosts, like *Anguilla*, may have small lens muscles and little or no accommodation. As fish grow the amplitude of accommodation movements increases (Fernald 1991).

Specializations in the fish cornea include 'spectacles', derived from a thickened dermal layer. In lampreys, flattening of the cornea by contraction of a *cornealis* muscle attached to the outer cornea or spectacle (Walls 1942; Schwab 2012) results in lens movement and the eyeball can also lengthen to facilitate focusing on the retina.

12.3.8 Optical system adaptations to habitat

Diurnal fish in shallow habitats (e.g. cichlids, Muntz 1975) may have yellow corneas (due to carotenoids, Kondrashev et al. 1986) and/or yellow lenses. Theories about the role of these yellow filters (Fernald 1993; Douglas et al. 1998b) include the reduction of short wavelengths resulting in less chromatic aberration and 'glare', and countering bioluminescent camouflage of animals above them against the daylight (Herring 2002). A deep-sea fish *Malacosteus niger* residing at 400–1000 m has a bright yellow lens (13.4.3) which may improve vision in the dim light, mainly long wavelength bioluminescence, of this deep environment, by filtering out short wavelengths (Somiya 1982; Herring 2002). In some species xanthophores, with perikarya at the edge of the cornea and processes which extend over it, can respond to change it from yellow in light to colourless in dark and may protect the retina from bright light (Kondrashev et al. 1986).

Fish inhabiting the deeper waters have a variety of optical system adaptations (Collin and Partridge 1996; Herring 2002; Warrant and Locket 2004) to the low illumination levels, additional to retinal specializations (12.3.12). Many deep-sea fish such as *Bathylagus*, *Zenion* and *Epigonus* have very large eyes, enlarged pupils and large lenses. The increased eye size in eleven families of fish (Marshall 1971) is accompanied by the development of a tubular shape, as in *Opisthoproctus* and *Scopelarchus*, facing upwards towards the brighter surface water, against which will be silhouettes of prey. At the upper end of the tubular eye is a large-aperture lens and a small area of retina at the lower end on which the image is focused. An image is also formed on an accessory retina in a diverticulum on the side of the tubular eye which extends the visual field in other directions (Locket 1977). In the spook-fish, *Dolichopteryx longipes*, the diverticulum, which faces ventrally, includes reflective plates (derived from a tapetum, 12.3.5, 12.3.12) orientated at varying angles to form a concave mirror reflecting light to form a focussed image on the lateral diverticular retina (Wagner et al. 2009). Accommodation is either absent or restricted. Deep-sea *Ipnops* lacks a lens; the eyes are large dorsal structures ('cephalic organs') with a flat retina containing rods, beneath transparent layers including cranial elements and the cornea (Munk 1959; Young 1981).

Deep-sea gadid species, from depths of 250–4000 m, display structural specializations in the cornea which may be associated with the high pressure of the water column, low illumination and low temperature. Included are structural alignments of the corneal dermis endoplasmic reticulum in the shallower water species and a thicker dermis in species at greater depths (Collin and Collin 1998). They also suggest that iridescence by the corneal structure in these fish suppresses reflectance to camouflage and thus protect the eyes from predators.

In complete contrast to the optical system adaptation to deep-sea habitats are the adaptations to aerial conditions in some fish species. In air, the curved surface of the cornea has a greater refractory power than the lens. The four-eyed top minnow *Anableps anableps* from South and Central America, when swimming at the water surface has the dorsal part of the eyes above water and the ventral part submerged. The eye has a dorsoventrally elongated lens and its short axis, together with the curved upper cornea, forms an aerial image on the ventral surface of the retina. Aquatic images are formed on the dorsal surface of the retina after transmission of the light through the elongated axis of the lens. This dual optical system can effectively provide simultaneous aerial and aquatic vision. Mud skippers (*Periopthalmus*) which frequently lie on exposed tidal banks have raised eyes and corneas with pronounced curvature, facilitating good retinal focus in air.

The cornea in the Australian sandlance, *Limnichthyes fasciatus*, is unusually thick for fish and has a high refractive index as well as a capacity for lens-like focusing (Pettigrew and Collin 1995; Pettigrew et al. 1999).

12.3.9 The retina and its neuronal circuitry

The retina is a photosensitive neuroepithelium with a neuronal organization in fish (Figure 12.2) similar to other vertebrates and which transforms the projected image into neural activity. It is multilayered, the terminology for the layers reflects their position or their composition with perikarya of the bipolar cells forming the inner nuclear layer and the photoreceptors (rods and cones, 12.3.10) the outer nuclear layer. The photoreceptors (outer nuclear layer) are distally located in the retina, abutting a pigmented epithelium. The retina is hence an 'inverted' retina and light passes through the neuron layer to reach it.

From the photoreceptor cells impulses are conducted vertically by bipolar cells (inner nuclear layer), neurons which synapse with rods or cones (photoreceptors for 'dark' or 'light' conditions respectively), and ganglion cells (ganglion cell layer) with axons to the optic tract (Kandel et al. 1991). Conduction also occurs laterally through horizontal cells (outer plexiform layer) which synapse with the rods and cones, and by amacrine cells (inner plexiform layer) which synapse with the axons of the bipolar cells and with ganglion cell dendrites. The amacrine cells represent part of a neuropil in the retina (Young 1957). The two layers of horizontal cells will increase the complexity and relative importance of retinal neural processing in visual acuity.

Overall the retinal neuronal circuitry is one of the determiners of visual sensitivity and its regulatory transmitters include melatonin and dopamine (Li and Maaswinkel 2007) which can function as neuromodulators. Dopamine concentrations are high when those for melatonin are low and vice versa and involve circadian rhythm activity (cyclical endogenous changes over a 24-hour period). Dopamine apparently regulates changes in sensitivity between rod and cone dominance which displays a circadian rhythm. It has been demonstrated (Korenbrot and Fernald 1989) that in the cichlid *Haplochromis burtoni* there is both circadian and photic determination of the level of expression of rod opsin mRNA.

The terminal nerve (11.8.1) modulates retinal neurotransmission (Li and Maaswinkel 2007). Electroretinogram recordings demonstrate that amino acid stimulation of teleost nostrils increases visual sensitivity to light and decreases the threshold of retinal ganglion cells. Also retinal dopamine release is reduced by olfactory stimulation. The terminal nerve has projections to both the fish retina and optic tectum.

Axons from the retinal ganglion cells leave the eye through the optic tract (Figures 12.1, 12.2). The decussation of the optic tracts at the optic chiasma (11.6.2) results in the ganglion cells of each retina projecting on to the superficial layer of the contralateral (opposite side) optic tectum (11.6.2), forming several bands of axon terminals in an ordered pattern or retinotopic map (von Bartheld and Meyer 1987; Li and Maaswinkel 2007). There is also projection of the retinal ganglion cells to form large terminal areas in the thalamus (11.6.2), but the physiology of these centres is not well understood (Schellart 1990).

12.3.10 Retinal photoreceptors: rods and cones

Rods and cones are elongated cells (Figure 12.2) with distinct outer and inner segments which may have a contractile myoid component in teleosts (12.3.11) and a synaptic terminal linking to bipolar cells. Whereas rods have a uniform width in the outer segment cones do not (see Young 1970). The photopigment is located on ordered lamellae in the outer segment where the light transduction process occurs. The lamellae of the rods are internal membrane discs, not continuous with the plasma membrane although formed from it. The cone lamellae remain as infoldings of the plasma membrane. The inner segment of both photoreceptors contains the cell's nucleus and organelles for biosynthesis. Rods are highly sensitive and can detect a single photon. They contain more photopigment than cones and function well under low illumination such as nocturnal conditions; they are not usually associated with colour vision although deep-sea fish may have selective wavelength discrimination using rods. There are different types of cones (e.g. double cones, Allison et al. 2004) with

variation in sensitivity and pigment/opsin content, giving the capacity for diurnal colour vision; they are less sensitive than rods. In teleosts cones can be arranged in mosaics of cone subtypes with different spectral characteristics, an organization which may improve colour detection (Fernald 1993; Raymond and Barthel 2004; Raymond et al. 2014).

Visual pigments

Visual pigments are composed of a chromophore, a carotenoid which is a derivative of vitamin A_1 (11-cis retinal) or vitamin A_2 (3-dehydroretinal), bound to a protein, opsin (Carleton et al. 2000). There are five main types of opsins, a rod opsin and four cone opsins each with peak sensitivity (or maximum absorbance measured by microspectrophotometry of single rods or cones) to different wavelengths. The peak sensitivity of each pigment depends on its unique amino acid sequence (Chang et al. 1995; Yokoyama and Radlwimmer 1998), with specific sites critical for their function. One of the cone opsins is sensitive to ultraviolet as demonstrated in African cichlids (Carleton et al. 2000).

Anableps has eyes 'split' (12.3.8) into two areas: upper (aerially specialized) and lower (aquatically specialized). Middle wavelength-sensitive opsins are prevalent in the lower aquatically specialized retina whereas long wavelength-sensitive opsins are confined to the upper aerially specialized retina (Owens et al. 2011).

The energy transduction begins with absorption of light by the retinal causing an isomer change from an 11-cis to an all-trans form through rotation within the retinal, causing conformational changes in the opsin component (Kandel et al. 1991). This leads to hyperpolarization (increasing the membrane potential) of the plasma membrane, that is, changing the receptor potential in response to light, reducing the release of neurotransmitter from the synaptic terminal of the photoreceptor, to modify postsynaptic neuronal activity. A steady release of neurotransmitter which occurs in darkness is altered in response to light.

12.3.11 Retinal photomechanical activity

Melanin movement in the pigmented epithelium and changes in the length of the myoids of the rods and cones is termed 'photomechanical activity'. It can occur, in some fish, in response to light-level changes and controls the light reaching the photo-sensitive cells (i.e. rods and cones). Raymond et al. (2014) describe this region as glial Müller cells which invest the rod and cone photoreceptors with processes in which pigment movement can occur.

An extensive literature on the photomechanical activity in the fish retina has been reviewed by Walls (1942), Brett (1957) and Fernald (1993). Such movements occur in teleosts which lack control of the pupillary aperture to alter the retinal light intensity (12.3.6). During darkness melanin migrates into the pigment epithelium (Figure 12.2) perikarya, but in light the migration is into their processes. Illumination also results in elongation of the myoids of the rods to move their outer segments in the opposite direction to that of the melanin so that the outer segments are shielded from the light. These movements in the rods reverse in darkness, the myoids contracting. Movements of the outer segment of the cones is counter to that in rods so that they do not become completely shaded by the melanin.

Both neural (Arey 1916) and humoral (Ali 1964) influences have been proposed in the regulation of these retinal photomechanical responses, but Muntz (1971) considered that this control was not understood. More recent recognition of melatonin synthesis in the fish retina by, for example, Falcón and Collin (1991), Zaunreiter et al. (1998) and Besseau et al. (2006), additional to its pineal synthesis, raises new possibilities for the control of photomechanical activity. This suggestion is prompted by pigment movements in fish skin melanophores which can be sensitive to diurnal cycles of pineal melatonin secretion (reviewed by Fujii and Oshima 1986; Fujii 2000; Burton 2010, 2011).

12.3.12 Retinal specializations

The retina of deep-sea fish has specializations related to the low level of illumination (Locket 1977). Cones are absent and rods are longer than usual, lengthening the light path through the rhodopsin and increasing photon capture, which is further enhanced by a high pigmentation density. A mirror-like tapetum (reflecting layer) of guanine or oil droplets in the choroid (Figure 12.1) reflects back light, providing a second opportunity for

its absorption by the photoreceptors (Douglas et al. 1998b). Many deep-sea fish have only a small number of neurons synapsing with the photoreceptor layer, several receptors converging on each neuron thereby increasing transduction countering the low frequency of photon capture. Another specialization is the occurrence of multiple layers or multibanks of rods, increasing the distance travelled by light through the photoreceptors and its absorption.

The deep-sea fish *Malacosteus niger* (12.3.8, 13.4.3) of the family Stomiidae (the dragonfishes) uniquely possesses both long-wave bioluminescence (invisible to other deep-sea animals) to illuminate a small trophic search area, as well as a retinal sensitizer (derived from a bacteriochlorophyll lacking magnesium) allowing long-wave visual sensitivity (Douglas et al. 1998c; Sutton 2005). Long-wave (greater than 700 nm) red bioluminescence is emitted by photophores ventral to the eyes which compares with 460–490 nm, blue, emitted by other photophores in this species and other deep-sea fish. The chlorophyll photosensitizer absorbs longer wavelengths and transfers their associated energy to visual pigments with innate shorter wavelength sensitivity. The presence of the photosensitizer is probably associated with a large number of copepods in the diet (4.4.1 and Sutton 2005).

In some fish the retina has acute zones, with a high neuronal density, known as *areae retinae* (Wagner et al. 1998). An area of this type, but associated with a pit, is a *fovea*. These have not been described in fish universally and usually cones are more abundant in the fish *foveae* described (Munz 1971), as occurs in other vertebrates. A pure-rod *fovea* has, however, been described in several families of deep-sea fish. Such a *fovea* occurs in the slickhead *Baiacalifornia drakei*; it contains up to 28 banks of rods (Locket 1985) which presumably increase the sensitivity of this area of the retina.

12.3.13 Perception of polarized light

Some fish can detect polarized light and may use this capacity for orientation and navigation.

Transmission of light through the atmosphere, water surface reflection and scatter by waterborne particles can all polarize light (Waterman 1954), confining wave motion to a single plane, which can be detected by some animals and used as a cue in navigation. In fish there is evidence that detection of polarized light may be associated with cones carrying ultraviolet-sensitive pigments, as in brown trout *Salmo trutta* (Bowmaker and Kuntz 1987). They are lost during smoltification and seaward migration of Pacific salmonids and regenerated at sexual maturation on return to fresh water for spawning (Beaudet et al. 1997). Rainbow trout can orientate to the plane of polarized light (Hawryshyn et al. 1989), but only if the light field includes ultraviolet radiation. Rainbow trout previously trained to orientate under a polarized light field retain this orientation when subsequently released in a circular tank under polarized light, but this ability is impaired by filtering out the ultraviolet component.

It is not known how many fish species use ultraviolet polarized light for their orientation, but they include goldfish (Bowmaker et al. 1991) as well as salmonids. More recently Flamarique (2011) has demonstrated that in northern anchovy *Engraulis mordax*, retinal ultrastructure reveals some cones with longitudinally oriented outer segment lamellae. These cones are sensitive to the polarization of axially incident light representing a basis in this species for segregated colour and polarization detection channels.

12.4 Chemoreceptors: taste, smell and internal stimuli

The capacity to react to specific chemicals can occur in response to external or internal stimuli, which is important in fish behaviour (14.7).

Internal chemoreception may have a role in regulating blood vascular responses to varying oxygen or carbon dioxide tensions and ventilation (6.5.10) or parameters involved in food digestion or perhaps detoxification, but little is known about this aspect of fish function. External chemoperception is important in feeding, predator avoidance and reproduction vital to fish survival and is considered the most ancient of the sensory systems (Hara 1993). In aquatic habitats the solubility of compounds and their capacity for diffusion in water are determiners of their role as chemical signals. To what extent can fish taste and smell in the aqueous environment be considered distinct sensations? Hasler (1957) answers this question by interpreting odour perception in fish as chemically induced sensations relayed to the brain from the olfactory sac; taste being relayed to the brain from gustatory nerve endings in the mouth, and over the body surface in some species.

Taste (gustation) and smell (olfaction) are well-developed senses in fish; additionally some species have solitary epidermal chemoreceptor cells (Kotrschal 1996). Smell is used in distance sensation, taste is a more local sense. Fish are sensitive to a narrower range of chemical stimuli than are mammals which may be a function of relative solubilities and volatility in respective aquatic and aerial habitats (Sorensen and Caprio 1998). Of 32,000 or more known extant species of fish, chemosensory capacities have been studied in only a few, mainly freshwater, and it is not known how representative these are of the range of such abilities. The possibility of a 'common chemical sense' has been considered as a role of the solitary chemosensory cells (Whitear 1971b). It was first described as a 'relatively high threshold' and general chemosense (Parker 1912 cited by Kotrschral et al. 1993). However, Kotrschal (1991) questions as to whether there is such a sense, suggesting that as our knowledge advances specific components will most likely become recognized. If the concept of a common chemical sense is retained, Kotrschal believes it should refer only to sensations associated with free spinal nerve or trigeminal (cranial nerve V) nerve endings.

12.4.1 Chemosensory cells

Chemoreceptor cells possess cilia or microvilli on their outer (sensory) plasma membrane (Figure 12.3). Electron microscope studies demonstrate that taste cells can have a single club-shaped microvillus or several smaller microvilli; olfactory cells can be ciliated or have microvilli as in other vertebrates (Sorensen and Caprio 1998). Whereas the receptor cells of the taste buds are epithelial the olfactory cells are receptors with the characteristics of bipolar neurons (11.3.2, 12.2 and Figure 12.3).

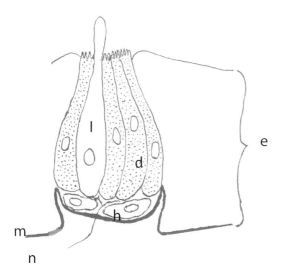

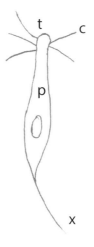

Figure 12.3 Fish chemosensory cells. Left, representative types of cells in a fish tastebud. Right, ciliated olfactory cell. c = cilium; d= dark (electron-dense) taste cell with several apical microvilli. e = epidermis; l = light (less electron-dense) taste cell with single club-shaped microvillus. m = basal membrane; n = gustatory nerve fibres; p=cylindrical dendrite; t = terminal knob; x = axon; b = basal cell. Based on ultrastructure descriptions by Sorensen and Caprio (1998).

The sequence of intracellular events in chemoreceptor cells leading to the generation of the receptor potential is not fully understood. In the case of olfactory sensory cells, available evidence suggests that the signalling (stimulus) chemical binds to G-proteins (Bruch and Teeter 1989; Ngai et al. 1993; Zielinski and Hara 2007) which influences the activity of the secondary messengers cyclic adenosine monophosphate (cAMP) and inositol triphosphate (IP3), causing an inflow of calcium ions through ion channels in the sensory cell plasma membrane, depolarizing it to generate the receptor potential. It is proposed (Brand et al. 1991; Caprio et al. 1993; Zielinski and Hara 2007) that G-proteins, cAMP and IP3 may also have roles in ion channel activation and membrane depolarization in taste receptor cells.

12.4.2 Taste (gustation)

Food discrimination is mediated by gustation; fish taste/gustatory receptors respond to dilute solutions in contrast to these receptors in terrestrial animals (Hara 2007).

12.4.3 Location and neural connections of taste receptor cells

Literature on the distribution of fish taste receptor cells is reviewed by Hasler (1957), Kotrschal

(1991) Whitear (1992) and Hara (1993, 2007). They are specialized epithelial cells synapsing with gustatory nerve fibres from cranial nerve VII, IX or X and in some cases with spinal nerves. Thirty to one hundred taste cells may be grouped into tastebuds (Figure 12.3), elevated as epidermal hillocks or level with the epidermis, or occur as solitary sense cells. They can occur on the skin and barbels additional to the mouth and oesophagus. There may be somatotopy (Finger 1982) in which localized tastebuds correspond with the spatial organization of their axon terminals within the CNS. The tastebuds include at least three types of gustatory cells as seen under the electron microscope: dark (electron dense), light (less electron dense) and basal cells (Figure 12.3). It is not clear whether they are different receptor types but the basal cells may act as neuromodulators rather than as precursors of the other cell types, unlike in mammals (Hara 2007).

12.4.4 Gustatory stimuli

The wide diversity of trophic relationships within foodwebs associated with fish habitats in marine and freshwater environments will have associated interspecific variability in taste discrimination. Training experiments and electrophysiological recording (Marui and Caprio 1992) indicate that fish are able to taste salt, sweet, sour and bitter substances, as well as amino acids, nucleotides, organic acids, ammonium compounds and small peptides. Such compounds are water-soluble, have relatively low molecular weights and occur in living tissues and are released on their breakdown providing an important means to locate food sources. As well as interspecific variability in taste discrimination within this range of compounds the skin and oropharyngeal taste buds, and associated innervation, also vary in responsiveness to specific chemicals and this should be taken into account in experiments on taste stimuli in fish.

It is suggested (Ogawa and Caprio 2010) that in channel catfish (*Ictalurus punctatus*) extraoral tastebuds innervated by cranial nerve VII may be involved in appetite behaviour whilst oropharyngeal tastebuds, innervated by cranial nerves IX and X, may be associated with consuming behaviour, the two being sensitive to different proportions of amino acids. Dorsal fin solitary sensory cells of rocklings are sensitive to heterospecific mucus and 'may be involved in predator avoidance' (Kotrschal 1991).

12.4.5 Smell (olfaction)

Smell in jawed fish is dependent on receptors in an olfactory epithelium lining two olfactory sacs, and connected to the CNS at the olfactory bulbs via the olfactory nerves, and the bulbs are linked to the rest of the CNS by the olfactory tracts (Figure 11.5).

12.4.6 The olfactory system

The olfactory system of fish is sensitive to a wider range of stimuli than the taste receptors. These include food odours, pheromones (2.3.1, 14.7) and 'geographic location' markers, but both smell and taste receptors are sensitive to amino acids and nucleotides which are associated with tissue degradation or disruption. The olfactory system can be involved in homing and navigation (14.7). Agnathans have a single dorsal nostril, all other fish have paired olfactory organs (Zielinski and Hara 2007), each with an olfactory sac possessing an incurrent anterior naris (nostril) and an excurrent posterior naris (Figure 12.4). In extant lungfish, both sets of nares open in the mouth and the anterior nares are supplied with water from a ventral groove on the snout

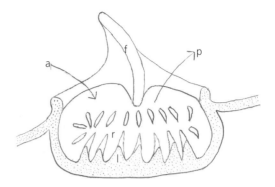

Figure 12.4 Diagram of longitudinal section through an olfactory organ of minnow (*Phoxinus*). a = anterior nares; f = fold of skin; l = lamellae of olfactory rosette; p = posterior nares; r = raphe of olfactory rosette. Arrows indicate direction of current. Based on von Frisch (1941b), cited by Hasler (1957).

(Schultze 1991). The interior location of the lungfish nares has been vigorously debated with reference to any homology with the tetrapod choana (glossary), the recent opinion being that the lungfish openings are not homologous with the tetrapod choana.

The olfactory sacs include an 'olfactory rosette' (Figure 12.4) lined by an olfactory epithelium with the receptor cells on its surface. Usually the rosette is characterized by a raphe, or olfactory ridge, with projecting folds or lamellae, increasing the area of the olfactory epithelium. In most fish, a current of water circulates through the olfactory sac via the nares as a result of swimming and breathing movements and/or cilia (Hasler 1957). Some fish, for example *Gasterosteus* and *Scophthalmus*, possess only a single naris on each side, through which incurrents and excurrents are created by breathing movements compressing the olfactory sac.

The strength of olfactory sensation depends on the degree of folding into lamellae and the density of receptor cells as well as being influenced by the central integrative processes. The extent of folding of the olfactory epithelium has been investigated in relation to the ecological habitat of 10 teleost species (Teichmann 1954 cited by Hara 1993) and the number of lamellae ranged from 2 in stickleback (*Gasterosteus aculeatus*) to 93 in eel (*Anguilla anguilla*) and up to 120 have been recorded more recently in the latter (Hara 1993) and 230 in the barred snapper, *Hoplopagrus guentheri* (Pfeiffer, 1964), the number increasing with the size of the fish. The large number of lamellae in this *Hoplopagrus* species was thought to be related to a nocturnal habit. Fishelson et al. (2010) have compared the morphology of the olfactory epithelium of 12 species of benthic predatory lizard fishes (Synodontidae) from widely separated geographical locations. The interspecific range in number of olfactory lamellae was from 5 to 22; the number increases with the growth of the fish. The length of the lamellae and approximate total number of olfactory receptor neurons (ORN) ranged from 3.5–4.0 mm (50,000 ORNs) to 8 mm (170,000 ORNs) between species. There is an inverse relationship between the number of lamellae and the importance of vision in behavioural activity (Teichmann 1954 cited by Hara 1993).

Receptor and non-receptor cells occur in the olfactory epithelium. Non-sensory cells do not have an axon and are ciliated or non-ciliated epidermal cells. The cilia have a ventilatory role. Mucus from goblet cells carries odorants to the receptors. The non-sensory cells also act as supporting cells and there are small, undifferentiated basal cells. The olfactory receptor cells (Figure 12.3) are bipolar neurons with cylindrical dendrites and an olfactory knob above the free surface of the epithelium, exposed to odorants. Olfactory receptor neurons occur in three forms in a single olfactory epithelium: ciliated, microvillous and crypt receptor cells, the latter with cilia in a crypt (Hansen and Finger 2000). In *Ictalurus punctatus* and rainbow trout, nucleotides are detected by the microvillous cells, bile salts by ciliated cells and amino acids by both ciliated and microvillous cells; a role for the crypt cells has yet to be determined (Sato and Suzuki 2001; Hansen et al. 2005).

There are interesting differences in the organization of the olfactory system in different groups of fish. In elasmobranchs, the olfactory sense is well developed, the nasal sacs and olfactory bulbs being large. These fish depend to a large extent on the olfactory sense in their search for food. Flatfish generally have a less well-developed olfactory system on their blind side (Prasada and Finger 1984); the flounder *Bothus robinsi* shows the extreme situation since the olfactory bulb and nerve do not develop at all on that side (Kobelkowsky 2004). The males in Lophiformes and Myctophiformes have large olfactory organs, a sexual dimorphism which

may be associated with the importance of pheromones at the depth of the habitats of these fish (Baird et al. 1990).

12.4.7 The olfactory bulbs

Unmyelinated axons from the receptor cells aggregate into bundles running posteriorly in the olfactory nerves to end in the olfactory bulb. The olfactory bulbs (Figure 11.5) are considered part of the prosencephalon of the brain (Gardner 1975) from which they are separated in fish by olfactory stalks/tracts, with considerable interspecific variations in their relative length (Sorensen and Caprio 1998). The olfactory bulb is a relay centre in which neurons from the incoming olfactory nerve synapse with mitral cells which are the outgoing neurons, their myelinated axons forming the olfactory tracts to the cerebral hemispheres of the telencephalon. The olfactory bulb is organized into four layers (Satou 1990; Sorensen and Caprio 1998): a superficial olfactory nerve layer, a deeper glomerular nerve layer, a mitral cell layer with perikarya and an internal cell layer. The glomeruli are clusters of synapses between olfactory nerve fibres and dendrites of mitral cells. The bulb also includes inhibitory granule cells and ruffed cells. Using tungsten microelectrodes, Zippel et al. (2000) have recorded extracellularly from the individual relay neurons in goldfish (*Carassius auratus*), demonstrating that ruffed cell stimulation can activate granule cells thereby inhibiting mitral cell activity. Such interactions can modulate centrally transmitted information. The olfactory bulb is more complex than would be anticipated for the relay of information alone and the organization of this bulb may facilitate odour discrimination.

In *Gadus morhua*, the ratio of olfactory receptor cell axons to mitral cells is 1000:1 (Døving 1986). Granule cells, with perikarya in the internal cell layer, synapse with the mitral cells and may inhibit spontaneous activity of the mitral cells. These granule cells also synapse with ruffed cells and with centrifugal fibres from the telencephalon (Satou 1990) facilitating regulation within the olfactory bulb by higher centres in the brain. Electrophysiology and studies on behaviour indicate that in goldfish (Kyle and Peter 1982) and sockeye salmon (Satou et al. 1984), medial tracts within the olfactory tract project to brain centres implied in the control of sexual behaviour. It is suggested that medial tracts of the olfactory tract carry pheromonal information whilst lateral tracts carry food-odour information which may also be associated with organizational differences in the olfactory bulb in *Cyprinus carpio* (Satou et al. 1983). Fibres of the terminal nerve (11.8.1) are also found in these medial tracts.

Wilson (2004) suggests that contrasting physicochemical attributes of different olfactory stimuli are enhanced by the neuronal circuitry in the olfactory bulb. The glomeruli are small areas of dense neuropil which may respond to different characteristics of various odours. Their spatial distribution within the olfactory bulb may result in a specific pattern of neuronal activity for individual odourants (Nikonov and Caprio 2001). Responses in individual large mitral cells to olfactory stimuli have been recorded and his co-workers (Nikonov and Caprio 2004) identified olfactory bulb neurons selectively responding to biologically relevant basic, neutral or acidic amino acids which can be distinguished further by side-chain characteristics. Miklavc and Valentinčič (2012) have reported that zebrafish, *Danio rerio*, respond more actively to amino acids they have previously been conditioned to (i.e. an experience-modified response), compared with non-conditioned amino acids, although they have difficulty discriminating between those amino acids evoking similar chemotopic patterns of activity in the olfactory bulb. Hansen and others (2005) concluded that in *I. punctatus* and *Carasssius auratus* distinct types of olfactory receptor neurons terminate in defined areas in the olfactory bulb.

12.4.8 Olfactory stimuli

Olfactory stimulants are similar to those for taste in fish, being very small water-soluble molecules often associated with metabolic processing and signalling food or other biologically important information. Comparative interspecific studies recognize amino acids, gonadal steroids (as water-soluble conjugates), prostaglandins and bile acids/salts as major olfactory stimuli. The potency of various odorants has been investigated using electroolfactogram

(EOG) recording (Hara 1993) of summated generator potentials of the receptor cells represented by voltage changes in the water over the olfactory epithelium, and by electroencephalogram recordings from the olfactory bulb, nerve and tract. Different stimuli can be discriminated by individual relay neurons which vary in their responses recorded by microelectrode (Zippel et al. 2000), with further discrimination in the CNS.

12.5 Mechanoreceptors: introduction

Vertebrates have an array of mechanoreceptors responsive to pressure associated with touch and motion as well as special 'baroreceptors' registering pressure change in blood vessels. These mechanoreceptors may be relatively simple or incorporated in a complex organ or system including ears (hearing, equilibrium) and lateral lines detecting water motion. Fish lack terrestrial vertebrate features such as external ears and Pacinian corpuscles. Some vertebrates, notably mammals, have relatively large Pacinian corpuscles (specialized pressure receptors) which can be readily observed in histological preparations of the skin and subcutaneous tissue (Gardner 1975) and in mesenteric tissue (Patton 1965).

Simple receptors may include Merkel cells (12.5.1), baroreceptors (12.5.2) and nerve endings acting as proprioceptors (12.5.3).

In the lateral lines and ear, the receptor cells are known as sensory hair cells. Their ultrastructure (Wersäll 1961, reviewed by Popper and Platt 1993; Flock and Wersäll 1962; Lowenstein et al. 1964; Flock 1965, 1971) reveals that the receptor apex of each hair cell has a ciliary bundle with a single long, true cilium, the kinocilium, and a cluster of smaller microvilli, the stereocilia, decreasing in length with increasing distance from the kinocilium (Figure 12.5). The kinocilium has a basal body with a laterally projecting foot on the side distal to the stereocilia (Lowenstein et al. 1964).

12.5.1 Merkel cells

Merkel cells which are tactile in vertebrates occur in the epidermis (including barbels and fin-rays) of some teleosts (Lane and Whitear

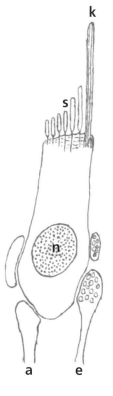

2μm

Figure 12.5 Left: Generalized structure of a vertebrate mechanosensory hair cell (multiple sources). Right: Scanning electron micrograph of a hair cell from the ear of a skate. Reproduced with permission from V.C. Barber and C.J. Emerson (1980). Scanning electron microscopic observations on the inner ear of the skate. *Cell and Tissue Research* 205, 199–215. © by Springer-Verlag 1980. a = afferent nerve fibre; e = efferent nerve fibre; k = kinocilium; n = nucleus; s = stereocilium.

1977; Zaccone 1984; Whitear 1986a,b) and lung-fish (Fox et al. 1980) but were not found in the chondrichthyan *Raja clavata* (Whitear and Moate 1998). Merkel cells are small and round, include membrane-bound granules, possess projecting microvilli and synapse with nerve fibres (Lane and Whitear 1977). Lucarz and Brand (2007) have questioned the function of these cells, suggesting alternate roles.

12.5.2 Baroreceptors

Internal baroreceptors, known to be important in mammals for monitoring blood pressure, may function in fish ventilation, particularly ram ventilation (6.5.11). Histologically mammalian cardiovascular baroreceptors appear to be terminal ramifications of nerve fibres in the blood vessel wall of the carotid sinus and aortic arch (reviewed by Scher 1965). A comparative review of baroreceptors by Bagshaw (1985) includes physiological evidence supporting their occurrence in fish, gill arches and the pseudobranch being cited as possible locations. Olson

(1998) cautions that apparent baroreceptor activity in fish cannot always be separated from responses to blood oxygenation levels.

12.5.3 Proprioceptors

Other internal mechanoreceptors, such as proprioceptors, providing feedback information on muscle tension, may occur in fish, but their identity has been difficult to establish; muscle spindles (mammalian proprioceptors) have not been described in fish. Sensory nerve fibres in myosepta of lamprey (*Petromyzon*) could be proprioceptors (Young 1981). In the elasmobranch *Scyliorhinus canicula* there is evidence that corpuscles of Wunderer under the skin of the body and fins may have mechanoreceptors with a proprioceptor role (Lowenstein 1956; Roberts 1969), based on electrophysiological recordings. Simple sensory nerve endings have been described in the *stratum compactum* (2.3.2) above the myotomes of the pickerels *Esox niger* and *E. americanus* and may have a proprioceptor function (Ono 1982). Williams et al. (2013) studied bluegill sunfish

(*Lepomis macrochirus*) and found evidence for proprioception associated with the pectoral fin-rays.

12.6 Water movement perception and equilibrium systems

Fish have complex systems of pressure receptors operating to detect hydrodynamic changes (water movements and pressure changes) including those associated with sound waves as well as changes in position relative to the water column.

The buoyancy resulting from the closer densities of water and fish body tissues provides fish with a high degree of support, facilitating their swimming over a range of depths. This freedom from substrate support necessitates a highly tuned equilibrium sensory and effector system which has evolved along different pathways in fish taxa.

The velocity of sound in water is 1457 m/s at 20 °C, which is about 4.4 times that in air, but can be influenced by water depth and salinity. Water resists compression and sound propagation, involves molecular displacement, a longitudinal oscillation or vibration, with frequencies measured in Hertz (Hz = cycles per second). In air, sound particle displacement attenuates over a short distance, but in water it persists for greater distances up to several wavelengths, metres, from the source. Accounts of hearing often distinguish between an acoustic near field and an acoustic far field (reviewed by Popper and Platt 1993). The near-field effect is caused by flow motions of the water resulting directly from sound source movements; the far-field water displacement is particle oscillation generated by the variable pressure component of the sound source. Both field effects may be significant in fish hearing.

The term 'acousticolateralis system' is used to refer collectively to the fish hearing and water movement sensory systems, and more rarely the term 'octavolateralis system' (McCormick 1981) is used to include the 'balance/postural' component of pressure reception. The auditory and lateral line systems are sensitive to water-borne vibrations; detected directly by lateral line receptors, but only after additional transmission through body tissues in the case of the auditory system which lacks an outer ear and eardrum in fishes. The equilibrium sensory system is sensitive to changes in the position of the body axes of fish. In all three systems (hearing, lateral line and equilibrium) the sensory cells are ciliated and bathed by fluid in which the ciliary bundles may bend in response to relative movement; this bending initiates transduction and nerve impulses.

At the base of the hair cells are presynaptic vesicles. The ciliary bundles are bent by fluid movements around them. The mechanoelectric transduction, associated with this bending, into a receptor potential is only partly understood. In the teleost *Lota lota* and the amphibian *Necturus maculotus*, intracellular electrophysiological recording of responses to mechanical stimulation demonstrate that bent ciliary bundles alter the membrane potential of the sensory hair cell (Harris et al. 1970). It is suggested (Lowenstein 1971) that bending of stereocilia may move the base of the kinocilium in the direction of its basal foot to initiate an excitatory depolarization and receptor potential which would be inhibited by hyperpolarization caused by the opposite movement in the ciliary bundle. This generation of the receptor potential may involve membrane gating, governing ionic currents (Flock 1971). The electrical charge represented by the receptor potential regulates the amount of neurotransmitter released in a graded manner which determines the frequency of action potentials generated in the afferent nerves to the central nervous system. The bases of the hair cells also synapse with nerve terminals, which can inhibit the impulses in the afferent nerve (Russell 1968).

Fish sensory hair cells have been considered to be Type II hair cells, that is, cylindrical with a basal innervation, as in Figure 12.5, in contrast to Type I which are flask-shaped and enveloped by a 'basket' of nerve fibres. Yan et al. (1991), however, found evidence supporting the presence of Type I cells in some fish.

In the skate *Raja ocellata*, gender differences have been reported (Barber et al. 1985) for the number of hair cells and associated nerve fibres stimulated by far field vibrations, the number being larger in females. Barber and others (1985) suggest that this difference may be related to mate detection or prey location. Morphological patterns of hair cell polarization (differences in the position of the kinocilia relative to the sterocilia) have been observed in the various regions of the elasmobranch ear in scanning electron micrographs (Lowenstein and Wersäll 1959; Barber and Emerson 1980) as well as in bony fish (reviewed by Popper and Platt 1993). Such patterns of polarization in different regions of the ear may facilitate recognition of sound direction (Platt and Popper 1981; Schellart and Wubbels 1998).

12.7 The lateral line system: perception of low-frequency vibration

Bleckmann (2007) provides a comprehensive review of lateral line morphology and function including behavioural and CNS physiological aspects. The lateral line (Figure 12.6) which is well developed in most fish (and some amphibians) enables them to detect water currents, including vibrations caused by the presence of other moving animals. The lateral line system is thus a hydrodynamic receiver or detector of low-frequency water movements.

A lateral line system is usually well developed in species continually exposed to water movement but is secondarily reduced in relatively inactive and benthic fish (Lowenstein 1971). Gobies (Family Gobiidae), Southern smelts (Family Retropinnidae) and Southern graylings (Family Prototroctidae) are among those fish which do not have a lateral line (Allen et al. 2002). Lamprey, *Lampetra*, has 'very simple' lateral line organs, sunk in rows of pits open to the exterior rather than linked in canals (Young 1981). Lateral lines of the shovel nose rays *Rhinobatos typus* and *Aptychotrema rostrata* include a variety of morphological types, some with tubules, some without (Wueringer and Tibbets 2008). The lateral lines can be 'pored' with a single terminal pore (endpore) and may have branched tubules with individual pores.

12.7.1 Neuromasts and neural connections

The water movements are perceived by mechanosensory hair cells grouped into neuromasts (Figure 12.6). They can occur as sensory hillocks embedded in the outer layer of epidermis with their kinocilia and stereocilia projecting above it within a gelatinous cupula which can move in response to water movements. In many fish the neuromasts and cupulae are sunk into pits or into the lateral line canals (Figure 12.6) which open to the surrounding water by pores through the scales. Those in pits have been described as 'pit organs', with lateral line innervation but obscure role and reviewed by Peach and Marshall (2000) in the case of elasmobranchs.

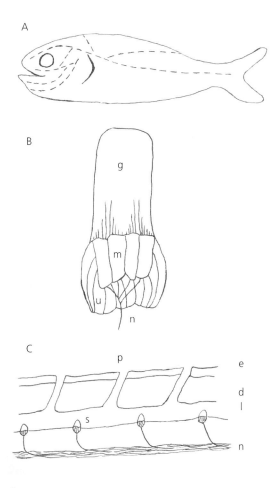

Figure 12.6 Fish lateral line system. (a) Generalized distribution of lateral line canals. (b) Diagram of a neuromast. (c) Representative longitudinal section through the main lateral line of the body. d = dermis (scales not included); e = epidermis; g = gelatinous cupula; l = lateral line canal; m = mechanosensory cell; n = sensory nerve; p = pore of lateral line canal; s = neuromast; u = supporting cell. (a) and (c) adapted from Goodrich (1930), (b) based on accounts by Lowenstein (1957) and Popper and Platt (1993).

The continuous lateral lines, linking neuromasts, commonly occur along each flank, and on the head nerves. Young (1981) suggested that lateral lines were innervated by branches of cranial nerve VIII, running with cranial nerves V, VII, IX and X (11.8). However, in bowfin, *Amia calva*, separate lateral line sensory nerves have been described (McCormick 1981). Details of lateral line innervation are discussed in Northcutt et al. (2000).

12.7.2 Neuromast function

Electrophysiological recording from lateral line organs in catfish, *Ameiurus nebulosus* (Hoagland 1933), and skate, *Raja spp.* (Sand 1937), demonstrate that the neuromasts are tonic receptors, spontaneously generating impulses. Experimentally forcing water currents along the lateral line canal changes the frequency of these impulses (Sand 1937). A water current flowing tailwards inhibits the discharge whilst a flow in the opposite direction increases the frequency of discharge indicating antagonistic neuromasts (Sand 1937) aligned with different directional sensitivities. However, recent work on goldfish (*Carassius auratus*) shows that 'lateral line fibres do not code bulk water flow direction in turbulent flow' (Chagnaud et al. 2008a). Stationary goldfish were exposed to running water and lateral line discharges recorded. Increasing responses to faster flow rates occurred even after reversal of flow direction.

Lateral line neuromasts are responsive between 10 and 200 Hz (Coombs et al. 1991) and to 'near-field effect'. Lateral lines detect hydrodynamic forces generating local water movements affecting areas of the fish. This may be important in schooling (Popper and Platt 1993) and in detecting moving prey or as a warning leading to rapid movements to avoid a predator (Blaxter et al. 1981). The wake created by swimming fish includes vortex rings generated by the caudal fin; the patterns and turbulence may be detected by fish predators via their lateral lines (Pohlmann et al. 2001, 2004). For blind cavefish the generation of water motion with fluctuations reflected from objects in the water and perceived by the lateral line are particularly important for navigation (von Campenhausen et al. 1981; Weissert and von Campenhausen 1981). Bleckmann (2007) discusses possible natural lateral line stimuli including self-generated (e.g. from the fish's own movements) and external background (e.g. from running water) 'noise'. Central processing cancelling noise stimulation is not yet understood. Knowledge of the circuitry and possible filtering functions of the various lateral line central nuclei, including the possible role of an extensive arborization of fibres from the lateral line nerves, is sparse. Mid-brain and fore-brain lateral line centres may also respond to other stimuli, for example visual stimuli, by summation, facilitation or inhibition.

12.8 The ear: hearing and gravity perception

Fish have no external ear, and no 'middle ear' in contrast to tetrapods, thus there is no external evidence of the ear such as the pinna of most mammals and the eardrum membrane of amphibians. The ear of fish, functionally equivalent to the tetrapod inner ear, has auditory and postural, or orientation, sensory roles. It is enclosed in the cranium and pressure differences are transmitted to it through the body of the fish, which has a similar density and acoustic characteristics as the surrounding water.

The auditory and postural components of the fish ear are not as clearly separated as in terrestrial vertebrates (Popper and Platt 1993). The fish ear does not have a cochlea (or its rudiments), which differs from all other vertebrates (Lowenstein 1971). It is from the fish lagena that the coiled cochlea, or sound receptor of tetrapods, is evolutionarily derived. Popper and Platt (1993) discuss a possible functional division within the brain between auditory and vestibular (postural) inputs from the ear.

In jawed fish each ear consists of three semicircular canals and three chambers; the otolith organs or (upper) utriculus, (lower) sacculus and small posterior lagena (Figure 12.7). Lampreys, however, only have two semicircular canals, and a single elongated chamber, a 'macula communis' rather than the utriculus, sacculus and lagena of jawed fish (Avallone et al. 2005). Hagfish (*Myxine*) has an even simpler structure with only a single semicircular canal (illustrated by Young 1981, after Retzius 1881).

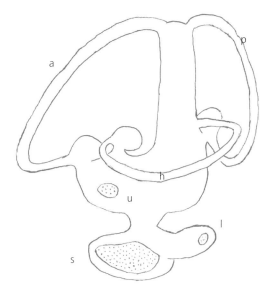

Figure 12.7 General plan of fish otolith organs and semicircular canals. a = anterior vertical canal; h = horizontal semicircular canal; l = lagena (including astericus); p = posterior semicircular canal; s = sacculus (including saggita); u = utriculus (including lapillus). The three otoliths are stippled. Based on von Frisch (1936), Hoar (1966) and Lowenstein (1971).

12.8.1 Semicircular canals

The semicircular canals lie at right angles to one another (Figure 12.7) and in jawed fish can thus discriminate between three planes of movement as in most other vertebrates. Each semicircular canal contains fluid, endolymph, and has an ampulla at one end (Figure 12.7), but with one at each end in *Myxine*. Ampullae have an internal ridge, the crista, of sensory and supporting cells. The ciliary hairs of the sensory cells are embedded in a gelatinous cupula (similar to that of the neuromasts of the lateral line) which can be deflected by relative movement of the fluid (endolymph) and the semicircular canal.

12.8.2 Otolith organs

The utriculus, sacculus and lagena (or otolith organs) have an otolith (ear stone) known respectively as the lapillus, sagitta and astericus (Figure 12.7), as well as a sensory epithelium, a macula, with sensory hair cells, innervated by branches of cranial nerve VIII, and supporting cells. Agnathan and elasmobranch otoliths consist of a gelatinous matrix

containing statoconia, with small granules of a calcium salt. Teleosts have solid otoliths with species specific shapes for each otolith organ and may have annual growth rings, which can be used to assess the age of fish (1.4.1, 3.3, 7.13.5). The teleost otolith consists of calcium salts and is rigid and hard in texture. It is coupled to its macula by a gelatinous otolithic membrane (Platt and Popper 1981) suspending the otolith in endolymph. The cilia of the sensory hair cells of the macula project into the narrow space between it and the otolith. The otoliths are denser than the endolymph and any movements of the two will be out of phase and will result in deflection of the sensory cell cilia relative to the otolith (Schellart and Wubbels 1998).

12.8.3 Semicircular canal and otolith organ functions

The cupulae and sensory hair cells of the semicircular canals function as gravity and angular acceleration receptors. Movement of the canal relative to its endolymph causes deflection of the cupulae, as a result of angular acceleration, leading to compensatory movement of fins and myotomal contractions. Electrophysiological recordings from branches of the VIIIth cranial nerve of dogfish *Scyliorhinus* (Lowenstein and Sand 1936) and *Raja* (Lowenstein and Sand 1940a,b) demonstrated spontaneous resting discharge from each ampulla, angular accelerations increasing or decreasing the frequency of the impulses. The vertical canals respond to rotation about all three axes; the horizontal canal to rotation around the vertical axis. The spontaneous resting discharges will probably be involved in maintaining muscle tone. In addition to the semicircular canals and otolith organs, the eyes may have considerable involvement in maintaining posture and equilibrium in fish (Lowenstein 1971).

Fish otolith organs can perceive both gravity and sound. Electrophysiological analysis of impulses from the macula of the utriculus, sacculus and lagena in *Raja clavata* indicate sensitivity to positional change and may be used to control equilibrium responses (Lowenstein and Roberts 1949), although this cannot be resolved by such experiments. An acoustic role for the otolith organs of *Raja* is also indicated by electrophysiological experiments which record sound-generated action potentials from the macula of the utriculus and sacculus, but not the lagena (Lowenstein and Roberts 1951). Bass and Lu (2007) suggest that, in sleeper goby, the orientation of each otolith organ is a determiner in direction hearing, the saccule being most important. The otolith organs respond to linear accelerations which would include centrifugal and rotational gravitational stimuli as well as vibrations and acoustic stimuli (Lowenstein 1971).

12.8.4 Direct and indirect sound transmission

Sound transmission from the water to the ear of fish can involve different routes. The fish body follows the particle movements of a water-transmitted sound and the cilia of the sensory hair cells are deflected by movement of the denser otoliths relative to the endolymph; that is, they respond to a 'direct' stimulation (Popper and Platt 1993). An additional route for soundwave pressure oscillations can involve the swim bladder producing an 'indirect' stimulation. The pressure oscillation component of soundwaves produce vibrations in the swim-bladder wall which are then transmitted to the sensory hair cells of the ear. Sonar is used in commercial fishing to detect scatter of the signal by swim bladders in a school (Schellart and Wubbels 1998). Indirect stimulation increases the range of frequencies detected and the sensitivity to lower thresholds. Popper and Platt (1993) have reviewed adaptations which improve fish hearing. In freshwater ostariophysans (cyprinids, silurids, characins, etc.) there are hinged Weberian ossicles, which are parts of vertebrae connecting the anterior end of the swim bladder to the inner ears, between the two saccules. This arrangement makes possible the transmission of the swim-bladder wall vibrations to the endolymph of the ear. These ossicles further increase the sound discrimination in these fish. In many species, such as in the herrings (Clupeidae) and some squirrelfish (Holocentridae), there is direct contact between the swim bladder and the auditory bulla (Nelson 1955). In mormyrids and anabantids the head has a bubble of gas in contact with the ear (reviewed by Popper and Platt 1993).

Fish with an indirect ear stimulation, involving Weberian ossicles or other acoustic coupling of the swim bladder to the ear, are known as hearing specialists (Popper and Platt 1993; Ladich 2000; Amoser and Ladich 2005). Popper and Platt (1993) compare graphically the ranges of sensitivity and frequency perception of representative hearing specialists and non-specialists (without hearing enhancing adaptations and species without a swim bladder). Hearing specialists possess superior sensitivity and frequency range perception (up to about 3000 Hz). Hearing non-specialists are characterized by high-intensity thresholds and hearing over relatively narrow frequency ranges (only up to about 1000 Hz). However, Popper and Fay (2011) recommend a new approach to this type of classification of hearing capacity on the grounds that the two categories are too simplistic and, instead, there is a range across species that is a continuum, but they retain the term 'specialization' for fish with anatomical structures such as the Weberian ossicles (1.3.8, 3.5.2) found in otophysans.

12.8.5 Hearing in vocally communicating fish

Some fish, are able to produce sound which may be involved in conspecific communication (13.7).

Bass and Lu (2007) review a series of studies on hearing in vocally communicating fish, and distinguish between mechanisms of sound communication and directional hearing in sonic (sound producing) and non-sonic species (13.6). Nesting male midshipmen fish and toadfish (Order Batrachiformes) produce vocal signals attracting gravid females, which make choices based on frequency, duration and intensity of the sounds. During the breeding season midshipmen fish and toadfish nest in shallow water in which 'upper harmonics' of sound are heard at a greater distance. In female midshipman, fish steroids produce seasonal changes in nerve impulse frequency generated in the sacculus epithelium. Oestrogenic receptor mRNA has been identified associated with the saccular nerve close to the hair cell epithelium in midshipman fish, further supporting a steroid modulatory role. Bass and Lu (2007) review evidence for mechanisms which decrease auditory sensitivity of the ears and lateral lines of an individual to self-generated vocalization.

12.9 Detecting electric fields and discharges

Fish can possess the capacity for 'electroreception' to detect either poor or good electrical conductors in their neighbourhood and this enables such fish to navigate or identify objects by a process termed electrolocation. Electroreception depends on specialized electroreceptors which occur in the coelacanth, Chondrichthyans (e.g. elasmobranchs), *Polyodon* (paddlefish) and some teleosts (Table 12.1).

Both fresh water and seawater contain ions of dissolved salts making both media good electrical conductors. Nerve impulses and muscle contraction in aquatic animals generate small bioelectrical changes (11.3.3 and Appendix 11.1) which generate electrical fields (spatial force associated with electrical charges) external to the animal. These small fields can be detected by the special elecetroreceptors.

12.9.1 Electrolocation

The ability to generate and/or use electrical fields to navigate and detect animate or inanimate objects in the surrounding water is a feature of fish function which has only fairly recently become apparent. It is a capacity which appears to have polyphyletic origins although there is some possibility of its early occurrence in the precursor agnathan groups, and persistence through some lineages. Moreover, it is not a capacity restricted to fish as it has been reported in other taxa including the platypus, the egg-laying monotreme mammal (Pettigrew 1999). Some fish have electric organs and are thus termed electogenic (13.8 and Table 12.1); in the case of Mormyridae (elephantfishes), the electric organs produce a weak electric field important in navigation and communication.

The osteoglossomorph family, the Mormyridae, which live in turbid African rivers, appear to be exceptionally adept in electrolocation using disturbances of the self-generated field. Several

Table 12.1 Electrogenic[†] and electroperceptive fish (not comprehensive).

Taxon	Genus	Electrogenic	Electroperceptive	Reference
Coelacanths	*Latimeria*	N	Y	Berquist et al. 2015
Chondrichthyes	*Torpedo*	Y +	?	Nelson 1976
	Raja	Y	Y	Bennett 1971b, Kalmijn 1971
	Scyliorhinus	N	Y	Bennett 1971b, Kalmijn 1971
	Hydrolagus	N	Y	Fields et al. 1993
Acipenseriformes	*Polyodon*	N	Y	Chagnaud et al. 2008b
Teleosts				
Osteoglossomorpha	*Gymnarchus*	Y	Y	Lissman 1958
	Xenomystus	N	Y	Bullock & Northcutt 1982
Ostariophysi				
Siluriformes	*Malapterurus*	Y +	?	Howes 1985
	Clarias	Y	Y	Bullock 1999[††]
Gymnotids	*Electrophorus*	Y +	Y	Bauer 1979
	Gymnotus	Y	Y	Schuster 2002
	Eigenmannia	Y	Y	Tan et al. 2005
	Sternarchus	Y	Y	Bennett 1971a, Szabo 1965
Acanthopterygii	*Uranoscopus*	Y	N	Alves-Gomes 2001

Y+ = capable of strong (high-voltage) electric organ discharges (EOD).
[†]Literally 'electrogenerative'.
[††]Bullock believes that all siluriformes are electroperceptive, but only a few electrogenic.

elasmobranchs, with a very different marine habitat, also have been shown to have extremely sensitive electrolocation, detecting the small fields associated with potential live prey. The various taxa which have electroperception may include members which generate powerful electric organ discharges, EODs (13.8 and Table 12.1), as an offensive or defensive strategy, thus siluroid catfish include the family Malapteruridae with two such genera, which are well known. The capacity of many of the other siluroids (Howes 1985) to be electroperceptive, but without a prominent electrogenic organ, has not been given so much attention until recently (Bullock 1999).

12.9.2 Electroreceptors

The presence of electroreceptors was first investigated in *Gymnarchus niloticus* (Lissman 1958, 1963; Lissman and Machin 1958), an osteoglossomorph taxonomically close to the mormyrids. Electroreceptors have since been described as passive or active (Bennett 1967; von der Emde 1998, 2007), the former detecting signals originating externally, the latter being sensitive to self-emitted signals from the same fish.

Electroreceptors are modified lateral line neuromasts (Bennett 1971b), are found in the coelacanth (Berquist et al. 2015) and can occur in both teleosts and elasmobranchs on the head and body. They are innervated by branches of the lateral line nerves and are closely associated with mechanoreceptors and their nerve centres (von der Emde 1998). The electroreceptor fibres are received by centres in the medulla oblongata of the brain, with neural connections to more anterior centres (von der Emde 1998, 2007). Compared with the optic, olfactory and auditory systems, electroreceptors are not so enclosed by elaborate accessory tissue structures. Electroreceptor cells are typically clustered at the bases of epidermal chambers, with an apical surface protruding from the lumen and the basal membrane synapsing chemically with afferent nerve fibres to the electrosensory lateral line nerves. Transduction of negative and positive external voltage changes into neural activity is by depolarization and hyperpolarization of the apical and basal membranes to regulate neurotransmitter release (von der Emde

1998). Electroreceptors have a very high level of sensitivity. The direct current, low-frequency fields generated by normal bodily activity of aquatic animals enable sharks to locate and attack prey, even without other sensory cues, and they will locate flounder buried in sand (Kalmijn 1966, 1971). Possession can have some negative effects with human interaction. Thus elasmobranchs may pick up metal objects found in harbours and may also be prone to confusion in the vicinity of underwater cables. The latter problem has recently been the subject of a United States Government report (Normandeau et al. 2011).

12.9.3 Types of electroreceptors

There are two major types of electroreceptors: ampullary (passive) receptors are connected to the external water by a canal, and respond to low frequencies, below 50 Hz, and tuberous (active) receptors without a duct, but with a plug of loose cells, sensitive to high frequencies, having large intercellular spaces with low electrical impedance. The 'active' tuberous receptors occur only in the weakly electric fish such as mormyriformes that electrolocate via self-emitting signals while the 'passive' type detect electric fields which are not self-generated. The histology and ultrastructure of both types of receptor are reviewed by Bennett (1971b). Ampullary receptors are rhythmically active and give tonic or long-lasting responses to low-frequency and direct current stimuli whereas tuberous receptors are phasic, responding only briefly to voltage changes and not to direct current stimuli (Bennett 1971b). The ducts of ampullary receptors have high resistance walls and these receptors occur generally in electrosensitive fish. In marine species (e.g. elasmobranchs) the duct is filled with a Ca^{2+}-rich jelly with a high conductance. In freshwater electrosensitive fish the ducts are shorter and their jelly is K^+-rich.

Ampullary electroreceptors occur in elasmobranchs as ampullae of Lorenzini, as well as in all electrosensitive teleosts according to von der Emde (1998, 2007). The ampullae of Lorenzini occur in capsules on the dorsal and ventral surfaces of the head region and have ducts distributed to external pores, often distant from the capsule, over the surface of the body (Murray 1960). The ampullae of Lorenzini are extraordinarily sensitive to weak electric fields, being able to detect voltage gradients as low as 0.01 µV/cm (Bennett 1971b). They facilitate direction finding from ducts oriented in the direction of the field (Young 1981). The ampullae of Lorenzini have an interesting distribution in the shovel-nose rays *Rhinobatis typus* and *Aptychtrema rostrata* (Wueringer and Tibbets 2008) with many more on the lower than on the upper surface, particularly in *R. typus*. In both species, ampullae of Lorenzini are more numerous on the rostral (snout) region, suggesting that its elongation may be an adaptation to enhance sensory input. Both species feed largely on benthic prey, *R. typus* occurring in both marine and freshwater habitats, *A. rostrata* being marine, and it is probable that electrolocation and mechanoreception are used in prey detection.

Tuberous electroreceptors occur in the weakly electric Mormyriformes (e.g. *Gymnarchus niloticus*) and Gymnotiformes (e.g. *Eigenmannia*) in which electric organs in the tail region rhythmically generate EODs. These EODs, either self-emitted or from neighbouring fish, are detected by the tuberous receptors, many of which are highly sensitive to the discharge frequency of the species, and often the sex, facilitating identification of competitors and possible mates (Hopkins 1988). At least eight tuberous electroreceptor types have been recognized in various teleost species differing physiologically and morphologically (von der Emde 1998). According to Heiligenberg (1993) there are two major categories of tuberous electroreceptors, one encoding the precise timing of an EOD, the other encoding its amplitude, both occurring in individual fish. Each type has separate neural pathways to the brain. The EODs are continually monitored by the electroreceptors which are most densely distributed anteriorly. Amplitude and timing changes caused by nearby objects can produce an electrical image for further processing in the mid-brain and cerebellum (Heiligenberg 1993).

The mormyriform tuberous electroreceptors are mormyromasts and Knollenorgans, the first perceiving self-emitted EODs, the second being part of a neural system processing input only from neighbours (Heiligenberg 1993). Mormyromasts are amplitude detectors and Knollenorgans are timing detectors (von der Emde 1998).

Mormyromasts possess 'A' and 'B' sensory cells (Szabo and Wersäll 1970) with separate neural pathways to distinct regions in the electrosensory lobe of the medulla oblongata (Bell 1990) and probably facilitate distinction between capacative and resistive characteristics of an object (von der Emde 1990, 2007). Capacitance is associated with living objects and can distort EOD waveforms, the B cells being sensitive to such distortions (reviewed by von der Emde 2007) which facilitate prey and predator detection. Most gymnotiforms and all mormyriforms except *Gymnarchus* have tuberous organs sensitive to pulse EODs. Experimentally simulated EOD pulses differing in waveform but not in amplitude can be discriminated by a pulse-type gymnotid *Hypopomus artedi*, electroperception matching its EODs (Heiligenberg and Altes 1978). *Gymnarchus* and some gymnotiformes, such as *Eigenmannia*, produce continuous wave-like EODs.

12.9.4 Electric organ discharge variation

Fugère and Krahe (2010) observe that EODs vary widely between species and therefore are appropriate for species recognition.

12.9.5 Jamming avoidance

As part of the very complex system some species notably *Gymnarchus* and *Eigenmannia* have jamming avoidance (Bullock et al. 1972; Heiligenberg 1975) so that the emitted signal can be adjusted in the presence of competing signals, for example from conspecifics (13.8).

12.10 Magnetoreceptors

Recent information supports the possibility of magnetoreception in animals including birds and fish (Walker et al. 2007), while passive magnetotaxis does occur in some bacteria (Blakemore 1975), that is, some bacteria move directionally with reference to a magnetic field.

In their review of fish magnetic reception, Walker et al. (2007) emphasize that its analysis is at a very early stage, but accumulating information is suggestive of a role in navigation.

Teleosts may have intracellular magnetite crystals in magnetosomes (Mann et al. 1988; Walker et al. 1988) within an organic matrix as seen in ultrastructure studies on ethmoid tissue from sockeye salmon *Onchorhynchus nerka*. Walker et al. (2007) report it in multi-lobed magnetoreceptor cells near the basal lamina of the olfactory epithelium in rainbow trout. Elasmobranchs are unique in that their electroreceptors are sufficiently sensitive to electric currents induced by movement across magnetic field lines that they are without a need for specialized magnetoreceptors (Kalmijn 1984, 1987).

12.11 Nociceptors and thermoreceptors

Numerous nerve fibres, seen with the electron microscope, occur in the epidermis of chondrichthyes (Whitear and Moate 1998) and teleosts (Whitear 1971a), the majority probably terminating as free nerve endings, which may be sensory including possible nociceptors (14.11.1) and thermoreceptors (14.12), which generally lack distinguishing characteristics, as is also suggested for fish proprioceptors (12.5.3).

12.12 Overview

Fish sensation involves sensory cells or nerve endings, often with accessory structures varying in complexity. Tetrapod morphology may be recognizable in fish sensory structures but the lateral lines and electrolocation systems are largely exceptions, although lateral lines occur in amphibians and the duck-billed platypus has electrolocation. Lissman (1963) regards electrolocating fish as 'living in a world totally alien to man'.

Bibliography

Albert, J.S. and Crampton, W.G.R. (2002). Electroreception and electrogenesis. In: D.H. Evans (ed.). *The Physiology of Fishes*, 3rd edn. Boca Raton, FL: CRC Press, pp. 431–72.

Collin, S.P., Kempster, R.M., and Yopak, K.E. (2016). How elasmobranchs sense their environment. In: R.E. Shadwick, A.P. Farrell, and C.J. Brauner (eds). *Fish Physiology Vol. 34, Part A: Physiology of Elasmobranch Fishes: Structure and Interaction with Environment*. Amsterdam: Elsevier Inc. 19–99.

Diebel, C.E., Proksch, R., Green, C.R., Nielson, P. and Walker, M.M. (2000). Magnetite defines a vertebrate magnetoreceptor. Nature, 406, 299–302.

Douglas, R.H. and Djamgoz, M.B.A. (eds) (1990). *The Visual System of Fish*. London: Chapman and Hall.

Dusenbery, D.B. (1992). *Sensory Ecology: How Organisms Acquire and Respond to Information*. New York, NY: Freeman.

Hara, T.J. (ed.). (1992). *Fish Chemoreception*. London: Chapman and Hall.

Lamb, D.T. (2009). Evolution of vertebrate retinal photoreception. *Philosophical Transactions of the Royal Society B*, 364: 2911–24.

Sneddon, L. (2007). Nociception. In: T.H. Hara and B. Zielinski (eds). *Sensory Systems Physiology: Fish Physiology, Vol. 25*. New York, NY: Academic Press/Elsevier, pp. 153–78.

Tavolga, W.N., Popper, A.N., and Fay, R.R. (1981). *Hearing and Sound Communication in Fishes*. New York, NY: Springer-Verlag.

CHAPTER 13

Special adaptations

13.1 Summary

Extant fish have a huge variety of adaptations linked to the depth, warmth, salinity and stability of their environment as well as interactions with other conspecifics and availability and accessibility of potential food.

Describing the extent of the differences between taxa can be rendered into habitat differences and special types of responses to the environment including sound production and electrogenesis.

While the oceans represent vast habitats, layered by depth and very cold at the poles, some fish have survived in minute habitats, fairly stable in the case of caves but extremely unstable in desert springs and temporary pools.

In external appearance, particular fish species have acquired different shapes from long to spherical, with polymorphisms possible, and there are even a few parasitic species.

13.2 Introduction

Many fish live in habitats remote from humans and are therefore unfamiliar, or are in other ways extraordinary, and have evolved special adaptations which encompass unexpected structures, behaviour and senses or spectacular appearance quite different from the more usual fish. Fish diversity was briefly outlined in Chapter 1, and Chapters 2 to 12 dealt with the various ways in which fish achieve the major functions of vertebrate life, with most reported work on common fish species. Studying fish functioning in major attributes like nutrition or reproduction has indeed necessarily concentrated on the common, commercially important or readily accessible fish species, the 'typical' fish of Chapter 15. The present chapter deals with the less usual attributes of some fish, which may have evolved, in response to particular habitats or selective pressures, adaptations facilitating display, communication, defence, camouflage or many other features enabling advantage for survival, resource utilization or reproduction.

Some habitats, notably in warm, shallow waters with high productivity, have multiple niches and intense competition for space and food resources with many fish species interacting. In such conditions speciation has been promoted, producing scores if not hundreds of different genera or species, and a bewildering array of structures, like differing dentition in cichlids or colour patterns in coral reef fish.

By contrast, in dark, aphotic habitats like the deep ocean there may have been convergence, with dark integuments prevalent. Perhaps surprisingly, however, lack of pigmentation (i.e. albinism) is usual in caves. Deep-ocean fish are not easily accessible and tend not to be found in large numbers though the habitat itself is very large and extensive; those species that have been described have a high incidence of photophores, with self-produced light, bioluminescence, found also in some mid-water species. Other characteristics of deep-ocean fish include arrays of structures maximizing chances for food capture and reproduction in conditions of scarcity.

Cave-fish live in food-poor conditions with very limited capacity for growth and reproduction and are usually in a single top predator position per species per cave system, very different from the multi-species systems of the African Great Lakes or coral reefs, or indeed the deep oceans. Other very limited habitats include desert springs, temporary pools and the polar regions. A few fish species are essentially terrestrial, while others can tolerate drought.

Fish may form aggregations when schooling or spawning and communicate by various mechanisms including the use of colour and sight, sound and via electroperception.

Electric fish can use high- or low-voltage discharges for various purposes. Electrolocation was considered in 12.9 for Mormyridae, for example, but other taxa have electrogenic organs for predation or defence. Defensive structures include spines and toxins as well as electric discharges.

There are many strange fish shapes, from extreme elongations to the compressed or spherical. Fin extensions occur in fish which 'fly', or have leaf-like camouflage, or long auxiliary support structures or fins elongated parallel to the body axis. Unusual body positions also occur in some taxa, notably the flatfish.

Hermaphroditism (9.4) and sexual dimorphism (9.3) or polymorphisms occurs in several taxa; anglerfish are particularly noted for having dwarf males, which may be 'parasitic' on females. True parasitism is very rare in jawed fish, but other symbioses occur.

13.3 Special adaptations linked to habitat: warm, shallow water

Speciation and a high degree of diversity of fish morphology, colour and function is typical of warm, shallow water with a multitude of niches whether in fresh water or inshore marine areas.

The large African lakes and the many marine coral reefs are particularly noted for fish diversity.

Freshwater adaptive radiation in the warm waters of Africa and South/Central America is well illustrated by the two very speciose teleost groups, the Family Cichlidae and the Suborder Characoidea:

13.3.1 Cichlids

Cichlids are widespread in warm fresh water, occurring in South America and Central America up to Texas, the West Indies, Africa, Madagascar, Syria and coastal India (Nelson 1976). This family (Cichlidae) has constituted the major fish fauna of the African Great Lakes (Lake Victoria, etc.) and has been intensively studied there and elsewhere. Within the African Great Lakes there has been enormous speciation (Table 13.1) and adaptive radiation, with the cichlids showing remarkable specializations, such as fin-biting (4.3.4), and dentition and behaviour modified correspondingly. Mouthbrooding is a common reproductive strategy, protecting the young, and brood parasitism has been reported (13.13). Nelson (1976) gives the following numbers of species.

Unfortunately, human interference, introducing the Nile Perch to Lake Victoria, has caused multiple problems, apparently eliminating many of the species previously described there (Witte et al. 1992). By 1992 these species were described as having included over 300 endemic haplochromine cichlids, with approximately two-thirds lost or threatened.

13.3.2 Characoids

Characoids are ostariophysans (best known by the South American piranhas), with suborder status, Characoidea (Nelson 1976), and 6 to 16 families. The largest of these families, the characins, have, somewhat like the cichlids, speciated in warm, fresh water in Africa as well as South and Central America and Mexico. There are over 1000, generally small in size, characoid extant species and these include blind cavefish, the potentially very large *Brycon* (up to 90 cm) and three genera (Family Gasteropelecidae) possibly capable of 'active propulsion' (but see 13.10.1) when airborne (i.e. flight), which is an interesting contrast to the marine Exocoetidae best known as 'flying fish'. The Amazonian piranhas (e.g. *Serrasalmus*) are feared as though fairly small, they have a reputation for attacking prey *en masse*, and such prey can include mammals at watering points.

Table 13.1 Numbers of fish species African Great Lakes (after Nelson 1976).

Name of Lake	Cichlid species (#)	Non-cichlid species (#)
Lake Malawi	> 200 (most endemic)	42
Lake Victoria	170 (most endemic)	38
Lake Tanganyika	126 (all endemic)	67

13.3.3 Coral reef fish

These fish also show enormous speciation, though based on many families, and adaptive radiation, with a spectacular array of colours and patterns which may change with age, but in general do not show quick changes with background (although there may be some exceptions, e.g. surgeonfish and trumpetfish). It could be anticipated that colour might be influenced diurnally by degree of illumination as in the freshwater neon tetra (*Paracheiodon innesi*) which is bright during daytime and dull and camouflaged at night (Lythgoe and Shand 1983). This phenomenon was first noticed by Verrill (1897). Spalding et al. (2001) estimate 1500 to 2000 fish species in the Australian Great Barrier Reef system, representing about 17 per cent of the global reef area. Some genera and families (Table 13.2) have a wide distribution.

The four families listed (Tables 13.2, 13.3) are fairly typical of reef fish in being acanthopterygians, many with bright colours.

The Pomacentridae includes a set of genera (e.g. *Premnas, Amphiprion, Dascyllus*) which, popularly termed clownfish or anemonefish, form commensal relationships with particular species of sea-anemones, darting into the tentacles for protection.

Table 13.2 Four reef families with a wide distribution (after Nelson 1976).

Family	Common name	Genera (#)	Species (#)
Pomacentridae	Damselfishes	23	230
Scaridae *	Parrotfishes	11	68
Labridae	Wrasses	58**	400
Acanthuridae	Surgeonfishes	9	75

*Recent work indicates this as a subfamily (Scarinae) in the family Labridae (Westneat and Alfaro 2005).
**Some occur in cool Northern waters

Table 13.3 The four families from Table 13.2 as found on Caribbean reefs (after Stokes 1980).

Family	Common name	Genera (#)	Species (#)
Pomacentridae	Damselfishes	5	16
Scaridae	Parrotfishes	4	14
Labridae	Wrasses	8	16
Acanthuridae	Surgeonfishes	1	3

The anenomefish are not harmed (Appendix 13.1) by the stinging cells (nematocysts) of the anemones, and can pick up scraps of food from the anemone. The anemonefish are noted for their broad bright stripes which persist; other damselfish are bright as young but may be dull grey as adults, a situation which complicates identification.

The parrotfish (Scaridae) have characteristic 'beaks' (4.4.5, Figure 4.10) and most species are brightly coloured, particularly as adults.

The wrasses (Labridae) are more streamlined; many species are very brightly coloured and some are very small, being 'cleanerfish', picking parasites off other species. The family is very widespread with two genera reaching into cold northern waters, one (*Tautogolabrus adspersus*) occurs off Newfoundland, where it may become torpid over the winter (Green and Farwell 1971).

The surgeonfish (of the Family Acanthuridae which includes unicornfish and sawtails) are distinguished by the presence of a sharp spine on each side of the caudal peduncle. This spine can lie flat or be extended. Two Caribbean genera (Stokes 1980) have dark and pale phases (perhaps linked to background). The juveniles of the group are commonly noted for different colouration and in the case of the 'Mimic' surgeonfish (*Acanthurus pyroferus*), for very different shape and colours, resembling one or other angelfish (yellow *Centropyge heraldi* or greyish *C. vrolikii*), noteworthy for speed and therefore less likely to be prey (Kuiter 1996). This is reminiscent of the situation with juvenile two-colour parrotfish (*Cetoscarus bicolor*), which as shown in Kuiter 1996 have some resemblance to clownfish, being very different from the adults.

From a different and very small family (Aulostomidae), trumpetfish (*Aulostomus maculatus*) may be 'capable of many colour changes' (Stokes 1980) but this does not confuse identification as the fish have other distinctive characters.

Perhaps the most beautiful coral reef fish are the basslets (Family Grammidae). These small, slim fish may have completely different coloration on the anterior region, a feature which may also occur in the Pseudochromidae (Kuiter 1996). Thus the fairy basslet (*Gramma loreto*) is reported to be purple (anterior) and orange (posterior) in part of its range (Kaplan 1982), while the heliotrope basslet (*Lipogramma klayi*) has a pink head and yellow body (Stokes 1980). Confusingly, the 'orange basslet' (*Pseudanthias squamipinnis*) off Australia has orange females and purple males (Kuiter 1996). The yellow-back basslet (*Pseudanthias bicolor*), also off Australia, is more appropriately named; it is yellow dorsally and pink ventrally, though it also has a pink dorsal fin.

The array of colour and pattern in reef fish is amazing, and it is difficult to summarize briefly. One feature that reappears in several groups is a 'false eye', for example posteriorly, or a stripe obscuring the eye (2.6, Appendix 2.1, Plates 4, 5, 6 and 7). Another feature, linked to the complex behaviour of many reef fish is the change of colour with dominance or with sex change (9.4); thus 'terminal males' may have a distinct colour form, as is illustrated for several parrotfish species in Stokes (1980).

The bright colours of many reef fish, although often marked by stripes or spots tend to be fairly uniform or solid (Plates 4, 5, 6 and 7) in contrast to the scintillation of some freshwater fish like neon tetras or, more spectacularly the rainbow fish (Plate 15), possibly due (in the latter) to iridophores (2.6.5).

13.4 Special adaptations linked to habitat; cold, deep-water systems

Deep water, whether marine or fresh water, can provide a stable, cool, food-poor habitat.

13.4.1 Lakes

Lakes are rarely very deep or very big and may be fairly isolated and species poor in northern regions. The large Canadian lakes include the five linked 'Great Lakes' bordering the United States and the northerly Great Slave Lake and Great Bear Lake.

The deepest lake is Lake Baikal, near Irkutsk. Lake Baikal has a maximum depth of about 1600 m (Kozhov 1963), a depth equivalent to 'abyssal' (Table 13.4); its deepest region (deeper than 500 m) is characterized by a stable cool temperature year-round (3.3–3.6 °C), no light and a scant invertebrate community of gammarids and oligochaetes. Fish species reported at this depth are restricted to five cottoids, typically with reduced eyes but well-developed lateral lines. Overall, 50 fish species (nine families) are reported from Lake Baikal, with one

Table 13.4 Major marine regions.

Region	Depth		Other characteristic
	Metric (m)	Imperial (feet)	
Coastal (continental slope)	To 2000	To 6600[1]	
Offshore	To 11000	To 36000	
Photic			External source of light
All wavelengths	To 100	To 330[2]	
Strong blue	To 500	To 1650[2]	
Trace (blue-violet)	To 1000	To 3300[2]	
Pelagic			In the water column
Epipelagic	0–200[3]		Photic region
Mesopelagic	200–1000[3]		Mid-water
Bathypelagic	1000–6000[3]		Close to the deep bottom
Benthic			On the bottom
Abyssal	> 1000 3000–6000[5]	> 3300[4]	Vast plains
Hadal	6000–11000	20000–36000[6]	The rare deep trenches

[1]From Marshall (1971). [2]Derived from Hardy (1956), from the Sars expedition. [3]Herring 2002, [4]Abercrombie et al. (1980). [5]Jamieson et al. (2009). [6]Converted and approximated from Wolff (1961).

family (Comephoridae) endemic. Pankhurst et al. (1994) analysed the visual adaptations of this family and concluded that there were fewer adaptations for vision than in marine fish at similar depths. They pointed out that freshwater adaptations for vision could be problematic because of turbidity though that is not so much of an issue in 'clear deep lakes'.

Similarly, Great Slave Lake and Great Bear Lake of Canada's North West Territories, have a small number ('at least' 22 for Great Slave Lake, Rawson 1951) of fish species reported, notably sticklebacks (*Culea inconstans* for Great Slave Lake and *Pungitius pungitius* for both lakes), cottids (*Cottus ricei* in Great Slave Lake) and walleye (*Stizostedion vitreum*). Salmoniforms present include lake trout (actually a char, *Salvelinus namaycush*) and grayling (*Thyllamus arcticus*) as well as coregonids (lake whitefish

Coregonus clupeaformis and inconnu *Stenodus leucich-thys*) and an esocid (Northern pike, *Esox lucius*), with little obvious adaptive radiation/local speciation in these northerly taxa which are widely distributed in Canada. The maximum depth of Great Slave Lake is around 600 m (Rawson 1951), which does not have an 'abyssal' component but the degree of ice cover and lack of light in winter provides an environment somewhat similar to the deeper layers of other waters. Fish in these northerly lakes may be slow growing, with the larger predators (pike, lake trout) growing to large sizes and potentially long-lived (more than 20 years, Miller and Kennedy 1948). The relict opossum shrimp *Mysis relicta* and plankton provide an important food source for young fish (e.g. lake trout), which become piscivorous later. Reproduction may be intermittent; for example, for inconnu it may be once per two to four years (Scott and Crossman 1973) and for lake trout once every two to three years (Miller and Kennedy 1948). In colour the fish tend to be silvery.

13.4.2 The deep sea

Deep-sea fish provide a contrast to the shallow, warm-water fish and the colder lake fish; typically they are dark with large eyes and mouths. Like the deep lakes their environment tends to be light-less, food-poor and stable, certainly in terms of tempera-ture unless close to hydrothermal vents. They have shown considerable specialization, and apparently have been well separated from surface or shallow wa-ter species, with allopatric speciation enabling con-siderable differentiation. It was only quite recently, with major deep-water explorations undertaken (e.g. the Challenger Expedition 1872–76), that it was un-derstood that deep oceans could be inhabited. Previ-ous doubts were based on the high pressures, lack of light, etc. expected in the deep oceans and some early sampling as deep-sea dredging by Edward Forbes in the 1840s had temporarily confirmed the doubts. However, subsequently a deep cable hauled up from 1000 fathoms (6000 feet) was encrusted with organisms and the age of deep-sea exploration en-sued. It became evident that deep-water organisms, including fish, had special adaptations.

Overall the seas comprise the productive shal-low coastal waters and the other very large and potentially very deep offshore area. The various re-gions/zones are summarized in Table 13.4. Deeper fish tend to share characteristics by zone: their col-ours tend to be silvery in the upper layers, red in the 'twilight' zone and dark in the aphotic (dark) regions; they may have modified eyes (large or very small) and a large number have light producing or-gans (photophores, 13.4.7).

In some areas of the world covered by ocean there are very deep trenches; there may also be un-derwater volcanoes (and thermal vents, 13.4.5) or their remains as seamounts, and ridges dividing the underwater territory. Seamounts can be associated with high abundance of fish including some larval fish like *Sebastes* (Dower and Perry 2001).

The deepest recorded ocean depth, the Marianus trench, is deeper than the height of Mount Everest and recorded as approximately 36,000 feet deep, about seven miles (Soule 1970). Soule states that the seven major deepest trenches are in the Pacific, while there are four big trenches in the Indian Ocean. The Puerto Rican trench, the deepest area in the Atlantic, is about five miles deep and 450 miles long.

Depending on the turbidity of the water sun-light may not penetrate very far, but even with clear water longer wavelength light (red–green) does not reach much below 300 feet, while some light of the blue–violet wavelengths persists to depths up to and sometimes exceeding 3000 feet (Hardy 1956). Photosynthesis (primary pro-duction) is thus limited to surface waters and the foodwebs are skewed; the ocean depths are food-poor, depending to some extent on detritus, 'snow' sinking from the surface. The presence of bioluminescence in the oceans, however, means that there may be faint light in the depths where sunlight does not penetrate.

The combination of pressure, darkness and pau-city of potential food in the deep oceans meant initially a prediction of no life at depths; however slowly we have learnt, since the late nineteenth cen-tury, that there are organisms in the deep sea which actually may hold some advantages for them (e.g. fairly stable conditions), including temperature and less seasonality than in surface waters.

Although there may be barriers to dispersion, some deep-sea species may be widespread but sparse numerically. It has been very difficult to study them because they are so remote and may be damaged on retrieval. Robot and manned mini-submarines have been useful in obtaining information (Soule 1970) but they are expensive and their presence may interfere with behaviour. In 1960, an American vessel made a dive into the depths of the Marianus trench where they recorded (Soule 1970), at about 35,000 feet, an 'ivory-coloured flat-fish', but this identification has been challenged (Wolff 1961) on various grounds: the animal may have been an invertebrate (holothurian). Trawls to this depth had not found fish before, though 'a brotulid' had been recovered from about 23,000 feet (Soule 1970); Nelson (1976) states that *Bassogigas* (a brotula) was dredged from the Sunda trench, off Java, being the 'deepest definite record of a captured fish' at about 23,000 feet. This record (further referred to by Nielsen and Munk 1964) having been achieved by the Galathea 1951–52 expedition under Bruun, who later (1956) suggested the term

'hadal' for extreme depths, the region below about 6000 m (20,000 feet) and extending to 11,000 m (36,000 feet). Nielsen (1977) studied 11 hadal individuals, obtained from several collections, and described them as a new genus/species *Abyssobrotula galateae* 'the deepest living species of fish known'; he had examined a '*Bassogigas*' (previously known as 'the deepest living fish ever caught', studied by Staiger (1972), recovered from 8370 m (27,500 feet approx.) and reclassified it as an *Abyssobrotula*.

Recently Jamieson et al. (2009) listed 20 hadal (ultra-deep) and 'near-hadal' species. They believe that fish recorded from extreme depths may only make forays into the trenches; to-date no fish have been recorded from the deepest trench bottoms which are a relatively small proportion of the earth's surface (0.56 per cent) as compared to about 50 per cent for coverage by the abyssal region (depths 10,000–20,000 feet; 3000–6000 m). Representative teleosts recorded from deep ocean waters are shown in Table 13.5.

Table 13.5 Teleosts from deep ocean habitats.

Superorders based on early Nelson, prior to 2006, Appendix 1.2a.

Superorder	Family	Common name	Example	Depth (m) adult	Habit
Protacanthopterygii	Opisthoproctidae	spookfish	*Opisthoproctus*	200–600[1]	MP
	Bathylagidae	deepwater smelt	*Bathylagus*	500–1500[2]	BP
Stenopterygii	Gonostomaditae	bristlemouths	*Cyclothone acclinidens*	300–1200[3]	BP
	Chauliodontidae	viperfish	*Chauliodus*	400[4]	MP?
	Sternoptychidae	marine hatchetfishes	*Sternoptyx*	500–1500[5]	MP
	Malacosteidae	loose-jaws	*Malacosteus australis*	> 500[6]	BP
Cyclosquamata	Ipnopidae		*Ipnops*	1500–5400[7]	B
Scopelomorpha	Myctophidae	lanternfishes	*Myctophum*	300–800[8]	MP
Paracanthopterygii	Brotulidae		*Bassogigas*	7000[5]	H
	Macrouridae	grenadiers	*Coryphaenoides yaquinae*	7000[8]	H
	Ceratiidae	seadevil (angler)	*Ceratius holboelli*	1500–2500[5]	H
Acanthopterygii	Liparidae	snailfish	*Pseudoliparis amblystomopsis*	7000[9]	H
	Trachichthyidae	orange roughy	*Hoplostethus atlanticus*	750–1800[10]	MP/BP D

[1]= Ruby and Morin 1978. [2]= Busch et al. 2008. [3]= DeWitt 1972. [4]= Dalyan and Eryilmaz 2008. [5]= Nelson 1976. [6]= Kenaley 2007. [7]= Okiyama and Ida 2010. [8]= Hulley 1992. [9]= Jamieson et al. 2009. [10]= Bell et al. 1992.
B = benthic; BP = bathypelagic; D = demersal; H = hadal; MP = mesopelagic

13.4.3 Deep-sea taxa

It is clear from various accounts that the concept of 'deep sea' is not consistent between authors, sometimes inclusive of mesopelagic forms to be found at depths just above the aphotic zone. Interestingly some mesopelagic fish appear to show considerable vertical migrations, the 'cookie-cutter shark' *Isistius brasiliensis* has been found at 3500 m but has also been recorded in surface waters, at night (Widder 1998). Another family of sharks which may be 'deep sea' is the Etmopteridae including *Etmopterus spinax*, collected at 180–250 m by long-line (Claes and Mallefet 2009a). Holocephalans (chimaeras) may be habitually fairly deep: 'the majority of chimaeroids occupy deep-sea habitats' (Lisney 2010), most of the species being found below 500 m.

Information on deep-sea fish is basically still at the stage of description and functions like reproduction may be very poorly known. It has become apparent that some pelagic fish migrate vertically and diurnally. The Myctophidae (lanternfish) may behave like this and the grenadiers (Macrouridae), which occur world-wide, appear to have the capacity to move fairly actively in the water column, being bathypelagic. The family is speciose and fairly abundant. The Malacosteidae, with huge mouths, are depicted as fairly active swimmers. Some deep-sea fish are apparently truly benthic however, remaining on the bottom, as ambush predators, as with many anglerfish families. The 'spider-' or 'tripod'-fish (i.e. *Bathypterois grallator*) are also well adapted for the benthos, being capable of resting above the sediment with their long appendages raising the main body off the bottom.

Fish from the deep oceans are generally characterized as dark (black) with large gapes, distensible stomachs and abdomens; teeth are usually numerous and very sharp. The capacity to engulf large prey enables deep-sea fish to tackle a great variety of other organisms. A photograph of *Chiasmodon* (Herring 2002) illustrates abdominal distension whereas *Eurypharynx* (Tchernavin 1947b) shows flexion of the head to enable swallowing. Strenzke (1957) figures the huge gape of *Malacosteus niger* (Figure 4.5) which may take large or small prey (Sutton 2005).

Sensory structures may be pronounced, with vision usually well adapted for low light (bioluminescence of specific wavelengths, 13.4.7) and large eyes which may be tubular (Herring 2002 and 12.3.8). However, some deep-sea fish such as Bathypteroidae and Ipnopidae may have very small eyes or, in the case of *Ipnops*, large lensless eyes (Nelson 1976). Marshall (1971) differentiates between mesopelagic species with good sight and bathypelagic species with 'small or regressed eyes'. Deep-sea fish may be elongate or even spherical (anglers) and are typically not very large as adults. Surprisingly they may be soft to the touch, with a tendency to disintegrate on handling.

Marshall (1971) when comparing features of mesopelagic and bathypelagic fish, included the usual presence of the swim bladder in the former and the lack of it in the latter, not unexpected. However, some 'deep-sea fish' have features which belie apparent reality. Thus *Malacosteus niger,* which has a huge gape and formidable teeth (Figure 4.5), has the reputation of being a top predator (Sutton 2005) but turns out to consume a lot of zooplankton, although it lacks gill rakers, inconsistent with expected features of planktivory (4.3.3). This fish has a 'unique visual system', including the use of a form of chlorophyll and a yellow lens (12.3.8, 12.3.12). As with many other deep-sea fish, *M. niger* has multiple rows of rods in its retina (a multibank retina) which should achieve greater light sensitivity (12.3.12); for the mesopelagic *Chauliodus* the number of rows increases as the fish grows (Locket 1980) but this only occurs during initial growth (to about half maximum size) with three bathybenthic species described by Fröhlich and Wagner (1998).

13.4.4 Anglerfish

The deep-sea anglerfish from the paracanthopterygian order Lophiiformes have been ascribed to 10 of 15 families (Nelson 1976), some of which are very poorly known. These 10 families are grouped as a sub-order (Ceratioidei), and are bathypelagic at about 4500–7500 feet (1500–2500 m). Anglerfish are dorso-ventrally flattened and typically have a 'fishing lure', the extended first dorsal spine (illicium) with a terminal 'bait' (esca) which can be dangled over the mouth. In ceratioids the male does not have an illicium, and is dwarfed, either swimming free in the water column or 'parasitic'

(13.12) on the female which occurs in 'at least' 4 of the 10 families updated to 5 of 11 families (Pietsch 2005). As many as eight dwarf males have been recorded attached to one female. Commonly the esca is luminous (13.4.7), and is very long (longer than the fish) in the gigantactinids. Females from each of the 11 families are illustrated by line-drawings in Pietsch (2005). It is not clear how many of these are naturally black though they usually appear so in photographs, and this would be typical of fish at these depths.

13.4.5 Hydrothermal vents and occasional visitors

Hydrothermal vent sites found in volcanically active sub-marine regions and which are characterized by high concentrations of sulphurous compounds and very specialized invertebrates, may have visits, from deep-sea fish as 'opportunistic predators' (Sancho et al. 2005), but it has been very difficult to identify such fish from photographs. Cohen et al. (1990) summarize the situation at the Galapagos rise thermal vents where 20 species were reported (Cohen and Haedrich 1983). Cohen et al. (1990) report the capture of a 'gravid' female of a new *Bythites* species (a brotulid, 13.4.2, Table 13.5) from the same region. Recently, Sancho et al. (2005) have investigated the prey of a vent zoarcid, *Thermarces cerberus*, which seems to specialize in vent predation, and for which analysis 27 specimens were collected by a submersible. Also, recently Munroe and Hashimoto (2008) described a new species of tonguefish, *Symphurus thermophilus*, in association with Western Pacific thermal vents.

13.4.6 The colour of deep-sea fish

In contrast to shallow-water fish, oceanic fish tend to have convergent skin colour, or, at the bottom in maximum depths, very little. Fish in the upper photic zone are often silver (lanternfish, hatchetfish), in the lower photic zone may be red (e.g. redfish *Sebastes*), and in the aphotic zone (e.g. bathypelagic species, Marshall 1971) tend to be black. All of these situations will tend to camouflage the fish; the black would be difficult to see by the faint light of bioluminescence,

red would not be evident as red when only blue–violet wavelengths penetrate, and silver would confuse a predator as light reflects off the potential prey. Depending on the direction from which predators might attack, counter-shading (counter-illumination) may benefit some fish in the deep photic zone, and this is described for some small shark species with small ventral photophores (13.4.7, Plate 14), a bright lower surface might not be perceived, either by a ventrally situated predator or, potential prey (e.g. in the case of the 'cookie cutter' shark, *Isistius*, which takes bites out of large fish) as viewed against a lighter photic area, from below.

The deepest recorded fish may have some pigment, thus the grenadier *Coryphaenoides yaquinae* is fairly dark (typical of macrourids) and liparids may be white or pinkish as seen by artificial light and recorded by camera (Jamieson et al. 2009). The brotula (*Abyssobrotula galataea*), trawled from the Puerto Rican trench at 8370 m (approx. 27,500 feet), was pale yellowish (Nielsen 1977). Benthic invertebrates found as deep are pale (Herring 2002).

13.4.7 Bioluminescence

Photophores, sites of light production, generally by symbiotic bacteria, occur in many deep-sea fish (Claes and Mallefet 2009a). It is estimated that about 70 per cent of 'deep-sea' Osteichthyes have photophores, though only about 6 per cent of deep-sea Chondrichthyans have them and they have not been recorded in chimaeras, being restricted to elasmobranchs. It is not clear what definition of 'deep-sea' is applied here but there is no doubt that many mesopelagic and bathypelagic teleosts as well as deep-water benthic teleosts have the capacity to emit light from luminescent organs called photophores, which may have a restricted location (e.g. on the esca of female anglerfish), or form distinctive species-specific rows and patterns as occurs in the mesopelagic lanternfish, Myctophidae (Figure 1.21). In the few sharks (all 12 species of 2 families) with bioluminescence the photophores (Plate 14) are tiny and ventral, forming a field of 'counter-shading' in *Etmopterus spinax* (Claes and Mallefet 2009a,b) and *Isistius brasiliensis* (Widder 1998). A single batoid ray species has a pattern of photophores outlining its dorsal periphery.

According to Ruby and Morin (1978), more than 40 teleost families have luminescence (in photophores), and in many cases the light is produced by mutualistic bacteria. These bacteria, of which there were six known species (three genera: *Benecka, Vibrio, Photobacterium*) for surface waters, occur as single species per fish and are species-specific. In deeper, cooler water macrourids and *Opisthoproctus* yielded only the cool-water-adapted luminous bacterium *Photobacterium phosphoreum*. However, Widder (2010) is not certain about the bacterial identity. This author also states that the bacteria have to be newly acquired from the environment by each generation. Some fish species have more than one source of their 'luciferin' (the general term for a light-producing chemical). Thus *Linophryne coronata* (anglerfish) females have a bacterial source in their lure and a different (currently unidentified) source in the chin barbel. Ostracods may be a source of a luciferin for some fish (Haddock et al. 2010). Coelenterazine as a light source has been reported from mesopelagic fish species (Mallefet and Shimomura 1995); these authors believe the coelenterazine to be diet-derived and, having found it in fish eggs, support the idea of maternal transfer, somewhat counter to Widder's (2010) report that (bacterial) luciferin has to be freshly acquired.

The physiological control of fish bioluminescence is reviewed by Claes and Mallefet (2009b) being neural in Osteichthyes and hormonal in Chondrichthyes. Adrenalin and noradrenalin can stimulate bioluminescent response *in vitro* in Osteichthyes. In the chondrichthyans (Claes and Mallefet 2009ab; Claes et al. 2012) photophores do not respond to neurotransmitters *in vitro* but do luminesce in response to melatonin (Plate 14) which is inhibited by prolactin and α-MSH. The secondary intracellular messenger cAMP is also probably concerned as forskolin inhibits the bioluminescent response to melatonin. It is suggested (Claes et al. 2012) that in bioluminescent sharks (Dalatiidea and Etmopteridae) the control of the luminosity appears to depend on responses of melanophores overlying the photophores.

Fish which consume bioluminescent organisms may have a screen, in the form of a dark peritoneum or melanophores in the gut wall, which could prevent the light in the gut, particularly the stomach, informing potential predators (Fishelson et al. 1997, 2012). However, a dark peritoneum, as found in the freshwater fathead minnow *Pimpehales promelas* (Stauffer et al. 1995), may have other functions (as yet unknown).

Although it is usual to associate fish with photophores with deeper water habitats (i.e. mesopelagic, bathypelagic or benthic), at least four families (Anomalopidae, Leiognathidae, Apogonidae, Monocentridae) with bioluminescence have members which inhabit or access relatively shallow water (based on Wada et al. 1999; Haddock et al. 2010); Wada et al. (1999) state that bioluminescence is known in many shallow-water marine species. Shallow-water species can be accessed and held in captivity, enabling experimentation. Thus the ponyfishes (Leiognathidae) have been reported to be able to transfer luminous bacteria between generations (Wada et al. 1999) but only post-hatch, when adults were kept with juveniles. The shallow-water Anomalopidae (flashlight or lanterneye fish) are remarkable for being able to shutter their anterior photophores, which can thus be used for prey capture, or be hidden from potential predators (Morin et al. 1975; McCosker 1977).

Bioluminescence produced by fish in the photophores is usually blue or blue-green but may be red (Herring and Cope 2005), as in three dragonfish genera which emit both colour ranges in separate photophores. The source of the light may be bacterial symbionts as mentioned earlier (*vide* Ruby and Morin 1978), but for particular species establishing this may prove controversial. Thus Foran (1991) claimed the involvement of bacteria in bioluminescence of myctophids and stomiiforms but this was disputed by Haygood et al. (1994). The light produced can be modified spectrally by filters (Denton et al. 1985) and can be varied in duration and intensity; light emitted from isolated photophores of hatchetfish can give a fast, weak response or a longer, strong light (Krönström et al. 2005), apparently via nitric oxide modulation of adrenergic nerves.

Morin et al. (1975) discussed the major proposed functions of bioluminescence, which included prey capture and predator evasion. A possible role for conspecific recognition for schooling or mate selection is also possible. McCosker (1977) considered that most bioluminescent fish use their photophores for one of several possibilities, including improved visibility, prey capture or luring, predator avoidance and communication, but that flashlight fish use theirs for all of these.

13.5 Limited habitats

Some fish species occur in very small numbers in severely constrained and small-volume locations; such species are very vulnerable to local extinctions. Somewhat less vulnerable are species living in habitats at the extremes of temperature or pressure or those that venture onto land. Limited habitats also include the very small-volume desert springs and temporary pools, and, in a different sense, the polar regions of large volume but physiologically problematic.

13.5.1 Cavefish

Cavefish present an interesting comparison to deep-sea fish. Caves are dark and dependent on secondary production and detritus. They may have a fairly stable temperature range but can be subject to flash-floods. Population density is necessarily low and fish usually represent the top predators, and in any one cave system there may be only one fish species, although two different genera exist in one location in Iran (Smith 1990). In contrast to deep-sea fish, cavefish lack melanin and are usually pink (Plate 13), attributable to the presence of haemoglobin, and typically have very poor eyesight; photophores are generally not found in cavefish but may be present in cave invertebrates.

The general subject has achieved quite a lot of attention, with the journal *Stygologia* launched in 1985, devoted to cave fauna and also including non-biological features of caves (e.g. hydrology). Cavefish may be referred to as troglobitic (cave-dweller/cave-restricted) hypogean (subterranean) and/or stygobitic (blind, skin unpigmented). The tendency for eye reduction in cavefish has been debated since the time of Darwin (1859). Recently this condition has been reviewed (Tian and Price 2005; Protas et al. 2007); the cost of maintenance of a relatively unusable structure has been evoked as a pressure for eye regression (Protas et al. 2007).

The major taxa represented in caves are ostari-ophysans (e.g. loaches, catfish and characins) as well as the Amblyopsidae (the family termed 'cavefish') and some gobiids (Hoese and Kottelat 2005). Perhaps the largest cavefish species is the Australian blind cave eel, *Ophisternon candidum* (Plate 13), which reaches a length of 40 cm (Allen et al. 2002).

Proudlove (2006), as described in Wilkens (2007), lists over 135 cavefish species, with a wide distribution only excepting Europe and the Antarctic. It is confirmed that co-occurrence is rare. Probably the best-known cavefish is a blind albino form of *Astyanax mexicanus*, a characin, which is available in pet shops. It has a widespread distribution in caves from Brazil through Mexico and into the south-west United States (Nelson 1976). The blind form and the ordinary form can interbreed. A catfish, *Trichomycterus itacarambiensis*, from Brazil, has enormous variability; one-third were reported as albinos by Trajano (1997). The situation with *A. mexicanus* and *T. itacarambiensis* may be unusual for cavefish, though apparently there are approximately 100 pigmented and sighted species which have been regarded (perhaps unreasonably) as 'accidentals' (Romero 2008). For those species which are regarded as true cavefish, having diverged further from their origin, many of them are isolated and threatened. An extreme example is *Speoplatyrhinus poulsoni* (Cooper and Kuehne 1974), an amblyopsid described as from one cave in Alabama, for which fewer than 100 individuals were estimated (Kuhajda and Mayden 2001). Quite contrary to the situation for deep-sea species it is (at least theoretically) relatively easy to gauge the size of many of the populations and confirm the tenuous status, made more precarious by limited food and low reproduction rate, so that recovery even if possible is very slow. Thus Trajano (1997) reports, of her study, that 50 per cent of females of *Trichomycterus itacarambiensis* appeared to be reproductive and observed this is high for troglobites.

13.5.2 Desert springs

The desert springs of North America are associated with the occurrence of 'desert pupfish', generally *Cyprinodon macularias*. This small fish can tolerate very high salinities and temperatures (Hendrickson and Romero 1989) but has been under severe pressure from habitat fragmentation, competition with introduced species and development of reservoirs and similar projects (Echelle et al. 2000). Divergent species include a very small population of *C. diabolis* at Devils Hole, Nevada (Andersen and Deacon 2001) and a more abundant species *C. eremus* close by the Colorado River delta (Echelle et al. 2000).

13.5.3 Polar fish

Polar fish tolerate very cold conditions year round and may thrive to the point of having quite large populations and fairly large individuals. Survival and success in the frigid waters close to either pole has elicited the production of 'antifreezes' which may be glycerol as in smelts or, as is more usual, variations of antifreeze proteins and glycoproteins (7.4) and excretion of these may be limited by the loss of glomeruli from the kidneys (7.4). One suborder of the acanthopterygians, Nothenoidei, has specialized for the Antarctic environment; the Family Channichthyidae includes genera which lack erythrocytes (5.13), an adaptation which reduces viscosity, higher blood viscosity being a potential problem in cold water, and is possible because the cold water has a fairly high level of dissolved oxygen.

13.5.4 Temporary pools

'Temporary pools' can occur in many different locations and may be associated with ephemeral fauna which can grow and reproduce quickly, leaving fertilized eggs capable of withstanding desiccation. There are some fish species that have this type of life cycle. They have been described, notably from South America, as 'annual' fish (Breder and Rosen 1966; Sterba 1962; Greenwood 1975). Fish described as 'annual' include six species of *Cynolebias*, two species of *Rachovia*, and *Austrofundulus*, all of which are cyprinodonts like the desert pupfish. Thomerson and Turner (1973) added the Venezuelan *Rivulus stellifer*. Sterba (1962) and Greenwood (1975) also list some African cyprinodonts (*Aphysemion, Epiplatys, Micropanchax, Nothobranchius, Pachypanchax*) as annual or possibly annual, with Sterba more tentative than Greenwood. 'Annual' fish are necessarily small and fast-growing.

Rivulus marmoratus (now *Kryptolebias marmoratus*) responds to drying of its aqueous habitat by becoming emergent and seeking refuge in logs or other damp places (Taylor et al. 2008). Ong et al. (2007) report, based on Abel et al. (1987), that this species can survive emerged for more than one month.

13.5.5 Terrestrial fish

Despite the usual expectations, there are fish which can air-breathe (6.6) and even spend extended periods of time out of water. Terrestrial fish are few but this kind of habit has considerable interest in view of the evolution of tetrapods and the physiological adaptations which have arisen in several lineages.

Littoral fish inhabit shores and some of them, besides being highly tolerant of varying temperature and salinity, can remain out of water for several hours without harm. Some, indeed, can even be deemed 'terrestrial' or amphibious because the majority of their time is spent on land. Perhaps the best known terrestrial fish is the goby *Periopthalmus* (the mudskipper), which has protruding eyes enabling a good field of vision; it can haul its small body around on mud, using its paired fins (notably the pectorals) as props and limbs. *Gillichthys mirabilis* is a large marine-shore goby from the Californian coast. It can tolerate being out of water for several hours. Blennies of the genus *Alticus* such as *A. monochrus, A. kirki* found in the splash zone of islands in the Indian and Pacific Oceans and by the Red Sea, spend most of the time out of water. *Anabas*, the 'climbing perch' as its name suggests, is noted for land excursions, and *Anableps* (the 'four-eyed' fish), although not terrestrial has the habit of cruising the water surface with its eyes protruding into the air. The upper-half of each eye is different from the lower, each half being suited for the different refractive index encountered (12.3.8). The freshwater archerfish *Toxotes* can remain in water but extends its predatory field by 'spitting' at insects resting on land (Figure 4.7).

The capacity to survive terrestrial conditions temporarily may be very important for fish in river systems that flood and recede periodically, as occurs in the Amazon basin. Kramer et al. (1978) reported several taxa from the Amazon that might be capable of overland movements, and observed several catfish species moving effectively when placed on land. Freshwater eel (anguillids) are traditionally known to move between water bodies, thus enabling either escape or invasion (as described for *Callichthys*, Lüling 1973, quoted by Kramer et al. 1978).

Recently there has been particular interest in *Kryptolebias* (formerly *Rivulus*) *marmoratus* (13.5.4), a small cyprinodont which has been described as amphibious. If its aqueous habitat dries it can emerge, with an apparently unique jumping/flipping locomotion, and seek damp shelter, and is capable of surviving out of water for more than one month. It is also a self-fertilizing hermaphrodite (9.4). These fish are thought to use the skin surface as a gas-exchange surface when emerged (Ong et al. 2007).

13.5.6 Burrowing

Burrowing fish include congrid eels, notably the garden eels, Subfamily Heterocongrinae, which are popular subjects for marine aquaria. Each eel has its individual vertical burrow from which its anterior end emerges (Plate 12) and into which it withdraws rapidly if disturbed. Synbranchids (swamp-eels) have a burrowing habit; they have a wide distribution from South America, Africa and Asia, with two genera in Australia where they include some cave-dwelling populations (notably of *Ophisternon candidum*, Plate 13). The swamp-eels generally lack fins, have a single ventral gill opening (6. 5.2) and small eyes.

The Japanese goby, *Luciogobius*, burrows in gravel sediment (Yamada et al. 2009) and the flatfish family Pleuronectidae are generally characterized by their habit of lying in the substrate with just their eyes protruding, which is very typical of the soles that favour soft sandy bottoms. One pleuronectid, the 'Greenland halibut' (*Reinhardtius hippoglossoides*), is much more active, tending to be pelagic, actively swimming in the water column as a fish predator. Rather than actively burrowing, some fish may utilize available shelters, even tin cans, as has been described for the nesting toadfish, *Opsanus tau* (Fine 1978). The New Zealand torrent fish *Cheimarrichthys fosteri*, which inhabits braided river systems, is thought to burrow during the day, emerging to feed at night (Glova et al. 1987).

Sandlance (*Ammodytes*), a small fish of the North Atlantic (and Pacific), is reported to burrow in sand or gravel, for example at night, rising into the water column to feed in the evening and remaining close to the bottom during the day (Scott 1985). The Australian sandlance (*Limnichthyes fasciatus*) is also a burrower with special corneal adaptations (12.3.8).

13.6 Fish communication

Schooling fish (14.6.1) may have excellent coordination and communication as a group. In non-schooling fish also there may be various inter-individual communications possible, depending on sight, sound or even electric fields.

Fish communication has recently been extensively reviewed (Ladich et al. 2006). It includes aggressive displays and mate recognition and courtship and depends on a variety of perceptions (Chapter 12) including electroperception, visual responses and less obviously, sound production and response.

Mate recognition may depend on shape, colour patterns and movements (thus males may court gravid females as occurs with sticklebacks, Wootton 1984) or, more rarely, sounds (13.7) produced by the male.

13.7 Sound production

Sound production by fish has been reported in many species (at least 700, Stuart et al. 2009), including paracanthopterygians (gadids, carapids, toadfish) and characins (piranha) and of course the 'grunts' (Pomadasyidae), grunters (Teraponti-dae) and 'drums'/'croakers' (Sciaenidae) as well as some mormyrids, triglids, croaking gouramis and others (Table 13.6).

Sound production may depend on 'drumming' (sonic) muscles adjacent to the swim bladder or, in the case of 'grunts', on the grinding of pharyngeal teeth with amplification by the swim bladder (Nelson 1976); catfish may use both drumming muscles and the stridulation of pelvic fin spines (Amorim 2006). The croaking gourami uses its pectoral fins (Ladich et al. 1992).

Table 13.6 Taxa with sound production.

Subdivisions as for Table 1.3.

Subdivision	Family	Species example	Common name
Osteoglossomorpha	Mormyridae	*Gnathonemus petersii*	elephantfish
Ostarioclupeomorpha	Callichthyidae	*Corydoras paleatus*	armored catfish
	Characidae	*Pygocentrus nattereri*	red-bellied piranha
Euteleostei	Gadidae	*Gadus morhua*	Atlantic cod
	Carapidae	*Carapus acus*	pearlfish
	Batrachoididae	*Opsanus tau*	toadfish
	Terapontidae	*Bidyanus bidyanus*	silver perch (grunters)
	Sciaenidae	*Pogonias cromis*	black drum (drums, croakers)
	Pomadasyidae (now Haemulidae)	*Haemulon plumieri*	grunts
	Belontidae	*Trichopsis vittata*	croaking gourami

Piranhas apparently use sound in food competition displays (Millot et al. 2011) whereas male toadfish use it for attracting females and repelling males. Female toadfish also produce sounds when disturbed and overall the sound communication system is quite complex (Thorson and Fine 2002), although the most obvious feature has been the 'advertisement' call of the male, termed the 'boatwhistle' (Fine 1978; Thorson and Fine 2002). The various sounds emitted by fish have been described as boops (toadfish, Thorson and Fine 2002), thumps, hoots, pops, knocks, grunts, moans, growls and hums with some of these categories recognized as a prolongation and fusion of shorter sound pulses (Amorim 2006). Thus some species are capable of several different sounds and individuals within a species may also show variation. Pearlfish (carapids), parasitic or commensal within invertebrates, can communicate with sound from within their hosts (Parmentier et al. 2006a).

Drumming by male cod has been studied in terms of variable drumming muscle mass, with higher mass linked to mating success (Rowe and Hutchings 2008). The drumming muscles of male

weakfish (a sciaenid) triple in mass at the reproductive season (Connaughton et al. 2002).

Most drumming muscles can be categorized as exceptionally fast (though not in carapids), being the fastest vertebrate muscle known (Parmentier et al. 2006b), with a very fast relaxation rate, and many structural and biochemical specializations, including a helical organization of the myofibrils in carapids, a prominent sarcoplasmic reticulum and protein isoforms. Intrinsic drumming muscles in toadfish and triglids attach directly to the swim bladder, whereas extrinsic muscles in sciaenids, carapids, holocentrids (squirrelfish) are linked less directly, via ribs and ligaments (Parmentier et al. 2006b).

13.8 Electric fish

Fish with electroperception (12.9) are capable of detecting potential prey by the presence of small electric fields generated round nerve and muscle due to the essential action and muscle potentials of the prey even when resting or hidden in sediment.

Some fish are also capable of generating much larger fields or discharging large or small voltage impulses for electrolocation, communication, aggression or defence.

Electric fish have special adaptations for navigation, prey capture and defence, and/or awareness of conspecifics. They can generate (depending on the species) high- and/or low-voltage discharges (electric organ discharges, EODs) from groups of special muscle or nerve cells (electrocytes). In some cases, such as the African elephantfish, mormyrids ('about ten genera' Nelson 1976), the field generated is used to navigate in the turbid water and to recognize conspecifics. These fish can detect and interpret electrical field disturbances.

There is a distinction between strong electric organs and weak electric organs (Lissman 1958), with Bullock (1999) describing individual species of electric fish as either strongly electric or weakly electric although *Electrophorus* is both. Bullock (1999) expects more weakly electric taxa to be described, especially as some 'episodic' fish discharge irregularly.

'Strong' electric fish include the marine electric rays (Torpedinidae, nine genera Nelson 1976) as well as two stargazer genera (*Astroscopus* and

Uranoscopus) and two freshwater genera: the African catfish *Malapterurus* (with three other catfish genera showing only weak EODs, Alves-Gomes 2001) and the South American electric eel (not a true eel) *Electrophorus electricus* (see also Table 12.1). The high-voltage discharges (up to 600 volts, Albert et al. 2008) of *Electrophorus*, used for predation (Bauer 1979) are sufficient to stun or kill small to medium-sized mammals at a watering point. This fish has both high- and low-voltage discharges (Heiligenberg 1993), achieved by three different electric organs on each side (Keynes and Martins-Ferreira 1953), two (the main organ and the posterior Hunter's organ) of which are capable of high-voltage discharges, whereas the posterior Sachs' organ produces low-voltage (millivolts, Albert et al. 2008) low-frequency EODs irregularly (Bauer 1979). The catfish *Malapterurus* apparently has high-voltage discharges (Nelson 1976). The elasmobranch marine electric rays (Family Torpedinae), of which there are nine genera (Nelson 1976), have high-voltage discharges; the Atlantic stingray *Dasyatis sabina* is reported (Sisneros and Tricas 2002) to have 'ampullary electroreceptors' (i.e. ampullae of Lorenzini, 12.9.3) to detect weak fields produced by prey and potential mates apparently due to muscle and nerve activity.

Most electric fish produce low-voltage pulses, which may be only a fraction of a volt as in *Gymnotus carapo* of South (and Central) America. It is only relatively recently that such weak electrical discharges have been recorded (Lissmann 1951, 1958) although there had been earlier morphological evidence for possible electric organs (Keynes 1957).

Weakly electric organs can be a component of an electrosensory system (12.9.1) for detecting objects, conspecifics or prey, which can distort the electric field created by the electric organ, facilitating electrolocation.

Weakly electric fish occur in the elasmobranch Rajidae and teleost Mormyridae, Gymnarchidae, Gymnotidae, Stenarchidae, Sternopygidae and Ramphichthyidae (see also Table 12.1). Electrogenic fish which are weak emitters tend to have similar (convergent) shapes with long median fins providing propulsion, which is expected to minimize interference. The term 'knife-fish' (culteriform shape) has been applied to several weak electrogenic species (e.g. the ostariophysans *Gymnotus*, *Stenarchus*,

Eigenmannia) and one member of the osteoglossomorph Notopteridae (*Xenomystus*) which is 'electroreceptive' (Bullock and Northcutt 1982).

Collectively electrogenic fish have a wide geographical distribution with marine and freshwater species.

13.8.1 Bioelectrogenesis

The generation of electricity by live organisms (i.e. 'bioelectrogenesis') includes the generation of action potentials in nerve and muscle which can be summed to achieve higher-voltage discharges as in electric fish. Questions central to electric organ discharge are how does the summation occur and how is self-electrocution avoided in strongly electric fish? Basically the electric organs are derived from either muscle or nerve tissue and have structure facilitating electrical summation.

The macro- and micro-structure of electric organs is quite variable, is species-specific in location but can occur at any point of the body, from the head region (Uranoscopidae), or the main body (many species), to the tail (Rajidae). They are usually formed from banks of modified muscle cells and originate from localized eye muscles in *Astroscopus*, more extensive branchial muscles in *Torpedo* and from much of the posterior body and/or caudal fin musculature in most other electric fish (Keynes 1957; Bennett 1971b). In *Electrophorus* most of the posterior body and caudal fin musculature is electrogenic, the three electric organs on each side collectively representing a large proportion of the fish.

The individual electrogenic cells are generally derived from embryonic muscle cells, that is they are myogenic in most electric fish. An exception is the electric organs in the family Stenarchidae, which are neurogenic, derived from nerve fibres. The individual cells of myogenic electric organs are usually flat 'plates' and are known as electroplaques or electroplates (Keynes and Martin-Ferreira 1953; Keynes 1957) but the term 'electrocyte' has been introduced by Bennett (1971b), taking into account the interspecific variability in the shape of these cells. Myogenic electrocytes are multinucleate, like striated muscle cells (Keynes 1957). They form columns embedded in a clear jelly-like extracellular matrix, which distinguishes electric organs from surrounding muscle.

Electrocytes typically have one smooth innervated face and an opposite (folded) non-innervated face and are oriented in the same direction in any species, for example the smooth face is posterior in *Electrophorus* but dorsal in *Astroscopus,* ventral in *Torpedo* and anterior in *Raia/Raja* (Keynes 1957). In *Electrophorus* the main electric organ has up to 6000 electrocytes arranged in series and the number increases in longer fish. The electrical resistance of the smooth and folded faces was measured by Keynes and Martin-Ferreira (1953), that of the folded face being lower. The large folds increase the surface area which decreases the electrical resistance of this face (the resistance of an electrical conductor being inversely proportional to its diameter), thereby facilitating discharge of the electrocyte.

Microelectrodes have been used to record the electrical potentials across the electrocyte membranes of *Electrophorus*, revealing that the innervated face is essentially a normal muscle excitable membrane (Appendix 11.1) in this species. Electrocyte discharge is due to a response asymmetry between the smooth and folded faces. Acetylcholine from the nerve endings increases the smooth face Na^+ and K^+ permeability to generate a post-synaptic potential. This generates an action potential (11.3.3) on the smooth face with a reversal of membrane polarity, but not on the folded face, resulting in a current flow across the cytoplasm and the low resistance folded face.

The voltage change is additive between electrocytes, like batteries arranged in series, resulting, in *Electrophorus*, in the head region becoming positive relative to the tail (Keynes and Martin-Ferreira 1953), and in a current in the external medium associated with the electrocyte discharge. With up to 6000 electroplaques in the main electric organ of *Electrophorus*, these additive individual voltage changes can result in an overall discharge of up to 600– and 850 volts. In the case of marine electrogenic fish the resistance of seawater is less than that of fresh water and the innervated electrocyte face is not excitable; the post-synaptic potentials (11.3.4), other than action potentials, generate the electrocyte discharges (Keynes 1957; Bennett 1971b).

Synchronization of electrocyte discharges is essential to achieve a maximum additive voltage. This is attained partly by variable axon diameters and conduction velocities and through direct and circuitous axonal pathways varying the conduction distances. In *Electrophorus*, electrogenic organ nerves originate from large perikarya in the spinal cord, exiting through the ventral roots of the spinal nerves (Keynes 1957).

In the weakly electric knifefish of the family Sternarchidae the electrocytes are neurogenic rather than myogenic, and are derived from spinal nerve neurons. The electric organ lies longitudinally along the fish, ventral to the vertebral column. Axons from segmental spinal nerves run anteriorly in the organ, then fold back posteriorly. The axons have dilations about 100 µm in diameter. It is suggested (Bennett 1971b) that this electrogenic organ lost its myogenic electrocytes and that the spinal nerves supplying them became enlarged. The axons have two types of nodes of Ranvier (11.3.5). Type I nodes are morphologically normal and produce action potentials, whereas Type II nodes do not. The Type II nodes are wide (50 and their axon membrane is folded extensively, acting as capacitors, transforming action potential electrical change into external discharge signals (Waxman et al. 1972; Waxman 1997) enhanced by the increased axon surface area.

This account has concentrated on two bioelectric organs demonstrating the myogenic and neurogenic differences. However, there is considerable diversity in the morphology of bioelectric organs (Bennett, 1971b). Also, the evolution of bioelectrogenesis in teleosts is reviewed by Alves-Gomes (2001) who states that 'electric organs have evolved eight times independently in teleosts'.

13.8.2 How is self-electrocution prevented?

The magnitude of the electrical discharges generated by strongly electric fish has prompted questions about how they avoid electrocuting themselves. Keynes (1957) and Young (1981) state that there is no complete answer to this question, although it is known that electric fish are protected against externally applied electric shocks. It has been suggested (Keynes 1957) that internal lines of current flow are at right angles to excitable structures which reduces their stimulation. Also, fats are good insulators, and the spinal cord and myotomes are invested with thick coverings of these compounds in *Electrophorus*

(Keynes 1957). The innermost layer of fish skin is the hypodermis (2.3.3) which is an adipose tissue which can form a relatively thick fatty layer around the body of fish. This invites a carefully planned comparative histological study of the hypodermis of electric and non-electric fish to determine the degree of its development, which could have an important role as an electrical insulator. Other attributes of the integument, such as its thickness and extent of junctions between epidermal cells, also require further attention.

13.9 Defence

Fish can be actively or passively defensive, although assigning these clear categories may not be satisfactory as, for example, spines could be regarded in either mode as could camouflage if it entails change in colour and/or pattern appropriate to background (2.9). Active defence includes escape reactions such as 'tail-flips' achieved via the very fast conducting Mauthner fibres (11.6.6). Mimicry, a situation when animals resemble other animals (or behave misleadingly), has been reported from coral reef fish (e.g. the 'Mimic' surgeonfish, 13.3.3).

Passive camouflage, depending mainly on morphology may be a very effective defence. Perhaps its most impressive form is in the leafy seadragon (13.10.1) whose shape disguises its presence very effectively when amongst weed.

Active camouflage, when fish change colour to match their background (2.9), is complex, depending on morphology (different chromatophores, colour, shape, size and number in different locations) and the ability to slowly change numbers of these cells and/or rapidly change colour when pigment bodies (chromatosomes) move within chromatophores (2.9).

Defensive behaviour, apart from the use of strong electric pulses (13.8), also includes roles for spines and toxins. Studies of sticklebacks as prey show (4.3.2) that their spines can be effective deterrents. The elasmobranch stingrays, and some teleosts, such as weevers and the *Noturus* catfish (stonecats and madtoms), are notorious for their toxins; the most toxic fish reported are the puffers (fugu) which synthesize tetraodotoxin, found in various body organs.

13.10 Special adaptations linked to morphology

13.10.1 Fin shapes and modifications

Electric fish with high- or low-voltage discharges for prey detection, defence or navigation tend to have elongated shape and fins that minimally interfere with discharge or with the field effect. *Electrophorus* and its kindred (the Superfamily Gymnotoidea; 'about 16 genera' Nelson 1976 have long median fins which can achieve locomotion by undulation, so that lateral body movement is minimized.

Fin modification takes many forms in other fish. Fish bodies may be compressed or elongated and the fins may take many roles, besides the apparently primary one of steering swimming or support. Fin extensions can achieve elongations along or away from the main body axis and whole fins or individual spines may be involved, to achieve a number of different functions including camouflage.

The leafy seadragon, *Phyllopteryx eques*, a seahorse from the Australian coast, has an amazing set of protruberances resembling seaweed; it is exceptionally difficult to see this fish *in situ*. Other fish with similar protruberances achieving camouflage include cottids. The sea moths (Pegasidae), as their name suggests, have a very strange shape, with extremely large pectoral fins, which may be splayed, giving the appearance of wings.

Elongations of single spines may produce lures as with the anglerfishes (13.4.4), or they may serve as channels forming toxic defence structures (weavers, stonefish).

The flyingfishes, for example *Exocoetus*, with seven other genera, have very long, paired fins which can achieve air-gliding after a leap, apparently evasive. Some genera are two-winged (long pectorals) and some four-winged (long pectorals and pelvics). The pectoral fins of the 'flying' characoid Gasteropelecids (13.3.2) are not as pronounced, the short 'flights' apparently being dependent on the large pectoral region muscles which enable leaps (Wiest 1995); it is doubtful whether active flight is actually accomplished though observers have believed it occurs.

The anal fin of male live-bearing poeciliids (e.g. sail-fin molly, *Poecilia latipinna*) is modified as an intromittent structure for internal fertilization.

13.10.2 Fish shape

Fish shape was briefly reviewed in 1.4.3. Fish can be tubular or flat but most fish are indeed 'fish-shaped' with a roughly oval body and prominent fins. Additional projections such as sensory barbels (gadids, catfish) or lures (anglerfish) are well known. Smaller defensive or offensive projections such as spines which may be protracted (surgeonfish) are found in a number of families (Gasterosteidae, Cottidae). Puffers (Tetraodontiformes) can inflate themselves as a defensive behaviour.

At least two families have 'necks', a most unusual feature for fish: the genus *Lepidogalaxias* (a protacanthopterygian) from south-west Australia cannot move its eyes but has large gaps between the anterior vertebrae enabling head flexibility (Allen et al. 2002), and the ostariophysan *Raphiodon vulpinus* from the Amazon has a hinge between the fourth and fifth vertebra, enabling it to bend the head upwards (Lesiuk and Lindsey 1978).

Extreme elongation occurs in the many eel families and in the oarfish as well as in ostariophysans like the 'electric eel'. This fish shape has recently been analysed by Ward and Mehta (2010).

Hatchetfish, angelfish and sunfish are flattened laterally but symmetrically whereas 'flatfish' (Order Pleuronectiformes) are flattened laterally but asymmetrically, lying on one side (13.11) during development. The elasmobranch skates, rays and torpedoes are flattened dorsoventrally and symmetrically.

13.11 Special adaptations: positional

Some fish, notably members of the catfish family Mochiokidae (Synodidae) regularly swim 'upside-down' (Blake and Chan 2007), a feature somewhat reminiscent of the mammalian sloth. The shrimpfish (family Centriscidae) swim with their heads vertically down, often as a group. The tube-eye *Stylephorus chordata* swims vertically head-up, and has the capacity to move its tubular eyes forward or upward (Nelson 1976). The seahorses tend to swim vertically, with their head upwards, and may coil their tail round vegetation, maintaining an upright position.

The flatfish (the basically marine Order Pleuronectiformes) begin development as normal-looking fry/juveniles swimming actively but, at metamorphosis, undergo an extraordinary transition in which they lie on one side with one eye migrating from the lower to the upper surface. Depending on the taxon (Family) these fish lie on their right side (eyes sinistral, e.g. the lefteye flounders Bothidae), or on their left side (eyes dextral, e.g. the righteye flounders Pleuronectidae) or on either side (Psettodidae, i.e. eyes sinistral or dextral). Flatfish spend most of their time on the bottom or partially buried in the sediment but at least one species (the Greenland halibut *Reinhardtius hippoglossoides*) has resumed a more pelagic habit, as an active predator.

The oceanic sunfish (*Mola mola*) which is flattened laterally and symmetrically has been reported to float at the water surface, on its side, but it is not clear whether such individuals are moribund.

13.12 Hermaphroditism and polymorphism

In association with sexual reproduction fish may develop particular, externally visible differences.

Hermaphroditism (9.4), in which an individual is both male and female either simultaneously or sequentially, occurs in some fish (as in many invertebrates) but not in tetrapods. Changes associated with sequential hermaphroditism may include colour changes which, once made, may persist, appropriate for display and courtship. Some reef species, notably wrasses and parrotfish, may have a complex social structure and different colour forms and sizes, amounting to 'polymorphism'. A form of polymorphism may also occur in salmonids which can have dwarf and/or land-locked forms coexisting with larger sea-run fish.

Sexual differences in colour, size, structure or shape produce temporary or more permanent sexual dimorphism (9.3). Temporary changes occur, for example with pre-spawning lumpfish, sticklebacks and salmonids, in which the male develops different colouration (notably bright pink all over for male lumpfish and blue for females, or red ventrally for male sticklebacks and salmonids). Large pre-spawning male salmonids also develop a jaw hook, the kype (Witten and Hall 2003), during their migration upriver. In some salmonids, moreover, there are freshwater resident dwarf (sneaker) males or jacks in addition to the 'normal' size males, and sneakers may be successful in achieving fertilization of some eggs when the regular pair spawns. It is quite common to find external signs of sexual difference during the reproductive period; generally males are more brightly coloured and may have elaborate displays (e.g. *Betta splendens*, the Siamese fighting fish) reminiscent of bird courtship. More permanent differences between males and females include size differences, with males often being reproductive at a smaller size than females (9.3) or males with claspers (Chondrichthyes) or less permanently, special tubercles (goldfish) or spawning ridges (capelin) for holding the female. Some flatfish males such as *Bothus* may have their eyes widely separated compared to the females (Greenwood 1975; Kobelkowsky 2004). Another flatfish genus, *Arnoglossus*, can have males with very elongated antero-dorsal spines (e.g. *A. tapeinosoma*, Norman 1934). The most bizarre dimorphism occurs with anglers (13.4.4) which have very small males, which, in some families of the deep-sea anglers (ceratioids) become permanently attached and fused to the females as 'parasites'.

13.13 Symbioses

Symbioses, in which organisms become dependent on relationships with live organisms of other species, include parasitism, an unequal situation between two species, with a usually larger host and a smaller species which harms the host usually by feeding on it externally (ectoparasite) or internally (endoparasite). Other symbiotic relationships include mutualism (both species benefit) and commensalism (one species benefits, the other is not harmed). Understanding symbiotic relationships and categorizing them may require lengthy observations in which a situation thought to be mutualistic may turn out to be parasitic.

Close interspecific relationships in fish are fairly rare, though fish may be host to various parasites parasitic fish are unusual, although the lampreys (the major component of the extant jawless fish) are parasitic, as adults, on jawed fish.

Commensalism or mutualism may exist with invertebrates such as anenomes or sea-urchins.

Brood parasitism has been reported in the African Great Lakes; Sato (1984) reported a catfish, *Synodontis multipunctatus*, parasitic on mouth-brooding cichlids in Lake Tanganyika and Stauffer and Loftus (2010) found a clariid catfish, *Bathyclarias nyasensis*, parasitic on the bagrid catfish, *Bagrus meridionalis*, in Lake Malawi. In these cases the parasite fish accomplish insertion of their fertilized eggs into the brood of the host (as occurs with cuckoo birds) and, in the case of *S. multipunctatus*, the eggs hatch before the hosts and after yolk-sac absorption feed on the fry of the host, as they hatch.

True parasitism is very rare in jawed fish. The remoras (seven genera, including *Remora*, Nelson 1976) are not parasites but they have a large sucker dorsally on the head and can attach to large fish, turtles or mammals and can pick up food scraps. They are thus 'commensal symbionts' as are the anemone fish (13.3.3) which get both protection and food from their hosts. In commensalism the host does not benefit but is not harmed either. A similar 'protective' relationship occurs between the Banggai cardinal (*Pterapogon kaudneri*) and anemones or the long-spined sea urchin *Diadema setosum* (Vagelli 2004); the fish can take refuge amongst the spines of the urchin. Clingfish (Gobiosoceidae) may also have close relationships with other species including urchins; the New Zealand urchin clingfish (*Dellichthys morelandi*) are found beneath the urchin *Evichinus chloriticus* which may provide 'shelter and food' (Dix 1969). Similarly the clingfish *Diademichthys deversor* is found amongst the spines of *Diadema*, and tubefeet, presumably grazed from the urchin, have been reported in the gut of the clingfish. At least two species of clingfish (*Cochleoceps bicolor* and *C. orientalis*) are commonly known as 'cleanerfish': cleanerfish (including some small wrasse species, Tables 13.2 and 13.3), have a mutualistic relationship with their hosts from which they remove ectoparasites. Pearlfish (carapids) also enter live invertebrates and some species are described as parasitic in that they feed on organs of the sea cucumbers they have entered (Nelson 1976; Parmentier et al. 2016).

The Trichomycteridae is a large family of mostly small catfishes (27 genera, about 185 spp., Nelson 1976; 41 genera, 207 spp., Ferraris 2007) that includes obligate parasitic fish. Thus 17 genera from 2 subfamilies, Vandelliinae; 4 genera, 9 spp. and

Stegophilinae; 13 genera, 31 spp. (Fernández and Schaefer 2009) are parasites on gills (the vandelliines) or skin (stegophilines) of fish. These fish are feared by humans in their South American range because *Vandellia cirrhosa* (candiru) may enter human orifices, although this has not been substantiated (Spotte 2002), with the possible exception of one old first-hand account of removal of a fish from a vagina and a recent somewhat questionable record of male urethral penetration. The deep-sea eel *Simenchelys parasiticus* is also believed to be parasitic, as its name shows. However, 'little is known of its feeding habits' (Nelson 1976), but a recent observation (Caira et al. 1997) led to the conclusion that this fish is a facultative endoparasite.

13.14 Conclusion

The considerable diversity in form and function which is found in fish is appropriate for such an ancient and large group with extensive niches in the aquatic environment. Often fish studies are based on one or two 'representative types' (15.6) to demonstrate the basic biology and some differences between major taxonomic groups. Beyond these limitations students of fish biology can expect variability comparable in scale to that in tetrapod adaptations to the diverse niches and habitats of the terrestrial environment.

Appendix 13.1 Anemone fish and nematocysts

The ability of some anemone fish (Pomacentridae) (13.3.3, 13.13) to live amongst the tentacles of coelenterates (anenomes, corals) has been the subject of conjecture. Normally any fish brushing against, for example, anemone tentacles would be adversely affected by discharge of the stinging cells called nematocysts. It is suspected that the pomacentrids which have this successful commensal relationship with such potentially dangerous hosts have the capacity to secrete special mucus or acquire mucus from their hosts (Mebs 2009; Balamurugan et al. 2014). Nelson et al. (2016) state that nematocyst discharge is inhibited.

Bibliography

Krahe, R. and Fortune, E. S. (eds). (2013). *Electric fishes: Neural Systems, Behaviour and Evolution*. Special issue, *Journal of Experimental Biology*, 216.

CHAPTER 14

Fish behaviour

14.1 Summary

Behaviour is a facet of animal life generally associated with movement. In vertebrates, it is based on the reflex arc and chains of events which may be continuous (ventilation), intermittent or even singular (some forms of reproduction being terminal).

Fish are capable of learning, with short- and long-term memories, some due to early imprinting.

Many fish functions involving interaction with the environment, including other fish, are directly dependent on locomotion. Signalling systems between fish include pheromones.

Reproduction usually involves a sequential set of behaviours that may include courtship and parental care.

Fish can recognize and discriminate topography, sometimes using electroreception; they can also recognize predators, kin and potential mates and (cleanerfish) potential clients.

Fish neuroanatomy has similarities with that of other vertebrates but differs from mammals in many features including, for fish, the lack of a cerebral cortex, important for mammalian learning and memory, and the presence in fish of both Mauthner fibres and chromatophores (not found in mammals).

14.2 Introduction

With the possession of muscles (Chapter 3), nerves (Chapter 11) and sensory receptors (Chapter 12), fish undergo many activities which can be regarded as behaviour. Animal behaviour generally refers to a set of reactions to environmental stimuli, often manifesting as changes in position or appearance but which may also involve the facility for learning and memory (detected by particular actions like associating objects or, for example, sounds with places or sights). Behaviour is characteristic not only of multicellular animals but also of those other organisms which are motile (e.g. *Paramoecium*).

Some movements either of whole organisms or their cells are classified as taxes or kineses, the former being directional, in response to a specific stimulus, for example chemotaxis in response to a chemical stimulus. Kineses are locomotory too but involve increase in speed or turning frequency (Abercrombie et al. 1980) in response to a stimulus and are not directional with regard to the origin of the stimulus. These kinds of responses (taxes, kineses) have been studied intensively for some protozoa or invertebrates; they may be very relevant to reproduction in fish, with regard to mate attraction or sperm movement.

Plants are usually excluded from the category of 'behaving' though they can show local movements in response to environmental change, phototropism, for example, being a directional response to light, when plant parts move, not with the use of muscles but by, for instance, pressure (turgor) change.

The study of behaviour (ethology) has generated widespread interest but has also involved some controversy in interpretation of observations. One problem in understanding fish behaviour is the risk of an anthropocentric approach, attributing human attitudes to fish, which may be inappropriate. Moreover, some fish behaviours such as electrolocation lie well outside human experience and this may make understanding difficult. The idea of 'intent' sometimes pervades terminology applied to fish behaviour, thus the words 'strategy' and 'tactics'

Essential Fish Biology: Diversity, Structure and Function. Derek Burton & Margaret Burton.
© Derek Burton & Margaret Burton 2018. Published 2018 by Oxford University Press.
DOI 10.1093/oso/9780198785552.001.0001

(Potts and Wootton 1984) have been applied to certain facets of fish reproduction. This usage seems to be accepted but if these terms are interpreted as voluntary behaviours they should be retained with caution as 'intent' is difficult to establish and falls within the possible criticism of being teleological and subjective (see discussion in Thorpe 1956). Currently the use of fish 'models' to understand vertebrate behaviour, including the medical/psychiatric implications, have been accompanied by acceptance of terms like 'anxiety' and 'fear' in this context.

As well as the science of 'ethology' neuroscience has had a particular interest in animal behaviour. Analyses of factors causing or influencing behaviour have dealt with a number of possibilities which include instinct, learning, intelligence and memory, items easier to study in mammals (rats, humans, chimpanzees) than in fish.

However, a large number of fish behaviours have been studied, some in detail, some, such as 'play', tentatively or at initial stages. Recent studies on zebrafish (*Danio rerio*) have shown that these small aquatic vertebrates can be used advantageously particularly as historic work with rats and mazes (Lashley 1950) have proved to be too complex for their 'purpose', involving too many sensory systems. Lashley was attempting to evaluate the presence of engrams as bases for memory and some progress has been made with this idea (14.4), and the occurrence of long-term potentiation (LTP) of neural circuits is becoming recognized as an important basis for learning and memory (14.4).

Understanding particular fish behaviours is frequently dependent on a sound knowledge of physiology and anatomy, as well as taxonomy; lacking such knowledge risks mis-interpretation of results. Thus fish may be ultrasensitive to changes in conditions that an experimental setup does not exclude. Flatfish, for example, can react to a very small amount of light that a human may not detect. If working with captive fish, taken from the wild, it is very important to double-check their comparability, that is, for size, known sex and, of course, species. An additional problem can be the capacity of small tropical fish to hybridize. The previous chapters have included reference to many different behaviours of fish, including colour change (2.9, 11.10.6), locomotion (3.8) and feeding (Chapter 4). The act

of reproduction (Chapter 9) can incorporate various behaviours such as mate identification, courtship and nest-building for some as well as the surprising mouth-brooding (Table 9.7). Migration is also a common behaviour associated with reproduction as well as changes in the physical environment (e.g. temperature). All these behaviours and many others depend upon the ability of the fish to respond actively; to recognize and react to environmental stimuli, whether animate or purely physical.

14.3 Neural pathways and behaviour

Behaviour in multicellular animals ultimately depends on individual nerve cells (neurons), their internal organization and their relationship with other nerve cells. Neurons are very complex in structure with processes (11.3.2) linking them to neighbouring neurons, and sensory structures (receptors) and effectors (e.g. muscles). Although we know that communication between neurons, receptors and effectors is achieved chiefly via chemical neurotransmitters (11.3.4, Appendix 14.1) and electrical pulses (e.g. action potentials), very little is known about how memory and learning (cognitive functions) is attained.

However, it is understood that learning is a matter of association, that accumulating/storing/recalling information from receptors must occur. Studying nerve tracts (multiple nerve processes, grouped together, and linking regions of the central nervous system) has been possible, so that regions of association can be identified. Within a brain region each individual neuron may have many synapses ('an average of 50,000' per neuron in the human cerebral cortex, Withers 1992).

Fish can be shown to recognize and associate certain signs and signals, for example colour, shape, sound, but it is not yet possible, for any animal, to present a coherent account of what animal memory is due to, at the cellular level. Neurons may show extremely complex organisation in ultrastructure, with Nissl substance forming many granular rosettes between parallel 'lines' of rough endoplasmic reticulum in mammals (Bloom and Fawcett 1975). The Nissl substance granules are essentially ribonucleic acid (RNA) and may just facilitate the production of enzymes necessary for neurotransmitter synthesis for example, rather than being any basis for information storage.

At the macroscopic level it has been shown, in fish, that the brain not only has centres receiving specific sensory input (in order from anterior to posterior: olfactory, visual, auditory) but also has 'behavioural' centres posteriorly. These controlling facets of 'innate' behaviour are the respiratory centre (6.5.10) and the paling centre (2.9, 11.6.2) both in the medulla oblongata (hind brain). Fajardo et al. (2013) report, in zebrafish, 'control of a specific motor program by a small brain area'. The centre(s) for learning in fish is not known. Arguing from the mammalian situation and the importance of olfaction, cases have been made that it would be in the fore-brain, whereas the optic tectum (mid-brain) has also been suggested (11.6.2). In any vertebrate learning centre a very large number of neurons in intimate association would be predicted.

Reactions of animals to stimuli range widely in their complexity but often take the form of a series of relatively simple 'reflex' responses; the stimulus evokes a train of action potentials (11.3.3) in the sensory nerve and processes of this nerve link (synapse), usually via an inter-neuron, with the motor nerve; activity in this nerve then reaches the 'effectors' (e.g. muscles which contract to produce movement). This reflex arc (11.5, Figure 11.4) can be modified by input from inter-neurons and as most behaviours are described, there may be a series of set responses initiated and then maintained as a chain of events. Such responses, though they can be modified, are usually regarded as innate or instinctive; they would be likely to include some feeding behaviours initiated when prey is sighted or courtship when a mate is identified (perhaps visually). A degree of complexity is present if a chemical stimulus first attracts a fish followed by visual identification/confirmation. Further complexity ensues if the fish with previous experience recognizes a (small?) conspecific as potential prey. Possible confusion concerning 'mate or prey' may be deflected if feeding ceases during the reproductive phase.

A set of behaviours with a very obvious endpoint, as occurs in feeding or courtship, may be called a 'drive' and an animal interrupted mid-way may resume at the first opportunity. Some animals, when interrupted, may not be able to resume and the concept of 'displacement activity' has been applied to subsequent inappropriate movements.

Young (1960, 1991) considered that memory in the invertebrate (octopus) cephalopods involves complex neural circuits, and Withers (1992) states that transient synaptic changes or reverberatory circuits may be involved in short-term memory whilst 'structural changes' involving synthesis of new protein or RNA is a more likely basis for long-term memory. Prolonged reverberatory activity in neuronal circuitry with many impulses in a short time may induce a growth change, a structural basis for information storage (Thorpe 1956). Although there is considerable knowledge about the physiology and biophysics of single neurons (11.3.3, Appendix 11.1) and their ultrastructure (e.g. Nissl substance), less was known about the functioning and anatomy of neural circuits in cognitive function, particularly in fish. However, considerable progress has been made recently in vertebrate studies, with the concepts of engrams, reverberation, Hebb's synapses, long-term potentiation and the role of specific receptors like N-methyl-D-aspartate receptors, NMDAR (14.4, Appendix 14.1). A range of modern multidisciplinary methodologies currently being used to investigate neuronal circuitry in zebrafish (*Danio rerio*) has been reviewed by Sumbre and Polavieja (2014). Roberts et al. (2013) list behaviours available for further study, referring to pioneer work by Hinz et al. (2012) whose investigations (using larval zebrafish) are expected to provide information on molecules involved in long-term memory.

14.4 Learning, memory and play

Learning in fish is dealt with briefly in 11.11. Any capacity to learn depends upon associative neural pathways in the brain. The utility of learning can be modified by the duration of memory, which may be months (or even years) in some fish species. Actually, a very durable memory of a single event may be less advantageous than a shorter memory (as in humans); the recognition of predators appropriate for a particular (perhaps very small) stage in the life cycle would have time limitations. Conversely, some memories are extremely durable as occurs with versions of vertebrate 'imprinting' when an association established early in a life persists and is relatively unalterable.

An intermediate duration of learning is represented by that of the sculpin (Green 1971) which can find its home territory six months after being removed from it; an example of spatial as opposed to social learning.

The capacity of fish to learn or acquire modified behaviours has been queried on the basis of their lack of certain brain areas associated with learning in mammals. This may be an inappropriate interpretation of morphology relative to function; rather, the presence of a neuropil may indicate a neural region capable of 'memory', and some fish have extensive neuropil in the brain (11.3.6, 11.6.2). The most abundant fish neuropil may be in the optic tectum (up to 15 layers, Wullimann 1998) with some layering (lamination) also found in the olfactory bulb (12.4.7). Layering in the optic tectum of striped bass is figured by Groman (1982); Northcutt (2002) provides a comprehensive table of tectal layering with chondrichthyans tending to fewer than teleosts. Ito and Yamamoto (2009) discuss the lack of layering in the teleost telencephalon in comparison to the six layers of the mammalian fore-brain (pallium/neocortex). Anatomically the teleost telencephalon receives input from most sensory systems (Wulliman 1998). The fore-brain has a regulatory role over other parts of the teleost brain and a role in schooling behaviour (11.6.2).

Mammalian cognitive function is associated with the fore-brain; the archipallium (hippocampus, amygdala) and very large neopallium (neocortex). Recently there have been attempts to compare the mammalian hippocampus (and perhaps the amygdala) with sites in the teleost telencephalon, with the implication (or stronger) that such areas would be involved in memory, etc. A hippocampal area was figured in cyprinids (Bernstein 1970) and Rodriguez et al. (2002) consider the lateral telencephalic pallium of fish to be homologous with the mammalian hippocampus, important in spatial memory. Braithwaite (2006) reports that the medial pallium of teleosts is likely homologous to the mammalian pallial amygdala and thus involved in learning. Rink and Wullimann (2004) state that comparabilities between teleosts and other vertebrates have been difficult because whereas in teleosts the pallial

structures are everted, they are evaginated in other vertebrates. This is not always reflected in figures provided but Young (1950) clearly shows the situation in cod, with the third ventricle at the surface of the fore-brain and also shows comparisons based on the work undertaken by Kappers et al. (1936). The anatomical situation, briefly, is that the 'roof' of the fore-brain in tetrapods (and lungfish) is thick whereas it is ultrathin in teleosts.

Eccles (1990) provides a possible anatomical analysis of mammalian cerebral cortex function in terms of neuronal morphology based on dendrons (11.3.2) of pyramidal cells as 'receptive units' interacting with 'psychons' (a unit representing a unique mental experience, e.g. thinking). Eccles describes dendrons as *possessing* apical dendrites which merge together as an afferent bundle to the perikaryon. These dendrites have 'spiny' projections which synapse with incoming axons, there being up to 100,000 such synapses on each dendron which can be excited by psychon interaction.

Engrams have been defined (Withers 1992) as a change in the nervous system that corresponds to a memory or 'the physical representation of memory', the 'memory trace', (Josselyn 2010) and have recently been revived (Liu et al. 2012; Hsiang et al. 2014) as a reasonable explanation for memory. The review by Thorpe (1956) considers an engram as an association of neurons with many alternative pathways involving increased synapses and usage. Ryan et al. (2015) state that memory consolidation, the transformation of a new unstable memory into stable long-term memory, involves specific patterns of engram connections. Using learning-dependent cell labelling on mice they also identified increased synaptic potentiation and increased density of cerebral cortex pre-synaptic terminals and they suggest that after memory consolidation the role of such synaptic plasticity may be to facilitate recall by reactivation of the perikarya (soma) of engram cells. Mayford (2014) reviews the search (in mammals) for a hippocampal engram.

Recently the concept of reinforcement of a neural circuit, long-term potentiation (LTP), has been increasingly explored as a basis for memory rather than the development of totally new structure. Collingridge (2003) has explained studies of the likelihood of NMDA receptors as the basis for synaptic activity in establishing LTP (in mammals). Bliss et al. (2004) consolidated an overview of LTP to that point, and a further review of synaptic plasticity (Bliss et al. 2014) refers to LTP's relevance to learning and memory. The underlying suggestion of reverberatory circuits which could be strengthened is attributed to Hebb (1949) and the current exploration of LTP depends to some extent on the 'Hebb synapse' (Bliss and Collingridge, 1993). These authors considered that 'LTP of synaptic transmission in the (mammalian) hippocampus is the primary model for investigating the synaptic basis of memory and learning in vertebrates'. It would seem that this view prevails, and the field is gaining momentum. It should be noted that the hippocampus may not occur in all vertebrates; its occurrence in fish is debatable. Our developing knowledge of the functions of neuroglia and tripartite synapses (11.3.1, 11.3.4) in neural tissue invites questions about the possible role of these cells in cognitive function.

Judging by brain size relative to body size some fish have impressively large brains (11.6.5). This feature occurs in chondrichthyans and in mormyrids (teleost osteoglossomorphs; elephantfish). Indeed, popular literature points out that the mormyrids have a superior relative brain size in comparison to humans. However, mormyrids also have a very well-developed electrolocation system which involves detection of other fish as well as 'mapping' the physical environment; much of the brain is expected to be involved with memory intricately related to identification of objects in the turbid water around these fish.

The concept of 'animal play' is discussed by Fagen (1981) and Burghardt (2011). For fish it may imply emotions, a situation which is difficult to reconcile with the recent and prevailing human attitude to fish. Accounts of 'play' in fish are usually anecdotal; video recordings of possible play behaviour will help to resolve some of the questions concerning interpretation. Whereas a shark rolling may be an aggressive action, it has also been linked to movements on its body surface of sharksuckers (Ritter 2002); the movement does not seem to have been interpreted as play although similar movements in aquatic mammals like otters or porpoises have readily been accepted as such.

The idea of 'play' in fish is contentious (Burghardt et al. 2015). However, Sterba (1962) states that mormyrids may play for a long time with a small ball of foil, and Burghardt et al. (2015) record individuals of one species of cichlid as spending time charging a bottom-weighted thermometer and even relocating it, while ignoring other objects (such as plastic plants). This 'play' appears to be species-specific, other fish species did not engage.

14.5 Behaviour centred on movement

14.5.1 Motile activity

Movement from place to place is locomotion, which for fish is usually swimming (3.8), dependent on the white and red muscles or variants. Regarded as a behaviour, locomotion can be seen to have various opposing attributes, like movement based on fins or body undulations and the capacity to steer or brake as well as altering speed. Locomotion is necessarily involved in other behaviour like predation, mate acquisition and migration, and may even be essential for oxygenation at the gills (ram ventilation, 6.5.11). Some fish are in more or less constant movement remaining in the water column as pelagics (14.5.2) whereas others spend most of their life on the bottom as benthics. Pelagic fish may also congregate (14.6.1), forming shoals or schools in which complex behaviour results in the fish group moving together synchronously. The capacity to 'hold station', hovering in one place is characteristic of many reef fish. The capacity to take other positions, like routine inversion occurs with the 'upside-down' catfish (Family Mochodidae, about 150 species, Nelson 1976).

14.5.2 Pelagic fish

A distinctive group of large fish occur in the world's oceans. These are large, active fish capable of swimming long distances and, with the exception of the huge planktivorous basking and whale sharks, are generally predators of other pelagic fish. Taxonomically this group includes the elasmobranch lamnid sharks and the large pelagic teleosts such as tunas, sailfish, marlins and swordfish, which are sought by sports anglers and for food.

However, although the lamnid sharks and the sailfish group share some characteristics of shape and tail structure which facilitate fast movement and a tendency to have heterothermy (7.4) with internal red muscle regions improving capacity to maintain high activity, they are very different in reproductive strategies. The sharks produce few young and are very slow in completing a reproductive cycle (Gilmore 1993). The sailfish, etc. tend to have very high fecundity and fairly continuous output (Arocha 2002, Poisson and Fauvel 2009).

Many pelagic fish species form groups (14.6.1) which may consist of a few or very many (hundreds or even thousands) of fish. This has made some such species (anchovy, sardines, herring) vulnerable to overexploitation but may have had advantages in confusing predators. Some species like shrimpfish (1.4. 3) move together synchronously with great precision and maintain this behaviour in small groups when in captivity. Although schooling behaviour (14.6.1) is usually associated with small fish species, it may also occur pre-spawning for much larger fish.

14.5.3 Feeding

Movement achieving predation has been analysed in much detail for various stages of the life cycle, with larval feeding behaviour (Puvanendran et al. 2004) developing when the yolk supply is near exhaustion and the mouth becomes operational. Timing and acquisition of predatory or grazing behaviour is associated with, or recognition and selection of, appropriate food; captive fish can learn to accept fish pellets or other artificial material (e.g. flakes) as food.

Feeding behaviour may be suspended for short or long periods; this cessation of behaviour may be important during or subsequent to the spawning period, while long periods of starvation can occur in cold or hot seasons, when fish enter an inactive state. This low activity (14.5.4) may be equivalent to hibernation (in the cold) or aestivation (in drought) or be mild torpor when a fish may move but rarely does so.

14.5.4 Torpor and sleep

Long periods of inactivity can occur in fish, the record perhaps held by the African lungfish (*Protopterus*) which can enter extended states of aestivation when, in response to adverse conditions, the individual fish burrows into the mud and ceases most activities; this torpor can last for several years (Hochachka and Guppy 1987) but is usually entered into seasonally, at the onset of the dry season, for the West African species *P. annectens* and *P. dolloi* (Greenwood 1975). These fish form a cocoon around themselves with mucus and a 'breathing hole' near the mouth; metabolism is maintained at a very low rate, utilizing muscle as an energy source, so that when fish emerge at the end of the aestivation period they are described as 'emaciated'.

Greenwood (1975) describes a number of observations which indicate that various fish species can have a non-active 'sleep' phase. This applies to fish which rest on the bottom during the night, which can be observed in tank-held individuals. He also reports accounts of schooling fish which rest on the bottom at night and resume schooling if disturbed. Presumably fish that cannot reach the bottom—that is, they are above deep water—must rest while in the water column. Conversely, some bottom-dwelling fish are reported to float at night.

Recently zebrafish (*Danio rerio*) have been used for sleep studies; it is regarded as a useful experimental model providing some insight into human sleep and sleep disorders. Studies have concentrated on understanding the neural basis and the endocrines (e.g. melatonin, 10.11.2) involved and the regulation and ontogeny of cycles controlling sleep and arousal (Chiu and Prober 2013; Sombes et al. 2013). The mammalian sleep–wake cycle is controlled through the suprachiasmatic 'nucleus' (a group of neurons) in the hypothalamus (Elbaz et al. 2013) and this can be found in fish too.

14.5.5 Territoriality

Non-pelagics, particularly fish that live in complex and/or productive ecosystems with many competitors, may set up and defend territories for feeding or for reproduction. This type of behaviour is noted where a male (like the stickleback or the lumpfish) constructs and patrols a nest which may be intricate (stickleback) or a simple scrape (lumpfish). In the case of anemone fish (Pomacentridae, 13.3.3) the territory is its anemone.

Defence of a territory (or 'home range', Welsh et al. 2013) has been studied in sticklebacks in which the males establish a nest and defend it against intruding males. It is more difficult to apply the concept to pelagic species which may, however, become more benthic when spawning. Thus salmonids, which may feed in the water column, construct and patrol a nest site as a scrape or 'redd' in the substrate. Defensive or aggressive behaviour, also known as 'agonistic', may take the form of display with particular positional changes. In sharks, a rolling movement has been reported (Ritter and Godknecht 2000). Presumably it must be clearly differentiated from courtship and minimize the risks of actual contact or injury.

14.5.6 Migration

Migrations occur in large number of fish species, often as a behaviour that precedes reproduction (9.14).

Some fish undergo long, once in a lifetime, migration prior to their final, and only, reproduction. For some fish this behaviour depends upon returning to their natal stream and it has been concluded that this occurs when early imprinting enables the fish to correctly identify the area from which it originally came. This type of 'homing' behaviour is found in some Pacific salmon species which move from seawater to fresh water (anadromous, 9.14), swimming upstream against the current.

The very long and arduous reproduction-linked migrations which occur with semelparous species such as some salmonids and anguillid eels are rare. Shorter movements prior to reproduction occur however, with many other species including cod off Russia and Norway and capelin which may move onshore to spawn. Their beach-spawning activity (Plate 11) is expected to terminate with death for the males, whereas the females can survive (Burton and Flynn 1998; Flynn et al. 2001). In both these cases (cod, capelin) the physiological stress of this behaviour does not involve accommodating to markedly different salinity. Conversely salmonids and anguillids, in passing to different salinities (*to* seawater for adult anguillids, *from* seawater for adult salmonids) have to make considerable physiological adjustments. Similarly the juveniles have

to make the opposite adjustments and salmonids go through a silvering process during preparation (smoltification, 2.10, 7.9, 9.14) for entering seawater. The migration of eels is also accompanied by morphological and colour shifts.

Some freshwater fish have brief pre-reproduction migration, for example from stream to lake; this incurs minimum stress, and presumably allows the eggs and newly hatched young to have a stable environment and stay in one locality until able to cope with flowing water. North American lake trout (*Salvelinus namycush*) disperse after spawning (usually in October for Canadian lake trout) and may move as far away as 100 miles from the reproduction site (Scott and Crossman 1973).

14.6 Social behaviour

Interaction with other fish of the same species (i.e. conspecifics) constitutes social behaviour. It includes group activities like schooling (14.6.1) and reproduction (14.8), the latter itself constituting many different and sequential activities. Signalling between fish includes the use of colour (2.6), bioluminescence (13.4.7), sound (13.7), electric fields (13.8) and pheromones (14.7).

14.6.1 Fish aggregations

Fish can form large or small groups, sometimes as 'schools' or 'shoals', which may be more or less temporary; pre-spawning aggregations occurs in some species, which can facilitate mating or fertilization if gametes are broadcast. Other advantages include the possibility of predator confusion and food location (Shaw 1962; Magurran 1990). Surgeon fish may form mixed species schools and schooling is generally restricted to adults (Lawson et al. 1999); it is thought to improve food access when in competition with other species.

Some schooling fish move and turn with great precision, indicating very specific and swift communication between the fish, with a role for lateral line near-field perception (12.7) and vision in schooling maintenance. Denton and Nicol (1966) mention the likely role of a black spot in enabling visual contact between silvery schooling fish and the fact that normally schooling fish may disperse at night. Partridge and Pitcher (1980) concluded, from experiments on saithe (*Pollachius virens*) involving surgical section of the lateral lines and blindfolding, that distance between neighbouring fish in a school is maintained by visual and lateral line stimulation. These authors consider that vision appears to have a role in position maintenance in the school whereas lateral lines are probably sensitive to the school's swimming speed and direction. The importance of vision in schooling is discussed by Denton and Nicol (1966).

Large groups of small fish may attract predators, other fish, birds and humans included, and in this situation the small fish may be termed forage fish or, as fished commodities, be rendered into fishmeal or used as bait. The North Atlantic capelin (*Mallotus villosus*) form pre-spawning aggregations which attract seabirds, whales and other fish like cod. Anchovies and anchovetas have formed huge schools off Peru and California where they were the subject of large fisheries. The bristle mouths (e.g. *Cyclothone*) and lanternfish (Myctophidae) may school in the mesopelagic regions. In fresh water, pre-spawning aggregations of trout and salmon can attract predation; spawning runs are notoriously unpredictable in numbers.

In fresh, brackish or inshore water there may be an abundance of small schooling fish; often termed 'minnows' by Westerners, in which case the fish are generally cyprinid ostariophysans or cyprinodonts like the killifish (*Fundulus heteroclitus*, Carson 1941). A list of Wisconsin minnows (24 species within the headwaters basin) compiled by Klosiewski and Lyons (no date) shows 'shiner' spp. (*Notropis*), and three other 'shiner' genera with a single species each. Other 'minnows' listed include species referred to as varieties of 'chub' and 'dace' with the original minnow genus *Phoxinus* (*P. neogaeus*) represented in Europe by *Phoxinus phoxinus*.

Delcourt and Poncin (2012) have reviewed the definitions of shoaling and schooling and the tendency of animals, particularly fish, to form cohesive groups, with schooling characterized by the habit of individuals to orient similarly, whereas shoaling is considered to be defined by the proximity of members. The two situations are not necessarily mutually exclusive but schools are considered to be more tightly structured than shoals. Bricard et al.

(2013) presented a physical analysis of 'coordinated motion', in relation to small particles but believed to extend to 'stable directed motion' of live groups including bacteria and animals.

14.7 Chemoreception and pheromones in fish behaviour

Some of the behaviour patterns involving chemoreception are responses to pheromones, feeding, orientation and homing, and social interactions including reproduction (14.9 and Chapter 9). Olfaction (12.4.5) is involved in all of these behaviours whilst gustation (12.4.2) is more restricted. Pheromones (2.3.1, 9.5.2, 9.18,) are substances released into the environment for signalling between conspecific individuals; they represent an interesting but not very well understood area of fish behavioural physiology. An injured minnow (*Phoxinus phoxinus*) releases an alarm substance (Schreckstoff) causing a school to flee or seek cover (von Frisch 1938, 1941a). This substance is produced in Ostariophysi, including cyprinids, in large club cells in the mid epidermis (2.3.1). Surgery has established that the olfactory organ mediates the response (Smith 1992), which can also occur in some other fish groups. Fish olfactory organs can also detect sex pheromones which trigger endocrinological change or behavioural change (9.18). Steroid pheromones and bile acids/salts (Li et al. 2002) can have an important role in fish reproduction. The sensitivity of elasmobranchs to bile salts has recently been described (Meredith et al. 2012).

Encoded olfactory chemosensory information within the central nervous system is a common feature of hypotheses attempting to explain the homing migrations of salmonid species, which remain incompletely understood. One hypothesis (Hasler and Wisby 1951) suggests that juvenile salmon learn or imprint chemical cues of the home river, migrating adults using the encoded information to return to the same river. Synthetic chemicals (e.g. morpholine, Hasler et al. 1978), additional to normal home-river odours, have been used successfully to imprint coho salmon smolt (Hasler and Scholz 1983). Natural imprinted odours are probably a mixture of conspecific and plant odours, although the chemical composition is unknown (Hasler and Scholtz 1983).

Yamamoto and Ueda (2007) propose, from electro-olfactogram (EOG) recordings and behavioural experiments, that amino acids facilitate the homing migration of salmonids. EOG recording suggests that sockeye salmon, *Oncorhynchus nerka*, may be imprinted experimentally by a single amino acid. An alternative hypothesis is that adult salmon home to population specific pheromone trails released by juveniles resulting in a constant population odour (Nordeng 1977). Skin mucus, bile and faeces could be possible routes for the release of these pheromones (Nordeng 1971; Stabell 1992). This hypothesis is not supported by any demonstration that homing depends on members of the population residing in the vicinity (Quinn 1990). The use of morpholine to imprint smolts and ensure their return as adults to a specific location may be challenged by the report that morpholine can act as a non-specific attractant for salmonids (Mazeaud 1981). The sea lamprey, *Petromyzon marinus*, spawns in rivers with many conspecific larvae (Teeter 1980; Li et al. 1995) and with no evidence of a home river. The ammocoete larvae may spend several years in fresh water and the incoming adults may possibly being guided by bile acids (which can act as a pheromone in salmonids, Døving et al. 1980), the olfactory sense in sea lamprey being acutely sensitive to two unique bile acids found only in the larvae (Li and Sorensen 1997).

14.8 Reproductive behaviours

Reproduction may be preceded by a gathering of mature individuals in a historically established area and the taking and defending of territories as well as the construction of nests.

This behaviour allows the mature fish to be very close as a preliminary to successful fertilization and involves a number of different behaviours to achieve mate location.

Reproduction may also be accompanied by profound morphological changes, not just in the gonads and developing gametes but in the skin and even the body shape (capelin, male Pacific salmon). Capelin males acquire pronounced lateral ridges and male Pacific salmon have extreme rostral development, forming a kype (13.12). Colour changes are common pre-spawning and enable identification of male or female partners (14.9).

The reproductive behaviour of some fish, particularly small freshwater species, has been studied intensively and the 'model' for understanding facets of fish reproduction has been the three-spined stickleback (Tinbergen 1952; Wootton 1984).

14.8.1 Mate location

In sequence, the mature fish may locate the reproductive site and potential mate using a set of sensory structures including chemoreceptors and photoreceptors (vision). It is expected that the neural pathways be innate (genetically set) with specific pheromones released, for example in the urine, attracting individuals to each other. Sound (e.g. drumming, 13.7) may also be important in attracting mates.

14.8.2 Courtship

Courtship behaviour establishes a brief period when two or more fish can come close together and synchronize their activities for successful egg and sperm contact. Courtship movements have been categorized into different phases (which are usually sequential) such as 'approach', 'quivering', 'tail crossing', etc. (Poncin 1994).

Once mature individuals are relatively close then vision or further chemical cues (or perhaps for electric fish, electric signals) secure identification of one or more potential mates (14.9) and the fish may pair up or form other small groupings (for capelin, often two males per female, one each side). For many fish species nest-building occurs and the male stakes out a territory based on the nest. The courtship, which may be very brief, if successful, culminates in mature oocytes, 'eggs' (9.10, Figures 9.9 and 9.13) being deposited, for example on a substrate and/or in a nest and fertilized for oviparous species, or fertilized and held in the oviduct for ovo- or viviparous fish (e.g. elasmobranchs, guppies).

14.8.3 Fertilization

If cued appropriately the male will deposit sperm near or on the 'eggs' and fertilization should ensue.

For fish with internal fertilization (chondrichthyans, *Poecilia*, *Gambusia*, etc.), the process must be very precise, with morphological adaptations to ensure the sperm locates appropriately close to the eggs; for chondrichyans the males have 'claspers' (Figures 1.7, 1.8, 1.9, 9.5) which help stabilize the transfer. For elasmobranchs there are two pelvic claspers. Chimaeras have an additional small cephalic clasper (tentaculum), positioned dorsally on the head; it seems fragile and it is not clear what role it has in sperm transfer, although the expectation is that claspers enable a fish pair to maintain contact during sperm transfer. Males of the small freshwater ovoviviparous (live-bearing) fish like mollies and guppies have an intromittent structure formed from modified pelvic fins.

14.8.4 Egg transfer

Subsequent to fertilization eggs may be picked up by a parent and transferred to a safer place, like the male's pouch for seahorses (*Hippocampus*, etc.) or the mouth (many cichlids). Mouth-brooding seems an extraordinary behaviour as it means the brooder has to stop eating/swallowing. Mouth-brooding may continue after hatching; the parent continuing to protect the young which will venture out under conditions recognized as 'safe'. This form of parental care is expensive for the brooder and has also co-evolved with a predatory habit where fish have developed a behaviour of ramming the brooder so that eggs or young are dropped (4.3.4).

14.8.5 Nest-building

Fish such as non-mouth-brooding cichlids and salmonids alter the substrate for egg deposition by producing a depression (scrape) sometimes in a mound built by the fish (some cichlids). Other fish actually construct a nest of bubbles (anabantoids, Greenwood 1975) or material is collected and assembled into a nest as occurs with sticklebacks. For these fish the male constructs the nest, bringing small pieces of vegetation to one location and glueing them together to form a muff-like nest with two openings. The glue, called spiggin, is secreted by the kidneys (Rushbrook et al. 2007 and 9.5) under the influence of androgens. Wunder (1930, cited by Thorpe 1956) reported that sticklebacks prefer green material for the nest; perhaps this reflects durability as compared to brown (and perhaps decaying) items.

14.8.6 Parental care

Once fertilized the eggs may be pelagic and attract no further attention from the parents or may be deposited and left to develop without parental involvement (salmonids). However, in many species one or both parents are actively involved in protecting, defending or fanning the brood.

About 20 per cent of teleost fish families show some form of parental care (Rüber et al. 2004, based on Blumer 1982); in particular some egg-laying (i.e. oviparous) fish have parental care, either by the male (sticklebacks, lumpfish, seahorses) or female (ocean pout, tilapia) or, more rarely, both parents (some cichlids). Blumer (1982) provides a comprehensive list of caring behaviours, some of which have been studied in detail. In the case of sticklebacks such as *Gasterosteus aculeatus*, the male constructs a muff-like nest, courts one or more females which deposit eggs and after fertilizing them he fans the nest and guards it and the young when they hatch (Wootton 1984). Lumpfish (*Cyclopterus lumpus*) males also fan and guard the eggs (Scott and Scott 1988) in a rudimentary 'nest' while the male seahorse (*Hippocampus* spp.) has a brood pouch for receipt of the eggs. The ocean pout (*Macrozoarces americanus*) has internal fertilization but the fertilized eggs are deposited and cared for by the female who coils her body around the egg mass (Yao and Crim 1995). Mouth-brooding occurs in 'at least eight families and 53 genera' (Mrowka and Schierwater 1988), but unfortunately some of these cichlids as species if not genera may now be extinct (16.5). Nine mouth-brooding families are listed by Blumer (1982): updated information (Appendix 9.2) shows at least ten families with some mouth-brooding species. Tilapia and many other cichlids are mouth-brooders, commonly females hold the developing eggs in their mouths; in cardinal fish (e.g. *Apogon*) it is the males that hold the broods (Fishelson et al. 2004). Biparental mouth-brooding is also reported; the parents may split the brood (Balshine-Earn and Earn 1998) or brood sequentially (Morley and Balshine 2003). At least one species of cichlid (*Sarotherodon galilaeus*) has extreme flexibility in mouth-brooding (Balshine-Earn and Earn 1998) with female, male or biparental brooding possible. For both sticklebacks and tilapia the young hatchlings may return to nest or mouth if alarmed.

Developing eggs are extremely vulnerable even to predation by conspecifics and the various guarding and defence behaviours are very important in promoting survival. It is possible that certain extreme behaviours like beach spawning (9.14) have evolved in response to predation, with some predator avoidance achieved by depositing eggs at high tide levels. Certainly eggs may be favoured food for conspecifics as well as other fish or invertebrates; avoiding of some conspecific or even cannibalistic egg consumption may be attained by cessation of feeding during the spawning period.

Mouth-brooding is successful in providing protection but it has a high cost; the number which can be brooded is obviously limited and the brooder cannot eat during brooding.

14.9 Recognition

Fish have the capacity to recognize conspecifics or other fish species.

The ability to identify other fish, particularly conspecifics, is obviously important for successful reproduction which depends on a series of cues; these may include colour and shape; enabling identification, for reproduction, of a potential mate which is ready to court and mate. Conversely, recognition of predators is also important, involving cues which promote avoidance.

In the case of reproduction, colour changes can allow identification of the right sex and the appropriate condition; thus male lumpfish (*Cyclopterus*) acquire some red colouration as they reach spawning-readiness while females are blueish (Goulet et al. 1986). Similarly males of both Pacific salmon (*Oncorhynchus* spp.) and three-spined sticklebacks (*Gasterosteus aculeatus*) become partially red pre-spawning. Thus male sticklebacks can recognize conspecific males (red abdomen) and gravid females (pale, swollen abdomen).

Unsuccessful identification can occur; thus Poncin (1994) reports observations in a Belgian river involving a grayling (*Thymallus thymallus*) which attempted to mate with a barbel (*Barbus barbus*), apparently because feeding movements of the barbel, disturbing gravel, simulated spawning acts of grayling.

Crane and Ferrari (2015) discuss the capacity of a small fish (*Pimephales promelas*) to observe the behaviour of other fish and assess predation risk, apparently judging the information received and favouring conspecifics' behaviour as a reliable source.

Various accounts suggest or confirm that fish can be quite good at identifying and differentiating conspecifics and even kin.

14.9.1 Kin recognition

The capacity to recognize related individuals is regarded as potentially beneficial (Olsen et al. 1998) and there is evidence that fish can do this (Olsen et al. 1998; Gerlach et al. 2008). For kin recognition it seems that odours are important and that imprinting for such odours may have a short window of time during which this can occur amongst siblings. Olsen et al. (1998) determined that for juvenile Arctic char (*Salvelinus alpinus*) the major histocompatibility type had an influence on kin identification. For zebrafish (*Danio rerio*), Gerlach et al. (2008) found a 24-hour 'window' shortly after hatching and six days post-fertilization, during which kin recognition was established (olfactory imprinting). This type of imprinting and kin recognition system presumably is only possible when hatching occurs in close proximity; it does not apply to marine fish with pelagic eggs and larvae, for example.

14.9.2 Interspecific communication: cleanerfish

Successful recognition of particular fish species is very important for the mutualistic symbiosis between cleanerfish and their clients, a relationship reported from coral reefs where many different fish species are in visual contact.

Cleanerfish (4.3.4, 13.3 3, 13.13) are small fish (e.g. some wrasses) which obtain all or part of their food by removing ectoparasites from 'clients', other fish species willing to cooperate. This relationship is mutualistic (13.13) but may be abused in that 'cleaners' could browse on scales or gills and/or mucus, and 'clients' could eat 'cleaners'. The occurrence of 'mimics' which successfully exploit the normal mutualism of cleaners was reported by Cheney and Côté (2005); the relationship between exploitation and cooperation was discussed by Gingins et al. (2013).

Cleanerfish, in order to function effectively, must be acceptable to the fish they service. They must come into neither predator nor prey or potential mate categories, the literature suggests that their identification depends partly on location; that is, there are 'cleaning stations' in an identifiable and semi-permanent topography. Recognition of cleanerfish by their potential clients apparently depends on visual cues; cleanerfish commonly have distinct longitudinal stripes, notably brilliant blue, associated with yellow or black (Cheney et al. 2009).

14.10 Innate behaviour

Some forms of behaviour are absolutely innate or nearly so. These behaviours include colour change (2.9) and ventilation (6.5.10, 14.10.2). Some phases of other linked behaviours may individually be close to innate and cannot be stopped once started.

14.10.1 Colour, crypsis and mimicry

Fish appearance can be adjusted by changing colour, either for crypsis (camouflage) or mimicry (simulating other species) or communication/recognition (14.9). Both crypsis and mimicry can be utilized in defence or predation.

A fish capable of changing colour or pattern does so by neural or hormonal processes, elicited by incident light and/or background or stress and reproductive condition and social status. A fish can have neural and endocrine control of the chromatophores and in both cases the fish cannot, as far as is currently known, be trained to change colour.

The neural and endocrine regulation of fish colour changes has been described (2.9, 11.10.6). Höglund et al. (2000) demonstrated experimentally a social behavioural role for melanophore-stimulating hormone (MSH) in Arctic char, *Salvelinus alpinus*. Subordinate fish in a group can display grades of skin darkening as a result of social stress. Data obtained in this work demonstrated that the different darkening levels were accompanied by elevated plasmatic levels of α-MSH. Stress-induced increases of noradrenalin recorded in the CNS could increase pituitary release of α-MSH. The skin darkening could be a social signal to reduce aggressive behaviour or it could be cryptic, making subordinates less visible. Adrenalin release from chromaffin cells (10.11.3, 11.9) in response to stress can also change fish colour by direct action on the melanophores. Excitement pallor or excitement darkening results dependent on whether alpha- or beta-adrenoceptors predominate on the melanophores (Fujii 1993).

Mimicry has been described in reef fish (13.3.3). In the case of the dusty dottyback (*Pseudochromis fuscus*), it involves colour changes (2.9) to resemble any one of several species (Cortesi et al. 2015). This increases the success of predation on juveniles of the mimicked species and also reduces predation of dottybacks. The dottybacks are also able to undergo cryptic colour change.

14.10.2 Ventilation

Fish respiratory ventilation may be active (6.5.10), with pharyngeal muscle action creating a current of water, with a regulatory centre in the medulla oblongata (11.6.2) or ram ventilation (6.5.11) dependent on rapid swimming with an open mouth.

Some fish (6.5.11) are capable of switching from active to ram ventilation and vice versa but it would be anticipated that change of ventilation type is not readily modified by cues other than swimming speed or oxygen demand, or internal signals related to oxygen demand or carbon dioxide load. Feeding interacts with ventilation; for example, during non-feeding, the gill ventilation volume, frequency, stroke volume and gape are all increased in comparison with non-feeding in Sacramento blackfish *Orthodon microlepidotus* (Cech and Massingill 1995). High temperatures, especially when accompanied by feeding (4.8), may correlate with increased ventilation to a point that is lethal, as increased muscle activity increases oxygen demand.

14.11 Fish behaviours and humans

The tendency of some fish species to form schools and/or pre-spawning aggregations has been utilized to maximize fishing effort at such times and places. This has sometimes meant that fish have been extremely vulnerable to overexploitation, and the design of legislation to protect fishes and to establish conservation areas has partly depended on identifying the risks involved, and balancing the trade-offs of different commercial and conservation goals.

The behaviour of fish in their natural environment and the capacity to study it, and in particular the capacity to study large fish, has been a challenge, partly answered by diving (Ritter 2002). However, the presence of divers may itself interfere with natural behaviour.

The growing concern about climate change has prompted investigations of fish behaviour with relation to changes in water temperature and also recommendations for ongoing observations of fish in their natural environment. Aguzzi et al. (2015) have discussed the need for coastal observatories to monitor fish behaviour with reference to the use of the Western Mediterranean observatory platform (seafloor observatory, OBSEA, off Barcelona) with video cameras and collecting physical data simultaneously.

It is obviously a great concern that fish behaviours will be adversely affected by change in water quality, particularly as some behaviours are relatively fixed, 'innate' or 'instinctive' and dependent on a set of cues that may be disrupted by large-scale changes in the environment.

14.11.1 Do fish experience pain?

The occurrence of pain in fish has been debated intensely. Sneddon and co-workers (2003) identified nociceptors on the head of rainbow trout *Oncorhynchus mykiss* by their responses to pressure, high temperature and acetic acid, which were similar to those of tetrapods. These authors also suggest that in fish, pain may occur in association with noxious stimulation. Rose (2002) considers that fish are unlikely to experience pain since they lack the cortical regions of the cerebral hemispheres necessary for the sensation of pain in humans. In this context the Cambridge Declaration of Consciousness (Low et al. 2012) includes the observation 'the neural substrates of emotions do not appear to be

confined to cortical structures'. Although there is only rudimentary development of a cerebral cortex in fish, except in lungfish, their optic tectum has a complex layered structure and includes a neuropil (11.3.6, 14.4). The optic tectum has been compared structurally with the mammalian cortex and is not solely concerned with vision (Healey 1957): 'In short the mesencephalon of fishes exhibits many of the functions which are associated with the cortex of higher vertebrates.' Human pain involves complex perceptual and emotional elements and anthropocentric considerations and the establishment of universally acceptable experimental criteria have added to the complexity of the debate. In fish, pain is possibly a different unpleasant sensation of discomfort than that experienced by humans (Sneddon et al. 2003).

In humans, noxious stimuli have a potential to cause tissue damage which can activate nociceptors directly or by release of chemicals which stimulate them (Gardner 1975), and in animals generally may elicit a reflex withdrawal (Sneddon 2015). High pressure, temperature extremes and some chemicals are noxious stimuli and fish have nociceptors which are preferentially sensitive to them. Fish possess substance P, *N*-methyl-D-aspartate and opioids all generally associated with nociception. The trigeminal nerve of rainbow trout (*Oncorhynchus mykiss*) conveys oral and facial nociceptive information and includes nerve fibres with diameters similar to those in mammalian nociception and with comparable electrophysiological properties. Mettam et al. (2012) demonstrated electrophysiologically that chemoreceptors innervated by the trigeminal nerves in rainbow trout are nociceptive, responding only to irritant stimuli including CO_2 in carbonated water. Changes in water quality producing low pH or high CO_2 concentration may stimulate the nociceptive system of fish, which generally swim away from such stimuli.

Recent reviews of the concept of pain in fish (Rose et al. 2014; Key 2015) doubt the position taken by other groups in support of its occurrence. Rose et al. (2014) claim that previous demonstrations of fish pain have been 'unsubstantiated' whereas Key (2015) makes the definite statement that 'fish do not feel pain'. The subject remains controversial.

14.12 Temperature sensitivity

Fish are able to migrate to water at temperatures at which they are physiologically most effective, reducing any requirement for temperature-related compensation (Hazel 1993), and such behaviour indicates the possession of thermoreceptors. Vertebrates possess thermoreceptors in the skin (cutaneous) and internally, for example in the hypothalamus, frequently being naked nerve terminals (Withers 1992). Some function in long-term responses and are tonic, others being more rapid, are phasic. Increasing heat, up to 50 °C, applied locally elicits behavioural responses in goldfish (*Carassius auratus*) mediated by activation of thermally sensitive nociceptors rather than by thermoreceptors (Nordgreen et al. 2009). The possible involvement of nociceptors was demonstrated by the effect of morphine treatment on the behavioural responses. It is suggested that thermosensitive nocireceptors in fish would have an important role in escape from lethal water temperatures (Nordgreen et al. 2009).

14.13 Habitat and behaviour

In broad terms it is obvious that the immediate environment is a fundamental influence on fish behaviour while morphology limits the behaviours possible. Fish living in the deep sea (13.4.3) share many characteristics, as do pelagic fish, fish that live in caves, etc., and taxonomically elasmobranchs (without exception) have internal fertilization. Whereas all fish (without exception) show locomotory movement it is pelagic fish that tend to cruise and benthic fish that tend to rest on the substrate.

In order to comprehend the complex interrelationships of organisms within their community some biologists have adopted the concept of 'guilds' (Bain et al. 1988; Aarts and Nienhuis 2003; Elliott et al. 2007; Mourão et al. 2014), species dependent on the same habitat, which in the case of animals may involve different behaviours to exploit a shared area, and may refer to a similar behaviour used by different species, such as occurs with cleanerfish and their mimics. Behaviour appropriate to a particular habitat often results in convergence, when taxonomically distant fish have, under singular selection pressures, evolved similar structures, for example photophores in deep-sea fish. Similarly, in a dark environment, cave-fish (13.5.1) tend to have lost sight, and perhaps surprisingly have not shown any tendency to develop photophores, presumably because they are often the top predator in a small ecosystem and have other means of communication.

14.14 Overview

Establishing an understanding of fish behaviour has required careful investigation of sensory perception (Chapter 12) and the links through the nervous system (Chapter 11) to the effectors enabling appropriate reactions. It has also been important to recognize that some hormones (Chapter 10) have influence on fish behaviour, notably reproduction, colour change and stress responses.

As with other physiological areas, the study of fish behaviour has benefitted from work on mammalian subjects like rats; it has required observation and experiment, and correlation with information on structure gained from both dissection and microscope. Many jurisdictions currently carefully control the use of live vertebrates, including fish, in experiments and major interventions like hypophysectomy (removal of the pituitary) and nerve sectioning or brain lesioning may not be allowed. Any person wishing to undertake live fish studies may have to go through a rigorous approval process which can take several months to complete.

Currently much progress has been made in rational studies of fish behaviour but much more remains to be done. Structurally although fish behaviour shares many features with, for example, mammals there are also profound differences The fish brain (11.6.2) has the basic regional organization (fore-, mid- and hind-brain) of other vertebrates but lacks the laminated cerebral cortex associated in mammals with many cognitive (learning, memory) functions. Very rapid responses involving movement are possible in fish (but not in mammals) because of the special (large diameter) fast-conducting Mauthner fibres (11.6.6) which pass down the spinal cord. Fish apparently lack the kind of lateral specialization found in the human fore-brain and may also lack the kind of commissures (corpus callosum of humans) which provide links enabling a high degree of association between different activities which promote learning and memory. However, fish undoubtedly can learn, for example, to recognize other fish as kin, predators, etc., and can remember locations. That they can be deceived is indicated by the various forms of mimicry described, for instance, for some tropical fish (14.10.1) with some juveniles mimicking fish less desirable as prey.

However, a degree of lateralization in fish is reported (Braithwaite 2005) in association with the capacity of fish to operate information collection separately with their lateral eyes; for example, one eye may be watching shoal movements while the other detects possible predators.

Some features of fish behaviour are shared with other vertebrates, though not necessarily mammals. Thus chromatophores, capable of internal pigment movement (2.9), the basis for colour change in fish, amphibia and reptiles, do not occur in birds or mammals, the integument of which is generally obscured by feathers or fur. The importance of signs and signals in eliciting behaviour is common throughout the vertebrates. Some behaviours are elicited by quite simple cues, such as colour or shape as shown by Tinbergen, quoted by Carthy (1966). Other behaviours are innate and may not be intermittent in response to cues but ongoing: a fish that ventilates actively often does so for an uninterrupted lengthy time period and a fish capable of colour change is always in some colour phase, dependent on structurally stable chromatophores.

Predator avoidance is a highly significant behaviour in most animals; it may dependent on passive or active defence mechanisms the latter including a range of behaviours. Predator avoidance requires identification of a potential predator followed by appropriate responses such as defence with raised spines or escape, perhaps with the rapid action of Mauthner fibres innervating white muscles of the myotomes.

Locomotion is a basic feature of almost all animals, exceptions being those sessile invertebrates like anenomes or barnacles. Even these have motile larvae enabling dispersal and settlement. A very few fish species (deep-sea anglers) have males that permanently adhere (and fuse) to the females and are thus essentially sessile. The morphological basis for body movement in fish includes the use of red or white muscles and the various roles of median and paired fins. Fins may be used for display and for fanning eggs, intermittent uses quite independent of basic locomotion.

There have been some advances in studying learning and memory, particularly the possible role of engrams and LTP. Fish studies may help, zebrafish having the advantages of small size, transparency (as larvae) and cutaneous breathing (as larvae); the latter meaning no opercular movement that could interfere with recording activity (Sumbre and Polarvieja 2014). The occurrence of a neuropil/lamination in the optic tectum and the olfactory bulbs of fish has invited comparison with its presence in the neocortex of the mammalian brain. The presence of neuropil and laminae provides the basis for a vast number of neuronal microcircuits, selective and/or enhanced activity within such circuits and their interconnections may be an important attribute of memory, learning and intelligence.

Fish behaviour has been studied for many years. Generally, the studies started with observations including the underlying morphology and subsequently included an understanding of the role of physiology. Suggestions as to why fish behaved in a particular manner or questioning if they could learn and remember have been tackled via experiments. Parallel understanding of sensory capacity in fish (Chapter 12) has underpinned the deeper understanding of behaviour or even allowed description of previously unsuspected behaviour (electrolocation). That fish are capable of complex behaviour including 'cognitive' behaviour (Braithwaite 2005) is certain; that some fish show parental care is also certain, but that they 'choose' certain behaviour or that they 'play' may be contentious.

The interactions of ethologists, neuroscience and psychology sometimes leads to conflicted attitudes by their protagonists, using words like 'intent' or 'stress' in anthropocentric modes and prompting resistance by some scientists. Although it is now well accepted that physiological stress (7.4, 7.12.1, 7.13.3, 10.12.7) occurs in fish, the study of fish stress was resisted previously. The resulting behaviour of stressed fish, however, is yet to be understood; it may indeed give rise to darkening, which in turn acts as a social signal (Höglund et al. 2000). Furthermore stress can be associated with increased levels of a number of hormones including MSH which causes darkening and adrenalin which can pale or darken the fish, depending on the predominance of either α- or β-adrenoceptors respectively (2.9 and Appendix 2.3). So fleeting or more permanent changes in the skin colour can involve a set of other behaviours, comprising avoidance of conflict. More permanent changes in skin structure can occur in response to captivity (2.10). It is certain that behaviourists should be wary of atypical responses of captive fish, but interactions with humans provided the basis for suspecting that fish can hear, with anecdotes associating fish feeding with sound as well as sight.

Appendix 14.1 Neurotransmitters and behaviour

Apart from electrical synapses (11.3.4), neurally controlled behaviour will incorporate neurotransmitter action. Chemical neurotransmitters (11.3.4, 11.9.4) which enable communication between neurons, sensory receptors and effectors can be acetylcholine, biogenic amines and amino acids. Donald (1998) reviews a number of additional putative transmitters.

Acetylcholine is involved in skeletal muscle control. Biogenic amines are the catecholamines (adrenalin, noradrenalin and dopamine) as well as serotonin (5-hydroxy tryptamine).

The occurrence of different molecular receptors enables a range of responses to individual neurotransmitters, adrenoceptors (with major sub-types alpha and beta), are responsive to adrenalin.

The amino acids GABA (gamma aminobutyric acid) and glutamic acid also function as neurotransmitters in fish; the latter functioning via glutaminergic receptors (e.g. NMDAR).

Putative neurotransmitters include neuropeptides such as substance P. Individual neurons may have more than one transmitter and may release cotransmitters together at a single nerve terminal (e.g. ATP and noradrenalin or GABA and dopamine, 11.3.4, 11.9.4). Neuromodulators induce changes in synaptic function and may be co-transmitters. Possible neuromodulators include GABA, serotonin, noradrenalin, enkephalins and opiates. There is still much to learn about co-transmission and neuromodulation and their influence on behaviour.

Winberg and Nilsson (1993) have reviewed roles of brain biogenic amine neurotransmitters in fish aggressive, defensive and stress behaviours and the formation of dominance hierarchies which can all be important in aquaculture. Dopaminergic neurons appear to stimulate aggressive and competitive behaviour and increased dopaminergic activity in dominant Arctic char *Salvelinus alpinus*, in a dominance hierarchy. The brain serotinergic neurons appear to inhibit aggressive behaviour and decrease swimming. Also, increased serotinergic activity occurs in fish in low positions in a dominance hierarchy.

Interest is developing in the possible effect on neurotransmitter activity and fish behaviour of predicted CO_2 change levels associated with climate change. Larval coral fish can exhibit olfactory and behavioural abnormalities when exposed to high CO_2 (Nilsson et al. 2012). The olfactory abnormalities cause these larval fish to be attracted to odours they normally avoid, such as those associated with predators and unsuitable habitats, which affects fish mortality. These behaviours can be reversed by blocking GABA-A neurotransmitter receptors indicating that high CO_2 interferes with neurotransmitter activity (Nilsson et al. 2012). *Pangasianodon hypophthalmus*, a fish occurring in hypercapnic conditions (high CO_2) in the Mekong river was used (Regan et al. 2016) to demonstrate a reversal of the above situation. When exposed experimentally to normocapnia these fish display hyperactivity which the GABA-A receptor blocker gabazine reverses. This further indicates that ambient CO_2 can alter GABA neurotransmitter activity in fish behaviour.

The participation of glutaminergic neurotransmitters in learning and memory has recently been gathering considerable support. In particular, NMDA receptors (NMDAR) are thought to be involved in memory encoding (Morris 2013) although this view is somewhat modified with the work of Taylor et al. (2013) who found NMDARs to be important for behavioural inhibition but not for 'associative' spatial memory. Moreover, as Bliss et al. (2014) report, the 'canonical' status of the NMDAR in long-term potentiation (LTP), thought to be a basis for memory, is somewhat diluted by the recognition of varying forms of LTP associated with different types of synapse, which may not be NMDAR-dependent.

Bibliography

Abrahams, M. The physiology of antipredator behaviour: what you do with what you've got. In: K.A. Sloman, R.W. Wilson, and S. Balshine (eds) (2005). *Fish Physiology, Vol. 25: Behaviour and Physiology of Fish.* Amsterdam: Elsevier, Amsterdam,pp. 79–108.

Braithwaite, V. Cognitive ability in fish. In: K.A. Sloman, R.W. Wilson, and S. Balshine (eds) (2005). *Fish Physiology, Vol. 25: Behaviour and Physiology of Fish.* Amsterdam: Elsevier, Amsterdam, pp. 1–37.

Breder, C.N. and Rosen, D.E. (1966). *Modes of Reproduction in Fishes.* New York, NY: American Museum of Natural History.

Brown, C., Laland, K., and Krause, J. (eds) (2011). *Fish Cognition and Behavior.* Hoboken, NJ: Wiley-Blackwell.

Fagen, R.M. (1981). *Animal Play Behaviour.* Oxford:Oxford University Press.

Johnssen, J.I., Winberg, S., and Sloman, K.A. Social interactions. In: K.A. Sloman, R.W. Wilson, and S. Balshine (eds) (2005). *Fish Physiology, Vol. 25: Behaviour and Physiology of Fish.* Amsterdam: Elsevier, Amsterdam, pp. 151–96.

Magnhagen, C. Braithwaite, V.A., Forsgren, E., and Kapoor, B.G. (eds) (2008). *Fish Behaviour.* Boca Raton, FL: CRC Press.

Pellegrini, A.D. (ed.) (2011). *Oxford Handbook of the Development of Play.* Oxford: Oxford University Press.

Sloman, K.A., Wilson, R.W., and Balshine, S. (eds) (2005). *Fish Physiology, Vol. 25: Behaviour and Physiology of Fish.* Amsterdam: Elsevier, Amsterdam.

Rosenthal, G.G. and Lobel, P.S. Communication. In: K.A. Sloman, R.W. Wilson, and S. Balshine (eds) (2005). *Fish Physiology, Vol. 25: Behaviour and Physiology of Fish.* Amsterdam: Elsevier, Amsterdam, pp. 39–78.

Smith, P.K. (ed.) (1984). *Play in Animals and Humans.* Oxford: Basil Blackwell Inc.

Stacey, N. and Sorensen, P. (2005). Reproductive pheromones. In: K.A. Sloman, R.W. Wilson, and S. Balshine (eds). *Fish Physiology, Vol. 24: Behaviour and Physiology of Fish.* Amsterdam: Elsevier, Amsterdam, pp. 359–412.

CHAPTER 15

Obtaining information

The typical fish

15.1 Summary

Managing the vast amount of information accumulated about fish has become a major enterprise, with large numbers of scientific papers published every year.

However, experimental design or data collection has often depended on very practical considerations such as size of holding tanks for live fish, accessibility of the fish under consideration and expense.

Very often the outcome has been the use of a very limited number of fish species based on their commercial value or their ease of accommodation to confinement.

There has been a tendency to concentrate on a very few species for study, with some being used more for experiments and others for anatomy/histology. Indeed, dogfish was often used as the basic fish for study although the bony fish are far more speciose and occupy a much wider range of habitats than elasmobranchs, with divergence in many features, both physiological and morphological.

Extrapolation from a limited number of type species has sometimes ensued and may give rise to problems.

15.2 Introduction

The term fish has been loosely applied to any animal living in water, hence starfish, shellfish, cuttlefish and so on, but biologists prefer to limit the term to aquatic vertebrates with gills as adults and fins rather than limbs (1.2).

In communicating information about vertebrate fish the starting point conceptually may be a constructed generalization, the 'typical fish', based on the most familiar species such as those fished for food or sport or kept for interest or pleasure, basically as pets. These familiar species are likely to be bony fish, and if fished species, generally seen dying or dead, whereas the 'pets' will usually be small and bright coloured and needing fairly warm water. However, whereas most studies of live fish are also based on bony fish, it has been common for students to start studying fish anatomy using dogfish (cartilaginous fish).

The general view of fish is also influenced by the exploitation of fish species commercially. This view is accompanied by economic pressures and the collection of large amount of data, with equations and rules set out for analysis of growth, reproduction, fishing effort and sustainability, and a certain amount of extrapolation and generalization. Fished species can be regarded, sometimes unrealistically, as sharing a number of characteristics such as rapid growth and high fecundity, representing another view of the typical fish.

Communication channels include formal reports and papers available in the scientific literature. It is usual for scientists to record their data in a format that can be transmitted to other scientists and held for future use. Availability of and access to such data varies but it is generally held in some version of printed documents. Conventions of expression for these communications are fairly well established: most publications of original data include an abstract, a summary which is brief and has a wider circulation than the full document or 'paper'.

The format of biological 'papers' when presented formally for publication is fairly standard. It includes a title, an abstract, an introduction, materials and methods, results, discussion and a list of references (relevant previous published work). The references

Essential Fish Biology: Diversity, Structure and Function. Derek Burton & Margaret Burton.
© Derek Burton & Margaret Burton 2018. Published 2018 by Oxford University Press.
DOI 10.1093/oso/9780198785552.001.0001

are either in alphabetic order (by first author) or by number, based on consecutive appearance in the text.

The length of papers has been highly variable, with a recent tendency to brevity. With the very large number of fish species and increased activity in fish research it has been difficult to manage the amount of information that has accumulated. Perhaps in response to this problem there has been an inclination to limit references to the very recent published material; this may mean that perfectly valid older work becomes effectively 'lost' and the foundation and basic understanding of a topic may be obscured. In the case of morphology, 'old' work was frequently very well executed although adjustment of nomenclature may be appropriate.

15.3 Choosing fish to study

For the study of live fish and function many, if not most, fish species are inappropriate to obtain or maintain, and so there has been a tendency to concentrate on a few bony fish (teleost) species which are relatively easy to keep. As fresh water is usually more accessible and large fish are hard to hold, requiring too much space, the preferred fish for study are often small freshwater species tolerant of fluctuating conditions. Such fish, though used as representative, may actually be fairly atypical of other fish species, many of which are marine.

The acanthopterygians (1.3.8), taxonomically the most speciose group, are very diverse with numerous marine and freshwater forms, some of which are easily held in aquaria. Two commonly used larger fish, goldfish and 'trout', in Western laboratories, however, are not acanthopterygian; goldfish are ostariophysan (1.3.8) and 'trout', whether rainbow or brown, are protacanthopterygians (1.3.8). These fish, goldfish and 'trout', are fairly tolerant of a wide temperature range, distinguishing them from more temperature-sensitive species, even those close to their own taxonomic grouping. Thus, char (*Salvelinus alpinus*) does best in relatively cold water whereas rainbow trout (previously *Salmo gairdneri* now *Oncorhynchus mykiss*, Table 15.1) is notably tolerant of warm water and thus easier to hold in captivity. An unfortunate result of this species-selective approach for experimental design may be an over-optimistic view of temperature tolerance or other potential stressors in fish.

15.4 Sources of background information

In planning fish research or more general study, background information available may include reports and refereed journal articles. Meetings of scientists, with subsequent publications of abstracts or whole proceedings allow current work to be discussed and become accessible quickly.

15.4.1 Reports

A very large amount of material has accumulated on fish and fisheries. Some of it takes the form of internal reports linked to institutions and it may be difficult to find or to use, particularly as some governments prohibit the use of such material without prior permission. Raw, untreated data are rarely accessible but was previously published as appendices, particularly in some government documents such as Wallace (1914), Graham (1923) or records from marine laboratories (Bagenal 1957).

A difficulty with treated data can be that 'outliers' may have been discarded, these being data that do not fit an expected pattern or hypothesis. Such specimens may later prove to represent examples of a previously unpredicted aspect of the phenomenon under study. Moreover, it is possible for the raw data to be altered either by mistake or intentionally, in the latter case a fish put in a particular category initially may be reassigned on what may, or may not, be valid objection to the first categorization (Burton 1999).

Before widespread computerization some information was published as compilations of abstracts and grouped by areas of interest or by authors.

Government reports are generally regarded as nonrefereed: no external scientist has read or criticized the document. The information contained in the report represents a history of the data collected, sometimes with considerable detail provided; such information can be very valuable. It is to be expected that the records provided are valid and may be useful for comparative purposes. (Burton 1999).

Particularly valuable reports are those of the various deep-sea expeditions such as those undertaken by the Challenger, which include copious figures, meticulously produced.

15.4.2 Refereed journals

Journal publications include highly specialized material through to publications which accept a wide range of submissions, the latter including *Science* and *Nature*, which generally publish brief accounts often as 'letters' which are deemed to be current and of wide interest.

Most journal 'papers' are checked over, not only by an editor but usually by at least one, usually more, scientists with expertise in the field covered. This may mean alteration of the text in line with the opinions/criticisms of the referee and the editor.

Some journals are published by a group of scientists forming a Society and may represent the material already presented to a meeting of that society. The Royal Society of London has had publications of *Proceedings* for a long time and it has been regarded as a valuable source of scientific data and opinion. Royal Society presentations were previously sponsored by an individual from the Society and it is presumed that such material had been carefully checked, the background of the author(s) having been vouched for by the sponsor. Nowadays this procedure has been discontinued and papers are reviewed, as described earlier, for other journals.

In Canada, the government-sponsored *Journal of the Fisheries Research Board of Canada* subsequently (with the demise of the Research Board) transmuted to the *Canadian Journal of Fisheries and Aquatic Science*. A wider publication, accepting scientific papers on animals, including fish, is the *Canadian Journal of Zoology*, published monthly and previously administered through government offices.

With increasing scientific activity in the last decades new journals of the classical format have emerged (including the *Journal of Fish Biology* and *Reviews of Fish and Fisheries*) as well as, more recently, online material.

15.4.3 Scientific (specialist) meetings

As an extension of the type of regular meetings held by scientific societies such as the Royal Society, a plethora of specialist groups have been formed which may have regularly scheduled or occasional scientific meetings. Some of these are organized nationally and may have an annual meeting at different locations (the Canadian Society of Zoology) but many are international in reach and may have meetings every three to five years. The Society for Experimental Biology has a number of specialist meetings annually. Such meetings may take the form of 'symposia' and be subsequently published as books.

15.4.4 Books

Books may be generated from edited symposia or from government interest in particular topics, such as Scott and Crossman (1973), Scott and Scott (1988), Morrison (1987, 1988, 1990, 1993), these last being a valuable contribution, via government, to the literature.

Otherwise, books may be the result of the author's or publisher's perception of a niche which remains unexplored and/or which is of public or scientific interest.

15.4.5 Language of communication

Over the centuries the language used to transmit scientific information shifted from Latin (and classical Greek) to vernaculars, notably German (e.g. von Frisch 1911), Spanish/French (e.g. Ramón y Cajal 1952), French (e.g. Grassé 1958, 1997) and English, the latter now being predominant. Most scientific publications are currently in English or have an English abstract. For a while, Russian science was avidly translated so that it became available to Western scientists. This was particularly relevant for fisheries work and made available some important material that might otherwise have been totally neglected (see Messiatzeva 1932; Nikolsky 1963; Fedorov 1971; Oganesyan 1993). Nevertheless some interesting material remained available in English only as an abstract (Shirakova 1969).

The terminology used for biological work is extensive and complex and often based on Latin and Greek; an understanding of some of the vocabulary (and plurals for nouns) is helpful for navigating dense texts and taxonomy (Appendix 15.1).

15.5 Species studied

15.5.1 Publications

In the twentieth century, the ability to keep abreast of the widening fields associated with studying form and function in fish as well as taxonomy resulted in collation of journal paper titles and abstracts into special publications so that it is possible to count the numbers of papers in particular subject areas.

Publications on fish were approximately 40,000 items from 1916 back to pre-Linnaean times (Dean 1916). Subsequently, in the second half of the twentieth century (Cvancara 1989, 1996), the journal publications' number annually was around 5000 (4792 for 1989, 5136 for 1996), numbers which suggest we should know a lot about fish. However, examination of the species studied also indicates a skew towards particular fish, which has increased in the time considered (Tables 15.1 and 15.2).

The overall impression from Tables 15.1, 15.2 and 15.3 is that salmonids are favourite subjects for study (which may reflect their intensive use in aquaculture), though research on commercial subjects may be under-reported in compilations, often being communicated in government documents not included as journal papers.

An interesting trend seen in Table 15.1 is the increase in zebrafish publications, something that has accelerated recently (15.6.4).

Regional accessibility can influence subjects for study; studies of live fish in Western Europe and North America may have concentrated on easily held, cheap, freshwater fish, like non-native goldfish and brown or rainbow trout, while similar studies in Asia have utilized native carp or smaller warm-water 'tropical' fish, and analyses may show bias towards papers published in English. Analyses of published papers in particular years (Tables 15.1 and 15.2) show a marked skew towards particular species which may shift over time if new protocols or concerns develop. As with the use of rapidly maturing, small-size species in genetics (fruit-fly, the plant *Arabidopsis*), success will promote the utilization of species for tank studies (see *Brachydanio rerio*, Table 15.1), a situation which may threaten a species if it is wild-caught and not breeding to replacement

Table 15.1 Numbers of journal papers published on particular freshwater teleosts in different years (1969, 1989, 1996).

Species	Superorder	1969[†]	1989[††]	1996[††]
	Ostariophysi			
Goldfish (*Carassius auratus*)		173	178	102
Carp (*Cyprinus carpio*)		216	207	150
Zebrafish (*Brachydanio*, or *Danio, rerio*)		9	13	27
	Protacanthopterygii			
Rainbow trout* as *Salmo gairdneri* as *Oncorhynchus mykiss*		277	207 17	150 404
Atlantic salmon (*Salmo salar*)		100	166	178
Brook trout (*Salvelinus fontinalis*)		91	49	49
Char (*Salvelinus alpinus*)		18	37	55
	Acanthopterygii			
Yellow perch (*Perca flavescens*)		30	34	15
Eurasian perch (*Perca fluviatilis*)		64	29	31
Three-spined stickleback (*Gasterosteus aculeatus*)		92	47	45
Sailfin molly (*Poecilia* spp.)		13	2	1
Swordtail (*Xiphophorus helleri*)		16	2	6
Siamese fighting-fish (*Betta splendens*)		12	3	2

[†]= Source Atz 1969.
[††]= Source Cvancara 1989, 1996.
*Binomial change from *S. garidneri* to *O.mykiss*.

levels. Zebrafish, however, are easy to breed in captivity so this caveat does not apply to them.

However, the most noteworthy trend is the increase in papers published on salmonids and the decrease for most other groups (Tables 15.1, 15.2

Table 15.2 Numbers of journal papers published on particular marine or catadromous* teleosts in different year (1969, 1989, 1996).

Species	Subdivision (Superorder)	1969[†]	1989[††]	1996[††]
Freshwater eel (*Anguilla* spp.)	Elopomorpha	52	45	46
Atlantic tarpon (*Megalops atlantica*)		13	0	2
Atlantic herring (*Clupea harengus*)	Ostarioclupeomorpha	139	44	39
Capelin (*Mallotus villosus*)	Euteleostei (Protacanthopterygii)	9	5	4
	(Paracacanthopterygii)			
Atlantic cod (*Gadus morhua*)		91	69	98
Pollack (*Pollachius pollachius*)		2	0	2
Haddock (*Melanogrammus aeglefinus*)		37	8	7
	(Acanthopterygii)			
Plaice (*Pleuronectes platessa*)		37	12	17
Winter flounder** As *Pseudopleuronectes americanus*		15	22	8
As *Pleuronectes americanus*				10
Sunfish (*Mola mola*)		6	0	0

*Catadromous species breed in salt water but may spend a long growth period in fresh water.
**Genus change from *Pseudopleuronectes* to *Pleuronectes*, since reversed.
[†]Atz 1969.
[††]Cvancara 1989, 1996.

Table 15.3 Numbers of journal papers published on chondrichyans in different years (1969, 1989, 1996).

Species	Subclass(Order)	1969[†]	1989[††]	1996[††]
	Holocephali			
Chimaera (*Chimaera monstrosa*)		10	0	0
	Elasmobranchi			
Spiny dogfish (*Squalus acanthias*)	(Squaliformes)	71	14	14
(*Scyliorhinus canicula*)	(Lamniformes)	35	12	10
Great white shark (*Carchorodon carcharias*)		3	0	3
Basking shark (*Cetorhinus maximus*)		9	1	0
	(Rajiformes)			
Winter skate (*Raja ocellata*)		0	1	2
Little skate (*Raja erinacea*)		7	7	5

[†]Source Atz 1969.
[††]Source Cvancara 1989, 1996.

and 15.3). The change in numbers of papers published is not due to the change in source from Atz (1969) to Cvancara (1989, 1996) because the decline for some species continues from Cvancara 1989 to Cvancara 1996. It seems as if many species that used to be studied, including commercially important gadids like haddock, are now neglected in favour of the salmonids, for which research funds may be more readily available, and which can be more readily cultured. Herring studies appear to have plummeted but this may be a reflection of the journals used for compilations, although it may also derive from the uncertainty of fishing the highly variable, relatively low-value, stocks. Interestingly cod studies have held fairly steady perhaps due to attempts to understand the recent (late 1980s and early 1990s) drastic decline in this commercially valuable species.

15.5.2 Reference works

Anatomy guides and laboratory manuals for fish dissection (Table 15.4) in the Western scientific community have been prevalently on elasmobranch dogfish (either *Squalus acanthias* in North America or *Scyliorhinus canicula* in Western Europe) as these, being large and cheap, were regarded as good targets for student dissection,

Table 15.4 Reference works for anatomy, dissection and histology, based on 'types'.

Subject	Types	Author	Date
Anatomy	Dogfish	Marshall and Hurst	1930
	Lamprey, dogfish, cod, lungfish	de Beer	1951
	Dogfish	Grove and Newell	1953
	Skate, trout, whiting roach	Saunders and Manton	1951, 1979
	Dogfish, skate, haddock	Ritchie	1953
	Lamprey, dogfish	Weichert	1961
	Channel catfish	Grizzle and Rogers	1976
Neuroanatomy	Zebrafish	Wulliman et al.	1996
Dissection guides	Dogfish	Whitehouse and Grove	1947
	Dogfish	Rowett	1953
	Dogfish	Wells	1961
	Dogfish	Gans and Parsons	1964
	Dogfish	Ashley	1969
	Perch	Peachey	1989
Histology and Anatomy	Zebrafish	Holden et al.	2012
Histology	Trout	Anderson and Mitchum	1974
	Rainbow trout	Hibaya	1982
	Striped bass	Groman	1982
	Rainbow trout	Yasutake	1983
	Cod	Morrison	1987, 1988

with interesting anatomy and fairly easily cut cartilage for brain access. Teleosts, being bony and often laterally flattened, may be much more difficult to manipulate on a dissecting board. Many Western students therefore have been reared on the anatomy of elasmobranchs, even though these are relatively uncommon and atypical. Perch (North America) are sometimes used as an addition or alternate to dogfish; herring and skate have been used in Europe.

Elementary texts on zoology such as Grove and Newell (1953) used to concentrate principally

or even exclusively on dogfish, with a two-year syllabus, including a year on vertebrates, only using dogfish as the fish example for dissection (Wells 1961). Currently, however, such intensive use of dogfish has diminished and teleosts such as herring may be used briefly for some anatomical observation. For the microscopic study of fish tissues (histology), a few manuals have been available (Table 15.4), mostly concentrating on one teleost species; that is, there have been no general guides on dogfish histology to match the anatomy guides even though dogfish were commonly used as sources of material for student histological studies of cartilage and striated (skeletal) muscle. The available teleost guides should be used with great caution because of the diversity of fish species.

15.6 Type fish

The concept of using 'types' is well established though it remains somewhat problematic. For fish, the huge diversity means that no one fish species can be truly representative. The main type species—dogfish for anatomy, sticklebacks, goldfish and zebrafish for experiments (the latter for genomics), trout for many purposes and gadids, perch and flatfish for culture and experiments—have special attributes (Table 15.5) which influence the generalizations which can be drawn from each study.

15.6.1 Dogfish

Dogfish (*Scyliorhinus* and *Squalus*) are not typical because most fish are not cartilaginous, and the anatomy of cartilaginous fish differs in many ways from that of other fish, not just in the composition of the skeleton (i.e. cartilage not bone). Dogfish have exposed gill slits not covered with an operculum. They also have a spiracle on each side. Their tail is heterocercal and their pectoral fins have a broad attachment to the body. Their pelvic fins are far posterior and in males bear claspers for mating, with internal fertilization. The eggs are large and few. The male reproductive system is complex, the thyroid, pancreas, supra-renal and inter-renal glands are conspicuous.

Table 15.5 Anatomical features compared in 'type' fish.

Fish type	Gills	Airbladder	Pancreas	Gut	Ovary
Cartilaginous					
Dogfish*	No cover	None	Obvious	Spiral valve No caeca	No wall
Bony					
Goldfish**	Cover	Present	Not visible	No spiral valve Caeca present	No Wall
Zebrafish	Cover	Present	Not visible	No spiral valve No caeca No stomach	No wall
Trout	Cover	Present	Not visible	No spiral valve Caeca present	No wall
Sticklebacks	Cover	Present	Not visible	No spiral valve No caeca?	Wall
Cod	Cover	Present	Not visible	No spiral valve Caeca present	Wall
Perch	Cover	Present	Not visible	No spiral valve Caeca present	Wall
Flatfish	Cover	None	Not visible	No spiral valve Caeca present	Wall

*The only 'type' fish listed here with ampullae of Lorenzini (12.9.3).
**The only 'type' fish listed here with Weberian ossicles (12.8.4).

The intestine has a spiral valve. The olfactory region is very well developed and there is a unique system of ampullae of Lorenzini which achieve electrolocation.

15.6.2 Sticklebacks

Sticklebacks (notably *Gasterosteus*) of the Family Gasterosteidae have been the subjects of over 2000 publications (Mattern 2006) with particular interest in their behaviour, including the work of Nobel Laureate Tinbergen (1952) and that of Wootton (1976, 1984), with a recent monograph (Ŏstland-Nilsson et al. 2006). As sticklebacks are widely distributed in the cooler regions of the Northern hemisphere and often occur in large numbers in shallow water, fresh, brackish or salt, they are fairly easy to obtain, and cheap. Physiologically they represent small euryhaline fish but have some unusual morphological features such as lateral scutes rather than scales and a

high degree of intra-specific variability in structures often regarded as more stable (e.g. scute and spine number). At least one species (*Gasterosteus aculeatus*) is also prone to a high level of parasitic infestation (Scott and Crossman 1973) which can affect reproductive physiology including secondary sexual characteristics (Rushbrook et al. 2007) and could complicate replication in experiments. These fish are reputed to be short-lived (maximum 3.5 y, Scott and Crossman 1973) and therefore not suitable for longer-term studies.

15.6.3 Goldfish

Goldish (*Carassius auratus*) are easily kept in small tanks of freshwater, breed in captivity, and are very tolerant of temperature fluctuations. They are cheap to buy and keep, as well as potentially long-lived and easy to obtain throughout the year in most locations. They are, therefore, ideal for many purposes including experiments on a wide range of functions. However, as members of the Ostariophysi they have some special characteristics which differentiate them from other fish; notably they have relatively good hearing because they have special Weberian ossicles (1.3.8, 12.8.4) linking the swim bladder to the ear, and have club cells in the skin (2.3.1). They are physostomatous (the swim bladder opening to the gut, 3.8.5, 6.6.2). Their colour is unusual and their diet (like most of the cyprinid and silurid ostariophysans) is herbivorous unlike the majority of fish (4.3).

15.6.4 Zebrafish

Zebrafish (*Brachydanio* or *Danio rerio*) have recently become exceedingly popular for behavioural and neurophysiological work and experiments because they are easily kept and bred, with frequent egg production possible and rapid development with, as Parker et al. (2013) express, *ex utero* 'embryos' making them increasingly useful. Perry et al. (2010) devoted a volume of the *Fish Physiology* series to zebrafish. An early use of zebrafish for endocrinological studies with both *in vivo* and *in vitro* approaches was a series of papers by Van Ree including Van Ree et al. (1977) and later Detrich et al. (2009) presented a miscellany of methodologies used with this species.

In 2014, an entire volume of *Frontiers in Neural Circuits* was devoted to a collection of papers on this species. As Sumbre and de Polavieja (2014) explain, the zebrafish genome-sequencing programme began in 2001 and various zebrafish mutant lines have been established.

15.6.5 Salmonids

Salmonids, though not so easy to obtain and keep as goldfish, can be held for extended periods in fresh water. They tolerate highly acidic conditions well and may thus be kept in peaty areas (parts of North America and Europe). Rainbow trout are particularly tolerant of temperature fluctuations and feed, grow and reproduce well in captivity. Rainbow trout can be migratory (anadromous, 9.14) but may accomplish all their life cycle in fresh water. As members of the Protacanthopterygii they have an adipose fin dorsally and anterior to the tail. They are active swimmers, their eggs are fairly large and deposited in gravel with an elaborate courtship possible. Colour changes preparatory to migration (silvering to smolts) and prior to reproduction. They can have high commercial value, their tank culture has been promoted and they may therefore be fairly easy to obtain in some regions of Western Europe and North America. *In vitro* fertilization can be achieved and DNA transfer is relatively easy because the eggs are large, microinjection through the micropyle being possible (Du et al. 1992).

15.6.6 Gadids

Gadids (e.g. cod *Gadus morhua*) can be held in captivity but need seawater and large tanks. They feed and grow well but therefore need even more space. They also reproduce well in captivity. They are actually inclined to over-eat and if temperatures rise the competing demands of digestion and respiration may produce anoxia and death (Specific Dynamic Effect, 4.8). Anatomically they are more 'typical' perhaps than flatfish or dogfish, having the generalized fish shape and fins as well as obvious gill covers.

15.6.7 Perch

Perch (*Perca*) are used because they are common in Europe, northern United States and southern Canada. As they are acanthopterygians these fish could be appropriate representatives of freshwater bony fish for both anatomical and experimental studies. However, they have little commercial value which may have affected research support; publications (by number) for both the American species (*Perca flavescens*) and the European species (*P. fluviatilis*) declined recently (Table 15.1). They are easy to obtain and cheap and have become popular subjects for elementary dissection in North America, but preserved specimens are not easy to dissect for details of the nervous or vascular system; they are best used for examining major organs in the abdominal cavity. The colour of preserved *P. flavescens* (yellow perch) is often more uniformly yellow than would be seen in live specimens which have dark stripes, similar to parr marks on salmonids.

15.6.8 Flatfish

Flatfish (notably pleuronectids) are often quite easy to keep if there is a good supply of sea water. An incentive to keep these fish is also their potential commercial value (as with trout). Species held in tank culture have included *Pseudopleuronectes americanus* (winter flounder, off the North Atlantic coast of North America), *Pleuronectes platessa* (plaice, North Sea, Irish Sea) and *Hippoglossus* species (halibut, North Pacific and North Atlantic). Flatfish can be fairly tolerant of a wide temperature range, may feed and accept handling and can be caught without much obvious stress. It is easy to take repeat blood samples and they are also very easy to sex: if the fish can be held up against a light source gonads may be seen by transparency; testes are shorter than ovaries. *In vitro* fertilization has been achieved for several species although handling large halibut can be difficult. Obviously the shape is not typical as is the fact that juveniles undergo a metamorphosis when one eye migrates. Flatfish genera are usually either dextral (right-side up) or sinistral (left-side up) but one genus (*Psettodes*) has no side predominance. Occasionally a wrong-side up flatfish may occur and captive bred young may show partial metamorphosis with incomplete eye migration, large white patches on the upper surface and other colour abnormalities (2.8). Captivity may also lead to other anatomical differences; for example, the epidermis may thicken up (2.10).

15.7 Fish habitat and availability/suitability for study

Benthic (bottom-living) fish may be easier to study than those which swim actively in the water column (pelagic, like herring or mackerel), although this problem does not seem to apply to salmonids.

Some pelagic fish are difficult to hold in captivity because they may depend on ram ventilation (6.5.11 and Roberts 1975) or may swim into tank sides and damage themselves as occurs with some open-water schooling fish like capelin.

Deep-sea fish are notoriously difficult to obtain alive, even the mid-water lantern fish do not do well when netted, so functional studies of such fish, including their bioluminescence, have been hampered. Their skin and muscles are soft and tend to disintegrate on handling.

15.8 Captive fish and sources

Studies of, and experiments with, live fish will rely on either wild-caught or captive-bred fish and both situations require careful consideration. Wild-caught fish may need an extensive period of acclimation before work can commence and matching wild fish for group studies can be difficult if there is a range of size and maturity available. Moreover tank-held fish can begin to show considerable diversity, even of structure (2.10) with time, and different conditions such as light, temperature and feeding. More domesticated captive-bred fish may also show differences from wild fish especially if they have been selected for particular traits either deliberately or inadvertently; therefore the extrapolation of results obtained from such fish, to wild fish, may be inappropriate. In using tank-held fish for extracting information the prior history of the fish may not be clear; trans-shipments and mixture of groups may have occurred so that they do not form a useful coherent group on which to base a study.

15.9 Conclusion

It is possible for scientists who have had their student fish studies based on a limited number of 'types' such as dogfish and/or trout and goldfish, to be very surprised if they try to extrapolate from these. Particular anatomical features very visible in one group as with the pancreas of elasmobranchs may appear to be totally absent in teleosts (Table 15.5). Comparison between Tables 15.1, 15.2 and 15.3 show that studies on rainbow trout, measured as numbers of journal publications, prevail but at least until recently macroscopic anatomical studies (Table 15.4) aimed primarily at undergraduates have concentrated on elasmobranchs.

During the Second World War, a novel view of zoology (*Animal Biology*, Grove and Newell 1942) was prefaced by the then Cambridge University Reader (later Professor) in Zoology, Dr Pantin, the man who discovered terrestrial nemertines on a forest walk, amongst many other contributions to science. Pantin wrote 'the old method of approaching Zoology by the detailed study of the anatomy of a few types has much to recommend it. It suffers however from two disadvantages'. He then refers to the 'vast array' of different animals 'so different from each other' as a problem when doing type-based work. He also considered earlier type studies to be neglectful of function, that is, physiology. However, realistically, fish are so many and various, now as then, that 'type' studies have to occur; it is impossible to be truly comprehensive. In using types though it is necessary to be wary of extrapolation and of generalization, and also of allowing too much research to concentrate on too few types, a situation that seems to have been worsening.

It is particularly interesting that some large fish 'well-known' from the point of popular science and marketing, such as the great white shark and the sunfish *Mola mola* (Navarro and Snodgrass 1995), show few journal publications (Atz 1969, Cvancara 1989, 1996). Even capelin, a North Atlantic forage fish at a critical position in many food webs, supporting birds, whales and other fish like cod, is not frequently studied (Atz 1969, Cvancara 1989, 1996). Although rainbow trout research has produced much interesting information it is surely appropriate now to diversify. To quote Pantin again, later in his life (1960) he observed, concerning one of his favourite animals, 'the genus is not of the slightest economic importance', and he added 'indeed in these days when the scientist is naturally expected to bend his energies to really useful things, like shooting rockets at the moon, some explanation is perhaps required for time spent on creatures of such obscurity'. His point is important; of course, whatever the immediate and obvious possible economic benefits from some research, it is unwise to restrict acquisition of knowledge to a very narrow focus. New knowledge, appreciated initially by only a few, may finally acquire its full significance later as other developments occur. At the very basic level we don't know how many fish species there are. Recently international efforts have been attempting to understand the number of marine species overall, not just fish, with estimates that only about 1 per cent of benthic species are known (Snelgrove 1999); it is likely there are still many unknown fish species in the Amazon–Rio Negro complex (Val et al. 1996). And just recognising the existence of a species may not tell us much about its function. Who could have predicted the existence of fish without red blood cells (5.13) or fish with 'parasitic' males (9.3)? There are even fish with 'necks' (*Lepidogalaxias* and *Raphiodon*, 13.10.2). However helpful trout and goldfish have been as type species, we need the stimulus of the unusual and the unexpected. Fish give us the opportunity to study a rich diversity, sufficiently like ourselves to pique our curiosity and sufficiently unlike mammals to give us a fresh view of vertebrate possibilities—the fish version of life.

Appendix 15.1 Examples of Latin and Greek basics for biological use

Much of the terminology used for fish structure and classification is Greek- or Latin-based. The following gives examples of the way these languages are used (see also Appendix 1.4). Various prefixes or suffixes added at the beginning (prefix) or end (suffix) of a word are particularly informative.

Prefixes:

Amount: hypo = less; hyper = more; iso = the same; di-two; tri = three; poly = many.

Salinity: hal(ine).

Validity: pseudo = false.

Colour: leuco = white; erythro = red.

Position: bathy = bottom, supra = above.

Xeno = strange, photo = light.

Pre = before, post = behind, endo = within, exo = without, meso = middle.

Somato = body, chondro = cartilage, osteo = bone.

Suffixes:

-cyte = cell; -phore = 'bearing', -crine = secretion, -trophic = feeding, -ase indicates an enzyme, -ose indicates a sugar.

Either prefix or suffix: coel(e) meaning cavity.

Plurals can give difficulty:

A word which has been latinized may have the ending 'us': the plural for this may end 'i'; for example, 'glomerulus' becomes 'glomeruli' and 'ramus' becomes 'rami'. Alternatively, and more rarely, the ending may be '-ora'; for example, 'corpus' becomes 'corpora'.

The ending 'um' has the plural form 'a'; that is, 'caecum' becomes 'caeca'; 'cilium' becomes 'cilia', 'oogonium' becomes 'oogonia'; 'ovum' becomes 'ova'.

A word with the ending 'a'; for example, 'larva', has the plural form 'ae'; that is, 'larvae'.

A word ending 'on' has the plural form 'a'; that is, mitochondrion becomes mitochondria, spermatozoan becomes spermatozoa.

A word ending 'is' has the plural form 'es', that is 'mitosis' becomes 'mitoses'.

A latinized word ending 'e' may take the plural form 'ia'; that is, 'rete' becomes 'retia'.

CHAPTER 16

Conservation and fish function

The future for fish

16.1 Summary

The future for fish species, both numerically and in prevalence, has become a subject for pessimism.

Pressures on fish survival include fishing and degradation of the environment as well as human interference with the balance of ecosystems by the direct use of water (the Aral Sea disaster) and the introduction of non-native species.

Contamination of coastal and inland waters, for example by pharmaceuticals and mine tailings, affects fish negatively as does change in temperatures.

The continued rise in human populations and expanding industrialization are matters for concern. The optimistic views of fish abundance held in historic times and still politically extant require modification in view of the realities of stock depletion and individual extinctions.

16.2 Introduction

Fish function well under a very wide range of environments but individual fish may require very specific conditions in which to thrive. Over a very long time range fish species are capable of change (evolution) with adaptations to particular environmental changes but individuals may also be capable of physiological changes enabling them to survive in shifts of conditions; such shifts may be 'normal' and part of the daily, seasonal or life cycle. Other changes in the environment may be less usual and provoked by singular events or by an ongoing situation such as leaching into the water of 'foreign' substances, for example non-normal material, ions or changes in temperature beyond the usual fluctuations.

In physiological parlance organisms which tolerate change well may be differentiated from those that do not; for aquatic animals it has been common to use the term euryhaline for those that tolerate salinity changes and stenohaline for those that do not (7.9). Similarly, animals that tolerate temperature changes well are 'eurythermal' while those that do not are 'stenothermal'. However, living organisms have a range of responses to temperature changes that encompass a wider range of possibilities; in the case of fish they may be capable of moving considerable distances to avoid temperature changes and they may also be capable of torpor in hot or cold conditions. In addition, their normal physiological responses to cold may include seasonal production of 'antifreeze'. Whereas some vertebrates (notably birds and mammals) regulate their internal temperature to a set point, most fish do not, but some large fish are able to maintain some parts of their bodies at high temperatures with a form of heterothermy (7.4).

Currently, and indeed since wide-scale industrialization and human population growth, fish and all other organisms are increasingly constrained by habitat change and habitat destruction. Their ability to survive in these circumstances has been remarkable but it is important to understand their capacity to withstand change and to avoid continuing deleterious conditions.

Direct utilization of fish as a human food source has not always been realistic. It was common to regard marine fisheries as inexhaustible with the view that fish produce thousands if not millions of eggs per annum, and that there is compensation in survival because of reduced competition (Appendix 16.1).

Essential Fish Biology: Diversity, Structure and Function. Derek Burton & Margaret Burton.
© Derek Burton & Margaret Burton 2018. Published 2018 by Oxford University Press.
DOI 10.1093/oso/9780198785552.001.0001

Studying fish reproductive function (9.16) reveals that they may be more conservative in output than initially expected and this warrants considerable care in prediction of future abundance.

16.3 Conservation physiology

Currently, a new field of study is emerging, with a new journal, *Conservation Physiology*, including an introductory article (Cooke et al. 2013). The subject was recently reviewed at a symposium organized by the Society for Integrative and Comparative Biology (Madliger and Love 2015). The aim of this new field is the application of information from a wide range of physiological sources to understanding processes associated with organismal responsiveness to environmental changes, including climate change, and to stressors. This field should stimulate new theoretical (modelling) and experimental initiatives and will also utilize existing physiological methodologies and information. The impacts of most foreseeable environmental perturbations (climate change, overpopulation, food shortages, pollution) will most likely influence physiological and cellular processes at an early stage, providing warnings.

16.4 Abundance

More and more fish species are being reported as extant, but the number of individuals per species may be very limited. Some fish species occupy very small niches, and there would never be very many individuals. In other cases, a particular species may number many individuals (1.4.5). It is obvious that fish with highly specific physiological requirements as to temperature or food are exceptionally vulnerable to changes, but we may not know the natural food or temperature limits for a species. The fact that some fish can be domesticated or tank-held does not necessarily mean that they can reproduce successfully in captivity or be capable of reintroduction to their usual habitat.

Most of the fish species that have been reported as extremely abundant are pelagic and oceanic, capable of moving considerable distances. Such fish are also prone to large-scale exploitation, the masses of

fish attracting commercial interest, such as anchovies off California (Table 1.6) and off Peru (Idyll 1973). Other fish species believed to be abundant may appear so because they form pre-spawning aggregations. A question arises as to the capacity of a species to withstand removal, before reproduction, of its mature individuals. Studying reproductive function (9.15, 9.16) suggests that individual females and, perhaps surprisingly, also males can be very sensitive to feeding levels prior to reproduction and down-regulate output. The rate at which they may up-regulate can be quite slow because cold-water species may have very slow oogenesis, reported to take several years in some species for any one putative cell to become yolky and reach spawning capacity. There can, however, be several cohorts of oocytes (immature eggs) within an ovary, with successive reproductions possible if conditions are optimal.

Recently it has become clear that not only do fish down-regulate their reproductive output (e.g. by a condition for females, known as atresia, when a few or more developing eggs are reabsorbed), but individual fish (male or female) may omit reproduction altogether, skipped spawning (9.16). Such reproductive omission can be related to feeding and feeding season, particularly relevant for fish that may have a very limited feeding season (4.8, 9.16).

In the case of some fish, notably sharks, their capacity to produce multiple young in any one year is severely limited, a situation that has to be allowed for in any exploitation.

16.5 Extinction

For fish which have never been abundant, or are currently very limited in habitat range and their capacity to relocate, absolute extinction in the relatively near future is a likely outcome; for fish which are commercially desirable, commercial extinction is also likely, particularly if calculations of sustainability are erroneous. Extinctions may also occur with change in habitat produced by human or natural activity as with changes in watercourses or use. Some fish, for example the relict fish (1.3.6), appear to have withstood change over a long time period, perhaps by inhabiting relatively isolated niches thus reducing competition with teleosts, although this does not apply to *Amia* which may have persisted due to a combination of hypoxic and temperature tolerance as well as behavioural traits such as aggression and winter breeding (Ilves and Randall 2007). However, some of the relict fish, like the Australian lungfish and the coelacanth (two species), do inhabit restricted areas, a low-volume river system for *Neoceratodus* (the lungfish), two widely separated but very small marine areas for the two species of *Latimeria* (the coelacanth), and are in danger of permanent extirpation.

Burkhead (2012) studied extinctions in North American freshwater fishes, the geographical area including Canada, the United States (with Alaska) and Mexico excluding Qintana Roo and Campeche. He found 81 extinctions in the twentieth century for world freshwater fish species with 57 taxa lost in North America. He emphasizes that exact dates of extinction are problematic but can be known when a habitat is destroyed. A species currently under threat and possibly extinct is the slender chub (*Erimystax cahni*) of the upper Tennessee River, which has not been observed since 1996 in spite of intensive searches. Burkhead suggests that the recent extinction rates for freshwater fishes are very high, approximately 877 times greater than the background extinction rate, a number that indicates a crisis.

The slender chub (Plate 16) is one of 12 freshwater fish species of the south-eastern United States, which have been listed as most imperilled for the region (Kuhajda et al. 2009); one of these, the Alabama cavefish (*Speoplatyrhinus polsoni*), is endangered because its habitat is a single cave system and it has a very small population, with no more than 100 individuals estimated, only 10 observed at one time.

Fish numbers may be very difficult to ascertain, unlike the situation for birds or mammals the former being particularly accessible and highly visible with range and brood size fairly easily delimited. Thus, for fish, the risks of extinction within or throughout a range, are perhaps higher than generally realized. The setting-up of risk lists attempts to address this problem by flagging species at risk. The International Union for Conservation of Nature (IUCN) has established eight categories in their 'Red list' (Appendix 16.2) with, for example, cod evaluated as 'vulnerable' in 1996, 'facing a high risk of extinction in the wild in the medium-term future', a situation which may have improved somewhat (2016).

16.6 The physiology of pollution

The term pollution implies human mistake or intentional contamination of the environment; that is, for fish, the water which they depend on for almost every phase of their existence, to swim, obtain oxygen, feed and in which to reproduce.

16.6.1 Thermal pollution

Direct use of water for cooling in industrial processing may lead to local upswings or sustained increase in water temperature affecting a region around the effluent system. The Swedish government has monitored the water temperature affects around the Forsmark Nuclear Power station on the Swedish coast north of Stockholm; a comprehensive study beginning in 1969. The cooling water is the combined product from two rivers and the warm discharge runs into an enclosed 'biotest' basin, which reaches 8–10 °C above the surrounding water. This basin is brackish and supports 12 different teleost species which form an important part of the ecosystem. One concern has been the possible increase in disease prevalence with temperature rise (Forsmarks Kraftgrupp 1982).

16.6.2 Chemical pollution

Aquatic systems are vulnerable to the leaching of mine 'tailings' and the discharge of many different chemicals from industrial processes, some of which include such 'natural' substances as are found within wood pulp. The latter can include endocrine disruptors (8.3.10) which mimic or interfere with the action of natural steroids. Effluent from domestic conurbations may, since the introduction of the contraceptive pill, also include steroids. In both the latter cases, fish can be affected as fish may share the same or very similar steroids with mammals; mammalian steroids can affect fish reproduction. An excess of female-type steroids in an aquatic system can feminize fish. Fish respond to chemical stress by avoiding it or secreting copious mucus at the gills or skin; the latter response may gain time to swim out of danger.

Metal toxins and other contaminants may induce a suite of changes in the liver (7.11) inducing enzymes which have some capacity to detoxify. Wood and co-workers (7.11, 8.3.9) have studied the physiology of interactions with metals in the aquatic environment.

Oil can have at least two major effects on fish; the physical problem may be a coating on the water and/or the gills which diminishes gas exchange. A chemical effect can be a systemic reduction in androgens (male steroids), and this in turn can decrease the expression of secondary male characteristics (e.g. skin thickness, Burton et al. 1985) which would be expected to reduce reproductive effectiveness.

Sumpter (2009), as part of a theme issue (Templeton et al. 2009), reviewed the widespread nature of chemical contamination and the lack of understanding of the problem.

16.6.3 Eutrophication

Run-off from agriculture or waste water containing detergents with phosphates caused a problem in freshwater when the increased nitrogen or phosphorus promoted algal growth, producing 'blooms'. These used a lot of oxygen which led to oxygen depletion (hypoxia) and fish kills. Some of these problems have been addressed by modification of detergent ingredients and by monitoring/controlling effluents including manure slurries reaching water.

Boutilier (1998) says that eutrophication is a 'growing problem' inshore, affecting the shallow nursery areas important for fish such as cod. Freshwater eutrophication in Britain is reviewed by Giles (1994), with the comment that phosphate levels in some British rivers are 'in many cases orders of magnitude above natural background levels'.

16.7 Habitat destruction

Wetlands have been subject to intense pressure because of direct water use or indirectly through pollution while marshy areas have been drained ('reclaimed') for agriculture and mangroves supplanted by aquaculture ventures. Deforestation around river margins has contributed to flood events and direct losses of wetland habitats occur as human building activities encroach.

Water quality may deteriorate to the point it is uninhabitable, a situation which may develop gradually with accumulation of unfavourable constituents. In the last several decades concerns about acidification were widespread; 'acids' (e.g. as produced by sulphur dioxide) emitted from industrial areas could precipitate as 'acid rain' in otherwise pristine areas. This was particularly damaging for regions which already had peaty water which was naturally acidic. Although some fish like trout (*Salmo* spp.) can tolerate fairly acid conditions, increased acidity stresses them, before lethal levels are reached.

Water use in countries like New Zealand, with its intense agriculture, and other regions, semi-desert or arid areas, generally includes irrigation and water diversion, the latter also occurring with construction of dams and hydroelectric schemes. Efforts have often been made to mitigate the effects of river dams when 'fish ladders' are provided to aid migratory species, notably salmon. For those fish which are imprinted (14.5.6) to home to natal headwaters, water diversions can be particularly serious.

A massive disaster occurred when Soviet bureaucrats decided that the region around the Aral Sea should grow cotton with irrigation derived directly from this inland (freshwater) sea which subsequently shrank to a vestige of its former size. Over many years and directly consequent on the Industrial Revolution many British rivers became uninhabitable by fish or most other living organisms.

Even a small amount of water use or diversion can seriously affect shallow water areas or braided rivers which have multiple small, shallow channels.

Wetland losses in Britain and the United States are discussed by Giles (1994).

16.8 Rehabilitation

Awareness of pollution issues has led to efforts, in some regions, to cut down contaminants of fresh water in particular, and some 'rehabilitations' have been deemed successful, for example, the River Thames in England. However, the ongoing controversy about water as a marketable commodity tends to treat any freshwater system as disposable. It is not uncommon for bureaucracies to regard entire ecosystems as saleable and their inhabitants as transferable. This has been the attitude for mammals like water voles in the Lee Valley near London in which, at an initial stage of planning, the rehabilitated river and lakes were regarded as candidates for the 2012 Olympic Games' water-sports venues (perhaps for rehabilitation, again, later). The fact that it would be very difficult to relocate the invertebrates such as dragon-flies did not seem to be addressed, and the fish population would have been negatively affected by the increased activity and reduced biodiversity. The very fact that 'rehabilitation' seems achievable in some instances may encourage degradation of areas in the future with the notion that harm can be reversed.

The loss of haplochromine cichlids in Lake Victoria during the 1980s was attributed to the introduction of Nile perch, *Lates niloticus* (16.10.5). With subsequent removal/reduction of the Nile perch population there has been some improvement in the haplochromines abundance (Witte et al. 2000) which had declined from over 90 per cent to less than 1 per cent of demersal 'ichthyomass'.

16.9 Disease

As the Swedish research group (Forsmarks Kraftgrupp 1982) have noted, pollution, particularly thermal pollution, can increase the risk/incidence of disease. Indeed some salmonids, notably char, are prone to increased disease as water temperatures rise, even to levels which would not be deleterious for other salmonids.

With the increase in fish culture there been concomitant concern with the incidence of disease, which may takes several forms including deficiency diseases associated with artificial diets. Diseases are also caused by a variety of pathogens and their severity/contagion also increases with high stocking density.

Disease-causing organisms can be categorized as endo- (internal) or ecto- (external) parasites, the latter including fish lice (sea lice) which thrive on salmonids in crowded conditions. Fish lice can be readily detected but other pathogens may be microscopic and less readily diagnosed. They include viruses such as cause infectious pancreatic necrosis (IPN), a serious infection of salmonids, as well as bacteria and fungi. Fin rot (ulcerative dermatitis) is a slowly progressive external condition of flatfish (e.g. winter flounder) associated with bacterial infection (Moore et al. 2004). A 'white-spot' (ich or ick) disease of freshwater fish is caused by a protozoan ectoparasite which is very contagious and can cause high mortality.

Wild fish may harbour a variety of multicellular parasites including copepods like the ectoparasitic fish lice, some genera of which may attach to the gills of, for example, cod, and cause extensive damage. Stickleback populations may be infested with tapeworms (e.g. *Schistocephalus*) which, when they occur in the body cavity, encroach on the space available for the viscera, affecting the animal's capacity for reproduction (MacNab et al. 2011).

Prevention of fish diseases in aquaculture can be attempted by the development of vaccines (Yanong 2009; Gudding et al. 2014). There have also been attempts to use bacteriophage viruses to control bacterial disease such as salmon furunculosis (Verner-Jeffreys et al. 2007). Treatment of skin diseases in fish may be accomplished with appropriate short-term immersion including the use of saline solutions for freshwater species. Antibiotics (e.g. oxytetracycline added to the feed) may be appropriate for any fish species thought to be infected by bacteria (Bowker 2010). However, the waste water may then contaminate the local water with potential for a wide range of harmful effects.

Fish lice, which have been problematic in fish farming, may be controlled biologically, using 'cleanerfish' such as wrasse which, however, do not do well in cold temperatures. Recently, the more cold-tolerant lumpfish (*Cyclopterus*) have been investigated and appear to be promising in this role (Imsland et al. 2014, 2015).

16.10 Fishing

Fish conservation is inexorably linked to the fishery and the possibility of over-fishing of wild stocks as well as the chance of rehabilitation through rein-troductions after captive breeding, and the benefits and problems of aquaculture.

16.10.1 The 'wild' fishery

Small-scale fishing has a very long history. It may now be referred to as 'artisanal' and depends on lo-cal fishers using specific gear, such as small nets or baited lines, operated from very small boats. Fish caught in excess of domestic needs may be mar-keted locally. Such fishing is not likely to pose a risk to most fish species.

Recent commercial fishing, however, has used gear which can cause much destruction to non-desired species inadvertently caught ('by-catch') and to the ocean bottom if draggers are used. The capacity to locate fish has improved and means that a very large proportion of a population can be found and removed.

Particularly destructive, wasteful fishing occurs when sharks are 'finned' (16.10.4) and nets lost in the water may be so tough and non-degradable that they continue killing for months if not years (ghost nets).

The physiology of fish is such that some of the recent moves toward conservation may have been ineffective. Thus, the rules that determine the re-turn of some particular fish (e.g. halibut) to the wa-ter may be pointless if the fish is severely stressed or injured and perhaps unlikely to survive. The prohibition of small mesh nets with legislation aimed at stopping the capture of under-sized fish may be completely disregarded and has given rise to some altercations at the national level (Spain versus Newfoundland with turbot, see Appendix 16.3). Conserving small fish is indeed desirable but wholescale removal of large adults may be a problem too if the large females (e.g. cod, May 1967) give many more eggs (up to about 11.5 mil-lion) than several small fish (300,000 each). Previ-ously (1958) Powles reported a large (140 cm) cod female with 12 million 'ripening eggs'. On this basis perhaps big females should not be regarded as trophy fish but returned unharmed if possible. The same or similar reservations may apply to big males. Unlike many terrestrial vertebrates, fish can reproduce and continue growing over a very long time period. The fact that 50 per cent of a species males may be capable of sperm production at a small size does not necessarily mean that they con-tribute to the current season's reproductive effort, but they may. The plasticity in some fish life cycles is considerable; small 'sneaker' males (jacks) may actually compete with full-size adults in salmonid populations (9.3). The presence of hermaphrodit-ism (9.4) as a regular and normal situation with some coral-reef fish requires considerable under-standing of colour phasing if even selective fishing occurs on reefs.

16.10.2 Problems with the wild fishery

Over-fishing appears to have been a massive prob-lem that occurred partly because of ignorance, partly because of optimism and partly because fish were sometimes regarded as equal, invariable units, 'the assumption that all codfish are alike' (Boutilier 1998). Fisheries mathematics dealt with fish as enti-ties which could grow and reproduce but was re-sistant to ideas that incorporated de-growth (7.13.4) or intermittent reproduction. Off Newfoundland, cessation of the cod fishery was legislated in 1992 (cod moratorium) for most of the fishing regions and a recovery of the stocks has been very slow; a full-scale fishery has not yet been possible (as at 2016) though big fish are now being reported, caught through a limited 'food fishery' for New-foundlanders, not for commerce.

16.10.3 Fish farming versus the wild fishery

In the case of fished species, fish farming has some-times replaced the wild fishery. However, fish farm-ing as an intensive activity may itself give problems in that nitrogenous waste and parasites (e.g. fish lice) may access, and negatively impinge on, wild species nearby (Krkošek et al. 2007/2008). The widespread development of aquaculture has been accompanied by attempts to increase productivity; some of these activities have encroached on habi-tat such as mangroves important as fish 'nursery' areas. Also the development of biotechnology in aquaculture has included genetically modified or-ganisms (more recently) and the use of hormones, for example, which have been a potential source of local objections. A recent review of aquaculture bio-technology is provided by Fletcher and Rise (2012).

16.10.4 Shark-fin fishery

The selective fishery for shark fins can deplete elasmobranchs rapidly, particularly as they are very slow in replacement capacity, with long gestations and few young produced. In this fishery sharks are killed just for their fins; the de-finned carcass being discarded. This 'wasteful' practice is discussed by Klimley (2013).

16.10.5 Non-endemics

Attempts to increase output from particular areas have sometimes been very damaging to endemic species, particularly when non-native species (e.g. Nile perch in the African Great Lakes) are introduced (Witte et al. 1992) as preferred in terms of growth or marketability. In Lake Titicaca, high in the Andes, introductions of non-native species such as trout have been held responsible for the probable extinction of *Orestias cuvieri*, a relatively large endemic cyprinodont. Lake Alaotra in Madagascar has suffered from the introduction of various species including *Gambusia* which was supposed to consume mosquito larvae but is reported to have preferred fish larvae (Mutschler 2003), contributing to severe problems with the fishery there. Problems associated with introductions were reviewed by Allendorf (1991).

16.10.6 Catch and post-catch handling

In historic times fish caught were preserved for future use by smoking or salting and drying but this practice has given way to refrigeration with risks of spoilage in power cuts. The capacity to refrigerate a catch prior to processing has led to some extraordinary journeys for fish, for example from Alaska to China for pollack, thence to Eastern Canada for sale. This is 'long-commute' food indeed and does not seem appropriate.

Fishing practices have also shifted from long-lining to gill-netting and trawling; that is, it has gone from selective to broad-spectrum capture with environmental damage suspected or proven and low-quality product too when fish is squashed in a trawl or suffocated and left in a gill net for hours. Over the years it is certain that fish caught for food has not always been optimally used, but instead dumped or rendered into fish meal for pet food or fertilizer.

Legal requirements for some fisheries stipulate that some species obtained as by-catch (i.e. caught inadvertently) be returned to the water, technically unharmed, as an extension of inland fishing's 'catch-and-release' programmes. Unfortunately many fish which have been caught by current commercial methods are not going to survive the process so returning them to the water may be a futile act.

Thus, attempts to conserve fish have included substituting 'catch-and-release' instead of terminal angling for sports fishing and, similarly, return of non-targeted species when netted commercially. Unfortunately fish may not survive such experiences and the whole idea of what constitutes 'stress' and its negative effects is an area which probably needs more consideration (see Barton 2002, also 7.4, 10.12.7).

Fish caught as pets or decor enhancers may also suffer considerable losses in transport or subsequent premature deaths in aquaria. Fish suppliers do not always seem to know which species they are dealing with and it is possible to order and pay for a species that turns out to be a different species from a different country than the one intended. Hybrids may also occur in captivity and, as with tiger conservation, the idea of re-stocking from captivity may not work very well if the genetics and place of origin have become confused.

16.11 Conservation and the fisheries

16.11.1 Marine protected areas (MPAs)

There have been massive attempts to designate large areas as 'protected', with the aim of providing refuges from intense human activity and to allow recovery of damaged regions. The United States has issued an inventory of such regions within its jurisdiction and has operated a 'National Marine Protected Areas Center'. The World Bank in 1995 issued a list of 1,306 subtidal global MPAs. However, the establishment of such MPAs, though initially comforting, may not have been as positive as would be hoped. Rogers and Beets (2001) state that the occurrence of over 100 MPAs in the Caribbean gives a 'misleading impression' regarding the actual protection value. Existing problems included anchor damage and the continuation of some commercial fishing, even though it is not supposed to occur. There continued to be 'loss of spawning aggregations', smaller sizes and decreased abundance. Agardy et al. (2003) reviewed some of the problems with MPAs, some political, some due to lack of knowledge of the ecosystems. Progress on this situation has been announced by John Kerry (American Secretary of State) for October 2016, with the establishment of the Antarctic Ross Sea as a new MPA, a highly significant event, and hopefully not reversible.

16.11.2 Predictability of fish numbers

Historically the prevalence of fish has been regarded as enduring. Particularly in the vast oceans, it has been common to regard the productivity as virtually limitless and unchanging. Received wisdom has it that 'there are more fish in the sea than ever came out of it', and eminent scientists thought that fish stocks were so large they could not be depleted (Pauly 1993). In this mood the regulation of fish removal has been sporadic or non-existent until the last few decades. The notion of inexhaustible supply has now been partially supplanted, but the idea of compensation (numerical, not financial) persists when numbers increase if a stock declines, a phenomenon expected to produce less competition and more food (Appendix 16.1). This idea had become firmly established (Ricker 1954; Beverton and Holt 1957) and it may take some time before fisheries mathematics allows adequately for the possibility of depensation (Appendix 16.1) when fish numbers do not respond well to a decline but instead decline further.

Part of the problem in estimating fish numbers arises from an over-optimistic view of fish reproductive capacity in the case of bony fish. There is certainly no basis for optimism in the case of elasmobranchs which have suffered from a negative image (depicted as man-eaters in the film 'Jaws', etc.); it is possible that the popular view in some areas would be that the elimination of all sharks

would be desirable. This view most likely extends to the large planktivorous sharks (basking shark, whale shark) which may, by their very size, appear threatening. Non-elasmobranchs, such as sturgeon and teleosts, may indeed have the capacity to produce very large numbers of eggs but it is now established that they do not necessarily do so. It is clear that adult female fish may regularly or occasionally omit egg production in a calendar year, and even if they do produce eggs they may not do so at the optimum number possible. Similarly, males may omit sperm production and 'skip' reproduction. For both males and females these omissions can be nutritionally (Burton 1991) or possibly temperature linked if conditions are unusual ('turbot' off Russia, Federov 1971). Moreover, changes in the environment may have an effect several years later if oocyte growth of cold water fish takes two or more years, with the maximum output established early.

Predictions for fish species in terms of numbers, 'renewable resource' and future expectations are much more difficult to make than for more visible species like large mammals or birds, generally terrestrial vertebrates with parental care and fairly accessible and knowable litter or brood sizes. For both mammals and birds, an optimum production for a specified number of adult females can be fairly confidently known. This is not the case for fish which may produce large numbers (even millions) of eggs per annum; a DFO (Department of Fisheries and Oceans, Canada) poster (Northern Cod Science Program) claims the 'odds on cod' are 'one in a million' surviving to adulthood (seven-years old). There has, however, been a prevailing idea that as stocks deplete, more of the larvae survive to adulthood due to diminished competition (Appendix 16.1). This viewpoint has led to the belief that stocks can easily recover from a small number of adults.

Like terrestrial vertebrates, wild fish can be marked or ringed to obtain data on their location and survival but information obtained through this procedure is rather limited. It has been used to establish post-spawning survival for capelin and may be useful to determine migratory routes for salmonids, for example. For very small fish, marking/tagging may become a hazard and may increase mortality.

Predictions on the future of fish species may include the effects of both climate and hydrological changes. Currently (2015) it seems that unstable conditions (as predicted by some climate change models) are common globally, but overall warming appears to be occurring. If Arctic/Greenland ice melts and the Labrador Current and Gulf Stream shift position, the effect on both marine and freshwater fish species may be profound. Similarly, changes (increases) in ice melt in the Antarctic may destabilize the marine fauna which have been adapted to a fairly stable situation in the previous centuries. Fish generally do not regulate their body temperature (7.4) but may respond to unfavourable temperatures by migrating; hence winter flounder are so-called because they are observed mainly in the winter months off New England. In response to changing conditions (light, temperature, food availability), fish may also undergo vertical migrations, more appropriate in the oceans and deep freshwater bodies than in shallower regions. However, in many aquatic environments there can be obstacles to relocation whether physical (dams) or chemical (oxygen paucity), or a combination of both when volcanic eruptions occur.

16.12 Responsibility: the tragedy of the commons

It has been suggested that the 'tragedy of the commons' (Hardin 1968) may apply to the plight of fish populations, meaning that nobody (no community) has taken adequate responsibility for the resource because the wild fish in general have no 'owners'. Where groups of interested (i.e. maritime) nations have formed to discuss problems, it is not clear that science has won out over politics. There is a danger that international trade deals may involve non-sustainable sales of fish to obtain a market for another commodity. Some species have been labelled 'under-utilized' at times with no justification or erroneously; one unfortunate species was apparently labelled under-utilized under one name when it already was 'utilized' under another. Confusion with nomenclature should have been avoided by using the correct Latin binomials at all times though this counsel is somewhat undermined by frequent changes in the binomials, sanctioned by

International Taxonomic bodies, desirous of reflecting phylogeny or cladistics.

Fish like some of the salmonids, flatfish and tilapia have undergone name changes which have confused the literature and non-scientists including political bureaucrats drafting legislation. Thus the common name 'turbot' is used for two quite different flatfish, one of which (*Scophthalmus maximus*) has high value. The other 'turbot' (*Reinhardtius hippoglossoides*) is also known as Greenland halibut, whereas the true halibut (*Hippoglossus hippoglossus*) is a much larger and higher-value fish. These obfuscations contribute to marketing and legal problems, as does the name 'sole' which in the North American market seems to be attributed to any flatfish which is not halibut, whereas in the European (e.g. British) market two high-value fish are recognized as sole: the Dover sole (*Solea vulgaris*) and the lemon sole (*Microstomus kitt*).

16.13 The role of physiology in conservation

Physiology is obviously important in determining the effect or even the presence of adverse environmental conditions such as studied in connection with toxicology. Problems considered have been the effects of xenobiotics such as pharmaceuticals in fresh water and excess of natural substances from oil extraction as well as contaminants from mining and industrial processes (7.11, 8.3.10).

Other areas of physiology which are very important in conservation issues include reproduction and the problems associated with temperature variation particularly with the risks of global warming.

16.13.1 Reproductive capacity

Physiological studies on reproduction have demonstrated that in marine fish with the capacity for annual reproduction, not all mature individuals participate always each year. This is the case for cod and winter flounder (Burton and Idler 1984; Rideout and Burton 2000). There are varying proportions of reproductive and non-reproductive adults in the population, which probably depends on the feeding conditions the previous year (Burton 1991). This situation applies to both females and males, the latter being probably under-reported when non-reproductive, as the emphasis on data collection for wild fish of commercial values has often been the fecundity of females.

To maintain a sustainable (fishable) stock, the reproductive cycles of fish need to be understood, in particular the habit of assigning fish to categories such as 'immature' or 'mature' or 'post-spawned' on the basis of weak or insufficient information should be revised. In the past some large fish (e.g. halibut) could be termed 'immature' when they were most likely 'non-reproductive', that is adults which had spawned in previous years and could spawn the following year, but were currently not undergoing active gametogenesis or holding gametes (or vitellogenic oocytes, 9.10.2).

It has also been observed that fishing pressure has led to decrease in fish size, which will reduce the number of oocytes produced and future stock size. This is another factor which should be incorporated in consideration of conservation and fishing practice.

16.13.2 Temperature changes

It is clear that fish inhabiting large bodies of water, whether lakes, rivers or seas, have at least an option of moving either horizontally or vertically to avoid adverse conditions. As eggs or larvae such an option does not exist or, for larvae, is very limited in scope. Some fish are fairly sedentary in habit and/or have a limited home territory, with little inclination to move far, or no ability to do so in the case of isolated bodies of water as in deserts or caves.

Some fish species seem quite tolerant of a large range of temperature whereas others (e.g. char) require a cool environment. Although some species can survive severe or abrupt temperature change (*Gillichthys mirabilis*), others, although they can survive temperature shifts, may respond by aborting reproduction (Fedorov 1971). Fish kills ascribed to temperature change or contact with ice have sometimes been reported, as occurred with cunner *Tautogolabrus adspersus* off Newfoundland (Green 1974). It is noteworthy that cunners are the most northerly species of a family (Labridae) that are more usually associated with tropical reefs (13.3.3). Cod are reported to use thermally defined 'highways' with

'great precision' (DeBlois and Rose 1995), indicating great sensitivity to temperature change.

Currently there is concern that climate change will produce global warming with shifts in temperatures that will adversely affect survival. Fish are likely to be particularly susceptible to warming and also to local cooling as ice melts. However, marine fish like cod and capelin 'shifted north very quickly' under conditions of warming (1920–40), supporting the inference that some fish can react appropriately and promptly to adverse conditions (Rose 2005).

Cheung et al. (2013a) state that ocean warming over the past four decades has already had effects on global fisheries; a particular example being rapid increases in a warm-water species, the red mullet (*Mullus barbatus*), in catches around the United Kingdom. These authors report (agreeing with Rose 2005) that in response to ocean warming, marine species, including fish, migrate to higher latitudes and deeper water, which contributes to 'tropicalization' of catches; that is, increasing dominance of warm-water fish. Pörtner and Knust (2007) indicate that ocean warming can 'affect marine fish through oxygen limitation of thermal tolerance' which Cheung et al. (2013a) suggest can reduce their growth, reproduction and abundance. In accordance with the idea of shrinking growth, Cheung and co-workers (2013b) present information suggesting overall reduction in size for marine fish. Brander et al. (2013) accept the prediction of decreased size but suggest that the rapidity of decline as forecast (Cheung et al. 2013b) is 'remarkable'. It is obvious that the subject of climate change is fraught with contention and debate at this time but, unfortunately there is a growing body of evidence that global warming is taking place and little doubt that fish will be adversely affected.

16.14 Conclusion

In spite of the ability of some marine fish species to tolerate, adapt and move, the future for fish, as for other vertebrates including man, looks somewhat doubtful. It is possible that the number of humans is at risk of exceeding the 'carrying capacity' of the world and it also seems likely that fish populations will continue to decline as demand for food goes up and environments appropriate for fish degrade.

It is not clear whether redemption (rehabilitation) of degraded environments will continue to be practicable but it is good that it has been possible in some instances. Reduction of long-commute products and local processing should help somewhat as would careful collation of data and sincere attempts to reduce over-fishing and wastage. Regulations on net size (to limit by-catch of juveniles) have been in position for decades in some jurisdictions including Canada but the regulations have not always been taken note of either by fishers or regulators.

Recently attempts to ameliorate the decline in fish and other species by some positive actions have included establishment of MPAs, and it is hoped that widespread application of similar conservation measures in fresh water and marine regions will help recovery. The partial rehydration of the Aral Sea area has been reported and though it is very limited in scope it does provide some hope, showing return of a few fish species following the previous calamitous collapse caused by overly optimistic irrigation. Support of conservation measures may depend on understanding the aims of fisheries closures and the establishment of other limits to human activities, and adequate compensation to local people temporarily deprived of food and employment. Perhaps most important is that education and knowledge-based decisions should be firmly established and that our collective understanding, in the case of fish and their habitats, should be widely available to legislators and politicians as well as the general human population. In particular, there have been problems with interpretation of science, and political compromises may have occurred when they were inappropriate; responsible decisions should be made on sound, undiluted data. To strike an optimistic note, it is hoped that increasing awareness and understanding will enable individuals and governments to take appropriate action.

Appendix 16.1 Compensation and depensation

Fisheries science went through an optimistic phase when it was believed, and stated, that fished species were capable of responding to fishing pressure by producing more young, or at least, by having more young survive. The latter was argued on the basis that there would be less competition amongst the young produced and less cannibalism. This view of fish reproduction was termed 'compensation' and was a reason not to reduce fishing pressure; in other words, fishing was supposed to increase fishable biomass: the more taken, the more there would be. Unfortunately this view, which supported the large-scale fishing efforts of the post-Second World War era, may have been erroneous. It is possible that once a reproductive population declines below a certain point (not necessarily known), fewer and fewer young are produced or survive, which is called 'depensation'. It could be reasoned that this would not occur when a large number of fish are involved in a mass spawning event, in which case the more fish present perhaps the greater likelihood of success. This could occur with capelin (*Mallotus villosus*), for which courtship seems to be minimal and very brief (seconds?), if it can be said to happen at all. The males tend to wait around close to shore and when the females arrive move in with them as waves break on the shore. It seems impossible that much mate selection occurs. Any female may spawn with a male on each side of her, but the event is very rapid; it seems like the contact grouping is almost fortuitous. In other fish which form spawning aggregations it seems that a large number of both males and females could increase the overall success. It seems certain that the reduction of a spawning biomass to a few individuals is always undesirable, bearing in mind the possibility of oil spills and the like which could kill or disable the entire local stock.

Appendix 16.2 Categories of vulnerability

The International Union for Conservation of Nature (IUCN, established 1948) maintains a list (the Red List) of the status of species of animals, plants and fungi. Any particular species is assigned a category based on the information available, or not; the list including two 'unknown' categories, 'not evaluated' and 'data deficient'. The remaining categories range from species of 'least concern' to 'extinct in the wild' with intermediate categories such as 'vulnerable' and 'critically endangered'. Some species, such as cod, plunged in numbers from super-abundant to 'vulnerable' and have remained in an unsatisfactory state off Newfoundland even after a fishing moratorium was declared in 1992. The general opinion (2016) is that though stocks have recovered somewhat, the cod numbers remain far below recent historic levels.

Appendix 16.3 The turbot war

In the 1990s (1994–96), Canada and Spain disputed the rights of Spanish vessels to take turbot (*Reinhardtius hippoglossoides*) in waters off Newfoundland. At the height of the dispute a Canadian vessel fired a shot across the bows of a Spanish trawler on the grounds that the Spanish trawler was involved in illegal activity. It transpired that the vessel was using a small-mesh net liner and therefore catching 'under-sized' fish, at least by Canadian law. Apparently this activity was legal by European standards.

This illustrates the desirability for more effective liaison between countries on the need for conservation and rules to achieve it. It can be compared with previous problems in clearly identifying fish in need of protection and the confusion arising not only from using different languages but the use of the same name ('turbot', 'halibut', 'sole') for different species.

Bibliography

Crook, D.A., Douglas, M.M., King, A.J., and Schnierer, S. (eds) (2016). Special issue. *Reviews in Fish Biology and Fisheries*, 26 (4), *Indigenous participation and partnerships in aquatic research and management*.

Woo, P.T.K. (ed.) (2006). *Fish Diseases and Disorders, Vol.1: Protozoan and Metazoan Infections*. 2nd edn. Wallingford: CAB Int.

Glossary

Nouns usually used:

abr. = abbreviation; acr. = acronym, initials (capitals) used instead of full words; adj. = adjective; aka: also known as; q.v. = literally 'which see'; see also (please check alternate or additional term given)

Abducens nerve cranial nerve VI; paired efferent nerves, each from brain to eye muscle (posterior rectus).

Abyssal (adj.) deep-water dwellers, from around 1000 m to 6000 m.

Acclimation the process of physiological adjustment to a change in external conditions, in the laboratory.

Acclimatization the process of physiological adjustment to a change in external conditions.

Acellular bone this was described as forming some fish scales but it is possible this is not bone at all.

Acetylcholine a neurotransmitter secreted from the terminals of some neurons (cholinergic); for example, motor neurons, preganglionic autonomic neurons and post-ganglionic cranial autonomic neurons.

Acidophil a cell which binds acid stains such as eosin and orange G.

Acousticolateralis system fish hearing and water vibration sensory systems.

ACTH (acr.) adrenocorticotrophic hormone (q.v.).

Action potential transient change in voltage across a neuron/muscle cell membrane, achieving conduction of an electrical impulse.

Activin a polypeptide paracrine which promotes gametogenesis/maturation.

Adelphagy *in utero* consumption of siblings (occurs in the shark *Carcharias taurus*).

Adipose fin a small fatty protuberance dorsally, just before the caudal fin, or ventrally, anterior to the anus. Found in eight orders of fish, including Salmoniformes, Charciformes, Siluriformes and Myctophiformes, as well as Percopsiformes, Osmeriformes, Aulopiformes and Stomiiformes. It is thought to have a hydrodynamic function (Reimchen and Temple 2004).

Adnexa eyelids, including the nictitating membrane (third eyelid).

Adrenergic (adj.) used for neurons or fibres for which catecholamines (excluding dopamine) are neurotransmitters.

Adrenoceptor cell membrane receptor binding adrenalin and noradrenalin.

Adrenocorticotrophic hormone the pituitary hormone controlling secretion of corticosteroids.

Aestivation a condition of torpor during adverse conditions particularly drought.

Afferent branchial artery a major blood vessel in fish, of which there are five pairs supplying the gills with blood directly (and therefore at fairly high pressure), from the heart via the short ventral aorta.

Afferent fibres/nerves conduct impulses to the CNS (= sensory fibres/nerves).

AFGP (acr.) antifreeze glycoprotein.

AFP (acr.) antifreeze protein.

Aglomerular (adj) applied to a kidney lacking glometruli, and therefore without ultrafiltration.**Agonistic:** intraspecific defensive or aggressive behaviour.

AHH (acr.) aryl hydrocarbon hydroxylase.

Alarm substance cells (acr. ASC) club cells (q.v.).

Albedo the ratio of incident to reflected light.

Alevin a newly hatched salmonid larva, with the remains of yolk in a ventral sac.

Altricial (adj.) refers to young animals, hatched or born at a relatively undeveloped state. The opposite of 'precocious'.

Amacrine cells retinal neurons possessing only denditic processes.

Aminopeptidase one of the two types of exopeptidase (q.v.), the term used for an enzyme which removes an amino acid from the amino terminal of a polypeptide/protein.

Ammocoete the larval form of lampreys.

Ammonotelic (adj.) an animal (usually aquatic) for which the principle nitrogenous excretion is via ammonia.

Amphidromy a migratory life cycle either to or from fresh water (diadromy) in which very young forms migrate, with most growth occurring after the migratory phase (McDowall 1992).

Amphistylic jaw suspension the arrangement of jaw support found in some sharks. with two points of attachment, the posterior one including the hyomandibular.

Ampulla a swelling on a duct or an elongated structure, differentiated from lobules which are subdivisions of organs. In general terms, a structure resembling a swelling. Swollen region (otolith organ) in each semicircular canal of the ear, containing a sensory structure (crista) detecting movements of the endolymph (q.v.).

Ampullae of Lorenzini occur in the skin of elasmobranchs as visible pores to the exterior, being a set of small ampullae incorporating electroreceptors and an underlying system of canals. The electroreceptors are extremely sensitive to weak electric fields and are used for prey detection.

Ampullary electroreceptors these are connected by a canal to the external water and are responsive to relatively low-frequency electrical activity.

Amnion an extra-embryonic membrane in reptiles, birds and mammals.

Amniote a vertebrate with an amnion in the embryonic stage; that is, reptiles, birds and mammals.

Amylase the term used for an enzyme which cleaves starch ('amylum') by hydrolysis.

Anadromy (adj. anadromous) refers to fish which migrate from the sea to fresh water to spawn.

Anamniote a vertebrate (e.g. fish) lacking an amnion in the embryonic stage.

Androgens steroids with 19 carbon atoms such as testosterone, 11-ketotestosterone (11KT), and which are essential in both males and females for reproduction. In females, testosterone is the precursor for oestrogen and in male fish, 11KT is the principle circulating steroid that stimulates phases of spermatogenesis, etc.

Annuli rings formed in body tissue (e.g. cleithrum, otoliths, scales) by seasonal deposition of circuli (q.v.), and used to assess age.

Annual fish short-lived fish which can survive drought as resistant eggs.

Anorexogenic conditions limiting feeding.

ANP atrial natriuretic peptide.

ANS autonomic nervous system.

Anthropogenic (adj.) produced by human activity.

Antibody a specific protein (immunoglobulin) evoked by the presence of an antigenic 'foreign' substance, within an animal.

Antifreeze protein a special (usually plasmatic) protein which helps to protect animals from freezing.

Aorta fish have a ventral aorta transporting blood towards the gills from the heart and a dorsal aorta transporting blood from the gills as a major distribution vessel to the periphery.

Aphakic space a region in the eye which enables light to bypass the lens, mainly produced by a mismatch in size of the lens and pupil.

Apoptosis programmed cell death.

Appendicular skeleton the structures supporting fins.

Aquaporins special molecules facilitating water movement across membranes.

APUD amine content and/or amine precursor uptake and decarboxylation.

Aquatic surface respiration a relatively rare situation in which fish access the highest, best-oxygenated water.

Areae retinae (pl.) very sensitive zones of the retina with a high neuronal density, for example fovea.

Aqueous humor fluid in the eyeball between the cornea and lens.

Arcualia embryonic rudiments of vertebrae.

Arteriole a small diameter blood vessel linking an artery to capillaries.

Artery a large diameter blood vessel transporting blood away from the heart.

ASC alarm substance cells.

ASR (acr.) aquatic surface respiration.

Astrocyte large star-shaped neuroglial cell, found in the CNS including the optic tracts. Can contribute to neuromodulation.

Asynchrony in oogenesis; occurs when individual oocytes show a continuum of development, with immature oocytes recruited into vitellogenesis one by one. This enables precise allocation of resources, with fecundity up-regulated as conditions permit.

Atresia resorption of developing oocuytes.

Atrium the heart chamber (one in fish) that receives venous blood from the sinus venosus.

Auditory nerve: cranial nerve VIII; paired short sensory nerves, each from the ear to the brain.

Auerbach's (myenteric) plexus autonomic network, between longitudinal and circular muscle layers of the gut wall.

Auricle synonym for atrium (heart chamber).

Autonomic nervous system regulates the (involuntary) function of smooth muscle of internal organs (e.g. the gut, blood vascular system, glands) and also skin chromatophores in some fish.

Autonomic outflow pathways of autonomic nerve fibres from the CNS, localized within specific spinal and cranial nerves.

Autostylic jaw suspension a jaw suspension type not found in fish, but typical of, for example, mammals, with the link between the upper jaw and the cranium achieved via the palatoquadrate element.

Axon process conducting nerve impulses away from the neuron perikaryon.

Balbiani body (yolk nucleus) a dense cytoplasmic structure found in pre-vitellogenic oocytes at the peri-nucleolar (large immature) stage.

Barbels elongated sensory structures usually projecting from the head as with 'catfish'.

Baroreceptors pressure receptors.

Basibranchial the lowest part of each branchial arch.

Basipterygium the skeletal element to which radials or fin-rays are attached.

Basophil a cell which binds basic (i.e. high pH) stains (e.g. haematoxylin).

Basophilic (adj.) a cell or region of a cell which binds basic stains. The nucleus of most cells is basophilic, due to the presence of DNA.

Batch spawning (serial or multiple spawning) occurs when a group of oocytes undergoes vitellogenic development at the same rate, and then is subdivided for deposition over days or weeks.

Bathypelagic deep pelagic (i.e. swimming in deep water, 1000 m to 6000 m).

BBB (acr.) blood–brain barrier.

BCF (acr.) body, caudal fin propulsion.

Benthic (adj.) bottom-living, for example flatfish, anglers.

Bilateral symmetry a structure which can be bisected into two equal (right and left) halves. Vertebrates including fish exhibit this, but imperfectly, as the gut and associated organs may have a lateral bias.

Bile fluid secreted by the liver and stored in the gall bladder.

Bile acids for example cholic acid; synthesized in the liver as precursors of bile salt (q.v.) formation. Synthesized from cholesterol (with a specific four-ring structure), which may be conjugated, for example, to taurine and are generally present as salts.

Bile pigment(s) biliverdin and bilirubin, formed from haemoglobin in the liver.

Bile salt(s) for example sodium taurocholate, with detergent (emulsifying) action in the gut, when released into the duodenum from the gall bladder via the bile duct.

Binomial the two-part Latinized (i.e. Latin-based) name which differentiates each species, for example '*Carassius auratus*' for goldfish, *Solea vulgaris* for sole.

Bioaccumulation the tendency of an organism to concentrate materials in different organs, which may be harmful.

Bioelectrogenesis the capacity of an organism to sum small voltages for discharge.

Bioluminescence the production of light by a live organism.

Blackwater fresh water coloured by, for example, tannins due to run-off from bogs, and tending to acidity, as with many tropical river systems. Alternatively, some rivers may be coloured by mud or silt as in the River Blackwater (Essex, United Kingdom).

Blast cells immature cells capable of recruitment; examples are chondrobasts which become cartilage-forming chondrocytes (Chapter 3), osteoblasts which become bone-forming osteocytes (Chapter 3) and erythroblasts which become erythrocytes (Chapter 5).

Blood–brain barrier some substances (e.g. large molecules) and blood cells cannot pass from the circulation to the brain; selective transport of some solutes occurs between the vascular system and the neural tissue of the brain across the choroid plexus.

BOD (acr.) biological oxygen demand.

Bohr effect increased carbon dioxide (or acidity) decreases haemoglobin's affinity for oxygen thus causing more oxygen to be released in conditions of high CO_2. Compare Haldane effect.

Bombesin a peptide isolated from amphibian skin, a variant stimulates gastrin secretion.

Bouquet similar to synzesis, when chromosomes cluster close to the nuclear membrane in meiotic prophase.

Bowman's capsule the first part of the kidney nephron in vertebrates.

Bradykinin a substance, which may be paracrine, producing vasodilation, tending to lower blood pressure.

Braided river in which the river divides into numerous riverlets or small channels which form a network.

Brain stem medulla oblongata and floor of mid-brain in fish (the term also includes optic tectum and diencephalon in mammals).

Branchial basket the collective term for the supports of the gill area.

Brockmann bodies a concentration of pancreatic islet tissue found in some teleosts.

Brood parasitism some fish which brood their eggs in their mouths are taken advantage of by another species which inserts its own eggs into the location of the original brood which are subsequently eaten by the intruders.

Bulbus arteriosus the most anterior heart region in teleosts.

Caducous (adj.) of a scale capable of being shed (see deciduous).

Calcitonin a hypocalcaemic hormone.

cAMP cyclic adenosine monophosphate.

Capillary the smallest diameter blood vessel with very thin (one cell thick) walls.

Carbonic anhydrase the enzyme catalyzing the reversible conversion of CO_2 to carbonic acid.

Carboxypeptidase one of the two types of exopeptidase (q.v.), the term used for an enzyme which removes an amino acid from the carboxy terminal of a polypeptide.

Cardiac output the volume of blood expelled by a heart in unit time.

Cardiomyocyte heart muscle cell.

Carnivore literally 'meat eater', an animal which consumes other animals.

Carrying capacity (K), the ecological term which expresses the idea that any organism has a finite maximum number of individuals within a given appropriate ecosystem. The competition for resources beyond that point as well as the toxic effect of wastes produced would ensure premature deaths of many individuals and/or elimination of the population.

Cartilage skeletal material characterized by the form of matrix containing a large amount of carbohydrate.

Cartilage bone see endochondral bone.

Catadromy (adj. catadromous) refers to fish which migrate from fresh water to the sea to spawn.

Catecholamines include ANS and CNS neurotransmitters and hormones; that is, adrenalin (epinephrine), dopamine and noradrenalin (norepinephrine). Catecholamine secretion includes that by chromaffin cells and adrenergic postganglionic spinal autonomic neurons.

Caudal neurosecretory system also known as the urohypophysis.

Caudal peduncle the posterior region of the tail from which the caudal (tail) fin arises.

CCK (acr) cholecystokinin.

CCS (acr.) cardiac conducting system.

Cdc2 (abr.) a gene (or the gene's product) controlling the formation of a kinase (cdk) which when bound to a cyclin comprises MPF, and in variants controls the passing of checkpoints in the cell cycle (mitosis and meiosis).

Cdc (abr.) cell division cycle.

Cdk (abr.) cell division kinase.

Cell cycle cells pass through a number of stages, including growth before reaching the 'cell division' stage or a resting point. The onward passage of cells through the cycle is controlled by various forms of MPF which phosphorylate target molecules for progression through the cycle.

Cellulase an enzyme digesting cellulose.

Central canal small tubular space within the spinal cord, contiguous with the brain ventricles and containing cerebro-spinal fluid.

Central nervous system the brain and spinal cord.

Cepahlization development of complex head structures.

Cerebellum large subregion of the hind-brain associated with balance/position.

Cerebral hemispheres paired outgrowth of fore-brain, well developed in tetrapods (but not in most fish) predominant in coordinating their functions.

Cerebro-spinal fluid clear fluid (basically acellular) filling the brain ventricles and spinal cord central canal.

CF condition factor.

cGMP cyclic GMP, guanosine 3^I5^I cyclic monophosphate; acts as an intracellular secondary messenger between many target cell receptors and response pathways to hormones and some neurotransmitters.

Chaperon(e) or chaperonin(e) a protein, for example an Hsp, capable of protecting other proteins from denaturation.

Chemoreception perception of chemical stimuli' for example taste and smell.

Chemotopy the concept that a spatial pattern for specific odours can be mapped in the brain, an example of somatotopy, being a correspondence between body position of a receptor and the area of central nervous system associated with it.

Chimaeras the holocephali are commonly referred to as chimaeras, rabbitfish or ratfish, all with the same theme of a blend of features (hence 'chimaera' from classical mythology) reminiscent externally of a mammal and fish joined together, as the head of these fish, with large eyes and a snout-like anterior, does have the appearance of a rabbit or rat.

Chitinase an enzyme digesting chitin.

Chloride cells also known as mitochondria-rich cells, are found on the gill filaments (and elsewhere, for example opercular epithelium), particularly clustered at the base, where they are involved in ion regulation traditionally, and especially active movement of chloride.

Choana the internal nasal opening, into the mouth, of tetrapods.

Cholinergic adjective used for neurons or fibres for which acetylcholine is the neurotransmitter.

Chondrichthyes the major fish taxon including the holocephali (chimaeras or rabbitfish) and elasmobranchs (sharks and rays, etc.). The term indicates 'cartilaginous fish' (chond = cartilage) but it excludes the 'Chondrostei' (q.v.) which also have cartilaginous skeletons.

Chondroblast an immature cartilage-forming cell.

Chondrocyte a cartilage-forming cell.

Chondrostei a term for fish like sturgeon with a cartilaginous skeleton, but which are taxonomically separated from the other major group with cartilaginous skeletons: the rabbitfish, sharks and rays, which are termed chondrichthyans.

Chorion when used in association with fish eggs means a layer around the ovulated egg.

Choroid plexus thin, folded vascular structure in roof of fore- and hind-brain which filters substances entering the CNS.

Chromaffin cells have a high catecholamine content detected by reaction with chromates; found in sympathetic ganglia and separate islets of 'chromaffin tissue' in the head kidney of teleosts or suprarenal glands of elasmobranchs.

Chromatic aberration colour distortion of light by a lens, producing separate colours as fringes around an image.

Chromatophore cells bearing pigment (or reflective platelets) in inclusions which can move intracellularly to cause colour change.

Chylomicrons small spheres formed intracellularly by the intestinal epithelium, and consisting of lipid coated by a phospholipid/protein membrane.

Chrysopsin a gold-coloured visual pigment, a rhodopsin found in retinas of deep-sea fish.

Chymotrypsin a vertebrate gut protease, an endopeptidase formed from the precursor chymotrypsinogen.

Ciliary bundle at the apex of sensory cells in the lateral line and ear, consisting of a single kinocilium (q.v.) and several stereocilia (q.v.); these latter bend in relation to the kinocilium in response to movements of the surrounding fluid.

Ciliary photoreceptors light-sensitive cells with outer and inner segments connected by a ciliary structure, for example rods and cones, with photopigments on flattened lamellae in the outer segment.

Cilium a microscopic projection of some cells, capable of moving or being moved. It has the same ultrastructure as a flagellum, containing 11 microtubules in a 9 + 2 arrangement, the '2' being surrounded by the 9.

Circadian an approximate 24-hour cycle.

Circadian rhythm cyclical endogenous changes over a 24-hOUR period, revealed by experimental constant exposure to an environmental factor such as light or darkness.

Circuli rings formed by changing patterns of growth in bone. Not to be confused with 'lamellae' of Haversian bone.

Cis-trans a change in molecular configuration/shape in which specific atoms or groups on one side of a molecule become reoriented to its opposite side.

Cladistics a recent view of speciation whereby one (usually structural or more rarely physiological) event separates species, rather than the old view of a suite of features changing more or less simultaneously to render a divergence.

Cleanerfish small fish which remove ectoparasites from other fish as their food supply.

Cleithrum a bone of the anterior pectoral region, dorsally situated, which may link to the skull.

Club cells epidermal cells in some taxa, notably cyrinids, which may release a warning pheromone on injury, the warning benefitting non-injured members of the shoal.

CNP (acr.) cardiac natriuretic peptide, secreted by the atrium of the heart.

CNS (acr.) central nervous system.

Coagulation the process of producing a blood clot.

Coeliac (celiac) ganglion sympathetic ganglion giving rise to anterior splanchnic nerve.

Collagen fibres supporting protein structures important as constituents of connective tissue and blood vessel walls.

Commensalism a type of mutualism where two species are very closely interacting in such a way as to benefit one and not harm the other.

Commissure neural tract linking right and left sides of CNS.

Common chemical sense a possible non-specific chemical sense, which may be associated with solitary chemosensitive cells in the skin of some fish.

Compatible osmolyte = counteracting osmolyte.

Complement a complex protein, found in blood plasma, important in developing and maintaining immunity.

Condition factor a measure of weight (mass) relative to length, for example, $W \times 100/L^3$ in which high CF indicates a prospering organism.

Cones elongated photoreceptors which taper distally and occur in the retina, and have the capacity for colour detection.

Conodonta fossil chordates with reversible mouth apparatus, markedly different from gnathostome jaws and therefore considered to be 'without jaws' (Aldridge and Donahue 1998). Until recently only known from their teeth ('conodonts').

Conspecific belonging to the same species.

Conus arteriosus the most anterior heart chamber of some fish, for example elasmobranchs.

Corpus cavernosum a blood-filled tissue in the gill of elasmobranchs.

Corpuscles of Stannius small, pale, round structures found in some teleosts close to the kidney.

Corpuscles of Wunderer non-encapsulated dermal proprioceptors.

Cortex (of spleen) outer region.

Cortical alveoli (sing. cortical alveolus) these develop in the periphery of oocyte cytoplasm, before yolk is accreted. They do not stain with routine dyes such as haematoxylin and eosin. They have been termed endogenous yolk, which may be misleading as they are precursor material associated with the cortical reaction occurring on fertilization much later, at the time of meiotic completion after ovulation.

Cortical reaction exocytosis (release) of the contents of the cortical alveoli into the perivitellinespace (between the plasma membrane and chorion) at fertilization. The result promotes blockage of the micropyle and prevents polyspermy (fertilization by more than one sperm).

Cortisol a steroid secreted during stress in vertebrates.

Cosmine the matrix of cosmoid scales (q.v.), with branched channels.

Cosmoid (adj) scales typical of (extinct) crossopterygians.

Counteracting osmolyte a molecule (e.g. TMAO) which protects proteins from the destabilizing effects of urea.

Countercurrent a system in which two fluids pass each other with opposing flow, a situation which can improve exchange between them.

Countershading a common colour pattern in animals, including fish, in which the lower part of the body is paler than the upper regions, so that the lower part blends with the bright surface of the water as seen from below, the upper part with the substrate and the water as seen from above. Also, shading of the pale lower part by the upper part produces some uniformity of hue.

Courtship behaviour with which two or more fish can synchronize their activities to ensure fertilization.

Cranial nerve specialized nerves directly associated with the brain.

Cranial autonomic system autonomic nerves with cranial outflows (see parasympathetic nervous system).

Creatine phosphate a source of energy for fish movement.

Crenation the shrunken state of red blood cells after being exposed to high salinity.

Cryoprotectant a substance which helps to prevent or delay freezing.

C-start fast starts typical of lunge carnivores.

Ctenoid (of scale) these scales found in teleost groups like acanthopterygians have small spiny projections on the outer surface.

Culteriform knife shape as in 'knife-fish'.

Cupula a gelatinous structure attached to cilia of mechanosensory cells of the lateral line and semicircular canal systems. Movement of the cupula relative to the cilia achieves detection of movement (e.g. due to vibration) of the surrounding fluid.

Cuticle an acellular layer on the skin surface.

Cyanophore a rare chromatophore type with blue pigment.

Cyclins proteins involved in control of the cell cycle; when a cyclin becomes bound to the *cdc2* (cell division cycle) product (another protein) the complex acts as a kinase enzyme promoting transfer of phosphate from ATP (adenosine triphosphate) to other proteins. This phosphorylation is important in the progress of cell division and the cell cycle.

Cycloid (adj. of scale) a round thin scale, typical of many modern teleosts.

Cystovarian (adj) describes a walled ovary which has a lumen, into which ovulation occurs. A completely walled ovary is found in paracanthopterygians and acanthopterygians.

Cysts structural and functional entities enclosing, in male fish, the clonal product of divisions by a single primary spermatogonium. The cyst wall structure may be homologous to the Sertoli cells of other vertebrates.

Cytochrome P450 a ubiquitous (common!) enzyme system which is evoked /increased under conditions such as stress.

Cytokine a small protein involved in intercellular communication, operating in larger quantities than hormones.

Cytoprotectant a substance which protects cytoplasm from destabilizing effects of urea.

Deciduous (adj.) easily rubbed-off scales.

Dendrite process of the neuronal cell body/perikaryon, increasing surface area for interaction of neurons, receives impulses from axons of other neurons.

Dendron long process (longer than dendrites) conducting nerve impulses towards the sensory neuronal cell body/perikaryon.

Dentine a hard material, with a channelled matrix, similar to the major (core) material of mam malian teeth, and expected to form the major part of elasmobranch scales or spines.

Dermal denticle synonym for placoid scales of elasmobranchs, and resembling small teeth in, for example, dogfish.

Dermis the inner cellular layer of the skin, contains numerous collagen fibres.

Degree-days for any particular species post-fertilization, the time to hatch can be calculated and predicted based on the multiple of days passed and temperature.

Depauperate (adj.) describes an ecosystem with few species.

Depensation in fisheries science it has been usual to consider that a downward change in numbers produces a compensatory increase in survival (due to less competition). However, the reverse situation, depensation, may occur; for example, reproduction may require the stimulus of a spawning aggregation.

Dermal bone see membrane bone.

Determinate spawning the egg number is set early, and additions do not occur (see also up- and down-regulation, atresia, indeterminate spawning).

Diadromy migration to (anadromy) or from (catadromy) fresh water for reproduction. See also amphidromy.

Diakinesis the last stage of meiotic prophase I.

Diapedesis the process whereby leukocytes exit the main circulation and enter, for example, connective tissue.

Dibasic aminoacid an aminoacid with more than one basic group per molecule.

Diencephalon subregion of fore-brain.

Dimorphism two distinctive structural types within one species.

Diphycercal tail a symmetrical tail typical of crossopterygians.

Diploid two chromosome sets (per cell).

Diploidization a process of returning to a diploid state from polyploidy.

Diplotene a stage of meiotic prophase I.

Dipsogenic effect promoting intake of water by drinking.

Displacement activity an inappropriate activity when an animal receives two incompatible stimuli simultaneously or the drive response to a stimulus is interrupted.

Distal convoluted tubule part of the nephron closest to the collecting duct and farthest from the Bowman's capsule.

Dorsal horns bilateral projections of the grey matter into the white matter of the dorsal spinal cord.

Dorsal root base of a cranial or spinal nerve carrying sensory (afferent) fibres. Emerging from the spinal cord, the spinal 'root' unites with the ventral root to form the spinal nerve.

Drive a sequence of behaviours with a satisfaction of some obvious endpoint.

EDCs endocrine-disrupting chemicals (see endocrine disruption).

Efferent branchial artery one of a series of arteries leaving the gills, with oxygenated blood, and passing to the dorsal aorta.

Efferent fibres/nerves conduct impulses from the CNS (= motor fibres/nerves).

Egg membrane is used loosely. It should probably have been restricted to the cell membrane or oolemma, but it is also used to refer to the follicle (q.v.) which is a cellular layer around the developing oocyte.

Electrolocation the capacity of some fish to generate and/or use electrical fields to navigate and detect objects including other fish.

Electroreceptors detect electrical changes in the external environment such as disturbances of electrical fields.

Eleuthero-embryo the post-hatch stage which still depends on yolk.

Elasmodine a tissue similar to isopedine (q.v.).

Elasmoid scale said to be typical of teleosts, the matrix contains collagen fibres (see elasmodine).

Elastic fibres protein structures able to stretch and resume shape, found in connective tissue and the walls of arteries. Their presence can be revealed in sections by special stains.

Electrocyte a cell capable of electrogenesis (generation of a potential difference which can be summed), as a group forming an electric organ.

Electrogenesis see electrocyte and bioelectrogenesis.

Electroplaque a flat electrocyte, also known as an electroplate.

Electric organ discharge an electrical event or series of events (pulses) produced by electric organs.

EM (acr.) electron microscope.

Enamel a very hard material associated with the outer covering of vertebrate teeth and placoid scales.

Endochondral bone cartilage bone; bone which replaces an ontongenetically earlier cartilage model.

Endocrine disruption an array of EDCs (endocrine-disrupting chemicals, McLachlan 2001) may interfere with hormones, notably sex-steroid function, by either enhancing or blocking at the receptor level.

Endogenous (of digestion) used to refer to digestive enzymes secreted by the fish.

Endolymph the fluid in the semicircular canals and otolith organs of the ear.

Endopeptidase a protease which cleaves proteins within the chain rather than removing a terminal amino acid.

Endorphins endogenous morphines, produced in the pituitary and having tranquillizing effects.

Endostyle an organ found in the base of the pharynx in some protochordates and in the larvae of lampreys.

Endothelium an inner (cellular) lining, for example of a blood vessel.

Engram a memory trace; a neural change corresponding to a memory.

Enteric system subdivision of the ANS controlling gut motility and possibly glandular secretions of the mucosa.

Enterocyte a columnar epithelial cell lining the gut cavity.

Enterokinase activates chymotrypsinogen and trypsinogen.

EOD (acr.) electric organ discharge.

EOG (acr.) electro-olfactogram, recording of electrical changes associated with olfactory stimulation.

Epiphysis the pineal.

Endothelium an inner (cellular) lining, for example of a blood vessel.

Eosinophilic granular cells distinctive cells which may occur in fish epidermis and may be identical to similar cells in the circulation (i.e. granular leucocytes).

Epidermis the outer cellular layer of the skin.

EROD (acr.) ethoxy resorufin-O-deethylase, a form of cytochrome P450, which can be involved in detoxification.

Erythrocyte red (actually individually straw-coloured under the light microscope) blood cell; pigmentation due to haemoglobin.

Erythrophore the type of chromatophore which is red due to carotenoid in the inclusions.

Erythropoiesis erythrocyte formation.

Esca lure found in anglerfish. The lure may have photophores. It projects dorsally or is suspended over the mouth by a long illicium derived from a dorsal spine.

Estivation see aestivation.

Estrogens also 'oestrogens', the female steroid hormones primarily involved with establishing oogenesis, particularly vitellogenesis in fish.

Eurypterid an extinct taxon of large marine invertebrates, allied to scorpions, and which may have preyed on fish.

Euryhaline an organism which tolerates a range of salinities, as in an estuary or the seashore.

Eurythermal an organism capable of surviving in a range of temperatures in contrast to more sensitive stenothermal organisms which have a more restricted temperature range.

Eutrophication a situation where nutrients promote excessive growth of, for example, algae and therefore produces competition for oxygen.

Excitable membrane cell membrane with gated channels (i.e. neuron and muscle cells).

Exocrine a gland with a duct.

Exogenous (of digestion) used to refer to digestive enzymes of microbial origin.

Exopeptidase a protease which cleaves a terminal amino acid from a protein.

Exogenous (of digestion) used to refer to digestive enzymes of microbial origin.

Exopeptidase a protease which cleaves a terminal amino acid from a protein.

Facial nerve cranial nerve VII; paired nerves with both afferent and efferent nerve fibres associated with facial sensations and jaw movements.

Falciform process structure in the eye which may have a nutritive role.

Fecundity refers to an estimate of egg production (as numbers of eggs).

Feedback an engineering term applied to physiological systems, in which information received controls the product; see 'positive' and 'negative' feedback.

Ferritin an iron-containing product of haemoglobin breakdown, which may be stored in the spleen.

Fibre/fiber general term includes, in the nervous system, afferent and efferent long nerve processes such as axons or dendrons. Term also used more widely in biology to describe elements of connective tissue (e.g. collagen fibres) and plant cells.

Fibroblast a cell type capable of secreting fibres such as collagen or elastin.

Fibrocyte a mature fibroblast.

Fibrin a protein fibre, formed from fibrinogen, the main constituent, with blood cells, of a blood clot.

Fibrinogen the soluble protein, the circulating precursor of fibrin.

Fick's principle the understanding that removal of some substances from the blood at the periphery can reveal certain features of the blood flow, for example the output from the heart.

FL (acr.) fork length.

Follicle in females, the cellular layers enclosing a developing oocyte. In males it has occasionally been used to refer to structures in the testis more usually known as lobules (teleosts), ampullae (elasmobranchs) or cysts/spermatocysts. In general it is a histological term describing a sphere of cells, usually 'hollow', for example with a lumen such as is found in thyroid follicles.

Follistatin a binding protein for activin.

Forage fish fish which because of their abundance and small size tend to be favoured prey items. Examples in the North Atlantic include the sand lance (*Ammodytes*) and the capelin (*Mallotus*).

Fork length the length (in cm) of a fish taken from the tip of the snout to the fork of the tail. N.B. the tail is not always forked so other measures may be used, see SL.

Forskolin agent used to experimentally manipulate intracellular cAMP activity.

Fovea an area retinae associated with a pit in the retina.

Frugiverous fruit-eating.

Fry (pl.) young hatchlings, which initially may have a ventral yolk sac.

FSH (acr.) a gonadotropin, the follicle-stimulating hormone, in females, chemically equivalent to ICSH in males.

Funny current 'mediates autonomic regulation of heart rate' (see DiFrancesco 2010).

G instantaneous growth rate = specific growth rate.

Gall bladder the round bladder-like organ which acts as a vessel for bile secreted from the liver, and has a bile duct emptying into the anterior intestine.

Gametogenesis the sequence of cell divisions and cell growth that produces gametes, either sperm by spermatogenesis or eggs by oogenesis.

Ganglion (pl. ganglia) a concentration of nerve cell bodies (perikarya) outside the CNS.

Ganoid scale a scale type rare in extant forms, and containing ganoin.

Ganoin(e) material formed by one or more layers of enamel or enameloid.

Gastric evacuation stomach emptying.

Gastrin a hormone released by the stomach and controlling stomach secretions (i.e. hydrochloric acid and pepsinogen release), as well as increasing gastric motility. Gastrin release is promoted by food entering the stomach.

Gated ionic channels membrane ionic channels which open by protein conformational change in response to changes in membrane voltage or chemical neurotransmitter enabling transmembrane movement of ions.

GBD (acr.) gas bubble disease.

Genus (pl. genera) the first unit of the Latin binomial used to name a species. Hence *Gadus morhua* (the Atlantic cod) shares the generic name 'Gadus' with Gadus ogac (the Greenland cod.

Germinal vesicle breakdown (acr. GVBD) the term seems archaic, it refers to the nucleus (germinal vesicle) of an oocyte. 'Breakdown' means the nucleus is not so easily seen, and it indicates loss of the nuclear membrane as meiosis, previously halted (i.e. 'arrested' in early prophase), resumes to complete oogenesis.

GH growth hormone.

Ghrelin a peptide hormone controlling appetite and growth.

Giemsa the name of a blood stain used on vertebrate blood smears. It contains methylene blue, methylene azure and eosin.

Gill arch the support for gills, formed by bilateral perforations of the pharynx.

Gill filament = primary lamella.

Glacial relicts organisms stranded in isolated ecosystems in cold conditions to which they are well adapted, and which are regarded as left *in situ* after ice has retreated, following a climate change of an interglacial episode.

Glomerulus (sing), glomeruli (pl.) 1. a small knot of blood vessels arising from, and giving rise to, small arterioles; enclosed by the Bowman's capsule of a nephron. 2. A cluster of synapses in the olfactory bulb.

Glossopharyngeal nerve cranial nerve IX; paired nerves containing afferent and efferent fibres associated with branchial movements and taste.

Glucagon regulates blood sugar, is hyperglycaemic (q.v.).

Glucosidase an enzyme capable of cleaving glycosidic bonds between two monosaccharides.

Glutaminergic receptors a group which share a similarity in chemical structure of the transmitter; that is, responding to the structure of or similar to glutamate or aspartate (excitatory aminoacids).

Glycocalyx the cuticle (of skin).

Gmax the upper limit of G.

GnRH (acr.) gonadotropin-releasing hormone.

Goblet cell mucous cell.

Gonadosomatic index (acr. GSI) reflects the gonad mass (practically as gram weight) as a percentage of the total body mass (g weight).

Gonadotropin A hormone controlling reproduction, see also FSH and LH.

Gonochorist any animal with separate sexes.

G-protein guanosine triphosphate binding protein, involved in transmembrane signal transduction.

Granulocyte a leucocyte containing cytoplasmic granules.

Granulosa a cellular layer (one cell thick) forming the inner layer of a follicle round an oocyte. The granulosa cells synthesize steroids, initially estradiol from testosterone formed in the outer layer of the follicle, the theca.

Grey matter the CNS tissue containing perikarya (neuron cell bodies) and dendrites, unmyelinated fibres and neuroglia, centrally located in the spinal cord, more superficial in the brain. Compare/contrast with white matter q.v.

Grey ramus (communicans) short neural connection between a sympathetic ganglion and a spinal nerve, enabling some post-ganglionic non-myelinated autonomic fibres to re-enter the spinal nerve from a sympathetic ganglion.

Group synchrony of females (not just fish or vertebrates) which show a reserve of immature oocytes with a subsection undergoing growth and vitellogenesis to form eggs for the next season.

GSI (acr.) gonadosomatic index.

GTH gonadotrophic hormone.

Guild species dependent on the same habitat, usually further defined; for example, 'fish guild'

Gular plates Support the throat region of some fish, reduced to branchiostegal rays (teleosts).

GVBD (acr.) germinal vesicle breakdown.

Gymnovarian literally 'naked ovary', no muscular wall round the organ, ovulation occurs outward to the abdominal cavity. Typical of many vertebrates including mammals/humans.

Gynogenesis the production of all-female progeny without male genetic involvement, sperm only activate.

Hadal found in the deepest ocean trenches.

Haem(at)opoiesis blood-cell formation.

Haldane effect deoxygenated blood carries relatively more carbon dioxide than the oxygenated state.

Haploid one chromosome set per cell as found in gametes post-meiosis.

Harmonics sound sources usually generate compound waves with a fundamental (low) frequency and a series of harmonics which are integer multiples of this frequency.

Haversian bone structure typical of mammalian compact bone in which bone cells (osteocytes) are arranged in layers (lamellae) concentrically around a central vascular channel, forming a series of units, each called an osteon.

Head kidney the anterior part of the kidney in fish.

Hemibranch one of the two rows of filaments on a gill arch.

Haemosiderin an iron-containing product of haemoglobin breakdown.

Hepatocyte the typical liver cell.

Herbivore plant eater.

Hermaphroditism the situation found in both some fish species and some invertebrates (e.g. earthworm) whereby an individual has both male (testes) and female (ovaries) organs, which may be active sequentially or simultaneously.

Heterocercal the asymmetrical tail condition with a larger upper lobe.

Heterodimer a molecular structure which includes two different subunits.

Heterodont having teeth of different shapes, specialized for cutting or chewing, for example.

Histocompatibility the degree to which 'non-self' elicits (or does not) adverse tissue reactions; that is, when tissues from two individuals come into contact. One group of proteins important in this regard is the 'major histo-compatibility complex' (MHC).

Holobranch the two rows of filaments on a gill arch.

Holocephali the chimaeras or rabbitfish, a small group of marine fish with an operculum rather than external gill slits, but otherwise sharing features with the sharks and rays (elasmobranchs), thus placing them together in the 'Chondrichthyes'.

Holostei the term used for gar and bowfins, currently not part of the formal classification.

Holostylic jaw suspension the upper jaw is fused to the cranium.

Homeobox a set of genes responsible for the successive ordering of body parts.

Homocercal the term used for a caudal fin which has two equal lobes as viewed externally.

HSI hepatosomatic index, a measure of relative liver size, as a percentage of body weight.

Hsp heat-shock protein.

HUFA (acr.) highly unsaturated fatty acids; polyunsaturated fatty acids with 20 or more carbon atoms.

Humoral (adj.) applied to control by hormones, in contrast/comparison to the neural control of the nervous system.

Hyaloin(e) a material found covering catfish scutes. It has no collagen and is highly mineralized.

Hydration fish oocytes may undergo a rapid water uptake prior to ovulation.

Hydrolysis splitting of a molecule such as a polypeptide, with the addition of the elements of water, facilitated by fairly specific enzymes.

Hyoid second gill arch.

Hyomandibular enlarged epibranchial of the hyoid arch, may be important in jaw support.

Hyostylic jaw suspension the anatomical situation where the second gill arch plays a major role in the attachment of the lower jaw.

Hypercapnia more than 'normal' carbon dioxide.

Hypercarbia as hypercapnia.

Hypercalcaemic (adj.) raises blood calcium.

Hyperglycaemic (adj.) raises blood glucose.

Hyperoxia more than 'normal' oxygen availability.

Hypodermis the region beneath the dermis, between the muscle blocks and the dermis.

Hypocalcaemic (adj.) lowers blood calcium; for example, calcitonin, stanniocalcin, teleocalcin.

Hypocalcin = stanniocalcin, lowers circulating calcium, similar in effect to calcitonin.

Hypogean (adj.) subterranean.

Hypoglycaemic lowers blood glucose.

Hypophysectomy surgical removal of the pituitary.

Hypophysis the pituitary.

Hypothalamic–hypophyseal portal system a special portal system (q.v.) found in some vertebrates (e.g. mammals). which enables a short circuit of the blood and thus rapid transit between the hypothalamus and the pituitary.

Hypothalamus ventral area of the diencephalon of the brain; links to the pituitary.

Hypoxia less than 'normal' oxygen.

Hz (abr.) Herz; cycles per second.

ICSH interstitial cell-stimulating hormone, the gonadotrophin which is expected to control early gametogenesis, the title refers to its role in males; the same molecule is termed FSH relative to females.

IGF (acr.) insulin-like growth factor.

Imbricated (adj.) overlapping (of scales).

Immature (adj.) a term applied either to a fish that has never reproduced or which has gonads not currently either in active gametogenesis or not holding fully formed gametes ready for reproduction.

Immunoglobulins large protein molecules circulating in the blood plasma, important in defence against bacteria, for example.

Imprinting an association established early in life which remains relatively unalterable.

Indeterminate spawning for fish, this term recognizes the ability of some fish to recruit oocytes over a long period, up-regulating numbers as conditions allow.

Inhibin a polypeptide paracrine which inhibits gametogenesis/maturation.

Innominate artery the 'unnamed' artery; it is paired and the most anterior branchial artery on each side, coming off the ventral aorta and splitting into the first and second branchial arteries.

Instantaneous growth rate a value ascribed to growth at a single and usually short time period.

Instinct an innate or internally driven behaviour with a set pattern, for example those in courtship.

Insulin regulates blood sugar, is hypoglycaemic, and antagonistic to glucagon.

Integument the outer cellular layer of a vertebrate, has protective and shape-retaining functions amongst others.

Intermediate filaments intracellular protein filaments (keratin in the epidermis), intermediate in diameter between microtubules and microfilaments (such as the actin and myosin of muscle).

Interneurons CNS neurons relaying (passing) information between other neurons.

Internuncial neurons interneurons.

Interrenal cells steroidogenic (corticosteroid synthesis) cells found either in the head kidney of teleosts or between the kidneys (in interrenal glands) in elasmobranchs.

Ionic channels enable ions to pass across a membrane, and includes 'gated' ionic channels q.v.

Ionocyte an epidermal cell capable of ion regulation and very similar to gill chloride cells.

Ionotropic receptor molecular receptor responding to a neurotransmitter directly by opening gated ionic channels.

Iridophore a chromatophore containing light reflective plates and responsible for silver or iridescent colouring found in the integument or peritoneum.

Isopedine a tissue, described in scales from both extant and extinct fish, characterized by the inclusion of parallel collagen fibres, but not readily distinguished from elasmodine.

Isospondyli a now-discarded term for some bony fish, previously differentiated/associated on the basis of soft fins, posteriorly placed pelvic fins and, if present, swim bladders physostomatous (i.e. linked to the gut) swim bladders.

Isozymes similar enzymes influencing the rate of a specific reaction.

Iteroparity (adj. iteroparous) repeat reproduction. Notably from year to year.

IUCN International Union for Conservation of Nature.

JG cells juxtaglomerular cells.

Juxtaglomerular cells found close to the anterior end of a nephron.

K carrying capacity of the environment, the maximum sustainable population.

Kelt post-spawned salmonid, with poor condition but regarded as capable of spawning again another year.

11-Ketotestosterone The dominant androgen in male fish.

Kin related individuals.

Kinocilium single long cilium (with 9 + 2 microtubules) in the ciliary bundle at the apex of sensory hair cells of the lateral lines and ear.

Kisspeptins peptides which may have a role in controlling reproduction in some vertebrates.

Knollenorgans mormyrid tuberous electroreceptors (q.v.) sensitive to conspecific EODs (q.v.).

kPA kiloPaschal, a metric unit of measure for pressure, has recently tended to replace units such as Torr (mmHg) or 'cm H_2O' which reflected the visual measurement of a column of fluid supported by the application of a pressure system.

Kupffer cell a phagocytic cell in the endothelium of the liver sinusoids.

Kype the jaw hook, developed prior to spawning, on the lower jaw of male salmonids, may also involve enlargement or development of canine teeth.

Lagena the most posterior of the otolith organs of the fish ear, sensitive to sound as well as gravitational stimuli.

Lamella (sing.), lamellae (pl) primary = gill filament, secondary = smaller 'platelets' (Gray 1954).

Lamellar bone bone in which layers can be discerned, a noted feature of mammalian Haversian bone, not associated with fish.

Larva (pl. larvae) newly hatched animals.

Lateral line system detector of low-frequency water movements enabling fish to identify presence of other moving animals as well as reflection of water from static objects.

Lateralization primarily associated with the human cerebral hemispheres in which functional areas are not symmetrical, with specialization of the right and left sides. Some lateralization has been reported in fish.

LCI liver condition index, = HSI.

Learning modification of behaviour by experience and is dependent on a memory.

Lecithotrophy dependent on yolk.

Lepidology the study of scales.

Lepidophagy scale eating.

Leptin a hormone characterized from mammalian adipose tissue and secreted proportionally, which may occur in fish.

Leptocephalous larva thin leaf-shaped larva characteristic of elopomorphs (eels, etc.).

Leptotene a stage of meiotic prophase I.

Leuc(k)ocyte a 'white' (actually non-pigmented) blood cell, with several subtypes.

Leucopiesis leucocyte formation.

Leucophore a chromatophore type collectively producing a white effect due to guanine in intracellular inclusions.

LH luteinizing hormone, a gonadotropin, originally named for its role in maintaining the corpus luteum in mammalian pregnancy. It is expected to have a different role in fish reproduction and is steroidogenic, promoting oestrogen synthesis and thence vitellogenesis.

Lipotropin produced in the pituitary, may be a prohormone in that the beta form can be cleaved to a smaller form and an endorphin.

Lipovitellin a component of yolk.

Littoral (adj.) of the shore.

Lobules a general histological term referring to organs divided by connective tissue into fairly large interconnecting compartments. Many bony fish have testes with lobular structure.

Loop of Henle a long mid-segment of the nephron associated with water regulation, and not found in fish.

L-start same as 'C-start'.

LTP long-term potentiation; the reinforcing of a neural circuit.

Luciferin a light-producing chemical.

Lymph tissue fluid which, in some vertebrates, has a defined circulation separate from, but draining into, the blood circulation.

Lymphocyte a leucocyte type, associated with immune responses.

Macula sensory epithelium in each otolith organ of the inner ear.

Macula communis the single elongated otolith organ occurring in lamprey.

Magnetosomes intracellular magnetite particles.

Malphigian cell epidermal cell which retains its mitotic capacity.

Malphigian corpuscle the structure formed by the Bowman's capsule enclosing a glomerulus.

Mature a term applied to a fish that is or has been reproductive and/or with gamete development or holding gametes.

Matrotrophy nutrition provided to the developing embryo by the mother; that is, viviparity and also situations where yolk nutrition is supplemented from maternal sources post-fertilization.

Maturation promoting factor (MPF) is a complex of a kinase and cyclin and, in variant forms, controls the cell cycle.

Mauthner fibres paired very large diameter (efferent) fibres, running ventrally in the spinal cord of fish and some amphibia, have very rapid conduction, and enable escape reactions by means of tail flips.

MCH melanophore- (or melanin-) concentrating hormone.

Medulla oblongata subregion of the hind-brain merging with the spinal cord.

Meissner's plexus autonomic nerves in the connective tissue beneath the gut epithelium.

Melanophore/melanin-concentrating hormone Produces paling in fish, may also control appetite.

Melanophore-stimulating hormone responsible for darkening in some fish and amphibia.

Melano-macrophage centres large patches of tissue, appearing black/brown which may be important in sequestering some potentially toxic particles. These aggregates may be found in the spleen.

Melanophore the chromatophore type which is brown or black due to melanin in melanosomes which can move intracellularly.

Melatonin pineal hormone; aggregates melanosomes in melanophores, is associated with circadian rhythms of light and dark and activity.

Membrane bone also termed dermal bone, formed superficially and does not have a prior cartilage model.

Membrane pumps molecular systems in cell membranes which actively transport substances such as sodium and potassium ions, against their concentration gradients,

Memory ability to retain neural information and subsequently recall it.

Meninx primitiva thin tissue surrounding CNS of fish, equivalent of meninges (pl. of meninx) of tetrapods.

Meristics countable traits; for example fin-ray number is expected to be species-specific for a particular fin.

Merkel cells small, round epidermal cells which are tactile in vertebrates, but their role in fish is controversial.

Mesencephalon mid-brain.

Mesonephros the elongated 'middle kidney' which forms the major part of the kidney in fish.

Mesopelagic mid-water fish.

Metabotropic channel molecular receptor responding to a neurotransmitter via a secondary intracellular messenger.

Metameric segmentation = metamerism.

Metamerism antero–posterior organization of an animal into repeating segments (metaneres) each interdependent in the activity of the whole animal; formally described as the serial repetition of homologous parts. Shown by vertebrates but somewhat obscured, for example, by cephalization (head development).

Metanephros the compact kidney of the tetrapods particularly mammals.

Metencephalon cerebellum.

MFO mixed function oxidase.

Micelles tiny aggregates of lipid products formed in the intestinal lumen.

Microglia small neuroglial cells with a few short processes; have an immune function.

Micropyle one or more specific openings (which may have complex structures, e.g. 'sperm-guides') in the egg covering (chorion) to enable sperm entry. No equivalent in mammals.

Microspectrophotometry technique enabling measurement of spectral absorption by individual morphologically distinct photoreceptors, such as rods/cones.

Migration movement between different habitats. Often associated with reproduction and can involve physiological changes.

Mimicry deceptive resemblance of a species to another for protection or to facilitate predation.

MIS maturation-inducing steroid (see progestogens).

Mitochondria-rich cells also known as chloride cells, are involved in ionic regulation.

Mitral cells neurons in the olfactory bulb, forming the olfactory tracts to the cerebral hemispheres.

Mixed function oxidase enzyme (s) associated with detoxification.

Mixed nerve peripheral nerve (e.g. spinal nerve) with both sensory (afferent) and motor (efferent) fibres.

MMC melanomacrophage centres.

Molal (of a solution) the gram molecular weight of a substance dissolved in a kilogram of solvent.

Molar (of a solution) the gram molecular weight of a substance dissolved in a litre of solvent.

Molecular receptor receptors at this cellular level are essential to the specific response/recognition mechanism which enables neurotransmission or hormone reaction in contrast to 'sensory receptors' which vary from cells to organs in picking up stimuli such as pressure, light, etc.

Monocyte a relatively large white blood cell with a rounded or bean-shaped nucleus.

Mormyromasts mormyrid tuberous electroreceptors processing self-generated EODs (q.v.).

MPA marine protected areas.

MPF 1. median, paired fin propulsion. 2. maturation-promoting factor, also used for 'mitosis-promoting factor'.

MR cells = mitochondria-rich cells.

MRG metal-rich granules.

MSH melanophore- (or melanin-) stimulating hormone, also called intermedin.

Muller fibres relatively large diameter, motor-coordinating fibres in the white matter of the spinal cord.

Mutualism a close relationship between two species.

Muscularis mucosa in the gut, a layer of smooth muscle cells lying beneath the epithelial 'mucosa' and able to move it.

Myelencephalon medulla oblongata.

Myelin an insulating layer (with a high lipid content) around axons and dendrons of some nerve cells.

Myenteric plexus Auerbach`s plexus q.v.

Myocardium the collective term for the heart musculature.

Myocyte muscle cell.

Myoid a contractile component of photoreceptors in the eye of some fish.

Myogenic literally 'determined or generated by muscle', refers to hearts which have an intrinsic and regular contraction of the cardiac muscle.

Nanoparticles very small particles, for example in the size range 10 nm to 1600 nm (see Scown et al. 2010).

Nares anterior and posterior openings of the olfactory sacs.

Natriuetic peptides secreted in, for example, specific regions of the heart and affecting sodium excretion at the gills or rectal gland.

Negative feedback detection of specific items as increasing (incoming information) decreases output, release or acquisition of these items, achieving regulation and a 'normal' level, that is, homeostasis.

Nephridia structures achieving excretion in some invertebrates; for example, earthworms.

Nephron the kidney tubule which forms the base unit of the kidney, in which the tubules are specially oriented to collect blood for filtration.

Nerve fibres nerve axons and dendrons.

Neural crest embryonic cells, identifiable as a distinct region in the late gastrula/neurala of vertebrates; this tissue gives rise to a variety of specialized tissues including those of the nervous system, and the chromatophores.

Neurocranium The region of the skull immediately surrounding the brain.

Neuroeffector junction physiological junction between a neuron and effector such as striated muscle.

Neuroglia cells in neural tissue with a variety of physiological supporting roles (e.g. myelin formation, neuromodulation); can undergo mitosis in contrast with neurons. There are three types; astrocytes, oligodendoglia and microglia.

Neuromast a group of mechanosensory 'hair' cells in lateral line systems.

Neuromodulator substances acting as neurotransmitters can also alter the release and or action of other transmitters (Burnstock 2009).

Neuron(e)s nerve cells which are specialized for conduction of electrical impulses.

Neuropil neural tissue with densely interlacing nerve cell processes which be associated with integration and processing of information.

Neurotransmitter chemical messenger enabling neural signals to pass between neurons (e.g. at a synapse) and between neurons and effectors.

Neurovascular unit anatomical association between an astrocyte and a capillary.

NMDA N-methyl-D-aspartate.

NMDAR the NMDA receptor, found in the brain; one of several glutaminergic receptors.

NO nitric oxide.

Noci (o)ceptors receptors sensitive to noxious stimuli.

Nodes of Ranvier microscopic gaps in the myelin sheath which facilitate saltatory (fast) conduction.

NOS nitric oxide synthase.

Notochord a supporting rod of ensheathed vacuolated cells, in the mid-dorsal position and typical of vertebrates in which it is replaced in adults by the vertebral column. Both the notochord and the vertebral column prevent telescoping when the lateral muscles (myotomes) contract.

Octavolateralis system term equivalent to 'acousticolateralis system' used to include a set of nerves and receptors achieving balance and posture.

Oculomotor muscles these move the eyeball in its orbit; they comprise six distinct muscles each side (four rectus, two oblique).

Oculomotor nerve cranial nerve III, paired motor nerve from the brain to four eye muscles; three rectus: anterior, inferior and superior rectus and the inferior oblique muscle.

Odontode the embryonic structure from which a scale (or tooth?) develops.

Odontoid synonym for placoid scale.

Olfactory bulb a centre receiving fibres from the olfactory nerve which synapse with mitral cells. Each bulb contains small areas of neuropil which may be responsive to characteristic differences between odours.

Olfactory nerve cranial nerve I; paired nerve from olfactory organ to the brain, enabling smell.

Olfactory rosette characteristic folding increasing the area of sensory olfactory epithelium within fish olfactory sacs.

Oligodendroglia myelin forming neuroglia in the CNS.

OM L oxygen minimum layer.

Omnivore an animal that is not very discriminating with regard to food, ingesting animals and/or plants, etc.

Oocyte stage in oogenesis when an oogonium enters meiosis.

Oogenesis the process of cell division and cell growth, starting with mitoses of diploid oogonia, proceeding to entry into meiosis and meiotic arrest of oocytes, growth usually with yolk.

Oogonium (pl. oogonia) the diploid 'stem-cell' capable of entering mitoses preparatory to transformation to an oocyte per oogonium when meiosis begins. Each oogonium is small and has a prominent nucleus and nucleolus.

Oolemma the cell (plasma) membrane of the cell series culminating in ova.

Oophagy consumption of eggs. Occurs in lamnoid sharks *in utero*; that is, young sharks eat surrounding eggs before hatching.

Operculum gill cover, a hard structure screening the gills, so that water does not pass directly to the exterior after passing over the gills in fish possessing it; for example, holocephali, teleosts, etc. It can be moved in and out (i.e. away from or towards the posterior head region) with a hinge anteriorly. Its presence forms an opercular cavity, which becomes part of the ventilation system.

Opsin the protein component of visual pigments.

Opsonins proteins that can coat antigens and thus promote their targeting by, for example, phagocytes.

Optic lobes paired regions of the mesencephalon; receive optic nerves/tracts.

Optic nerves/tracts cranial nerve II; paired nerve (tract) from the eye to the brain, enabling sight.

Optic tectum roof of the mesencephalon, considered to be somewhat equivalent (functionally) to the tetrapod cortex.

Outflow see autonomic outflow.

Orexogenic conditions stimulating feeding.

ORN olfactory receptor neuron.

Ornithine cycle the biochemical pathway transforming ammonia to urea.

Osmoconformer an animal that has internal body fluids roughly identical in osmotic effect to the surrounding medium.

Osmolality solute concentration given as osmoles per kg solvent.

Osmolarity solute concentration given as osmoles per litre of solvent.

Osmole a mole of solute particles.

Osmolyte ions or molecules that, when in solution, can affect osmotic movement of water across a semipermeable membrane. Osmolytes include urea.

Osmoregulation the capacity to control internal conditions to minimize osmotic disturbance, see also osmosis.

Osmosis the tendency of water to move from a region of high water concentration to a region of low water concentration, across a semipermeable membrane such as a cell membrane. This tendency, with water moving, for example, from a region of low salinity to a region of high salinity or equivalently effective molecular composition (see osmolytes) can be regarded as a special kind of diffusion. This process can lead to cell volume changes, increases if surrounded by freshwater, decreases if surrounded by a saline higher than the cytoplasm's normal constitution.

Osmotic fragility the tendency of some cells to lyse under certain osmotic constraints, for example when placed in water, in which case the cell membrane can rupture as water passes into the cell by osmosis. Cells with cell walls (plants, bacteria) or an effective internal osmotic regulation via contractile vacuoles (*Paramecium*) or amino acid/taurine regulation (fish red blood cells) can avoid rupture if exposed to water or dilute saline.

Osteichthyes bony fish, a major taxon, which may however exclude some fish with bone which are considered to be closer to tetrapods.

Osteoblast an immature bone-forming cell.

Osteocyte a bone-forming cell.

Otoconia calcareous granular structures in the gelatinous matrix of the otoliths of elasmobranchs and agnathans.

Otoliths solid, hard, calcareous structure, (which may have annual growth rings), in the teleost otolith organs, which are coupled by gelatinous membranes to the maculae.

Oval found in the swim bladder of some fish, capable of gas resorption.

Ovipary the female deposits eggs before or after fertilization.

Ovovivipary (aplacental vivpary) after internal fertilization eggs may be retained and hatch in the reproductive tract where they may obtain some form of nourishment from maternal fluids or from eating sibling eggs (oophagy) or young (adelphagy); that is, without placentation.

Ovulation the emergence of a late-stage oocyte from its follicle preparatory to fertilization.

Ovum (pl. ova) the large haploid cell produced as the terminal point of oogenesis.

Oxynticopeptic cells cells which are capable of releasing both HCl and pepsin, unlike mammals which have two separate cell types for these functions.

Pacemaker 1. a region in the vertebrate heart which determines overall heart rate. 2. any region achieving rhythm as can occur for the breathing/respiratory centre in the hind-brain of vertebrates, controlling ventilation rate.

Pachytene a stage in meiotic prophase I.

Paedophagy 'young' eating, for example consumption of larvae.

Paedomorphosis a situation, whereby a larval form reproduces, also referred to as occurring by paedogenesis.

PAH polyaromatic hydrocarbon(s).

Paling centre autonomic centre, in the medulla oblongata, controlling rapid colour changes in some teleosts.

Paracellular a pathway between cells.

Paracrine a substance with hormone-like qualities except it has a local effect.

Parasite a form of symbiosis harming one participant.

Parasitism an intimate and lasting relationship between organisms of two different species, in which one, the host, is harmed.

Parasympathetic nervous system subdivision of the mammalian autonomic nervous system; not fully developed in fish (see cranial autonomic system).

Parenchyma (of spleen) inner region containing red and white pulp.

Parental care a diverse range of caring behaviours, displayed by parents following egg fertilization.

Parr juvenile salmonids typically with bilateral vertical grey/black and short stripes or bars (parr marks).

Parr marks a series of black/brown roundish patches along the side of young salmonids, which disappear on smoltification.

Pars anterior the anterior region of the pituitary.

Pars intermedia a region of the pituitary, inappropriately named in fish because it is actually posterior in position.

Pauciglomerular a small number of glomeruli.

Pavement cell 1. Malphighian cell of epidermis (q.v.). 2. The epithelial cells of the gill external surface.

Pelagic swimming animals; in fish would refer to active swimmers like herring, mackerel and tuna, swimming in the water column.

Pepsin the main (or only) stomach protease enzyme.

Pepsinogen the inactive precursor form of pepsin.

Perikaryon (pl. perikarya) cell body containing the nucleus in cells possessing projecting processes; for example, neurons, neuroglia and melanophores.

Perimeningeal tissue fatty tissue around spinal cord peripheral to the meninx primitive.

Periphyton a mat of mixed small organisms (e.g. algae) and detritus on aquatic substrates. Also known as 'Aufwuchs'.

Peritoneum the thin epithelium lining the major body cavities and covering the abdominal organs.

pH The measurement of acidity/alkalinity, formally defined as minus the logarithm to the base 10 of the hydrogen ion concentration. Neutrality is indicated by pH 7 (10^{-7} g hydrogen ions per litre), whereas < 7 is acidic and > 7 is alkaline. In reality, though water would be expected to be neutral, it rarely is. Distilled water is often acidic (as constituents may accompany the condensation process) and fresh water may vary considerably from very acidic to alkaline depending on the composition of rocks and vegetation leaching into the system. Because the pH scale is logarithmic, particular care needs to be used in its interpretation and graphical representation.

Phagocytes motile cells which can engulf (and destroy) particles such as other cells or cell fragments.

Phasic receptors sensory receptors manifesting a rapid decline (adaptation) in impulse frequency when a stimulus is maintained.

Pharyngeal jaws Structures derived from the branchial skeleton, capable of holding or manipulating food.

Pharynx a major gut component, between the buccal cavity and the oesophagus, pierced bilaterally by the gill slits.

Pheromone a substance released into the environment that influences behaviour of conspecifics.

Phosphocreatine see creatine phosphate.

Phosvitin a component of yolk.

Photophore a biological structure which emits light.

Photoreception sensitivity to photons of light energy.

Physoclystous a closed swim bladder; that is, it has no patent connection to the gut.

Physostomatous a swim bladder with a duct open to the anterior gut.

Phytoestrogens plant-produced molecules which mimic steroids, and can accumulate in pulp-mill effluents.

Pineal a light-sensitive region dorsal to the brain which has an endocrine role in producing melatonin.

Pinealocyte a photoreceptor cell of the pineal gland/organ.

Pituitary a complex endocrine gland beneath the hypothalamus of the brain.

Pillar cells in the gill, separating gill epithelia in the gill lamellae.

Piscivore fish eater.

PKD proliferative kidney disease.

Placoid scale the basic scale type of elasmobranchs.

Planktivore plankton eater, usually by a form of filter-feeding through the gill slits for adult fish.

Plankton floating organisms including algae, jellyfish, fish eggs and fish larvae.

Planktophagy plankton eater.

Plasma the fluid portion of blood including fibrinogen (see also serum).

Platelet secondary lamella of the gill, not to be confused with the blood-clotting platelets of tetrapod blood.

Play behaviour patterns which are free from attaining an animal's needs and their specific goals and may involve exploratory manipulation of objects.

Plesiomorphic (adj.) features regarded as ancient and relatively unchanged; see also 'primitive'.

Pleiotropic affecting different characteristics.

Plexus (pl. plexi) interweaving neural or vascular network.

Podocyte a special cell within Bowman's capsule; a group of these cells facilitate ultrafiltration.

Polar body the very small discarded cell(s) of oogenesis, formed during meiosis, during which time only one cell per division receives the yolk and almost all of the other cell constituents except the chromosomes which are divided equally. The first polar body is discarded after meiosis I and the second polar body after meiosis II.

Polypeptide a linked chain of aminoacids, which are termed proteins when the number reaches double digits; up to ten aminoacids in a chain may be referred to by the number involved; for example, octapeptide for eight, decapeptide for ten.

Polyploid more than two chromosome sets per cell.

Polyspermy fertilization by more than one sperm, and generally undesirable. Blocked by the 'cortical reaction'.

POMC proopiomelanocortin.

Positive feedback Physiologically, incoming information, identifying a particular substance as present or increasing leads to further increase, which may occur in growth or reproduction.

Post-ganglionic fibre/neuron non-myelinated nerve process/neuron between autonomic ganglion and effector.

Post-ovulatory follicle (acr. POF) the remains of the follicle left in the ovary after ovulation.

Precocious refers to young animals born or hatched at a well-developed and relatively independent stage.

Preganglionic fibre/neuron myelinated fibre/neuron linking CNS and autonomic ganglion.

Primary growth phase occurs when an oogonium becomes an oocyte and enlarges prior to the laying down of yolk.

Primitive a now somewhat archaic (or arcane) view of characteristics of living species which are also attributed to fossil, or very much older, lineages. See also plesiomorphic.

Primordial germ cells the distinct cell line which gives rise to the gametes, initially multiplying by mitoses to produce 'gonia', either spermatogonia (males) or oogonia (females).

Progestogens steroids with 21 carbon atoms (see also corticosteroids), and which have a role in females at 'egg' maturation time. There are several progestogens which are believed to have important roles in fish reproduction; these include 17α 20β dihydroprogesterone (DHP) also referred to as an MIS.

Prolactin the vertebrate hormone eliciting lactation in mammals, which occurs in fish, but its function therein is debatable, but may be hypercalcaemic.

Pronephros the most anterior part of the kidney in agnatha and, perhaps, ontogenetically during embryonic development).

Pro-opiomelanocortin also known as pro-opiocortin, large precursor molecule, synthesized in the pituitary, can cleave to form endorphin, MSHs and ACTH and other small peptides (see Goldsworthy et al. 1981).

Proprioceptors tension (not stress!) detectors, which may be important in maintaining muscle tone.

Prosencephalon fore-brain.

Protandry the hermaphrodite condition when a male stage precedes a female stage.

Protogyny the hermaphrodite condition when a female stage precedes a male stage.

Proton pump may occur in cell membranes, driving H+ across them to create an electrical gradient which can result in transmembrane movement of other ions.

Proximal (convoluted) tubule a major part of the nephron, that region occurring close behind the renal corpuscle region (or its position in aglomerular fish).

Proximal pars distalis a region of the anterior pituitary.

Pseudobranch a gill-like structure in the spiracle of elasmobranchs or within the operculum of teleosts. Its function is not known but it is unlikely to act as a gill (hence its name).

Psychon a unit of inner sense, enabling memory and other mental processes such as learning and thought in mammals.

PUFA polyunsaturated fatty acids.

Purkinje cells neurons in the cortex of the cerebellum with large flask-shaped perikarya and an elaborate dendritic system.

Pyloric caeca (ceca) small finger-like processes extending from the pyloric (posterior) region of the stomach in teleosts.

Rabbitfish chimaeras, the SubClass Holocephali.

Ramus communicans (pl. rami communicantes) connection between spinal nerve and sympathetic ganglion. Preganglionic fibres are in white rami and postganglionic fibres in grey rami (reflecting presence or absence of myelin).

RAS renin angiotensin system.

Receptor potential an electrical depolarization of a sensory cell membrane resulting in release of neurotransmitter.

Receptors structures sensitive to specific stimuli; the term is used to describe receptor cells/tissues/organs and molecular membrane receptors.

Recruits young fish large enough to contribute to a fishery, or first-time spawners.

Rectal gland a small finger-like blind-ending tube attached to the rectum in elasmobranchs and, for example, the coelacanth.

Red heart found in fish from relatively anoxic habitats, the colour being due to large amounts of myoglobin. (see 'white heart').

Red muscle striated muscle containing a large proportion of myoglobin (hence red), generally found as a layer beneath the hypodermis of fish skin, and responsible for slow, continuous swimming movements.

Red pulp a region of the spleen, red because of the prevalence of erythrocytes.

Relict fish a few extant fish species which have retained characteristics of very ancient lineages; that is they are plesiomorphic, formerly termed primitive.

Renal corpuscle aka Malphigian corpuscle, the usual start point for formation of urine but not present in all fish. It comprises a knot of blood vessels (the glomerulus) partly surrounded by Bowman's capsule which is continuous with the kidney tubule.

Renal portal vein returns blood to the kidney from the caudal vein.

Reproductive scope the area in which or time during which reproduction occurs (Potts et al. 2014).

RES reticuloendothelial system.

Residual egg an 'egg' (a late-stage oocyte) which has remained in the female post-ovulation.

Residual sperm sperm remaining in the testes ducts after the reproductive season.

Respiratory centre groups of neurons in the medulla oblongata controlling respiratory movements enabling ventilation of the gills, that is buccopharyngeal cavity volume change and opercular position in teleosts.

Rete (mirabile) literally 'wonderful net', a closely packed array of small arteries and veins (sometimes termed capillaries), usually disposed in a regular pattern as seen in cross-section, with venules running counter to arterioles and thus appropriate for countercurrent and rapid exchange of materials or maintaining higher temperature.

Retes anglicized plural for 'rete'.

Retia latinized plural for 'rete'.

Retina a light-sensitive neural tissue in the vertebrate eye.

Retinomotor movements retinal pigment movements in response to changes in illumination, occurs in teleosts.

Rhombencephalon hind-brain.

Root effect in some fish increased acidity has more effect in promoting oxygen release (Bohr effect) than in other vertebrates.

Rods elongated photoreceptor cells, with an outer segment uniform in width, in the retina; function well in dim light.

Rostral pars distalis a region of the anterior pituitary.

Sacculus one of the three otolith organs sensitive to sound and gravity in the fish ear.

Saccus vasculosus structure with folded vascularized walls attached posteriorly to the hypothalamus, present in some fish. Its function is not known.

Saggital a mid-longitudinal section.

Saltatory conduction rapid impulse conduction caused by myelin sheath restricting action potential voltage changes to the non-insulated membrane at the nodes of Ranvier.

Scale pockets a term used for either the insertion point of (e.g. teleost) scales or the intucking of skin between overlapping scales.

School a cohesive (close) group of swimming animals, usually of the same species.

Schreckstoff the name given by Von Frisch to the 'alarm substance' released by club cells.

Schwann cells cells wrapped around axons and dendrons of peripheral nerves, producing myelin insulating layers. Compare oligodendroglia of the CNS.

SCN suprachiasmatic nucleus (in this case 'nucleus' is a concentration of nerve cell perikarya).

Scute a bony plate, part of a series embedded in the skin of some fish.

SDA specific dynamic action (same as SDE).

SDE specific dynamic effect (same as SDA).

Seamount a large subsea projection from the seafloor upwards as a cone derived from volcanic activity.

Secondary growth phase the stage during which yolk deposition (vitellogenesis) occurs in oocytes.

Secondary messenger intracellular control molecule, for example cAMP, which is in turn controlled by an extracellular hormone or neurotransmitter.

Secretogogue a substance evoking enzyme secretion/release in the gut.

Semelparity the condition in which an animal life cycle incorporates just one possibility for reproduction, as the termination.

Semicircular canals three curved tubes (on each side) at right-angles to one another in the ear, sensitive to movement in particular planes. Extant Agnatha have fewer; two in lamprey, one in hagfish, on each side.

Serum the fluid portion of blood minus fibrinogen/fibrin; that is, obtained after coagulation.

Sessile an organism that remains in one location and cannot move from it.

Shoal a group of swimming animals, usually of the same species.

Sinusoid a very small blood-filled space, typically lined by endothelium.

Sinus venosus the first chamber of the fish heart, thin-walled and therefore distensible.

Siphon in male chondrichthyans a sac (paired) which assists flushing of sperm down the claspers to the female.

SL 1. standard length. 2. somatolactin.

Sleep a period of inactivity under the neural control of the suprchiasmatic nucleus of the hypothalamus in mammals, which is also present in fish.

Smolt the first stage of a seaward migration in the life cycle of anadromous salmonids. Smolt are silvery and pre-adapted for increased salinity.

Smoltification the process of transformation from a 'parr' to a smolt.

Sneaker also known as a jack or dwarf male, a precocious male capable of fertilizing eggs of a normal size courting pair.

Soma the body of an animal or the central part of a cell; for example, the perikaryon of a neuron.

Somatic nervous system motor nerves controlling striated muscles (usually 'voluntary' responses) achieving movement and support.

Somatolactin a pituitary hormone chemically similar to growth hormone and prolactin.

Somatotopy the correspondence between peripheral nerve endings/sensory receptors and the spatial organization of their connections within the CNS. See also chemotopy.

Sp one species.

Spp the expression used to indicate the inclusion of more than one species, but not necessarily every species in a genus.

Specific dynamic action or effect increased consumption of oxygen associated with digestion.

Specific growth rate growth rate determined for a specified time interval, for example, percentage body weight gain per day.

Species the basic unit for classification, representing a group of potentially interbreeding organisms, and known by an internationally agreed binomial q.v. (see also genus).

Spermatids in males, the initial product of meiosis. A spermatid has no flagellum. Each can become a spermatozoan through the process of spermiogenesis.

Spermatocyte the meiotic cell which, by completion of two successive divisions, form haploid spermatids.

Spermatogonium (pl. spermatogonia) 'stem-cells' in males, which form the reserve for sperm formation. After mitoses a secondary spermatogonium can enter meiosis as a spermatocyte. Spermatogonia strongly resemble oogonia, the female equivalent.

Spermatophore sperm packets, used for transfer to the female in elasmobranchs.

Spermatozeugmata (pl.) specialized sperm packets found in the Poeciliidae.

Spermatozoon (pl. spermatozoa, abr. sperm) the male gametes, haploid and potentially motile. Most fish sperm have just one flagellum, though a bi-flagellated condition occurs in pout.

Spermiation the release of sperm into the duct system.

Spermiogenesis the formation of flagellated sperm from spermatids.

Spinal autonomic system autonomic nerves with outflows from the spinal cord.

Spinal nerve a pair of 'mixed' nerves for each body segment, passing from the spinal cord on each side to the periphery.

Spinal outflow the term used, together with and in contrast to, 'cranial outflow', to describe the pathways of autonomic fibres from the CNS to the periphery.

Spiracle a reduced form of gill opening, usually round with minimal or no gill tissue. If present, for example in elasmobranchs, usually bilateral; dorsal for skates/rays.

Splanchnic nerve autonomic (anterior and posterior) postganglionic nerves to the gut.

Spleen a dark red-coloured organ, normally found close to the stomach in vertebrates. It has several functions, including erythropoiesis.

Squamation collective term for the scales.

StAR steroidogenic acute regulatory protein.

Standard length the length (cm) from the tip of the snout to the mid-point of the end of the caudal peduncle (q.v.).

Stanniectomy surgical removal of corpuscles of Stannius (not easy to do?).

Stanniocalcin also known as hypocalcin.

STC-1 stanniocalcin.

Stenohaline an organism not well adapted to living in different salinities.

Stenopeic The condition in elasmobranchs in which the pupil is narrowed to form clear images of both near and distant objects.

Stereocilia microvilli in a cluster in a 'ciliary' bundle at the apex of sensory 'hair' cells of the lateral line and ear; in contrast the kinocilium of the ciliary bundle is a true cilium with a 9 + 2 microtubular ultrastructure.

Steroidogenesis the process of steroid synthesis.

Steroidogenic (adj.) the term applied to cells or tissues capable of synthesizing steroids.

Stomach a major region of the gut, serving to collect food and start protein digestion; is present in most, but not all, fish.

S-start escape movements, for example tail-flips.

Stratum argenteum a subdermal layer, typical of pelagic fishes, which contains numerous guanine crystals.

Stratum compactum the major inner layer of the fish dermis.

Stratum spongiosum the major outer layer of the fish dermis.

Stress a physiological state evoked by an unusual environment or, for example, incipient predation, and accompanied by increased heart rate, etc., and the increased circulatory presence of adrenalin and some steroids, for example cortisol.

Stroke volume the blood emitted from the heart per systolic contraction cycle.

Stygobitic living in the dark, but not used for marine habitats, and implies (for fish) lack of melanophores and poor sight.

Substance P a tachykinin, found in fish but its role is uncertain.

Substantia gelatinosa Rolandi region of neural tissue in each of the dorsal horns of the spinal cord grey matter.

Suction feeding the ability to create negative pressure in the buccal cavity, drawing in water (and prey).

Supersaturation a condition where the amount of gas dissolved in water is more than usual for the temperature.

Suprachiasmatic nucleus region of the hypothalamus which may be involved in control of circadian rhythms.

Surfactant emulsifier, rendering water-insoluble fatty substances into smaller droplets.

Suspensorium the mechanism of attachment of the jaws.

SV saccus vasculosus.

Swim bladder a dorsal structure between the kidney region and the gut, which may have an opening to the anterior gut (physostomatous) or be closed (physoclystous). Usually functions as a hydrostatic organ.

Symbiosis close relationship between two species.

Synchrony in females, when all oocytes develop together at the same pace, typical of semelparous fish.

Syncytium a group of cell-like units which are not completely separate, and act in unison.

Synzesis the zygotene stage in meiosis I when chromosomes aggregate in one place close to the nuclear membrane.

Sympathetic chain series of connected ganglia, on each side of the vertebral column, which can fuse into a single chain posteriorly in some teleosts.

Sympathetic ganglia neural swellings on the sympathetic chain, where synapses occur between preganglionic and postganglionic neurons. The ganglia contain the perikarya of the postganglionic neurons and some chro-

maffin cells. Uninterrupted 'preganglionic' axons also pass through these ganglia.

Sympathetic nervous system subdivision of the autonomic nervous system in mammals; nowadays termed spinal autonomic nervous system in fish.

Synapse physiological junction between neurons, transmission of signals across the synapse can be chemical (neurotransmitter) or electrical (electrotonic).

Systole the contraction of a heart or individual heart chambers.

T3 triiodothyronine.

T4 tetraiodothyronine.

Tachykinin substances (which may be paracrine) which produce vasoconstriction, tending to raise blood pressure.

Tapetum a reflecting layer in the choroid.

Taurine 2-aminoethylsulfonic acid.

Taxon (pl. taxa) any unit of taxonomy including genus, family, phylum.

Taxonomy the formal study of biological classification.

Tegmentum base of the mid-brain, has a role in motor functions.

Teleocalcin a hypocalcaemic factor found in corpuscles of Stannius.

Telencephalon cerebral hemispheres.

Teleosts the term used taxonomically and generally to refer to most bony fish, and therefore currently the most speciose and abundant fish group.

Telomere a region at the end of each chromosome.

Tentaculum one of three structures (pl. tentacula) on male chimaeras, the anterior cephalic and two pelvic, believed to be important for holding the female in position during copulation. The anterior tentaculum is also termed a clasper, in line with terminology for the pelvic claspers of male chondrichyans.

Terminal nerve cranial nerve 0: associated with the olfactory tract but function is an issue for conjecture.

Tessarae plate-like, close-fitting units of cartilage.

Tetraiodothyronin (e) thyroxine.

Tetraodotoxin a neurotoxin found in organs of puffer fish.

Tetrapods terrestrial (usually) vertebrates, with four limbs, or descended from four-limbed forms; that is, Amphibian, Reptiles, Birds, Mammals.

Thalamus middle region of the diencephalon, relays sensory signals.

Theca the outer layer of the follicle surrounding an oocyte; it is a single layer of flat cells, expected to be steroidogenic, capable of androgen (testosterone) synthesis.

Thermoreceptors temperature receptors.

Thrombocyte a cell actively important in coagulation.

Thrombopoiesis thrombocyte formation.

Thymus a gland involved in white cell production in vertebrates particularly young mammals.

Thyroid the endocrine tissue found, as follicles close to the bifurcation of the branchial arteries in elasmobranchs and more diffuse in teleosts. Its principle secretion is thyroxin, which increases metabolism.

Thyrotropin the pituitary hormone that controls the release of thyroxin.

Thyroxin tetraiodothyronine, the major thyroid hormone.

TL total length.

TMAO trimethylamine oxide.

Tonic receptors Sensory receptors manifesting no decline, or a slow one, in impulse frequency when a stimulus is maintained.

Torr a measure of pressure, which can also be expressed as 'mm of mercury'.

Total length the length of a fish from its most anterior to its most posterior point, which may be off-centre; that is, the top or bottom lobe of the caudal fin.

Tragedy of the commons the view that lack of specific ownership leads to neglect.

Transduction of energy transformation of one form of energy into another form. Such changes represent the basis for stimulus perception by sensory receptors.

Trigeminal nerve cranial nerve V; paired mixed nerves each including multiple sensory pathways and controlling jaw movements.

Trochlear nerve cranial nerve IV; paired short motor nerves, efferent, each from brain to eyeball (superior oblique muscle).

Troglobitic (adj.) cave dwelling.

Trophic ecology the study of food chains and webs.

Trophic level refers to the position of an organism on the 'food chain'.

Trypsin a protease associated with the intestine.

Trypsinogen the precursor of trypsin.

TSH thyroid-stimulating hormone, synonym for thyotropin (q.v.).

Tuberous electroreceptors electroreceptors lacking an open connection to the exterior water, are sensitive to relatively high-frequency electrical activity.

Tunica elastica a region of a blood vessel wall, rich in elastic fibres.

Tunica externa the outermost layer of a blood vessel wall.

Tunica intima the innermost layer of a blood vessel.

Tunica media the middle layer of a blood vessel wall.

Ultimobranchial glands or bodies small anterior concentrations of endocrine tissue secreting hypocalcaemic factor(s).

Ultrafiltration the process in Bowman's capsule which retains large molecules in the blood but allows formation of urine from small molecules and ions.

Urea the small nitrogenous molecule which enables nitrogenous excretion in terrestrial vertebrates and some fish, where it may also act to boost osmotic pressure.

Ureogenic producing urea.

Ureosmotic an animal using urea to boost osmotic pressure.

Ureotelic an animal excreting urea as a major excretory product.

Uric acid a minor excretory component in fish, the major nitrogenous excretion of birds.

Urohypophysis/urophysis is found in the caudal region of the spinal cord of actinopterygians. Named in relation to the anterior hypophysis (the pituitary). Also known as the caudal neurosecretory system.

Urotensin a product of the urohypophysis.

Utriculus the most anterior of the three otolith organs, sensitive to sound and gravity, in the fish ear.

Vagus nerve cranial nerve X, paired long mixed nerve with motor function controlling branchial movements and a sensory branch to the lateral line on each side.

Vas deferens (pl. vasa deferentia) the main male duct, conducting sperm from each testis to the genital pore.

Vas efferens (pl. vasa efferentia) a tributary duct from the testis to the vas deferens, several are seen on each side in elasmobranchs.

Vasoconstriction a change in blood vessel dimensions, giving a smaller diameter and thus raising blood pressure, achieved by smooth muscle contraction.

Vasodilat(at)ion a change in blood vessel dimensions, giving a wider diameter and lowering blood pressure, achieved by smooth muscle relaxation.

Vasomotor the constriction or dilation of blood vessels via the action of smooth muscles in the walls.

Vasopressive factors affect the blood pressure, for example through altering the diameter of blood vessels (see vasoconstriction/vasodilation.

Vein a thin-walled blood vessel, returning blood to the heart.

Venous return the mechanism(s) enabling blood to move from the periphery to the heart.

Ventral horns bilateral projections of the grey matter into the white matter of the ventral spinal cord.

Ventral root base of a spinal or cranial nerve carrying motor (efferent) fibres. Emerging from the spinal cord, this 'root' unites with the dorsal root to form the spinal nerve.

Ventricle 1. the major pumping chamber of the heart. 2. interconnected cavity in system containing cerebro-spinal fluid within brain neural tissue, extends into the spinal cord central canal. The ventricles may be numbered using Latin numerals; for example, ventricle III, IV.

VIP vasoactive intestinal peptide.

Visceral nervous system autonomic nervous system.

Visceral skeleton an alternate term for the branchial skeleton.

Visual pigment a combination of opsin protein and a vitamin A derivative.

Vitellogenin the yolk precursor, a large and complex molecule synthesized in the liver, released to the blood, taken up by the oocytes.

Vitellogenesis either a) the synthesis of vitellogenin in the liver or b) the synthesis of yolk in oocytes following the uptake and cleavage of vitellogenin.

Vitreous humor gelatinous medium between the lens and retina of the eye.

Vivipary retention of fertilized eggs and development of the young dependent on a placenta.

Viviparous (adj.) briefly 'live-bearing', but see also 'vivipary', above.

Weberian ossicles parts of modified vertebrae connecting the swim bladder to the ear on each side, in Ostariophysans, for example goldfish, increasing sound discrimination.

White heart hearts with small amounts of myoglobin are pale, and termed 'white'.

White matter an area of the nervous system containing a high concentration of nerve fibres, particularly myelinated fibres producing the white effect.

White muscle the major portion of fish striated muscle, contains little myoglobin (hence 'white'), anaerobic and responsible for sudden (e.g. 'escape') movements such as a tail-flip.

White pulp part of the spleen, divided into subareas; white because of the prevalence of lymphocytes.

White ramus communicans short neural connection enabling pre-ganglionic myelinated autonomic fibres to enter a sympathetic ganglion from a spinal nerve.

Whitewater fresh water with high levels of suspended particles, high pH, high calcium and magnesium.

Winterkill mass deaths due to, for example, hypoxia caused by ice cover.

Woven bone bone with collagen fibres appearing to be interlaced; that is, with a more random structure than lamellar bone (q.v.).

Xanthophore the chromatophore type which is yellow or orange due to carotenoid inclusions (xanthosomes).

Xenobiotics 'foreign substances'; that is, man-made chemicals which can have biological effects.

Yolk nutrient material, deposited as discrete inclusions (globules/vesicles/droplets) or forming a coalesced mass, in each developing oocyte in its secondary growth phase.

Yolk nucleus see Balbiani body.

Yolk sac as the embryo develops the yolk forms a separate structure surrounded by a membrane, with blood vessels passing to the embryo. The contents of the yolk sac diminish with time but it may persist for some days post-hatching.

Zona radiata a term used for the complex membrane surrounding a vitellogenic oocyte; in its fully developed form it has strong radial markings and a close relationship with the granulosa cells of the surrounding follicle.

Zymogens enzyme precursor molecules; for example, pepsinogen.

Zygotene a stage of meiotic prophase I.

References

Aarts, B.G.W. and Nienhuis, P.H. (2003). Fish zonations and guilds as the basis for assessment of ecological integrity of large rivers. *Hydrobiologia*, 500, 157–78.

Abe, H. and Oka, Y. (2007). Neuromodulatory functions of terminal nerve Gn RH neurons. In: T.J. Hara and D.J. Zielinski (eds). *Fish Physiology, Sensory Systems Neuroscience, Vol. 25*. New York, NY: Academic Press, Elsevier Inc., pp. 455–503.

Abe, N. and Maehara, T. (2013). Molecular characterization of kudoid parasites (Myxozoa: Multivalvulida) from somatic muscles of Pacific bluefin (*Thunnus orientalis*) and yellowfin (*T. albacores*) tuna. *Acta Parasitologica*, 58, 226–30.

Abel, D.C., Koenig, C.C., and Davis, W.P. (1987). Emersion in the mangrove forest fish *Rivulus marmoratus*: a unique response to hydrogen sulfide. *Environmental Biology of Fishes*, 18, 67–72.

Abercrombie, M., Hickman, C.J., and Johnson, M.L. (1961, 1980). *A Dictionary of Biology*, 3rd revision of the 1st edn, 1951. Chicago, IL: Aldine Publishing Co.

Abrahamsson, T., Holmgren, S., Nilsson, S., and Petersson, K. (1979). Adrenergic and cholinergic effects on the heart, the lung and the spleen of the African lungfish, *Protopterus aethiopicus*. *Acta Physiologica Scandinavica*, 107, 141–7.

Adrian, E.D. and Buytendijk, F.J. (1931). Potential changes in the isolated brain stem of the goldfish. *Journal of Physiology, London*, 71, 121–35.

Agardy, T., Bridgewater, P., Crosby, M.P., Day, J. Dayton, P.K., Kenchington, R., Laffoley, D., McConney, P., Murray, P.A., Parks, J.E., and Peau, L. (2003). Dangerous targets? Unresolved issues and ideological clashes around marine protected areas. *Aquatic Conservation: Marine and Freshwater Ecosystems*, 13, 353–67.

Agius, C. and Roberts, R.J. (2003). Melano-macrophage centres and their role in fish pathology. *Journal of Fish Diseases*, 26, 499–509.

Aguzzi, J., Doya, C., Tecchio, S., De Leo, F.C., Azzurro, E., Costa, C., Sbragaglia, V., Del Rio, J., Navarro, J., Ruhl, H.A. Company, Favali, P., Purser, A., Thomsen, L., and Catalán, I.A. (2015). Coastal observations for monitoring of fish behaviour and their responses to environmental changes. *Reviews in Fish Biology and Fisheries*, 25, 463–83.

Akazome, Y., Kanda, S., Okubo, K., and Oka, Y. (2010). Functional and evolutionary insights into vertebrate kisspeptin systems from studies of fish brain. *Journal of Fish Biology*, 76, 161–82.

Albert, J.S., Zakon, H.H., Stoddard, P.K., Unguez, G.A., Holmberg-Albert, S.K.S., and Sussma, M.R. (2008). The case for sequencing the genome of the electric eel *Electrophorus electricus*. *Journal of Fish Biology*, 72, 331–54.

Aldman, G., Jönsson, A.C., Jensen, J., and Holmgren, S. (1989). Gastrin/CCK-like peptides in the spiny dogfish, *Squalus acanthias*; concentrations and actions in the gut. *Comparative Biochemistry and Physiology C*, 92, 103–8.

Aldridge, R.J. and Donoghue, P.C.J. (1998). Conodonts: a sister group to hagfishes? In: J.M. Jørgensen, J.P. Lomholt, R.E. Weber, and H. Malte (eds). *The Biology of Hagfishes*. London: Chapman and Hall, 15–31.

Alexander, D.R., Kerekes, J.J., and Sabean, B.C. (1986). Description of selected lake characteristics and occurrence of fish species in 781 Nova Scotia lakes. *Proceedings of the Nova Scotian Institute of Science*, 36, 63–106.

Alexander, R.M. (1970). *Functional Design in Fishes*, London: Hutchison University Library.

Alexander, R.M. (1972). The energetics of vertical migration by fishes. In: C. Sleigh, M.A. and A.G. MacDonald (eds). The effect of pressure on organisms. *Symposium of the Society for Experimental Biology*, 26, Cambridge: Cambridge University Press, 273–94.

Ali, M.A. (1964). Retinomotor responses of the goldfish (*Carassius auratus*) to unilateral photic stimulation. *Revue Canadienne de Biologie*, 23, 45–53.

Aliç, T.Z., Oray, I.K., Karakulak, F.S., and Kahraman, A.E. (2012). Age, sex ratio, length-weight relationships and reproductive biology of Mediterranean swordfish, *Xiphias gladius* L. 1758, in the eastern Mediterranean. *African Journal of Biotechnology*, 11, 3673–880.

Allen, G.R., Midgley, S.H., and Allen, M. (2002). *Field Guide to the Freshwater Fishes of Australia*. Perth: Western Australian Museum.

Allendorf, F.W. (1991). Ecological and genetic effects of fish introductions: synthesis and recommendations. *Canadian Journal of Fisheries and Aquatic Science*, 48 (Supplement 1), 178–81.

Allison, W.T. Haimberger, J., Hawryshyn, C.W., and Temple, S.E. (2004). Visual pigment shifting in zebrafish: Evidence for a rhodopsin–porphyropsin interchange system. *Visual Neuroscience*, 21, 945–52.

Almeida-Val, V.M.F. (1999). Phenotypic plasticity in Amazon fishes: water versus air-breathing. In: A.K. Mittal, F.B. Eddy, and J.S. Datta Munshi (eds). *Water/Air Transition in Biology*. Enfield, NH: Science Publishers Inc.

Altman, P.L. (1961). *Blood and Other Body Fluids*. Washington, DC: Federation of American Societies for Experimental Biology.

Alvarez-Pellitero, P. (2008). Fish immunity and parasite infections: from innate immunity to immunoprophylactic prospects. *Veterinary Immunology and Immunopathology*, 126, 171–98.

Alves-Gomes, J.A. (2001).The evolution of electroreception and bioelectrogenesis in teleost fish: a phylogenetic perspective. *Journal of Fish Biology*, 58, 1489–511.

Amelio, D., Garofalo, F., Wong, W.P., Chew, Y.K., Ip, Y.K., Cerra, M.C., and Tota, B. (2013). Nitric oxide synthase-dependent 'On/Off' switch and apoptosis in freshwater and aestivating lungfish, *Protopterus annectens*: Skeletal muscle versus cardiac muscle. *Nitric Oxide*, 32, 1–12.

Amorim, M.C.P. (2006). Diversity of sound production in fish. In: F. Ladich, S.P. Collin, P. Moller, B.G. Kapoor, and N.H. Enfield (eds). *Fish Communication, Vol. 1*. Enfield, NH: Science Publishers Inc.

Amoser, S. and Ladich, F. (2005). Are hearing sensitivities of freshwater fish adapted to the ambient noise in their habitats? *Journal of Experimental Biology*, 208, 3533–42.

Andersen, M.E. and Deacon, J.E. (2001). Population size of Devils Hole pupfish (*Cyprinodon diabolis*) correlates with water level. *Copeia*, 2001, 224–8.

Anderson, B.G. and Mitchum, D.L. (1974). *Atlas of Trout Histology*. Wyoming, CO: Wyoming Game and Fish Department.

Anderson, W.G. (2015). Endocrine systems in elasmobranchs. In: R.E. Shadwick, A.P. Farrell, and C.J. Brauner (eds). *Physiology of Elasmobranch Fishes: Internal Processes*. Amsterdam: Elsevier, pp. 457–530.

Anderson, W.G., Takei, Y., and Hazon, N. (2001). The diposgenic effect of the rennin-angiotensin system in elasmobranch fish. *General and Comparative Endocrinology*, 124, 300–7.

Andrew, W. (1959). *Textbook of Comparative Histology*. Oxford: Oxford University Press.

Angelone, T., Gattuso, A., Imbrogno, S., Mazza, R., and Tota, B. (2012). Nitrite is a positive modulator of the Frank–Starling response in the vertebrate heart. *American Journal of Physiology, Regulatory, Integrative and Comparative Physiology*, 302, R1271–R1281.

Anon. (1984). Atlantic salmon scale reading, *Report of the Atlantic salmon scale reading workshop*, Aberdeen, Scotland, 23–28 April 1984. Copenhagen: International Council for the Exploration of the Sea.

Anon. (2012). Gar Family (Lepisosteidae). *Indiana Division Fish and Wildlife Information Series*. 2012-MLC. Indianapolis, IN: Indiana Government.

Araque, A. (1999). Tripartite synapses: glia, the unacknowledged partner. *Trends in Neurosciences*, 22, 208–15.

Ardelli, B.F. and Woo, P.T.K. (2006). Immunocompetent cells and their mediators in fin fish. In: P.T.K. Woo (ed.). *Fish Diseases and Disorders, Vol. 1*. Protozoans and metazoan infections. Cambridge, MA, CAB International, pp. 702–24.

Arey, L.B. (1916). The movements in the visual cells and retinal pigment of the lower vertebrates. *Journal of Comparative Neurology*, 26, 121–201.

Arinç, E., Zinck, M.E., and Willis, D.E. (2000). Cytochrome P4501A and associated mixed function oxidase in fish as a biomarker for toxic carcinogenic pollutants in the aquatic environment. *Pure and Applied Chemistry*, 72, 985–94.

Arocha, F. (2002). Oocyte development and maturity classification of swordfish from the north-western Atlantic. *Journal of Fish Biology*, 60, 13–27.

Aronson, L.R. (1963). The central nervous system of sharks and bony fishes. With special reference to sensory and integrative mechanisms. In: P.W. Gilbert (ed.). *Sharks and Survival*. Boston, MA: Heath, pp. 165–241.

Artedi, P. (1738). *Ichthyologia sive Opera omnia de Piscibus*. In: C. Linnaeus (ed.). *Five Parts in One*. Leiden (Lugdunum Batavorum): Linnaeus.

Ashley, L.M. (1969). *Laboratory Anatomy of the Shark*. Dubuque, IO: Wm.C. Brown Company Publishers.

Atz, J.W. (1969). *Dean Bibliography of Fishes*. New York, NY: The American Museum of Natural History.

Atz, J.W. (1972). Review on Hardisty and Potter 1971. *Science*, 176 (4042), 1409.

Au, W.W.L., Benoit-Bird, K.J., and Kastelerlein, R.A. (2007). Modelling the detection range of fish by echolocating bottlenose dolphin and harbour porpoise. *Journal of the Acoustical Society of America*, 121, 3954–62.

Avallone, B., Fascio, U., Senatore, A., Balsamo, G., Bianco, P.G., and Marmo, F. (2005). The membraneous labyrinth during larval development in lamprey (*Lampetra planeri*, Bloch, 1784). *Hearing Research*, 201, 37–43.

Axelsson, M., Farrell, A.P., and Nilsson, S. (1990). Effects of hypoxia and drugs on the cardiovascular dynamics of the Atlantic hagfish, *Myxine glutinosa*. *Journal of Experimental Biology*, 151, 297–316.

Azpeleta, C., Martinez-Alvarez, R.M., Delgado, M.J., Isoma, E., and De Pedro, N. (2010). Melatonin reduces locomotor activity and circulating cortisol in goldfish. *Hormones and Behaviour*, 57, 323–9.

Bagenal, T.B. (1957). The breeding and fecundity of the long rough dab *Hippoglossoides platessoides* (Fabr.) and the associated cycle in condition. *Journal of the Marine Biological Association of the United Kingdom*, 36, 339–75.

Bagenal, T.B. (1965). The fecundity of long rough dab in the Clyde sea area. *Journal of the Marine Biological Association of the United Kingdom*. 45, 599–606.

Bagnara, J.T., Fernandez, P.J., and Fujii, R. (2007). On the blue colouration of vertebrates. *Pigment Cell Research*, 20, 14–26.

Bagshaw, R.J. (1985). Evolution of cardiovascular baroreceptor control. *Biological Reviews*, 60, 121–62.

Bai, J.P.F. (1994). Distribution of brush-border membrane peptidases along the rat intestine. *Pharmaceutical Research*, 11, 897–900.

Bain, M.B.H., Finn, J.T., and Booke, H.E. (1988). Stream flow regulation and fish community structure. *Ecology*, 69, 382–92.

Baird, H.B., Krueger, C.C., and Josephson, D.C. (2002). Differences in incubation period and survival of embryos among brook trout strains. *North American Journal of Aquaculture*, 64, 233–41.

Baird, R.C., Jumper, G.Y., and Gallaher, E.E. (1990). Sexual dimorphism and demography in two species of oceanic midwater fish (Stomiformes: Sternoptychidae) from the eastern Gulf of Mexico. *Bulletin of Marine Science*, 47, 561–6.

Baker, B.I. (1994). Melanin-concentrating hormone updated functional considerations. *Trends in Endocrinology and Metabolism*, 5, 120–6.

Balamurugan, J., Kumar, T.T.A., Kannan, R., and Pradeep, H.D. (2014). Acclimation behaviour and bio-chemical changes during anemonefish (*Amphiprion sebae*) and sea anemone (*Stichodactylus haddoni*) symbiosis. *Symbiosis*, 64, 127–38.

Baldwin, C.C. (2013). The phylogenetic significance of colour patterns in marine teleost larvae. *Zoological Journal of the Linnean Society*. 168, 496–563.

Balfour, F.M. (1878). *A Monograph on the Development of Elasmobranch Fishes*. London: Macmillan.

Ball, J.N. (1965). Partial hypophysectomy in the teleost *Poecilia*: separate identities of teleostean growth hormone and teleostean prolactin-like hormone. *General and Comparative Endocrinology*, 5, 654–61.

Ball, J.N. and Baker, B.I. (1969). The pituitary gland: anatomy and histophysiology. In: W.S. Hoar and D.J. Randall (eds). *Fish Physiology, Vol. II. The Endocrine System*. New York, NY: Academic Press, pp. 1–11.

Ballantyne, J.S. (1997). Jaws: The inside story. The metabolism of elasmobranch fishes. *Comparative and Biochemical Physiology*, 118B, 703–42.

Ballantyne, J.S. (2016). Metabolism of elasmobranchs (Jaws II). In: R.E. Shadwick, A.P. Farrell, and C.J. Brauner (eds). Fish Physiology 34, B. *Physiology of Elasmobranch Fishes: Internal Processes*. Amsterdam: Elsevier, pp. 395–456.

Balment, R.J., Lu, W., Weybourne, E., and Warne, J.M. (2006). Arginine vasotocin a key hormone is fish physiology and behavior: a review with insights from mammalian models. *General and Comparative Endocrinology*, 147, 9–16.

Balon, E.K. (1984). Patterns in the evolution of reproductive style in fishes. In: G.W. Potts and R.J. Wootton (eds). *Fish Reproduction: Strategies and Tactics*. London: Academic Press, 35–53.

Balon, E.K. (1991). Probable evolution of the coelacanth's reproductive style: lecithotrophy and orally feeding embryos in cichlid fishes and in *Latimeria chalumnae*. *Environmental Biology of Fishes*, 32, 249–65.

Balon, E.K., Bruton, M.N., and Fricke, H. (1988). A fiftieth anniversary reflection on the living coelacanth, *Latimeria chalumnae*: some new interpretations of its natural history and conservation status. *Environmental Biology of Fishes*, 23, 241–80.

Balshine-Earn, S. and Earn, D.J.D. (1998). On the evolutionary pathway of parental care in mouth-brooding cichlid fish. *Proceedings of the Royal Society of London B*, 265, 2217–22.

Barber, V.C. and Emerson, C.J. (1980). Scanning electron microscope observations on the inner ear of the skate, *Raja ocellata*. *Cell and Tissue Research*, 205, 195–215.

Barber, V.C., Yake, K.I., Clark, V., and Pungur, J. (1985). Quantitative analyses of sex and size differences in the macula neglecta and ramus neglectus in the inner ear of the skate, *Raja ocellata*. *Cell and Tissue Research*, 241, 597–605.

Bardack, D. (1998). Relationships of living and fossil hagfishes. In: J.M. Jørgensen, J.P. Lomholt, R.E. Weber, and H. Malte (eds). *The Biology of Hagfishes*. London: Chapman and Hall, pp. 3–14.

Baroiller, J.F. and D'Cotta, H. (2001). Environment and sex determination in farmed fish. *Comparative Biochemistry and Physiology C*, 130, 399–409.

Barr, W.A. (1963). The endocrine control of the sexual cycle in the plaice (*Pleuronectes platessa* L.) I. Cyclical changes in the normal ovary. *General and Comparative Endocrinology*, 3, 197–204.

Barrington, E.J.W. (1957). The alimentary canal and digestion. In: M.E. Brown (ed.). *The Physiology of Fishes, Vol. 1*. New York, NY: Academic Press, pp. 109–62.

Barrington, E.J.W. and Matty, A.J. (1954). Seasonal variation in the thyroid gland of the minnow *Phoxinus* L., with some observations on the effect of temperature. *Proceedings of the Zoological Society of London*, 124, 89–95.

Barrington, E.J.W. and Sage, M. (1963). On the responses of iodine-binding regions of the endostyle of the larval lamprey to goitrogens and thyroxine. *General and Comparative Endocrinology*, 3, 669–79.

Barrington, E.J.W. and Sage, M. (1966). On the response of the endostyle of the hypophysectomised larval lamprey to thiourea. *General and Comparative Endocrinology*, 7, 463–74.

Bartelmez, G.W. (1915). Mauthner's cell and the nucleus notorious tegmenti. *Journal of Comparative Neurology*, 25, 87–128.

Bartheld, C.S. von and Meyer, D.L. (1987). Comparative neurology of the optic tectum in ray-finned fishes: patterns of lamination formed by retinotectal projections. *Brain Research*, 420, 277–88.

Barton, B.A. (2002). Stress in fishes: A diversity of responses with particular reference to changes in circulating corticosteroids. *Integrative and Comparative Biology*, 42, 517–25.

Baruscotti, M., Bucchi, A., and DiFrancesco, D. (2005). Physiology and pharmacology of the cardiac ('funny') current. *Pharmacology and Therapeutics*, 107, 59–79.

Bass, H.W. (2003). Telomere dynamics unique to meiotic prophase: formation and significance of the bouquet. *Cellular and Molecular Life Sciences*, 60, 2319–24.

Bass, A.H. and Lu, Z. (2007). Neural and behavioural mechanisms of audition. In: T.H. Hara and B. Zielinski (eds). *Fish Physiology Vol. 25. Sensory Systems Neuroscience*. New York, NY: Academic Press/Elsevier, pp. 377–410.

Bateson, W. (1889). Contractility of the iris in fishes and in cephalopods. *Journal of the Marine Biological Association of the United Kingdom*, 1, 215–16.

Bateson, W. (1890). The sense organs and perceptions of fishes: with remarks on the supply of bait. *Journal of the Marine Biological Association of the United Kingdom*, 1, 228–56.

Battle, H. (1944). The embryology of the Atlantic salmon (*Salmo salar* Linnaeus). *Canadian Journal of Research*, 22, 105–25.

Bauer, M.P., Bridgham, J.T., Langenau, A.L. Johnson, A.L., and Goetz, F.W. (2000). Conservation of steroidogenic acute regulatory (StAR) protein structure and expression in vertebrates. *Molecular and Cellular Endocrinology*, 168, 119–25.

Bauer, R. (1979). Electric organ discharge (EOD) and prey capture behaviour in the electric eel, *Electrophorus electricus*. *Behavioral Ecology and Sociobiology*, 4, 311–19.

Bayliss, W.M. and Starling, E.H. (1902). The mechanism of pancreatic secretion. *Journal of Physiology*, 28, 125–38.

Beamish, F.W.H., Dougan, J.L., Thomson, C.E., Healey, P.J., and Dunn, L.R. (1989). *The Cyclostomata—An Annotated Bibliography.* Supplement 1984–1988. Chicago, IL: Great Lakes Fishery Commission.

Beaudet, L., Flamarique, L., and Hawryshyn, C.W. (1997). Cone receptor topography in the retina of sexually mature Pacific salmonid fishes. *Journal of Comparative Neurology*, 383, 49–59.

Becker, W.M., Kleinsmith, L.J., and Hardin, J. (2000). *The World of the Cell*, 4th edn. San Francisco, CA: Benjamin/Cummings Publishing Company.

Bedford, J.J., Harper, J.L., Leader, J.P., Yancey, P.H., and Smith, R.A.J. (1998). Betaine is the principal counteracting osmolyte in tissues of the elephant fish, *Callorhincus millii* (Elasmobranchii, Holocephali). *Comparative Biochemistry and Physiology B*, 119, 521–6.

Beeman, J.W., Vendetti, D.A., Morris, R.G., Gadomski, D.M., Adams, B.J., Vanderkooi, S.P., Robinson, T.C., and Maule, A.G. (2003). Gas bubble disease in resident fish below Grand Coulee Dam. *Report Western Fisheries Research Center.* Cook, WA:Columbia River Research Laboratory.

Beer, T. (1894). Die Akkommodation des Fischauges. *Pflügers Archivfuer die Gessammt Physiologie des Menschen und der Tiere*, 58, 523–650.

Bejda, A.J. and Phelan, B.A. (1998). Can scales be used to sex winter flounder? *Biological Bulletin*, 96, 621–3.

Bell, C.C. (1981). Central distribution of octavolateral afferents and efferents in a teleost (Mormyridae). *Journal of Comparative Neurology*, 195, 391–414.

Bell, C.C. (1990). Mormryomast electroreceptor organs and their afferent fibres in mormyrid fish. III Physiological differences between two morphological types of fibres. *Journal of Neurophysiology*, 63, 319–32.

Bell, G.R. (1964). A guide to the properties, characteristics, and uses of some general anaesthetics for fish. *Fisheries Research Board of Canada Bulletin*, No. 148. Ottawa: Fisheries Research Board of Canada.

Bell, G.H., Davidson, J.N., and Scarborough, H. (1957). *Textbook of Physiology and Biochemistry*. London: E. and S. Livingstone Ltd.

Bell, J.D., Lyle, J.M., Bulman, C.M., Graham, K.J., Newton, G.M., and Smith, D.C. (1992). Spatial variation in reproduction, and occurrence of non-reproductive adults, in orange roughy, *Hoplostethus atlanticus* Collett (Trachichthyidae), from south-eastern Australia. *Journal of Fish Biology*, 40, 107–22.

Bendell-Young, L.I., Bennett, K.E., Crowe, A., Kennedy, C.J., Kermode, A.R., Moore, M.M., Plant, A.L., and Wood, A. (2000). Ecological characteristics of wetlands receiving industrial effluent. *Ecological Applications*, 10, 310–22.

Benedet, S., Björnsson, B.T., Taranger, G.L., and Andersson, E. (2008). Cloning of somatolactin alpha, beta forms and the expression profile in pituitary and ovary of maturing female broodstock. *Reproductive Biology and Endocrinology*, 6, 42. doi: 10. 1186/140 77-7827-6-42.

Benedet, S., Johansson, V., Sweeney, G., Galay-Burgos, M., and Björnsson, B.T. (2005). Cloning of two Atlantic salmon growth hormone receptor isoforms and *in vitro* ligand-binding response. *Fish Physiology and Biochemistry*, 31, 315–29.

Bennett, M.G. and Conway, K.W. (2010). An overview of North America's diminutive freshwater fish fauna. *Ichthyological Exploration of Freshwaters*, 21, 63–72.

Bennett, M.V.L. (1967). Mechanisms of electroreception. In: P. Cahn (ed.). *Lateral Line Detectors*. Bloomington, IN: Indiana University Press, pp. 313–93.

Bennett, M.V.L. (1971a). Electroreception. In: W.S. Hoar and D.J. Randall (eds). *Fish Physiology, Vol. V Sensory Systems and Electric Organs*. New York, NY: Academic Press, pp. 493–574.

Bennett, M.V.L. (1971b). Electric organs. In: W.S. Hoar and D.J. Randall (eds). *Fish Physiology, Vol. V. Sensory Systems and Electric Organs*. New York, NY: Academic Press, pp. 347–491.

Berenbrink, M., Koldkjær, P., Kepp, O., and Cossins, A.R. (2005). Evolution of oxygen secretion in fishes and the emergence of a complex physiological system. *Science*, 307, 1752–7.

Bergmann, O., Bhardwaj, R.D., Bernard, S., Zdunek, S., Barnabé-Heider, F., Walsh, S., Zupicich, J., Alkass, K., Buchholz, B.A. Druid, H., Jovinge, S., and Frisén, J. (2009). Evidence for cardiomyocyte renewal in humans. *Science*, 324, 98–102.

Berkowitz, E.G. (1956). Functional properties of the spinal pathways in the carp, *Cyprinus carpio* L. *Journal of Comparative Neurology*, 106, 269–89.

Bern, H.A. and Takasugi, N. (1962). The caudal neurosecretory system of fishes. *General and Comparative Endocrinology*, 2, 96–110.

Bernstein, J.J. (1970). Anatomy and physiology of the central nervous system. In: W.S. Hoar and D.J. Randall (eds). *Fish Physiology, Vol. IV. The Nervous System, Circulation, and Respiration*. New York, NY: Academic Press, pp. 1–90.

Bernstein, J.J. and Streicher, E. (1965). The blood–brain barrier of fish. *Experimental Neurology*, 11, 464–73.

Berquist, R.M., Galinsky, V.L. Kajiura, S.M., and Frank, L.R. (2015). The coelacanth rostral organ is a unique low-resolution electro-detector that facilitates the feeding strike. *Scientific Reports*, 5, 8962. doi: 10.1038/srep08962

Berra, T.M. (2007). *Freshwater Fish Distribution*. Chicago, IL: University of Chicago Press.

Berra, T.M., Sever, D.M., and Allen, G.R. (1989). Gross and histological morphology of the swimbladder and lack of accessory respiratory structures in *Lepidogalaxias salamandroides*, an aestivating fish from Western Australia. *Copeia*, 1989, 850–6.

Berwein, M. (1941). Beobachtungen und Versuche über des gesellige Leben von Elritzen. *Zeitschrift fuer vergleichende Physiologie*, 28, 402–20.

Besseau, L., Benyassi, A., Møller, M., Coon, S.L., Weller, J.L., Boeuf, G., Klein, D.C., and Falcón, J. (2006). Melatonin pathway: breaking the 'high-at-night' rule in trout retina. *Experimental Eye Research*, 82, 620–7.

Beukema, J.J. (1970). Angling experiments with carp: decreased catchability through one trial learning. *Netherlands Journal of Zoology*, 20, 81–92.

Beverton, R.J.H. and Holt, S.J. (1957). *On the Dynamics of Exploited Fish Populations*. London: Chapman and Hall.

Beyenbach, K.W. (2004). Kidneys sans glomeruli. *American Journal of Physiology*, 212, 2856–63.

Bhikajee, M. and Green, J.M. (2002). Behaviour and habitat of the amphibious blenny, *Alticus monochrus*. *African Zoology*, 37, 221–230.

Billard, R. Solari, A., and Escaffre, A-M. (1974). Mêthode d'analyse quantitative de la spermatogenèse des poissons téléostéens. *Annales de biologie animale, biochimie, biophysique*, 14, 87–104.

Bjenning, C. and Holmgren, S. (1988). Neuropeptides in the fish gut. An immunohistochemical study of evolutionary patterns. *Histochemistry*, 88, 155–63.

Black, D. and Love, R.M. (1986). The sequential mobilisation and restoration of energy reserves in tissues of Atlantic cod during starvation and refeeding. *Journal of Comparative Physiology B*, 156, 469–79.

Blackmore, C.G., Varro, A., Dimaline, R., Bishop, L., Gallacher, D.V., and Dockray, G.J. (2001). Measurement of secretory vesicle pH reveals intravesicular alkinisation by vesicular monoamine transporter type 2 resulting in inhibition of prohormone cleavage. *Journal of Physiology*, 531, 605–17.

Blackstock, N. and Pickering, A.D. (1980). Acidophilic granular cells in the epidermis of the brown trout, *Salmo trutta* L. *Cell and Tissue Research*, 210, 359–69.

Blake, R.W. (1983). *Fish Locomotion*. Cambridge: Cambridge University Press.

Blake, R.W. (2004). Fish functional design and swimming performance. *Journal of Fish Biology*, 65, 1193–222.

Blake, R.W. and Chan, K.H.S. (2007). Swimming in the upside down catfish *Synodontis nigriventris*: it matters which way is up. *Journal of Experimental Biology*, 210, 2979–89.

Blakemore, R.P. (1975). Magnetotactic bacteria. *Science*, 140, 377–9.

Blaxter, J.H.S., Denton, E.J., and Gray, J.A.B. (1981). Acoustico-lateralis systems in clupeid fishes. In: W.N. Tavolga, A.N. Popper, and R.R. Fay (eds) *Hearing and Sound Communication in Fishes*. New York, NY: Springer Verlag, pp. 39–59.

Blaxter, J.H.S., Wardle, C.S., and Roberts, B.L. (1971). Aspects of the circulatory physiology and muscle systems of deep-sea fish. *Journal of the Marine Biological Association of the United Kingdom*, 51, 991–1006.

Bleckmann, H. (2007). The lateral line system of fish. In: T.H. Hara and B. Zielinski (eds). *Fish Physiology Vol 25. Sensory Systems Neuoscience*. New York, NY: Academic Press/Elsevier, pp. 411–53.

Bliss, T.V.P. and Collingridge, G.L.G. (1993). A synaptic model of memory: long-term potentiation in the hippocampus. *Nature*, 361, 31–9.

Bliss, T., Collingridge G., and Morris, R (2004). *LTP: Long-Term Potentiation, Enhancing Neuroscience for 30 Years*. Oxford: Oxford University Press.

Bliss, T.V.P., Collingridge, G.L., and Morris, R.G.M. (2014). Synaptic plasticity in health and disease: introduction and overview. *Philosophical Transactions of the Royal Society B*, 369. doi: 20130129

Bloom, G., Östlund, E., Euler, U.S.V. Lishajko, F., Ritzén, M., and Adams-Ray, J. (1961). Studies on catecholamine-containing granules of specific cells in cyclostomes' hearts. *Acta Physiologica Scandinavica*, 53 (Suppl. 185), 1–34.

Bloom, W. and Fawcett, D.W. (1975). *A Textbook of Histology*. Philadelphia, PA: W.B. Saunders and Co.

Blumer, L.S. (1982). A bibliography and categorization of bony fish exhibiting parental care. *Zoological Journal of the Linnean Society*, 76, 1–22.

Bogevik, A.S., Tocher, D.R., Waagbø, R., and Olsen, R.E. (2008). Triacylyglycerol, wax ester- and sterol ester- hydrolases in midgut of Atlantic salmon (*Salmo salar*). *Aquaculture Nutrition*, 14, 93–8.

Bolotovsky, A.A. and Levin, B.A. (2011). Influence of development rate on pharyngeal teeth formula in *Abramis brama* (L.) bream: experimental data. *Russian Journal of Developmental Biology*, 42, 141–7.

Bond, C.E. (1996). *Biology of Fishes*, 2nd edn. London: Saunders College Publishing.

Bone, Q. (1972). Buoyancy and hydrodynamic functions of integument in the castor oil fish, *Ruvettus pretiosus* (Pisces: Gempyidae). *Copeia*, 78–87.

Bone, Q. (1978). Locomotor muscle. In: W.S. Hoar and D.J. Randall (eds). *Fish Physiology, Vol. VII. Locomotion*. New York, NY: Academic Press, pp. 361–424.

Borlangan, I.E. and Coloso, R.M. (1993). Requirements of juvenile milkfish (*Chanos chanos* Forsskal) for essential amino acids. *Journal of Nutrition*, 123, 125–32.

Boulenger, G.A. (1904). *Fishes (Systematic account of Teleostei)*. In: S.F. Harmer and A.E. Shipley (eds). *The Cambridge Natural History, Vol. VII. Fishes, Ascidians, etc.* London: Macmillan and Co., pp. 541–727.

Boutilier, R.G. (1998). Physiological ecology in cold ocean fisheries: a case study in Atlantic cod. In: H.O. Pörtner and R.C. Playle (eds). *Cold Ocean Physiology*. Society for Experimental Biology Symposium Seminar Series 66. Cambridge: Cambridge University Press.

Bowker, J.D. (2010). The efficacy of terramycin® 200 for fish (oxytretracycline dehydrate) type A medicated article administered in feed to control mortality caused by susceptible pathogens of freshwater-reared finfish. *Protocol OTC-10-EFFU.S. Fish and Wildlife Service Aquatic animal Drug Approval Patrtnership*. Washington, DC.

Bowker, J.D. Trushenski, J.T., Glover, D.C., Carty, D.G., and Wandelear, N. (2015). Sedative options for fish research: a brief review with new data on sedation of warm-, cool-, and coldwater fishes and recommendations for the drug approval process. *Reviews of Fish Biology and Fisheries*, 25, 147–63.

Bowmaker, J.K. and Kuntz, Y.W. (1987). Ultraviolet receptors, tetrachromatic colour vision and retinal mosaics in the brown trout (*Salmo trutta*) age dependent changes. *Vision Research*, 27, 2101.

Bowmaker, J.K. and Wagner, H-J. (2004). Pineal organ of deep-sea fish: photopigments and structure. *Journal of Experimental Biology*, 207, 2379–87.

Bowmaker, J.K., Thorpe, A., and Douglas, R.H. (1991). Ultraviolet-sensitive cones in the goldfish. *Vision Research*, 31, 349–52.

Boyd, C.A.R. (2001). Amine uptake and peptide hormone secretion: APUD cells in a new landscape. *Journal of Physiology*, 531, 581.

Boyle, K.S and Horn, M.H. (2006). Comparison of feeding guild structure and ecomorphology of intertidal fish assemblages from central California and central Chile. *Marine Ecology Progress Series*, 319, 65–84.

Bracewell, P., Cowx, I.G., and Uglow, R.F. (2004). Effects of handling and electrofishing on plasma glucose and whole blood lactate of *Leuciscus cephalus*. *Journal of Fish Biology*, 64, 65–71.

Brainerd, E.L. and Ferry-Graham, L.A. (2006). Mechanics of respiratory pumps. In: R.E. Shadwick and G.V. Lauder (eds). *Fish Physiology, Vol. 23: Fish Biomechanics*. New York, NY: Academic Press, pp. 1–28.

Braithwaite, V. (2006). Cognitive ability in fish. In: K.A. Sloman, R.W. Wilson, and S. Balshine (eds). *Fish Physiology Vol. 24, Behaviour and Physiology of Fish*. Elsevier, Amsterdam, pp. 1–37.

Brand, J.G., Teeter, J.H., Kumazawa, T., Huque, T., and Bayley, D.C. (1991). Transduction mechanisms for the taste of amino acids. *Physiology and Behaviour*, 49, 899–904.

Brander, K., Neuheimer, A., Andersen, K.H., and Hartvig, M. (2013). Overconfidence in model projections.

International Council for the Exploration of the Sea, Journal of Marine Science, 70, 1063–8.

Braum, E. and Junk, W.J. (1982). Morphological adaptation of two Amazonian characoids (Pisces) for surviving in oxygen deficient water. International Review of Hydrobiology, 67, 869–86.

Brauner, C.J., Seidelin, M. Madsen, S.S., and Jensen, F.B. (2000). Effects of freshwater hyperoxia and hypercapnia and their influences on subsequent seawater transfer in Atlantic salmon (Salmo salar) smolts. Canadian Journal of Fisheries and Aquatic Science, 57, 2054–64.

Brauner, C.J., Wang, T., Wang, Y., Richards, J.G., Gonzalez, R.J. Bernier, N.J., Xi, W., Patrick, M., and Val, A.L (2004). Limited extracellular but complete intracellular acid-base regulation during short-term environmental hypercapnia in the armoured catfish, Liposarcus pardalis. Journal of Experimental Biology, 207, 3381–90.

Breder, C.M. and Rasquin, P. (1950). A preliminary report on the role of the pineal organ in the control of pigment cells and light reactions in recent teleost fishes. Science, 111, 10–12.

Breder, C.M. and Rosen, D.E. (1966). Modes of Reproduction in Fishes. New York, NY: American Museum of Natural History.

Brett, J.R. (1957). The sense organs: The Eye. In: M.E. Brown (ed.). The Physiology of Fishes, Vol. 2: Behaviour. New York, NY: Academic Press, pp. 121–54.

Brett, M.T. and Müller-Navarra, D.C. (1997). The role of highly unsaturated fatty acids in aquatic foodweb processes. Freshwater Biology, 38, 483–99.

Bricard, A., Caussin, J-P., Desreumaux, N., Dauchot, O., and Bartolo, D. (2013). Emergence of macroscopic directed motion in populations of motile colloids. Nature, 503, 95–8.

Bridge, T.W. (1904). Fishes (Exclusive of the Systematic account of Teleostei). In: N.F. Harmer and A.E. Shipley (eds.). The Cambridge Natural History, Vol. VII: Fishes, Ascidians, etc. London: Macmillan and Co. 141–537.

Bridges, C.R., Berenbrink, M., Müller, R., and Waser, W. (1998). Physiology and biochemistry of the pseudobranch: an unanswered question? Comparative Biochemistry and Physiology, 119A, 67–77.

Brito, P.M., Meunier, F.J., Clement, G., and Geffard-Kuriyama D. (2010). The histological structure of the calcified lung of the fossil coelacanth Axelrodichthyes arapensis (Actinistia: Mawsoniidae). Palaeontology, 53, 1281–90.

Brix, O., Grüner, R., Rønnestad, I., and Gemballa, S. (2009). Whether depositing fat or losing weight fish maintain a balance. Proceedings of the Royal Society B, 276, 3777–82.

Brönmark, C., Hulthén, K., Nilsson, P.A., Hansson, L.A., Brodersen, J., and Chapman, B.B. (2014). There and back again: migration in freshwater fishes. Canadian Journal of Zoology, 92, 467–79.

Brown, A.R., Gunnarsson, L., Kristiansson, E. and Tyler, C.R. (2014). Assessing variation in the potential susceptibility of fish to pharmaceuticals, considering evolutionary differences in their physiology and ecology. Philosophical Transactions of the Royal Society B, 369, 20130576.

Brown, C., Laland, K., and Krause, J. (2011). Fish Cognition and Behaviour. Oxford: Wiley-Blackwell.

Brown, J.A., Thonney, J-P., Holwell, D., and Wilson, W.R. (1991). A comparison of the susceptibility of Salvelinus alpinus and Salmo salar ouananiche to proliferative kidney disease. Aquaculture, 96, 1–6.

Brown, M. E. (1957). Experimental studies on growth. In: M.E. Brown (ed.). The Physiology of Fishes, Vol. 1. New York, NY: Academic Press, pp. 361–400.

Brown, S.B., Adams, B. A., Cyr, D. G., and Eales, J.G. (2004). Contaminant effects on the teleost fish thyroid. Environmental Toxicology and Chemistry, 23, 1680–701.

Bruch, R.C. and Teeter, J.H. (1989). Second-messenger signalling mechanisms in olfaction. In: J. Brand, J.H. Teeter, R. Cagan, and M.R. Kare (eds). Chemical Senses, Vol. 1: Receptor Events and Transduction in Taste and Olfaction. New York, NY: Marcel Dekker, pp. 283–98.

Bruun, A.F. (1956). The abyssal fauna: its ecology, distribution and origin. Nature, 177, 1105–8.

Bucke, D. (1971). The anatomy and histology of the alimentary tract of the carnivorous fish the pike Esox lucius L. Journal of Fish Biology, 3, 421–31.

Buddington, R.K. and Diamond, J.M. (1986). Aristotle revisited: The function of pyloric caeca in fish. Proceedings of the National Academy of Science USA, 83, 8012–14.

Budgett, J.S. (1903). Notes on the spiracles of Polypterus. Proceedings of the Zoological Society of London, 73, 10–11.

Bueno, E.M. and Glowacki, J. (2011). Biologic Foundation for Skeletal Tissue Engineering. San Rafael CA: Morgan and Claypool Publishers.

Buhler, R., Lindros, K.O., Nordling, A., Johansson, I., and Ingelmann-Sundberg, M. (1992). Zonation of cytochrome P450 isozyme expression and induction in rat liver. European Journal of Biochemistry, 204, 407–12.

Bull, H.D. (1957). Behaviour and conditioned responses. In: M.E. Brown (ed.). The Physiology of Fishes, Vol. 2. New York, NY: Academic Press, pp. 211–28.

Bullock, T.H. (1999). The future of research on electroreception and electrocommunication. Journal of Experimental Biology, 202, 1455–8.

Bullock, T.H. and Northcutt, R.G. (1982). A new electroreceptive teleost: Xenomystus nigri (Osteoglossiformes: Notopteridae). Journal of Comparative Physiology, 148, 345–52.

Bullock, A.M. and Roberts, R.J. (1975). The dermatology of marine fish. I. The normal integument. Oceanography and Marine Biology: An Annual Review, 13, 383–411.

Bullock, T.H. Hamstra, R.H., and Scheich, H. (1972). The jamming avoidance response of high frequency electric fish. I. General features. *Journal of Comparative Physiology*, 77, 1–22.

Bureau of Marine Fisheries (1929). *The Commercial Fish Catch of California for the Years 1926 and 1927*. Division of fish and game of California. Fish Bulletin No. 15. Sacramento, CA: California State Printing Office.

Bureau of Marine Fisheries (1941). *The Commercial Fish Catch of California for the Years 1936–1939 Inclusive*. Division of fish and game of California. Fish Bulletin No. 57. Sacramento, CA: California State Printing Office.

Bureau of Marine Fisheries (1942). *The Commercial Fish Catch of California for the Year 1940*. Division of fish and game California. Fish Bulletin No. 58. Sacramento, CA: California State Printing Office.

Burger, J.W. (1962). Further studies on the function of the rectal gland in the spiny dogfish. *Physiological Zoology*, 35, 205–17.

Burger, J.W. (1965). Roles of the rectal gland and kidneys in salt and water excretion in the spiny dogfish. *Physiological Zoology*, 38, 191–6.

Burger, J.W. and Hess, W.N. (1960). Functions of the rectal gland in spiny dogfish. *Science*, 131, 670–1.

Burghardt, G.M. (2011). Defining and recognising play. In: A.D. Pellegrini (ed.). *Oxford Handbook of the Development of Play*. Oxford: Oxford University Press, pp. 9–18.

Burghardt, G.M., Dinets, V., and Murphy, J.B. (2015). Highly repetitive object play in a cichlid fish (*Tropheus duboisi*). *Ethology*, 121, 38–44.

Burkhardt-Holm, P., Schmidt, H., and Meier, W. (1998). Heat shock protein (hsp 70) in brown trout epidermis after sudden temperature rise. *Comparative Biochemistry and Physiology A*, 120, 35–41.

Burkholder, T.J. and Lieber, R.L. (2001). Sarcomere length operating range of vertebrate muscles during movement. *Journal of Experimental Biology*, 204, 1529–36.

Burkhead, N.M. (2012). Extinction rates in North American freshwater fishes, 1900–2010. *Bioscience*, 62, 798–808.

Burnett, J., O'Brien, L., Mayo, R.K., Darde, J., and Bohan, M. (1989). Finfish maturity sampling and classification schemes use during Northeast Center bottom trawl surveys, 1963–1989. NOAA Technical memorandum NMFS-F/NEC–76. Woods Hole, MA: Northeast Fisheries Center.

Burnstock, G. (1959). The innervation of the gut of brown trout. *Quarterly Journal of Microscopical Science*, 100, 199–220.

Burnstock, G. (1969). Evolution of the autonomic innervations of visceral and cardiovascular systems in vertebrates. *Pharmacological Reviews*, 21, 247–324.

Burnstock, G. (2007). Physiology and pathophysiology of purinergic neurotransmission. *Physiological Reviews*, 87, 659–797.

Burnstock, G. (2009). Autonomic neurotransmission: 60 years since Sir Henry Dale. *Annual Review of Pharmacology and Toxicology*, 49, 1–30.

Burnstock, G. and Costa, M. (1975). *Adrenergic Neurons*. London: Chapman and Hall.

Burrow, C.J., Newman, M.J., Davidson, R.G., and den Blaauwens, J.L. (2011). Sclerotic plates or circumorbital bones in early jawed fishes. *Palaeontology*, 54, 207–14.

Burrows, A.S., Fletcher, T.C., and Manning, M.J. (2001). Haematology of the turbot, *Psetta maxima* (L.): ultrastructural, cytochemical and morphological properties of peripheral blood leucocytes. *Journal of Applied Ichthyology*, 17, 77–84.

Burton, D. (1964a). The anatomy of the spinal cord of the minnow Phoxinus phoxinus L. with particular reference to the nerve fibres controlling colour change. Ph.D. Thesis, University of London.

Burton, D. (1964b). The spinal pigmentomotor tract of the minnow *Phoxinus phoxinus* (L). *Nature*, 201, 1149.

Burton, D. (1966). Effect of spinal lesions on the colour change of the minnow (*Phoxinus phoxinus* L.). *Nature*, 211, 992–3.

Burton, D. (1969). The colour changes of the minnow, *Phoxinus phoxinus*, with particular reference to the effect of spinal lesions. *Journal of Zoology, London*, 157, 169–85.

Burton, D. (1978). Melanophore distribution within the integumentary tissues of two teleost species, *Pseudopleuronectes americanus* and *Gasterosteus aculeatus* form *leiurus*. *Canadian Journal of Zoology*, 56, 526–35.

Burton, D. (1979). Sexual dimorphism in the integumentary tissues of three-spine stickleback, *Gasterosteus aculeatus* form *leiurus*. *Copeia*, 1979, 533–5.

Burton, D. (1981).The physiological responses of melanophores and xanthophores of hypophysectomised and spinal winter flounder *Pseudopleuronectes americanus* Walbaum. *Proceedings of the Royal Society, London*, 213B, 217–31.

Burton, D. (1988a). Melanophore comparisons in different forms of ambicolouration in the flatfish *Pseudopleuronectes americanus* and *Reinhardtius hippoglossoides*. *Journal of Zoology London*, 214, 353–60.

Burton, D. (1988b). A comparison of the effects of temperature on melanophore *in vitro* responses to noradrenalin in two teleost species adapted to cold ocean and tropical conditions. *Journal of Fish Biology*, 32, 433–42.

Burton, D. (1993). Should MSH be XSH or is it really MSH? *Bulletin of the Canadian Society of Zoologists*, 24, 40.

Burton, D. (2008). A physiological interpretation of pattern changes in a flatfish. *Journal of Fish Biology*, 73, 639–49.

Burton, D. (2010). Flatfish (Pleuronectiformes) chromatic biology. *Reviews in Fish Biology and Fisheries*. 20, 31–46.

Burton, D. (2011). Coloration and Chromatophores in Fishes. In: A.P. Farrell (ed.). *Encyclopedia of Fish Physiology: From Genome to Environment, Vol. 1*. San Diego, CA: Academic Press, pp. 489–96.

Burton, D. and Everard, B.A. (1991a). A comparison of the effect of captivity on the epidermis of pre-spawning and postspawned winter flounder *Pseudopleuronectes americanus*. Walbaum. *Journal of Zoology, London*, 223, 1–7.

Burton, D. and Everard, B.A. (1991b). The effect of androgen treatment on the epidermis of post-pawned winter flounder, *Pseudopleuronectes americanus* (Walbaum). *Journal of Fish Biology*, 38, 73–80.

Burton, D. and Fletcher, G.L. (1983). Seasonal changes in the epidermis of the winter flounder, *Pseudopleuronectes americanus*. *Journal of the Marine Biological Association of the United Kingdom*, 63, 273–87.

Burton, D. and O'Driscoll, M.P. (1992). 'Facilitation' of melanophores responses in winter flounder, *Pseudopleuronectes americanus*. *Journal of Experimental Biology*, 168, 289–99.

Burton, D., Burton, M.P., and Idler, D.R. (1984). Epidermal condition in post-spawned winter flounder *Pseudopleuronectes americanus* Walbaum maintained in the laboratory and after exposure to crude petroleum. *Journal of Fish Biology*, 25, 593–606.

Burton, D., Burton, M.P.M., Truscott, B., and Idler, D.R. (1985). Epidermal cellular proliferation and differentiation in sexually mature male *Salmo salar* with androgen levels depressed by oil. *Proceedings of the Royal Society B*, 225, 121–8.

Burton, M.P.M. (1991). Induction and reversal of the non-reproductive state in winter flounder (*Pseudopleuronectes americanus* Walbaum by manipulating food availability. *Journal of Fish Biology*, 39, 909–10.

Burton, M.P.M. (1994). A critical period for nutritional control of gametogenesis in female winter flounder *Pleuronectes americanus* (Pisces: Teleostei). *Journal of Zoology*, 233, 405–15.

Burton, M.P.M. (1999). Notes on potential errors in estimating spawning biomass. *Journal of Northwest Atlantic Fisheries Science*, 25, 205–13.

Burton, M.P.M. (2008). Reproductive omission and 'skipped spawning': Detection and importance. *Cybium*, 32 (No. 2 supplement), 315–16.

Burton, M.P.M. and Flynn, S. (1998). Differential post-spawning mortality among male and female capelin (*Mallotus villosus* Müller) in captivity. *Canadian Journal of Zoology*, 76, 588–92.

Burton, M.P. and Idler, D.R. (1984). The reproductive cycle in winter flounder, *Pseudopleuronectes americanus* (Walbaum). *Canadian Journal of Zoology*, 62, 2563–7.

Burton, M.P. and Idler, D.R. (1987). A comparison of testicular function in mature salmonids (*Salmo* spp.) and winter flounder (*Pseudopleuronectes americanus*) with particular reference to putative sites of steroidogenesis. Third International Symposium on Reproductive Physiology of Fish. St. John's NL: Memorial University of Newfoundland, pp. 249–249.

Busch, M.W., Klimpel, S., Sutton, T., and Piatkowski, U. (2008). Parasites of the deep-sea smelt *Bathylagus europs* (Argentiniformes: Microstomatidae) from the Charlie-Gibbs Fracture Zone. *Marine Biology Research*, 4, 313–17

Cailliet, G.M., Love, M.S., and Ebeling, A.W. (1986). *Fishes. A Field and Laboratory Manual on their Structure, Identification, and Natural History*. Belmont, CA: Wadsworth Inc.

Caira, J.N., Benz, G.W., Borucinska, J., and Kohler, N.E. (1997). Pugnose eels, *Simenchelys parasiticus* (Synaphobranchidae) from the heart of a shortfin mako, *Isurus oxyrhincus* (Lamnidae). *Envonmental Biology of Fishes*, 49, 139–44.

Calow, P. and Woollhead, A.S. (1977). Locomotory strategies in freshwater triclads and their effects on the energetic of degrowth. *Oecologia*, 27, 353–62.

Camp, A.L., Konow, N., and Sanford, C.P.J. (2009). Functional morphology and biomechanics of the tongue-bite apparatus in salmonid and osteoglossomorph fishes. *Journal of Anatomy*, 214, 717–28.

Campbell, C.M. and Idler, D.R. (1976). Hormonal control of vitellogenesis in hypophysectomised winter flounder (*Pseudopleuronectes americanus* Walbaum). *General and Comparative Endocrinology*, 28, 143–50.

Campbell, C. M., Walsh, J.M., and Idler, D.R. (1976). Steroids in the plasma of the winter flounder (*Pseudopleuronectes americanus* Walbaum). A seasonal study and investigation of steroid involvement in oocyte maturation. *General and Comparative Endocrinology*, 29, 14–20.

Campbell, K.S.W. and Barwick, R.E. (1985). An advanced massive dipnorhynchid lungfish from the early Devonian of New South Wales, Australia. *Records of the Australian Museum*, 37, 301–16.

Cánepa, M.M., Pandolfi, M., Maggese, M.C., and Vissio, P.G. (2006). Involvement of somatolactin in background adaptation of the cichlid fish *Cichlasoma dimerus*. *Journal of Experimental Zoology*, 305A, 410–19.

Canfield, J.G. and Rose, G.J. (1993). Activation of Mauthner neurons during prey capture. *Journal of Comparative Physiology A*, 172, 611–18.

Capréol, S.V. and Sutherland, L.E. (1968), Comparative morphology of juxtaglomerular cells in fish. *Canadian Journal of Zoology*, 46, 249–56.

Caprio, J., Brand, J.G., Teeter, J.H., Valentincic, T., Kalinoshi, D.L., Kohbara, J., Kumazawa, T., and Wegert, S. (1993). The taste system of the channel catfish: from biophysics to behaviour. *Trends in Neuroscience*, 16, 192–7.

Cardoso, J.C.R., Viera, F.A., Gomes, A.S., and Power, D.M. (2010). The serendipitous origin of chordate secretin peptide family members. *Evolutionary Biology*, 10, 135–54.

Carey, F.G., Kanwisher, J.W., and Stevens, E.D. (1984). Bluefin tuna warm their viscera during digestion. *Journal of Experimental Biology*, 109, 1–20.

Carey, J.R. and Judge, D.S. (2000). *Longevity Records, Life Spans of Mammals, Birds, Amphibians, Reptiles and Fish.* Odense: Odense University Press.

Carleton, H.M. and Leach, E.H. (eds) (1949). *Schafer's Essentials of Histology*, 15th edn. London: Longmans, Green and Co.

Carleton, K.L., Hárosi, F.I., and Kocher, T.D. (2000). Visual pigments of African cichlid fishes: evidence for ultraviolet vision from microspectrophotometry. *Vision Research*, 40, 879–90.

Carlson, J.K., Goldman, K.J., and Lowe, C.J. (2004). Metabolism, energetic demand, and endothermy. In: J.C. Carrier, J.C. Musick, and M.R. Heithaus (eds). *Biology of Sharks and Their Relatives.* Boca Raton, FL: CRC Press, pp. 203–24.

Carrera, I., Sueiro, C., Molist, P., Holstein, G.R., Martinelli, G.P., Rodrigez-Moldes, I., and Anadón, R. (2006). GABAergic system of the pineal organ of an elasmobranch (*Scyliorhinus caniculа*): a developmental immunocytochemical study. *Cell and Tissue Research*, 323, 273–81.

Carroll, S.M. Bridgham, J.T., and Thornton, J.W. (2008). Evolution of hormone signaling in elasmobranchs by exploitation of promiscuous receptors. *Molecular Biology and Evolution*, 25, 2643–52.

Carson, R. (1941). *Under the Seawind.* New York, NY: Oxford University Press.

Carthy, J.D. (1966). *The Study of Behaviour. Studies in Biology 3.* New York, NY: St. Martin's Press Inc.

Casselman, J.M. (1987). Determination of age and growth. In: A.H. Weatherley and H.S. Gill (eds). *The Biology of Fish Growth.* London: Academic Press, pp. 209–42.

Casselman, J.M. (2007). Dr. E.J. (Ed) Crossman's scientific contributions on muskellunge: celebrating a fishing legacy. *Environmental Biology of Fish*, 79, 5–10.

Catton, W.T. (1951). Blood cell formation in certain teleost fishes. *Blood*, 6, 39–60.

Cech, J.J. and Massingill, M.J. (1995). Tradeoffs between respiration and feeding in Sacramento blackfish, *Orthodon microlepidotus. Environmental Biology of Fishes*, 44, 157–63.

Çek, S. Bromage, N., Randall, C., and Rana, K. (2001). Oogenesis, hepatosomatic and gonadosomatic indexes, and sex ratio in rosy barb (*Puntius conchonius*). *Turkish Journal of Fisheries and Aquatic Sciences*, 1, 33–41.

Chagnaud, B.P., Bleckmann, H., and Hofmann, M.H. (2008a). Lateral line nerve fibres do not code bulk water flow direction in turbulent flow. *Zoology*, 111, 204–17.

Chagnaud, B.P. Wilkens, L.A., and Hofmann, M.H. (2008b). Receptive field organization of electrosensory neurons in the paddlefish (*Polyodon spathula*). *Journal of Physiology*, 102, 246–55.

Chambers, J. (1987). The cyprinodontiform gonopodium, with an atlas of the gonopodia of the fishes of the genus *Limia. Journal of Fish Biology*, 30, 389–418.

Chang, B.S.W., Crandall, K.A., Carulli, J.P., and Hartl, D.L (1995). Opsin phylogeny and evolution: a model for blue shifts in wavelength regulation. *Molecular Phylogenetics and Evolution*, 4, 31–43.

Chao, L.N. (1978). A basis for classifying Western Atlantic Sciaenidae (Teleostei: Perciformes). *National Oceanic and Atmospheric Administration Technical Report, National Marine Fisheries Service, Circular* 415. Washington, DC: National Oceanic and Atmospheric Administration, National Marine Fisheries Service.

Cheney, K.L. and Côté, I.M. (2005). Frequency-dependent success of aggressive mimics in a cleaning symbiosis. *Proceedings of the Royal Society B*, 272, 2635–9.

Cheney, K.L., Grutter, A.S., Blomberg, S.P., and Marshall, N.J. (2009). Blue and yellow signal cleaning behaviour in coral reef fishes. *Current Biology*, 19, 1283–7.

Chester Jones, I., Bellamy, D., Chan, D.K.O., Follett, B.K., Henderson, I.W., Phillips, J.G., and Smart, R.S. (1972). Biological actions of steroid hormones in nonmammalian vertebrates. In: D.R. Idler (ed.). *Steroids in Nonmammalian Vertebrates.* New York, NY: Academic Press, pp. 414–80.

Cheung, W.W.L. Watson, R., and Pauly, D. (2013a). Signature of ocean warming in global fisheries catch. *Nature*, 497, 365–9.

Cheung, W.W.L., Sarmiento, J.L. Dunne, J., Frölicher, T.L., Lam, V.W.Y., Palomares, M.L.D., Watson, R., and Pauly, D. (2013b). Shrinking of fishes exacerbates impacts of global ocean changes on marine ecosystems. *Nature Climate Change*, 3, 254–8.

Chezik, K.A., Lester, N.P., and Venturelli, P.A. (2014). Fish growth and degree-days I: selecting a base temperature for a within-population study. *Canadian Journal of Fisheries and Aquatic Science*, 71, 47–55.

Chiasson, R.B. (1974). *Laboratory Anatomy of the Perch.* Dubuque, IO: Wm.C. Brown Company.

Ching, B., Jamieson, S., Heath, J.W., and Hubberstey, A. (2010). Transcriptional differences between triploid and diploid Chinook salmon (*Oncorhynchus tshawytscha*) during live *Vibrio anguillarum* challenge. *Heredity*, 104, 224–34.

Chiu, C. and Prober, D.A. (2013). Regulation of zebrafish sleep and arousal states: current and prospective approaches. *Frontiers in Neural Circuits*, 7, Article 58.

Cho, G.K. and Heath, D.D. (2000). Comparison of tricaine methanesulphonate (MS222) and clove oil anaesthesia effects on the physiology of juvenile chinook salmon *Oncorhynchus tschawytscha* (Walbaum). *Aquaculture Research*, 31, 537–46.

Christiansen, J.S., Dalmo, R.A., and Ingebrigtsen, K. (1996). Xenobiotic excretion in fish with aglomerular kidneys. *Marine Ecology Progress Series*, 136, 303–4.

Christiansen, J.S., Praebel, K., Siikavuopio, S.I., and Carscadden, J.F. (2008). Facultative semelparity in capelin *Mallotus villosus* (Osmeridae) – an experimental test of a life history phenomenon in a subarctic fish. *Journal of Experimental Marine Biology and Ecology*, 360, 47–55.

Claes, J.M. and Mallefet, J. (2009a). Ontogeny of photophore pattern in the velvet belly shark, *Etmopterus spinax*. *Zoology*, 112, 433–41.

Claes, J.M. and Mallefet, J. (2009b). Hormonal control of luminescence from lantern shark (*Etmopterus spinax*) photophores. *Journal of Experimental Biology*, 212, 3684–92.

Claes, J.M., Ho, H-C., and Mallefet, J. (2012). Control of luminescence from pygmy shark (*Squaliolus aliae*) photophores. *Journal of Experimental Biology*, 215, 1691–9.

Clement, A.M. (2012). A new species of long-snouted lungfish from the late Devonian of Australia, and its functional and biogeographical implications. *Paleontology*, 55, 51–71.

Cobb, J.L.S. (1974). Gap junctions in the heart of teleost fish. *Cell and Tissue Research*, 154, 131–4.

Cohen, D.M. and Haedrich, R.L. (1983). The fish fauna of the Galapagos thermal vent region. *Deep-Sea Research*, 30, 371–9.

Cohen, D.M. Rosenblatt, R.H., and Moser, H.G. (1990). Biology and description of a bythitid fish from deep-sea thermal vents in the tropical eastern Pacific. *Deep-Sea Research*, 37, 267–83.

Cohen, L., Dean, M., Shipov, A., Atkins, A., Monsonego-Oman, E., and Shahar, R. (2012). Comparison of structural, architectural and mechanical aspects of cellular and acellular bone in two teleost fish. *Journal of Experimental Biology*, 215, 1983–93.

Cole, A.M., Weis, P., and Diamond, G. (1997). Isolation and characterization of pleurocidin, an antimicrobial peptide in the skin secretions of winter flounder. *Journal of Biological Chemistry*, 272, 12008–13.

Collin, S.P. and Collin, H.B. (1998). The deep-sea cornea: a comparative study of gadiform fishes. *Histology and Histopathology*, 13, 325–36.

Collin, S.P. and Partridge, J.C. (1996). Retinal specializations in the eyes of deep-sea teleosts *Journal of Fish Biology*, 49 (Suppl. A), 157–74.

Collingridge, G.L. (2003). The induction of N-methyl-D-aspartate receptor-dependent long-term potentiation. *Philosophical Transactions of the Royal Society B*, 358, 635–41.

Collins, L.E. and Waldeck, R.F. (2006). Telencephalic ablation results in decreased startle responses in goldfish. *Brain Research*, 1111, 162–5.

Colombe, L., Fostier, A., Bury, N., Pakdel, F., and Guiguen, Y. (2000). A mineralocorticoid-like receptor in the rainbow trout, *Oncorhynchus mykiss*: cloning and characterization of its steroid binding domain. *Steroids*, 65, 319–28.

Compagno, L.J.V. (1990). Alternative life-history styles of cartilaginous fishes in time and space. *Environmental Biology of Fish*, 28, 33–75.

Compagno, L.J.V. (1999). *FAO species guide for identification*, Vol. 3. Rome: The Food and Agriculture Organization of the United Nations.

Compagno, L.I.V., Dando, M., and Fowler, S. (2005). *Sharks of the World*. Princeton, NJ: Princeton University Press.

Company, R., Astola, A., Pendon, C., Valdiva, M.M., and Pérez-Sánchez, J. (2001). Somatotrophic regulation of fish growth and adiposity: growth hormone (GH) and somatoloactin (SL) relationship. *Comparative Biochemistry and Physiology C*, 130, 435–45.

Conde-Sieira, M., Alvarez, R., López-Patiño, Miguez, J.M., Flik, G., and Soengas, J.L. (2013). ACTH-stimulated cortisol release from head kidney of rainbow trout is modulated by glucose concentration. *Journal of Experimental Biology*, 216, 554–67.

Conlon, J.M. (1999). Bradykinin and its receptors in non-mammalian vertebrates. *Regulatory Peptides*, 5, 71–81.

Connaughton, M.A., Fine, M.L., and Taylor, M.H. (2002). Weakfish sonic muscle: influence of size, temperature and season. *Journal of Experimental Biology*, 205, 2183–8.

Connell, J.J. and Howgate, P.F. (1959). Studies on the proteins of fish skeletal muscle. 6. Amino acid composition of cod fibrillar proteins. *Biochemical Journal*, 71, 83–6.

Conover, D.O. and Heins, S.W. (1987). The environmental and genetic components of sex ratio in *Menidia menidia* (Pisces: Atherinidae). *Copeia*, 1987, 732–43.

Conrath, C.L. (2005). Reproductive biology. In: J.A. Musick and R. Bonfil (eds). *Management Techniques for Elasmobranch Fisheries. FAO Fisheries Technical paper* T474. Part 7.

Cooke, S.J., Sack, L., Franklin, C.E., Farrell, A.J., Beardall, J. Wikelski, M., and Chown, S.L. (2013). What is conservation physiology? Perspectives on increasingly integrated and essential science. *Conservation Physiology*, 1, 1–23.

Coombs, S., Janssen, J., and Montgomery, J.C. (1991). Functional and evolutionary implications of peripheral diversity in lateral line systems. In: D.B. Webster, R.R. Fay, and A.N. Popper (eds). *The Evolutionary Biology of Hearing*. New York, NY: Springer Verlag, pp. 267–94.

Cooper, J.E. and Kuehne, R.A. (1974). *Speoplatyrhinus poulsoni*, a new genus and species of subterranean fish from Alabama. *Copeia*, 1974, 486–93.

Cortesi, F., Feeney, W.E., Ferrari, M.C.O., Waldie, P.A., Phillips, G.A.C., McClare, E.C. Sköld, H.N., Salzburger, W., Marsall, N.J., and Cheney, K.L. (2015). Phenotypic plasticity confers multiple fitness benefits to a mimic. *Current Biology*, 25, 949–954.

Cott, H.B. (1940). *Adaptive Coloration in Animals*. London: Methuen.

Cowey, C.B. (1994). Amino acid requirements of fish: A critical appraisal of present values. *Aquaculture*, 124, 1–11.

Cowey, C.B. and Sargent, J.R. (1979). Nutrition. In: W.S. Hoar and D.J. Randall (eds). *Fish Physiology, Vol. VIII*. New York, NY: Academic Press, pp. 267–94.

Cragg, B.G. (1955). A physical theory on the growth of axons. *Journal of Cellular and Comparative Physiology*, 4533–60.

Craig-Bennett, A. (1931). The reproductive cycle of the three-spined stickleback, *Gasterosteus aculeatus* Linn. *Philosophical Transactions of the Royal Society*, CCXIX B, 466, 197–279.

Crane, A.L. and Ferrari, M.C.O. (2015). Minnows trust conspecifics more than themselves when faced with conflicting information about predation risk. *Animal Behaviour*, 100, 184–90.

Crawford, C.M. (1986). Development of eggs and larvae of the flounders *Rhombosolea tapirina* and *Ammotretis rostratus* (Pisces: Pleuronectidae). *Journal of Fish Biology*, 29, 325–34.

Currey, J.D. (2002). *Bones: Structure and Mechanics*. Princeton, NJ: Princeton University Press.

Cushing, D. (1995). *Population Production and Regulation in the Sea: A Fisheries Perspective*. Cambridge: Cambridge University Press.

Cushing, D.H. and Walsh, J.J. (eds) (1976). *The Ecology of the Seas*. Toronto: W.B. Saunders Co.

Cutler, C.P. and Cramb, G. (2000). Water transport and aquaporin expression in fish. In: E.S. Hohmann and S. Nielsen (eds). *Molecular Biology and Physiology of Water and Solute Transport*. New York, NY: Kluwer Academic/ Plenum Publishing.

Cutler, C.P. and Cramb, G. (2002). Branchial expression of an aquaporin 3 (AQP-3) homologue is downregulated in the European eel *Anguilla anguilla* following seawater acclimation. *Journal of Experimental Biology*, 205, 2643–51.

Cutler, C.P., Martinez, A.S., and Cramb, G. (2007). The role of aquaporin 3 in teleost fish. *Comparative Biochemistry and Physiology A*, 148, 82–91.

Cvancara, V. (1989). *Current References in Fish Research*. Chippeaw Falls, WI: Cvancara.

Cvancara, V. (1996). *Current References in Fish Research*. Chippeaw Falls, WI: Cvancara.

Dabrowski, K. (1990). Review of *The Biology of Fish Growth* by A.H. Weatherley and H.S. Gill. *Copeia*, 1990, 887–9.

Daget, J., Gayet, M., Meunier, F.J., and Sire, J-Y. (2001). Major discoveries on the dermal skeleton of fossil and recent Polypteriformes: A review. *Fish and Fisheries*, 2, 113–24.

Dahlgren, U. (1898). The maxillary and mandibular breathing valves of teleost fishes. *Zoological Bulletin*, 2, 117–24.

Dalyan, C. and Eryilmaz, L. (2008). A new deepwater fish, *Chauliodus sloani* Bloch and Schneider, 1801 (Osteichthyes: Stomiidae), from the Turkish waters of levant Sea (Eastern Mediterranean). *Journal of the Black Sea/Mediterranean Environment*, 14, 33–7.

Dammerman, K.W. (1910). Der saccus vasculosus der Fische ein Tieforgan. *Zeitschrift fuer wissenschaftliche Zoologie*, 96, 654–726.

Danulat, E. and Hochachka, P.W. (1989). Creatine turnover in the starry flounder *Platichthys stellatus*. *Fish Physiology and Biochemistry*, 6, 1–9.

Danylchuk, A.J. and Tonn, W.M. (2003). Natural disturbances and fish: Local and regional influences on winterkill of fathead minnow in boreal lakes. *Transactions of the American Fisheries Society*, 132, 289–98.

Danylchuk, A.J., Tonn, W.M., and Paszkowski, C.A. (2011). Fathead minnow, In: *Fishes of Wisconsin*. Available at: http:/ infotrek.er.usgs.gov/wdnr_fishes/account.jsp? species_param=1290.

D'Août, K. and Aerts, P. (1999). A kinematic comparison of forward and backward swimming in the eel *Anguilla anguilla*. *Journal of Experimental Biology*, 202, 1511–21.

Darias, M.J., Mazurais, D., Koumoundouros, G., Cahu, C.L., and Zambonino-Infante, J.L. (2011). Overview of vitamin D and C requirements in fish and their influence on the skeletal system. *Aquaculture*, 315, 49–60.

Darwin, C. (1859). *On the Origin of Species by Means of Natural Selection, or The Preservation of Favoured Races in the Struggle for Life*. London: John Murray.

Davenport, J., Lonning, S., and Kjorsvik, E. (1986). Some mechanical and morphological properties of the chorions of marine teleost eggs. *Journal of Fish Biology*, 29, 289–301.

Davies, D.T. and Rankin, J.C. (1973). Adrenergic receptors and vascular responses to catecholamines of perfused dogfish gills. *Comparative and General Pharmacology*, 4, 139–47.

Davis, A.M., Pusey, B.J., Thorburn, D.C., Dowe, J.L. Morgan, D.L., and Burrows, D. (2010). Riparian contributions to the diet of terapontid grunters (Pisces: Terrapontidae) in wet-dry tropical rivers. *Journal of Fish Biology*, 76, 862–79.

Dawley, E.M. (1996). Evaluation of the effects of dissolved gas supersaturation on fish and invertebrates downstream from Bonneville, Ice Harbor, and Priest Rapids Dams, 1993 and 1994. *National Oceanic and Atmospheric Administration. Seattle, Coastal Zone and Estuarine*

Studies Division, Northwest Fisheries Science Center. Seattle, WA: Northwest Fisheries Science Center.

Dean, B. (1916). *Bibliography of Fishes, Vol. 1*. New York, NY: American Museum of Natural History.

Dean, M.N. and Summers, A.P. (2006). Mineralised cartilage in the skeleton of chondrichthyan fishes. *Zoology*, 109, 164–8.

de Beer, G.R. (1951). *Vertebrate Zoology*. London: Sidgwick and Jackson Ltd.

DeBlois, E.M. and Rose, G.L. (1995). Effect of foraging activity on the shoal structure of cod (*Gadus morhua*). *Canadian Journal of Fisheries and Aquatic Science*, 52, 2377–87.

De Groot, S.J. (1969). Digestive system and sensorial factors in relation to the feeding behaviour of flatfishes (Pleuronectiformes). *International Council for the Exploration of the Sea Journal du Conseil*, 32, 385–95.

Delcourt, J. and Poncin, P. (2012). Shoals and schools: back to the heuristic definitions and quantitative references. *Reviews in Fish Biology and Fisheries*, 22, 595–619.

Della Torre, C., Corsi, L., Alcaro, L., Amato, E., and Focardi, S. (2006). The involvement of cytochrome P450 in the fate of 2,4,6-trinitrotoluene (TNT) in European eel [*Anguilla anguilla* (Linnaeus, 1758)]. *Biochemical Society Transactions*, 34, 1228–30.

de Lussanet, M.H.E. and Muller, M. (2007). The smaller your mouth, the longer your snout: predicting the snout length of *Syngnathus acus*, *Centriscus scutatus* and other pipette feeders. *Journal of the Royal Society, Interface*, 4, 561–73.

de Luze, A. and Leloup, J. (1984). Fish growth hormone enhances peripheral conversion of thyroxine to triiodothyronine in the eel (*Anguilla anguilla* L.). *General and Comparative Endocrinology*, 56, 308–12.

Demski, L.S. and Northcutt, R.G. (1983). The terminal nerve: a new chemosensory system in vertebrates? *Science*, 220, 435–7.

Demski, L.S. and Schwanzel-Fukuda, M. (eds) (1987). The terminal nerve (nervus terminalis) structure, function and evolution. 37 articles. *Annals of the New York Academy of Science*, 519, 1–468.

Denton, E.J. and Marshall, N.B. (1958). The buoyancy of bathypelagic fishes without a gas-filled swimbladder. *Journal of the Marine Biological Association of the United Kingdom*, 37, 753–67.

Denton, E.J. and Nicol, J.A.C. (1966). A survey of reflectivity in silvery teleosts. *Journal of the Marine Biological Association of the United Kingdom*, 46, 685–722.

Denton, E.J., Herring, P.J., Widder, E.A., Latz, M.F., and Case, J.F. (1985). The roles of filters in the photophores of oceanic animals and their relation to vision in the oceanic environment. *Proceedings of the Royal Society B*, 225, 63–97.

de Souza, P.C. and Borilla-Rodriguez, G.O. (2007). Fish hemoglobins. *Brazilian Journal of Medical and Biological Research*, 40, 769–78.

Dethloff, G.M. and Schmitt, C.J. (2000). Condition factor and organo-somatic indices. In: C.J. Schmitt, and G.M. Dethloff (eds). *Biomonitoring of Environmental Status and Trends (BEST) Program: Selected methods for monitoring environmental contaminants and their effects in aquatic ecosystems*. U.S. Geological Survey Information and Technology Report USGS/BRD 2000-0005 (2000). Columbia, MO:U.S. Geological Survey, pp. 13–18.

Detrich, H.W., Westerfield, M., and Zon, L.I. (eds) (2009). *Essential Zebrafish Methods, Cell and Developmental Biology*. Oxford: Academic Press/Elsevier.

Devlin, A.J., Danks, J.A., Faulkner, M.K., Power, A.V.M. Martin, T.J., and Ingleton, P.M. (1996). Immunochemical detection of parathyroid hormone-related protein in the saccus vasculosus of a teleost fish. *General and Comparative Endocrinology*, 101, 83–90.

DeVries, A.L. and Eastman, J.T. (1981). Physiology and ecology of notothenioid fishes of the Ross Sea. *Journal of the Royal Society of New Zealand*, 11, 329–40.

DeWitt, Jr., F.A. (1972). Bathymetric distribution of two common deep-sea fishes, Cyclothone acclinidens and C. signata, off Southern California. *Copeia*, 1972, 88–96.

Diamond, J. (1968). The activation and distribution of gaba and L-glutamate receptors on goldfish Mauthner neurons: an analysis of dendritic remote inhibition. *Journal of Physiology*, 194, 669–723.

Diamond, J. (1971). The Mauthner cell. In: W.S. Hoar and D.J. Randall (eds). *Fish Physiology, Vol. V.Sensory systems and sense organs*. New York, NY: Academic Press, pp. 265–346.

Di Bernardo, M., Bellomonte, D., Castagnetti, S., Melfi, R., Oliveri, P., and Spinelli, G. (2000). Homeobox genes and sea urchin development. *International Journal of Developmental Biology*, 44, 637–43.

Dibner, C., Schibler, U., and Albrecht, U. (2010). The mammalian circadian timing system: organization and coordination of central and peripheral clocks. *Annual Review of Physiology*, 72, 517–49.

Dietz, T.J. and Somero, G.N. (1992). The threshold induction temperature of the 90kDa heat shock protein is subject to acclimatization in eurythermal goby fishes (genus *Gillichthys*). *Proceedings of the National Academy of Science of the USA*, 89, 3389–93.

DiFrancesco, D. (2010). The role of the funny current in pacemaker activity. *Circulation Research*, 106, 434–46.

Dix, T.G. (1969). Association between the echinoid *Evechinus chloroticus* (Val.) and the clingfish *Dellichthys morelandi* Briggs. *Pacific Science*, 33, 332–6.

Dodd, J.M. (1983). Reproduction in cartilaginous fishes (Chondrichthyes). In: W.S. Hoar, D.J. Randall, and E.M. Donaldson (eds). *Fish Physiology, Vol. IX, Part A. Reproductive tissues and hormones.* New York, NY: Academic Press, pp. 31–95.

Domeneghini, C., Arrighi, S., Radaelli, G., Bosi, G., Berardenelli, P., Vaini, F., and Mascarrello, F. (1999). Morphological and histochemical analysis of the neuroendocrine system of the gut in *Acipenser transmontanus. Journal of Applied Ichthyology*, 15, 81–6.

Domeneghini, C., Radaelli, G., Arrighi, S., Mascarello, F., and Veggetti, A. (2000). Neurotransmitters and putative neuromodulators in the gut of *Anguilla anguilla* (L.). Localizations in the enteric nervous and endocrine systems. *European Journal of Histochemistry*, 44, 295–306.

Domenici, P. and Blake, R.W. (1997). The kinematics and performance of fish fast-start swimming. *Journal of Experimental Biology*, 200, 1165–78.

Donald, J.A. (1998). The autonomic nervous system. In: D.H. Evans (ed.). *The Physiology of Fishes*, 2nd edn. Boca Raton, FL: CRC Press, pp. 407–39.

Donoghue, P.C.J. and Sansom, I.J. (2002). Origin and early evolution of vertebrate skeletonization. *Microscopical Research and Technique*, 59, 352–72.

Donoghue, P.C.J., Forey, P.L., and Aldridge, R.J. (2000). Conodont affinity and chordate phylogeny. *Biological Reviews*, 75, 191–251.

Douglas, R.H., Harper, R.D., and Case, J.F. (1998a). The pupil response of a teleost fish, *Porichthys notatus*: description and comparison to other species. *Vision Research*, 38, 2697–3710.

Douglas, R.H. Partridge, J.C., and Marshall, N.D. (1998b). The eyes of deep-sea fish I: Lens pigmentation, tapeta and visual pigments. *Progress in Retinal and Eye Research*, 17, 597–636.

Douglas, R.H., Partridge, J.C., Dulai, K., Hunt, Mullineaux, C.W., Tauber, A.Y., and Hynninen, P.H. (1998c). Dragon fish see using chlorophyll. *Nature*, 393, 423–4.

Døving, K. (1986). Functional properties of the fish olfactory system. In: D. Ottoson (ed.). *Progress in Sensory Physiology* 6. Berlin: Springer Verlag, pp. 39–104.

Døving, K., Selset, R., and Thommesen, G. (1980). Olfactory sensitivity to bile acids in salmonid fishes. *Acta Physiologica Scandinavica*, 108, 123–31.

Dowding, A.J. and Scholes, J. (1993). Lymphocytes and macrophages outnumber oligodendrocytes in normal fish spinal cord. *Proceedings of the National Academy of Sciences of the USA*, 90, 10183–7.

Dower, J.F. and Perry, R.I. (2001). High abundance of larval rockfish over Cobb Seamount, an isolated seamount in the Northeast Pacific. *Fisheries Oceanography*, 10, 268–74.

Driedzic, W. (1983), The fish heart as a model for the study of myoglobin. *Comparative Biochemistry and Physiology A*, 764, 87–93.

Driedzic, W.R. and Stewart, J.M. (1982). Myoglobin content and the activities of enzymes of energy metabolism in red and white hearts. *Journal of Comparative Physiology*, 149, 67–73.

Driedzic, W.R., Clow, K.A., Short, C.E., and Ewart, K.V. (2006). Glycerol production in rainbow smelt (Osmerus mordax) may be triggered by low temperature alone and is associated with the activarion ofglycerol-3-phosphate dehydrogenase and glycerol-3-phsophatase. *Journal of Experimental Biology*, 209, 1016–23.

Drucker, E.G. and Jensen, J.S. (1991). Functional analysis of a specialized prey processing behaviour: winnowing by surfperches (Teleostei: Embiotocidae). *Journal of Morphology*, 210, 267–87.

Du, S.J., Gong, Z., Fletcher, G.L., Shears, M.A., King, M.J., Idler, D.R., and Hew, C.L. (1992). Growth enhancement in transgenic Atlantic salmon by the use of an 'all fish' chimeric growth hormone gene construct. *Biotechnology (New York)*, 10, 176–81.

Dumont, E.R. (2010). Bone density and the lightweight skeletons of birds. *Proceedings of the Royal Society B.* doi: 10.1098/rspb.2010.0117

Dunn, R.SA. (1970). Further evidence for a three-year oocyte maturation time in the winter flounder (*Pseudopleuronectes americanus*). *Journal of the Fisheries Research Board of Canada*, 27, 957–60.

Dyck, M., Warkentin, P.H., and Treble, M.A. (2007). A bibliography on the Greenland halibut, *Reinhardtius hippoglossoides* (a.k.a. Greenland turbot) 1936–2005. *Canadian Technical Report of Fisheries and Aquatic Sciences*, 2683. Winnipeg, MB: Fisheries and Oceans Canada, pp. 1–309.

Earley, R.L, Hanninen, A.F., Fuller, A., Garcia, M.J., and Lee, E.A. (2012). Phenotypic plasticity and integration in the mangrove rivulus (*Krptolebias marmoratus*): A prospectus. *Integrative and Comparative Biology*, 52, 814–27.

Eales, J.G. (1985). The peripheral metabolism of thyroid hormones and regulation of thyroidal status in poikilotherms. *Canadian Journal of Zoology*, 63, 1217–31.

Eales, J.G. (1988). The influence of nutritional state on thyroid function in various vertebrates. *American Zoologist*, 28, 351–62.

Eales, J.G. and Brown, S.B. (1993). Measurement and regulation of thyroidal status in teleost fish. *Reviews in Fish Biology and Fisheries*, 3, 299–347.

Eastman, J.T. (1990). The biology and physiological ecology of notothenioid fishes. In: O. Gon and P.C. Heemstra (eds). *Fishes of the Southern Ocean.* Grahamstown, South Africa: J.L.B. Smith Institute of Ichthyology, pp. 34–51.

Eastman, J.T. and Underhill, J.C. (1973). Intraspecific variation in the pharyngeal tooth formulae of some cyprinid fishes. *Copeia*, 1973, 45–53.

Eastman, J.T. Boyd, R.B., and De Vries, A.L. (1987). Renal corpuscle development in boreal fishes with and without antifreezes. *Fish Physiology and Biochemistry*, 4, 89–100.

Eastman, J.T., DeVries, A.L. Coalson, R.E., Nordquist, R.E., and Boyd, R.E. (1979). Renal conservation of antifreeze peptide in Antarctic eelpout, *Rhigophila dearborni*. *Nature*, 282, 217–18.

Easy, R.H. and Ross, N.W. (2006). Proteomic analysis of the epidermal mucus of Atlantic salmon (*Salmo salar*): biological indicators of stress. *VIIth International Congress on the Biology of Fish*, St. John's, NL. Bethesda MD: American Fisheries Society Abstract p., programme guide, pp. 90–90.

Easy, R.H. and Ross, N.W. (2010). Changes in Atlantic salmon (*Salmo salar*) epidermal mucus following short- and long-term handling stress. *Journal of Fish Biology*, 77, 1616–31.

Eaton, R.C., Nissanov, J., and DiDomenico, R. (1991). The role of the Mauthner cell in sensorimotor integration by the brainstem escape network. *Brain, Behaviour and Evolution*, 37, 272–85.

Ebbeson, S.O. and Northcutt, R.G. (1976). Neurology of anamniotic vertebrates. In: R.B. Masterton, C.B.G. Campbell, M.E. Bitterman, and N. Hotton (eds). *Evolution of Brain and Behaviour in Vertebrates*. Hillsdale, NJ: Erlbaum, pp. 115–46.

Ebel, W.J. and Raymond, H.L. (1976). Effect of atmospheric gas supersaturation on salmon and steelhead trout of the snake and Columbia Rivers. *Marine Fisheries Review*, 1191, 1–14.

Eberhardt, L.L. (1977). Relationship between two stock-recruitment curves. *Journal of the Fisheries Research Board of Canada*, 34, 425–8.

Eccles, J. (1990). A unitary hypothesis of mind-brain interaction in the cerebral cortex. *Proceedings of the Royal Society, London B*, 240, 433–51.

Eccles, J.C. (1992). Evolution of consciousness. *Proceedings of the National Academy of Sciences of the USA*, 89, 7320–4.

Echelle, A.E., van den Bussche, R.A., Malloy, T.P., Haynie, M.L., and Mickley, C.O. (2000). Mitochondrial DNA variation in pupfishes assigned to the species *Cyprinodon macularius* (Atherinomorpha: Cyprinodontidae): taxonomic implications and conservation genetics. *Copeia*, 2000, 353–64.

Eddy, F.B. and Handy, R.D. (2012). *Ecological and Environmental Physiology of Fishes*. Oxford: Oxford University Press.

Edgar, G.J. and Shaw, C. (1995). The production and trophic ecology of shallow-water fish assemblages in southern Australia. II. Diets of fishes and trophic relationship between fishes and benthos at Western Port, Victoria. *Journal of Experimental Marine Biology and Ecology*, 194, 83–106.

Edinger, L. (1899). *The Anatomy of the Central Nervous System of Man and of Vertebrates in General*. Trans. from 5th German edn by W.S. Hall, P.L. Holland, and E.P. Carlton.Philadelphia, PA: F.A.Davis Co.

Eduardo, J. Bicudo, P.W., and Johansen, K. (1979). Respiratory gas exchange in the airbreathing fish, *Synbranchus marmoratus*. *Environmental Biology of Fishes*, 4, 55–64.

Edwards, S.L. and Gross, G.G. (eds) (2015). *Hagfish Biology*. Boca Raton, FL: CRC Press.

Einarsson, S. and Davies, P.S. (1966). On the localisation and ultrastructure of pepsinogen, trypsinogen and chymotrypsinogen secreting cells in the Atlantic salmon, *Salmo salar* L. *Comparative Biochemistry and Physiology*, 114B, 295–301.

Einarsson, S., Davies, P.S., and Talbot, C. (1996). The effect of feeding on the secretion of pepsin, trypsin and chymotrypsin in the Atlantic salmon, *Salmo salar* L. *Fish Physiology and Biochemistry*, 15, 439–46.

Ekström, P. and Meissl, H. (1997). The pineal organ of teleost fishes. *Reviews in Fish Physiology and Fisheries*, 7, 199–284.

Elbaz, I., Foulkes, N.S., Gothiff, Y., and Appelbaum, L. (2013). Circadian clocks, rhythmic symaptic plasticity and the sleep-wake cycle in zebrafish. *Frontiers in Neural Circuits*, 7P, Article 9. doi.10.3389/fncir.2013.0009

Elger, M., Wahlquist, L., and Hentschel, H. 1984 Ultrastructure and adrenergic innervation of preglomerular arterioles in the euryhaline teleost *Salmo gairdneri*. *Cell and Tissue Research*, 237, 451–8.

Elliott, D. (2000). Early vertebrates and fish. In: P.L. Hancock, B.J. Skinner, and D.L. Dineley (eds). *The Oxford Companion to the Earth*. Oxford: Oxford University Press.

Ellis, A.E. (1977). The leucocytes of fish: A review. *Journal of Fish Biology*, 11, 453–91.

Elliott, M., Whitfield, A.K., Potter, I.C., Blaber, M., Cyrus, D.P. Nordlie, F.G., and Harrison, T.D. (2007). The guild approach to categorising estuarine fish assemblages: a global review. *Fish and Fisheries*, 8, 241–68.

Endler, J.A. (1990). On the measurement and classification of colour in studies of animal colour patterns. *Biological Journal of the Linnean Society*, 41, 315–52.

Engel, D.W., Angelovic, J.W., and Davis, E.M. (1966). Effects of acute irradiation on the blood constituents of pinfish *Lagodon rhomboids*. *Chesapeake Science*, 7, 90–9.

Engin, K. and Carter, C.G. (2001). Ammonia and urea excretion rates of juvenile Australian short-finned eel (*Anguilla australis australis*) as influenced by dietary protein level. *Aquaculture*, 194, 123–36.

Epstein, F.H. and Silva, P. (2005). Mechanisms of rectal gland secretion. *Bulletin of the Mount Desert Island Laboratory*, 44, 1–5.

Erickson, J.D. (2002). Folic acid and prevention of spina bifida and anencephaly. *Morbidity and Mortality Weekly Report*, 51 (RR13), 1–3.

Erlandsson, A. and Ribbink, A.J. (1997). Patterns of sexual size dimorphism in African cichlid fishes. *South African Journal of Science*, 93, 498–508.

Escobar, S. Servill, A. Espigares, F., Guegen, M-M., Brocal, I., Felip, A., Gómez, Carrillo, M., A., Zanuy, S., and Kah, O. (2013). Expression of kisspeptins and kiss receptors suggests a large range of functions for kisspeptin systems in the brain of the European sea bass. *PloS One*, 8(7). doi: org/10.1371/journalpone. 0070177.

Esteban, M.Á. (2012). An overview of the immunological defenses in fish skin. *International Scholarly Research Notes Immunology*, 1–29.

Evans, D.H. and Piermarini, P.M. (2001). Contractile properties of the elasmobranch rectal gland. *Journal Experimental of Biology*, 204, 59–67.

Evans, D.H. Piermarini, P.M., and Choe, K.P. (2005). The multifunctional fish gill: Dominant site of gas exchange, osmoregulation, acid-base regulation, and excretion of nitrogenous waste. *Physiological Review*, 85, 97–177.

Evans, D.L. Carlson, R.L., Graves, S.S., and Hogan K.T. (1984). Nonspecific cytotoxic cells in fish (*Ictalurus punctatus*). IV. Target cell binding and recycling capacity. *Developmental and Comparative Immunology*, 8, 823–33.

Fabra, M., Raldúa, D., Power, D.M., and Cerdà, J. (2005). Marine fish egg hydration is aquaporin-mediated. *Science*, 307, 545.

Fagen, R.M. (1981). *Animal Play Behaviour*. Oxford: Oxford University Press.

Fahay, M.P. (1983). *Early Stages of Fishes in the Western North Atlantic Ocean (Davis Strait, Southern Greenland and Flemish Cap to Cape Hatteras), Vol. 1: Acipenseriformes through Syngnathiformes*. Dartmouth, Nova Scotia: North Atlantic Fisheries Organization.

Fajardo, O., Zhu, P., and Friedrich, R.W. (2013). Control of a specific motor program by a small brain area in zebrafish. *Frontiers in Neural Circuits*, 7, Article 67, 19 pp. doi: 10.3389/fncir.2013.00067

Falcón, J. and Collin, J.P. (1991). Pineal-retinal relationships: rhythmic biosynthesis and immunocytochemical localization in the retina of the pike (*Esox lucius*). *Cell and Tissue Research*, 265, 601–9.

Falcón, J., Migaud, H., Muňoz-Cueto, J.A., and Carillo, M. (2010). Current knowledge on the melatonin system in teleost fish. *General and Comparative Endocrinology*, 165, 469–82.

Fänge, R. and Fugelli, K. (1962). Osmoregulation in chimaeroid fishes. *Nature*, 196, 689.

Fänge, R. and Fugelli, K. (1963). The rectal salt gland of elasmobranchs, and osmoregulation in chimaeroid fishes. *Sarsia*, 10, 27–34.

Fänge, R. and Nilsson, S. (1985). The fish spleen: structure and function. *Cell and Molecular Life Sciences*, 41, 152–8.

Farmer, C. (1997). Did lungs and the intracardiac shunt evolve to oxygenate the heart in vertebrates? *Palaeobiology*, 23, 358–72.

Farina, S.C., Near, T., and Bemis, W.E. (2015). Evolution of the branchiostegal membrane and restricted gill openings in actinopterygian fishes. *Journal of Morphology*, 276, 682–94.

Farrell, A.P. (1993). Cardiovascular system. In: D.H. Evans (ed.). *The Physiology of Fishes*. Boca Raton, FL: CRC Press, pp. 219–50.

Farrell, A.P., Thorarensen, H., Axelsson, M., Crocker, C.E., Gamperl, A.K., and Cech, J.J. (2001). Gut blood flow in fish during exercise and severe hypercapnia. *Comparative Biochemistry and Physiology A*, 128, 551–63.

Fasulo, S., Tagliafierro, G., Contini, A., Ainis, L., Ricca, M.B., Yanihara, N., and Zaccone, G. (1993). Ectopic expression of bioactive peptides and serotonin in the sacciform gland cells of teleost skin. *Archives of Histology and Cytology*, 56, 117–25.

Feder, M.E. and Hofmann, G.E. (1999). Heat-shock proteins, molecular chaperones, and the stress response: evolutionary and ecological physiology. *Annual Review of Physiology*, 61, 243–82.

Fedorov, K.Y. (1971). The state of the gonads of the Barents Sea Greenland halibut [*Reinhardtius hippoglossoides* (Walb.)] in connection with failure to spawn. *Journal of Ichthyology*, 11, 673–82.

Fenwick, J.C. (1970). The pineal organ. In: W.S. Hoar and D.J. Randall (eds). *Fish Physiology, Vol. IV. The nervous system, circulation and respiration*, New York, NY: Academic Press, pp. 91–108.

Fenwick, J.C. (1976). Effect of stanniectomy on calcium activated adenosine triphosphatase activity in the gills of fresh water adapted North American eel *Anguilla rostrata* LeSouer. *General and Comparative Endocrinology*, 29, 383–7.

Fernald, R.D. (1991). Teleost vision: seeing while growing. *Journal of Experimental Zoology*, 5, 167–80.

Fernald, R.D. (1993). Vision. In: D.H. Evans (ed.). *The Physiology of Fishes*. Boca Raton, FL: CRC Press, pp. 161–89.

Fernández, L. and Schaefer, S.A. (2009). Relationships among the neotropical candirus (Trichomycteridae, siluriformes) and the evolution of parasitism based on analysis of mitochondrial and nuclear gene sequences. *Molecular Phylogenetics and Evolution*, 52, 416–23.

Fernandes, M.N., da Cruz, A.L. da Costa, O.T.F., and Perry, S.F. (2012). Morphometric partitioning of the respiratory surface area and diffusion capacity of the gills and swim bladder in juvenile Amazonian air-breathing fish, *Arapaima gigas*. *Micron*, 43, 961–70.

Ferraris, C. J. (2007). *Checklist of Catfishes, Recent and Fossil (Osteichthys: Siluriformes) and Catalogue of Siluriform Primary Types*. Auckland: Magnolia Press.

Fields, R.D., Bullock, T.H., and Lange, G.D. (1993). Ampullary sense organs; peripheral, central and behavioural electroreception in chimaeras (*Hydrolagus*; Holocephali; Chondrichthyes). *Brain, Behaviour and Evolution*, 41, 269–89.

Fine, M.L. (1978). Seasonal and geographical variation of the mating call of the oyster toadfish *Opsanus tau* L. *Oecologia*, 36, 45–57.

Finger, T.E. (1982). Somatotopy in the representation of the pectoral fin and free fin rays in the spinal cord of the searobin, *Prionotus carolinus*. *Biological Bulletin*, 163, 154–61.

Finn, R.N. (2007). Vertebrate yolk complexes and the functional implications of phosvitins and other subdomains in vitellogenesis. *Biology of Reproduction*, 76, 926–35.

Finn, R.N., Fyhn, H.J., Norberg, B., Munholland, J., and Reith, M. (2000). Oocyte hydration as a key feature in the adaptive evolution of teleost fishes to seawater. In: B. Norberg, O.S. Kjesbu, G.L. Taranger, E. Andersson, and S.O. Stefansson (eds). *Proceedings of the Sixth International Symposium on the Reproductive Physiology of Fishes*. Bergen: Institute of Marine Research and University of Bergen., pp. 289–291.

Fishelson, L., Delarea, Y., and Zverdling, A. (2004). Taste bud formation and distribution on lips and in the oropharyngeal cavity of cardinal fish species (Apogonidae, Teleostei), with remarks on their dentition. *Journal of Morphology*, 259, 316–27.

Fishelson, L., Golani, D., Galil, B., and Goren, M. (2010). Comparison of the nasal olfactory organs of various species of lizard fishes (Teleostei: Aulopiformes: Synodontidae) with additional remarks on the brain. *International Journal of Zoology*, Article ID 807913.

Fishelson, L., Golani, D., Russell, B., Galil, B., and Goren, M. (2012). Melanization of the alimentary tract in lizardfishes (Teleostei, Aulopiformes, Synodontidae). *Environmental Biology of Fishes*, 95, 195–200.

Fishelson, L., Goren, M., and Gon, O. (1997). Black gut phenomenon in cardinal fishes (Apogonidae, Teleostei). *Marine Ecology Progress Series*, 161, 293–8.

Fisher, R.A., Fritzsche, R.A., and Hendrickson, G.L. (1985). Histology and ultrastructure of the 'jellied' condition in Dover sole, *Microstomus pacificus*. *Proceedings of the Fifth Congress of European Ichthyologists. Stockholm 1985*. 345–50. Stockholm: Museum of Natural History.

Fives, J.M. (1970). *Fiches d'identification des oeufs et larves de poisons. Fiche No.3 Blenniidae of the North Atlantic*. Conseil international pour l'exploration de la mer.

Flamarique, I.N. (2011). Unique photoreceptor arrangements in a fish with polarized light discrimination. *Journal of Comparative Neurology*, 519, 714–37.

Flammang, B.E. (2010). Functional morphology of the radialis muscle in shark tails. *Journal of Morphology*, 271, 340–52.

Fleming, I.A. and Ng, S. (1987). Evaluation of techniques for fixing, preserving, and measuring salmon eggs. *Canadian Journal of Fisheries and Aquatic Science*, 44, 1957–62.

Fletcher, G.L. (1975). The effects of capture, 'stress', and storage of whole blood on the red blood cells, plasma proteins, glucose, and electrolytes of the winter flounder (*Pseudopleuronectes americanus*). *Canadian Journal of Zoology*, 53, 197–206.

Fletcher, G.L. (1977). Circannual cycles of blood plasma freezing point and Na + and Cl- concentrations in Newfoundland winter flounder (*Pseudopleuronectes americanus*): correlation with water temperature and photoperiod. *Canadian Journal of Zoology*, 55, 789–95.

Fletcher, G.L. and King, M.J. (1978). Seasonal dynamics of Cu^{2+}, ZN^{2+}, Ca^{2+}, and Mg^{2+} in gonads and liver of winter flounder (*Pseudopleruronectes americanus*): evidence for summer storage of Zn^{2+} for winter gonad development in females. *Canadian Journal of Zoology*, 56, 284–90.

Fletcher, G.L. and Rise, M.L. (eds) (2012). *Aquaculture Biotechnology*. Oxford: Wiley-Blackwell.

Fletcher, G.L., Hew, C.L., and Davies, P.L. (2001). Antifreeze proteins of teleost fishes. *Annual Review of Physiology*, 63, 359–90.

Fletcher, G.L., King, M.J., Kao, M.H., and Shears, M.A. (1989). Antifreeze proteins in the urine of marine fish. *Fish Physiology and Biochemistry*, 6, 121–7.

Flik, G., Labedz, T., Lafeber, F.P.J.G., Wendelaar Bonga, S.E., and Pang, P.K.T. (1989). Studies on teleost corpuscles of Stannius: Physiological and biochemical aspects of synthesis and release of hypocalcin in trout, goldfish and eel. *Fish Physiology and Biochemistry*, 7, 343–9.

Flock, Å. (1965). Transducing mechanisms in lateral line canal organ receptors. *Cold Spring Harbor Symposium*, 30, 133–45.

Flock, Å. (1971). The lateral line organ mechanoreceptors. In: W.S. Hoar and D.J. Randall (eds). *Physiology of Fishes, Vol. V: Sensory Organs*. New York, NY: Academic Press, pp. 241–63.

Flock, Å. and Wersäll, J. (1962). A study of the orientation of the sensory hairs of the receptor cells in the lateral line organ of fish with special reference to the function of the receptors. *Journal of Cell Biology*, 15, 19–27.

Flynn, E.J., Trent, C.M., and Rawls, J.F. (2009). Ontogeny and nutritional control of adipogenesis in zebrafish (*Danio rerio*). *Journal of Lipid Research*, 50, 1641–52.

Flynn, S.R. and Benfey, T.J. (2007). Effects of dietary estradiol-17β in juvenile shortnose sturgeon, *Acipenser brevirostrum*, Lesuer. *Aquaculture*, 270, 405–12.

Flynn, S.R. and Burton, M.P.M. (2003). Gametogenesis in capelin, *Mallotus villosus* (Müller), in the northwest Atlantic Ocean. *Canadian Journal of Zoology*, 82, 1511–23.

Flynn, S.R., Matsuoka, M., Reith, M., Martin-Robichaud, D.J., and Benfey, T.J. (2006). Gynogenesis and sex determination in shortnose sturgeon, *Acipenser brevirostrum*. *Aquaculture*, 253, 721–7.

Flynn, S.R., Nakashima, B.S., and Burton, M.P.M. (2001). Direct assessment of post-spawning survival of female capelin, *Mallotus villosus*. *Journal of the Marine Biological Association of the United Kingdom*, 81, 307–12.

Fong, A.Y., Zimmer, M.B., and Milsom, W.K. (2009). The conditional nature of the 'Central Rhythm Generator' and the production of episodic breathing. *Respiratory Physiology and Neurobiology*, 168, 179–87.

Foote, C.J., Brown, G., and Wood, C.C. (1997). Spawning success of males using alternative mating tactics in sockeye salmon, *Oncorhynchus nerka*. *Canadian Journal of Fisheries and Aquatic Science*, 54, 1785–95.

Foran, D. (1991). Evidence of luminous bacterial symbionts in the light organs of myctophid and stomiiform fishes. *Journal of Experimental Zoology*, 259, 1–8.

Forbes, J.M. (2001).Consequences of feeding for future feeding. *Comparative Biochemistry and Physiology Part A*, 128, 463–70.

Forbes, S.A. (1907). On the local distribution of certain Illinois fishes: an essay in statistical ecology. *Bulletin of the Illinois State Laboratory of Natural History*, VII, Art. VIII.

Forsmarks Kraftgrupp (1982). *The Biotest Basin at Forsmark—A Giant Research Aquarium*. The National Swedish Environment Protection Board.

Foster, G.D., Zhang, J., and Moon, T.W. (1993). Carbohydrate metabolism and hepatic zonation in the Atlantic hagfish, *Myxine glutinosa* liver: effects of hormones. *Fish Physiology and Biochemistry*, 12, 211–19.

Fox, H., Lane, E.B., and Whitear, M. (1980). Sensory nerve endings in receptors in fish and amphibians. In: Spearman R.I.C. and Riley P.A. (eds). *The Skin of Vertebrates.* Linnean Society Symposium, 9. New York, NY: Academic Press, pp. 271–81.

Fox, S.I. (2002). *Human Physiology*, 7th edn. Boston, MA: McGraw-Hill.

Franz, V. (1913). *Schorgan In Oppels Lehrbuch der Vergleichenden Mikrosopischen Anatomie der Wirbeltiere*. Jena: Fischer.

Frick, N.T. and Wright, P.A. (2002a). Nitrogen metabolism and excretion in the mangrove killifish *Rivulus marmoratus* II. Significant ammonia volatilization in a teleost during air-exposure. *Journal of Experimental Biology*, 205, 91–100.

Frick, N.T. and Wright, P.A. (2002b). Nitrogen metabolism and excretion in the mangrove killifish *Rivulus marmoratus* 1. The influence of environmental salinity and external ammonia. *Journal of Experimental Biology*, 205, 79–89.

Fridberg, G. and Bern, H.A. (1968).The urophysis and the caudal neurosecretory system of fishes. *Biological Reviews*, 43, 175–99.

Fritsch, G. (1878). *Untersuchangen über den feineren Bau des Fischgehirns mit besonderer Berücksichtigung der Homologien bie anderen Wirbelthierklassen*. Berlin: Verlag Gutmann'sche Buchhandlung.

Fritzche, R.A. (1978). *Development of fishes of the Mid-Atlantic Bight. An Atlas of Egg, Larval and Juvenile Stages, Vol. V: Chaetodontidae through Ophididae*. Washington, DC: U.S. Department of the Interior Fish and Wildlife Service. (FWS/OBS–78/12).

Frölich, E. and Wagner, H.J. (1998). Development of multibank rod retinae in deep-sea fishes. *Visual Neuroscience*, 15, 477–83.

Frost, W.E. (1954). The food of the pike, *Esox lucius* L. in Windermere. *Journal of Animal Ecology*, 23, 339–60.

Fryer, G. (1959). The trophic interrelationships and ecology of some littoral communities of Lake Nyasa with especial reference to the fishes, and a discussion of the evolution of a group of rock-frequenting Cichlidae. *Proceedings of the Zoological Society of London*, 132, 153–281.

Fryer, G. and Iles, T.D. (1972). *The Cichlid Fishes of the Great Lakes of Africa*. Edinburgh: Oliver and Boyd.

Fudge, D.S. and Stevens, E.D. (1996). The visceral *retia mirabilia* of tuna and sharks: an annotated translation and discussion of the Eschricht and Müller 1835 paper and related papers. *Guelph Ichthyology Reviews* 4. 1–92

Fudge, D.S. Levy, N., Chiu, S., and Gosline, J.M. (2005). Composition, morphology and mechanics of hagfish slime. *Journal of Experimental Biology*, 208, 4613–25.

Fugère, V. and Krahe, R. (2010). Electric signals and species recognition in the wave-type gmnotiform fish *Apteronotus leptorhynchus*. *Journal of Experimental Biology*, 213, 225–36.

Fuhrman, F.A. (1967). Tetrodotoxin. *Scientific American*, 217 (no. 2), 60–71.

Fujii, R. (1993). Colouration and chromatophores. In: D.H. Evans (ed.). *The Physiology of Fishes*. Boca Raton, FL: CRC Press, pp. 535–562.

Fujii, R. (2000). The regulation of motile activity in fish chromatophores. *Pigment Cell Research*, 13, 300–19.

Fujii, R. and Miyashita, Y. (1976). Beta adrenoceptors, cyclic AMP and melanosome dispersion in guppy melanophores. In: V. Riley (ed.). *Pigment Cell, Vol. 3*. Basel: S. Karger, pp. 336–44.

Fujii, R. and Oshima, N. (1986). Control of chromatophore movements in teleost fishes. *Zoological Science*, 3, 13–47.

Fujii, R. and Taguchi, S. (1969). The responses of fish melanophores to some melanin-aggregating and dispersing agents in potassium-rich medium. *Annotationes, Zoologicae Japonenses/Nihon*, 42, 176–82.

Fujita, I., Sorensen, P.W., Stacey, N.E., and Hara, T.J. (1991). The olfactory system, not the terminal nerve, functions as the primary chemosensory pathway mediating responses to sex pheromones in male goldfish. *Brain, Behavior and Evolution*, 38, 313–21.

Furness, J.B. and Costa, M. (1987). *The Enteric Nervous System*. Edinburgh: Churchill Livingstone.

Galigher, A.E. and Kozloff, E.N. (1964). *Essentials of Practical Microtechnique*. Philadelphia, PA: Lea and Febiger.

Gambaryan, S.P. (1994). Microdissectional investigation of the nephrons in some fishes, amphibians, and reptiles. *Journal of Morphology*, 219, 319–39.

Gans, C. and Parsons, T.S. (1964). *A Photographic Atlas of Shark Anatomy. The Gross Anatomy of Squalus acanthias*. Chicago, IL: University of Chicago Press.

Garcia-Berhou, E., Carmona-Catot, G., Merciai, R., and Ogle, D.H. (2012). A technical note on seasonal growth models. *Reviews in Fish Biology and Fisheries*, 22, 635–40.

Garcia-Romeu, F., Cossins, A.R., and Motais, R. (1991). Cell volume regulation by trout erythrocytes: characteristics of the transport systems activated by hypotonic swelling. *Journal of Physiology*, 440, 547–67.

Gardner, E. (1975). *Fundamentals of Neurology*, 6th edn. Philadelphia, PA: W.B. Saunders Co.

Gauldie, R.W. (1993). Polymorphic crystalline structure of fish otoliths. *Journal of Morphology*, 218, 1–28.

Gaye-Siessegger, J., Focken, U., Abel, H., and Becker, K. (2007). Influence of dietary non-essential amino acid profile on growth performance and amino acid metabolism of Nile tilapia, *Oreochromis niloticus* (L.). *Comparative Biochemistry and Physiology A. Molecular and Integrative Physiology*, 146, 71–7.

Gemballa, S. and Bartsch, P. (2002). Architecture of the integument in lower teleostomes: functional morphology and evolutionary implications. *Journal of Morphology*, 253, 290–309.

Gendron, R.L. Armstrong, E., Paradis, H., Haines, L., Desjardins, M., Short, C.E., Clow, K.A., and Driedzic, W.R. (2011). Osmotic pressure-adaptive responses in the eye tissues of rainbow trout (*Osmerus mordax*). *Molecular Vision*, 17, 2596–604.

Genicot, S., Feller, G., and Gerday, C. (1988). Trypsin from Antarctic fish (*Paranotothenia magellanica* Forster) as compared with trout (*Salmo gairdneri*) trypsin. *Comparative Biochemistry and Physiology*, 90B, 601–9.

Gerlach, G. (2006). Pheromonal suppression of reproductive success in female zebrafish: female suppression and male enhancement. *Animal Behaviour*, 72, 1119–24.

Gerlach, G., Hodgins-Davis, A., Avolio, C., and Schunter, C. (2008). Kin recognition in zebrafish: a 24-hour window for olfactory imprinting. *Proceedings of the Royal Society B*, 275, 2165–70.

German, D.P., Horn, M.H., and Gawlicka, A. (2004). Digestive enzyme activities in herbivorous and carnivorous prickleback fishes (Teleostei: Stichaeidae): ontogenetic, dietary, and phylogenetic effects. *Physiological and Biochemical Zoology*, 77, 789–804.

Geven, E.J.W., Nguyen, N-K., van den Boogaart, M., Spanings, F.A.T., Flik, G., and Klaren, P.H.M. (2007). Comparative thyroidology: thyroid gland location and iodothyronine dynamics in Mozambique tilapia (*Oreochromis mossambicus* Peters) and common carp (*Cyprinus carpio* L.). *Journal of Experimental Biology*, 210, 4005–15.

Gibbins, I. (1994). Comparative anatomy and evolution of the autonomic nervous system. In: S. Nilsson and S. Holmgren (eds). *Comparative Physiology and Evolution of the Autonomic Nervous System*. Chur, Switzerland: Harwood pp. 1–67.

Gibbins, I.L. (1989). Co-existence and co-function. In: S. Holmgren (ed.). *The Comparative Physiology of Regulatory Peptides*. London: Chapman and Hall, pp. 308–43.

Gilbert, R.C. (1993). Evolution and Phylogeny. In: D.H. Evans (ed.). *The Physiology of Fishes*, 2nd edn. Boca Raton, FL: CRC Press, pp. 1–45.

Gilbert, P.W. and Heath, G.W. (1972). The clasper-siphon sac mechanism in *Squalus acanthias* and *Mustelus canis*. *Comparative Biochemistry and Physiology*, 42A, 97–119.

Giles, N. (1994). *Freshwater Fish of the British Isles*. Shrewsbury: Swan Hill Press.

Gillis, G.B. (1996). Undulatory locomotion in elongate aquatic vertebrates: anguilliform swimming since Sir James Gray. *Amererican Zoologist*, 36, 656–65

Gilmore, R.G. (1993). Reproductive biology of laminoid sharks. *Environmental Biology of Fishes*, 38, 95–114.

Gilmour, K.M. (1998). Gas exchange. In: D.H. Evans (ed.). *The Physiology of Fishes*, 2nd edn. Boca Raton, FL: CRC Press, pp. 101–27.

Gilmour, K.M., Perry, S.F., Bernier, N.J., Henry, R.P., and Wood, C.M. (2001). Extracellular carbonic anhydrase in the dogfish *Squalus acanthias*: a role in CO_2 excretion. *Physiological and Biochemical Zoology*, 74, 477–92.

Gingerich, W.H. (1982). Hepatic toxicology of fishes. In: L.J. Weber (ed.). *Aquatic Toxicology*. New York, NY: Raven Press.

Gingins, S., Werminghausen, J., Johnstone, R.A. Grutter, A.S., and Bshary, R. (2013) Power and temptation cause shifts between exploitation and cooperation in a cleaner wrasse mutualism. *Proceedings of the Royal Society B*, 280 (issue 1761, 6), 20130553.

Ginzburg, A.S. (1961). The block to polyspermy in sturgeon and trout with special reference to the role of cortical granules (alveoli). *Journal of Embryology and Experimental Morphology*, 9, 173–90.

Ginzburg, A.S. (1972). *Fertilization in Fishes and the Problem of Polyspermy*. Trans. Russian. Academy of Sciences of the USSR, Institute of Developmental Biology. Jerusalem: Israel Program for Scientific Translations.

Gislason, H. and Thorarensen, H. (2006). Gill surface area and heart size of different strains of Arctic charr (*Salvelinus alpinus*). VIIth International Congress of Biology of Fish, St. John's NL, Bethesda MD: American Fisheries Society, pp. 108–108.

Glova, G.J., Sagar, P.M., and Docherty, C.R. (1987). Diel feeding periodicity of torrentfish (*Cheimarrichthys fosteri*) in two braided rivers of Canterbury, New Zealand. *New Zealand Journal of Marine and Freshwater Research*, 21, 555–61.

Glover, C.N., Bucking, C., and Wood, C.M. (2011). Adaptations to *in situ* feeding: novel nutrient acquisition pathways in an ancient vertebrate. *Proceedings of the Royal Society B*, 278, 3096–101.

Goda, M. and Fujii, R. (1995). Blue chromatophores in two species of callionymid fish. *Zoological Science*, 12, 811–13.

Goetz, F.W. and Bergman, H.L. (1978). The effects of steroids on final maturation and ovulation of oocytes from brook trout (*Salvelinus fontinalis*) and yellow perch (*Perca flavescens*). *Biology of Reproduction*, 18, 293–8.

Goldsworthy, G.J., Robinson, J., and Mordue, W. (1981). *Endocrinology*. New York, NY: John Wiley and Sons.

Good, N.E., Winger, G.D., Winter, W., Connolly, T.N., Izawa, S., and Singh, R.M.M. (1966). Hydrogen ion buffers for biological research. *Biochemistry*, 5, 467–77.

Goodrich, E.S. (1901). On the pelvic girdle and fin of *Eusthenopteron*. *Quarterly Journal of Microscopical Science*, 45, 311–24.

Goodrich, E.S. (1909). *The Vertebrate Craniata, Vol. IX: Treatise on Zoology*. London: Adam and Charles Black.

Goodrich, E.S.a (1930, 1958). *Studies on the structure and development of vertebrates*. London: Macmillan. (Republished 1958, New York, NY: Dover).

Goodrich, E.S.b (1930, 1958). Excretory organs and genital ducts. In: *Studies on the Structure and Development of Vertebrates*, *Vol. II*. New York, NY: Dover Publications Inc., pp. 657–719.

Gonzales, T.T., Katoh, M., and Ishimatsu, A. (2008). Respiratory vasculature of the inter-tidal air-breathing eel goby, *Odontamblyopsis lacepedii* (Gobiidae: Amblyopinae). *Environmental Biology of Fishes*, 82, 341–51.

Goss, G.G., Perry, S.F., Wood, C.M., and Laurent, P. (1992). Mechanisms of ion and acid-base regulation at the gills of freshwater fish. *Journal of Experimental Zoology*, 263, 143–59.

Goto, D. and Wallace, W.G. (2010). Metal intracellular portioning as a detoxification mechanism for mummichogs (*Fundulus heteroclitus*) living in metal-polluted salt marshes. *Marine Environmental Research*, 69, 163–71.

Goulet, D., Green, J.M., and Shears, T.M. (1986). Courtship, spawning and parental care behaviour of the lumpfish, *Cyclopterus lumpus* L., in Newfoundland. *Canadian Journal of Zoology*, 64, 1320–5.

Graham, H.T. and O'Leary, J.L. (1941). Fast control fibres in fish, properties of Mauthner and Müller fibres of medullospinal system. *Journal of Neurophysiology*, 4, 224–42.

Graham, J.B. (1997). *Air-breathing Fishes: Evolution, Diversity and Adaptation*. San Diego: Academic Press.

Graham, M. (1923). The annual cycle in the life of the mature cod on the North Sea. *Fisheries Investigations Series II*, VI(6), 1–74.

Graham, M.S., Haedrich, R.L., and Fletcher, G.L. (1985). Hematology of three deep-sea fishes: a reflection of low metabolic rates. *Comparative Biochemistry and Physiology*, 80A, 79–84.

Grassé, P-P. (1958, 1997). *Traité de Zoologie* Tome XIII, Facscule 1,2 Agnathes et poisons, Anatomie, éthologie, systèmatique. Paris: Masson.

Grau, E.G., Prunet, P., Gross, T., Nishioka, R.S., and Bern, H. (1984). Bioassay for salmon prolactin using hypophysectomised *Fundulus heteroclitus*. *General and Comparative Endocrinology*, 53, 78–85.

Gray, E.G. (1955). An asymmetrical response of teleost melanophores. *Nature*, 175, 642–3.

Gray, E.G. (1956). Control of the melanophores of the minnow (*Phoxinus phoxinus* L.). *Journal of Experimental Biology*, 33, 448–59.

Gray, I.E. (1954). Comparative study of the gill area of marine fishes. *Biological Bulletin*, 107, 219–25.

Gray, J. (1928). The growth of fish II. The growth rate of the embryo of *Salmo fario*. *Journal of Experimental Biology*, 6, 110–24.

Gray, J. (1933a). Studies in animal locomotion I. The movement of fish with special reference to the eel. *Journal of Experimental Biology*, 10, 88–104.

Gray, J. (1933b). Studies in animal locomotion II. The relationship between waves of muscular contraction and the propulsive mechanism of the eel. *Journal of Experimental Biology*, 10, 386–90.

Gray, J. (1933c). Studies in animal locomotion III. The propulsive mechanism of the whiting (*Gadus merlangus*). *Journal of Experimental Biology*, 10, 391–400.

Gray, J. (1968). *Animal Locomotion*. London: Weidenfeld and Nicolson.

Green, J.M. (1971). High tide movements and homing behaviour of the tidepool sculpin *Oligocottus maculosus*. *Journal of the Fisheries Research Board, Canada*, 28, 383–9.

Green, J.M. (1974). A localised mass winter kill of cunners in Newfoundland. *Canadian Field Naturalist*, 88, 96–7.

Green, J.M. and Farwell, M. (1971). Winter habits of the cunner, *Tautogolabrus adspersus* (Walbaum 1792), in Newfoundland. *Canadian Journal of Zoology*, 49, 1497–9.

Green, J.M., Mlewa, C., and Dunbrack, R. (2006). Are marbled lungfish obligate air-breathers? Rates of aerial respiration in *Protopterus aethiopicus*. VIIth International Congress of Biology of Fish, St. John's, Canada. Bethesda, MD: American Fisheries Society, pp. 112–13.

Greenwood, M.P., Flik, G., Wagner, G.F., and Balment, R.J. (2009). The corpuscles of Stannius, calcium-sensing receptor and stanniocalcin: response to calciomimetics and physiological challenges. *Endocrinology*, 150, 3002–10.

Greenwood, P.H. (1975). Revised 3rd Edition of J.R. Norman, *A History of Fishes*. London: Ernest Benn Ltd.

Greenwood, P.H., Rosen, D.E. Weitzman, S.H., and Myers, G.S. (1966). Phyletic studies of teleostean fishes, with a provisional classification of living forms. *Bulletin of the American Museum of Natural History*, 131, 339–456.

Greer-Walker, M. and Pull, G.A. (1975). A survey of red and white muscle in marine fish. *Journal of Fish Biology*, 7, 295–300.

Gregory, W.K. (1933, 1959). Fish skulls: a study of the evolution of natural mechanisms. *Transactions of the American Philosophical Society*, 23, 75–481.

Grier, H.R. and Lo Nostro, F. (2000). The teleost germinal epithelium as a unifying concept. In: B. Norberg, O.S. Kjesbu, G.L. Taranger, E. Andersson, and S.O. Stefansson (eds). *Proceedings of the 6th International Symposium on the Reproductive Physiology of Fishes*. Bergen, Bergen: Institute of Marine Research and University of Bergen, pp. 233–6.

Griffith, R.W., Umminger, B.L., Grant, B.F., Pang, P.K.T., and Pickford, G.E. (1974). Serum composition of the coelacanth. *Journal of Experimental Zoology*, 187, 87–102.

Grizzle, J.M. and Rogers, W.A. (1976). *Anatomy and Histology of the Channel Catfish*. Auburn, AL: Auburn University.

Grizzle, J.M. and Thiyagarjah A. (1987). Skin histology of *Rivulus ocellatus marmoratus*: apparent adaptation for aerial respiration. *Copeia*, 1987, 237–40.

Groman, D.B. (1982). *Histology of the Striped Bass*. American Fisheries Society Monograph Number 3. Bethesda, MD: American Fisheries Society.

Grosell, M. (2010). The role of the GI tract in salt and H_2O balance. In: M. Grosell, A.P. Farrell, and C.J. Brauner (eds). *Fish Physiology, Vol. 30. The multifunctional gut of fish*. New York, NY: Academic Press.

Grosell, M., Farrell, A.P., and Brauner, C.J. (2010). In: A.P. Farrell and C.J. Brauner (eds). *Fish Physiology, Vol. 30. The multifunctional gut of fish*. New York, NY: Academic Press.

Grove, D.J. (1994). Chromatophores. In: S. Nilsson and S. Holmgren (eds). *Comparative Physiology and Evolution of the Autonomic Nervous System*. Chur, Switzerland: Harwood, pp. 331–52.

Grove, D.J. and Holmgren, S. (1992a). Intrinsic mechanisms controlling cardiac stomach volume of the rainbow trout (*Oncorhynchus mykiss*) following gastric distension. *Journal of Experimental Biology*, 163, 33–48.

Grove, D.J. and Holmgren, S. (1992b). Mechanisms controlling stomach volume of the Atlantic cod (*Gadus morhua*) following gastric distension. *Journal of Experimental Biology*, 163, 49–63.

Grove, A.J. and Newell, G.E. (1942). *Animal Biology*, 1st edn. London: University Tutorial Press.

Grove, A.J. and Newell, G.E. (1953). *Animal Biology*, 4th edn. London: University Tutorial Press.

Grüter, C. and Taborsky, B. (2004). Mouthbrooding and biparental care: an unexpected combination but male brood care pays. *Animal Behaviour*, 68, 1283–9.

Grynfellt, E. (1909). Les muscles de l'iris chez les teleostiens. *Bibliotheca Anatomica*, 20, 265–332.

Gudding, R., Lillehaug, A., and Evensen, O. (eds) (2014). *Fish Vaccination*. Chichester: Wiley-Blackwell.

Guerrero-Estévez, S.M. and López-López, E. (2016). Effects of endocrine disruptors on reproduction in viviparous teleosts with intraluminal gestation. *Reviews in Fish and Fisheries*. doi: 10.1007/s11160-19443-0.

Guest, M.M., Bond, T.P., Cooper, R.G., and Derrick, J.R. (1963). Red Blood cells: change in shape in capillaries. *Science*, 142, 1319–21.

Günther, K. and Deckert, K (1950). Wunderweit der Tiefsee. Berlin: Herbig Verlags buchhandlung.

Guraya, S.S. (1986). *The Cell and Molecular Biology of Fish Oogenesis*. Basel: Karger.

Gutowska, M.A., Drazen, J.C., and Robison, B.H. (2004). Digestive chinolytic activity in marine fishes of Monterey Bay, California. *Comparative Biochemistry and Physiology A*, 139, 351–8.

Haddock, S.H.D., Moline, M.A., and Case, J.F. (2010). Bioluminescence in the sea. *Annual Review of Marine Science*, 2, 443–93.

Hadley, M. E. (1996). *Endocrinology*. Upper Saddle River, NJ: Prentice-Hall Inc.

Hagey, L.R., Møller, P.R., Hofmann, A.F., and Krasowski, M.D. (2010). Diversity of bile salts in fish and amphibians: evolution of a complex biochemical pathway. *Physiological and Biochemical Zoology*, 83, 308–21.

Hale, M.E. (2000). Startle responses of fish without Mauthner neurons: Escape behavior of the lumpfish (*Cyclopterus lumpus*). *Biological Bulletin*, 199, 180–2.

Hale, P.A. (1965). The morphology and histology of the digestive systems of two freshwater teleosts of two freshwater teleosts, *Poecilia reticulata* and *Gasterosteus aculeatus*. *Journal of Zoology*, 146, 132–49.

Hall, B.K. (2005). *Bones and Cartilage: Developmental and Evolutionary Skeletal Biology*. San Diego, CA: Elsevier Academic Press.

Halliday, R.G. (1970).Growth and vertical distribution of the glacier lanternfish, *Benthosema glaciale*, in the Northwestern Atlantic. *Journal of the Fisheries Research Board of Canada*, 27, 105–16.

Halpern, B.N. (1959). The role and function of the reticuloendothelial system in immunological processes. *Journal of Pharmacy and Pharmacology*, 11, 321–8.

Halver, J.E. (1978). The vitamins. In: *Aquaculture Development and Coordination Programme. Fish Feeding Technology*. FAO (UN). Corporate Document. Rome: United Nations Food and Agriculture Organization.

Halver, J.E. (1989). *Fish Nutrition*, 2nd edn. New York, NY: Academic Press.

Halver, J.E. and Hardy, R.W. (eds) (2002). *Fish Nutrition*. San Diego, CA: Academic Press.

Halver, J.E., Ashley, L.M., and Smith, R.R. (1969). Ascorbic acid requirements of coho salmon and rainbow trout. *Transactions of the American Fisheries Society*, 98, 762–71.

Ham, A.W. (1957). *Histology*, 3rd edn. London: Pitman Medical Publishing Co. Ltd.

Hammond, C.L. and Schulte-Merker, S. (2009). Two populations of endochondral osteoblasts with differential sensitivity to Hedgehog signalling. *Development*, 136, 3991–4000.

Hampton, J.A., McCuskey, P.A., McCuskey, R.S., and Hinton, D.E. (1985). Functional units in rainbow trout (*Salmo gairdneri*) liver. 1. Arrangement and histochemical properties of hepatocytes. *Anatomical Record*, 213, 166–75.

Hamre, K. (2011). Metabolism, interactions, requirements and functions of vitamin E in fish. *Aquaculture Nutrition*, 17, 98–115.

Hansen, A. and Finger, T.E. (2000). Phyletic distribution of crypt-type olfactory receptor neurons in fishes. *Brain, Behaviour and Evolution*, 55, 100–10.

Hansen, A., Rolen, S.H., Anderson, K., Morita, Y., Caprio, J., and Finger, T.E. (2005). Olfactory receptor neurons in fish, molecular and functional correlates. *Chemical Senses*, 30 (Suppl. 1), 311.

Hara, T.J. (1993). Chemoreception. In: D.H. Evans (ed.). *The Physiology of Fishes*, 2nd edn. Boca Raton, FL: CRC Press,pp. 191–218.

Hara, T.J. (2007). Gustation. In: T.H. Hara and B. Zielinski (eds) *Fish Physiology 25 Sensory Systems Neuroscience*, New York, NY: Elsevier,pp. 45–96.

Harden Jones, F.R. and Scholes, P. (1985). Gas secretion and resorption in the swimbladder of the cod *Gadus morhua*. *Journal of Comparative Physiology B*, 155, 319–31.

Harder, W. (1975). *Anatomy of Fishes*. Stuttgart: Schweizerbart'sche Verlagbuchhandlung.

Hardin, G. (1968). The tragedy of the commons. *Science*, 162, 1243–8.

Hardisty, M.W. (1979). *Biology of the Cyclostomes*. London: Chapman and Hall.

Hardisty, M.W. (2006). *Lampreys: Life without Jaws*. Ceredigion: Forrest Text.

Hardisty, M.W. and Potter, I.C. (1971). *The Biology of Lampreys, Vol. 1*. New York, NY: Academic Press.

Hardisty, M.W. and Potter, I.C. (1982). *The Biology of Lampreys. Vol. 4B*. New York, NY: Academic Press.

Hardy, A.C. (1956). *The Open Sea, The World of Plankton*. London: Collins.

Hardy, J.D. Jr. (1978a). *Development of Fishes of the Mid-Atlantic Bight. An Atlas of Egg, Larval and Juvenile Stages. Vol. II: Angullidae through Syngnathidae*. U.S. Department of the Interior, Fish and Wildlife Service. (FWS/OBS–78/12).

Hardy, J.D. Jr. (1978b). *Development of Fishes of the Mid-Atlantic Bight. An Atlas of Egg, Larval and Juvenile Stages. Vol. III: Aphredoderidae through Rachycentridae*. Washington, DC: U.S Department of the Interior, Fish and Wildlife Service. (FWS/OBS–78/12).

Harper, L., Golubovskaya, I., and Cande, W.Z. (2004). A bouquet of chromosomes. *Journal of Cell Science*, 1117, 4025–32.

Harrington, R.W. Jr. (1961). Oviparous hermaphroditic fish with internal self-fertilization. *Science*, 134, 1749–50.

Harrington, R.W. Jr. (1963). Twenty-four hour rhythms of internal self-fertilization and ovoposition by hermaphrodites of *Rivulus marmoratus*. *Physiological Zoology*, 36, 325–41.

Harrington, R.W. Jr. (1967). Environmentally controlled induction of primary male gonochorists from eggs of the self-fertilizing hermaphroditic fish *Rivulus marmoratus* Poey. *Biological Bulletin* 132, 174–99.

Harrington R.W. Jr. (1968). Delineation of the thermolabile phenocritical period of sex determination and differentiation in the ontogeny of the normally hermaphroditic fish, *Rivulus marmoratus* Poey. *Physiological Zoology*, 41, 447–60.

Harris, L. (1990). *Independent Review of the Northern Cod Stock*. Ottawa: Communication Directorate of the Department of Fisheries and Oceans.

Harris, G.G., Frishkopf, L., and Flock, Á. (1970). Receptor potentials from hair cells of the lateral line. *Science*, 167, 76–9.

Harrison, P., Zummo, G. Farina, F., Tota, B. and Johnston, I.A. (1991). Gross anatomy, myoarchitecture, and ultrastructure of the heart ventricle in the haemoglobinless icefish *Chaenocephalus aceratus*. *Canadian Journal of Zoology*, 69, 1339–47.

Hart, J.L. (1973). *Pacific Fishes of Canada*. Fisheries Research Board of Canada Bulletin No. 180. Ottawa: Fisheries Research Borad of Canada.

Hasler, A.D. (1957). The Sense Organs: Olfactory and Gustatory Senses of Fishes. In: M.E. Brown (ed.). *The*

Physiology of Fishes, *Vol. 2*. New York, NY: Academic Press, pp. 187–209.

Hasler, A.D. and Scholz, A.T. (1983). *Olfactory imprinting and homing in salmon. Investigation into the mechanism of the imprinting process*. Berlin: Springer Verlag.

Hasler, A.D. and Wisby, W.J. (1951). Discrimination of stream odors by fishes and its relation to parental stream behaviour. *The American Naturalist*, 85, 223–38.

Hasler, A.D., Scholz, A.T., and Horrall, R.M. (1978). Olfactory imprinting and homing in salmon: recent experiments in which salmon have been artificially imprinted to a synthetic chemical verify the olfactory hypothesis for salmon homing. *Scientific American*, 66, 347–55.

Hattingh, J. (1975). Heparin and ethylenediamine tetraacetate as anticoagulants for fish blood. *Pflüeger's Archiv European Journal of Physiology*, 355, 347–52.

Hawryshyn, C.W. (1998). Vision. In: D.H. Evans (ed.). *The Physiology of Fishes*, 2nd edn. Boca Raton, FL: CRC Press, pp. 345–74.

Hawryshyn, C.W., Arnold, M.G., Chiasson, D.J., and Martin, P.C. (1989). The ontogeny of ultraviolet light sensitivity in rainbow trout *Salmo gairdneri*. *Visual Neuroscience*, 2, 247–54.

Hayashi, H., Sugimoto, M., Shima, N., and Fujii, R. (1993). Circadian motile activity of erythrophores in the red abdominal skin of tetra fishes and its possible significance in chromatic adaptation. *Pigment Cell Research*, 6, 29–36.

Haygood, M.G., Edwards, D.B., Mowlds, G., and Rosenblatt, R.H. (1994). Bioluminescence of myctophid and stomiiform fishes is not due to bacterial luciferase. *Journal of Experimental Zoology*, 270, 225–31.

Haywood, G.P. (1974). A preliminary histochemical examination of the rectal gland in the dogfish *Poroderma africanum*. *Transactions of the Royal Society of South Africa*, 41, 203–11.

Hazel, R. (1993). Thermal Biology. In: D.H. Evans (ed.). *The Physiology of Fishes*, 2nd edn. Boca Raton, FL: CRC Press, pp. 427–67.

Hazelhoff, E.H. and Evenhuis, H.H. (1952). Importance of the 'counter current principle' for the oxygen uptake in fishes. *Nature*, 169, 77.

Hazon, N., Tierney, M.L., and Takei, Y. (1999). Renin-angiotensin system in elasmobranch fish: A review. *Journal of Experimental Zoology*, 284, 526–34.

Healey, E.G. (1957). The nervous system. In: M.E. Brown (ed.). *The Physiology of Fishes*, *Vol. 2*. New York, NY: Academic Press, pp. 1–119.

Hebb, D.O. (1949). *The Organization of Behaviour*. New York, NY: John Wiley.

Hebrank, J.H., Hebrank, M.R., Long, J.H., Block, B.A., and Wainwright, S.A. (1990). Backbone mechanics of the blue marlin *Makaira nigricans* (Pisces Istiophoridae). *Journal of Experimental Biology*, 148, 449–59.

Heiligenberg, W. (1975). Theoretical and experimental approaches to spatial aspects of electrolocation. *Journal of Comparative Physiology*, 87, 137–64.

Heiligenberg, W. (1993). Electrosensation. In: D.H. Evans (ed.). *The Physiology of Fishes*, 2nd edn. Boca Raton, FL: CRC Press, pp. 137–60.

Heiligenberg, W. and Altes, R.A. (1978). Phase sensitivity in electroreceptors. *Science*, 199, 1001–4.

Helfman, G.S., Collette, B.B., Facey, D.E., and Bowen, B.W. (2009). *The Diversity of Fishes. Biology, Evolution and Ecology*. Oxford: Wiley-Blackwell.

Hendersen, J.R. and Daniel, P.M. (1978). Portal circulations and their relation to countercurrent systems. *Quarterly Journal of Microscopical Science*, 63, 355–69.

Henderson, P.A. and Walker, I. (1986). On the leaf litter community of the Amazonian blackwater stream Tarmazinho. *Journal of Tropical Ecology*, 2, 1–16.

Hendrickson, D.A. and Romero, A.V. (1989). Conservation status of desert pupfish, *Cyrinodon macularius*, in Mexico and Arizona. *Copeia*, 1989, 478–83.

Hepher, B. (1988). *Nutrition of Pond Fishes*. Cambridge: Cambridge University Press.

Hernandez, L.P., Gibb, A.C., and Ferry-Graham, L. (2009). Trophic apparatus in Cyprinodontiform fishes: Functional specializations for picking and scraping behaviour. *Journal of Morphology*, 270, 645–61.

Herrero, M.J., Martinez, F.J., Miguez, J.M., and Madrid, J.A (2007). Response of plasma and gastrointestinal melatonin, plasma cortisol and activity rhythms of European sea bass (*Dicentrarchus labrax*) to dietary supplementation with tryptophan and melatonin. *Journal of Comparative Biochemistry and Physiology B*, 177, 319–26.

Herring, P. (2002). *The Biology of the Deep Ocean*. Oxford: Oxford University Press.

Herring, P.J. and Cope, C. (2005). Red bioluminescence in fishes: on the suborbital photophores of *Malacosteus*, *Pachystomias* and *Aristostomias*. *Marine Biology*, 148, 383–94.

Hibaya, T. (1982). *An Atlas of Fish Histology*. Tokyo: Kodansha Ltd.

Hickman, C.P. and Trump, B.F. (1969). The kidney. In: W.S. Hoar and D.J. Randall (eds). *Fish Physiology, Vol. I*. Excretion, ionic regulation and metabolism. New York, NY: Academic Press, pp. 91–239.

Hidalgo, M.C., Urea, E., and Sanz, A. (1999). Comparative study of digestive enzymes in fish with different nutritional habits. Proteolytic and amylase activities. *Aquaculture*, 170, 267–83.

Higuchi, M., Celino, F.T., Tamai, A., Miura, C., and Miura, T. (2012). The synthesis and role of taurine in the Japanese eel testis. *Amino Acids*, 43, 7734–81.

Hinton, D.E., Lantz, R.C., Hampton, J.A., McCuskey, P.R., and McCuskey, R.S. (1987). Normal versus abnormal structure: Considerations in morphologic responses of teleosts to pollutants. *Environmental Health Perspectives*, 71, 139–46.

Hinz, F.I., Dieterich, D.C., Tirrell, D.A., and Schuman, E.M. (2012). Nonconical amino acid labelling in vivo to visualize and affinity purify newly synthesized proteins in larval zebrafish. *ACS Chemical Neuroscience*, 3, 40–9.

Hjort, J. (1914). Fluctuations in the great fisheries of Northern Europe. *Conseil Permanent International pour L'exploration de la Mer, Rapports et Process-verbaux, Vol. XX.* 237 pp.

Hoagland, H. (1933). Electrical responses from the lateral line nerves of catfish I. *Journal of General Physiology*, 16, 695–714.

Hoar, W.S. (1957). Endocrine organs. In: M.E. Brown (ed.). *The Physiology of Fishes, Vol. 1: Metabolism.* New York, NY: Academic Press, pp. 245–86.

Hoar, W.S. (1966). *General and Comparative Physiology*. Englewood Cliffs, NJ: Prentice-Hall.

Hoar, W.S. (1983). *General and Comparative Physiology*. Englewood Cliffs, NJ: Prentice-Hall.

Hoar, W.S., Randall, D.J., and Farrell, A.P. (eds) (1992). *The Cardiovascular System*. New York, NY: Academic Press.

Hochachka, P.W. and Guppy, M. (1987). *Metabolic Arrest and the Control of Biological Time*. Cambridge, MA: Harvard University Press.

Hochachka, P.W. Moon, T.W., Bailey, J., and Hulbert, W.C. (1978). The osteoglossal kidney: correlations of structure, function and metabolism with transition to air breathing. *Canadian Journal of Zoology*, 56, No. 4 April (Part 2), 820–32.

Hodgkin, A.L. and Huxley, A.F. (1939). Action potential recorded from inside a nerve fibre. *Nature*, 144, 710.

Hoese, D.F. and Kottelat, M. (2005). *Bostrychus microphthalmus*, a new microophthalmic cavefish from Sulawesi (Teleostei: Gobiidae). *Ichthyological Exploration of Freshwaters*, 16, 183–91.

Hofer, R. and Sturmbauer, C. (1985). Inhibition of trout and carp α-amylase by wheat. *Aquaculture*, 48, 277–83.

Hoffman, R.A. (1970). The epiphyseal complex in fish and reptiles. *American Zoologist*, 10, 191–9.

Höglund, E., Balm, P.H.M., and Winberg, S. (2000). Skin darkening, a potential social signal in subordinate Arctic charr (*Salvelinus alpinus*): the regulatory role of brain monamines and pro-opiomelanocortin-derived peptides. *Journal of Experimental Biology*, 203, 1711–21.

Holden, J.A., Layfield, L.J., and Matthews, J.L. (2012). *The Zebrafish Atlas of Macroscopic and Microscopic Anatomy*. Cambridge: Cambridge University Press.

Holmes, R.L. and Ball, J.N. (1974). *The Pituitary Gland. A Comparative Account*. Cambridge: Cambridge University Press.

Holmgren, S. (1977). Regulation of the heart of a teleost, *Gadus morhua*, by autonomic nerves and circulating catecholamines. *Acta Physiologica Scandinavica*, 99, 62–74.

Holmgren, S. and Olsson, C. (2011a). Comparative physiology of the autonomic nervous system. *Autonomic Neuroscience: Basic and Clinical*, 165, 1, 1–148.

Holmgren, S. and Olsson, C. (2011b). Autonomic control of glands and secretion: A comparative view. *Autonomic Neuroscience: Basic and Clinical*, 165, 102–12.

Holmgren, U. (1959). *Studies on the pineal gland*. Ph.D. Thesis, Harvard University, Cambridge, MA.

Holzman, R., Day, S.W., Mehta, R.S., and Wainwright, P.C. (2008). Jaw protrusion enhances forces exerted on prey by suction feeding fishes. *Journal of the Royal Society Interface*, 5, 1445–57.

Hoogland, R., Morris, D., and Tinbergen, N. (1957). The spines of sticklebacks (*Gasterosteus* and *Pygosteus*) as means of defence against predators (*Perca* and *Esox*). *Behaviour*, 10, 205–36.

Hopkins, C.D. (1988). Neuroethology of electrocommunication. *Annual Review of Neuroscience*, 11, 497–535.

Hopkins, K.D. (1992). Reporting fish growth: A review of the basics. *Journal of the World Aquaculture Society*, 23, 173, 179.

Horbe, A.M.C. and da Silva Santos, A.G. (2009). Chemical composition of black-watered rivers in the Western Amazon region (Brazil). *Journal of the Brazilian Chemical Society*, 20, 1119–26.

Hori, M. (1993). Frequency-dependent natural selection in the handedness of scale-eating cichlid fish. *Science*, 260, 216–19.

Horton, J.M. and Summers, A.P. (2009). The material properties of acellular bone in a teleost fish. *Journal of Experimental Biology*, 212, 1413–20.

Howes, G.J. (1985). The phylogenetic relationships of the electric catfish family Malapteruridae (Teleostei: Siluoidei). *Journal of Natural History*, 19, 37–67.

Hsiang, H.L, Epp, J.R., van den Oever, Yan, C., Rashid, A.J., Insel, N., Ye, L., Niibori, Y., Deisseroth, K., Frankland, P.W., and Josselyn, S.A. (2014). Manipulating a 'cocaine engram' in mice. *Journal of Neuroscience*, 34, 14115–27.

Htun-Han, M. (1978). The reproductive biology of the dab *Limanda limanda* L. in the North Sea: seasonal changes in the ovary. *Journal of Fish Biology*, 13, 351–9.

Huang, P.L., Huang, Z., Mashimo, H., Bloch, K.D., Moskowitz, M.A., Bevan, J.A., and Fishman, M.C. (1995). Hypertension in mice lacking the gene for endothelial nitric oxide synthase. *Nature*, 377, 239–42.

Hubbs, C.L. and Lagler, K.F. (1958). *Fishes of the Great Lakes Region*. Cranbrook Institute of Science Bulletin No. 26, Bloomfield Hills, MI: Cranbrook Institute.of Science.

Hueter, R.E., Murphy, C.J., Howland, M., Sivak, J. G., Paul-Murphy and Howland, H.C. (2001). Refractive state and accommodation in the eyes of free-swimming versus restrained juvenile lemon sharks (*Negapriopn brevirostris*). *Vision Research*, 41, 1885–9.

Huggins, A.K., Skutsch, G., and Balwin, E. (1969). Ornithine-urea cycle enzymes in teleostean fish. *Comparative Biochemistry and Physiology*, 28, 587–602.

Hughes, G.M. (1960). A comparative study of gill ventilation in marine teleosts. *Journal of Experimental Biology*, 37, 28–45.

Hughes, G.M. (1963). *Comparative Physiology of Vertebrate Respiration*. Cambridge, MA: Harvard University Press.

Hughes, G.M. (1966). The dimensions of fish gills in relation to their function. *Journal of Experimental Biology*, 45, 177–95.

Hughes, G.M. (1982). An introduction to the study of gills. In: D.F. Houlihan, J.C. Rankin, and T.J. Shuttleworth (eds). *Gills*. Cambridge University Press, Cambridge, pp. 1–24.

Hughes, G.M. (1999). Fish gills: 'Old fourlegs' and air-breathing fishes. In: A.K. Mittal, F.B. Eddy, and J.S. Datta Munshi (eds). *Water/Air Transition in Biology*. Enfield, NH, Science Publishers Inc.

Hughes, G.M. and Munshi, J.S.D. (1979). Fine structure of the gills of some Indian air-breathing fishes. *Journal of Morphology*, 160, 169–94.

Hughes, G.M., Peyraud, C., Peyraud-Waitzenegger, M., and Soulier, P. (1982). Physiological evidence for the occurrence of pathways shunting blood away from the secondary lamellae of eel gills. *Journal of Experimental Biology*, 98, 277–88.

Hulbert, W.C., Moon, T.W., and Hochachka, P.W. (1978). The osteoglossid gill: correlations of structure, function, and metabolism with transition to air breathing. *Canadian Journal of Zoology*, 56, 801–8.

Hulley, P.A. (1992). Upper-slope distribution of oceanic lanternfishes (family: Myctophidae). *Marine Biology*, 114, 365–83.

Humason, G. (1967). *Animal Tissue Techniques*. London: W.H. Freeman and Co.

Hunn, J.B. (1967). Bibliography on the Blood Chemistry of Fishes. Washington, DC: U.S. department of the Interior, Fish and Wildlife Service. Bureau of Sport Fisheries and Wildlife. Research report 72. 950–2.

Hutchings, J.A. (2000). Collapse and recovery of marine fishes. *Nature*, 406, 862–5.

Huxley, A.F. (1957). Muscle structure and theories of contraction. *Progress in Biophysics and Biophysical Chemistry*, 7, 255–318.

Huxley, T.H. (1863). *Man's Place in Nature*. Ann Arbor, MI: University of Michigan Press.

Huysseune, A., Sire, J-Y., and Witten, P.E. (2009). Evolutionary and developmental origins of the vertebrate dentition. *Journal of Anatomy*, 214, 465–76.

Hyman, L.H. (1942). *Comparative Vertebrate Anatomy*. Chicago, IL: University of Chicago Press.

Hyodo, S., Bell, J.D., Healy, J.M., Kaneko, T., Hasegawa, S., Takei, Y., Donald, J.A., and Toop, T. (2007). Osmoregulation in elephant fish *Callorhinchus milii* (Holocephali), with special reference to the rectal gland. *Journal of Experimental Biology*, 210, 1303–10.

Icardo, J.M. (2006). Conus arteriosus of the teleost heart: dismissed, but not missed. *The Anatomical Record Part A*, 288A, 900–8.

Icardo, J.M., Wong, W.P., Coivee, E., Loong, A.M., and Ip, Y.K. (2012). The spleen of the African lungfish *Protopterus annectens*: freshwater and aestivation. *Cell and Tissue Research*, 350, 143–56.

Idler, D.R. and Truscott, B. (1966). 1-alpha-hydroxycorticosterone from cartilaginous fish: a new adrenal steroid in blood. *Journal of the Fisheries Research Board of Canada*, 23, 615–19.

Idler, D.R. and Truscott, B. (1972). Corticosteroids in fish. In: D.R. Idler (ed.). Steroids in Non-Mammalian Vertebrates. New York, NY: Academic Press, pp. 127–252.

Idler, D.R., Bitners, I., and Schmidt, P.J. (1961). 11-ketotestosterone: an androgen for sockeye salmon. *Canadian Journal of Biochemistry and Physiology*, 39, 1737–42.

Idler, D.R., Schmidt, P.J., and Ronald, A.P. (1960). Isolation and identification of 11-ketotestosterone in salmon plasma. *Canadian Journal of Biochemistry and Physiology*, 38, 1053–7.

Idyll, C.P. (1973). The anchovy crisis. *Scientific American*, 228, No. 6, 22–9.

Iger, Y., Abraham, M., Dotan, A., Fattal, B., and Rahamin, E. (1988). Cellular responses in the skin of carp maintained in organically fertilized water. *Journal of Fish Biology*, 33, 711–20.

Ihara, J., Yoshida, N., Tanaka, T., Mita, K., and Yamashita, M. (1998). Either cyclin B1 or B2 is necessary and sufficient for inducing germinal vesicle breakdown during frog (*Rana japonica*) oocyte maturation. *Molecular Reproduction and Development*, 50, 499–509.

Ilves, K.L. and Randall, D.J. (2007). Why have primitive fish survived? In: D.J. McKenzie and A.P. Farrell (eds). *Fish Physiology*, Vol. 26. Primitive fishes. Amsterdam: Elsevier/Academic Press, pp. 515–36.

Imbrogno, S., Tota, B., and Gattuso, A. (2011). The evolutionary functions of cardiac NOS/NO in vertebrates tracked by fish and amphibian paradigms. *Nitric Oxide*, 25, 1–10.

Imsland, A.K., Reynolds, P., Eliassen, G., Hanstad, T.A., Foss, A., Vikingstad, E., and Elvegård, T.A. (2014). The use of lumpfish (*Cyclopterus lumpus* L.) to control sea lice (*Lepeophtheirus samonis* Krøyer) infestations in intensively farmed Atlantic salmon (*Salmo salar* L.). *Aquaculture*, 424–425, 18–23.

Imsland, A.K., Reynolds, P., Eliassen, G., Hangstad, T.A., Nytrø, A.V., Foss, A., Vikingstad, E., and Elvegård, T.A. (2015). Feeding references of lumpfish *Cyclopterus lumpus* L.) maintained in open net-pens with Atlantic salmon (*Salmo salar* L.). *Aquaculture*, 436, 47–51.

Ingram, G.A. (1980). Substances involved in the natural resistance of fish to infection—a review. *Journal of Fish Biology*, 16, 23–60.

International Chicken Genome Sequencing Consortium (2004). Sequencing and comparative analysis of the chicken genome provide unique perspectives on vertebrate evolution. *Nature*, 432, 695–716.

International Pacific Halibut Commission (2012). Commission seeks information on mushy halibut syndrome. *2012 Halibut landing report No.6* Seattle WA: Internation Pacific Halibut Commission.

Ip, Y.K. and Chew, S.F. (2010). Ammonia production, excretion, toxicity, and defense in fish: a review. *Frontiers in Physiology*, 1, 134. doi.org/10.3389/fphys.2010.00134

Ip, Y.K., Loong, A.M., Ching, B., Tham, G.H.Y., Wong, W.P., and Chew, S.F. (2009).The freshwater stingray, *Potamotrygon motoro*, upregulates glutamine synythetase activity and protein abundance, and accumulates glutamine when exposed to brackish (15%) water. *Journal of Experimental Biology*, 212, 3828–36.

Irving, P.W. (1996). Sexual dimorphism in club cell distribution in the European minnow and immunocompetence signalling. *Journal of Fish Biology*, 48, 80–8.

Isaera, V.V., Pushkina, E.V., and Karetin, Y.A. (2004). The quasi-fractal structure of fish brain neurons. *Russian Journal of Marine Biology*, 30, 127–34.

Ito, H. and Yamamoto, N. (2009). Non-laminar cerebral cortex in teleost fishes? *Biology Letters*, 5, 117–21.

Iwata, K.S. and Fukuda, H. (1973). Central control of colour changes in fish. In: E.W. Chavin (ed.). *Responses of Fish to Environmental Changes*. Springfield, IL: Charles C. Thomas, pp. 316–41.

Iwata, E., Nagai, Y., Hyoudou, M and Sasaki, H. (2008). Social environment and sex differentiation in the false clown anemone fish *Amphiprion ocellatus*. *Zoological Science*, 25, 123–8.

Jain, S.L. (1985). Variability of dermal bones and other parameters in the skull of *Amia calva*. *Zoology, Journal of the Linnean Society*, 84, 385–95.

Jalabert, B. (2008). An overview of 30 years of international research in some selected fields of the Reproductive Physiology of Fish. *Cybium*, 32 supplement, 7–13.

Jalabert, B. and Finet, B. (1986). Regulation of oocyte maturation in the trout, *Salmo gairdneri*: role of cyclic AMP in the mechanism of action of the maturation inducing steroid (MIS), 17α-hydroxy, 20β-dihydroprogesterone. *Fish Physiology and Biochemistry*, 2, 65–74.

Jamieson, A.J., Fujii, T., Solan, M., Matsumoto, A.K., Bagley, P.M., and Priede, I.G. (2009). Liparid and macrourid fishes of the hadal zone: *in situ* observations of activity and feeding behaviour. *Proceedings of the Royal Society B*, 276, 1037–45.

Janech, M.G., Fitzgibbon, W.R. Ploth, D.W., Lacy, E.R., and Miller, D.H. (2006). Effect of low environmental salinity on plasma composition and renal function of the Atlantic stingray, a euryhaline elasmobranch. *American Journal of Renal Physiology*, 291, F770–F780.

Janovetz, J. (2005). Functional morphology of feeding in the scale-eating specialist *Catoprion mento*. *Journal of Experimental Biology*, 208, 4757–68.

Jansen, W.F. and van der Kamer, J.C. (1961). Histochemical analysis and cytological investigations on the coronet cells of the saccus vasculosus of the rainbow trout (*Salmo irideus*). *Zeitschrift fur Zellforschung und Mikroscopische Anatomie*, 55, 370–8.

Janvier, P. (1999). Catching the first fish. *Nature*, 402, 21–2.

Janvier, P. (2007). Living primitive fishes and fishes from deep time. In: D.J. McKenzie, A.P. Farrell, and C.J. Brauner (eds). *Fish Physiology, Vol. 26. Primitive fishes* Amsterdam: Elsevier/Academic Press, pp. 1–51.

Janvier, P. and Sansom, R.S. (2015). Fossil hagfishes, fossil cyclostomes, and the lost world of 'ostracoderms'. In: S.L. Edwards and G.G. Gross (eds). *Hagfish Biology*. Boca Raton, FL: CRC Press.

Janz, D.M. (2000). Gross functional anatomy. The endocrine system. In: G.K. Ostrander (ed.). *The Laboratory Fish*. London: Academic Press, pp. 189–217.

Janzen, W. (1933). Untersuchengen űber Grosshirnfunktionen des Goldfisches (*Carssius auratus*). *Zoologische Jahrbucher Abteilung fuer allgemeine Zoologie und Physiologie der Tiere*, 52, 591–628.

Jawad, L.A. (2005). Comparative scale morphology and squamation patterns in triplefins (Pisces: Teleostei: Perciformes: Tripterygiidae). *Tuhinga*, 16, 137–67.

Jensen, F.B. (2004). Red blood cell pH, the Bohr effect, and other oxygenation-linked phenomena in blood O_2 and CO_2 transport. *Acta Physiologica Scandinavia*, 182, 215–27.

Jensen, J. and Holmgren, S. (1994). The gastrointestinal canal. In: S. Nilsson and S. Holmgren (eds). *Comparative Physiology and Evolution of the Autonomic Nervous System*. Chur, Switzerland: Harwood, pp. 119–68.

Jewell, M.E. and Brown, H. (1924). The fishes of an acid lake. *Transactions of the American Microscopical Society*, XLIII, No. 2. 77–84.

Jørgensen, J.M., Lomholt, J.P., Weber, R.E., and Malte, H. (1998). *The Biology of Hagfishes*. New York, NY: Chapman and Hall.

Johansen, K. (1968). Air-breathing fishes. *Scientific American*, 219, 102–11.

Johansen, K. (1970). Air-breathing in fishes. In: W.S. Hoar and D.J. Randall (eds). *Fish Physiology, Vol. IV. The nervous system, circulation and respiration*. New York, NY: Academic Press, pp. 361–422.

Johnsen, S. (2014). Hide and seek in the open sea: pelagic camouflage and visual countermeasures. *Annual Review of Marine Science*, 369–92.

Johnson, D.S. (1968). Malayan blackwaters. In: R. Misra and B. Gopal (eds.) Ch. IV *Ecology of Aquatic communities. Proceedings of the Symposium on Recent Advances in Tropical Ecology.* Varanasi: The International Society for Tropical Ecology, pp. 303–310.

Johnson, G.D. (1978). *Development of fishes of the Mid-Atlantic Bight. An atlas of egg, larval and juvenile stages. Vol. IV: Carangidae through Ephippidae* U.S Department of the Interior Fish and Wildlife Service. (FWS/OBS–78/12).

Jóhnsson, G. (1982). Contributions to the biology of catfish (*Anarhichas lupus*) at Iceland. *Rit Fiskideildar*, 6, 2–26.

Johnsson, M. and Axelsson, M. (1996). Control of the systemic heart and the portal heart of *Myxine glutinosa*. *Journal of Experimental Biology*, 199, 1429–34.

Johnston, I.A. (1981a). Structure and function of fish muscles. In: M.H. Day (ed.). *Vertebrate Locomotion*. London: Academic Press, pp. 91–113.

Johnston, I.A. (1981b).Quantitative analysis of muscle breakdown during starvation in the marine flatfish *Pleuronectes platessa*. *Cell and Tissue Research*, 214, 369–86.

Johnston, I.A. Ward, P.S., and Goldspink, G. (1975). Studies on the swimming musculature of the rainbow trout. I. Fibre types. *Journal of Fish Biology*, 7, 451–8.

Jones, P.W Martin, F.D., and Hardy, J.D. Jr. (1978). Development of fishes of the Mid-Atlantic Bight. An atlas of egg, larval and juvenile stages. Vol. I: Acipenseridae through Ictaluridae. U.S Department of the Interior Fish and Wildlife Service. (FWS/OBS–78/12). Washington, DC: U.S. Epartment of the Interior, Fish and Wildlife Service.

Josselyn, S.A. (2010). Continuing the search for the engram: examining the mechanism of fear memories. *Journal of Psychiatry and Neuroscience*, 35, 221–8.

Jow, L.Y., Chew, S.F., Lim, C.B., Anerdon, P.M., and Ip, Y.K. (1999). The marble goby *Oxyeleotris marmoratus* activates hepatic glutamine synthetase and detoxifies ammonia to glutamine during air exposure. *Journal of Experimental Biology*, 202, 237–45.

Jucá-Chagas, R. and Bocardo, L. (2006). The air-breathing cycle of Hoplosternum littorale (Hancock 1828) (Siluriformes: Callichthyidae). *Neotropical Ichthyology*, 4, 371–3.

Jutfelt, F. (2006). The intestinal epithelium of salmonids. Transepithelial transport, barrier functions and bacterial interactions. Thesis, University of Göteberg, Sweden.

Kajiura, S.M. (2010). Pupil dilation and visual field in the spiked dogfish, *Squalus acanthias*. *Environmental Biology of Fish*, 88, 133–41.

Kalashnikova, M.M. (2000). Ultrastructure of fish and amphibian liver during catabolism of degenerating erythrocytes. *Bulletin of Experimental Biology and Medicine*, 129, 101–4.

Kalmijn, A.J. (1966). Electro-perception in sharks and rays. *Nature*, 212, 1232–3.

Kalmijn, A.J. (1971). The electric sense of sharks and rays. *Journal of Experimental Biology*, 55, 371–83.

Kalmijn, A. (1984). Theory of electromagnetic orientation: a further analysis. In: L. Bolis, R.D. Keynes, and S.H.P. Maddrell (eds). *Comparative Physiology of Sensory Systems*. Cambridge: Cambridge University Press, pp. 525–59.

Kalmijn, A. (1987). Detection of weak electric fields. In: J. Atema, R.R. Fa, A.N. Popper, and W.N. Tavolga (eds). *Sensory Biology of Aquatic Animals*. Berlin: Springer-Verlag, pp. 151–86.

Kandel, E.R., Schwartz, J.H., and Jessel, T.M. (1991). *Principles of Neural Science*, 3rd edn. New York, NY: Elsevier.

Kaneko, T and Hirano, T. (1993). Role of prolactin and somatolactin in calcium regulation in fish. *Journal of Experimental Biology*, 184, 31–45.

Kao, M.H., Fletcher, G.L., Wang, N.C., and Hew, C.L. (1986). The relationship between molecular weight and antifreeze polypeptide activity in marine fish. *Canadian Journal of Zoology*, 64, 578–82.

Kaplan, E.H. (1982). *Peterson Field Guides: Coral Reefs*. Boston, MA: Houghton Mifflin.

Kappers, C.U.A. (1926). The meninges in lower vertebrates compared with those of mammals. *Archives of Neurology and Psychiatry*, 15, 281–96.

Kappers, C.U.A., Huber, G.C., and Crosby, E. (1936). *The Comparative Anatomy of the Nervous System of Vertebrates Including Man*. New York, NY: Macmillan.

Kappers, J.A. (1971). The pineal organ: An introduction. In: G.E.W. Wolstenholme and J. Knight (eds) *The Pineal Gland*. London: Ciba Foundation Symposium Churchill Livingston, pp. 3–34.

Karnaky, K.J. (1986). Structure and function of the chloride cell of *Fundulus heteroclitus* and other teleosts. *American Zoologist*, 26, 209–24.

Kato, A., Chang, M-H, Kurita, T., Nakada, T., Ogoshi, M., Nakazato, T., Doi, H., Hirose, S., and Romero, M.F. (2009). Identification of renal transporters involved in sulfate excretion in marine teleost fish. *American Journal of Physiology, Regulatory, Integrative and Comparative Physiology*, 297, R1647–R1659.

Katz, S.L. (2002). Design of heterothermic muscle in fish. *Journal of Experimental Biology*, 205, 2251–66.

Keast, A. (1968). Feeding of some Great Lakes fishes at low temperatures. *Journal of the Fisheries Research Board of Canada*, 25, 1199–968.

Keats, D.W., Steele, D.H., and South, G.R. (1986). Atlantic wolfish (*Anarhichas lupus* L.; Pisces: Anarhichidae) predation on green sea urchins (*Strongylocentrotus drobachiensis* (O.F.Mull.) Echindermata: Echinoidea) in eastern Newfoundland. *Canadian Journal of Zoology*, 64, 1920–5.

Keenan, E. (1928). The phylogenetic development of the substantia gelatinosa Rolandi. Part I: Fishes. *Proceedings of the Royal Academy, Amsterdam*, 31, 837–56.

Kemp, A. (1999a). Ontogeny of the skull of the Australian lungfish *Neoceratodus forsteri* (Osteichthyes: Dipnoi). *Journal of Zoology, London*, 248, 97–137.

Kemp, A. (1999b). Anomalies in skull bones of the Australian lungfish, *Neoceratodus forsteri*, compared with aberrations in fossil dipnoan skulls. *Journal of Vertebrate Paleontology*, 19, 407–29

Kenaley, C.P. (2007). Revision of the stoplight loosejaw Genus *Malacosteus* (Teleostei: Stomiidae: Malacosteinae), with description of a new species from the temperate southern hemisphere and Indian Ocean. *Copeia*, 2007, 886–900.

Kennedy, B.P. and Davies, P.L. (1980). Acid-soluble nuclear proteins of the testis during spermatogenesis in the winter flounder. *Journal of Biological Chemistry*, 255, 2533–9.

Kennedy, W.A. (1953). Growth, maturity, fecundity and mortality in the relatively unexploited whitefish, *Coregonus clupeaformis*, of Great Slave Lake. *Journal of the Fisheries Research Board of Canada*, 10, 413–41.

Kerry, J. (2016). *On the new Marine Protected Area in Antarctica's Ross Sea*. Press Statement, Secretary of State. Washington, DC: US Government.

Ketola, H.G. (1982). Amino acid nutrition of fishes: requirements and supplementation of diets. *Comparative Biochemistry and Physiology*, 73B, 17–24.

Key, A. and Retzius, G. (1876). *Studien in der Anatomie des Nervensystems und des Bindegewebes*. Stockholm: Sampson and Wallin.

Key, B. (2015). Fish do not feel pain and its implications for understanding phenomenal consciousness. *Biology and Philosophy*, 30, 149–65.

Keynes, R.D. (1957). Electric organs. In: M.E. Brown (ed.). *The Physiology of Fishes, Vol. 2*. New York, NY: Academic Press, pp. 323–43.

Keynes, R.D. and Martin-Ferreira, H. (1953). Membrane potentials in the electroplates of the electric eel. *Journal of Physiology London*, 119, 315–51.

Keys, A. and Willmer, E.N. (1932). 'Chloride secreting cells' in the gills of fishes, with special reference to the common eel. *Journal of Physiology*, 76, 368–80.

Khani, S. and Tayek, J.A. (2001). Cortisol increases gluconeogenesis in humans: its role in the metabolic syndrome. *Clinical Science (London)*, 101, 739–47.

Kiceniuk, J.W. and Jones, D.R. (1977). The oxygen transport system in trout (*Salmo gairdneri*) during sustained exercise. *Journal of Experimental Biology*, 69, 247–60.

Kim, H.D. and Isaacks, R.E. (1978). The osmotic fragility and critical haemolytic volume of red blood cells of Amazon fishes. *Canadian Journal of Zoology*, 56, 860–2.

Kime, D.F. and Manning, N.J. (1982). Seasonal patterns of free and conjugated androgens in the brown trout *Salmo trutta*. *General and Comparative Endocrinology*, 48, 222–31.

King, W., Ghosh, S., Thomas, P., and Sullivan, C.V. (1997). A receptor for the oocyte maturation-inducing hormone 17α, 20β, 21-trihydroxy-4-pregnen-3-one on ovarian membranes of striped bass. *Biology of Reproduction*, 56, 266–71.

Kingsley, A.S. (1926). *Outlines of Comparative Anatomy of Vertebrates*. Philadelphia, PA: P. Blakiston's Son and Co.

Kinsey, W.H., Sharma, D., and Kinsey, S.C. (2007). Fertilization and egg activation in fishes. In: P.J. Nabin, J. Cerdà, and E. Lubzens (eds) *The Fish Oocyte: From Basic Studies to Biotechnological Applications*. Dordrecht, NL: Springer, pp. 397–410.

Klimley, A.P. (2013). *The Biology of Sharks and Rays*. Chicago, IL: University of Chicago Press.

Klok, M.D., Jakobsdottir, S., and Drent, M.L. (2007). The role of leptin and ghrelin in the regulation of food intake and body weight in humans: a review. *Obesity Review*, 8, 21–34.

Klosiewski, J. and Lyons, J.L. (no date). *Confusing Wisconsin Minnows*. Field Key. Department of Natural Resources, Madison, WI: Department of Natural Resources.

Knight, W. (1968). Asymptotic growth: An example of nonsense disguised as mathematics. *Journal of the Fisheries Research Board of Canada*, 25, 1303–7.

Knutson, M. and Wessling-Resnick, M. (2003), Iron metabolism in the reticuloendothelial system. *Critical Reviews in Biochemistry and Molecular Biology*, 38, 61–88.

Kobelkowsky, A. (2004). Sexual dimorphism of the flounder *Bothus robinsi* (Pisces: Bothidae). *Journal of Morphology*, 260, 165–71.

Kohler, A.C. (1967). Size at maturity, spawning season, and food of Atlantic halibut. *Journal of the Fisheries Research Board of Canada*, 24, 53–66.

Komourdjian, M. and Idler, D.R. (1977). Hypophysectomy of rainbow trout, *Salmo gairdneri*, and its effect on plasma sodium regulation. *General and Comparative Endocrinology*, 32, 536–42.

Kondrashev, S.L., Gamburtzeva, A.G., Gnjubkina, V.P., Orlov, O., Jun., and Pham, Thi My. (1986). Colouration of corneas in fish, a list of species. *Vision Research*, 26, 287–90.

Konow, N. and Sanford, C.P.J. (2008). Biomechanics of a convergently derived prey-processing mechanism in fishes: evidence from comparative tongue bite apparatus morphology and raking kinematics. *Journal of Experimental Biology*, 211, 3378–91.

Konow, N., Camp, A.L., and Sanford, C.P.J. (2008). Congruence between muscle activity ad kinematics in a convergently derived prey-processing behaviour. *Integrative and Comparative Biology*, 48, 246–60.

Korenbrot, J.I. and Fernald, R.D. (1989). Circadian rhythm and light regulate opsin mRNA in rod photoreceptors. *Nature*, 337, 454–7.

Kotrschal, K. (1991). Solitary chemosensory cells–taste, common chemical sense or what? *Reviews in Fish Biology and Fisheries*, 1, 3–22.

Kotrschal, K. (1996). Solitary chemosensory cells: why do aquatic vertebrates need another taste system? *Trends in Ecology and Evolution*, 11, 110–14.

Kotrschal, K., van Staaden, M.J., and Huber, R. (1998). Fish brain: evolution and environmental relationships. *Reviews in Fish Biology and Fisheries*, 8, 373–408.

Kotrschal, K., Whitear, M., and Finger, T.E. (1993). Spinal and facial innervation of the skin in the gadid fish *Ciliata mustela* (Teleostei). *Journal of Comparative Neurology*, 331, 407–17.

Kottelat, M., Britz, R., Hui, T.H., and Witte, K-E. (2006). *Paedocypris*, a new genus of Southeast Asian cyprinid fish with a remarkable sexual dimorphism, comprises the world's smallest vertebrate. *Proceedings of the Royal Society B*, 273, 895–9.

Koumans, J.T., Akster, H.A., Booms, G.H., Lemmens, C.J., and Osse, J.W. (1991). Numbers of myosatellite cells in white axial muscle of growing fish: *Cyprinus carpio* L. (Teleostei). *American. Journal of Anatomy*, 192, 418–24.

Kovacs, T.G., Martel, P.H., O'Connor, B.I., Hewitt, L.M., Parrott, J.L., McMaster, M.E. MacLatchy, D.L., Van Der Kraak, G.J., and van den Heuvel, M.R. (2013). A survey of Canadian mechanical pulp and paper mill effluents; insights concerning the potential affect to fish reproduction. *Journal of Environmental Science and Health Part A: Toxic/.Hazardous Substances and Environmental Engineering*, 48, 1178–89.

Kozhov, M. (1963). *Lake Baikal and its Life*. The Hague: Dr. W. Junk.

Krahe, R. and Fortune, E.S. (eds.) (2013). *Electric Fishes: Neural systems, Behaviour and Evolution*. Special issue, *Journal of Experimental Biology*, 216.

Kramer, D.L. and Bryant, M.J. (1995a). Intestine length in the fishes of a tropical stream: 1. Ontogenetic allometry. *Environmental Biology of Fishes*, 42, 115–27.

Kramer, D.L. and Bryant, M.J. (1995b). Intestine length in the fishes of a tropical stream: 1. Relationships to diet- the long and short of a convoluted issue. *Environmental Biology of Fishes*, 42, 129–41.

Kramer, D.L., Lindsey, C.C., Moodie, G.E.E., and Stevens, E.D. (1978). The fishes and the aquatic environment of the central Amazon basin, with particular reference to respiratory patterns. *Canadian Journal of Zoology*, 56, 717–29.

Krkošek, M., Ford, J.S. Morton, A., Lefe, S., Myers, R.A., and Lewis, M.A. (2007, corrected 19 December 2008). Declining wild salmon populations in relation to parasites from farm salmon. *Science*, 318, 1772–5.

Krogdahl, Å., Hemre, G.-I., and Mommsen, T.P. (2005). Carbohydrates in fish nutrition: digestion and absorption in postlarval stages. *Aquaculture*, 11, 103–22.

Krogh, A. (1941). *The Comparative Physiology of Respiratory Mechanisms*. Philadelphia, PA: University of Philadelphia Press. Reprinted by Dover Publications, New York, NY, 1968.

Krönström, J., Holmgren, S., Baguet, F., Salpietro, L., and Mallefet, J. (2005). Nitric oxide in control of luminescence from hatchetfish (*Argyropelecus hemigymnus*) photophores. *Journal of Experimental Biology*, 208, 2951–61.

Krossøy, C., Waagbø, R., and Ørnsrud, R. (2011). Vitamin K in fish nutrition. *Aquaculture Nutrition*, 17, 585–594.

Kuchnow, K.P. (1971).The elasmobranch pupillary response. *Vision Research*, 11, 1395–406.

Kuhajda, B.R. and Mayden, R.L. (2001). Status of the federally endangered Alabaman cavefish, *Speoplatyrhinus poulsoni* (Amblyopsidae), in Key Cave and surrounding caves, Alabama. *Environmental Biology of Fishes*, 62, 215–22.

Kuhajda, B.R., George, A.L., and Williams, J.D. (2009). The desperate dozen: Southeastern freshwater fishes on the brink. *Southeastern Fishes Council Proceedings*, No. 51. 10–30.

Kuiter, R.H. (1996). *Guide to the Sea Fishes of Australia*. Sydney: New Holland Publishers.

Kujawa, R., Kucharczyk, D., and Mamcarz, A. (2010). The effect of tannin concentration and egg unsticking time on the hatching success of tench (*Tinca tinca* (L.) larvae. *Reviews in Fish Biology and Fisheries*, 20, 339–43.

Kulczykowska, E., Warne, J.M., and Balment, R.J. (2001). Day–night variations in plasma melatonin and arginine vasotocin concentrations in chronically cannulated flounder (*Platichthys flesus*). *Comparative Biochemistry and Physiology A*, 130, 827–34.

Kumakura, S. (1927). Versuche an Goldfischen, denen beide Hemisphären des Grosshirns extirpiert worden waren. *Nagoya Journal of Medical Science*, 3, 19–24.

Kumazawa, T. and Fujii, R. (1984).Concurrent releases of norepinephrine and purines by potassium from adrenergic melanosome-aggregating nerve in tilapia. *Comparative Biochemistry and Physiology*, 78C, 263–6.

Kumazawa, T. and Fujii, R. (1986). Fate of adenylic cotransmitter released from adrenergic pigment-aggregating nerve to tilapia melanophore. *Zoological Science*, 3, 599–603.

Kuperman, B.I., and Kuz'mina, V.V. (1994). The ultrastructure of the intestinal epithelium in fishes with different types of feeding. *Journal of Fish Biology*, 44, 181–93.

Kuwamura, T., Suzuki, S., Karmo, K., and Nakashima, Y. (2007). Sex change of primary males in a diandric labrid *Halichoestes trimaculatus*: coexistence of protandry and protogyny with a species. *Journal of Fish Biology*, 70, 1898–906.

Kwan, L., Cheng, Y.Y., Rodd, F.H., and Rowe, L. (2013). Sexual conflict and the function of genitalic claws in guppies (*Poecilia reticulata*). *Biology Letters*, 9, 4pp. doi.org/10.1098/rsb.2013.0267

Kwong, R.W.M., Kumai, Y., and Perry, S.F. (2013). The role of aquaporin and tight junction proteins in the regulation of water movement in larval zebrafish. *PLoS One*, 8(8), e70764.

Kyle, A.L. and Peter, R.E. (1982). Effects of forebrain lesions on spawning behaviour in the male goldfish. *Physiology and Behaviour*, 28, 1103–9.

Laale, H.W. (1980). The perivitelline space and egg envelope of bony fishes: a review. *Copeia*, 1980, 210–26.

Lacalli, T.C. (2005). Protochordate body plan and the evolutionary role of larvae: old controversies resolved? *Canadian Journal of Zoology*, 83, 216–24.

Lacy, E.R. and Reale, E. (1985). The elasmobranch kidney I. Gross anatomy and general distribution of the nephrons. *Anatomy and Embryology*, 173, 23–34.

Ladich, F. (2000). Acoustic communication and the evolution of hearing in fishes. *Philosophical Transactions of the Royal Society of London B*, 355, 1285–8.

Ladich, F., Brittinger, W., and Kratochvil, H. (1992). Significance of agonistic vocalization in the croaking gourami (*Trichopsis vittatus*, Teleostei). *Ethology*, 90, 307–14.

Ladich, F., Collin, S.P., Moller, P., and Kapoor, B.G. (2006). *Communication in Fishes. Vol. I, Section I. Acoustic and Chemical Communication. Vol. II, Section II. Visual Communication. Section III. Electric Communication.* Enfield, NH: Science Publishers Inc.

Lagerström, M.C., Fredricksson, R. Bjarnadottir, T.K., and Schiöth, H.B. (2005). The ancestry of the prolactin-releasing hormone precursor. *Annals of the New York Academy of Sciences*, 1040, 368–70.

Lagios, M.D. (1975). The pituitary gland of the coelacanth *Latimeria chalumnae* Smith. *General and Comparative Endocrinology*, 25, 126–46.

Lake, J.S. (1967). Principal fishes of the Murray–Darling River system. In: A.H. Weatherley (ed.). *Australian Inland Waters and their Fauna*. Canberra: Australian National University Press.

Lambert, Y. and Dutil, J-D. (1997). Can simple condition indices be used to monitor and quantify seasonal changes in the energy reserves of Atlantic cod (*Gadus morhua*)? *Canadian Journal of Fisheries and Aquatic Science*, 54 (Suppl. 1), 104–12.

Lane, E.B. and Whitear, M. (1977). On the occurrence of Merkel cells in the epidermis of teleost fishes. *Cell and Tissue Research*, 182, 235–46.

Lang, F., Busch, G., Ritter, M., Volkl, H., Waldegger, S., Gulbins, E. , and Hàussinger, D. (1998). Functional significance of cell volume regulatory mechanisms. *Physiological Review*, 78, 247–306.

Langecker, T.G. and Longley, G. (1993). Morphological adaptations of the Texas blind catfishes *Trogloglanis pattersoni* and *Satan eurystomus* (Siluriformes: Ictaluridae) to their underground environment. *Copeia*, 1992, 976–86.

Langeland, A. and Nøst, T. (1995). Gill raker structure and selective predation on zooplankton byparticulate feeding fish. *Journal of Fish Biology*, 47, 719–32.

Langley, J.N. (1898). On the union of cranial autonomic (visceral) fibres with the nerve cells of the superior cervical ganglion. *Journal of Physiology, London*, 23, 240–70.

Lanzing, W.J.R. and van Lennep, E.W. (1971). The ultrastructure of the saccus vasculosus of teleost fishes III. Supporting tissue, innervations, vascularization. *Australian Journal of Zoology*, 19, 327–45.

Larsson, D.G. and Förlin, L. (2002). Male-biased sex ratios of fish embryos near a pulp mill: temporary recovery after a short-term shutdown. *Environmental Health Perspectives*, 110, 739–42.

Lashley, K.S. (1950). In search of the engram. In: *Physiological Mechanisms in Animal Behaviour, Society of Experimental Biology Symposium, No.4*. Cambridge: Cambridge University Press, pp. 454–82.

Lauder, G.V. (1979). Feeding mechanics in primitive teleosts and in the halecomorph fish *Amia calva*. *Journal of Zoology London*, 187, 543–78.

Lauder, G.V. (1980). Evolution of the feeding mechanism in primitive Actinopterygian fishes: A functional analysis of *Polypterus, Lepisosteus*, and *Amia*. *Journal of Morphology*, 163, 283–317.

Lauder, G.V. and Liem, K.F. (1981). Prey capture by *Luciocephalus pulcher*: implications for models of jaw protrusion in teleost fishes. *Environmental Biology of Fish*, 6, 257–68.

Lauder, G.V. and Liem, K.F. (1983). The evolution and interrelationships of the Actinopterygian fishes. *Bulletin of the Museum of Comparative Zoology*, 150, 95–197.

Lauder, G.V. and Tytell, E.D. (2004). Three Gray classics on the biomechanics of animal movement. *Journal of Experimental Biology*, 207, 1597–9.

Laurent, P. (1982). Structure of vertebrate gills. In: D.F. Houlihan, J.C. Rankin, and T.J. Shuttleworth (eds). *Gills*. Cambridge: Cambridge University Press, pp. 25–44.

Lawson, G.L., Kramer, D.L. , and Hunte, W. (1999). Size-related habitat use and schooling behaviour in two species of surgeonfish (*Acanthurus bahianus* and *A. coeruleus*) on a fringing reef in Barbados, West Indies. *Environmental Biology of Fishes*, 54, 19–33.

Leaman, B.M. (1991). Reproductive styles and life-history variables relative to exploitation and management of *Sebastes* stocks. *Environmental Biology of Fishes*, 30, 253–71.

Leatherland, J.F. (1994). Reflections on the thyroidology of fishes: from molecules to humankind. *Guelph Ichthyology Reviews*, No. 2, 67pp.

Le Douarin, N.M. and Kalcheim, C. (1999). *The Neural Crest*, 2nd edn. Cambridge: Cambridge University Press.

Lefevre, S., Bayley, M., McKenzie, D.J. and Craig, J.F. (2014). Air-breathing fishes. *Journal of Fish Biology*, 84, 547–53.

Leggett, W.C. and Power, G. (1969). Differences between two populations of landlocked Atlantic salmon (*Salmo salar*) in Newfoundland. *Journal of the Fisheries Research Board of Canada*, 26, 1585–96.

Leonard, G., Maie, T., Moody, K.N., Schrank, G.D., Blob, R.W., and Schoenfuss, H.L. (2012). Finding paradise: cues directing the migration of the waterfall climbing Hawaiian gobioid *Sicyopterus stimpsoni*. *Journal of Fish Biology*, 81, 903–20.

Lesiuk, T.P. and Lindsey, C.C. (1978). Morphological peculiarities in the neck-bending Amazonian characoid fish *Rhaphiodon vulpinus*. *Canadian Journal of Zoology*, 56 (4) Part 2: 991–7.

Lessard, J., Val, A., Aota, S. , and Randall, D.J. (1995). Why is there no carbonic anhydrase activity available to fish plasma? *Journal of Experimental Biology*, 198, 31–8.

Levesque, H.M., Short, C., Moon, T.W., Ballantyne, J.S., and Driedzic, W.R. (2005). Effects of seasonal temperature and photoperiod on Atlantic cod (*Gadus morhua*). 1. Morphometric parameters and metabolites. *Canadian Journal of Fisheries and Aquatic Science*, 62, 2854–63.

Levine, J.S., and MacNichol, F.F. Jr. (1979). Visual pigments in teleost fishes: effects of habitat, microhabitat and behaviour on visual system evolution. *Sensory Processes*, 3, 95–131.

Li, L. and Maaswinkel, H. (2007). Visual sensitivity and signal processing in teleosts In: T.H. Hara and B. Zielinski (eds). *Fish Physiology, Vol. 25. Sensory systems neuroscience*. New York, NY: Academic Press, Elsevier Inc., pp. 179–241.

Li, W. and Sorensen, P.W. (1997). Highly independent olfactory receptor sites for conspecific bile acids in the sea lamprey, *Petromyzon marinus*. *Journal of Comparative Physiology A*, 180, 429–38.

Li, W., Sorensen, P.W., and Gallaher, D. (1995). The olfactory system of migratory sea lamprey (*Petromyzon marinus*) is specifically and acutely sensitive to unique bile acids released by conspecific larvae. *Journal of General Physiology*, 105, 569–89.

Li, W., Scott, A.P. Siefkes, M.J., Yan, H., Liu, Q., Yun, S.S. , and Gage, S.A. (2002). Bile acid secreted by male sea lamprey that acts as a sex pheromone. *Science*, 296, 138–41.

Liao, Y-Y and Lucas (2000). Diet of the common wolfish *Anarhichas lupus* in the North Sea. *Journal of the Marine Biological Association of the United Kingdom*, 80, 181–2.

Lieber, R.L., Raab, R., Kashin, S., and Edgerton, V.R. (1992). Sarcomere length changes during fish swimming. *Journal of Experimental Biology*, 169, 251–54.

Liem, K.F. (1974). Evolutionary strategies and morphological innovations: cichlid pharyngeal jaws. *Systematic Zoology*, 22, 425–41.

Liem, K.F. and Greenwood, P.H (1981). A functional approach to the phylogeny of the pharyngognath teleosts. *American Zoologist*, 21, 83–101.

Liem, K.F. and Osse, J.W.M. (1975). Biological versatility, evolution and food resource exploitation in African cichlid fishes. *Amererican Zoologist*, 15, 427–54.

Lin, J.J. and Somero, G.N. (1995). Temperature-dependent changes in expression of thermostable and thermolabile isozymes of cytosolic malate dehydrogenase in the eurythermal goby fish *Gillichthys mirabilis*. *Physiological Zoology*, 68, 114–28.

Lin, Y., Dobbs III, G.H., and Devries, A.L. (1974). Oxygen consumption and lipid content in red and white muscles of Antarctic fishes. *Journal of Experimental Zoology*, 189, 379–86.

Lindsay, G.J.H. (1984). Distribution and function of digestive tract chitinolytic enzymes in fish. *Journal of Fish Biology*, 24, 529–36.

Lindsay, G.J.H. and Gooday, G.W. (1985). Chitinolytic enzymes and the bacterial microflora in the digestive tract of cod, *Gadus morhua*. *Journal of Fish Biology*, 26, 255–65.

Lisney, T.J. (2010). A review of the sensory biology of chimaeroid fishes (Chondrichthyes; Holocephali). *Reviews in Fish Biology and Fisheries*, 20, 571–90.

Lisney, T.J. and Collin, S.P. (2007). Relative eye size in elasmobranchs. *Brain, Behaviour and Evolution*, 69, 266–79.

Lissmann, H.W. (1951). Continuous electrical signals from the tail of a fish, *Gymnarchus niloticus* Cuv. *Nature*, 167, 201.

Lissmann, H.W. (1958). On the function and evolution of electric organs in fish. *Journal of Experimental Biology*, 35, 156–91.

Lissmann, H.W. (1963). Electric location by fishes. *Scientific American*, March 1963, 50–9

Lissmann, H.W. and Machin, K.E. (1958). The mechanism of object location in *Gymnarchus niloticus* and similar fish. *Journal of Experimental Biology*, 35, 451–86.

Liu, X., Ramirez, S., Pang, P.T., Puryear, C.B., Govindarajan, A., Deisseroth, K., and Tonegawa, S. (2012). Optogenetic stimulation of a hippocampal engram activates fear memory. *Nature*, 484, 381–5.

Locket, N.A. (1977). Adaptation to the deep-sea environment. In: F. Crescelli (ed.). *Handbook of Sensory Physiology, Vol. VII/5. The Visual System in Vertebrates*. Berlin: Springer-Verlag, pp. 67–192.

Locket, N.A. (1980). Variation of architecture with size in the multiple-bank retina of a deep-sea teleost *Chauliodus sloani*. *Proceedings of the Royal Society of London*, 208, 223–242.

Locket, N.A. (1985). The multiple bank rod fovea of *Baja-california drukei*, an alocephalid deep-sea teleost. *Proceedings of the Royal Society of London B*, 224, 7–22.

Lofts, B. and Bern, H.A. (1972). The functional morphology of steroidogenic tissues. In: D.R. Idler (ed.). *Steroids in Non-Mammalian Vertebrates.* New York, NY: Academic Press, pp. 37–126.

Logan, A.G., Moriarty, R.J., and Rankin, J.C. (1980). A micropuncture of kidney function in the river lamprey, *Lampetra fluviatilis*, adapted to fresh water. *Journal of Experimental Biology*, 85, 137–47.

Lorenzen, C.J. (1971). Continuity in the distribution of surface chlorophyll. *Journal du Conseil/Conseil Permanent International de l'Exploration de la Mer*, 34, 18–23.

Loretz, C.A. and Pollina, C. (2000). Natriuretic peptides in fish physiology. *Comparative Biochemistry and Physiology A, Molecular and Integrative Physiology*, 125, 169–87.

Lourie, S.A. and Randall, J.E. (2003). A new pygmy seahorse, *Hippocampus denise* (Teleostei: Syngnathidea) from the IndoPacific. *Zoological Studies*, 42, 284–91.

Love, R.M. (1960). Water content of cod (*Gadus callarius* L.) muscle Nature **185**, 692.

Love, R.M. (1970). *The Chemical Biology of Fishes.* New York NY: Academic Press.

Love, R.M. and Lavéty, (1977). Wateriness of white muscle: A comparison between cod (*Gadus morhua*) and jelly cat (*Lycichthys denticulatus*). Marine Biology **43**, 117–121.

Low, P., Pankscepp, J., Reiss, D. Edelman, D., Van Swinderen, B. and Kock, C. (2012). *The Cambridge Declaration on Consciousness.* Proceedings of the Francis Crick Conference on Consciousness in Humans and non-human Animals. 7 Juky, 2012. Cambridge: University of Cambridge.

Lowe-Jinde, L. and Nimi, A.J. (1983). Influence of sampling on the interpretation of haematological measurements of rainbow trout, *Salmo gairdneri. Canadian Journal of Zoology*, 61, 396–402.

Lowenstein, O. (1956). Pressure receptors in the fins of the dogfish, *Scyliorhinus canicula. Journal of Experimental Biology*, 33, 417–21.

Lowenstein, O. (1957). The sense organs: The acoustic-lateralis system. In: M.E. Brown (ed.). *The Physiology of Fishes, Vol. 2.* New York, NY: Academic Press, pp.155–86.

Lowenstein, O. (1971). The labyrinth. In: W.S. Hoar and D.J. Randall (eds). *Fish Physiology, Vol. V.* New York, NY: Academic Press, pp. 207–40.

Lowenstein, O. and Roberts, T.D.M. (1949). The equilibrium function of the otolith organs of the thornback ray (*Raja clavata*). *Journal of Physiology (London)*, 110, 392–415.

Lowenstein, O. and Roberts, T.D.M. (1951). The localization and analysis of the responses to vibration from the isolated elasmobranch labyrinth. A contribution to the problem of the evolution of hearing in vertebrates. *Journal of Physiology (London)*, 114, 471–89.

Lowenstein, O. and Sand, A. (1936). The activity of the horizontal semicircular canal of the dogfish, *Scyllium canicula. Journal of Experimental Biology*, 13, 416–28.

Lowenstein, O. and Sand, A. (1940a). The individual and integrated activity of the semicircular canals of the elasmobranch labyrinth. *Journal of Physiology (London)*, 99, 89–101.

Lowenstein, O. and Sand, A. (1940b). The mechanism of the semicircular canal. A study of the responses of single-fibre preparations to angular accelerations and rotations at constant speed. *Proceedings of the Royal Society of London B*, 129, 256–75.

Lowenstein, O. and Wersäll, J. (1959). A functional interpretation of the electron microscope structure of the sensory hairs in the cristae of the elasmobranch *Raja clavata* in terms of directional sensitivity. *Nature*, 184, 1807–10.

Lowenstein, O. Osborne, M.P., and Wersäll, J. (1964). Structure and innervation of the sensory epithelia of the labyrinth in the thornback ray (*Raja clavata*). *Proceedings of the Royal Society B*, 160, 1–12.

Lowerre-Barbieri, S.K. and Barbieri, L.R. (1993). A new method of oocyte separation and preservation for fish reproduction studies. *Fishery Bulletin*, 91, 165–70.

Lowry, D., Wintzer, A.P., Mattot, M.P. Whitenack, L.D. Dean, M. And Motta, P.J. (2005). Aerial and aquatic feeding in the silver arawama, *Osteoglossum bicirrhosum.* Environment Biology of Fishes, 73, 453–462.

Lubzens, E., Young, G., Bobe, J., and Cerdà, J. (2010). Oogenesis in teleosts: how fish eggs are formed. *General and Comparative Endocrinology*, 165, 367–89.

Lucarz, A. and Brand, G. (2007). Current considerations about Merkel cells. *European Journal of Cell Biology*, 86, 243–51.

Luer, C.A., Blum, P.C., and Gilbert, P.W. (1990). Rate of tooth replacement in the nurse shark, *Ginglymosta cirratum. Copeia*, 1990, 182–91.

Lüling, K.H. (1973). *Südamerikanische Fische und ihr Lebensraum.* Wuppertal: Engelbert Pfriem Verlag.

Lutz, P.L. and Robertson, J.D. (1971). Osmotic constituents of the coelacanth *Latimeria chalumnae* Smith. *Biological Bulletin*, 141, 553–60

Lythgoe, J.N. and Shand, J. (1983). Diel colour changes in the neon tetra *Paracheirodon innesi. Environmental Biology of Fish*, 8, 249–54.

MacDonald, J. and Waiwood, K.G. (1982). Rates of digestion of different prey in Atlantic cod (*Gadus morhua*) ocean pout (*Macrozarces americanus*), winter flounder (*Pseudopleuronectes americanus*), and American plaice (*Hippoglossoides platessoides*). *Canadian Journal of Fisheries and Aquatic Science*, 39, 651–9.

Macer, C.T. (1967). *Fiches d'identification des oeufs et larves de poissons* No.2. Ammodytidae. Conseil permanent international pour l'exploration de la mer. Copenhagen: International Council for the Exploration of the Sea.

MacFarlane, R.B. and Bowers, M.J. (1995). Matrotrophic viviparity in the yellowtail rockfish *Sebastes flavidus*. *Journal of Experimental Biology*, 198, 1197–206.

MacNab, V., Scott, A.P., Katsiadaki, I., and Barber, I. (2011). Variation in the reproductive potential of *Schistocephalus* infected male sticklebacks is associated with 11-ketotesterone titre. *Hormones and Behaviour*, 60, 371–9.

Mack, A.F. and Wolberg, H. (2006). Growing axons in fish optic nerve are accompanied by astrocytes interconnected by tight junctions. *Brain Research*, 1103, 25–31.

Maddock, D.M. and Burton, M.P.M. (1994). Some effects of starvation on lipid and muscle layers of winter flounder. *Canadian Journal of Zoology*, 72, 1672–9.

Madliger, C.L. and Love, O.P. (2015). The power of physiology in changing landscapes: considerations for the continued integration of conservation and physiology. *Society for Integrative and Comparative | Biology Annual Meeting Abstracts S2.1*, 205–205.

Magnuson, J.J. (1969). Digestion and food consumption by skipjack tuna. *Transactions of the American Fisheries Society*, 98, 379–92.

Magurran, A.E. (1990). The adaptive significance of schooling as an anti-predator defence in fish. *Annales Zoologici Fennici*, 27, 51–66.

Mallefet, J. and Shimomura, O (1995). Presence of coelenterazine in mesopelagic fishes from the Strait of Messina. *Marine Biology*, 124, 381–5.

Mann, K.D., Hoyt, C., Feldman, S., Blunt, L., Raymond, A., and Page-McCaw, P.S. (2010). Cardiac response to startle stimuli in larval zebrafish: sympathetic and parasympathetic components. *American Journal of Physiology— Regulatory, Integrative and Comparative Physiology*, 298, R1288–R1297.

Mann, S., Sparks, N.H.C., Walker, M.M., and Kirschvink, J.L. (1988). Ultrastructure, morphology and organisation of biogenic magnetite from sockeye salmon, *Oncorhynchus nerka*: Implications for magnetoreception. *Journal of Experimental Biology*, 140, 35–49.

Manning, A.J.M. and Crim, L.W. (1998). Mortality and interannual comparison of the ovulatory periodicity, egg production and egg quality of the batch –spawning yellow tail flounder. *Journal of Fish Biology*, 53, 954–72.

Maradonna, F., Batti, S., Marino, M., Mita, D.G., and Carnevalli, O. (2009). Tamoxifen as an emerging endocrine disruptor. Effects on fish reproduction and detoxification target genes. *Annals of the New York Academy of Sciences*, 163, 457–9.

March, B.E. (1993). Essential fatty acids in fish physiology. *Canadian Journal of Physiology and Pharmacology*, 71, 684–9.

Marshall, A.M. and Hurst, C.H. (1930). *A Junior Course of Practical Zoology*. London: John Murray.

Marshall, E.K., Jr. and Smith, H.W. (1930). The glomerular development of the vertebrate kidney in relation to habitat. *Biological Bulletin*, 59, 135–53.

Marshall, N.B. (1971). *Explorations in the Life of Fishes*. Cambridge, MA: Harvard University Press.

Marshall, W.S (1995). Transport processes in isolated teleost epithelia: opercular epithelium and urinary bladder. In: C.M. Wood and T.J. Shuttleworth (+). *Fish Physiology 14. Cellular and Molecular Approaches to Fish Ionic Regulation*, New York, NY: Academic Press, pp. 1–23.

Marshall, W.S. and Nishioka, R.S. (1980). Relation of mitochondria-rich chloride cells to active transport in the skin of a marine teleost. *Journal of Experimental Zoology*, 214, 147–56.

Martin, F.D. and Drewer, G.E. (1978). *Development of fishes of the Mid-Atlantic Bight. An atlas of egg, larval and juvenile stages. Vol. VI: Stromateidae through Ogcocephalidae*. U.S DepInt. Fish and Wildlife Service. (FWS/OBS–78/12).

Martin, L.M. (2015). *Beach Spawning Fish—Reproduction in an Endangered Ecosystem*. Boca Raton, FL: CRC Press.

Martinez, M., Guderley, H., Dutil, J-D., Winger, P.D., He, P., and Walsh, S.J. (2003). Condition, prolonged swimming performance and muscle metabolic capacities of cod *Gadus morhua*. *Journal of Experimental Biology*, 206, 503–11.

Martínez-Porchas, M., Martínez-Córdova, L.F., and Ramos-Enriquz, R. (2009). Cortisol and glucose: reliable indicators of fish stress? *Pan-American Journal of Aquatic Science*, 4, 158–78.

Martini, L. (1971). Chairman's Introduction. In: G.E.W. Wolstenholme and J. Knight (eds). *The Pineal Gland*. CIBA Foundation Symposium. Edinburgh: Churchill Livingstone, pp. 1–2.

Marui, T. and Caprio, J. (1992). Teleost gustation. In: T.J. Hara (ed.). *Fish Chemoreception*. London: Chapman and Hall, pp. 171–98.

Marza, V.D. (1938). *Histophysiologie d'ovogenèse*. Paris: Hermann.

Mathuru, A.S., Kibat, C., Cheong, W.F., Shui, G., Wenk, M.R., Friedrich, R.W. , and Jesuthasan, S. (2012). Chondroitin fragments are odorants that trigger fear behaviour in fish. *Current Biology*, 22, 538–44.

Mattern, M.Y. (2006). Phylogeny, systematics and taxonomy of sticklebacks. In: S. Östland, I. Mayer, and F.A. Huntingford (eds). *Biology of Three-Spined Sticklebacks*. Boca Raton, FL: CRC Press.

Matty, A.J. (1985). *Fish Endocrinology*. London: Croom Helm.

Mauck, P.E. and Summerfelt, R.C. (1971). Sex ratio, age of spawning fish, and duration of spawning in the carp, *Cyprinus carpio* (Linnaeus) in Lake Carl Blackwell, Oklahoma. *Transactions of the Kansas Academy of Science*, 74, 221–7.

May, A. (1967). Fecundity of Atlantic cod. *Journal of the Fisheries Research Board of Canada*, 24, 1531–51.

Mayford, M. (2014). The search for a hippocampal engram. *Philosophical Transactions of the Royal Society B*, 369, 20130161.

Mayo, D.J. and Burton, D. (1998a). β_2-adrenoceptors mediate melanosome dispersion in winter flounder (*Pleuronecxtes americanus*). Canadian Journal of Zoology, 76, 175–80.

Mayo, D.J. and Burton, D. (1998b). The *in vitro* physiology of melanophores associated with integumentary patterns in winter flounder (*Pleuronectes americanus*). *Comparative Biochemistry and Physiology*, A121, 241–7.

Mazeaud, F. (1981). Morpholine, a nonspecific attractant for salmonids. *Aquaculture*, 26, 189–91.

McAllister, D.E., Parker, B.J., and McKee, P.M. (1985). Rare, endangered, and extinct fishes in Canada. *Syllogeus*, 54, 1–192.

McBride, J.R. and van Overbeeke, A.P. (1971). Effects of androgens, estrogen and cortisol on the skin, stomach, liver, pancreas and kidney in gonadectomized adult sockeye salmon (*Oncorhynchus nerka*). *Journal of the Fisheries Research Board Canada*, 28, 485–90.

McClung, C.E. (1905). The chromosome complex of orthopteran spermatocytes. *Biological Bulletin*, 9, 304–40.

McCormick, C.A. (1981). Central projection of the lateral line and eighth nerves in the bowfin *Amia calva*. *Journal of Comparative Neurology*, 197, 1–15.

McCormick, S.D. (2001). Endocrine control of osmoregulation in teleost fish. *American Zoologist*, 41, 781–94.

McCormick, S.D., Regish, A., O'Dea, M.F., and Shrimpton, J.M. (2008). Are we missing a mineralocorticoid in teleost fish? Effects of cortisol, deoxycortisone and osmoregulation, gill Na^+, K^+-ATPase activity and isoform mRNA levels in Atlantic salmon. *General and Comparative Endocrinology*, 157, 35–40.

McCosker, J.E. (1977). Flashlight fishes. *Scientific American*, 236, 106–14.

McDonald, M.D. and Grosell, M. (2006). Maintaining osmotic balance with an aglomerular kidney, *Comparative Biochemistry and Physiology A. Molecular and Integrative Physiology*, 143, 447–58.

McDowall, R.M. (1992). Diadromy: Origins and definitions of terminology. *Copeia*, 1992, 248–51.

McDowall, R.M. (1999). Different kinds of diadromy: Different kinds of conservation problems. *International Council for the Exploration of the Sea Journal of Marine Science*, 56, 410–13.

McDowall, R.M. (2008). Jordan's and other ecogeographical rules, and the vertebral number in fish. *Journal of Biogeography*, 35, 501–8.

McEwan, M.R. (1938). A comparison of the retina of the mormyrids with that of various other teleosts. *Acta Zoologica*, 19, 427–65.

McGinnis, W., Garber, R.L., Wirz, J., Kuriowa, A., and Gehring, W. (1984). A homologuous protein-coding sequence in *Drosophila* homeotic genes and its conservation in other metazoans. *Cell*, 37, 403–8.

McKaye, K.R. and Kocher, T. (1983). Head ramming behaviour by three paedophagous cichlids in Lake Malawi, Africa. *Animal Behaviour*, 31, 206–10.

McKenzie, D.J., Farrell, A.P., and Brauner, C.J. (eds.) (2007). *Fish Physiology, Vol. 26. Primitive Fishes*. Amsterdam: Elsevier/Academic Press.

McLachlan, J.A. (2001). Environmental signaling: What embryos and evolution teach us about endocrine disrupting chemicals. *Endocrine Reviews*, 22, 319–41.

McLeese, J.M. and Moon, T.W. (1989). Seasonal changes in the intestinal mucosa of winter flounder, *Pseudopleuronectes americanus* (Walbaum) from Passamaquoddy Bay, New Brunswick. *Journal of Fish Biology*, 35, 381–93.

McNulty, J.A (1976). A comparative study of the pineal complex in the deep-sea fishes *Bathylagus wesethi* and *Nezumia liolepis*. *Cell and Tissue Research*, 172, 205–25.

McNulty, J.A. (1978a). Fine structure of the pineal organ in the troglobytic fish, *Typhlichthyes subterraneus* (Pisces: Amblyopsidae). *Cell and Tissue Research*, 195, 535–45.

McNulty, J.A. (1978b). The pineal of the troglophilic fish, *Chologaster agassizi*: an ultrastructural study. *Journal of Neural Transmission*, 43, 47–71.

McNulty, J.A. (1978c). A light and electron microscope study of the pineal in the blind goby, *Typhlogobius californiensis* (Pisces: Gobiidae). *Journal of Comparative Neurology*, 181, 197–212.

Mebs, D. (2009). Chemical biology of the mutualistic relationships of sea anemones with fish and crustaceans. *Toxicon*, 54, 1071–4.

Merchie, G., Lavens, P., and Sorgeloos, P. (1997). Optimization of dietary vitamin C in fish and crustacean larvae: a review. *Aquaculture*, 155, 165–81.

Meredith, T.L., Caprio, J., and Kajiura, S.M. (2012). Sensitivity and specificity of the olfactory epithelia of two elasmobranch species to bile salts. *Journal of Experimental Biology*, 215, 2660–7.

Messiatzeva, E. (1932). Chief results of the fishery research in the Barents Sea in 1930 by the GOIN (State Oceanographical Institute of U.S.S.R. *Rapports et process-verbaux des reunions* LXXXI Appendix 3. Copenhagen: International Council for the Exploration of the Sea.

Metcalfe, J.D. and Butler, P.J. (1982). Differences between directly measured and calculated values for cardiac output in the dogfish: a criticism of the Fick method. *Journal of Experimental Biology*, 99, 255–68.

Metcalfe, J.D. and Butler, P.J. (1984). On the nervous regulation of gill blood flow in the dogfish, *Scyliorhinus canicula*. *Journal of Experimental Biology*, 113, 253–67.

Metcalfe, J.D. and Butler, P.J. (1986). The functional anatomy of the gills of the dogfish (*Scyliorhinus canicula*). *Journal of Zoology, London, A*, 208, 519–30.

Mettam, J.J., McCrohan, C.R. , and Sneddon, L.U. (2012). Characterisation of chemosensory trigeminal receptors in the rainbow trout, *Oncorhynchus mykiss*: responses to chemical irritants and carbon dioxide. *Journal of Experimental Biology*, 215, 685–93.

Meunier, F.J. (1984). Spatial organization and mineralization of the basal plate of elasmoid scales in Osteichthyans. *American Zoologist*, 24, 953–64.

Meunier, F.J. (1991). The concept of bone tissue in Osteichthyes. *Seventh International Ichthyology Congress*. The Hague. Abstract, p. 48. Amsterdam: University of Amsterdam.

Meunier, F.J. and Huysseune, A. (1991). The concept of bone tissue in Osteichthyes. *Netherlands Journal of Zoology*, 42, 445–58.

Miklavc, P. and Valentinčič, T. (2012). Chemotopy of amino acids on the olfactory bulb predicts olfactory discrimination capabilities of zebrafish *Danio rerio*. *Chemical Senses*, 37, 65–75.

Miller, R.B. and Kennedy, W.A. (1948). Observations on the lake trout of Great Bear Lake. *Journal of the Fisheries Research Board of Canada*, 7, 176–89.

Millott, S., Vandewalle, P. , and Parmentier, E. (2011). Sound production in red-bellied piranhas (*Pygocentrus natteri*, Kner): an acoustic, behavioural and morphofunctional study. *Journal of Experimental Biology*, 214, 3613–18.

Mittal, A.K. and Banerjee, T.K. (1975). Histochemistry and the structure of the skin of a murrel, *Channa striata* (Bloch 1797) (Channiformes, Channidae). II Dermis and subcutuis. *Canadian Journal of Zoology*, 53, 844–52.

Mittal, A.K. and Whitear, M. (1979). Keratinisation of fish skin with special reference to the catfish *Bagarius bagarius*. *Cell and Tissue Research*, 202, 213–30.

Mohammad, M.G., Chilmonczyk, S., Birch, D., Aladaileh, S., Raftos, D., and Joss, J. (2007). Anatomy and cytology of the thymus in juvenile Australian lungfish, *Neoceratodus forsteri*. *Journal of Anatomy*, 211, 784–97.

Mölich, A., Waser, W., and Heisler, N. (2009).The teleost pseudobranch: a role for preconditioning of ocular blood supply? *Fish Physiology and Biochemistry*, 35, 273–86.

Moll, R., Divo, M. , and Langbein, L. (2008). The human keratins: biology and pathology. *Histochemistry and Cell Biology*, 129, 705–33.

Mommsen, T.P. (2001). Paradigms of growth in fish. *Comparative Biochemistry and Physiology B*, 129, 207–19.

Mommsen, T.P. and Walsh, P.J. (1991). Metabolic and enzymatic heterogeneity in liver of the ureogenic toadfish *Opsanus tau*. *Journal of Experimental Biology*, 146, 407–18.

Moncada, S. and Higgs, E.A. (2006). Nitric oxide and the vascular endothelium. *Handbook of Experimental Pharmacology*, 176, 213–54.

Moon, T.W., Busby, E.R., Cooper, G.A. , and Mommsen, T.P (1999). Fish hepatocyte glycogen phosphorylase—a sensitive indicator for hormonal modulation. *Fish Physiology and Biochemistry*, 21, 15–24.

Moore, M.J. Smolowitz, R., and Uhlinger, K. (2004). *Final report for flounder ulcer studies*. Massachusetts Water Resources Authority. Environmental quality Dept. Report ENQUAD 2004–2010. Duxbury MA: Environmental Quality Department.

Moore, R.A. (1933). Morphology of the kidneys of Ohio fishes. *Contributions, The Franz Theodore Stone Laboratory, Ohio University*, 5, 1–34.

Morgan, D.L. (2010). Fishes of the King Edward River in the Kimberley region, *Western Australia Records, Western Australia*, 25, 351–68.

Morgan, R.L., Wright, P.A., and Ballantyne, J.S. (2003). Urea transport in kidney brush-border membrane vesicles from an elasmobranch, *Raja erinacea*. *Journal of Experimental Biology*, 206, 3293–302.

Morin, J.G., Harrington, A., Nealson, K., Krieger, N., Baldwin, T.O. , and Hastings, J.W. (1975). Light for all reasons: Versatility in the behavioural repertoire of the flashlight fish. *Science*, 190, 74–6.

Morley, J.I. and Balshine, S. (2003). Reproductive biology of *Eretmodus cyanostictus*, a cichlid fish from Lake Tanganyika. *Environmental Biology of Fishes*, 66, 169–79.

Morris, R.G. (2013). NMDA receptors and memory encoding. *Neuropharmacology*, 74, 32–40

Morrison, C.M. (1987). *Histology of the Atlantic Cod, Gadus morhua. An Atlas. Part One. Digestive Tract and Associated Organs*. Canada: Canadian Special Publications of Fisheries and Aquatic Sciences 98. Ottawa: Department of Fisheries and Oceans.

Morrison, C.M. (1988). *Histology of the Atlantic Cod, Gadus morhua. An Atlas. Part Two. Respiratory System and Pseudobranch*. Canada: Canadian Special Publications of Fisheries and Aquatic Sciences 199. Ottawa: Department of Fisheries and Oceans.

Morrison, C.M. (1990). *Histology of the Atlantic Cod, Gadus morhua. An Atlas. Part Three. Reproductive Tract*. Canada: Canadian Special Publications of Fisheries and Aquatic Sciences 110. Ottawa: Department of Fisheries and Oceans.

Morrison, C.M. (1993). *Histology of the Atlantic cod, Gadus morhua. An Atlas. Part Four. Eleutheroembryo and Larva*. Canada: Canadian Special Publications of Fisheries and

Aquatic Sciences119. Ottawa: Department of Fisheries and Oceans.

Morriss-Kay, G.M. (2001). Derivation of the mammalian skull vault. *Journal of Anatomy*, 199, 143–51.

Moss, M.L. (1961a). Osteogenesis of acellular teleost fish bone *American Journal of Anatomy*, 108, 99–110.

Moss, M.L. (1961b). Studies of the acellular bone of teleost fish. 1. Morphological and systematic variations. *Acta Anatomica*, 46, 343–62.

Moss, M.L. (1963). The biology of acellular teleost fish bone. *Annals of the New York Academy of Science*, 109, 337–50.

Moss, M.L. (1977). Skeletal tissues in sharks. *American Zoologist*, 17, 335–42.

Mosse, P.R.L. and Hudson, R.C.L. (1977). The functional roles of different muscle fibre types identified in the myotomes of marine teleosts: a behavioural and histochemical study. *Journal of Fish Biology*, 11, 417–30.

Mott, J.C. (1957). The cardiovascular system. In: M.E. Brown (ed.). *The Physiology of Fishes, Vol. 1*. New York, NY: Academic Press, pp. 81–108.

Motta, P.J. (1984). Mechanics and function of jaw protrusion in teleost fishes: A review. *Copeia*, 1984, 1–18.

Moulton, L.L. and Penttila, D.E. (2001). *Field Manual for Sampling Forage Fish Spawn in Intertidal Shore Regions*. San Juan County forage fish assessment project. 23pp. LaConner, WA: Washington Department of Fish and Wildlife.

Mourão, K.R.M., Ferreira, V., and Lucena-Frédou, F. (2014). Composition of functional ecological guilds of the fish fauna of the internal sector of the Amazon Estuary, Pará, Brazil. *Annals of the Brazilian Academy of Science*, 86, 1783–800.

Moyle, P.B. and Cech, J.J. (1996). *An Introduction to Ichthyology*. Upper Saddle River, NJ: Prentice-Hall.

Mrowka, W. and Schierwater, B. (1988). Energy expenditure for mouthbrooding in a cichlid fish. *Behavioural Ecology and Sociobiology*, 22, 161–4.

Muir, B.S. (1970). Contributions to the study of blood pathways in fish gills. *Copeia*, 1970, 19–28.

Mulcahy, M.F. (1970). Blood values in the pike *Esox lucius* L. *Journal of Fish Biology*, 2, 203–9.

Munk, O. (1959). *The eyes of Ipnops murrayi Günther, 1878*. Galatea Report 3.

Munk, O. and Frederiksen, R.D. (1974). On the function of aphakic apertures in teleosts. *Videnskabelige Meddelelser fraDansk. Naturhistorisk Foring i Kobenhavn*, 137, 65–94.

Munroe, T.A. and Hashimoto, J. (2008). A new Western Pacific tonguefish (Pleuronectiformes: Cynoglossidae): The first pleuronectiform discovered at active hydrothermal vents. *Zootaxa*, 1839, 43–59.

Munz, F.W. (1971). Vision: visual pigments. In: W.S. Hoar and D.J. Randall (eds). *Fish Physiology, Vol. V: Sensory Systems and Electric Organs*. New York, NY: Academic Press, pp. 1–32.

Münz, H. and Claas, B. (1987). The terminal nerve and its development in the teleost fishes. In: L.S. Demski and M. Schwanzel Fukuda (eds). The terminal nerve (nervus terminalis) structure, function and evolution. *Annals of the New York Academy of Science*, 519, 50–9.

Muntz, W.R.A. (1975). The visual consequences of yellow filtering pigments in the eyes of fishes occupying different habitats. In: G.C. Evans, R. Bainbridge, and O. Rackham (eds). *Light as an Ecological Factor II*. Oxford: Blackwell Scientific Publications, pp. 271–87.

Murata, N. and Fujii, R. (2000). Pigment-aggregating action of endothelins on medaka xanthophores. *Zoological Science*, 17, 853–62.

Murphy, C.J. and Howland, H.C. (1991). The functional significance of crescent-shaped pupils and multiple pupillary apertures. *Journal of Experimental Zoology*, Supplement 5, 22–8.

Murray, H.M., Wright, G., and Goff, G.P. (1994). A study of the posterior esophagus in the winter flounder, *Pleuronectes americanus*, and the yellowtail flounder, *Pleuronectes ferruginea*: morphological evidence for pregastric digestion? *Canadian Journal of Zoology*, 72, 1191–8.

Murray, H.M., Hew, C.L. , and Fletcher, G.L. (2003). Spatial expression patterns of skin-type antifreeze protein in winter flounder (*Pseudopleuronectes americanus*) epidermis following metamorphosis. *Journal of Morphology*, 257, 78–86.

Murray, R.W. (1960). The response of the ampullae of Lorenzini of elasmobranchs to mechanical stimulation. *Journal of Experimental Biology*, 37, 417–24.

Mutschler, T. (2003). Lac Alaotra. In: S.M. Goodman and J.F. Banstead (eds.). The Natural History of Madagascar. Chicago: University Press, Chicago, pp. 1530–4.

Myers, R.A., Hutchings, J.A., and Barrowman, N.J. (1997). Why do fish stocks collapse? The example of cod in Atlantic Canada. *Ecological Applications*, 7, 91–106.

Nadkarni, V.B. and Gorbman, A. (1966). Structure of the corpuscle of Stannius in normal and radiothyroidectomized Chinook fingerlings and spawning Pacific salmon. *Acta Zoologica*, XLVII, 61–6.

Nakada, T., Westhoff, C.M., Kato, A., and Hirose, S. (2007). Ammonia secretion from fish gill depends on a set of Rh glycoproteins. *The Federation of American Societies for Experimental Biology Journal*, 21, 1067–74.

Nakada, T., Zandi-Nejad, K., Kurita, H., Kudo, H., Broumand, V., Kwon, C., Mercado, A., Mount, D.B. and Hirose, S. (2005). Roles of Slc13a1 and Slc26a1 sulfate transporters of eel kidney in sulfate homeostasis and osmoregulation in freshwater. *American Journal of Physiology, Regulatory, Integrative and Comparative Physiology*, 289, R575–R585.

Nakanishi, S., Inoue, A., Kita, T., Nakamura, M. Chang, A.C.Y., Cohen S.N., and Numa, S. (1979). Nucleotide sequence of cloned cDNA for bovine corticotrophin-β-lipotropin precursor. *Nature*, 278, 423–7.

Nash, D.M., Valencia, A.H., and Geffen, A.J. (2006). The origin of Fulton's condition factor- setting the record straight. *Fisheries*, 31, 236–8.

Navarro, D. and Snodgrass, B. (1995). *Ocean Encounter*. Monterey Bay, CA: Monterey Bay Aquarium.

N'Da, K. and Déniel, C. (1993). Sexual cycle and seasonal changes in the ovary of the red mullet *Mullus surmeletus* from the southern coast of Brittany. *Journal of Fish Biology*, 43, 229–44.

Nelson, E.M. (1955). The morphology of the swimbladder and auditory bulla in Holocentridae. *Fieldiana, Zoology*, 37, 121–37.

Nelson, J.S. (1976). *Fishes of the World*, 1st edn. New York, NY: John Wiley and Sons.

Nelson, J.S. (1994). *Fishes of the World*, 2nd edn. New York, NY: John Wiley and Sons.

Nelson, J.S. (2006). *Fishes of the World*, 4th edn. New York, NY: John Wiley and Sons.

Nelson, J.S., Grande, T.C., and Wilson, M.V.H. (2016). *Fishes of the World*, 5th edn. Hoboken, NJ: John Wiley and Sons Inc.

Neuhuber, W. and Schrödl, F. (2011). Autonomic control of the eye and the iris. *Autonomic Neuroscience: Basic and Clinical*, 165, 67–79.

Ngai, J., Dowling, M.M., Buck, L. Axel, R., and Chess, A. (1993). The family of genes encoding odorant receptors in channel catfish. *Cell*, 72, 657–66.

Nichols, J.H. (1971). *Fiches d'identification des oeufs et larves de poisons Fiches 4–6. Pleuronectidae*. Copenhagen: Conseil international pour l'exploration de la mer.

Nielsen, J.G. (1977). The deepest living fish *Abyssobrotula galateae*. A new genus and species of oviparous ophidioids (Pisces, Brotulidae). *Galatea Report*, 14, 41–8.

Nielsen, J.G. and Munk, O. (1964). A hadal fish (*Bassogigas profundissimus*) with a functional swimbladder. *Nature*, 204, 594–5.

Nikol'skii, G.V. (1961). *Special Ichthyology*. Trans. and published in Jerusalem by The Israel program for Scientific Translations.

Nikolsky, G.V. (1963). *The Ecology of Fishes*. Trans. I. Birkett. London: Academic Press.

Nikonov, A.A. and Caprio, J. (2001). Electrophysiological evidence for a chemotopy of biologically relevant odors

in the olfactory bulb of the channel catfish. *Journal of Neurophysiology*, 86, 1869–76.

Nikonov, A.A. and Caprio, J. (2004). Odorant specificity of single olfactory bulb neurons to amino acids in the channel catfish. *Journal of Neurophysiology*, 92, 123–4.

Nilsson, G.E. (1996). Brain and body oxygen requirements of *Gnathonemus petersii*, a fish with an exceptionally large brain. *Journal of Experimental Biology*, 199, 603–7.

Nilsson, G.E., Dixson, D.L., Domenici, P., McCormick, M.I., Sørensen, C., Watson, S. , and Munday, P.L. (2012). Near-future carbon dioxide levels alter fish behaviour by interfering with neurotransmitter functions. *Nature Climate Change*, 2, 201–4.

Nilsson, G.E., Löfman, C.O., and Block, M. (1995). Extensive erythrocyte deformation in fish gills observed by *in vivo* microscopy: apparent adaptations for enhancing oxygen uptake. *Journal of Experimental Biology*, 198, 1151–6.

Nilsson, S. (1976). Fluorescent histochemistry and cholinesterase staining in sympathetic ganglia of a teleost, *Gadus morhua*. *Acta Zoologica (Stockholm)*, 57, 69–77.

Nilsson, S. (1980). Sympathetic nervous control of the iris sphincter of the Atlantic cod, *Gadus morhua*. *Journal of Comparative Physiology A*, 138, 149–55.

Nilsson, S. (1983). *Autonomic Nerve Function in the Vertebrates*. Berlin: Springer-Verlag.

Nilsson, S. (1984). Innervation and pharmacology of the gills. In: W.S. Hoar and D.J. Randall (eds). *Fish Physiology, Vol. 10A. Gills anatomy, gas transfer, and acid-base regulation*. Orlando, FL: Academic Press, pp. 185–227.

Nilsson, S. (2011). Comparative anatomy of the autonomic nervous system. *Autonomic Neuroscience: Basic and Clinical*, 165, 3–9.

Nilsson, S. and Holmgren, S. (1993). Autonomic nerve functions. In: D.H. Evans (ed.). *The Physiology of Fishes*. Boca Raton, FL: CRC Press, pp. 279–313.

Nilsson, S. Abrahamsson, T. , and Grove, D.J. (1976). Sympathetic nervous control of adrenaline release from the head kidney of the cod, *Gadus morhua*. *Comparative Biochemistry and Physiology*, 55C, 123–7.

Nishiyama, T. (1963). Variations in the shape and blood supply of the saccus vasculosus in teleostean fishes. *Bulletin of the Faculty of Fisheries, Hokkaido University*, XIV, 47–61.

Noaillac-Depeyre, J. and Hollande, E. (1981). Evidence for somatostatin, gastrin and pancreatic polypeptide-like substance in the mucosa cells of the gut in fishes with and without a stomach. *Cell and Tissue Research*, 216, 193–203.

Nolte, W. (1933). Experimentelle Untersuchungen zum Problem der Lokalisation des Assoziationsvermögens im Fichgehirn. *Zeitschrift fuer vergleichende Physiologie*, 18, 255–79.

Nordeng, H. (1971). Is the local orientation of anadromous fishes determined by pheromones? *Nature*, 233, 411.

Nordeng, H. (1977). A pheromone hypothesis for homeward migration in anadromous salmonids. *Oikos*, 28, 155–9.

Nordgreen, J., Garner, J.P., Janczuk, M.J., Ranheim, B., Muir, W.M., and Horsberg, T.E. (2009). Thermonociception in fish: Effects of two different doses of morphine on thermal threshold and post-test behaviour in goldfish (*Carassius auratus*). *Applied Animal Behavior Science*, 119, 101–7.

Norman, J.R. (1934). *Monograph of the Flatfishes*. London: Trustees of The British Museum.

Normandeau, Exponent, Tricas, T., and Gill, A. (2011). *Effects of EMF, from Undersea Power Cables on Elasmobranchs and other Marine Species*. U.S. Department of the Interior, Bureau of Ocean EnergyManagement, Regulation and Enforcement, Pacific OCS Region, Camarillo, CA. OCS study BOEMRE 2011–2009. Camarillo, CA: Normandeau Associates Inc; Outer Continental Shelf Region Study.

Norris, D.O (1980). *Vertebrate Endocrinology*. Philadelphia, PA: Lea and Febiger.

Northcote, T.G., Arcifer, M.S., and Froehlich, O. (1987). Fin-feeding by the piranha (*Serrasalmus spilopleura* Kner): the cropping of a novel renewable resource. *Proceedings of the Vth Congress of European Ichthyologists, Stockholm 1985*, 133–45.

Northcutt, R.G. (2002). Understanding vertebrate brain evolution. *Integrative and Comparative Biology*, 42, 743–56.

Northcutt, R.G., Holmes, P.H., and Albert, J.S. (2000). Distribution and innervations of lateral line organs in the channel catfish. *Journal of Comparative Neurology*, 421, 570–92.

Northcutt, R.G., Neary, T.J., and Senn, D.G. (1978). Observations on the brain of the coelacanth *Latimeria chalumnae*: external anatomy and quantitative analysis. *Journal of Morphology*, 155, 181–92.

Ochi, H., Rossiter, A. , and Yanagisawa, Y. (2002). Paternal mouthbrooding bagrid catfishes in Lake Tanganyika. *Ichthyological Research*, 49, 270–73.

O'Connell, C.P. (1972). The interrelation of biting and filtering in the feeding activity of the Northern anchovy. *Journal of the Fisheries Research Board of Canada*, 29, 285–93.

Odum, E.P. (1993). *Ecology and Our Endangered Life-Support Systems*. Sunderland, MA: Sinauer Associates Inc.

Oganesyan, S.A. (1993). Periodicity of the Barents Sea cod reproduction. *International Council for the Exploration of the Sea*. Conference and Meeting, 1993/G:64. Demersal Fish Committee.

Ogawa, N. and Caprio, J. (2010). Major differences in the population of amino acid fiber types. *Journal of Neurophysiology*, 103, 2062–73.

O'Grady, S.M. and Maniak, P.J. (2005). Potassium secretion in winter flounder (*Pseudopleuronectes americanus*) intestine: Effects of K^+ channel blockers on electrogenic NaCl absorption. *The Bulletin, Mount Desert Island Biological Laboratory*, 44, 6–9.

Oguri, M. (1978). On the hepatosomatic index of holocephalian fish. *Bulletin of the Japanese Society of Scientific Fisheries*, 44, 131–4.

Oguri, M. (1989). Distribution of renal juxtaglomerular cells in the shortnosed sea horse *Hippocampus brevirostris*. *Nippon Suisan Gakkaishi*, 55, 1379–81.

Ojeda, J.L., Icardo, J.M., Wong, W.P., and Ip, Y.K. (2006). Microanatomy and ultrastructure of the kidney of the African lungfish *Protopterus dolloi*. *Anatomical Record*, 288A, 609–25.

Okiyama, M. and Ida, H. (2010). Record of *Ipnops* sp. (Ipnopidae: Aulopiformes) from northern Japan. *Ichthyological Research*, 57, 422–3.

Olsen, K.H. Grahnt, M., Lohmt, J., and Langeforst, A. (1998). MHC and kin discrimination in juvenile Arctic charr, *Salvelinus alpinus* (L.). *Animal Behaviour*, 56, 319–27.

Olsen, R.E., Myklebust, R., Kaino, T., and Ringo, E. (1999). Lipid digestibility and ultrastructural changes in the enterocytes of Arctic char (*Salvelinus alpinus* L.) fed linseed oil and soybean lecithin. *Fish Physiology and Biochemistry*, 21, 35–44.

Olson, K.R. (1996). Secondary circulation in fish: anatomical organization and physiological significance. *Journal of Experimental Zoology*, 275, 172–85.

Olson, K.R. (1998). The cardiovascular system. In: D.H. Evans (ed.). *The Physiology of Fishes*, 2nd edn. Boca Raton, FL: CRC Press, pp. 129–54.

Olsson, C. and Holmgren, S. (2001). The control of gut motility. *Comparative Biochemistry and Physiology A*, 128, 481–503.

Ong, K.J., Stevens, E.D., and Wright, P.A. (2007). Gill morphology of the mangrove killifish (*Kryptolebias aormoratus* is plastic and changes in response to terrestrial air exposure. *Journal of Experimental Biology*, 210, 1109–15.

Ono, R.D. (1982). Proprioceptive endings in the myotome of the pickerel (Teleostei: Esocidae). *Journal of Fish Biology*, 21, 525–35.

Osse, J.W.M. and van deb Boogaart, J.G.M. (1999). Dynamic morphology of fish larvae, structural implications of friction forces in swimming, feeding and ventilation. *Journal of Fish Biology*, 55 (Supplement A), 1565–74.

Östland-Nilsson, S., Mayer, I., and Huntingford, F.A. (eds) (2006). *Biology of the Three-spined Stickleback*. Boca Raton, FL : CRC Press.

Overnell, J. (1973). Digestive enzymes of the pyloric caeca and of their associated mesentery in the cod (*Gadus morhua*). *Comparative Biochemistry and Physiology B*, 46, 519–31.

Owens, D.W. and Lane, E.B. (2003). The quest for function of simple epithelial keratins. *BioEssays*, 25, 748–58.

Owens, G.L., Rennison, D.J., Allison, W.T., and Taylor, J.S. (2011). *Biology Letters*. doi: 10.1098/rsbt.0582

Oxley, A., Tocher, D.R., Torstensen, B.E. , and Olsen, R.E. (2005). Fatty acid utilisation and metabolism in caecal enterocytes of rainbow trout (*Oncorhynchus mykiss*) fed dietary fish or copepod oil. *Biochimica et Biophysica Acta*, 1737, 119–29.

Ozaka, C., Yamamoto, N., and Somiya, H. (2009). The aglomerular kidney of the deep-sea fish, *Ateleopus japonicus* (Ateleopodiformes: Ateleopodidae): Evidence of wider occurrence of the aglomerular condition in Teleostei. *Copeia*, 2009, 609–17.

Ozon, R. (1972). Estrogens in fishes, amphibians, reptiles and birds. In: D.R. Idler (ed.). *Steroids in Nonmammalian Vertebrates*. New York, NY: Academic Press, pp. 390–413.

Padhi, K.P. Akimento, M-A., and Ekker, M. (2006). Independent expansion of the keratin gene family in teleostean fish and mammals: An insight from phylogenetic analysis and radiation hybrid mapping of keratin genes in zebrafish. *Gene*, 368, 37–45.

Page, L.M. (1976). The modified midventral scales of *Percina* (Osteichthyes; Percidae). *Journal of Morphology*, 148, 255–64.

Palmer, A.R. (2010). Scale-eating cichlids: from hand(ed) to mouth. *Journal of Biology*, 9, 11–15.

Palstra, A.P., Blok, M.C., Blom, E., Tuinhof-Koelma, N. Dirks, R.P., Forlenza, M., and Blonk, R.J.W. (2014). In- and outdoor reproduction of first generation common sole under a natural photothermal regime: Temporal progression of sexual maturation determined by plasma steroids, genome sequencing and pituitary gonadotropin expression. *Proceedings of the 10th International Symposium on Reproductive Physiology of Fish, Alhao, Portugal*. p. 80.

Pankhurst, N.W., Sideleva, V.G., Pankhurst, P.M., Smirnova, O., and Janssen, J. (1994). Ocular amorphology of the Baikal oil-fishes, *Comephora baicalensis* and *C.dybowskii* (Comephoridae). *Environmental Biology of Fishes*, 39, 51–8.

Pandolfi, M., Cánepa, M.M., Meijidae, F.J., Alonso, F., Vázquez, G.R., Maggese, M.C., and Vissio, P.G. (2009). Studies on the reproductive and developmental biology of *Cichlosoma dimerus* (Perciformes, Cichlidae). *Biocell*, 33, 1–20.

Pang, P.K.T. (1971). Calcitonin and ultimobranchial glands in fishes. *Journal of Experimental Zoology*, 178, 88–99.

Pang, P.K.T. (1973). Endocrine control of calcium metabolism in teleosts. *American Zoologist*, 13, 775–92.

Pankhurst, N.W. (1985). Final maturation and ovulation of oocytes of the goldeye *Hiodon alosoides* (Rafinesque) *in vitro*. *Canadian Journal of Zoology*, 63, 1003–9.

Pantin, C.F.A. (1960). Presidential Address: 24 May, 1960. *Geonemertes*: A study in island life. *Proceedings of the Linnean Society of London*, 172.

Papoutsoglou, E.S. and Lyndon, A.R. (2005). Effect of incubation temperature on carbohydrate digestion in important teleosts for aquaculture. *Aquaculture Research*, 36, 1252–64.

Park, E-H. and Lee, S-H. (1988). Scale growth and squamation chronology for the laboratory-reared hermaphrodite fish *Rivulus marmoratus* (Cyprinodontidae). *Japanese Journal of Ichthyology*, 34, 476–82.

Parker, G.H. (1908). Structure and function of the ear of the squeteague. *Bulletin of the United States Bureau of Fisheries*, 28, 1211–24.

Parker, G.H. (1912). The relation of smell, taste and the common chemical sense in vertebrates. *Proceedings of the Academy of Natural Sciences of Philadelphia*, 15, 219–34.

Parker, G.H. (1948). *Animal Colour Changes and their Neurohumors*. Cambridge: Cambridge University Press.

Parker, M.O., Brock, A.J. Walton, R.T., and Brennan, C.H. (2013). The role of zebrafish (*Danio rerio*) in dissecting the genetics and neural circuits of executive function. *Frontiers in Neural Circuits*, 7. doi.10.3389/fncir.2013.0063 eCollection 2013

Parmentier, E., Fine, M., Vanderwalle, P., Ducamp, J-J., and Lagerdère J-P. (2006a). Sound production in two carapids (*Carapus acus* and *C. mourlani*) and through the sea cucumber tegument. *Acta Zoologica (Stockholm)*, 87, 113–19.

Parmentier, E., Lagardère, J-P., Braquegnier, J-P., Vanewalle, P., and Fine, M.L. (2006b). Sound production in carapid fish: first example with a slow sonic muscle. *Journal of Experimental Biology*, 209, 2952–60.

Parmentier, E., Lanterbecq, D., and Eeckhaut, I. (2016). From commensalism to parasitism in Carapidae (Ophidiiformes): heterochronic modes of development? *Peer Journal*, 4, 1786. DOI 10.7717/peerj.1786.

Partridge, B.L. and Pitcher, T.J. (1980). The sensory basis of fish schools: relative roles of lateral line and vision. *Journal of Comparative Physiology*, 135, 315–25.

Patshnik, M. and Groninger, H.S. Jr. (1964). Observations on the milky condition in some Pacific coast fishes. *Journal of the Fisheries Research Board of Canada*, 21, 335–46.

Patterson, S. and Goldspink, G. (1972). The fine structure of red and white myotomal muscle fibres of the coalfish

(*Gadus virens*). *Zeitschrift fuer Zellforschung und Microsko-pische Anatomie*, 133, 403–74.

Patterson, S., Johnston, I.A., and Goldspink, G. (1974). The effect of starvation on the chemical composition of red and white muscles in the plaice (*Pleuronectes platessa*). *Experientia*, 30, 892–3.

Patton, H.D. (1965). Receptor mechanism. In: Ruch, T.C. and Patton, H.D. (eds). *Physiology of Biophysics*, 19th edn. Philadelphia, PA: Saunders, pp. 95–112.

Patton, J.S., Nevenzel, J.C., and Benson, A.A. (1975). Specificity of digestive lipase in hydrolysis of wax esters and triglycerides studied in anchovy and other selected fish. *Lipids*, 10, 575–83.

Patzner, R.A. and Seiwald, M. (1985). The reproduction of *Blennius pava* Risso (Teleostei, Blenniidae). Secondary sexual organs and accessory glands of the testis during the reproductive cycle. *Proceedings of the Vth Congress of European Ichthyologists*, Stockholm 1985, pp. 293–8. Stockholm: Swedish Museum of Natural History.

Paulsen, D.F. (1993). *Basic Histology*, 2nd edn. Norwalk, CT: Appleton and Lange.

Pauly, D. (1980). A new methodology for rapidly acquiring basic information on tropical fish stocks: growth, mortality and stock-recruitment relationships. In: S.B. Saila and P. Roedel (eds). *Stock Assessment for Tropical Small-Scale Fisheries*. Proceedings of an International workshop held 19–21 September 1979. Kingston, University of Rhode Island, International Center for Marine Research Development, pp. 154–72.

Pauly, D. (1993). Foreword. In: *On the Dynamics of Exploited Fish Populations*. Reprint, Beverton, R.J.H. and Holt, S.J. (eds). London: Chapman and Hall.

Pauly, D., Christensen, V., Guénnette, S., Pitcher, T.J., Sumaila, U.R., Walters, C.J., Watson, R., and Zeller, D. (2002). Towards sustainability in world fisheries. *Nature*, 418, 689–95.

Peach, M.B. and Marshall N.J. (2000). The pit organs of elasmobranchs: a review. *Philosophical Transactions of the Royal Society of London B*, 355, 1131–4.

Peachey, G. (1989). *The Perch, Dissection Guide*. Kelowna: Biocam Communications.

Pearcey, W.G. (1961). Seasonal changes in osmotic pressure of flounder sera. *Science*, 134, 193–4.

Pearse, A.G.E. (1968). *Histochemistry Theoretical and Applied*. Boston, MA: Little, Brown and Co.

Pearse, A.G.E. (1969). The cytology and ultrastructure of polypeptide hormone-producing cells of the APUD series and the embryologic, physiologic and pathologic implications of the concept. *Journal of Histochemistry and Cytochemistry*, 17, 303–13.

Pearson, R. and Ball, J.N. (1981). *Lecture Notes on Vertebrate Biology*. Oxford: Blackwell Scientific Publications.

Penttila, J. and Dery, L.M. (eds). (1988). Age determination methodsfor Norhwest Atlantic species. National Oceanic and Atmospheric Administration, U.S. Department of Commerce, Technical Report National Marine Fisheries Service, 72. Springfield NA: National Marine Fisheries Service.

Perry, S.F. (1997). The chloride cell: structure and function in the gills of freshwater fishes. *Annual Review of Physiology*, 59, 325–47.

Perry, S.F., Ekker, M., Farrell, A.P., and Brauner, C.J. (eds) (2010). *Zebrafish, Fish Physiology, Vol. 29. Zebrafish*. Amsterdam: Elsevier Inc.

Petersen, R.H., Spinney, H.C.E., and Sreedharan, A. (1977). Development of Atlantic salmon (*Salmo salar*) eggs and alevins under varied temperature regimes. *Journal of the Fisheries Research Board of Canada*, 34, 31–43.

Petrino, T.R., Toussaint, G., and Lin, Y-W.P. (2007). Role of inhibin and activin in the modulation of gonadotropin- and steroid-induced oocyte maturation in the teleost *Fundulus heteroclitus*. *Reproductive Biology and Endocrinology*, 5, 21–30.

Pettigrew, J.D. (1999). Electroreception in monotremes. *Journal of Experimental Biology*, 202, 1447–54.

Pettigrew, J.D. and Collin, S.P. (1995). Terrestrial optics in an aquatic eye: the sandlance *Limnichthyes fasciatus* (Creediidae, Teleostei). *Journal of Comparative Physiology A*, 177, 397–408.

Pettigrew, J.D., Collin, S.P. , and Ott, M. (1999). Convergence of specialized behaviour, eye movements and visual optics in the sandlance (Teleostei) and the chameleon (Reptilia). *Current Biology*, 9, 421–24.

Pfeiffer, W. (1963). The fright reaction in North American fish. *Canadian Journal of Zoology*, 41, 69–77.

Pfeiffer, W. (1964). The morphology of the olfactory organ of *Haplopagrus guentheri* Gill 1862. *Canadian Journal of Zoology*, 42, 235–7.

Pfeiffer, W. and Fletcher, T.F. (1964). Club cells and granular cells in the skin of lamprey. *Journal of the Fisheries Research Board of Canada*, 21, 1083–8.

Phelps, Q.E. Edwards, K.R., and Willis, D.W. (2007). Precision of five structures for estimating age of common carp. *North American Journal of Fisheries Management*, 27, 103–5.

Pickering, A.D. (1977). Seasonal changes in the epidermis of the brown trout (*Salmo trutta* (L.) *Journal of Fish Biology*, 10, 561–6.

Pickering, A.D. (1989). Environmental stress and the survival of brown trout, *Salmo trutta*. *Freshwater Biology*, 21, 47–55.

Pickering, A.D. and Fletcher, J.M. (1987). Sacciform cells in the epidermis of the brown trout, *Salmo trutta*, and the Arctic char, *Salvelinus alpinus*. *Cell and Tissue Research*, 247, 259–65.

Pickford, G.E. and Atz, J.W. (1957). *The Physiology of the Pituitary Gland of Fishes*. New York, NY: New York Zoological Society.

Pickford, G.E. and Grant, F.B. (1967). Serum osmolarity in the coelacanth, *Latimeria chalumnae*: urea retention and ion regulation. *Science*, 155, 568–70.

Pieperhoff, S., Bennett, W., and Farrell, A.P. (2009). The intercellular organization of the two muscular systems in the adult salmonid heart, the compact and the spongy myocardium. *Journal of Anatomy*, 215, 536–47.

Pietsch, T.W. (2005). Dimorphism, parasitism, and sex revisited: modes of reproduction among deep-sea ceratioid anglerfishes (Teleostei: Lophiiformes). *Ichthyological Research*, 52, 207–36.

Pinkus, F. (1894). Über einen noch nicht beschriebenen Hinneren des *Proptopterus annectens*. *Anatomischer Anzeiger*, 9, 562–6.

Plante, S., Audet, C., Lambert, Y., and de la Noüe, J. (2003). Comparison of stress responses in wild and captive winter flounder (*Pseudopleuronectes americanus* Walbaum) broodstock. *Aquaculture Research*, 34, 803–12.

Platt, C. and Popper, A.N. (1981). Fine structure and function of the ear. In: W.N. Tavolga, A.N. Tavolga, A.N. Popper, and R.R. Fay (eds). *Hearing and Sound Communication in Fishes*. New York, NY: Springer Verlag, pp. 3–36.

Pohlmann, K., Atema, J., and Breithaupt, T. (2004). The importance of the lateral line in nocturnal predation of piscivorous catfish. *Journal of Experimental Biology*, 207, 2971–8.

Pohlmann, K., Grasso, F.W., and Breithaupt, T. (2001). Tracking wakes: the nocturnal predatory strategy of piscivorous catfish. *Proceedings of the National Academy of Sciences of the USA*, 98, 7371–4.

Poisson, F. and Fauvel, C. (2009). Reproductive dynamics of swordfish (*Xiphias gladius*) in the southwestern Indian Ocean (Reunion Island). Part 2: fecundity and spawning pattern. *Aquatic Living Resources*, 22, 59–68.

Poncin, P. (1994). Field observations on a mating attempt of a spawning grayling, *Thymallus thymallus* with a feeding barbel *Barbus barbus*. *Journal of Fish Biology*, 45, 904–6.

Poon, S-K., So, W-K., Xu, X., Liu, L., and Ge, W. (2009). Characterization of inhibin α subunit (*inha*) in the zebrafish: evidence for a potential feedback loop between the pituitary and ovary. *Reproduction Research*, 138, 709–19.

Popper, A.N. and Fay, R.R. (2011). Rethinking sound detection by fishes. *Hearing Research*, 273, 25–36.

Popper, A.N. and Platt, C. (1993). Inner ear and lateral line. In: D.H. Evans (ed.). *The Physiology of Fishes*. Boca Raton, FL: CRC Press, pp. 99–136.

Portavella, M. and Vargas, J.P. (2005). Emotional and spatial learning in goldfish is dependent on different telencephalic pallial systems. *European Journal of Neuroscience*, 21, 2800–6.

Porter, W.P. (1967). Solar radiation through the living body walls of vertebrates with emphasis on desert reptiles. *Ecological Monographs*, 37, 273–96.

Pörtner, H.O. and Knust, R. (2007). Climate change affects marine fishes through the oxygen limitation of thermal tolerance. *Science*, 315, 95–7.

Potter, I.C. and Robinson, E.S. (1991). Development of the testis during post-metamorphic life of the Southern hemisphere lamprey *Geotria australis* Gray. *Acta Zoologica*, 72, 113–19.

Pottinger, T.G. and Pickering, A.D. (1985). Changes in skin structure associated with elevated androgen levels in maturing male brown trout, *Salmo trutta* (L.). *Journal of Fish Biology*, 26, 745–53.

Potts, G.W. and Wootton, R. (eds) (1984). *Fish Reproduction: Strategies and Tactics*. London: Academic Press.

Potts, W.M., Booth, A.J., Richardson, T.J., and Sower, W.H.H. (2014). Ocean warming affects the distribution and abundance of resident fishes by changing their reproductive scope. *Reviews in Fish Biology and Fisheries*, 24, 493–504.

Powell, D.P. (2000). In: G.K. Ostrander (ed.). *The Laboratory Fish: The Handbook of Experimental Animals*. London: Academic Press, pp. 79–92.

Powles, P.M. (1958). Studies of reproduction and feeding of Atlantic cod (*Gadus callarias* L.) in the southwestern Gulf of St. Lawrence. *Journal of the Fisheries Research Board of Canada*, 15, 1383–402.

Prasada Rao, P.D. and Finger, T.E. (1984). Asymmetry of the olfactory system in the brain of the winter flounder *Pseudopleuronectes americanus*. *Journal of Comparative Neurology*, 225, 492–510.

Preston, E., McManus, C.D., Jonsson, A.C., and Courtice, G.P. (1995). Vasoconstrictor effects of galanin and distribution of galanin containing fibres in three species of elasmobranch fish. *Regulatory Peptides*, 58, 123–34.

Prosser, C.L. (ed.). (1952). *Comparative Animal Physiology*. Philadelphia, PA: W.B. Saunders Company.

Protas, M., Conrad, M., Gross, J.B., Tabin, C., and Borowsky, R. (2007). Regressive evolution in the Mexican cave tetra, *Astyanax mexicanus*. *Current Biology*, 17, 452–4.

Proudlove, G.S. (2006). *Subterranean Fishes of the World*. Moulis: International Society for Subterranean Biology.

Prozorovskaia, M.L. (1952). On the method of determining the fat content of the roach from the quantity of fat on the intestines. *Bulletin of the Institute of Fisheries*, USSR No. 1.

Pusey, B., Kennard, M., and Arthington, A. (2004). *Freshwater Fishes of North-Eastern Australia*. Collingwood: CSIRO Publishing.

Pütter, A. (1920). Studien über physiologische Ähnlichkeit. VI. Wachstumsähnlichkeiten. *Pflüger's Archiv-fuer die Gesamte Physiologie des Menschen und der Tiere*, 180, 298–340.

Puvanendran, V., Salies, K., Laural, B., and Brown, J.A. (2004). Size-dependent foraging of larval Atlantic cod (*Gadus morhua*). *Canadian Journal of Zoology*, 82, 1360–89.

Pyle, G and Couture, P. (2011). Nickel. In: C.M. Wood, A.P. Farrell, and C.J. Brauner (eds). *Fish Physiology, Vol. 31, Part A. Homeostasis and toxicology of essential metals* New York, NY: Academic Press, pp. 253–89.

Quinn, T.P. (1990). Current controversies in the study of salmon homing. *Ethology, Ecology and Evolution*, 2, 49–63.

Radhakrishnan, S., Stephen, M., and Nair, N.B. (1976). A study on the blood cells of a marine teleost fish, *Diodon hystrix*, together with a suggestion as to the origin of the lymphocytes. *Proceedings of the Indian National Science Academy*, 42, 212–26.

Raeder, I.L.U. Paulsen, S.M., Smalas, A.O., and Willassen, N.P. (2007). Effect of fish skin mucus on the soluble proteome of *Vibrio salmonicida* analysed by 2-D gel electrophoresis and tandem mass spectrometry. *Microbial Pathogenesis*, 42, 36–45.

Rahman, M.S., Takemura, A., Park, Y.J., and Takano, K. (2003). Lunar cycle in the reproductive activity in the forktail rabbitfish. *Fish Physiology and Biochemistry*, 28, 443–4.

Ramón-y-Cajal, S. (1952). *Histologie du système nerveux de l'homme et des vertébrés*. (Traduite de l'espagnol par L. Azoulay). 2 vols. Consejo superior de investigaciones cientifixas, Instituto Ramón-y-Cajal, Madrid.

Ramos, C.A., Fernandes, M.N., da Costa, O.T.F. , and Duncan, W.P. (2013). Implications for osmorespiratory compromise by anatomical remodeling in the gills of *Arapaima gigas*. *The Anatomical Record*, 296, 1664–75.

Randall, D.J. (1968). Functional morphology of the heart in fishes. *American Zoologist*, 8, 179–89.

Randall, D.J (1982). The control of respiration and circulation in fish during exercise and hypoxia. *Journal of Experimental Biology*, 100, 275–88.

Randall, D.J. (2014). Hughes and Shelton: the fathers of fish respiration. *Journal of Experimental Biology*, 217, 3191–2.

Randall, D.J. and Hoar, W.S. (1971). Special techniques. In: W.S. Hoar and D.J. Randall (eds). *Fish Physiology, Vol. VI. Environmental relations and behavior.* New York, NY: Academic Press, pp. 511–28.

Randall, D., Burggren, W., and French, K. (1997, 2002). *Eckert Animal Physiology Mechanisms and Adaptations*, 4th edn. New York, NY: W.H. Freeman and Co.

Rand-Weaver, M., Baker, B.I., and Kawauchi, H. (1991b). Cellular localization of somatolactin in the pars

intermedia of some teleost fishes. *Cell and Tissue Research*, 263, 207–15.

Rand-Weaver, M., Noso, T., Muamoto, K. , and Kawauchi, H. (1991a). Isolation and characterization of somatolactin, a new protein related to growth hormone and prolactin from Atlantic cod (*Gadus morhua*) pituitary gland. *Biochemistry*, 30, 1509–15.

Rang, H.P. and Dale, M.M. (1991). *Pharmacology*. Edinburgh: Churchill Livingstone.

Ranvier, L. (1872). Des étranglements annulaires et des segments interannulaires chez les *Raies* et les *Torpilles*. *Comptes Rendus, Academie des Sciences Paris*, 75, 1129–1132.

Raschi, W. and Tabit, C. (1992). Functional aspects of placoid scales: A review and update. *Australian Journal of Marine and Freshwater Research*, 43, 123–47.

Rasquin, P. (1956). Cytological evidence for a role of the corpuscles of Stannius in the osmoregulation of teleosts. *Biological Bulletin*, 111, 399–409.

Rauther, M. (1923). Zur vergleichender Anatomie der Schwimmblase der Fische. *Ergebnisse und Fortschritt der Zoologie*, 5, 1–66.

Rawson, D.S. (1951). Studies of the fish of Great Slave Lake. *Journal of the Fisheries Research Board of Canada*, 8, 207–40.

Raymond, J.A. and Driedzic, W.R. (1997). Amino acids are a source of glycerol in cold-acclimatized rainbow smelt. *Comparative Biochemistry and Physiology*, 118B, 387–93.

Raymond, P.A. and Barthel, L.K. (2004). A moving wave patterns the cone photoreceptor mosaic array in the zebrafish retina. *International Journal of Developmental Biology*, 48, 935–45.

Raymond, P.A., Colvin, S.M., Jabeen, Z., Nagashima, M., Barthel, L.K., Hadidjojo, J., Popova, L., Pejaver, V.R., and Lubensky, D.K. (2014). Patterning the cone mosaic array in zebrafish retina requires specification of ultraviolet-sensitive cones. *Plos One*, 9 (1). Doi.org/10.1371/journal.pone.0085325

Rebok, K., Jordanova, M., and Tavciovska-Vasileva, I. (2011). Spleen histology in the female Ohrid trout, *Salmo letnica* (Kar.) (Teleostei, Salmonidae) during the reproductive cycle. *Archives of Biological Sciences, Belgrade*, 63, 1023–30.

Rees, H. (1964). The question of polyploidy in the Salmonidae. *Chromosoma*, 15, 275–9.

Regan, M.D., Turko, A.J., Heras, A.J., Andersen, M.K., Lefevre, S., Wang, T., Bayley, M., Brauner, C.J., Huong, D.T.T., Phuong, N.T., and Nilsson, G.E. (2016). Ambient CO_2 fish behaviour and altered GABAergic neurotransmission: exploring the mechanism of CO_2-altered behaviour by taking a hypercapnia dweller down to low CO_2 levels. *Journal of Experimental Biology*, 219, 109–18.

Reid, S.D. (2011). Molybdenum and chromium. In: C.M. Wood, A.P. Farrell, and C.J. Brauner (eds). *Fish Physiology, Vol. 31, Part A Homeostasis and toxicology of essential metals*. New York, NY: Academic Press, pp. 375–415.

Reid, S.G., Sundin, L., Florindo, L.H., Rantin, F.T., and Milsom, W.K. (2003). Effects of afferent input on the breathing pattern continuum in the tambaqui (*Colosomma macropomum*). *Respiratory Physiology and Neurobiology*, 36, 39–53.

Reighard, J. E. (1903). *The Natural History of Amia calva Linnaeus*. G.H. Parker (ed.) Mark Anniversary Volume. Art IV. New York: Henry Holt and Company.

Reilly, B.D., Cramp, R.L., Wilson, J.M., Campbell, H.A. , and Franklin, C.E. (2011). Branchial osmoregulation in the euryhaline bull shark, *Carcharhinus leucas*: a molecular analysis of ion transporters. *Journal of Experimental Biology*, 214, 2883–95.

Reimchen, T.E. and Temple, N.F. (2004). Hydrodynamic and phylogenetic aspects of the adipose fin in fishes. *Canadian Journal of Zoology*, 82, 910–16.

Retzius, G. (1881). Das Gehörorgan der Wirbeltiere. *Morphologische-histologische Studien*. Stockholm: Samson und Wallin.

Ribbink, A.J. and Ribbink, A.C. (1997), Paedophagia among cichlid fishes of Lake Victoria and Lake Malawi/Nyasa. *South African Journal of Science*, 93, 509–12.

Ribeiro, M.L da S., DaMatta, R.A., Diniz, J.A.P., de Souza, W., Do Nascimento, J.L.M., and de Carvalho, T.M.U. (2007). Blood and inflammatory cells of the lungfish *Lepidosiren paradoxa*. *Fish and Shellfish Immunology*, 23, 178–87.

Ribeiro, V.M.A., Santos, G.B., and Bazzoli, N. (2007). Reproductive biology of *Steindachnerina insculpta* (Fernandez-Yépez) (Teleostei, Curimatidae) in Furnas reservoir, Minas Gerais, Brazil. *Revista Brasileira de Zoologia*, 24, 71–6.

Rice, A.N. and Lobel, P.S. (2002). Enzyme activities of pharyngeal jaw musculature in the cichlid *Tramitichromis intermedius*: implications for sound production in cichlid fishes. *Journal of Experimental Biology*, 205, 3519–23.

Richards, T.D.J., Fenton, A.L., Syed, R., and Wagner, G.F. (2012). Characterization of stanniocalcin-1 receptors in the rainbow trout. *ISRN Endocrinology Article ID 257841*, Vol. 2012. doi.10.5402/2012/257841

Richter, M. and Smith, M. (1995). A microstructural study of the ganoine tissue of selected lower vertebrates. *Zoological Journal of the Linnean Society*, 114, 173–212.

Ricker, W.E. (1954). Stock and recruitment. *Journal of the Fisheries Research Board of Canada* 11, 559–623.

Rideout, R.M. & Burton, M.P.M. (2000). Peculiarities in ovarian structure leading to multiple year delays in oogenesis and possible senescence in Atlantic cod, *Gadus morhua* L. *Canadian Journal of Zoology*, 78, 1840–1844.

Rideout, R.M. Rose, G.A., and Burton, M.P.M. (2005). Skipped spawning in female iteroparous fishes. *Fish and Fisheries*, 6, 50–72.

Rink, R.E. and Wullimann, M.F. (2004). Connections of the ventral telencepahlon (subpallium) in the zebrafish (*Danio rerio*). *Brain Research*, 1011, 206–20.

Ritchie, J. (1953). *Thomson's Outlines of Zoology*. London: Oxford University Press.

Ritter, E.K. (2002). Analysis of sharksucker, *Echeneis naucrates*, induced behaviour patterns in the blacktip shark, *Carcharhinus limbatus*. *Environmental Biology of Fishes*, 65, 111–15.

Ritter, E.K. and Godknecht, A.J. (2000). Agonistic displays in the black-tip shark (*Carcharhinus limbatus*). *Copeia*, 2000, 283–4.

Robb, A.P. (1982). Histological observations on the reproductive biology of the haddock, *Melanogrammus aeglefinus*. *Journal of Fish Biology*, 20, 397–408.

Roberts, A.C., Bill, B.R., and Glanzman, D.L. (2013). Learning and memory in zebrafish larvae. *Frontiers in Neural Circuits*, 7, Article 126. doi.10.3389/fncir.2013.00126

Roberts, B.L. (1969). The response of a proprioceptor to the undulatory movements of dogfish. *Journal of Experimental Biology*, 51, 775–85.

Roberts, J.L. (1975). Active branchial and ram gill ventilation in fishes. *Biological Bulletin*, 148, 85–105.

Roberts, R.J. and Bullock, A.M. (1981). Recent observations on the pathological effect of ultraviolet light on fish skin. *Fish Pathology*, 15, 237–9.

Roberts, R.J. and Ellis, A.E. (2012). The anatomy and physiology of the teleosts. In: R.J.Roberts (ed.). *Fish Pathology*, 4th edn. Oxford: Wiley-Blackwell, pp. 17–61.

Roberts, R.J. Young, H., and Milne, J.A. (1972). Studies on the skin of plaice (*Pleuronectes platessa* L.) 1. The structure and ultrastructure of normal plaice skin. *Journal of Fish Biology*, 4, 87–98.

Roberts, T.R. (1967). Tooth formation and replacement in characoid fishes. *Stanford Ichthyological Bulletin*, 8, 231–47.

Roberts, T.R. (1972). Ecology of fishes in the Amazon and Congo basins. *Bulletin of the Museum of Comparative Zoology*, 143, 117–47.

Robertson, J.D. (1975). Osmotic constituents of the blood plasma and parietal muscle of *Squalus acathias* L. *Biological Bulletin*, 148, 305–19.

Robins, C.R., Ray, G.C., and Douglas, J. (1986). *A Field Guide to Atlantic Coast Fishes North America*. Boston, MA: Houghton Mifflin Company.

Rockwell, H., Evans, F.G., and Pheasant, H.C. (1938). The comparative morphology of the vertebrate spinal column. Its form as related to function. *Journal of Morphology*, 63, 87–117.

Rodriguez, F., López, J.C., Vargas, J.P., Gómez, Y., Broglio, C., and Salas, C. (2002). Conservation of spatial memory function in the pallial forebrain of reptiles and ray-finned fishes. *Journal of Neuroscience*, 22, 2894–903.

Rodriguez, M.S., Cavallaro, M.R., and Thomas, M.R. (2012). A new diminutive species of *Loricaria* (Siluriformes: Loricariidae) from the Rio Paraguay system, Mato Grosso do Sul, Brazil. *Copeia*, 2012, 49–56.

Rogers, C.S. and Beets, J. (2001). Degradation of marine ecosystems and decline of fishery resources in marine protected areas in the US Virgin Islands. *Environmental Conservation*, 28, 312–22.

Rome, L.C. and Sosnicki, A.A. (1991). Myofilament change in swimming carp. II. Sarcomere length changes during swimming. *American Journal of Cell Physiology*, 260, C289–C296.

Romer, A.S. (1955). *The Vertebrate Body*, 2nd edn. London: W.B.Saunders Co.

Romer, A.S. and Parsons, T.S. (1986). *The Vertebrate Body*, 6th edn. Philadelphia, PA: Saunders College Publishing.

Romero, A. (2008). Typological thinking strikes again. *Environmental Biology of Fish*, 81, 359–63.

Roos, G., Van Wassenbergh, S., Herrel, A., and Aets, P. (2009). Kinematics of suction feeding in the seahorse *Hippocampus reidi*. *Journal of Experimental Biology*, 212, 3490–8.

Roosen-Runge, E.C. (1977). *The Process of Spermatogenesis in Animals*. Cambridge: Cambridge University Press.

Rose, G.A. (2005). On distributional responses of North Atlantic fish to climate change. International Council for the Exploration of the Sea. *Journal of Marine Biology*, 62, 1360–74.

Rose, J.D. (2002). The neurobehavioural nature of fishes and the question of awareness and pain. *Review Fisheries Science*, 10, 1–38.

Rose, J.D., Arlinghaus, R., Cooke, S.J., Diggles, B.K., Sawynok, W., Stevens, E.D., and Wynne, C.D.L. (2014). Can fish really feel pain? *Fish and Fisheries*, 15, 97–133.

Roth, W.W., Mackin, R.B., Speiss, J., Goodman, R.H. , and Noe, B.D. (1991). Primary structure and tissue distribution of anglerfish carboxypeptidase. *Molecular and Cellular Endocrinology*, 78, 171–8.

Rough, K.M., Nowak, B.F., and Reuter, R.E. (2005). Haematology and leucocyte morphology of wild caught *Thunnus maccoyi*. *Journal of Fish Biology*, 66, 1649–59.

Rousseau, K., Le Belle, N., Shaihi, M., Marchelidon, J., Schmitz, M., and Dufour, S. (2002). Evidence for a negative feedback in the control of eel growth hormone by thyroid hormones. *Journal of Endocrinology*, 175, 605–13.

Rowe, S. and Hutchings, J.A. (2008). A link between sound producing musculature and mating success in Atlantic cod. *Journal of Fish Biology*, 72, 500–11.

Rowett, H.G.Q. (1950). *Dissection Guides II. The Dogfish*. London: John Murray.

Rowett, H.G.Q. (1953). *Histology and Embryology*. London: John Murray.

Rowley, A.F., Hunt, T.C., Page, M., and Mainwaring, G. (1988). Fish. In: A.F. Rowley and N.A. Ratcliffe (eds). *Vertebrate Blood Cells*. Cambridge: Cambridge University Press, pp. 19–128.

Rüber, L., Britz, R., Tan, H.H., Ng, P.K.L., and Zardoya, R. (2004). Evolution of mouthbrooding and life-history correlates in the fighting fish genus *Betta*. *Evolution*, 58, 799–813.

Rubin, L.R. and Nolte, J. (1981). Autonomic innervation and photosensitivity of the sphincter pupillae muscle of two teleosts *Lophius piscatorius* and *Opsanus tau*. *Current Eye Research*, 1, 543–52.

Ruby, E.G. and Morin, J.G. (1978). Specificity of symbiosis between deep-sea fishes and psychrotrophic luminous bacteria. *Deep-Sea Research*, 25, 161–7.

Rüdeberg, C. (1971). Structure of the pineal organs of *Anguilla anguilla* L. and *Lebistes reticulatus* Peters (Teleostei). *Zeitschrift fuer Zellforschung und Mikoscopische Anatomie*, 122, 227–43.

Rusakov, Y.I., Kolychev, A.P., Bondareva, V.M., van der Schors, R.C., and Li, K.W. (2000). Primary structure of insulin of the Black Sea rockfish *Scorpaena porcus*. *Journal of Evolutionary Biochemistry and Physiology*, 36, 728–33.

Rushbrook, B.J., Katsiadaki, I., and Barber, I. (2007). Spiggin levels are reduced in male sticklebacks infected with *Schistocephalus solidus*. *Journal of Fish Biology*, 71, 298–303.

Rushton, W.A.H. (1951). A theory on the effects of fibre size in medullated nerve. *Journal of Physiology*, 115, 101–22.

Russell, I.J. (1968). Influence of efferent fibres on a receptor. *Nature*, 219, 177–8.

Rutherford, E.S., Rose, K., and Cowan, J.H. Jr. (2003). Evaluation of the Shepherd and Cushing (1980) model of density-dependent survival: a case study using striped bass (*Morone saxitilis*) larvae in the Potomac River, Maryland, USA. *International Council for the Exploration of the Sea, Journal of Marine Science*, 60, 1275–87.

Ryan, T.J., Roy, D.S., Pignatelli, M., Arens, A., and Tonegawa, S. (2015). Engram cells retain memory under retrograde amnesia. *Science*, 348, 1007–13.

Sadovy, Y. and Shapiro, D.Y (1987). Criteria for the diagnosis of hermaphroditism in fishes. *Copeia*, 1987, 135–56.

Sage, M. (1971). Respiratory, behavioural and endocrine responses of a teleost to a restricted environment. *Fishery Bulletin*, 69, 879–81.

Sagemehl, M. (1884). Beiträge zur vergleicheuden Anatomie der Fische. II. Einige Bemerkurgen über die Gehirnshäute der Knockenfische. *Morphologisches Jahrbuch*, 9, 457–74.

Saha, N. and Ratha, B.K. (2007). Functional ureogenesis and adaptation to ammonia metabolism in Indian freshwater air-breathing catfishes. *Fish Physiology and Biochemistry*, 33, 283–95.

Saint-Paul, U. (1988). Diurnal O_2 consumption at different O_2 concentrations by *Colossoma macropomum* and *Colossoma brachypomum* (Teleostei, Serrasalmidae). *Comparative Biochemistry and Physiology*, 89A, 675–82.

Sakamoto, T. and McCormick, S.D. (2006). Prolactin and growth hormone in fish osmoregulation. *General and Comparative Endocrinology*, 147, 24–30.

Sancho, G., Fisher, C.R., Mills, S. Johnson, G.A., Lenihan, H.S., Peterson, C.H., and Mullineaux, L.S. (2005). Selective predation by the zoarcid fish *Thermarces cerberus* at hydrothermal vents. *Deep-Sea Research*, 52, 837–44.

Sand, A. (1937). The mechanism of the lateral line organ of fishes. *Proceedings of the Royal Society of London B*, 123, 472–95.

Sandblom, E. and Axelsson, M. (2011). Autonomic control of circulation in fish: A comparative view. *Autonomic Neuroscience: Basic and Clinical*, 165, 127–39.

Sander, G.E. and Huggins, C.G. (1972). Vasoactive peptides. *Annual Review of Pharmacology*, 12, 227–64.

Sanford, C.P.J. and Lauder, G.V. (1990). Kinematics of the tongue-bite apparatus in osteoglossomorph fishes. *Journal of Experimental Biology*, 154, 137–62.

Santer, R.M. and Greer Walker, M. (1980). Morphological studies on the ventricle of teleost and elasmobranch hearts. *Journal of Zoology, London*. 190, 259–72.

Sargent, J., Bell, G., McEvoy, L., Tocher, D., and Estevez, A. (1999). Recent developments in the essential fatty acid nutrition of fish. *Aquaculture*, 177, 191–9.

Sargent, J.R., Gatten, R.R. and Merrett, N.R. (1983). Lipids of *Hoplostethus atlanticus* and *H.mediterraneus* (Beryciformes: Trachichthyoidae) from deep water to the west of Britain. *Marine Biology*, 74, 281–6.

Sasano, Y., Yoshimura, A., and Fukamachi, S. (2012). Reassessment of the function of somatolactin alpha in lipid metabolism using medaka mutant and transgenic strains. *BioMed Central Genetics*, 13, 64. doi. 10.1186/1471. 2156–13-64

Sasayama, Y. (1999). Hormonal control of Ca homeostasis in lower vertebrates: considering the evolution. *Zoological Science*, 16, 857–69.

Sasayama, Y., Matsubara, T., and Takano, K. (1999). Topography and immunohistochemistry of the ultimobranchial glands in some subtropical fishes. *Ichythology Research*, 46, 219–22.

Satchell, G.H. (1986). Cardiac function in the hagfish, Myxine (Myxinoidea: Cyclostomata). *Acta Zoologica (Stockholm)*, 67, 115–22.

Satchell, G.H. (1971,1991). *Physiology and Form of Fish Circulation*. Cambridge: Cambridge University Press.

Satchell, G.H. (1999). Circulatory system: Distinctive attributes of the circulation of elasmobranch fish. In: W.C. Hamlett (ed.). *Sharks, Skates and Rays. The Biology of the Elasmobranchs*. Baltimore, MD: Johns Hopkins University Press.

Sato, K. and Suzuki, N. (2001). Whole-cell response characteristics of ciliated and microvillous olfactory receptors to amino acids, pheromone candidates and urine in rainbow trout. *Chemical Senses*, 26, 1145–56.

Sato, T. (1984). A brood parasitic catfish of mouthbrooding cichlid fishes in Lake Tanganyika. *Nature*, 323, 58–9.

Satou, M. (1990). Synaptic organization, local neuronal circuitry, and functional segregation of the teleost olfactory bulb. *Progress in Neurobiology*, 34, 115–42.

Satou, M., Fujita, I., Ichikawa, M., Yamaguchi, K., and Ueda, K. (1983). Field potential and intracellular potential studies of the olfactory bulb in carp for a functional separation of the olfactory bulb into lateral and medial subdivisions. *Journal of Comparative Physiology*, 152A, 319–33.

Satou, M., Oka, Y., Kusonoki, M., Matsushima, T., Kato, M., Fujita, I., and Ueda, K. (1984). Telencephalic and preoptic areas integrate sexual behaviour in himé salmon (landlocked red salmon *Oncorhynchus nerka*): results of electrical brain stimulation experiments. *Physiology and Behaviour*, 33, 441–7.

Saunders, R.L. (1961). The irrigation of the gills in fishes: I. Studies of the mechanism of branchial irrigation. *Canadian Journal of Zoology*, 39, 637–53.

Saunders, J.T. and Manton, S.M. (1951). *A Manual of Practical Vertebrate Morphology*. Oxford University Press: Oxford.

Saunders, J.T. and Manton, S.M. (1979). Rev. Manton, S.M., and Brown M.E. *A Manual of Practical Vertebrate Morphology*. Oxford: Clarendon Press.

Sauter, R.M. (1994). Chromaffin systems. In: S. Nilsson and S. Holmgren (eds). *Comparative Physiology and Evolution of the Autonomic Nervous System*. Chur, Switzerland: Harwood, pp. 97–117.

Saville, A. (1964). Identification sheets on fish eggs and larvae Sheet 1. Clupeoidae. Conseil international pout l'exploration de la mer. Copenhagen: International Council for the Exploration of the Sea.

Scemes, E. and Giaume, C. (2006). Astrocyte calcium waves. *Glia*, 54, 716–25.

Schaffield, M., Knappe, M., Hunzinger, C., and Markl, J. (2003). cDNA sequences of the authentic keratins 8 and 18 in zebrafish. *Differentiation*, 71, 73–82.

Schär, M., Maly, I.P., and Sasse, D. (1985). Histochemical studies on metabolic zonation of the liver in the trout (*Salmo gairdneri*). *Histochemistry*, 83, 147–51.

Schellart, N.A.M. (1990). The visual pathways and central non-tectal processing. In: R.H. Douglas and M.B.A. Djamgoz (eds). *The Visual System of Fish*. London: Chapman and Hall, pp. 345–73.

Schellart, N.M. and Wubbels, R.J. (1998). The auditory and mechanosensory lateral line system. In: D.H. Evans (ed.). *The Physiology of Fishes*, 2nd edn. Boca Raton, FL: CRC Press, pp. 285–312.

Scher, A.M. (1965). Control of arterial blood pressure: measurement of pressure and flow. In: T.C. Ruch, and H.D. Patton (eds). *Physiology and Biophysics*, 19th edn. Philadelphia, PA: Saunders, pp. 660–83.

Schindler, I and Schmidt, J. (2008). *Betta kuehnei*, a new species of fighting fish (Teleostei, Osphronemidae) from the Malay Peninsula, *Bulletin of Fish Biology*, 10, 34–96.

Scholander, P.F. (1954). Secretion of gases against high pressures in the swimbladder of deep sea fishes. II. The rete mirabile. *Biological Bulletin*, 107, 260–77.

Scholander, P.F. and van Dam, L. (1953).Composition of the swimbladder gas in deep sea fishes. *Biological Bulletin*, 104, 75–86.

Schreck, C.B. (2010). Stress and fish reproduction: the roles of allostasis and hormesis. *General and Comparative Endocrinology*, 165, 549–56.

Schulz, R.W., de França, L.R, Lareyre, J-J., Legac, F., Chiarini-Garcia, H., Nobrega, R.H., and Miura, T. (2010). Spermatogenesis in fish. *General and Comparative Endocrinology*, 165, 390–411.

Schultze, H.-P. (1991). A comparison of controversial hypotheses on the origin of tetrapods. In: H.-P. Schultze and L. Trueb (eds). *Origins of the Higher Groups of Tetrapods. Controversy and Consensus*. New York, NY: Cornell University Press.

Schultze, H.-P. (1993). Patterns of diversity of jawed fishes. In: J. Hanken and B.K. Hall (eds) *The Skull, Vol 2: Patterns of Structural and Systematic Diversity*. Chicago, IL: University of Chicago Press.

Schultze, H.-P. and Arratia, G. (1986). Reevaluation of the caudal skeleton of Actinopterygian fishes: *Lepisosteus* and *Amia*. *Journal of Morphology*, 190, 215–41.

Schumann, D. and Piiper, J. (1966). Der Sauerstoffbedarf der Atmung beir Fischen nach Messungen an dernarkotisierten Schleie (*Tinca tinca*). *Pflügers Archiv*, 288, 15–26.

Schuster, S. (2002). Behavioural evidence for post-pause reduced responsiveness in the electrosensory system of *Gymnotus carpio*. *Journal of Experimental Biology*, 205, 2525–33.

Schwab, I.R. (2012). *Evolution's Witness. How Eyes Evolved*. Oxford: Oxford University Press.

Schwarzkopf, L. and Brooks, R.J. (1987). Nest-site selection and offspring sex ratio in painted turtles, *Chrysemys picta. Copeia*, 1987, 53–61.

Scott, A.P., Bye, V.J., and Baynes, S.M. (1980). Seasonal variations in sex steroids of female rainbow trout (*Salmo gairdneri* Richardson). *Journal of Fish Biology*, 17, 587–92.

Scott, A.P. and Canario, A.V.M. (1990). Plasma levels of ovarian steroids, including 17α, 21-dihydroxy–4-pregnene-3, 20-dione and 3α17α, 21-trihydroxy-5β-pregnan-20-one, in female plaice (*Pleuronectes platessa*) induced to mature with human chorionic gonadotrophin. *General and Comparative Endocrinology*, 78, 286–98.

Scott, D.P. (1962). Effect of food quantity on fecundity of rainbow trout, *Salmo gairdnieri. Journal of the Fisheries Research Board of Canada*, 19, 715–24.

Scott, P.D., Milestone, C.B., Smith, D.S., Maclatchey, D.L., and Hewitt, L.M. (2011). Isolation and identification of ligands for the goldfish testis and androgen receptor in chemical recovery condensates from a Canadian bleached kraft pulp and paper mill. *Environmental Science and Technology*, 45, 10226–34.

Scott, J.S. (1985). *Sand Lance*. Ottawa: Department of Fisheries and Oceans.

Scott, W.B. and Crossman, E.J. (1973). *Freshwater Fishes of Canada*. Ottawa: Fisheries Research Board of Canada Ottawa, Bulletin, 184.

Scott, W.B. and Scott, M.G. (1988). Atlantic fishes of Canada. *Canadian Bulletin of Fisheries and Aquatic Sciences*, 219. Toronto: University of Toronto Press.

Scown, T.M., Santos, E.M., Johnston, B.D., Gaiser, B., Baalousha, M., Mitov, S., Lead, J.R., Stone, V., Fernandez, T.F., Jepson, M., von Aerle, R., and Tyler, C.T. (2010). Effects of aqueous exposure to silver nanoparticles of different sizes in rainbow trout. *Toxicological Science*, 115, 521–34.

Sedmera, D., Reckova, M., de Almeida, A., Sedmerova, M., Biermann, M., Sarre, A., Raddatz, E., McCarthy, R.A., Gourdie, and Thompson, R.P. (2003). Functional and morphological evidence for a ventricular conduction system in zebrafish and *Xenopus* hearts. *American Journal of Physiology. Heart and Circulatory Physiology*, 284, 1152–60.

Seliger, H.H. (1962). Direct action of light in naturally pigmented musc*le* fibres. I Action spectrum for contraction in eel iris sphincter. *Journal of General Physiology*, 46, 333–42.

Sepulveda, C.A., Wegener, N.C., Bernal, D., and Graham, J.B. (2005). The red muscle morphology of the thresher sharks. *Journal of Experimental Biology*, 208, 4255–61.

Seternes, T., Dalmo, R.A., Hoffman, J., Bøgwald, J., Zykova, S., and Smedsrød, B. (2001). Scavenger-receptor-mediated endocytosis of lipopolysaccharide in Atlantic cod (*Gadus morhua* L.). *Journal of Experimental Biology*, 204, 4055–64.

Seth, H., and Axelsson, M. (2010). Sympathetic, parasympathetic enteric regulation of the gastrointestinal vasculature in rainbow trout (*Oncorhynchus mykiss*) under normal and postprandial conditions. *Journal of Experimental Biology*, 213, 3118–26.

Shaklee, J.B., Christiansen, J.A., Sidell, B.D. Prosser, C.L., and Whitt, G.S. (1977). Molecular aspects of temperature acclimation in fish: Contributions of changes in enzyme activities and isozyme patterns to metabolic reorganization in the green sunfish. *Journal of Experimental Zoology*, 201, 1–20.

Shaw, E. (1962). The schooling of fishes. *Scientific American*, 206, 128–38.

Shephard, K.L. (1994). Functions for fish mucus. *Reviews in Fish Biology and Fisheries*, 4, 401–29.

Shepherd, J.G. and Cushing, D.H. (1980). A mechanism for density-dependent survival of larval fish as the basis of a stock-recruitment relationship. *Journal Conseil International pour l'Exploration de la mer*, 185, 255–67.

Shepro, D., Belamarich, F.A., and Branson, R. (1966). The fine structure of the thrombocyte in the dogfish (*Mustelus canis*) with special reference to microtubule orientation. *Anatomical Record*, 156, 203–14.

Shiels, H.A., Calaghan, S.C., and White, E. (2006). The cellular basis for enhanced volume-modulated cardiac output in fish hearts. *Journal of General Physiology*, 128, 37–44.

Shier, D., Butler, J., and Lewis, R. (2002). *Hole's Human Anatomy and Physiology*, 9th edn. Boston, MA: McGraw-Hill.

Shirakova, M.Y. (1969). The sexual maturation rate of the generations of Baltic cod taken from 1961–1963. *Atlanticheskii nauchno-issledovatel'skogo institute rybnogo khozyaistva I Okeanograf II. Trudy*. Kaliningrad: Trudy Atlanticheskogo nauchno –Issledovatel skogo Instituta Rybnogo.

SICB (2015). Society for Integrative and Comparative Biology, Annual Meeting. West Palm Beach FL.

Silman, R.E., Leone, R.M., and Hooper, R.J.L. (1979). Melatonin, the pineal gland and human puberty. *Nature*, 282, 301–3.

Silva, A.G. and Martinez, C.B.R. (2007). Morphological changes in the kidney of a fish living in an urban stream. *Environmental Toxicology and Pharmacology*, 23, 185–92.

Silva, P., Spokes, K.C., and Kinne, R. (2012). Identification of aquaporin mRNA in the rectal gland of *Squalus acanthias*. *Bulletin of Mount Desert Island Laboratory*, 51, 1–3.

Silverthorn, D.U. (2001). *Human Physiology: An Integrated Approach*. Upper Saddle River, NJ: Prentice-Hall.

Simmons, P. and Young, O. (1999). *Nerve Cells and Animal Behavior*, 2nd edn. Cambridge: Cambridge University Press.

Sire, J-Y. and Akimento, M.A. (2004). Scale development in fish: a review with description of sonic hedgehog (shh) expression in the zebrafish (*Danio rerio*). *International Journal of Developmental Biology*, 48, 233–47.

Sire, J-Y. and Arnulf, I. (1990). The development of squamation in four teleostean fishes with a survey of the literature. *Japanese Journal of Ichthyology*, 37, 133–43.

Sire, J-Y., Donoghue, P.C.J., and Vickaryous, M.K. (2009). Origin and evaluation of the integumentary skeleton in non-tetrapod vertebrates. *Journal of Anatomy*, 214, 409–40.

Sisneros, J.A. and Tricas, T.C. (2002). Ontogenetic changes in the response properties of the peripheral electrosensory system in the Atlantic stingray (*Dasyatis sabina*). *Brain, Behaviour and Evolution*, 59, 130–40.

Sivak, J.G. (1978). The functional significance of the aphakic space of the fish eye. *Canadian Journal of Zoology*, 56, 513–16.

Skov, P.V. and Steffensen, J.F. (2003). The blood volumes of the primary and secondary circulation system in the Atlantic cod *Gadus morhua* L., using plasma bound Evans blue and compartmental analysis. *Journal of Experimental Biology*, 206, 591–9.

Slack, J.M.W. (1995). Developmental biology of the pancreas. *Development*, 121, 1569–80.

Smith, A. (1990). *Blind White Fish in Persia*. London: Penguin Books.

Smith, C. and Reay, P. (1991). Cannibalism in teleost fish. *Reviews in Fish and Fisheries*, 1, 41–64.

Smith, L.S. (1980). Digestion in teleost fishes. In: *Aquaculture Development and Coordination Programme. Fish Feed Technology*. Rome: Food and Agriculture Organization of the United Nations, pp. 1–19.

Smith, M.M. and Hall, B.K. (1990). Development and evolutionary origins of vertebrate skeletogenic and odontogenic tissues. *Biological Reviews*, 65, 277–373.

Smith, R.J.F. (1976). Seasonal loss of alarm cells in North American cyprinoid fishes and its relation to abrasive spawning behaviour. *Canadian Journal of Zoology*, 54, 1172–82.

Smith, R.J.F. (1982). The adaptive significance of the alarm substance-fright reaction system. In: T.J. Hara (ed.). *Chemoreception in Fishes*. Amsterdam: Elsevier, pp. 327–42.

Smith, R.J.F. (1992). Alarm signals in fish. *Reviews in Fish Biology and Fisheries*, 2, 33–63.

Smoot, J.C. and Findlay, R.H. (2000). Digestive enzyme and gut surfactant activity of detrivorous gizzard shad (*Dorosoma cepedianum*). *Canadian Journal of Fisheries and Aquatic Science*, 57, 1113–19.

Sneddon, L.U. (2015). Pain in aquatic animals. *Journal of Experimental Biology*, 218, 967–76.

Sneddon, L.U., Braithwaite, V.A., and Gentle, M.J. (2003). Do fish have nociceptors? Evidence the evolution of a vertebrate sensory system. *Proceedings of the Royal Society of London B*, 270, 1115–21.

Snelgrove, P.V.R. (1999). Getting to the bottom of biodiversity: sedimentary habitats. *Bioscience*, 49, 129–38.

Soengas, J.L. and Moon, T.W. (1998). Transport and metabolism of glucose in isolated enterocytes of the black bullhead *Ictalurus melas*: effect of diet and hormones. *Journal of Experimental Biology*, 201, 3263–73.

Solc, D. (2007). The heart and conducting system in the kingdom of animals. A comparative approach to its evolution. *Experimental and Clinical Cardiology*, 12, 113–18.

Sombes, A., Porsteinsson, H., Amardóttir, H., Jóhannes-dóttir, I.P., Sigurgeirsson, B. De Polavieja, G.G. and Karisson, K.Æ. (2013). The ontogeny of sleep-wake cycles in zebrafish: a comparison to humans. *Frontiers in Neural Circuits*, 7, Article 178. doi:10.3389/fncir.2013.00178

Somiya, H. (1982), 'Yellow lens' eyes of a stomiatoid deep-sea fish, *Malacosteus niger. Proceedings of the Royal Society of London B*, 215, 481–9.

Somiya, H. (1988). Guanine-type retinal tapetum of three species of mormyrid fishes. *Japanese Journal of Ichthyology*, 36, 220–6.

Sorensen, P.W. and Caprio, J. (1998). Chemoreception. In: D.H. Evans (ed.). *The Physiology of Fishes*, 2nd edn. Boca Raton, FL: CRC Press, pp. 375–405.

Soule, G. (1970). *The Greatest Depths*. Philadelphia, PA: Macrae Smith Co.

Spalding, M.D., Ravilious, C., and Green, E.P. (2001). *World Atlas of Coral Reefs*. Berkeley, CA: University of California Press.

Sparre, P. and Venema, S.C. (1998). Introduction to tropical fish stock assessment. *Food and Agriculture Organization of the United Nations, Fisheries technical paper 306/11.* Rome: Food and Agriculture Organization of the United Nations.

Spazier, E., Storch, V., and Braunbeck, T. (1992). Cytopathology of spleen in eel *Anguilla anguilla* exposed to a chemical spill in the Rhine River. *Diseases of Aquatic Organisms*, 14, 1–22.

Specker, J.L. and Schreck, C.B. (1980). Stress responses to transportation and fitness for marine survival in coho salmon (*Oncorhynchus kisutch*) smolts. *Canadian Journal of Fisheries and Aquatic Science*, 37, 765–9.

Speers-Roesch, B. and Treberg, J.R. (2010). The unusual energy metabolism of elasmobranch fishes. *Comparative Biochemistry and Physiology A*, 155, 417–34.

Spotte, S.H. (2002). *Candiru: Life and Legend of the Bloodsucking Catfishes*. Berkeley, CA: Creative Arts Book Co.

Springate, J.R.C. and Bromage, M.R. (1985). Effects of egg size on early growth and survival in rainbow trout (*Salmo gairdneri* Richardson). *Aquaculture*, 47, 163–72.

Srivastav, A.K., Srivastav, S.K., Sasayama, Y., and Suzuki, N. (1998). Salmon calcitonin induced hypocalcemia and hyperphosphatemia in an elasmobranch, *Dasyatis akajei. General and Comparative Endocrinology*, 109, 8–12.

Srivastav, A.K., Srivastav, S.K., Shasayama, Y., and Suzuki, N. (2003). Plasma magnesium levels in the stingray, *Dasyatis akajei* in response to administration of salmon calcitonin. *Asian Journal of Experimental Science*, 17, 17–21.

Srivastava, R.K. and Brown, J.A. (1991). The biochemical characteristics and hatching performance of cultured

and wild Atlantic salmon (*Salmo salar*) eggs. *Canadian Journal of Zoology*, 69, 2436–441.

Stabell, O.B. (1992). Olfactory control of homing behaviour in salmonids. In: T.J. Hara (ed.). *Fish Chemoreception*. London: Chapman and Hall, pp. 249–70.

Stacey, N. (2003). Hormones, pheromones and reproductive behaviour. *Fish Physiology and Biochemistry*, 28, 229–35.

Stacey, N.E. and Cardwell, J.R. (1995). Hormones as sex pheromones in fish: widespread distribution among freshwater species. *Proceedings of the Fifth International Symposium on the Reproductive Physiology of Fish, Austin 1995*, pp. 244–8. Austin: Fish Symposium 95.

Stacey, N. and Sorensen, P. (2005) Reproductive pheromones. In K.A. Sloman, R.W.Wilson and S.Balshine (eds.) *Fish Physiology 24: Behaviour and Physiology of Fish*. Amsterdam: Elsevier, pp. 359–412.

Stacey, N., Chojnacki, A., Narayanan, A., Cole, T., and Murphy, C. (2003). Hormonally derived sex pheromones in fish: exogenous cues and signals from gonad to brain. *Canadian Journal of Physiology and Pharmacology*, 81, 329–41.

Stacey, N.E., Kyle, A.L., and Liley, N.R. (1986). Fish reproductive pheromones. In: D. Duvall, D. Müller-Schwarze, and S.M. Silverstein (eds). *Chemical Signals in Vertebrates V*. New York, NY: Plenum Press.

Staiger, J.C. (1972). *Bassogigas profundissimus* (Pisces; Brotulidae) from the Puerto Rico trench. *Bulletin of Marine Science*, 22, 26–33.

Standen, E.M. (2010). Muscle activity and hydrodynamic function of pelvic fins in trout (*Oncorhynchus mykiss*). *Journal of Experimental Biology*, 213, 831–41.

Stanley, H., Chieffi, G., and Botte, V. (1965). Histological and histochemical observations on the testis of *Gobius paganellus. Zeitschrift fuer Zellforschung und Mikroscopische Anatomie*, 65, 350–62.

Stauffer, J.R. and Loftus, W.F. (2010). Brood parasitism of a bagrid catfish (*Bagrus meridionalis*) by a clariid catfish (*Bathyclarias nyasensis*) in Lake Malawi, Africa. *Copeia*, 2010, 71–4.

Stauffer, J.R., Boltz, J.M., and Whitem, L.R. (1995). The fishes of West Virginia *Proceedings of the Academy of Natural Sciences, Philadelphia*, 146, 1–389.

Steen, E.B. and Montagu, A. (1959). *Anatomy and Physiology, Vol. 1*. New York, NY: Barnes and Noble.

Steffensen, J.F. (1985). The transition between branchial pumping and ram ventilation in fishes: Energetic consequences and dependence on water oxygen tension. *Journal of Experimental Biology*, 114, 141–50.

Steffensen, J.F. and Lomholt, J.P. (1992). The secondary vascular system In: W.S. Hoar, D.J. Randall, and A.P. Farrell (eds). *Fish Physiology XIIA*. New York, NY: Academic Press, pp. 185–216.

Steffensen, J.F., Lomholt, J.P., and Johansen, K. (1981). The relative importance of skin oxygen uptake in the naturally buried plaice, *Pleuronectes platessa*, exposed to graded hypoxia. *Respiratory Physiology*, 44, 269–75.

Steinach, E. (1890). Utersuchungen zur vergleichenden Physiologie der Iris, Erste Mitteilung. *Pflügers Archiv; European Journal of Physiology*, 47, 289–340.

Stensiö, E. (1958). Les cyclostomes fossils ou Ostracodermes. In: P.P. Grassé (ed.). *Traité de zoologie*. 1st fascicle, Vol. 13. Paris: Masson et Cie.

Sterba, G. (1959, 1962). *Freshwater Fishes of the World*. Trans. and rev. by D.W. Tucker. London: Studio Vista.

Sterzi, G. (1901). Ricerche intorno all' anatomia comparata ed all'ondtogenesi delle meningi. *Atti del reale instituto veneto di scienze, lettere ed arti*, 60, 1101–361.

Stevens, B.R., Kaunitz, J.D., and Wright, E.M. (1984). Intestinal transport of amino acids and sugars: advances using membrane vesicles. *Annual Review of Physiology*, 46, 417–33.

Stevens, E.D. and Randall, D.J. (1967). Changes in blood pressure, heart rate and breathing rate during moderate swimming activity in rainbow trout. *Journal of Experimental Biology*, 46, 307–15.

Stevens, E.D., Lam, H.M., and Kendall, J. (1974) Vascular anatomy of the counter-current heat exchanger of skipjack tuna. *Journal of Experimental Biology*, 61, 145–53.

Stillwell, C.E. and Kohler, N.E. (1985). Food and feeding ecology of the swordfish *Xiphias gladius* in the western North Atlantic Ocean. *Marine Ecology Progress Series 22*, 239–47.

Stock, D.W. (2001). The genetic basis of modularity in the development and evolution of the vertebrate dentition. *Philosophical Transactions of the Royal Society of London B*, 356, 1633–53.

Stokes, F.J. (1980). *Collins Handguide to the Coral Reef Fishes of the Caribbean*. London: Collins.

Stoklosowa, S. (1966). Sexual dimorphism in the skin of sea-trout, *Salmo trutta*. *Copeia*, 1966, 613–14.

Storey, J.M. and Storey, K.B. (1985). Triggering of cryoprotectant synthesis by initiation of ice nucleation in the freeze tolerant frog, *Rana sylvatica*. *Journal of Comparative Physiology*, 156, 191–5.

Strenzke, F.J. (1957). *Ökologie der Wassertiere*. Konstanz: Akademische Verlagsgesellschaft Athenaion Dr Albert Hachfeld.

Striedter, G.F. (2004). Brain evolution. In: G. Paxinos, and J.K. Mai (eds). *The Human Nervous System*, 2nd edn. Amsterdam: Elsevier, pp. 3–121.

Striedter, G.F. and Norhcutt, R.G. (2006). Head size constrains forebrain development and evolution in ray-finned fishes. *Evolution and Development*, 8, 215–22.

Stuart, L., Smith, M., and Baumgartner, L. (2009). *Counting Murray–Darling fishes by validating sounds associated with spawning*. A final report to the Murray–Darling Basin Commission (now Murray–Darling Basin Authority). Canberra: Murray-Darling Basin Commission.

Sturm, A., Bury, N., Dengreville, L., Fagart, J., Flouriot, G., Rafetin-Oblin, M.E., and Prunet, P. (2005). 11-deoxycorticosterone is a potent agonist of the rainbow trout (*Oncorhynchus mykiss*) mineralocorticoid receptor. *Endocrinology*, 146, 47–55.

Sucré, E., Charmantier-Daures, M., Grousset, E., Charmantier, G., and Cucchu-Mouillet, P. (2010). Embryonic occurrence of ionocytes in the sea bass *Dicentrarchus labrax*. *Cell and Tissue Research*, 339, 543–50.

Sumbre, G. and de Polavieja, G.G. (2014). The world according to zebrafish: how neural circuits generate behaviour. *Frontiers in Neural Circuits*, 8, Article 91

Sumpter, J.P. (2009). Protecting aquatic organisms from chemicals: the harsh realities. *Philosophical Transactions of the Royal Society A*, 367, 3877–94.

Sumpter, J.P., Carragher, J., Potteinger, T.G., and Pickering, A.D. (1987). The interaction of stress and reproduction in trout. *Proceedings of the 3rd International Symposium on the Reproductive Physiology of Fish*, St. John's, NL, 299–302.

Sumpter, J.P., Dye, H.M., and Benfey, T.J. (1986). The effects of stress on plasma ACTH, α-MSH, and cortisol levels in salmonid fishes. *General and Comparative Endocrinology*, 62, 377–85.

Sumpter, J.P., Le Bail, P.Y., Pickering, A.D., Pottinger, T.G., and Carragher, J.F. (1991). The effect of starvation on growth and plasma growth hormone concentrations of rainbow trout, *Oncorhynchus mykiss*. *General and Comparative Endocrinology*, 83, 94–102.

Sumpter, J.P., Pickering, A.D., and Pottinger, T.G. (1985). Stress-induced elevation of plasma alpha-MSH and endorphin in brown trout, *Salmo trutta* L. *General and Comparative Endocrinology*, 59, 257–65.

Sundararaj, B.I. and Prasad, M.R.N. (1964). The histochemistry of the saccus vasculosus of *Notopterus chitala* (Teleostei). *Quarterly Journal of Microscopical Science*, 105, 91–8.

Sutton, T.T. (2005). Trophic ecology of the deep-sea fish *Malacosteus niger* (Pisces, Stomiidae): An enigmatic feeding ecology to facilitate a unique visual system. *Deep-Sea Research*, 52, 2065–76.

Sylvester, R.M. and Berry, C.R. Jr. (2006). Comparison of white sucker age estimates from scales, pectoral fin rays, and otoliths. *North American Journal of Fisheries Management*, 26, 24–31.

Swift, D.R. (1964). The effect of temperature and oxygen on the growth rate of the Windermere char (*Salvelinus alpinus* Willughby). *Comparative Biochemistry and Physiology*, 12, 179–83.

Szabo, T. (1965). Sense organs of the lateral line system in some electric fish of the Gymnotidae, Mormyridae and Gmynarchidae. *Journal of Morphology*, 177, 229–50.

Szabo, T. and Wersäll, J. (1970). Ultrastructure of an electroreceptor (mormyromast) in a mormyrid fish, *Gnathonemus petersi* II. *Journal of Ultrastructure Research*, 30, 473–90.

Szelistowski, W.A. (1989). Scale-feeding in juvenile marine catfishes (Pisces: Ariidae). *Copeia*, 1989, 517–19.

Szent-György, A. (1949). Free-energy relations and contraction of actomyosin. *Biological Bulletin*, 96, 140–61.

Taborsky, M. (1984). Broodcare helpers in the cichlid fish *Lamprologus brichardi*: Their costs and benefits. *Animal Behaviour*, 32, 1236–52.

Takagi, W., Kajimura, M., Bell, J.D., Toop, T., Donald, J.A., and Hyodo, S. (2012). Hepatic and extrahepatic distribution of ornithine urea cycle enzymes in holocephalan elephant fish (*Callorhincus milii*). *Comparative Biochemistry and Physiology B*, 161, 331–40.

Takahashi, T. and Koblmüller, S. (2011). The adaptive radiation of cichlid fish in Lake Tanganyika: A morphological perspective. *International Journal* of *Evolutionary Biology*, ID 620754.

Takemura, A., Rahman, S., Nakamura, S., Park, Y.J., and Takano, K. (2004). Lunar cycles and reproductive activity in reef fishes with particular attention to rabbit fishes. *Fish and Fisheries*, 5, 117–28.

Takemura, A., Rahman, M.S., and Park, Y.J. (2010). External and internal controls of lunar-related reproductive rhythms in fishes. *Journal of Fish Biology*, 76, 7–26.

Tam, S.J. and Watts, R.J. (2010). Connecting vascular and nervous system development: angiogenesis and the blood–brain barrier. *Annual Review of Neuroscience*, 33, 379–408.

Tan, E., Nizar, J. Carrera, G, E., and Fortune, E. (2005). Electrosensory interference in naturally occurring aggregates of a species of weakly electric fish, *Eigenmannia virescens*. *Behavioural Brain Research*, 164, 83–92.

Tarrant, H., Llewellyn, N., Lyons, A., Tattersall, N., Wylde, S., Mouzakitis, G., Maloney, M., and McKenzie, C. (2005). *Endocrine Disruptors in the Irish Aquatic Environment*. Final Report. Wexford, Ireland: Environmental Protection Agency.

Tavares-Dias, M. (2006). A morphological and cytochemical study of erythrocytes, thrombocytes and leucocytes in four freshwater teleosts. *Journal of Fish Biology*, 68, 1822–33.

Tavares-Dias, M., Ono, E.A., Pilarski, F., and Moraes, F.R. (2007). Can thrombocytes participate in the removal of cellular debris in the blood circulation of teleost fish? A cytochemical study and ultrastructural analysis. *Journal of Applied Ichthyology* 23, 709–712.

Taylor, A.M., Bus, T., Sprengel, R., Seeburg, P.H., Rawlins, J.N.P., and Bannerman, D.M. (2013). Hippocampal NMDA receptors are important for behavioural inhibition but not for encoding associative spatial memories. *Philosophical Transactions of the Royal Society B*, 369, 2013.0149.

Taylor, D.S., Turner, B.J., Davis, W.P., and Chapman, B.B. (2008). A novel terrestrial fish habitat inside emergent logs. *American Naturalist*, 171, 263–66.

Taylor, E.W., Leite, C.A.C., McKenzie, D.J., and Wang, T. (2010). Control of respiration in fish, amphibians and reptiles. *Brazilian Journal of Medical and Biological Research*, 43, 409–24.

Tchernavin, V. (1938). II. Changes in the salmon skull. *Transactions of the Zoological Society of London*, 24, 103–84.

Tchernavin, V.V. (1947a). Six specimens of Lyomeri in the British Museum (with notes on the skeleton of Lyomeri). *Journal of the Linnean Socety of London, Zoology*, 41, 287–353.

Tchernavin, V.V. (1947b). Further notes on the structure of the bony fishes of the Order Lyomeri (Eurypharynx). *Journal of the Linnean Society of London, Zoology*, 41, 377–93.

Teeter, J. (1980). Pheromone communication in sea lampreys (*Petromyzon marinus*): implications for population management. *Canadian Journal of Fisheries and Aquatic Science*, 37, 2132–3.

Teichmann, H. (1954). Vergleichende Untersuchungen am der Nase der Fische. *Zeitschrift fuer Morphologie und Oekologie der Tiere*, 43, 171–212.

Temple, S., Hart, N.S., Marshall, N.J., and Collin, S.P. (2010). A spitting image: specializations in archerfish eyes for vision at the interface between air and water. *Proceedings of the Royal Society B*, 277, 2607–15.

Templeman, W. and Andrews, G.L. (1956). Jellied condition in the American plaice *Hippoglossoides platessoides* (Fabricius). *Journal of the Fisheries Research Board of Canada*, 13, 147–82.

Templeman, W., Hodder, V.M., and Wells, R. (1978). Sexual maturity and spawning in haddock, Melanogrammus aeglefinus, of the Southern Grand Bank. *International Commission for the Northwest Atlantic Fisheries Organization Research Bulletin*, 13, 53–65.

Templeton, M.R., Graham, N., and Voulvoulis, N. (eds) (2009). Theme issue 'Emerging chemical contaminants in water and wastewater'. *Philosophical Transactions of the Royal Society A*, 367, 3893–5.

Te Winkel, L.E. (1935). A study of *Mistichthys luzonensis* with special reference to conditions correlated with reduced size. *Journal of Morphology*, 58, 463–535.

Tewksbury, H.T. and Conover, D.O. (1987). Adaptive significance of intertidal egg deposition in the Atlantic silverside *Menidia menidia*. *Copeia*, 1987, 76–83.

Thanuthong, T., Francis, D.S., Manickam, E. Senadheera, S.D., Cameron-Smith, D., and Turchini, G.M. (2011). Fish oil replacement in rainbow fish diets and total dietary PUFA content: II) Effects on fatty acid metabolism and *in vivo* fatty acid bioconversion. *Aquaculture*, 322/323, 99–108.

Thienpont, D. and Niemegeers, C.J.E. (1965). Propoxate (R7464): a new potent anaesthetic agent in cold-blooded vertebrates. *Nature*, 205, 1018–19.

Thomas, M.B., Thomas, W., Hornstein, M., and Hedman, S.C. (1999). Seasonal leukocyte and erythrocyte counts in fathead minnows. *Journal of Fish Biology*, 54, 1116–18.

Thomas, P.K. (1955). Growth changes in the myelin sheath of peripheral nerve fibres in fishes. *Proceedings of the Royal Society of London B*, 143, 380–91.

Thomas, P.K. (1956). Growth changes in the diameter of peripheral nerve fibres in fishes. *Journal of Anatomy, London*, 90, 5–14.

Thomas, P.K. and Young, J.Z. (1949). Internode lengths in the nerves of fishes. *Journal of Anatomy, London*, 83, 336–51.

Thomas, R.E. and Rice, S.D. (1981). Excretion of aromatic hydrocarbons and their metabolites by freshwater and seawater Dolly Varden char. In: J. Verberg, A. Calabrese, F.P. Thurberg, and W.B. Vernberg (eds). *Biological Monitoring of Marine Pollutants*. New York, NY: Academic Press, pp. 425–48.

Thomerson, J.E. and Turner, B.J. (1973). *Rivulus stellifer*, a new species of annual killifish from the Orinoco basin of Venezuela. *Copeia*, 1973, 783–787.

Thompson, D'Arcy W. (1961). *On growth and form*. Abridged edn. J.T. Bonner (ed.). Cambridge: Cambridge University Press.

Thompson, J.R. and Banaszak, L.J. (2002). Lipid-protein interactions in lipovitellin. *Biochemistry*, 41, 9398–409.

Thomson, D.A. (1969). Toxic secretions of the boxfish, *Ostracion meleagris* Shaw. *Copeia*, 1969, 335–52.

Thomson, K.S. (1966). Intracranial mobility in the coelacanth. *Science*, 153, 999–1000.

Thomson, K.S. (1969). The biology of the lobe-finned fishes. *Biological Reviews*, 44, 91–154.

Thomson, K.S. (1967). Mechanisms of intracranial kinetics in fossil rhipidistian fishes (Crossopterygii) and their relatives. *Journal of the Linnaean Society (Zoology)*, 46, 223–53.

Thorarensen, H., Gallaugher, P., and Farrell, A.P. (1996). Cardiac output in swimming rainbow trout, *Oncorhynchus mykiss*, acclimated to seawater. *Physiological Zoology*, 69, 139–53.

Thorpe, W.H. (1956). *Learning and Instinct in Animals*. London: Methuen.

Thorson, R.F. and Fine, M.L. (2002). Crepuscular changes in emission rate and parameters of the boatwhistle advertisement call of the gulf toadfish, *Opsanus beta*. *Environmental Biology of Fishes*, 63, 321–31.

Tian, N.M.M.-L. and Price, D.J. (2005). Why cavefish are blind. *Bioassays*, 27, 235–8.

Tiedemann, R., Moll, K., Paulus, K.B., and Schlupp, I. (2005). New microsatellite loci confirm hybrid origin (*Poecilia formosa*). *Molecular Ecology Notes*, 5, 586–9.

Tinbergen, N. (1952). The curious behaviour of the stickleback. *Scientific American*, 187, 22–6.

Tobey, N., Heizer, W., Yeh, R., Huang, T.I., and Hoffner, C. (1985). Human intestinal brush border peptidases. *Gastroenterology*, 88, 913–26.

Tolentino-Pablico, G., Bailly, N., Froese, R., and Elloran, C. (2008). Seaweeds preferred by herbivorous fishes. *Journal of Applied Phycology*, 20, 933–8.

Tomita, K. and Cooper, J.P. (2007). The telomere bouquet controls the meiotic spindle. *Cell*, 130, 113–26.

Tontisirin, K. and Clugston, G. (2004). *Vitamin and Mineral Requirements in Human Nutrition*. Geneva: World Health Organization and Food and Agriculture Organization of the United Nations.

Townshend, T.J. and Wootton, R.J. (1987). Seasonal patterns in the diets, gut lengths and feeding behaviour of three Panamanian cichlids. *Proceedings of the Vth Congress of European Ichthyologists, Stockholm 1985*, 127–32.

Trajano, E. (1997). Food and reproduction of *Trichomycterus itacarambiensis*, cave catfish from south-Eastern Brazil. *Journal of Fish Biology*, 51, 53–63.

Trautman, M.B. (1957). *The Fishes of Ohio*. Columbus, OH: Ohio State University Press.

Treberg, J.R., Speers-Roesch, B., Piermarini, P.M., Ip, Y.K., Ballatyne, J.S., and Driedzic, W.R. (2006). The accumulation of methylamine counteracting solutes in elasmobranchs with differing levels of urea: a comparison of marine and freshwater species. *Journal of Experimental Biology*, 209, 860–70.

Tripathi, P. and Mittal, A.K. (2010). Essence of keratin in lips and associated structures of a freshwater fish *Puntius sophore* in relation to its feeding ecology: histochemistry and scanning electron microscope investigation. *Tissue and Cell*, 42, 223–33.

Truscott, B. So, Y.P., Nagler, J.J., and Idler, D.R. (1992). Steroids involved with final maturation in the winter flounder. *Journal of Steroid Biochemistry and Molecular Biology*, 42, 351–356.

Truscott, B., Walsh, J.M., Burton, M.P., Payne, J.F., and Idler, D.R. (1983). Effects of acute exposure to crude petroleum in some reproductive hormones in salmon and flounder. *Comparative Biochemistry and Physiology*, 75C, 121–30.

Tsui, T.K.N., Hung, C.Y.C., Nawata, C.M., Wilson, J.M., Wright, P.A., and Wood, C.M. (2009). Ammonia transport in cultured gill epithelium of freshwater rainbow trout: the importance of Rhesus glycoproteins and the presence of an apical Na^+/NH_4 exchange complex. *Journal of Experimental Biology*, 212, 878–92.

Tsujii, T. and Seno, S. (1990). Melano-macrophage centers in the aglomerular kidney of the sea horse (Teleosts):

Morphologic studies on its formation and possible function. *Anatomical Record*, 226, 460–70.

Tsuneki, K. (1992). A systematic survey of the occurrence of the hypothalamic saccus vasculosus in teleost fish. *Acta Zoologica (Stockholm)*, 73, 67–77.

Tubbs, C., Tan, W., Shi, B., and Thomas, P. (2011). Identification of 17, 20beta, 21-trihydroxy-4-pregen-3-one (20beta-S) receptor binding and membrane progestin receptor alpha on southern flounder sperm (*Paralichthys lethostigma*) and their likely role in 20beta-S stimulation of sperm hypermotility. *General and Comparative Endocrinology*, 170, 629–39.

Twelves, E.L., Everson, I., and Leith, I. (1975). Thyroid structure and function in two Antarctic fishes. *British Antarctic Survey Bulletin*, 40, 7–14.

Tyler, A.V. (1973). Alimentary tract morphology of selected north Atlantic fish in relation to food habits. *Fisheries Research Board of Canada Technical Report*, 361, 23pp.

Uematsu, K. Holmgren, S., and Nilsson, S. (1989). Autonomic innervations of the ovary in the Atlantic cod *Gadus morhua*. *Fish Physiology and Biochemistry*, 6, 213–19.

Umminger, B.L. (1971). Osmoregulatory role of serum glucose in freshwater-adapted killifish (*Fundulus heteroclitus*) at temperatures near freezing. *Comparative Biochemistry and Physiology*, 38A, 141–45.

Urban, J. and Alheit, J. (1988). *Oocyte development cycle of plaice, Pleuronectes platessa, and North Sea sole, Solea solea*. International Council for the Exploration of the Sea. C.M. 1988/G.52. Demersal Fish Committee. Copenhagen, Denmark: International Council for the Exploration of the Sea.

Vagelli, A.A. (2004). Ontogenetic shift in habitat preference by *Pterapogon kaudneri*, a shallow water coral reef apogonid, with direct development. *Copeia*, 2004, 364–9.

Val, A., Almeida-Val, V.M., and Randall, D.J. (eds) (1996). 1. *Volume overview. In Physiology and Biochemistry of Fishes of the Amazon*. Manaus: INPA.

Van Dam (1938). On the utilization and regulation of breathing in aquatic animals. Gröningen University. Dissertation.

Van den Burg, E.H., Metz, J.R., Arends, R.J., Devreese, B., Vandenberghe, I., Van Beeumen, J., Wendelaar Bonga, S.E., and Flik, G. (2001). Identification of β-endorphins in the pituitary gland and blood plasma of the common carp. *Journal of Endocrinology*, 169, 271–80.

Van den Heuvel, M.R. Martel, P.H., Kovacs, T.G., MacLatchy, D.L., Van der Kraak, G.J. Parrott, J.L., McMaster, M.E., O'Connor, B.I., Melvin, S.D., and Hewitt, L.M. (2010). Evaluation of short-term fish reproductive bioassays for predicting effects of a Canadian bleached kraft mill effluent. *Water Quality Research Journal Canada*, 45, 175–86.

Van Eys, G.J.J.M. (1980). Structural changes in the pars intermedia of the cichlid teleost *Sarotherodon mossambicus* as a result of background adaptation and illumination. *Cell and Tissue Research*, 208, 99–110.

Van Noorden, S. and Patent, G.J. (1980). Vasoactive intestinal polypeptide-like immunoreactivity in nerves of the teleost fish, *Gillichthys mirabilis*. *Cell and Tissue Research*, 212, 139–46.

Van Oosten, J. (1957). The skin and scales. In: M.E. Brown (ed.). *The Physiology of Fishes*, *Vol. 1*. New York, NY: Academic Press, pp. 207–44.

Van Ree, G.E., Lok, D., and Bosman, G.J.G.M. (1977). In vitro induction of nuclear breakdown in oocytes of the zebrafish *Brachydanio rerio* (Ham.Buch.). Effects of the composition of the medium and of protein and steroid hormones. *Proceedings of the Koninklijke Nederlandse Akademie van Wetenschappen, Amsterdam, Series C*, 80, 353–71.

Van Wassenbergh, S., Roos, G. Aets, P., Herrel, A., and Adriaens D. (2011). Why the long face? A comparative study of feeding kinematics of two pipefishes with different snout lengths. *Journal of Fish Biology*, 78, 1786–98.

Velsen, F.P.J. (1980). Embryonic development in eggs of sockeye salmon, *Oncorhynchus nerka*. *Canadian Special Publication of Fisheries and Aquatic Sciences*, 49, pp. 1–24.

Verbost, P.M., Butkus, A., Willems, P., and Wendelaar Bonga, S.E. (1993). Indications for two bioactive principles in the corpuscles of Stannius. *Journal of Experimental Biology*, 177, 243–52.

Verkhratsky, A., Solovyeva, N., and Toescu, E.C. (2002). Calcium excitability of glial cells. In: A. Volterra, P. Magistretti, and P. Haydon (eds). *The Tripartite Synapse: Glia in Synaptic Transmission*. Oxford: Oxford University Press, pp. 99–109.

Vernberg, W.B. and Vernberg, F.J. (1972). *Environmental Physiology of Marine Animals*. New York, NY: Springer-Verlag.

Verner-Jeffreys, D., Algoet, M., Pond, M.J., Virdee, H.K., Bagwell, N.J., and Roberts, E.G. (2007). Furunculosis in Atlantic salmon (*Salmo salar* L.) is not readily controllable by bacteriophage therapy. *Aquaculture*, 270, 475–84.

Verrill, A.E. (1897). Nocturnal and diurnal changes in the colours of certain fishes and of the squid (*Loligo*), with notes on their sleeping habits. *Nature*, 55, 451.

Vicentini, R.N. and Araüjo, F.G. (2003). Sex ratio and size structure of *Micropogonius furneri* (Desmarest, 1823) (Perciformes, Sciaenidae) in Sepetiba Bay, Rio de Janiero, Brazil. *Brazilian Journal of Biology*, 63, 559–66.

Videler, J.J. (1993). *Fish Swimming*. New York, NY: Springer.

Villee, C.A., Walker, W.F. Jr., and Barnes, R.D. (1984). *General Zoology*. Philadelphia, PA: Saunders College Publishing.

Volkoff, H. and Peter, R.E. (2006). Feeding behaviour of fish and its control. *Zebrafish*, 3, 131–40.

Volkoff, H., Canosa, L.F. Unniappan, S., Cerdà,-Reverter, J.M., Bernier, N.J., Kelly, S.P., and Peter, R.E. (2005). Neuropeptides and the control of food intake in fish. *General and Comparative Endocrinology*, 142, 3–19.

Volkoff, H., Hoskins, L.J., and Tuziak, S.M. (2010). Influence of intrinsic signals and environmental cues on the endocrine control of feeding in fish: Potential application in aquaculture. *General and Comparative Endocrinology*, 167, 352–9.

Volterra, A. Magistretti, P., and Haydon, P. (eds) (2002). *The Tripartite Synapse: Glia in Synaptic Transmission*. Oxford: Oxford University Press.

Von Bartheld, C.S. and Meyer, D.L. (1987). Comparative neurology of the optic tectum in ray-finned fishes: patterns of lamination formed by retinotectal projections. *Brain Research*, 420, 277–88.

Von Bertalanffy, L. (1957). Quantitative laws in metabolism and growth. *Quarterly Review of Biology*, 32, 217–31.

Von Campenhausen, C., Riess, I. and Weissert, R. (1981). Detection of stationary objects by the blind cave fish *Anoptichthys jordani* (Characidae). *Journal of Comparative Physiology*, 143, 369–74.

Von der Emde, G. (1990). Discrimination of objects through electrolocation in the weakly electric fish *Gnathonemus petersi*. *Journal of Comparative Physiology*, 167, 413–21.

Von der Emde, G. (1998). Electroreception. In: D.H. Evans (ed.). *The Physiology of Fishes*, 2nd edn. Boca Raton, FL: CRC Press, pp. 313–43.

Von der Emde, G. (2007). Electroreception: object recognition in African weakly electric fish. In: T.H. Hara and B. Zielinski (eds). *Fish Physiology 25, Sensory Systems Neuroscience*. New York, NY: Academic Press, pp. 7–336.

Von Frisch, K. (1911). Beiträge zur Physiologie der Pigmentzellen in der Fischhaut. *Pflügers Archiv, European Journal of Physiology*, 138, 319–87.

Von Frisch, K. (1936). Über den Gehörsinn der Fische. *Biological Reviews*, 11, 210–46.

Von Frisch, K. (1938). Zur Psychologie des Fisch-Swarmes. *Naturwissenschaften*, 26, 601–6.

Von Frisch, K. (1941a). Über einen Schreckstoff der Fischhaut und seine biologische. *Bedeutung Zeitschrift für vergleichende Physiologie*, 29, 6–145.

Von Frisch, K. (1941b). Die Bedeutung des Geruchssinnes in Leben der Fische. *Naturwissenschaften*, 291, 321–33.

Wada, M., Azuma, N., Mizuna, N., and Kurokura, H. (1999). Transfer of symbiotic luminous bacteria from parental *Leiognathus nuchalis* to their offspring. *Marine Biology*, 135, 683–7.

Wagner, H-J., Douglas, R.H., Frank, T.M., Roberts, N.W., and Partridge, J.C. (2009). A novel vertebrate eye using both refractive and reflective optics. *Current Biology*, 19, 108–14.

Wagner, H-J., Fröhlich, E. Negishi, K., and Collin, S.P. (1998). The eyes of deep-sea fish II. Functional morphology of the retina. *Progress in Retinal and Eye Research*, 17, 637–85.

Wahlquist, I. and Nilsson, S. (1981). Sympathetic nervous control of the vasculature in the tail of the Atlantic cod *Gadus morhua*. *Journal of Comparative Physiology*, 144B, 153–6.

Wald, G. (1960). The distribution and evolution of visual systems. *Comparative Biochemistry*, 1, 311–45.

Wales, W. and Tytler, P. (1996). Changes in chloride cell distribution during early larval stages of *Clupea harengus*. *Journal of Fish Biology*, 49, 801–14.

Walker, M.M., Diebel, C.E., and Kirschvink, J.L. (2007). Magnetoreception. In: T.H. Hara and B. Zielinski (eds). *Sensory System Neuroscience: Fish Physiology 25*. New York, NY: Academic Press/Elsevier, pp. 337–76.

Walker, M.M., Quinn, T.P., Kirschvink, J.L., and Groot, C. (1988). Production of single-domain magnetite throughout life by sockeye salmon, *Oncorhynchus nerka*. *Journal of Experimental Biology*, 140, 51–63.

Walker, W.F. and Homberger, D.G. (1992). *Vertebrate Dissection*. Fort Worth, TX: Harcourt Brace.

Wallace, R.A. and Selman, K. (1981). Cellular and dynamic aspects of oocyte growth in teleosts. *Amererican Zoologist*, 21, 325–43.

Wallace, W. (1914). Report of the age, growth and sexual maturity of the plaice in certain parts of the North Sea. *Fisheries Investigations*, Series II, Vol. II, No.2.

Walls, G.L. (1942 Re-issued 1963). *The Vertebrate Eye and its Adaptive Radiation*. New York, NY: Hafner.

Walsh, P.J., Grosell, M., Goss, G.G., Bergman, H.L., Betman, A.N., Wilson, P., Laurent, P., Alper, S., Smith, C.P., Kamunde, C., and Wood, C.M. (2001). Physiological and molecular characterization of urea transport by the gills of the Lake Magadi tilapia (*Alcolapia grahami*). *Journal of Experimental Biology*, 204, 509–20.

Walton, K.M. and Dixon, J.E. (1993). Protein tyrosine phosphatases. *Annual Review of Biochemistry*, 62, 101–20.

Waltzek, T.B. and Wainwright, P.C. (2003). Functional morphology of extreme jaw protrusion in neotropical cichlids. *Journal of Morphology*, 25, 96–106.

Wang, H., Yan, T., Tan, J.T., and Gong, Z. (2000). A zebrafish 167 itellogenin gene (vg3) encodes a novel 167 itellogenin without a phosvitin domain and may represent a primitive vertebrate 167 itellogenin gene. *Gene*, 256, 303–10.

Ward, A.B. and Mehta, R.S. (2010). Axial elongation in fishes: Using morphological approaches to elucidate developmental mechanisms in studying body shape. *Integrative and Comparative Biology*, 50, 1106–19.

Warrant, F.J. and Locket, N.A. (2004). Vision in the deep sea. *Biological Reviews*, 79, 671–712.

Waterman, J.J. (2001). Measures, stowage rates and yields of fishery products. *Department of Scientific and Industrial Research, Torry Advisory Note No. 17*. 13 pp. London: Department of Scientific and Industrial Research.

Waterman, T.H. (1954). Polarization patterns in submarine illumination. *Science*, 120, 927.

Watson, W. and Walker, H.J. (2004). The world's smallest vertebrate *Schindleria brevipinguis*, a new pedomorphic species in the Family Schindleriidae (Perciformes: Gobioidei). *Records of the Australian Museum*, 56, 139–42.

Waxman, S.G. (1972). Regional differentiation of the axon: a review with special reference to the concept of the multiplex neuron. *Brain Research*, 47, 269–88.

Waxman, S.G. (1997). Axon-glia interactions: building a smart nerve fibre. *Current Biology*, 7, R406–R410.

Waxman, S.G., Pappas, G.D., and Bennett, M.V.L. (1972). Morphological correlates of functional differentiation of nodes of Ranvier in the neurogenic electric organs of the knife fish, *Sternarchus*. *Journal of Cell Biology*, 53, 210–24.

Weatherley, A.H. and Gill, H.S. (1987). *The Biology of Fish Growth*. London: Academic Press.

Webb, P.W. (1982). Locomotor patterns in the evolution of actinopterygian fishes. *American Zoologist*, 22, 329–42.

Webb, P.W. (1984). Body form, locomotion and foraging in aquatic vertebrates. *American Zoologist*, 24, 107–20.

Webb, P.W. (1998). Swimming. In: D.H. Evans (ed.). *The Physiology of Fishes*, 2nd edn. Boca Raton, FL: CRC Press, pp. 3–24.

Webb, R.C. (2003). Smooth muscle contraction and relaxation. *Advances in Physiology Education*, 27, 201–6.

Wegner, N.C. and Graham, J.B. (2010). George Hughes and the history of fish ventilation: From Du Verrney to the present. *Comparative Biochemistry and Physiology, A*, 157, 1–6.

Wegner, N.C., Sepulveda, C.A., Bull, K.B., and Graham, J.B. (2010). Gill morphometrics in relation to gas transfer and ram ventilation in high-energy demand teleosts: scombrids and billfishes. *Journal of Morphology*, 271, 36–49.

Weichert, C. K. (1961). *Representative Chordates*. New York, NY: McGraw-Hill.

Weihrauch, D., Wilkie, M.P., and Walsh, P.J. (2009). Ammonia and urea transporters in gills of fish and aquatic crustaceans. *Journal of Experimental Biology*, 212, 1716–30.

Weissert, R. and von Campenhausen, C. (1981). Discrimination between stationary objects by the blind cave fish *Anoptichthys jordani* (Characidae). *Journal of Comparative Physiology*, 143, 375–81.

Weitkamp, D.E. (2000). *Total Dissolved Gas Supersaturation in the Natural River Environment*. Report prepared for Chelan County Public Utility District No.1, Washington, DC: Chelan Couny Public Utility District.

Wells, R.M.G., Tetens, V., and Devries, A.L. (1984). Recovery from stress following capture and anaesthesia of Antarctic fish: haematology and blood chemistry. *Journal of Fish Biology*, 25, 567–76.

Wells, T.A.G. (1961). *Three Vertebrates*. London: Heinemann.

Welsh, J.Q., Goatley, C.H.R., and Bellwood, D.R. (2013). The ontogeny of home ranges: evidence from coral reef fishes. *Proceedings of the Royal Society B*, 280 (issue 1773). DOI: 10.1098/rspb.2013.2066.

Wendelaar Bonga, S.E. (1997). The stress response in fish. *Physiological Reviews*, 77, 591–625.

Wersäll, J. (1961). Vestibular receptor cells in fish and mammals. *Acta Oto-Laryngology*, Supplement 163, 25–9.

Westcott, S.W. and Burrows, P.M. (1991). Degree-day models for predicting egg hatch and population increase of *Criconemella xenoplax*. *Journal of Nematology*, 23, 366–92.

Westneat, M.W. and Alfaro, M.E. (2005). Phylogenetic relationships and evolutionary history of the reef fish family Labridae. *Molecular Phylogenetics and Evolution*, 36, 370–90.

Whitear, M. (1952).The innervations of the skin of teleost fishes. *Quarterly Journal of Microscopical Science*, 93, 567–84.

Whitear, M. (1970). The skin surface of bony fishes. *Journal of Zoology London*, 160, 437–54.

Whitear, M. (1971a). The free nerve endings in fish epidermis. *Journal of Zoology, London*, 163, 231–6.

Whitear, M. (1971b). Cell specialization and sensory function in fish epidermis. *Journal of Zoology, London*, 163, 237–64.

Whitear, M. (1986a). The skin of fish including cyclostomes, Epidermis. In: J. Bereiter-Hahn, A.G. Matoltsy, and K.S. Richards (eds). *Biology of the Integument*. Berlin: Springer-Verlag, pp. 8–38.

Whitear, M. (1986b). The skin of fish including cyclostomes, Dermis. In: J. Bereiter-Hahn, A.G. Matoltsy, and K.S. Richards (eds). *Biology of the Integument*. Berlin: Springer-Verlag, pp. 39–64.

Whitear, M. (1992). Solitary chemosensory cells. In: T.J. Hara (ed.). *Fish Chemoreception*. London: Chapman and Hall, pp. 102–25.

Whitear, M. and Moate, R. (1998). Cellular diversity in the epidermis of *Raja clavata* (Chondrichthyes). *Journal of Zoology, London*, 246, 275–85.

Whitear, M. Mittal, A.K., and Lane, E.B. (1980). Endothelial layers in fish skin. *Journal of Fish Biology*, 17, 43–65.

Whitehouse, R.H. and Grove, A.J. (1947). *Dissection of the Dogfish*. London: University Tutorial Press Ltd.

Whitledge, T.E. and Dugdale, R.C. (1972). Creatine in seawater. *Limnology and Oceanography* 17, 309–14.

Whitlock, K.E. (2004). Development of the Nervus Terminalis: Origin and migration. In: C.R. Wirsig-Wiechmann (ed.). Special Issue: The anatomy and function of the nervus terminalis. *Microscopy Research and Technique*, 65, 2–12.

Whittington, C.M. and Wilson, A.B. (2013). The role of prolactin in fish reproduction. *General and Comparative Endocrinology*, 191, 123–36.

Whyte, J.J. and Tillitt, D.E. (2000). EROD activity. In: C.J. Schmitt and G.M. Dethloff (eds). Biomonitoring of Environmental Status and Trends (BEST) Program: Selected Methods for Monitoring Chemical Contaminants and their Effects in Aquatic Ecosystems. Information and Technology Report U.S. Dept of the Interior/U.S. Geol. Survey Biological Resources Division USGS/BRD 2000-0005, pp. 5–8. Washington, DC: Department of the Interior.

Widder, E.A. (1998). A predatory use of counterillumination by the squaloid shark. *Isistius brasiliensis. Environmental Biology of Fishes*, 53, 267–73.

Widder, E.A. (2010). Bioluminescence in the ocean: Origins of biological, chemical and ecological diversity. *Science*, 328, 704–8.

Wiest, F.C. (1995). The specialized locomotory apparatus of the freshwater hatchetfish family Gasteropelecidae. *Journal of Zoology*, 236, 571–92.

Wilga, C.D. and Lauder, G.V. (2002). Function of the heterocercal tail in sharks: quantitative wake dynamics during steady horizontal swimming and vertical manoevering. *Journal of Experimental Biology*, 205, 2365–74.

Wilkens, H. (2007). Graham S. Proudlove, Subterranean fishes of the world. An account of the subterranean (hypogean) fishes described to 2003 with a bibliography 1541–2004. *Reviews in Fish Biology and Fisheries*, 71, 310–12.

Wilkie, M.P. (1997). Mechanisms of ammonia excretion across fish gills. *Comparative Biochemistry and Physiology*, 118A, 39–50

Wilkie, M.P. (2002). Ammonia excretion and urea handling by fish gills: present understanding and future research challenges. *Journal of Experimental Zoology*, 293, 284–301.

Wilkins, N.P. and Jancsar, S. (1979). Temporal variations in the skin of Atlantic salmon *Salmo salar* (L.). *Journal of Fish Biology*, 15, 299–307.

Williams, E.E. (1959). Gadow's arcualia and the development of tetrapod vertebrae. *Quarterly Review of Biology*, 34, 1–32.

Williams, R.IV, Neubarth, N., and Hale, M.E. (2013). The function of fin rays as proprioceptive sensors in fish. *Nature Communications*, 4, #1729. doi:10.1058/ncomms2751

Willoughby, F. (1686). *History of Fishes*. London: The Royal Society.

Wilson, D.A. (2004). Focus on odorant specificity of single olfactory bulb neurons to amino acids in the channel catfish. *Journal of Neurophysiology*, 92, 38–9.

Winberg, S. and Nilsson, G.E. (1993). Roles of brain monoamine neurotransmitters in agonistic behaviour and stress reactions, with particular reference to fish. *Comparative Biochemistry and Physiology*, 106C, 597–614.

Winemiller, K.O. (1989). Development of dermal lip protuberances for aquatic surface respiration in South American characid fishes. *Copeia*, 1989, 382–90.

Winemiller, K.O. (1990). Spatial and temporal variation in tropical fish trophic networks. *Ecological Monographs*, 60, 331–67.

Winemiller, K.O. and Yan, H.Y. (1989). Obligate mucus-feeding in a South American trichomycterid catfish (Pisces: Ostariophysi). *Copeia*, 1989, 511–14.

Wingstrand, K.G. (1956). Non-nucleated erythrocytes in a teleostean fish *Maurolicus mülleri* (Gmelin). *Zeitschrift fuer Zellforschung und Mikroscopische Anatomie*, 45, 195–200.

Winterbottom, R. and Emery, A.R. (1981). A new genus and two new species of gobiid fishes (Perciformes) from the Chagos Archipelago, Central Indian Ocean. *Environmental Biology of Fish*, 6, 139–49.

Winterbottom, R. and Southcott, L. (2008). Short lifespan and high mortality in the western Pacific coral reef goby *Trimma nasa. Marine Ecology Progress Series*, 366, 203–8.

Wirsig-Wiechmann, C.R. (2004). Introduction to the anatomy and function of the nervus terminalis. *Microscopy Research and Technique*, 65, 1.

Withers, P.C. (1992). *Comparative Animal Physiology*. Fort Worth, TX: Saunders.

Witte, F. Goldschmidt, A., Goudswaard, P.C., Ligtvoet, W., van Oijen, M.J.P., and Wanik, J.H. (1992). Species extinction and concomitant ecological changes in Lake Victoria. *Netherlands Journal of Zoology*, 42, 214–32.

Witte, F., Msuku, B.S., Wanink, J.H., Seehausen, O., Katunzi, E.F.B., Gouswaard, P.C., and Goldschmidt, T. (2000). Recovery of cichlid species in Lake Victoria: an examination of factors leading to differential extinction. *Reviews in Fish Biology and Fisheries*, 10, 233–41.

Witten, P.E. and Hall, B.K. (2003). Seasonal changes in the lower jaw skeleton in male Atlantic salmon (*Salmo salar* L.): remodelling and regression of the kype after spawning. *Journal of Anatomy*, 203, 435–50.

Witten, P.E. and Huysseune, A. (2009). A comparative view on the mechanisma and functions of skeletal remodelling in teleost fish, with special emphasis on osteoclasts and their functions. *Biological Reviews*, 84, 315–346.

Witten, P.E. and Villwock, W. (1997). Growth requires bone resorption at particular skeletal elements in a teleost fish with acellular bone (*Oreochromis niloticus*, Teleostei: Cichlidae). *Journal of Applied Ichthyology*, 13, 149–58.

Witten, P.E. Hansen, A., and Hall, B.K. (2001). Features of mono- and multinucleated bone resorbing cells of the zebrafish *Danio rerio* and their contribution to skeletal development, remodeling and growth. *Journal of Morphology*, 250, 197–207.

Witten, P.E., Sire, J-Y., and Huysseune, A. (2014). Old, new and new-old concepts about the evolution of teeth. *Journal of Applied Ichthyology*, 30, 636–42.

Wittenberg, J.B. and Haedrich, R.L. (1974). The choroid *rete mirabile* of the fish eye. II Distribution and relation to the pseudobranch and to the swimbladder rete mirabile. *Biological Bulletin*, 146, 137–56.

Wolf, K. (1963). Physiological salines for fresh-water teleosts. *The Progressive Fish Culturist* 25, 135–40.

Wolf, K. and Quimby, M.C. (1969). Fish cell and tissue culture. In: W.S. Hoar and D.J. Randall (eds). *Fish Physiology, Vol. II Reproduction and Growth, Bioluminescence, Pigments and Poisons.* New York, NY: Academic Press, pp. 253–306.

Wolfe, K.H. (2001). Yesterday's polyploids and the mystery of diploidization. *Nature Reviews Genetics*, 2, 333–41.

Wolff, T. (1961). The deepest recorded fishes. *Nature*, 190, 284.

Wood, C.M. (2011). Silver. In: C.M. Wood, A.P. Farrell, and C.J. Brauner (eds). *Fish Physiology, Vol. 31, Part B: Homeostasis and Toxicology of Non-Essential Metals.* New York, NY: Academic Press/Elsevier.

Wood, C.M. Farrell, A.P., and Brauner, C.J. (2011a). *Fish Physiology, Vol. 31A: Homeostasis and Toxicology of Essential Metals.* New York, NY: Academic Press.

Wood, C.M. Farrell, A.P., and Brauner, C.J. (2011b). *Fish Physiology, Vol. 31B: Homeostasis and Toxicology of Non-Essential Metals.* New York, NY: Academic Press.

Wood, C.M., Grosell, M., McDonald, M.D., Playle, R.C., and Walsh, P.J. (2010). Effects of waterborne silver in a marine teleost, the gulf toadfish (*Opsanus beta*): effects of feeding and chronic exposure on bioaccumulation and physiological responses. *Aquatic Toxicology*, 99 138–48.

Wood, C. M., Hopkins, T. E., Hogstrand, C., and Walsh, P. J. (1995). Pulsatile urea excretion in the ureogenic toadfish *Opsanus beta*: An analysis of routes and rates. *Journal of Experimental Biology*, 198, 1729–54.

Woodbury, L.A. (1942). A sudden mortality of fishes accompanying a supersaturation of oxygen in Lake Waubesa, Wisconsin. *Transactions of the American Fisheries Society*, 71, 112–17.

Wootton, R.J. (1976). *The Biology of the Sticklebacks.* London: Academic Press.

Wootton, R.J. (1984). *A Functional Biology of Sticklebacks.* Berkeley, CA: University of California Press.

Wourms, J.P. (1977). Reproduction and development in chondrichthyan fishes. *Amererican Zoologist*, 17, 379–410.

Wourms, J.P. (1981). Viviparity: the maternal fetal relationship in fishes. *American Zoologist*, 21, 473–515.

Wourms, J.P., Grove, B.D., and Lombardi, J. (1988). The maternal-embryonic relationship in viviparous fishes. In: W.S. Hoar and D.J. Randall (eds). *Fish Physiology, Vol. 11B.The Physiology of Developing Fish Viviparity and Posthatching Juveniles* New York, NY: Academic Press, pp. 1–134.

Wright, D.E. (1973). The structure of the gills of the elasmobranch, *Scyliorhinus canicula* (L.). *Zeitschrift fuer Zellforsche und Mikroscopische Anatomie*, 144, 489–509.

Wright, G. and Youson, J.H. (1980). Transformation of the endostyle of the anadromaous sea lamprey, *Petromyzon marinus* L., during metamorphosis. II. Electron microscopy. *Journal of Morphology*, 166, 231–57.

Wright, J.R., Bonen, A., Conlon, J.M., and Pohajdak, B. (2000). Glucose homeostasis in the teleost fish Tilapia: Insights from Brockmann body xenotransplantation studies. *American Zoologist*, 40, 234–45.

Wright, P.A., Felskie, A., and Anderson, P.M. (1995). Induction of ornithine-urea cycle enzymes and nitrogen metabolism and excretion in rainbow trout (*Oncorhynchus mykiss*) during early life stages. *Journal of Experimental Biology*, 198, 127–35.

Wu, G., Jaeger, L.A., Bazer, F.W., and Rhoads, J.M. (2004). Arginine deficiency in preterm infants: biochemical mechanisms and nutritional implications. *Journal of Nutritional Biochemistry*, 15, 442–51.

Wueringer, B.E. and Tibbetts, I.R. (2008). Comparison of the lateral line and ampullary systems of two species of shovel nose ray. *Reviews in Fish Biology and Fisheries*, 18, 47–64.

Wullimann, M.F. (1998). The central nervous system. In: D.H. Evans (ed.). *The Physiology of Fishes*, 2nd edn. Boca Raton, FL: CRC Press, pp. 247–82.

Wullimann, M.F., Rupp, B., and Reichert, H. (1996). *Neuroanatomy of the Zebrafish Brain: A Topological Atlas.* Basel: Birkhäuser Verlag.

Wunder, W. (1930). Experimentalle Untersuchungen am Dreistachligen Stickling (*Gasterosteus aculeatus*) während der Laichzeit. *Zeitschrift fuer Morphologie und Oekologie der Tiere*, 16, 453–98.

Xu, C., Schneider, D.C., and Rideout, C. (2013). When reproductive value exceeds economic value: an example from the Newfoundland cod fishery. *Fish and Fisheries*, 14, 225–33.

Yamada, T., Sugiyama, T., Tamaki, N., Kawakita, A., and Kato, M. (2009). Adaptive radiation of gobies in the interstitial habitats of gravel beaches accompanied by body elongation and excessive vertebral segmentation. *BioMedCentral Evolutionary Biology*, 9, 145–59.

Yamaguchi, K., Yasuda, A., Lewis, U.J. Yokoo, Y., and Kawauchi, H. (1989). The complete amino acid sequence of growth hormone of an elasmobranch, the blue shark (*Prionace glauca*). *General and Comparative Endocrinology*, 73, 252–9.

Yamamoto, Y. and Ueda, H. (2007). Physiological study on imprinting and homing related olfactory functions in salmon. *North Pacific Anadromous Fish Commission Technical Report No. 7*, 113–14.

Yamane, S. (1981). Sexual differences in ultrastructure of the ultimobranchial gland of mature eels (*Anguilla japonica*). *Bulletin of the Faculty of Fisheries, Hokkaido University*, 32, 1–5.

Yamanoue, Y., Settamarga, D.H.E., and Matsuura, K. (2010). Pelvic fins in teleosts: structure, function and evolution. *Journal of Fish Biology*, 77, 1173–208.

Yamasaki, F. (1972). Effects of methyltestosterone on the skin and gonads of salmonids. *General and Comparative Endocrinology*, Suppl. 3, 741–50.

Yan, H.Y., Saidel, W.M., Chang, J., Presson, J.C., and Popper, A.N. (1991). Sensory hair cells of a fish ear, evidence of multiple types based on ototoxity sensitivity. *Proceedings of the Royal Society London B*, 245, 133–8

Yancey, P.H. (2001). Water stress, osmolytes and protein. *American Zoologist*, 41, 699–709.

Yancey, P.H. (2005). Organic osmolytes as compatible, metabolic and counteracting cytoprotectants in high osmolarity and other stresses. *Journal of Experimental Biology*, 208, 2819–30.

Yancey, P.H., Blake, W.R., and Conley, J. (2002). Unusual organic osmolytes in deep-sea animals: adaptations to hydrostatic pressure and other perturbations. *Comparative Biochemistry and Physiology A*, 133, 667–76.

Yancey, P.H., Lawrence-Berry, R., and Douglas, M.D. (1989). Adaptations in mesopelagic fishes I. Buoyancy glycosaminoglycan layers in species without diel vertical migrations. *Marine Biology*, 103, 453–9.

Yáñez, J. and Anadón, R. (1998). Neural connections of the pineal organ in the primitive bony fish *Acipenser baeri*: a carbocyanine dye tract-tracing study. *Journal of Comparative Neurology*, 398, 151–61.

Yang, H. and Wright, J.R. (1995). A method for mass harvesting islets (Brockmann bodies) from teleost fish. *Cell Transplantation*, 4, 621–8.

Yanong, R.E. (2009). *Use of vaccines in finfish aquaculture*. Ruskin, FLA: University of Florida Institute of Food and Agricultural Sciences, extension.

Yao, Z. and Crim, L.W. (1995). Copulation, spawning and parental care in captive ocean pout. *Journal of Fish Biology*, 47, 171–3.

Yao, Z., Emerson, C.J., and Crim, L.J. (1995). Ultrastructure of the spermatozoa and eggs of the ocean pout (*Macrozoarces americanus* L.), an internally fertilizing marine fish. *Molecular Reproduction and Development*, 42, 58–64.

Yashpal, M., Kumari, U., Mittal, S., and Mittal, A.K. (2009). Morphological specialization of the buccal cavity in relation to the food and feeding habit of a carp *Cirrhinus mrigala*: A scanning electron microscope investigation. *Journal of Morphology*, 270, 71428.

Yasutake, W.T. (1983). *Microscopic Anatomy of the Salmonids. An Atlas*. Washington, DC: United States Department of the Interior Fish and Wildlife Service.

Yokoyama, S. and Radlwimmer, R. (1998). The 'five-sites' rule and the evolution of red and green colour vision in mammals. *Molecular Biology and Evolution*, 15, 560–7.

Yoneda, M., Kitano, H. Selvaraj, S., Matsuyama, M., and Shimizu, A. (2013). Dynamics of gonadosomatic index of fish with indeterminate fecundity between subsequent egg batches: application to Japanese anchovy *Engraulis japonicus* under captive conditions. *Marine Biology*, 160, 2733–41.

Yoshioka, S., Matsuhana, B., Tanaka, S., Inouye, Y., Oshima, N., and Kinoshita, S. (2011). Mechanism of variable structural colour in the neon tetra: quantitative evaluation of the Venetian blind model. *Journal of the Royal Society Interfacer*, 8, 56–66.

Yoshizawa, M. and Jeffery, W.R. (2008). Shadow response in the blind cavefish *Astyanax* reveals conservation of a functional pineal eye. *Journal of Experimental Biology*, 211, 292–9.

Young, B., Conti, D.V., and Dean, M.D. (2013). Sneaker 'jack' males outcompete dominant 'hooknose' males under sperm competition in Chinook salmon (*Oncorhynchus tshawytscha*). *Ecology and Evolution*, 3, 4987–97.

Young, B., Dunstone, R.L., Senden, T.J., and Young, G.C. (2013). A gigantic sarcopterygian (tetrapodomorph lobe-finned fish) from the Upper Devonian of Gondwana (Eden, New South Wales, Australia. *PLoS One*, 8(3), e5387.1

Young, J.Z. (1931a). On the autonomic nervous system of the teleost fish *Uranoscopus scaber*. *Quarterly Journal of Microscopical Science*, 74, 491–535.

Young, J.Z. (1931b). The pupillary mechanism of the teleostean fish *Uranoscopus scaber*. *Proceedings of the Royal Society of London B*, 107, 464–85.

Young, J.Z. (1933a). The autonomic system of selachians. *Quarterly Journal of Microscopical Science*, 75, 571–624.

Young, J.Z. (1933b). Comparative studies on the physiology of the iris. I. *Selachians*. *Proceedings of the Royal Society London B*, 112, 228–41.

Young, J.Z. (1933c). Comparative studies on the physiology of the iris. II. *Uranoscopus* and *Lophius*. *Proceedings of the Royal Society London B*, 112, 242–9.

Young, J.Z. (1936). The innervations and reactions to drugs of the viscera of teleostean fish. *Proceedings of the Royal Society of London B*, 120, 303–18.

Young, J.Z. (1945). History of the shape of a nerve fibre. In: W.E. Le Gros Clark (ed.) *Essays on Growth and Form, presented to D'Arcy Wentworth Thompson*. Oxford: Clarendon Press, pp. 41–94.

Young, J.Z. (1950). *The life of Vertebrates*. Oxford: Oxford University Press.

Young, J.Z. (1957). *The Life of Mammals*. Oxford: Clarendon Press.

Young, J.Z. (1960). *Doubt and Certainty in Science*. New York, NY: Oxford University Press.

Young, J.Z. (1978). *Programs of the Brain*. Oxford: Oxford University Press.

Young, J.Z. (1981). *The Life of Vertebrates*, 3rd edn. Oxford: Clarendon Press.

Young, J.Z. (1991). Computation in the learning system of cephalopods. *Biological Bulletin*, 180, 200–8.

Young, R.W. (1970). Visual cells. *Scientific American*, 223, 4, 81–91.

Youson, J.H. and Butler, D.G. (1988). Morphology of the kidney of adult bowfin, *Amia calva*, with emphasis on 'renal chloride cells' in the tubule. *Journal of Morphology*, 196, 137–56.

Yudkin, M. and Offord, R. (1975). *Biochemistry*. Oxford: Oxford University Press.

Yûki, R. (1957). Blood cell constituents in fish. *Bulletin of the Faculty of Fisheries, Hokkaido University*, VIII.I, 36–46.

Zaccone, G. (1979). Histochemistry of keratinization and epithelial mucus in the skin of the marine teleost *Muraena helena* (L.) (Anguilliformes, Pisces). *Cell and Molecular Biology*, 24, 37–50.

Zaccone, G. (1984). Immunochemical demonstration of neurospecific enolase in the nerve endings and skin receptors of marine eels. *Histochemical Journal*, 16, 1231–6.

Zapata, A. (1980). Splenic erythropoiesis and thrombopoiesis in elasmobranchs: an ultrastructural study. *Acta Zoologica (Stockholm)*, 61, 59–64.

Zapata, A., Diez, B., Cejalvo, T., Gutiérrez-de Frias, C., and Cortés, A. (2006). Ontogeny of the immune system of fish. *Fish and Shellfish Immunology*, 20, 126–36.

Zaunreiter, M., Brandstätter, R., and Goldschmid, A. (1998). Evidence for an endogenous clock B. in the retina of rainbow trout: I. Retinomotor movements, dopamine and melatonin. *Neuroreport*, 9, 1205–9.

Zelazowska, M., Kilarski, W. Billinski, S.M., Podder, D.D., and Kloc, M. (2007). Balbiani cytoplasm in oocytes of a primitive fish, the sturgeon *Acipenser gueldenstaedtii*, and its potential homology to the Balbiani body (mitochondrial cloud) of *Xenopus laevis* oocytes. *Cell and Tissue Research*, 329, 137–45.

Zhdanova, I.V., Wang, S.Y., Leclair, O.U., and Danilova, N.P. (2001). Melatonin promotes sleep-like state in zebrafish. *Brain Research*, 903, 263–26

Zielinski, B and Hara, T.J. (2007). Olfaction. In: T.J. Hara and B. Zielinski (eds). *Sensory Systems Neuroscience, Fish Physiology Vol. 25, Sensory Systems Neuroscience* New York, NY: Academic Press/Elsevier, pp. 1–43.

Zinkl, J.G., Cox, W.T., and Kono, C.S. (1991). Morphology and cytochemistry of leucocytes and thrombocytes of six species of fish. *Comparative Haematology International*, 1, 187–95.

Zippel, H.P., Gloger, M., Nasser, S., and Wilcke, S. (2000). Odour discrimination in the olfactory bulb of goldfish: Contrasting interactions between mitral cells and ruffed cells. *Philosophical Transactions of the Royal Society of London B*, 355, 1229–32.

Index